Probabilistic Methods in Structural Engineering

EDITOR'S FOREWORD

Dramatic innovations and developments have occurred in Civil and in Structural Engineering in recent years. Difficulties of analysis which appeared insurmountable only twenty years ago have largely disappeared with the advent of the mainframe computer and the finite elements method; new generation microcomputers now increasingly provide such analyses with greater convenience and economy. The Engineer today has more time to devize new forms of construction, to improve design details, and to allow for phenomena and data which were previously overlooked or approximated. Much of this new expertize has been used to improve the design of ships and aircraft, offshore platforms, subway systems, high-rise towers and buildings, and many other forms of construction previously designed by rules-of-thumb and simple codes of practice. There is now much more internationalism in Engineering too, with design methods and codes becoming more standardized, and large computers providing technical literature and patent information from all over the world. There is a need for these advances to be presented to an international audience by leading Engineers of international repute; this is the purpose of the new Civil Engineering Series by Chapman and Hall.

The second of the new Chapman and Hall Civil Engineering Series, entitled *Probabilistic Methods in Structural Engineering*, is by Professors G. Augusti, A. Baratta and F. Casciati of the universities of Florence, Naples and Pavia respectively. These authors have contributed to this subject via technical papers over many years; though their research continues apace they have nevertheless been pleased to produce a completely new exposition of this new and important subject. Professional engineers are already acquainted with the so-called semiprobabilistic methods which are used to incorporate limit-state philosophy in most of the new codes of practice; this book explains their theoretical basis. But probabilistic methods are of interest and importance in themselves and will no doubt be increasingly used in the future (just as reliability theory is increasingly applied in production engineering) – the educated engineer therefore needs to learn this subject now. These three experts are just the people to help!

The senior author, Professor G. Augusti, has already acquired a substantial international reputation and, among other things, sits on the organizing committees of the International Conferences on Structural Safety and Reliability (ICOSSAR) and International Conferences Applications of Statistics and Probability in Sai and Structural Engineering (ICASP). He has also chaired the 4th conference of the ICASP held in Florence in June 1983 and is at present the Dean of the Faculty of Engineering at Florence University. Professor F. Casciati heads the Structural Mechanics Department of Pavia University. Professor Baratta has remained at Naples University where he leads research in this field. Italy has emerged as a major source of development in this subject; these professors are three of her most accomplished exponents; their treatise should be welcomed as a major contribution to current knowledge and understanding.

E. Lightfoot *Oxford*

Probabilistic Methods in Structural Engineering

Giuliano Augusti
Professor, University of Florence

Alessandro Baratta
Professor, University of Naples

Fabio Casciati
Professor, University of Pavia

LONDON NEW YORK
Chapman and Hall

First published 1984 by
Chapman and Hall Ltd
11 New Fetter Lane, London EC4P 4EE
Published in the USA by
Chapman and Hall
733 Third Avenue, New York NY10017

Printed in Great Britain by
J. W. Arrowsmith Ltd, Bristol

ISBN 0 412 22230 2

British Library Cataloguing in Publication Data

Augusti, Giuliano
Probabilistic methods in structural engineering.
1. Safety factor in engineering 2. Structures, theory of—Statistical methods 3. Probabilities
I. Title II. Baratta, Alessandro
III. Casciati, Fabio
624.1'71 TA656.5

ISBN 0–412–22230–2

Library of Congress Cataloging in Publication Data

Augusti, Giuliano.
Probabilistic methods in structural engineering.

Bibliography: p.
Includes index.
1. Structural engineering—Statistical methods.
2. Engineering mathematics. 3. Probabilities
I. Baratta, Alessandro. II. Casciati, Fabio.
III. Title.
TA636.A93 1984 624.1'7'015192 83–7589
ISBN 0–412–22230–2

Contents

Introduction

'We apologize for the length of this **book**: we did not have time to make it short.' (after a letter attributed to G. B. Shaw)

'God keeps me from ever completing anything. This whole book is but a draught . . . Nay, but the draught of a draught!' (from H. M. Melville, *Moby Dick*, Chapter 32)

This is an ambitious book. Its ultimate scope is to fill what the authors (three structural engineers and teachers of structural mechanics with little – if any – formal education in probability and statistics) think is a significant gap in the available literature.

Indeed, in recent years probabilistic and statistical approaches and procedures are finding wider and wider applications in structural mechanics and engineering, and in more or less sophisticated forms are at the basis of many of the most recent versions of Structural Codes of Practice throughout the world. Detailed commentaries of these Codes have been written, and some of them (like the CIRIA Report, from which Chapter 10 of this book heavily draws) are true handbooks, quite satisfactory for any engineer who wishes not only to apply correctly, but also to understand, these new Codes.

On the other hand, although there are some very good books devoted to special aspects of the subject (see the list at the end of this Introduction), there is no place where a structural engineer or a student of these problems can find in one source the whole background of theory of probability, data collection and analysis, load and structural models, the only comprehensive text being perhaps Benjamin and Cornell's momentous book which is not, however, limited to structural problems.

This sort of text the authors set out to write, and tried to blend mathematical rigour and engineering approximations, and always to illustrate the theoretical principles with examples of concrete applications. In short, the objective was a self-contained reference book, suitable for use by structural engineers and structural engineering students without previous knowledge of probability and statistics. Although the book is not written as a textbook for immediate classroom use, it was also felt that it should provide sufficient matter for one or two courses at advanced undergraduate and/or postgraduate level.

Whether these ambitious objectives have been fulfilled, is left to the reader to say.

The book is divided in two parts, preceded by a short introductory chapter, which can be omitted by the reader who has some previous knowledge of the basic formulation of the relevant problems, and followed by a third part, containing the Appendices. In Part One, Chapter 2 contains definitions and formulae of probability theory which will be used in later chapters: a more formal mathematical presentation of the same subject matter, including many proofs, is presented in Appendices A1–A7. The basic ideas for setting up suitable models for structural reliability calculations conclude Chapter 2: it will be seen that these essentially consist in filtering an appropriate input (the *loads*) through an appropriate system (the *structure*), then in analysing the response with respect to some critical condition (or *limit state*).

The following three chapters deal, respectively, with models of the input loadings (Chapter 3) and with the two basic types of structural models, namely elastic (Chapter 4) and plastic (Chapter 5).

The final chapter of Part One, much longer than previous ones, tackles the engineering analysis of the response, i.e. the structural design problem. It therefore opens with a description of the limit states which are considered in the most recent formulations, then illustrates how to investigate them in a probabilistic context. The basic structural models of Chapters 4 and 5 must be completed (and complicated) to this purpose. The second half of Chapter 6 deals with structural design as a *decision* problem, and in particular illustrates (and criticizes) the *maximum utility* approach.

Part Two deals specifically with numerical and operative procedures. The first two chapters treat the analysis of statistical data for time-constant (Chapter 7) and time-varying (Chapter 8) quantities, respectively. Chapter 9 describes the computational procedures used for structural analysis in a probabilistic framework, and in particular the *simulation* procedures generically (and somewhat inappropriately) indicated as Monte Carlo methods.

Chapter 10 discusses how probabilistic approaches are being introduced in Structural Regulations and Codes of Practice, and the advantages thus obtained. Chapter 11, completing Part Two, deals with some open questions and basic problems about modelling of structural systems and the ensuing approximations, pointing out the improvements in practical results and in understanding that can be obtained thanks to the probabilistic viewpoint.

Finally, as anticipated, Part Three contains eight appendices. Appendices

A1–A7 cover the mathematical background needed for a full understanding of the subject matter of the whole book. However, rigorous proofs of many statements have been omitted, not only because of length, but mainly because they would have been beyond the scope of the book. Tables of the most used probability functions will be found in Appendix A8.

As for the choice of materials for classroom use, it may be noted that most chapters are almost self-contained, although some (Chapters 2, 7, 8) are intended only to furnish tools for successive use. Therefore, a few chapters might be used to complement more traditional courses (e.g. Chapter 4 for a course on Theory of Elastic Structures, Chapter 5 for a course of Plastic Limit Design, Chapter 9 for a course on Numerical Methods in Civil Engineering). A one-semester course on Probabilistic Structural Design, having the new Codes in mind, might follow the threadline of Chapters 6 and 10. These, however, are only some suggestions, because, as already stated, the main use of this book is not intended as a course textbook, but as a unified, self-contained reference book.

It is usual to close the Introduction to a book of this sort with a list of suggestions for further reading. Such a list always risks being criticized for its incompleteness, but all the more so in our case, because many interesting books, devoted in particular to either the more controversial aspects or the already established sectors of the field, have appeared in the last few years, and certainly more will appear between the time of writing and the time of actual publication. So we have decided to take the risk (we had to!), but present only a very short list of the books that we feel have been essential in our own study of the main aspects of our subject.

(a) On fundamental and methodological aspects

Benjamin, J. R. and Cornell, C. A. (1970) *Probability, Statistics and Decision for Civil Engineers,* McGraw-Hill, New York.

(b) On structural mechanics problems

Bolotin, V. V. (1969) *Statistical Methods in Structural Mechanics* (Translated from Russian), Holden-Day, San Francisco.

Lin Y. K. (1976) *Probabilistic Theory of Structural Dynamics,* Krieger, Huntington, New York (Reprint).

Clough, R. W. and Penzien, J. (1975) *Dynamic of Structures,* McGraw-Hill, New York.

(c) On structural engineering and structural design

Ferry-Borges, J. and Castanheta, M. (1971) *Structural Safety,* L.N.E.C., Lisbon.

Tichy, M. and Vorlicek, M. (1972) *Statistical Theory of Concrete Structures,* Irish University Press, Shannon, Ireland.

Construction Industry Research and Information Association (1977) *Rationalization of safety and serviceability factors in structural codes,* CIRIA, London.

Schuëller, G. I. (1981) *Einführung in die Sicherheit und Zuverlässigkeit von Tragwerken*, W. Ernst & Son, Berlin.

G.A., A.B., F.C.

ACKNOWLEDGEMENT

The Authors wish to acknowledge that many of the results included in this book were obtained in the course of researchers financially supported by the Italian National Research Council (C.N.R.), through the Research Groups for 'Advanced Problems in Structural Engineering' and for 'Steel Constructions'.

Thanks are also due to the publishers who have granted permission to reproduce material from copyrighted publications, which is acknowledged at appropriate places in the text.

Glossary of symbols and abbreviations

Symbols in brackets (and in general symbols in **bold** type) indicate vectors or matrices (as opposed to scalars); other symbols and/or meanings are specified where employed; [ ] = meaning assumed in Appendices. Symbols in this list reproduce exactly those used in the text, not in the figures

LOWER-CASE LETTERS

a	sample central moment [lower bound of an interval; threshold level]	a	ground acceleration	$\mathscr{a}$	pseudo-acceleration
b	bandwidth [upper bound of an interval]	b	estimator of $B(\ )$	$\mathscr{b}$	binomial law
c ($\mathbf{c}$)	real number; ratio between load components	c ($\mathbf{\mathsf{c}}$)	damping constant or matrix	$\mathscr{c}$	ratio
d ($\mathbf{d}$)	duration, generic quantity	d	distance	$\mathscr{d}$	duration (of an earthquake)
e ($\mathbf{e}$)	integral number; eccentricity; generic quantity	e	estimator of the coefficient of variation		
f	relative (statistical) frequeny	f ($\mathbf{\mathsf{f}}$)	modal coordinates	$\mathscr{f}$	[failure]
g ($\mathbf{g}$)	generic function; acceleration due to gravity	g	estimator of $G(\ )$		
h	height; real number	h ($\mathbf{h}$)	impluse response function	$\mathscr{h}$	snow height; cost of saved life
i	integral number; index				

j	integral number; index				
k	integral number; index				
l	length; integral number; index			ℓ	linear operator
m	integral number	m	sample mean	$\mathscr{m}$	median or mode
n	integral number; counting process	n	sample size		
		o ($\mathbf{o}$)	infinitesimal quantity		
p	probability density	p ($\mathbf{p}$)	probability of a given event	$\mathscr{p}$	pressure (wind pressure)
q ($\mathbf{q}$)	generic input/output variable or function generally with physical meaning	q	probability $(= 1 - \mathsf{p})$		
r	number (run number)	r	sample linear correlation coefficient; radius of gyration	$\mathscr{r}$	core radius of section
s	standard deviation	s	sample standard deviation	$\mathscr{s}$	[success]
t ($\mathbf{t}$)	time, independent variabile	t	Student's ratio		
u ($\mathbf{u}$)	output variable or function; displacement [real number]	u	standardized normal variable	$\mathscr{u}$	pseudo-random number
v	time interval	v ($\mathbf{v}$)	wind velocity	$\mathscr{v}$	number of lives; pseudo-velocity
w	load per unit area [time of occurrence]	w ($\mathbf{w}$)	active acceleration	$\mathscr{w}$	Fisher's function
x ($\mathbf{x}$)	generic mathematical variable or function; design parameter	x ($\mathbf{x}$)	abscissa; coordinate		
y ($\mathbf{y}$)	generic mathematical variable or function	y	ordinate; coordinate		
z ($\mathbf{z}$)	generic mathematical variable or function	z	coordinate	$\mathscr{z}$	sum of random variables

CAPITAL LETTERS

A	cross-sectional area	A ($\mathbf{A}$)	frequency response function	$\mathscr{A}$	decision region
B	(correlation) covariance function	B	bias [Borel set]	$\mathscr{B}$	family of Borel sets
				$\mathbb{B}$	benefit
C ($\mathbf{C}$)	generic quantity	C	transient live load	$\mathscr{C}$	boundary of the safe region
D ($\mathbf{D}$)	generic quantity; sample range	D ($\mathbf{D}$)	structural matrix	$\mathscr{D}$	space of integration
E	Young's modulus	E	event	$\mathscr{E}$	error
F	cumulative frequency	F	[F-distribution]	$\mathscr{F}$	[family subset of S]
G	spectral density function	G ($\mathbf{G}$)	elastic law coefficient		
H ($\mathbf{H}$)	cost; load effect	H	Heaviside function		
I	influence surface	I ($\mathbf{I}$)	identity matrix	$\mathscr{I}$	earthquake intensity
J	cross-sectional moment of inertia	J ($\mathbf{J}$)	Jacobian		
K ($\mathbf{K}$)	stiffness	K ($\mathbf{K}$)	generic quantity	$\mathscr{K}$	set of the system quantities
L	length	L ($\boldsymbol{L}$)	work; structural matrix, likelihood, losses	$\mathscr{L}$	non-linear operator
M ($\mathbf{M}$)	bending moment	M ($\mathbf{M}$)	mass	$\mathscr{M}$	earthquake magnitude
N ($\mathbf{N}$)	axial force	N	integral number		
P	cumulative probability distribution	P ($\mathbf{P}$)	probability of defined event		
Q ($\mathbf{Q}$)	parameter or function with physical meaning, point	Q ($\mathbf{Q}$)	compatibility term		
R	resistance	R	radius, distance	$\mathscr{R}$	sample description space of a numerical valued phenomenon
				$\mathbb{R}$	reliability
S ($\mathbf{S}$)	action effect homogeneous with R	S	response spectrum [set symbol]	$\mathscr{S}$	success (safe) region; [sample description space]

T	period; lifetime; length of record	T	return period	$\mathscr{T}$	stress field (set)
U ($\mathbf{U}$)	real number	U	generic quantity [response function of a FPP]	$\mathscr{U}$	unsafe region
				$\mathbb{U}$	utility
V ($\mathbf{V}$)	spectral distribution function	V ($\mathbf{V}$)	volume	$\mathscr{V}$	collapse mechanism (set)
W ($\mathbf{W}$)	load, action, input variable or function			$\mathscr{W}$ ($\boldsymbol{\mathsf{W}}$)	load impulse
X ($\mathbf{X}$)	generic mathematical quantity	Y ($\mathbf{Y}$)	reduced stiffness matrix		
Y ($\mathbf{Y}$)	generic mathematical quantitity	Z ($\boldsymbol{\mathsf{Z}}$)	reduced damping matrix	$\mathscr{Z}$	load domain
Z ($\mathbf{Z}$)	performance function; generic quantity				

GREEK LETTERS

α ($\boldsymbol{\alpha}$)	load factor
β	reliability index
γ ($\boldsymbol{\gamma}$)	safety factor
δ	coefficient of variation
ε ($\boldsymbol{\varepsilon}$)	strain; arbitrarily small quantity
ζ ($\boldsymbol{\zeta}$)	generic quantity
η	population parameter; Poisson probability law
θ ($\boldsymbol{\theta}$)	rotation, angle
λ	intensity of a Poisson process
λ_B	slenderness
λ_P ($\boldsymbol{\lambda}_P$)	plastic multiplier
μ ($\boldsymbol{\mu}$)	mean value
ν	frequency; mean crossing rate
ξ	damping ratio; envelope process
ρ ($\boldsymbol{\rho}$)	coefficient of correlation
σ ($\boldsymbol{\sigma}$)	stress
τ	time; time interval
υ	white-noise
ϕ ($\boldsymbol{\phi}$)	generic function; phase; load–deformation law; yield function
χ ($\boldsymbol{\chi}$)	chi for the χ square; strength parameters
ψ ($\boldsymbol{\psi}$)	combination factor
ω ($\boldsymbol{\omega}$)	angular frequency

Γ	gamma function
Δ ($\boldsymbol{\Delta}$)	variation
Θ ($\boldsymbol{\Theta}$)	generic quantity
Λ ($\boldsymbol{\Lambda}$)	generic matrix
Ξ ($\boldsymbol{\Xi}$)	[transition intensity]
Π ($\boldsymbol{\Pi}$)	probability
Σ ($\boldsymbol{\Sigma}$)	covariance matrix; correlation functions
Υ ($\boldsymbol{\Upsilon}$)	surface drag coefficient
Φ	normal distribution
Ψ ($\boldsymbol{\Psi}$)	diagonal damping matrix; [moment generating function]
Ω ($\boldsymbol{\Omega}$)	diagonal stiffness matrix; [characteristic function]

ABBREVIATIONS AND SPECIAL SYMBOLS

(a) *Abbreviations*

Prob[ ]	probability that	CDF	cumulative distribution function
freq[ ]	frequency of	PSD	power spectral density
r.v.	random variable	JPDF	joint probability density function
r.p.	random process	CPF	conditional probability of failure
p.p.	point process	FPP	filtered Poisson process
r.h.s.	right-hand side		
l.h.s.	left-hand side		
PMF	probability mass function		
PDF	probability density function		

(b) *Functions*

E[ ]	expected value	exp	exponential
Var[ ]	variance	ln	natural logarithm
Cov[ ]	covariance	log	logarithm
Det	determinant	Re	real part
tan	tangent function	Im	imaginary part
sin	sine function	erf	error function
cos	cosine function		

(c) *Accents*

~	random	·	derivation on time
\	sample value	"	estimated

(d) *Constants*

i	imaginary unit $= \sqrt{(-1)}$	$\Rightarrow$	implies
e	basis of ln = 2.718281	$\Leftrightarrow$	if and only if
π	3.14159	$\simeq$	approximately equal
:	such that		

(e) *Symbols*

$\in$	belonging to	$\dagger$	complex conjugate
$\notin$	not belonging to	$\sum$	summation
$\mid$	on condition that	$\prod$	product
$\subset$	subset of	d	differential
$\cup$	union	$\partial\ /\partial$	partial derivative
$\cap$	intersection	Δ	finite variation; increment
$\varnothing$	empty set	∂	small quantity; Dirac function; variation
^{T}	transpose	$\exists$	existence of
$*$	standardized		
^{c}	complement		

CHAPTER 1

Theory of structures and theory of probability

In this introductory chapter the most usual approaches and problems of structural analysis will be presented in a form that will allow further treatment from a *probabilistic* point of view

1.1 STRUCTURAL SYSTEMS

In very general terms, the 'structures' on which the engineers' analyses are developed can be regarded as *filtering systems*, and the quantities which appear in structural engineering broadly distinguished as three different types, namely:

(1) *input quantities* that act on the filter;
(2) *system quantities* that define the filter; and
(3) *output quantities* that measure the filter response.

A structural engineer would perhaps prefer to call these three types respectively:

(1) *actions* or *loads* (also indicated in the following as *environmental quantities*);
(2) *structural properties* (or sometimes *technological quantities*); and
(3) *structural response parameters.*

For instance, consider the cantilever bent in Fig. 1.1(a), which is loaded by a horizontal force in the corner point B and a vertical force in an arbitrary point D of the horizontal element BC: both forces and the point D can vary with time t. The *input quantities* (*actions*) are the intensities W_1, W_2 of the two forces and the distance $d = \mathrm{BD}$, which can be defined through a vector function $\mathbf{W} = \mathbf{W}(t)$, whose three components are W_1, W_2 and d.

The *system quantities* are the geometrical dimensions of the bent (l, h) and its cross-section(s), the restraint conditions, the mechanical properties of the material (modulus of elasticity E, yield stress σ_y, inelastic parameters, mass M, damping c, etc.), . . . : the set of these quantities will be in general indicated by $\mathscr{K}$.

The *output* (*response*) can be characterized through many component quantities, for instance (Fig. 1.1(b)) the bending moment in A, the horizontal displacement of B, the vertical displacement and rotation of C, etc. Also, these quantities can be collected into an n-component (vector) response function which will generally be indicated by $\mathbf{u} = \mathbf{u}(t)$; in the limit as $n \to \infty$, the response is measured by functions $u_k(x, t)$ of time t and a continuous geometrical parameter (abscissa) x. Throughout this book, for the sake of simplicity and uniformity, input, output and system quantities will in general be presented as discrete in space and continuous in time, i.e. in the already presented form of vector functions of time $\mathbf{W}(t)$ and $\mathbf{u}(t)$. The generalization to spatial continuity does not present any conceptual difficulty, but may involve formidable mathematical complications.

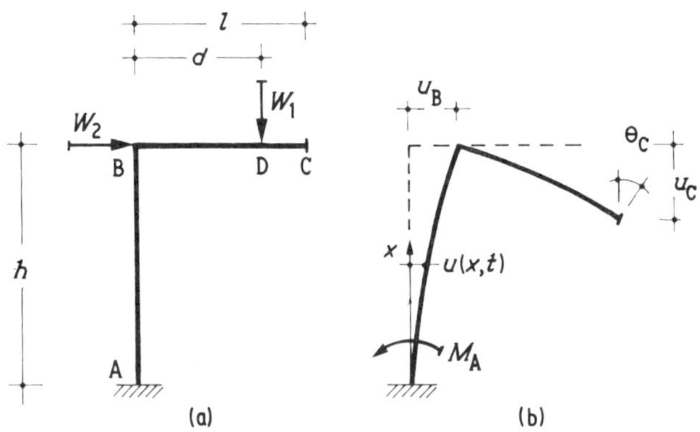

Fig. 1.1 Example structure.

In structural analysis, usually *input* and *system* quantities, $\mathbf{W}$ and $\mathscr{K}$, are considered as the *data* for the problem: the solution consists of the determination of the *response* (*output*). Symbolically,

$$\mathbf{u}(t) = \mathbf{u}[\mathbf{W}(t), \ \mathscr{K}(t)] \tag{1.1}$$

If the function on the right-hand side (r.h.s.) of equation (1.1) is defined analytically, the filter admits a 'closed-form' solution. Otherwise, (1.1) indicates a numerical procedure that allows the determination of the output from the data.

In some design problems, it is not possible to distinguish between *input* and *system* quantities as neatly as in the example of Fig. 1.1.

1.2 UNCERTAINTY AND RANDOMNESS

Structural analysis may be performed with reference to:

(1) existing structures; or
(2) structures being designed.

Both cases, although at first sight so different from each other, can be discussed together. In fact, in both cases, the 'knowledge' of the data (input and system quantities) is imprecise and uncertain to some extent.

The loads that are acting and/or will act on a structure are never known exactly. They can be estimated from records of loads on similar constructions, but two such records will never be identical. The uncertainty can be reduced, but never eliminated, by accurate measures of dead loading and severe controls on live loads (which are possible only in special cases and, even then, often only on the maximum values and not on the whole 'history').

Also, the system quantities are subject to uncertainties, although these may be smaller than for the inputs. In an existing structure, some geometrical quantities (like member lengths, angles, etc.) may perhaps be measured with sufficient accuracy, but others (like cross-sectional dimensions, placement of bars in reinforced concrete, etc.) can be measured only in a limited number of points, but may vary in between. The mechanical and 'strength' properties of the material, besides varying from point to point, are seldom prone to certain *in situ* measurements, but only to indirect 'estimates' (e.g. the resistance of concrete cubes or cylinders, cast in special forms, is assumed to be a reliable estimate of concrete strength in the actual structure).

A quantity that is subject to any of the uncertainties exemplified above is said to be a *random quantity*: usually, it is called a *random variable* (r.v.) if it does *not* vary in time, and a *random process* (r.p.) or *stochastic process* if it may vary with time (or with some other independent parameter). Thus, the *random variable* (which is an inappropriate, but much-used, locution) is a special *random process*. A random quantity cannot be defined by a single measure or function: in fact, every time it is recorded, a different *sample function* is obtained. Each sample function is a particular *realization* of the r.p.; the set of all possible sample functions constitutes the random process, which therefore can only be *described* through its *probability properties*, which allow estimates or forecasts within some degree of confidence.

In opposition to random quantities, a quantity whose random uncertainty is negligible is called *deterministic*. In this case, all *sample functions* are identical to each other.

If *input* and/or *system* quantities are random, their randomness is reflected, to a greater or lesser extent, in the *output* quantities. Therefore, *probabilistic structural analysis* consists of applying equation (1.1) to determine the probabilistic properties of the output *random process* $\tilde{\mathbf{u}}(t)$‡ from the probabilistic properties of

‡ Throughout this book, a superimposed tilde (˜) will indicate *randomness* of the relevant quantity. It is sometimes omitted for simplicity, if not confusing.

the data given as random quantities and from the values of the data given as deterministic quantities.

Another source of uncertainties in the output process lies in the formulation of the relationship symbolically represented by equation (1.1), which is a mathematical (analytical or numerical) model of reality, and therefore necessarily approximate. This problem will be ignored for the moment, and taken up again in Chapter 11.

1.3 LIMIT STATES AND RELIABILITY

Once the response of a structural system, be it deterministic or random, has been calculated through structural analysis, it must be judged from the point of view of performance and safety. This judgement may be expressed in different forms according to the type of response parameter under consideration. For instance, if a displacement or acceleration component is examined, it would be of interest to know how often and for how long it goes beyond any given value (and, in particular, beyond uncomfortably high values) during the lifetime of the structure.

It is evident that, if the response is a random process, the answer to this question can be given only in probabilistic terms, i.e. trying to quantify the time during which the relevant parameter is likely to overcome each given value. The description of a random process in this way is formally very cumbersome; this approach is therefore not dealt with in this book, except in special cases (see Chapter 6).

However, the response judgement procedure in structural engineering is often formulated in a somewhat simpler form, which will be accepted in the following. Namely, a structure is said to reach a *limit state* when a specific response parameter attains a *threshold* value. Several possible limit states can be defined (e.g. excessive displacement, first crack, plastic collapse, etc.). Some of these must absolutely be avoided, as they imply the ruin of the structure (and consequent danger to human lives), whereas others can be accepted although inconvenient.

To fix ideas in the context of this book, a specific example of a specific limit state will now be considered. Practical formulations and appraisals of alternative limit states is beyond our scope, but some general principles will be discussed later in Chapter 6.

Consider again the bent of Fig. 1.1. It survives if, and only if, the bending moment in the built-in section A (and, therefore, in all other sections) is smaller than an ultimate limit moment M_{Ay}:

$$W_1 d + W_2 h \leqslant M_{Ay} \tag{1.2}$$

The equality sign in inequality (1.2) corresponds to the attainment of the *collapse threshold* of the bent. In general, for each limit state, it is possible to formulate a *critical inequality* like (1.2) and identify, in the space of the relevant quantities, a *safe region* (or *success region*) $\mathscr{S}$, where the critical inequality holds, and an *unsafe region* (or *failure region*) $\mathscr{U}$, where it does not hold. In the specific simple example, *success region* and *unsafe region* can be represented as shown in Fig. 1.2 (a), where:

$$S = W_1 d + W_2 h$$
$$R = M_{Ay} \tag{1.3}$$

(it is assumed that $S \geqslant 0$).

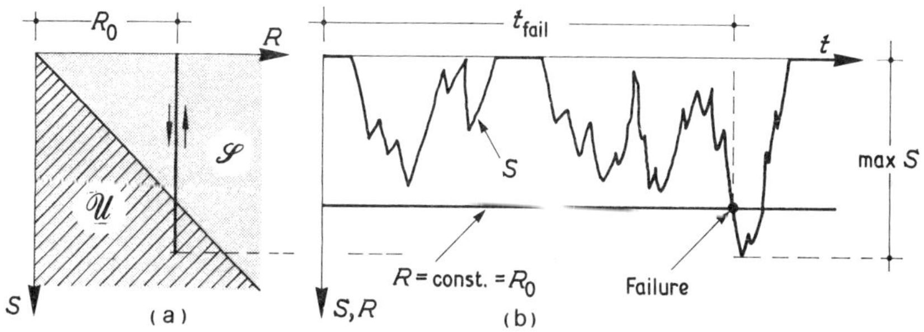

Fig. 1.2 (a) Example of safe region; (b) realization of $\tilde{S}(t)$.

Because of the variability and randomness of the input quantities, already discussed, $S = S(t)$ is in most cases a *random process*. The resistance R, which is calculated according to any appropriate assumption as to the material behaviour (and depends on yield stress and cross-sectional dimensions), is also in general a random quantity but in many cases can be considered constant in time (*random variable*).

Therefore, for each actual structure, the resistance takes up a constant, although uncertain, value R_0; and the representative point (R, S) moves in time up and down the solid line in Fig. 1.2 (a). In Fig. 1.2 (b) a possible realization of the random loading process $\tilde{S}(t)$ is shown: the limit state is attained when $S(t)$ violates the threshold R_0 (that is, when the point (R, S) reaches the boundary of the unsafe region in Fig. 1.2 (a)). The time to failure (with respect to the limit state under consideration) $\tilde{t}_{fail}$ would be the best measure of the reliability of the structure, and therefore the most appropriate descriptive property of the random process $\tilde{S}(t)$.

Some structural problems formulated in this way will be treated in Chapter 6. However, in structural problems of civil engineering, a formally different treatment is often adopted. Namely, for any time interval $(0, t)$, the critical inequality is checked only in the *worst possible* condition. This approach, which is particularly significant when the *design life* of the structure is predetermined, is sometimes necessary because of the form in which the data are available (e.g. as the *largest* load in a given time interval). It can be formulated in probabilistic terms by stating that the probability of failure (with respect to the limit state under consideration) P_{fail} and the complementary probability of success *reliability* $\mathbb{R} = P_{\text{suc}}$ coincide respectively with the probability that the critical inequality is violated at least once in the interval $(0, t)$, and with the probability that it holds at all instants in that interval. For the simple illustrative example,

$$\begin{aligned} P_{\text{fail}} &= \text{Prob}[\tilde{S} > \tilde{R} \text{ at least once in } (0,t)] = \text{Prob}[\max_t \tilde{S} > \tilde{R}] \\ P_{\text{suc}} &= 1 - P_{\text{fail}} = \text{Prob}[\tilde{S} \leqslant \tilde{R} \text{ always in } (0,t)] = \text{Prob}[\max_t \tilde{S} \leqslant \tilde{R}] \end{aligned} \tag{1.4}$$

where $\max_t \tilde{S}$ is a random variable, defined as the largest value attained by the random process $\tilde{S}(t)$ in the interval $(0, t)$.

The above formulations of the basic structural engineering problem of safety evaluation can be extended, without great conceptual difficulty, to more general cases. For example, the 'load effect' random process $\tilde{S}(t)$ may be undefined in sign, and in this case the possibility of violation of either a positive threshold $\tilde{R}^+$ or a negative threshold $\tilde{R}^-$ (of equal or different absolute value) must be considered. In other instances, the 'strength' may vary (usually, decrease) in time, $\tilde{R} = \tilde{R}(t)$, and this variation be independent (*aging*) or dependent (*fatigue*) on the loading process $\tilde{S}(t)$.

When all these possibilities are considered, and the structural reliability problem reduced to its essence, it is still seen that it can be tackled in either of the two formally different ways that have already been mentioned: namely, through the probabilistic description of the time to failure, or the evaluation of the probability that, in a space of random variables, a point $\tilde{\mathbf{Q}}$, which represents the significant input and system parameters, falls either in the *failure region* $\mathscr{U}$ or in the *success region* $\mathscr{S}$:

$$P_{\text{fail}} = \text{Prob}[\tilde{\mathbf{Q}} \in \mathscr{U}] \qquad P_{\text{suc}} = \text{Prob}[\tilde{\mathbf{Q}} \in \mathscr{S}] \tag{1.5}$$

1.4 ALTERNATIVE MODES OF FAILURE

A structure may in general fail according to several alternative 'modes'. For instance, the bent in Fig. 1.1 can collapse because the limit moment is reached in any of its sections (and if these are not assumed to be equal, each may be a 'dangerous' section) or also because of in-plane or out-of-plane instability. It is well known in structural engineering that all potentially dangerous modes of failure must be taken into consideration, and that, if some modes are overlooked, an unsafe result can be achieved. In the terminology of probabilistic structural analysis, the probability of failure P_{fail} is the probability that the considered structure fails in at least one of the relevant modes: if the event *failure according to the* i*th mode* is indicated by E_i, then (Appendix A1):

$$P_{\text{fail}} = \text{Prob}[E_1 \cup E_2 \cup E_3 \cup \ldots] \tag{1.6}$$

where the *union* operation indicated by the symbol $\cup$ must be performed on all relevant modes. Indeed, if one is not sure that all modes have been considered, the calculated P_{fail} is only a lower (i.e. *unsafe*) bound to the actual probability of failure. This point will be taken up again in Chapter 5, where it will be shown that, under some restrictive conditions, an upper (i.e. *safe*) bound to P_{fail} can also be calculated.

As for the actual determination of P_{fail}, note first that

$$P_{\text{fail}} \geqslant \max_i P_{\text{f}i} \tag{1.7}$$

where $P_{\text{f}i}$ is the probability calculated for the event E_i,

$$P_{\text{f}i} = \text{Prob}[E_i] \tag{1.7'}$$

Then, in accord with equation (A1.5),

$$P_{\text{fail}} = P_{\text{f}1} + P_{\text{f}2} + \ldots - \text{Prob}[\text{failure in two or more modes}] \leqslant \sum_{i=1}^{n} P_{\text{f}i} \tag{1.8}$$

Note explicitly that a structure cannot actually fail by two different modes, but two events E_i and E_j are not necessarily *mutually exclusive* in the sense used in Appendix A1. In fact, E_i and E_j are evaluated *a priori*, and therefore it is not to be excluded that the structure be prone to failure in both modes, i.e. that

$$\text{Prob}[E_i \cap E_j] > 0 \tag{1.9}$$

In general, the two bounds on P_{fail} given by inequalities (1.7) and (1.8) are too wide for practical applications. However, if all E_i are *stochastically independent* from each other, equation (A1.11) yields

$$\text{Prob}[E_i \cap E_j] = P_{\text{f}i} P_{\text{f}j} \tag{1.10}$$

If moreover all relevant probabilities of failure are very small, i.e.

$$P_{\text{fail}} \ll 1 \qquad \text{and} \qquad P_{\text{f}i} \ll 1 \qquad \text{for all } i \tag{1.11}$$

equation (1.8) becomes to good approximation

$$P_{\text{fail}} \simeq \sum_{i=1}^{n} P_{\text{f}i} \tag{1.12}$$

Since in practice many evaluations of structural safety are explicitly or implicitly based on equation (1.12), it is worth spending some time in noting its limitations. First of all, it must not be forgotten that *stochastic independence* of the failure modes is a prerequisite; secondly, that it is based on a *discrete* set of n failure modes.

A good illustrative example of equation (1.12) is the probability of failure, under a given load W, of the *chain* diagrammatically depicted in Fig. 1.3 (a), provided the n elements have been chosen at random in independent trials, so that their resistances can be assumed to be stochastically independent of each other. In this example, the n independent failure modes are the failures of each of the n chain links, and

$$P_{fi} = \text{Prob}[W > \sigma_{yi} A_i] \tag{1.13}$$

where σ_{yi} and A_i indicate the yield stress and cross-sectional area of the ith link. In particular, if all links have been taken from the same *population* P_{fi} is independent of i, and (1.12) becomes

$$P_{fail} = nP_{fi} \tag{1.14}$$

Remember, however, that, because of (1.11), equations (1.12) and (1.14) are valid only if n is sufficiently small: otherwise, the joint probability term in (1.8) is not negligible. This becomes evident if one notes that, no matter how small each P_{fi} is, equations (1.12) and (1.14) tend to infinity with n, while P_{fail} is a probability and therefore, by definition, cannot be larger than unity (cf. Appendix A1).

The chain in Fig. 1.3 (a) can represent a *discrete* (*finite element*) model of a continuous tie. In this case, problems may arise not only from the number n of elements, but also by the fact that σ_{yi} and A_i, although they may vary randomly along the tie axis, are certainly related to each other unless the sections considered are sufficiently far apart. Therefore, even this extremely simple example of probabilistic structural analysis appears to involve delicate engineering choices. The number of finite elements must be large enough so that the sum in equation (1.12) is not truncated too early (and, in more complicated problems, the structural solution is sufficiently approximate); but, on the other hand, each element must be so large that its resistance can be considered as stochastically independent of the resistance of the adjoining elements.

This sort of problem, although sometimes overlooked, can indeed be the most difficult point in probabilistic structural analyses, and can impair otherwise correct calculations.

The same problem discussed above with reference to the chain of Fig. 1.3 (a) can indeed be presented with reference again to the bent of Fig. 1.1. Assume that the vertical member AB, instead of having a constant cross-section as implicitly assumed so far, has been designed such that in each section X, the nominal limit moment is given by

$$M_y(x) = \gamma^0 [W_{10} d_0 + W_{20} (h - x)] \tag{1.15}$$

where (Fig. 1.3(b)) W_{10}, W_{20}, d_0 are reference values of the random variables $\tilde{W}_1$, $\tilde{W}_2$, $\tilde{d}$, and γ^0 is a constant (safety coefficient). In this case, each cross-section X is potentially dangerous to the same degree, and the relevant collapse modes are infinite.

In conclusion, one can say that two errors of opposite sign are introduced when using equation (1.12). First, neglecting the joint probability term in equation (1.8), one tends to overestimate P_{fail}: but the neglected term can be of the same order of magnitude as the calculated P_{fail} if the number of terms in the sum is large and/or the events E_i are strictly correlated. On the other hand, to limit the number of significant modes to n implies an underestimation of P_{fail}, which can be significant

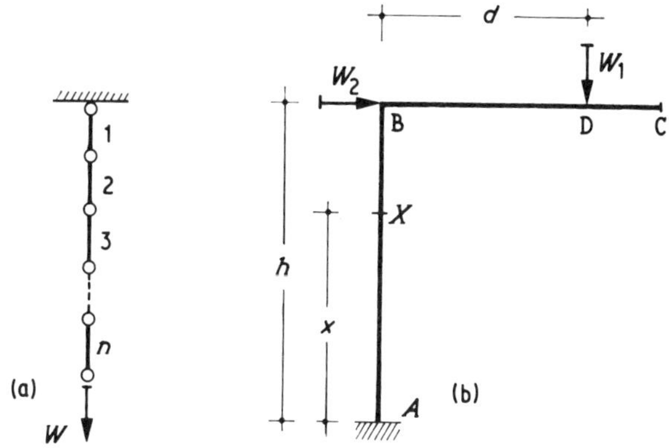

Fig. 1.3 Structural brittle behaviour.

if not all the *most dangerous* modes (i.e. the modes with the largest P_{fi}) have been considered. Sometimes these two errors compensate each other, and equation (1.12) yields quite satisfactory results: sometimes they do not, and (1.12) is just misleading.

1.5 'WEAKEST-LINK' AND 'FAILSAFE' STRUCTURES

Structures like those shown diagrammatically in Fig. 1.3 fail as soon as one of the elements (or sections) fails: this type of behaviour is usually defined as *weakest-link* behaviour, and the probability of failure is related to the probability of failure of each component by the additive formulae (1.8)–(1.14).

Weakest-link behaviour (in which terminology, 'link' may be understood to indicate any potential mode of failure) applies to the ultimate limit state of statically determinate structures, but also to other cases, such as brittle failure of any structure, or to serviceability limit states (deformation, cracking) which essentially coincide with the end of the linear behaviour. More generally, the additive formulae apply whenever failure occurs as soon as at least one of the potential modes of failure is activated. The care necessary when a continuous structure is discretized into finite elements has already been underlined in Section 1.4.

In other types of structures, usually denoted as *failsafe* structures, when the limit condition is reached in a point or element, a redistribution of stresses takes place, so that the structure is able to resist further loads. In these cases, contrary to (1.7),

$$P_{fail} \leqslant \max_i P_{fi} \qquad (1.16)$$

where P_{fi} now indicates the probability of the *i*th component reaching the limit condition.

Typical instances of failsafe behaviour are given by redundant ductile structures, a limit example being the bundle of ductile elements shown diagrammatically in Fig. 1.4(a): it is easy to show that the random variability of the limit load of the bundle decreases indefinitely with the number *n* of elements.

Fig. 1.4(b) shows a structure with one statical redundancy: the redistribution of stresses which takes place after, say, the compressed rod CE buckles, is obvious: the behaviour of redundant structures, and in particular of plastic structures, will be taken up again, in a more formal way, in Chapter 5. In this introductory chapter, suffice it to say that, in general, in *weakest-link* behaviour the overall probability of failure (and usually the random variability of the strength) is larger than the local values, while in failsafe behaviour the reverse may be true.

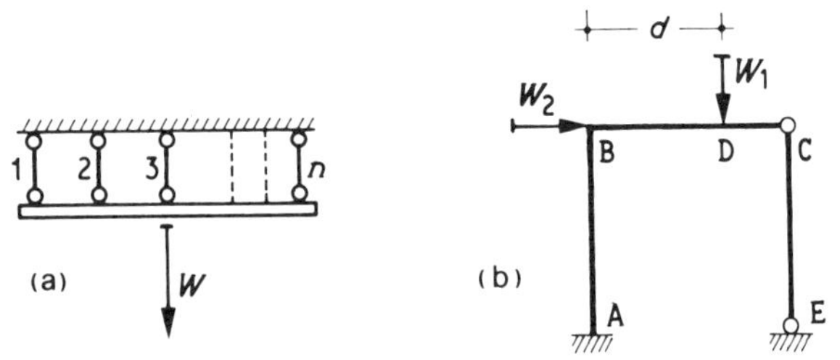

Fig. 1.4 Failsafe behaviour.

Sound design is often identified with provisions for failsafe behaviour. This is confirmed by the probabilistic approach, but it should not be forgotten that the same overall P_{fail}, which is after all the ultimate aim of the structural engineer, can be achieved in other ways than by increasing structural redundancy.

1.6 PROBABILITY DIAGRAMS AND 'DESIGN PARAMETER SPACE'

Let us consider a structure with given geometry and materials. To *design* this structure consists of choosing a set of (nominal) values for a number of quantities x_i, which can be denoted as *structural design parameters* (typically, the cross-sectional dimensions). The actual values of the design parameters are, in general, random variables; they must be chosen in such a way that the probability of failure of the structure is sufficiently low.

Consider the (real) space of the design parameters x_i, which by definition can assume only positive values (and are *system* quantities, according to the definition of Section 1.1). For each given load condition **W** there exists a surface $\mathscr{C}_{\mathbf{W}}$ separating the structures able to withstand the loads **W** from the structures failing under **W**, i.e. the *safe* and *unsafe* regions corresponding to **W**, say $\mathcal{S}_{\mathbf{W}}$ and $\mathcal{U}_{\mathbf{W}}$. The surface $\mathcal{C}_{\mathbf{W}}$ is, in general, an indefinite surface which, for obvious physical reasons, does not intersect the coordinate axes.

To obtain $\mathcal{C}_{\mathbf{W}}$ and $\mathcal{U}_{\mathbf{W}}$ is quite easy in the case of *weakest-link* behaviour, provided the surfaces $\mathcal{C}_i$ and regions $\mathcal{U}_i$ corresponding to each potential failure mode are known. In fact, from equations (1.5) and (1.6), as indicated diagrammatically in Fig. 1.5,

$$\mathcal{U}_{\mathbf{W}} = \mathcal{U}_1 \cup \mathcal{U}_2 \cup \ldots \cup \mathcal{U}_n \tag{1.17}$$

and $\mathcal{C}_{\mathbf{W}}$ is the envelope of all $\mathcal{C}_i$.

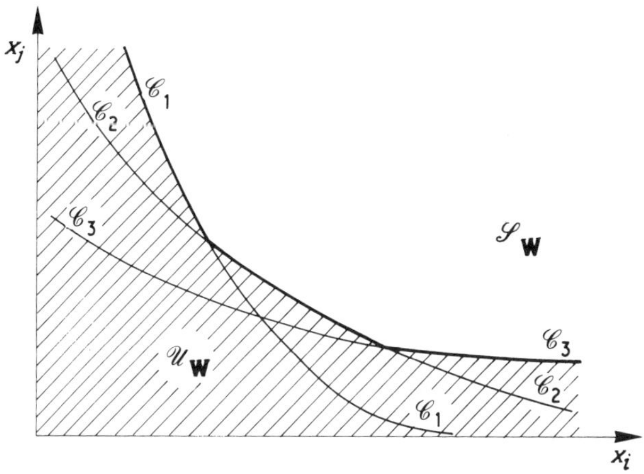

Fig. 1.5 Diagrammatic failure surface and unsafe region in design parameter space.

A self-explanatory example of such a diagram is shown in Fig. 1.6, which refers to the structure of Fig. 1.3(b), under the assumption that the only two possible failure modes correspond (1) to yielding of section B, and (2) to yielding of section A. As design parameters, the limit moments M_y^{AB} and M_y^{BC} of beams AB and BC (assumed constant along each beam) have been chosen.

Diagrams of the type shown in Figs 1.5 and 1.6 are useful in two respects. First, they put in evidence which is the relevant failure mechanism in each region of the design parameter space. Secondly, equation (1.17) makes inequalities (1.7) and (1.8) physically evident: in fact, indicating by **x** the *design point* of coordinates $\{x_i\}$ (cf. also the example in Fig. 1.6)

$$P_{fi} = \text{Prob}[\mathbf{x} \in \mathcal{U}_i] \tag{1.18}$$

and, from elementary probability (Appendix A1)

$$\text{Prob}[\mathbf{x} \in \mathcal{U}_i \quad \text{for all } i] \leqslant \text{Prob}[\mathbf{x} \in \mathcal{U}_{\mathbf{W}}] \leqslant \sum_{i=1}^{n} \text{Prob}[\mathbf{x} \in \mathcal{U}_i] \tag{1.19}$$

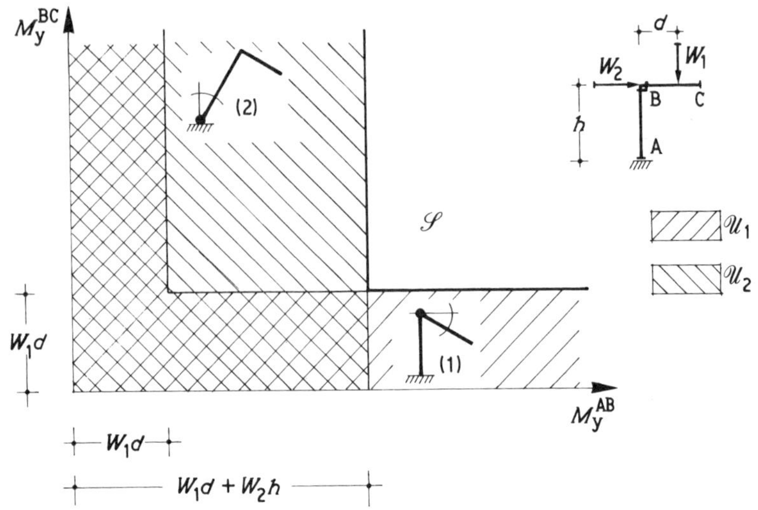

Fig. 1.6 Failure surfaces and unsafe region for example structure.

1.7 SIGNIFICANCE OF RELIABILITY CALCULATIONS IN STRUCTURAL ENGINEERING

It seems appropriate to conclude this chapter with some considerations of a general nature.

First, note that the acceptable probabilities of failure P_{fail} are, in civil engineering, very low: their order of magnitude, as rough quantitative indications, should be $P_{fail} = 10^{-3}$ for limit states which do not endanger lives, and $P_{fail} = 10^{-5}$–10^{-6} for disastrous limit states. The corresponding reliabilities P_{suc} would then be 99.9% or 99.9990–99.9999%.

These figures mean that, on average, out of 1000 nominally identical buildings one will crack or deform excessively, and out of 1 000 000 a number between 1 and 10 will collapse. Now, it is evident that in civil engineering '1000 identical buildings' (let alone 1 000 000) rarely occur . . . even neglecting the fact that a statistically significant test would require samples at least 10 or 20 times larger! Moreover, the determination of these low probabilities requires, as will become evident in the following, extrapolations (sometimes drastic) of statistical properties that are experimentally known (or can be obtained) only around the *central* values of the random quantities.

For these reasons, the 'probabilities of failure' which can be calculated or estimated in civil engineering have no real *statistical* significance (Section 2.1): rather, they are conventional, comparative values. Provided this point is clearly understood and accepted, probabilistic methods can play a very important role in making rational comparisons possible between alternative structural designs. Otherwise, they are vulnerable to all sorts of criticisms.

In this framework, *approximate* procedures that allow consistent comparisons of reliabilities appear just as valid as the procedures formally aimed at the calculation of reliability values: the choice between them is in most practical cases be determined by the relative simplicity of the calculations and by the availability of the necessary stochastic input data.

Two 'levels' of approximate reliability calculations have been formulated in recent years, mainly for use in structural codes of practice. In the latter context, they will be treated in detail in Chapter 10: however, their basic formulation will be presented earlier in Section 2.8, because many analytical procedures of probabilistic structural analysis have been developed with specific reference to them.

PART ONE

Analytical Methods

CHAPTER 2

Random quantities and structural reliability

After an introductory section on the relation between (empirical) frequency and (axiomatic) probability, the properties and characteristics of random processes and random variables, qualitatively introduced in Chapter 1, are briefly presented and discussed: thus, the stage is set for their use in structural problems. The chapter ends with the basic formulations of probabilistic safety analysis.

2.1 'FREQUENCY' AND 'PROBABILITY'

Probability theory (Appendix A1)[‡] is a self-consistent logical construction whose basic element (the *probability of an event*) is axiomatically defined and cannot be measured. Such an idealization would be useless in engineering or physical applications if no key were available to enter in it. The first idea for such a key is to introduce the experimental *statistical frequency* of the event (Chapter 7) as an empirical *measure* of its probability, measure that becomes more and more exact as the size n of the experimental *sample* increases. However, as it will be shown in Chapter 7 and Appendix A4, there is a proof that when $n \to \infty$ the probability of a difference between *statistical frequency* and *probability* becomes very small, but not that it vanishes. In other words, the identification of *frequency* and *probability* is not rigorously possible, not even in the limit for $n \to \infty$. Nevertheless, human experience teaches that a phenomenon with probability very close to one is almost certain to take place, while an event whose probability of occurrence is very close

‡ The theory of probability and random processes is systematically presented in Appendices A1 to A8 where many mathematical proofs are also presented. For the reader's convenience, throughout the book frequent citations refer to the section of Appendices where more detailed and/or rigorous treatment can be found, or to the relevant equations: these citations are easily recognized from the letter A.

to zero happens very rarely [1]. These *empirical* indications play a fundamental role in all actual applications of the theory of probability: in fact, they allow one to infer an almost certain coincidence, in the limit, between *frequency* and *probability*, and therefore justify one to *postulate* (from the mathematical point of view) their identification. Thus, without going further into tricky philosophical and logical questions, let us say only that empirical *frequency* becomes the key to enter into the analytical models of probability laws, and this assumption will be accepted in the following: the question of *statistical inference*, i.e. the actual treatment of experimental data to infer probability laws, will be tackled in Chapters 7 and 8.

2.2 TIME PROCESSES: CLASSIFICATION AND BASIC DEFINITIONS

In all engineering problems, actual phenomena must be assimilated to appropriate mathematical models. The purpose of this chapter is to present synthetically the models used for the random quantities (variables and processes) which have been introduced in Chapter 1: this presentation is limited to essential concepts and definitions. A more complete treatment can be found in the Appendices: therefore, many references to them will be found in this chapter. Also, more definitions of random quantities will be given in the following chapters, where relevant to the physical problem being presented. Note that, for simplicity's sake, the mathematical symbols used in the Appendices are sometimes simplified for use in the body of the book.

The most important types of *deterministic* time functions $q(t)^‡$ encountered in structural mechanics are collected and classified in Table 2.1, together with their *frequency spectrum* which measures the *frequency content* of the function. For periodic and almost-periodic functions $q(t)^§$, a discrete spectral representation $Q(\nu)$ holds, whereas for transient non-periodic functions, a continuous spectral representation can be obtained in most cases from a Fourier integral. In this case, the Fourier spectrum is generally a complex-valued function given by

$$Q(\nu) = |Q(\nu)|\mathrm{e}^{-\mathrm{i}\theta(\nu)} = \int_{-\infty}^{\infty} q(t)\mathrm{e}^{-\mathrm{i}2\pi\nu t}\,\mathrm{d}t \tag{2.1}$$

‡ All processes will be described in this chapter with reference to the independent variable 'time'. Of course, other independent variables can be considered, if appropriate.

§ As well known, a *periodic function* of period T is defined by

$$q(t) = q(t + nT) \qquad n = 1, 2, \ldots$$

Period T, *frequency* ν and *circular frequency* (or *pulsation*) ω are related by

$$T = 1/\nu = 2\pi/\omega$$

Note the different meaning of the *frequency* of a periodic function and the relative (statistical) frequency of an event, introduced in Section 2.1.

Table 2.1 Classification of deterministic time functions.

Category	*Type*	*Mathematical form*	*Time diagram*	*Frequency spectrum*	*Examples*
Periodic	Constant	$q(t) = Q = \text{const.}$	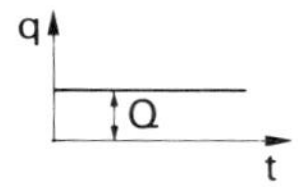		(a) Measured self-weight (b) Permanent displacement
	Sinusoidal	$q(t) = Q\cos(2\pi\nu_0 t)$	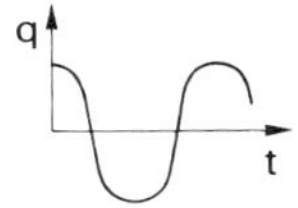	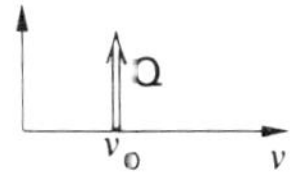	Free vibration of a deterministic single-degree-of-freedom linear undamped system
	General periodic	$q(t) = Q_0 + \sum_{n=1}^{\infty} Q_n \cos(2\pi n\nu_1 t)$	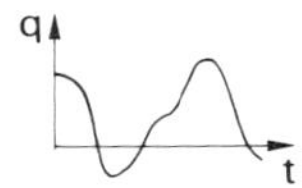	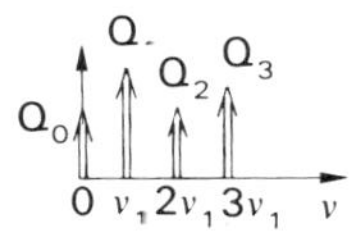	Vibration of a deterministic rigid–plastic single-degree-of-freedom system subject to a *n* sinusoidal force
Non-periodic	Almost periodic	$q(t) = Q_0 + \sum_{n=1}^{\infty} Q_n \cos(2\pi\nu_n \mathrm{t})$	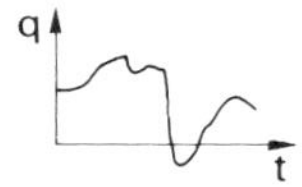	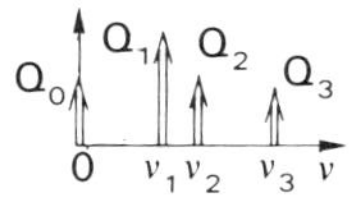	Free vibration of a deterministic multi-degree-of-freedom linear undamped system
	Transient	$Q(\nu) = \int_{-\infty}^{\infty} q(t)\mathrm{e}^{-i2\pi\nu t}\,\mathrm{d}t$ $= \lvert Q(\nu)\rvert \mathrm{e}^{-i\theta(\nu)}$		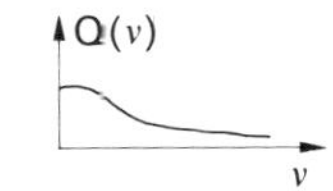	Free vibration of a deterministic damped mechanical system

with magnitude (equations (A5.39) and (A5.40))

$$|Q(\nu)| = |Q(\omega/2\pi)|$$
$$= \left[\left(\int_{-\infty}^{\infty} q(t)\cos(2\pi\nu t)\,\mathrm{d}t\right)^2 + \left(\int_{-\infty}^{\infty} q(t)\sin(2\pi\nu t)\,\mathrm{d}t\right)^2\right]^{1/2} \quad (2.1')$$

Of course, all combinations of these basic types are possible.

The classification of a *random process* $\tilde{q}(t)$ is less straightforward because, as already stated in Section 1.2, a different *sample function* $\grave{q}_k(t)$ is obtained every time the quantity is recorded (see e.g. Fig. 2.1). The mathematics of random processes can be found in Appendices A5 and A6: the characteristics that identify the basic types of r.p. are indicated in Fig. 2.2 and will be defined later in this chapter.

A continuous r.p. $\tilde{q}(t)$ is fully described by its *moments* and *joint moments*, defined as follows. Consider the set of sample functions $\grave{q}_k(t)$, $k = 1, 2, \ldots, n$ that

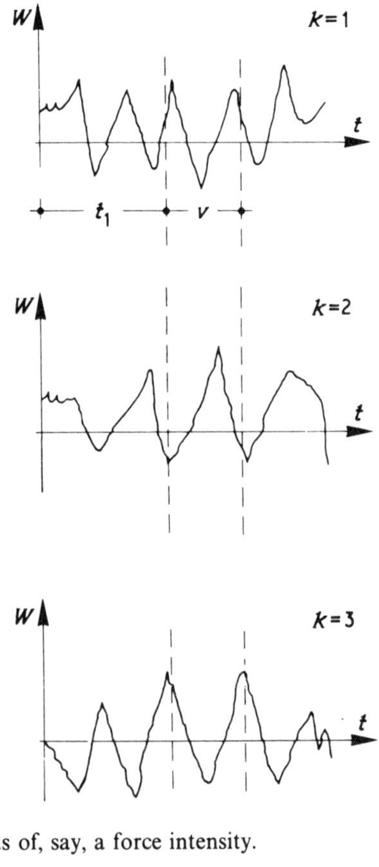

Fig. 2.1 Sample records of, say, a force intensity.

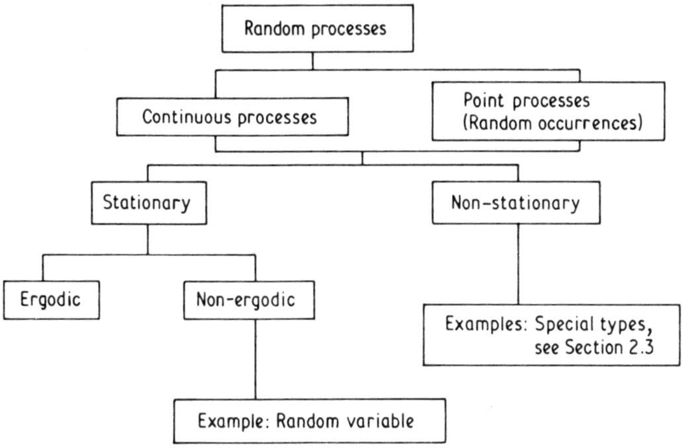

Fig. 2.2 Classification of random processes.

describe the random process $\tilde{q}(t)$. The *mean value* of the random process at any given time t_1 can be defined as (equations (A5.2), and (A2.11)[§]

$$\mu_{q(t)}(t_1) = \mu_q(t_1) = \lim_{n\to\infty} \frac{1}{n} \sum_{k=1}^{n} \grave{q}_k(t_1) \tag{2.2}$$

The *mean value* is also called the *average*, or *first moment* of the process, or also the (*mathematical*) *expectation* of its value at time $t = t_1$:

$$\mu_q(t_1) = \mathrm{E}[\tilde{q}(t_1)] \tag{2.2'}$$

The mean value, which is in general a function of time t, gives a first indication of what can be expected from a r.p. at that time.

The *interdependence* or, better, the *correlation* between the values of a r.p. at two different time instants, say t_1 and $t_2 = t_1 + v$, is measured by the *autocorrelation function* $\Sigma'_{q(t)}(t_1, t_2)$ defined by (equation (A5.3))

$$\Sigma'_{q(t)}(t_1, t_2) = \Sigma'_q(t_1, t_2) = \lim_{n\to\infty} \frac{1}{n} \sum_{k=1}^{n} \grave{q}_k(t_1)\grave{q}_k(t_2) \tag{2.3}$$

The autocorrelation function is the *joint moment* (of order 2) of the r.p. $\tilde{q}(t)$, while *moments* and *joint moments* of order r are defined by

$$\mathrm{E}[\tilde{q}^r(t_1)] \tag{2.4}$$

$$\mathrm{E}[\tilde{q}(t_1)\tilde{q}(t_2)\tilde{q}(t_3) \ldots \tilde{q}(t_r)] \qquad r \geqslant 2 \tag{2.4'}$$

Although most of the definitions given in this chapter apply both to continuous and point processes, they are written with specific reference to the former type.

[§] For simplicity of notation, tildes (˜) are omitted from subscripts and superscripts throughout the book (except in the Appendices).

Point processes will be treated in some detail in Section 2.7, with reference to the problem of threshold crossing.

2.3 STATIONARY AND NON-STATIONARY RANDOM PROCESSES

2.3.1 Stationary processes

A random process $\tilde{q}(t)$ is said to be *weakly stationary* or *stationary in the wide sense* in the special case in which $\mu_q(t_1)$ (equation (2.2)) and $\Sigma'_q(t_1, t_1+v)$ (equation (2.3)) do not vary as the time t_1 varies, i.e. the mean value of the r.p. is constant and the autocorrelation function depends only on the time interval $v = t_2 - t_1$ (Section A5.1)

$$\mu_q(t_1) = \mu_q \tag{2.2''}$$

$$\Sigma'_q(t_1, t_2) = B'_q(t_2 - t_1) = B'_q(v) \tag{2.3'}$$

In the case where all moments and joint moments, equations (2.4), are time invariant, the random process $\tilde{q}(t)$ is said to be *stationary in the strict sense* or *strongly stationary*.

2.3.2 Ergodic processes

While the already introduced mean value μ_q (2.2) and autocorrelation function (2.3) of a r.p. $\tilde{q}(t)$ are averages over the set of sample functions defining the process (*ensemble averages*), other averages (*time averages*) can be defined for each sample function $\grave{q}_k(t)$, as well as for any deterministic time function.

Namely, the (time) mean $m_q(k)$ and autocorrelation function $b'_q(v, k)$ of a (well defined) time function $\grave{q}_k(t)$ are by definition (equations (A5.6) and (A5.13))

$$m_q(k) = \lim_{T\to\infty} m_T(\grave{q}_k) = \lim_{T\to\infty} \frac{1}{T}\int_0^T \grave{q}_k(t)\,dt \tag{2.5}$$

$$b'_q(v, k) = \lim_{T\to\infty} b'_T(v, \grave{q}_k) = \lim_{T\to\infty} \frac{1}{T}\int_0^T \grave{q}_k(t)\grave{q}_k(t+v)\,dt \tag{2.6}$$

Consider a stationary random process: if $m_q(k)$ and $b'_q(v, k)$ do not differ when computed over different sample functions $\grave{q}_k(t)$, the random process is said to be *ergodic* (Section A5.1), and

$$m_q(k) = \mu_q \qquad \text{for any } k \tag{2.7}$$

$$b'_q(v, k) = B'_q(v) \qquad \text{for any } k \tag{2.8}$$

Therefore the properties of ergodic random processes can be determined by performing time averages over a single sample function.

It is worth underlining that the definition of ergodicity implies consideration of infinitely long samples. In practice, only finite time intervals are of interest: this allows to model as *ergodic* some processes that in rigour may have a different character. A good example of this case is the long-term wind velocity at a given meteorological station in a non-tropical region.

2.3.3 Non-stationary processes

The estimation of the properties of non-stationary random processes requires generally the determination of instantaneous averages at all times over the ensemble of sample functions forming the process. However, it is often not feasible to obtain a sufficient number of sample records and measure these properties with acceptable accuracy. Therefore special categories of non-stationary processes which simplify the measurement and analysis problem have been studied in detail.

The most important family is constituted by the *separable processes* in which non-stationarity can be separated from randomness, i.e. the process can be written in the form

$$\tilde{q}(t) = g(t)\tilde{q}_0(t) \tag{2.9}$$

where $g(t)$ is a *deterministic* function of time t and $\tilde{q}_0(t)$ a *stationary* (and possibly *ergodic*) random process. In this case, ensemble averaging is not necessary to describe the data (Section 8.7.2). For its comparative simplicity, equation (2.9) is often used, for instance, to model the ground acceleration during an earthquake (e.g. Section 9.2.4).

2.4 DESCRIPTIVE PROPERTIES OF RANDOM QUANTITIES

The most important functions used to describe the probabilistic properties of random quantities are [2]:

(1) the *mean value* $\mu_q(t)$, and
(2) the *autocorrelation function* $\Sigma'_q(t_1, t_2)$, already introduced in Section 2.2;
(3) the *mean square value* $E[\tilde{q}^2(t)]$ that gives rudimentary information on the intensity (or magnitude or 'energy') of the quantity;
(4) the *variance* $s_q^2(t)$ that gives the same information on its random variability;
(5) the *probability density function* (PDF) that furnishes a complete indication of the probabilistic properties of the process in the time domain; and
(6) the *spectral density functions* defined in the frequency domain that give an indication of the time variation of the quantity (as mean, mean square and autocorrelation function do in the time domain); of these, the *power spectral density* (PSD) will be presented in Section 2.4.4.

Other properties are also of importance in structural engineering and will be defined where used.

2.4.1 Mean, mean square, variance and standard deviation

While the mean value of a r.p. is the *first moment* already defined by (2.2), the *mean square value* is the *second moment*‡ given by (Section A2.2; cf. also equation (2.4))

$$E[\tilde{q}^2(t)] = \Sigma'_q(t_1, t_1) = \lim_{n\to\infty} \frac{1}{n} \sum_{k=1}^{n} \grave{q}_k^2(t_1) \tag{2.10}$$

while the *central second moment*⁺

$$\begin{aligned} s^2_{q(t)}(t_1) = s^2_q(t_1) &= E[\{\tilde{q}(t_1) - \mu_{q(t)}(t_1)\}^2] \\ &= E[\tilde{q}^2(t_1)] - E^2[\tilde{q}(t_1)] \\ &= \Sigma'_q(t_1, t_1) - \mu^2_q(t_1) \\ &= \Sigma_{q(t)}(t_1, t_1) = \Sigma_q(t_1, t_1) \end{aligned} \tag{2.11}$$

is called the *variance* of the r.p.

The positive square root $s_q(t_1)$ of the variance§

$$s_q(t_1) = +\sqrt{[\Sigma'_q(t_1, t_1) - \mu^2_q(t_1)]} = \sqrt{[\Sigma_q(t_1, t_1)]} \tag{2.12}$$

is the *standard deviation* of $\tilde{q}(t)$ at time $t = t_1$ and the ratio

$$\delta_q(t_1) = \frac{s_q(t_1)}{\mu_q(t_1)} \tag{2.12'}$$

is the relevant *coefficient of variation* (c.o.v.).

If $\tilde{q}(t)$ is an ergodic random process, $E[\tilde{q}^2(t)]$ does not depend on t and it is also given by

$$E[\tilde{q}^2(t)] = \lim_{T\to\infty} \frac{1}{T} \int_0^T \grave{q}_k^2(t)\,dt \tag{2.13}$$

2.4.2 Probability density function

The *probability density function* (PDF) $p_{q(t_1)}(q)$ (Section A2.1) of a random process $\tilde{q}(t)$, multiplied by dq, is the probability that, at time t_1, $q(t)$ assumes a value in the range $(q, q + dq)$

$$p_{q(t_1)}(q) = p(q, t_1) = \lim_{\Delta q\to 0} \{\mathrm{Prob}[q \leqslant \tilde{q}(t_1) < q + \Delta q]/\Delta q\} \tag{2.14}$$

The PDF is always a real-valued, non-negative function.

The probability that $\tilde{q}(t_1)$ is between any two values q_1 and $q_2 (: q_1 < q_2)$ is equal to the integral of the PDF between q_1 and q_2

$$\mathrm{Prob}[q_1 \leqslant q(t_1) < q_2] = \int_{q_1}^{q_2} p(q, t_1)\,dq \tag{2.15}$$

‡ The use of the term '*moment*' will be justified in Section 2.4.2.

§ Most statistical textbooks use the symbol σ for the standard deviation. This use should be avoided in an engineering book, because in solid mechanics σ usually denotes the stress.

Thus the probability that $\tilde{q}(t_1)$ is less than or equal to q is equal to the integral of the PDF from $-\infty$ to q. This function, usually indicated by $P_{q(t_1)}(q)$ or $P(q, t_1)$, is known as the *cumulative distribution function* (CDF)

$$P_{q(t_1)}(q) = P(q, t_1) = \text{Prob}[q(t_1) \leqslant q] = \int_{-\infty}^{q} p(\zeta, t_1)\,\mathrm{d}\zeta \tag{2.16}$$

Like any value of probability (Appendix A1)

$$0 \leqslant P(q, t_1) \leqslant 1 \qquad \text{for any } q, t_1 \tag{2.16'}$$

In terms of the PDF, the mean and mean square values of $\tilde{q}(t)$ at $t = t_1$ are given by (equation (A2.11))

$$\mu_q(t_1) = \int_{-\infty}^{\infty} q\, p(q, t_1)\,\mathrm{d}q \tag{2.17a}$$

$$\mathrm{E}[\tilde{q}^2(t_1)] = \int_{-\infty}^{\infty} q^2 p(q, t_1)\,\mathrm{d}q \tag{2.17b}$$

and the variance by (equation (A2.12))

$$s_q^2(t_1) = \int_{-\infty}^{\infty} (q - \mu_q(t_1))^2 p(q, t_1)\,\mathrm{d}q \tag{2.17c}$$

Therefore, they coincide respectively with the first, second and central second moments (moment of inertia) of the area below the curve $y = p(q, t_1)$ (Fig. 2.3).

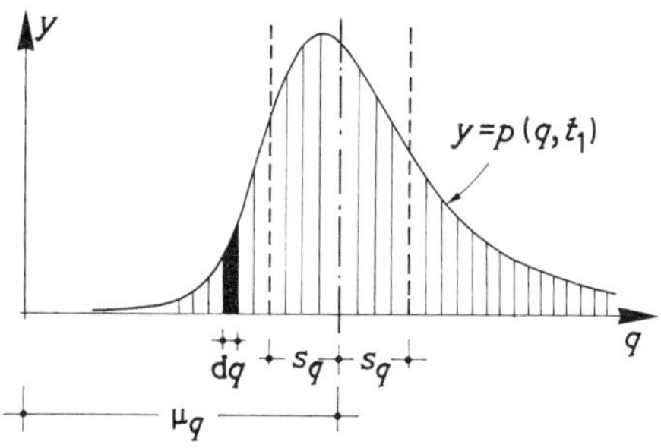

Fig. 2.3 Definition of PDF.

Moreover, since (cf. equation (2.15))

$$\int_{-\infty}^{\infty} p(q, t_1)\,\mathrm{d}q = 1 \tag{2.15'}$$

$\mu_q(t_1)$ and $s_q(t_1)$ coincide respectively with the abscissa of the centroid, and with the radius of gyration of the same area about the axis $q = \mu_q(t_1)$.

The most important forms of PDFs are collected in Table 2.2. Numerical tables

Table 2.2 Probability density functions.

Type		*Probability density function*
Discrete	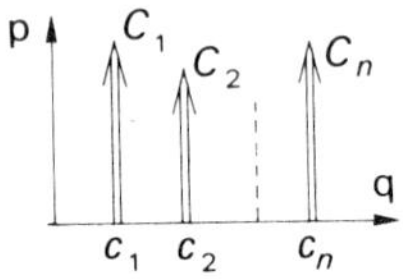	$p(q) = C_1\partial(q-c_1)+C_2\partial(q-c_2) + \dots + C_n\partial(q-c_n)$ with $C_1+C_2+\dots+C_n = 1$, ∂ = Dirac delta function
Uniform (rectangular)	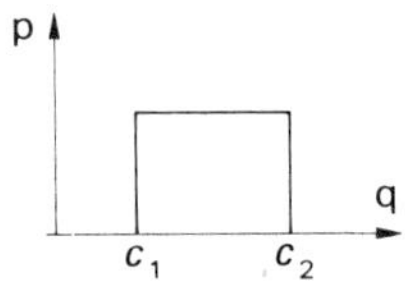	$p(q) = \begin{cases} (c_2-c_1)^{-1} & c_1 \leqslant q \leqslant c_2 \\ 0 & \text{otherwise} \end{cases}$
Gaussian (normal)	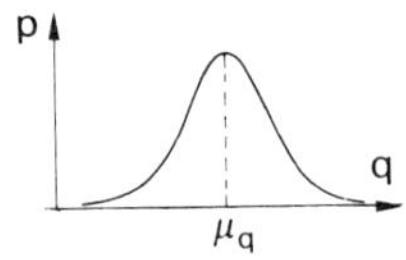	$p(q) = [s_q\surd(2\pi)]^{-1}\exp[-(q-\mu_q)^2/2s_q^2]$
Rayleigh	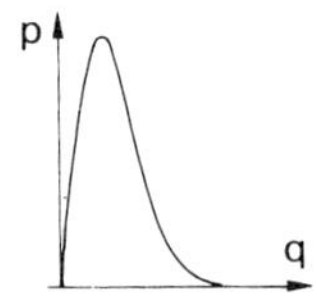	$p(q) = \begin{cases} (q/c^2)\exp[-q^2/2c^2] & q \geqslant 0 \\ 0 & q < 0 \end{cases}$
Maxwell	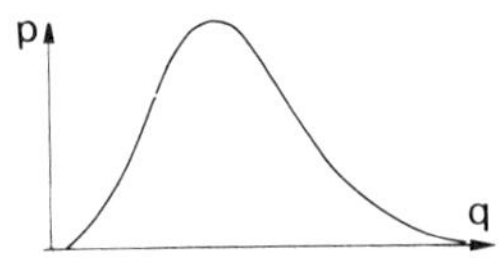	$p(q) = \begin{cases} [(q^2/c^3)\surd(2/\pi)]\exp[-q^2/2c^2] & q \geqslant 0 \\ 0 & q < 0 \end{cases}$
Extreme I (maxima) [Fisher-Tippet]	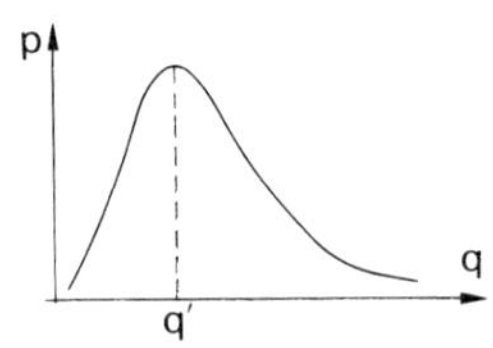	$p(q) = c\exp\{-c(q-q')-\exp[-c(q-q')]\}$
Extreme II (maxima) [Frechet]	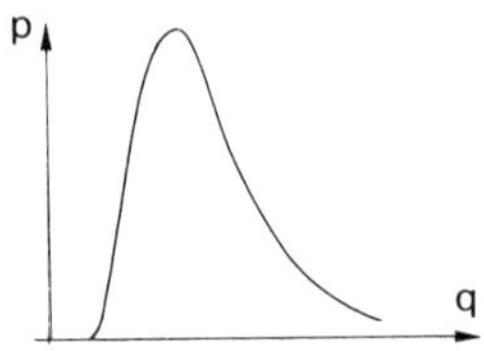	$p(q) = \begin{cases} \dfrac{c}{q'}\left(\dfrac{q'}{q}\right)^{c+1}\exp\left[-\left(\dfrac{q'}{q}\right)^{c}\right] & q \geqslant 0 \\ 0 & q < 0 \end{cases}$

Table 2.2 (*Contd.*)

Type		*Probability density function*
Truncated	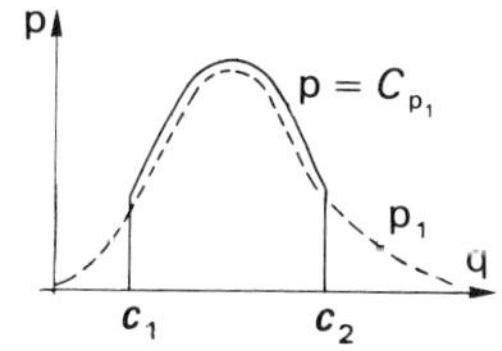	Let the original $p_1(q)$ be defined over $(-\infty, +\infty)$. Then $p(q)=\begin{cases} Cp_1(q) & c_1 \leqslant q \leqslant c_2 \\ 0 & \text{otherwise}\end{cases}$ with $C\int_{c_1}^{c_2} p_1(q)\,\mathrm{d}q = 1$
Clipped (censored)	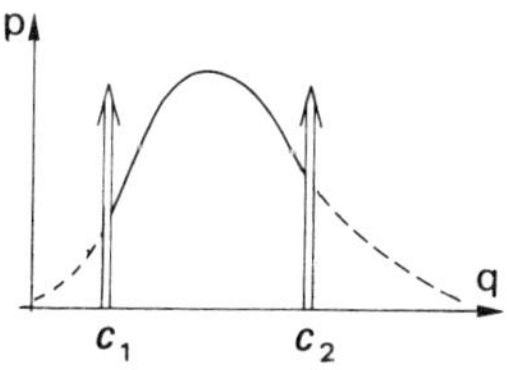	Let the original $p_1(q)$ be defined over $(-\infty, +\infty)$. Then $p(q)=\begin{cases} p_1(q) & c_1 < q < c_2 \\ C_1\partial(q-c_1) & q=c_1 \\ C_2\partial(q-c_2) & q=c_2 \\ 0 & \text{otherwise}\end{cases}$ with $\int_{c_1}^{c_2} p_1(q)\,\mathrm{d}q + C_1 + C_2 = 1$

of these functions can be found in textbooks of statistics: some are also given in Appendix A8 of this book.

If the r.p. is *stationary*, the PDF is independent of time and the argument t_1 can be dropped from all formulae. If it is also *ergodic*, (2.14) can be given the form

$$p(q) = \lim_{\Delta q \to 0} \left\{ \left[\lim_{T\to\infty} (t_q/T) \right] \Big/ \Delta q \right\} \tag{2.14'}$$

where t_q is the total amount of time during which $q(t)$ falls in the range $(q, q+\Delta q)$ (Fig. 2.4: $t_q = t_1 + t_2 + t_3 + \ldots$).

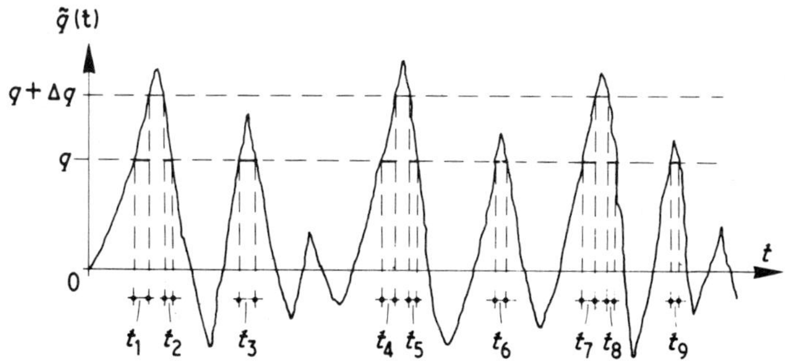

Fig. 2.4 PDF of ergodic r.p.

2.4.3 Autocorrelation and autocovariance functions

The influence of the value taken by $\tilde{q}(t)$ at time t_1 over the value taken at time t_2, can be measured either by the autocorrelation function, already defined by equation (2.3), or by the autocovariance function

$$\Sigma_q(t_1, t_2) = \Sigma'_q(t_1, t_2) - \mu_q(t_1)\mu_q(t_2) \tag{2.3''}$$

The following symmetry properties hold:

$$\Sigma'_q(t_1, t_2) = \Sigma'_q(t_2, t_1) \tag{2.18}$$

$$\Sigma_q(t_1, t_2) = \Sigma_q(t_2, t_1) \tag{2.18'}$$

The autocorrelation function and the mean and mean square values are related by

$$\mu_q(t_1) = \sqrt{[\Sigma'_q(t_1, \infty)]} \tag{2.19a}$$

$$\mathrm{E}[\tilde{q}^2(t_1)] = \Sigma'_q(t_1, t_1) \tag{2.19b}$$

It can also be shown that (equation A3.16)

$$|\Sigma'_q(t_1, t_2)|^2 \leqslant \Sigma'_q(t_2, t_2)\Sigma'_q(t_1, t_1) \tag{2.19'}$$

For a *stationary* r.p., the autocovariance function depends only on the difference $v = t_2 - t_1$, and, because of 2.18')

$$\Sigma_q(t_2 - t_1) = B_q(v) = B_q(-v) \tag{2.18''}$$

The *extinction interval* (or *extinction distance*) v_{ext} is defined as the smallest interval $t_2 - t_1$ beyond which the values $\tilde{q}(t_1)$ and $\tilde{q}(t_2)$ are independent of each other with sufficient approximation:

$$v_{\text{ext}} = \{\min v : B_q(v) \simeq 0\} \tag{2.20}$$

2.4.4 Power spectral density

The spectral density functions (or *spectra*) play for a random process $\tilde{q}(t)$ a role analogous to the Fourier analysis of a deterministic function and describe the frequency content of the process.

With reference to *stationary processes*, in which spectral analysis finds most applications, the *power spectral density* (PSD) is a real non-negative function defined as the Fourier transform of the autocorrelation function‡ (equations (A5.46′) and (A5.51))

$$G'^{\nu}_{q(t)}(\nu) = G'^{\nu}_q(\nu) = \int_{-\infty}^{\infty} B'_q(v)\,\mathrm{e}^{-\mathrm{i}2\pi\nu v}\,\mathrm{d}v \tag{2.21}$$

‡ The spectra of non-stationary processes are defined in Appendix A5, equations (A5.54) and (A5.55).

and vice versa

$$B'_q(v) = \int_{-\infty}^{\infty} G'^{\nu}_q(\nu)\, e^{i2\pi\nu v}\, d\nu \tag{2.22}$$

Analogously, the *spectral density function* $G^{\nu}_q(\nu)$ may be introduced starting from the autocovariance function $B_q(v)$ (equation (A5.44)), while, adopting the angular frequency ω instead of the frequency ν, one writes (equation (A5.51)

$$\begin{aligned} G'^{\nu}_q(\nu) &= 2\pi G'_q(\omega) \\ G^{\nu}_q(\nu) &= 2\pi G_q(\omega) \end{aligned} \tag{2.21$'$}$$

The PSD is a real-valued function for a stationary process, since as the imaginary part is the integral of the product of $B'(v)$, for which equation (2.18″) holds, with $\sin \omega v$, for which $\sin \omega v = -\sin(-\omega v)$. Hence equation (2.21) can also be written for a stationary process

$$G'^{\nu}_q(\nu) = 2\int_0^{\infty} B'_q(v)\cos(2\pi\nu v)\, dv = 2\pi G'_q(\omega) \tag{2.21$''$}$$

Since

$$G'^{\nu}_q(\nu) = G'^{\nu}_q(-\nu) \tag{2.21$'''$}$$

but negative values of the frequency ν have no real significance, it is customary to define and use the one-sided PSD and spectral density, given respectively by

$$\begin{aligned} G'^{o\nu}_q(\nu) &= 2G'^{\nu}_q(\nu) \qquad 0 \leqslant \nu < \infty \\ G^{o\nu}_q(\nu) &= 2G^{\nu}_q(\nu) \end{aligned} \tag{2.23}$$

The (one-sided) PSD $G'^{o\nu}_q(\nu)$ is related to the mean and mean square values of the (stationary!) r.p. by

$$\mu_q^2 = \lim_{\varepsilon \to 0} \int_{-\varepsilon}^{+\varepsilon} G'^{o\nu}_q(\nu)\, d\nu \tag{2.24}$$

and (equation (A5.49))

$$\begin{aligned} E[\tilde{q}^2(t)] &= \int_0^{\infty} G'^{o\nu}_q(\nu)\, d\nu \\ &= \mu_q^2 + \int_0^{\infty} G^{o\nu}_q(\nu)\, d\nu = \mu_q^2 + s_q^2 \end{aligned} \tag{2.25}$$

Therefore, for a r.p. with non-zero mean ($\mu_q \neq 0$), the PSD includes a Dirac delta-function at zero frequency (Fig. 2.5(a)), while the one-sided spectral density $G^{o\nu}_q(\nu)$ does not contain this information, $G^{\nu}_q(\nu)$ being the Fourier transform of $B_q(v) = B'_q(v) - \mu_q^2$. Equation (2.24) holds only if $G^{o\nu}_q(\nu)$ does not contain or reduce to a Dirac delta-function at zero frequency: in the latter case,

$$\lim_{\varepsilon \to 0} \int_{-\varepsilon}^{+\varepsilon} G'^{o\nu}(\nu)\, d\nu = \mu_q^2 + \lim_{\varepsilon \to 0} \int_{-\varepsilon}^{+\varepsilon} G^{o\nu}_q(\nu)\, d\nu = \mu_q^2 + s_q^2 \tag{2.24$'$}$$

Table 2.3 Autocorrelation and spectral density functions of some common random processes.‡

Type		*Autocorrelation function*		*One-sided power spectral density function*
Constant	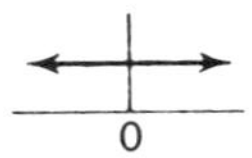	$B'_q(v) = c^2$		$G'^{ov}_q(\nu) = c^2\partial(\nu)$
Sine wave	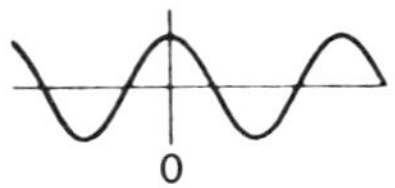	$B'_q(v) = \frac{1}{2}Q^2\cos(2\pi\nu_0 v)$	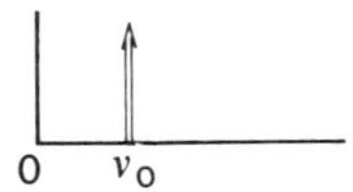	$G'^{ov}_q(\nu) = \frac{1}{2}Q^2\,\partial(\nu - \nu_0)$
White noise	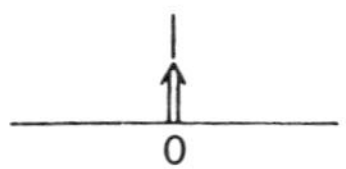	$B'_q(v) = \frac{1}{2}c\partial(v)$	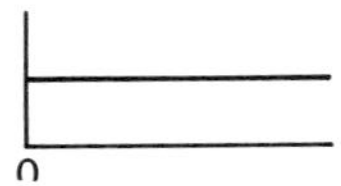	$G'^{ov}_q(\nu) = \begin{cases} c & \nu \geqslant 0 \\ 0 & \nu < 0 \end{cases}$
Low pass, white noise	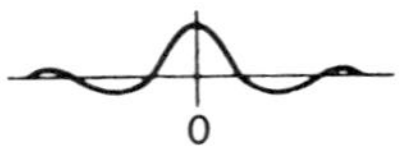	$B'_q(v) = cb\dfrac{\sin(2\pi bv)}{2\pi bv}$	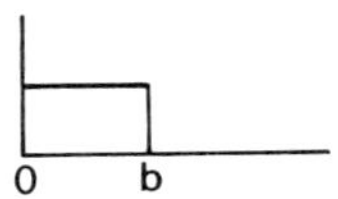	$G'^{ov}_q(\nu) = \begin{cases} c & 0 \leqslant \nu \leqslant b \\ 0 & \text{otherwise} \end{cases}$

Band-pass, white noise	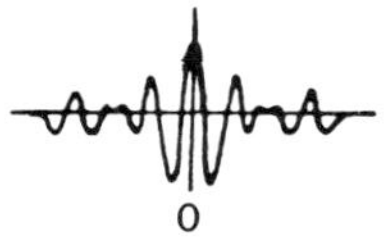	$B'_q(v) = cb\,\frac{\sin(\pi b v)}{\pi b v}\cos(2\pi\nu_0 v)$	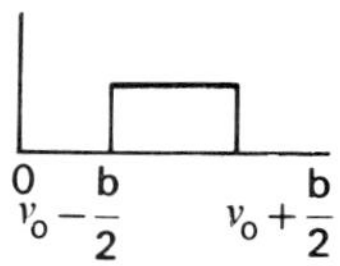	$G_q'^{o\nu}(\nu) = \begin{cases} c & 0 < \nu_0 - \frac{1}{2}b \leqslant \nu \leqslant \nu_0 + \frac{1}{2}b \\ 0 & \text{otherwise} \end{cases}$
Exponential	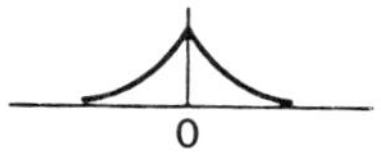	$B'_q(v) = e^{-c\lvert v\rvert}$		$G_q'^{o\nu}(\nu) = \frac{4c}{c^2 + 4\pi^2\nu^2}$
Exponential cosine	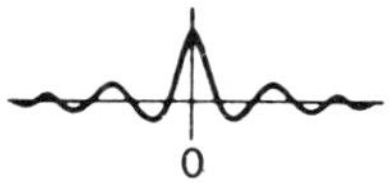	$B'_q(v) = e^{-c\lvert v\rvert}\cos(2\pi\nu_0 v)$	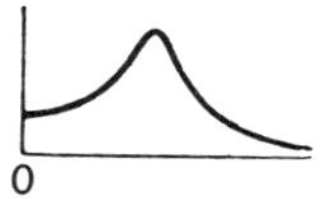	$G_q'^{o\nu}(\nu) = 2c\left(\frac{1}{c^2 + 4\pi^2(\nu+\nu_0)^2} + \frac{1}{c^2 + 4\pi^2(\nu-\nu_0)^2}\right)$
Exponential cosine, exponential sine	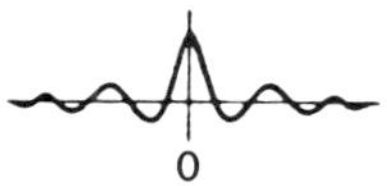	$B'_q(v) = e^{-c_0\lvert v\rvert}[c_1\cos(2\pi\nu_0 v) + c_2\sin(2\pi\nu_0\lvert v\rvert)]$	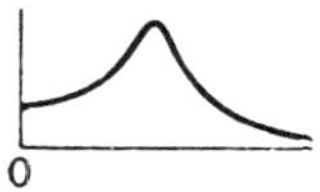	$G_q'^{o\nu}(\nu) = \frac{2c_0c_1 + 4\pi c_2(\nu+\nu_0)}{c_0^2 + 4\pi^2(\nu+\nu_0)^2} + \frac{2c_0c_1 - 4\pi c_2(\nu-\nu_0)}{c_0^2 + 4\pi^2(\nu-\nu_0)^2}$

‡ Adapted from Bendat and Piersol [2].

Equation (2.24′) holds for random processes whose realizations remain constant in time, i.e. random variables (Section 2.6).

Equation (2.25) follows easily from (2.22) and (2.23). An alternative proof gives us the possibility of establishing some basic relations. Remember first that the following terminology is often employed with reference to the fluctuations of a real or complex deterministic function $q_k(t)$ around the value $m_q(k)$ defined by equation (2.5) [3] (The value $m_q(k)$ is assumed to be zero without any loss of generality):

(1) $|q_k(\tau)|^2$ is the *power* of $q_k(t)$ at time τ;

(2) $\int_{t_1}^{t_1} |q_k(\tau)|^2 \, d\tau$ is the *energy* of $q_k(t)$ in the interval $[t_1, t_2]$;

(3) $\dfrac{1}{t_2 - t_1} \int_{t_1}^{t_2} |q_k(\tau)|^2 \, d\tau$ is the *mean power* of $q_k(t)$ in the interval $[t_1, t_2]$;

(4) $\int_{-\infty}^{\infty} |q_k(\tau)|^2 \, d\tau$ is the *total energy* of $q_k(t)$;

(5) $\lim\limits_{T \to \infty} \dfrac{1}{2T} \int_{-T}^{T} |q_k(\tau)|^2 \, d\tau$ is the *mean power* of $q_k(t)$.

The quantity $b'_q(0, k)$ defined by (2.6) denotes the mean power of a realization $\grave{q}_k(t)$ of a random process $\tilde{q}(t)$ that is equal to zero if $t \leqslant 0$ or $t \geqslant T$:

$$b'_q(0, k) = \lim_{T \to \infty} \frac{1}{T} \int_0^T \grave{q}_k^2(t) \, dt = \lim_{T \to \infty} \frac{1}{T} \int_{-\infty}^{\infty} \grave{q}_k^2(t) \, dt \tag{2.6'}$$

Consider now Parseval's theorem [3]

$$\int_{-\infty}^{\infty} \grave{q}_k^2(t) \, dt = \int_{-\infty}^{\infty} |\grave{q}_{kF}(\nu, \infty)|^2 \, d\nu \tag{2.25'}$$

which states that the total energy is equal to the integral over the whole frequency axis of the square modulus of the finite Fourier transform of the realization $\grave{q}_k(t)$ over a period $T = \infty$, the finite Fourier transform being defined by

$$\grave{q}_{kF}(\nu, T) = \int_0^T \grave{q}_k(t) \, e^{-i2\pi\nu t} \, dt \tag{2.25''}$$

From equations (2.6′) and (2.25′) it follows that

$$b'_q(0, k) = \lim_{T \to \infty} \frac{1}{T} \int_{-\infty}^{\infty} |\grave{q}_{kF}(\nu, T)|^2 \, d\nu = 2 \lim_{T \to \infty} \frac{1}{T} \int_0^{\infty} |\grave{q}_{kF}(\nu, T)|^2 \, d\nu \tag{2.6''}$$

By considering the expected value over all possible $\grave{q}_k(t)$, remembering that the order of integration and averaging can be changed, and introducing the relation

$$G_q^{\prime ov}(\nu) = 2 \lim_{T \to \infty} \frac{1}{T} \, \mathrm{E}[|\grave{q}_{kF}(\nu, T)|^2] \tag{2.26}$$

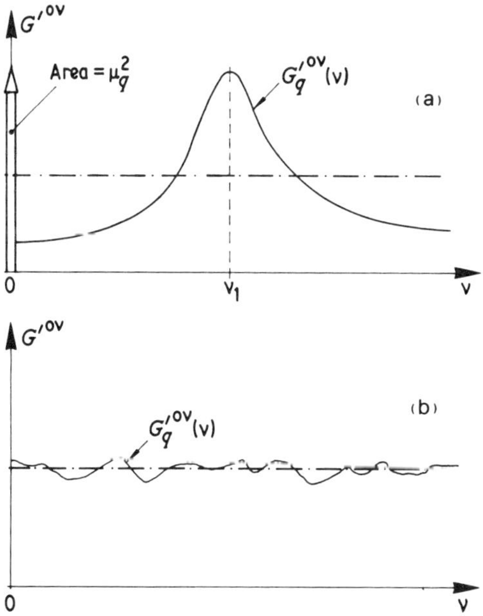

Fig. 2.5 Typical power spectral densities: (a) r.p. with non-zero mean μ_q and one prevailing frequency $\nu_1 = \omega_1/2\pi$; (b) (almost) white noise ($\mu_q = 0$; $G_q'^{ov} \simeq$ const.)

which is easily obtained from (2.25″) and (2.21), one derives (2.25). Equation (2.26) is often introduced to provide an estimate (cf. Chapter 8) of the PSD.

The power spectral density function supplies no new information with respect to the autocorrelation function, as the two are Fourier transform pairs. The two functions present the information in different forms, and either form is more desirable than the other for each specific application. With reference to stationary processes, some special autocorrelation and power spectral density functions are presented in Table 2.3.

2.5 JOINT PROPERTIES AND OPERATIONS ON RANDOM PROCESSES

Often, it is necessary to study two or more contemporary r.p.: for instance, this is the case for a structure on which a multi-parameter load acts, as in the example of Fig. 1.1‡. Therefore, *joint properties* have to be defined, among which the most important are:

‡ Alternatively, one can speak of a *vector-valued* r.p. $\tilde{\mathbf{q}}(t)$, whose components $\tilde{q}_i(t)$ coincide with each considered (scalar) r.p.

(1) the joint probability density function that supplies information in the amplitude domain;
(2) the cross-correlation function that supplies information in the time domain; and
(3) the cross-spectral density function that supplies information in the frequency domain.

These three functions are actually extensions of the basic formulation used to describe the properties of a single (scalar) random process.

2.5.1 Joint probability density functions

Consider two scalar r.p. $\tilde{q}(t)$ and $\tilde{y}(t)$. Their joint (second-order) PDF is the real-valued, non-negative function defined by (Section A2.4)

$$p_{q(t_1),y(t_2)}(q, y) = p(q, t_1; y, t_2)$$
$$= \lim_{\substack{\Delta q \to 0 \\ \Delta y \to 0}} \frac{\mathrm{Prob}[(q \leqslant \tilde{q}(t_1) < q + \Delta q) \cap (y \leqslant \tilde{y}(t_2) < y + \Delta y]}{\Delta q \Delta y} \tag{2.27}$$

where the *intersection* symbol $\cap$ can be read simply as *and* (Section A1.1).

The PDFs of each r.p. (*marginal* PDF) are related to the JPDF by

$$p_q(q, t_1) = \int_{-\infty}^{\infty} p(q, t_1; y, t_2)\,\mathrm{d}y \tag{2.28a}$$

$$p_y(y, t_2) = \int_{-\infty}^{\infty} p(q, t_1; y, t_2)\,\mathrm{d}q \tag{2.28b}$$

whence

$$\int_{-\infty}^{\infty}\int_{-\infty}^{\infty} p(q, t_1; y, t_2)\,\mathrm{d}q\,\mathrm{d}y = 1 \tag{2.29}$$

If both $\tilde{q}(t)$ and $\tilde{y}(t)$ are stationary processes, then

$$p(q, t_1; y, t_2) = p(q, t_1 + c; y, t_2 + c) = p(q0; y, v) \tag{2.30}$$

where c is an arbitrary constant, and $v = t_2 - t_1$.

Note also that according to the *conditional probability theorem* (equation (A2.43)), the PDF of $\tilde{q}$ at time t_1, given a value y of $\tilde{y}$ at time t_2, is

$$p_{q|y}(q, t_1 | y, t_2) = \frac{p(q, t_1; y, t_2)}{p_y(y, t_2)} \tag{2.31}$$

If

$$p(q, t_1; y, t_2) = p_q(q, t_1)\,p_y(y, t_2) \qquad \text{for any } t_1, t_2 \tag{2.32}$$

which, because of (2.31), is equivalent to

$$p_{q|y}(q, t_1 | y, t_2) = p_q(q, t_1) \tag{2.33a}$$

$$p_{y|q}(y, t_2 | q, t_1) = p_y(y, t_2) \tag{2.33b}$$

the two processes are *stochastically independent* of each other (equation (A2.39)). This synthetically means that the value taken by $\tilde{q}$ at time t_1 has no influence on the value taken by $\tilde{y}$ at time t_2 (or, better, on the PDF of $\tilde{y}$ at any time) and vice versa.

Stochastic independence, as it will be seen in the following, simplifies many complicated formulations. Because of this property, together with lack of knowledge on the joint properties of the relevant processes, stochastic independence is assumed in practical calculations much more often than proved or justified. A word of warning in this respect seems appropriate, because the assumption of independence is not always a *safe* one!

The previous definitions and formulae are easily extended to more than two r.p.

If all processes in the argument of $p(\quad)$ are the same (scalar) r.p. $\tilde{q}(t)$, the second- (and higher-) order PDFs of the r.p. $\tilde{q}(t)$ are obtained.

2.5.2 Cross-correlation and cross-covariance functions

The cross-correlation function $\Sigma'_{q,y}(t_1, t_2)$ of two random processes $\tilde{q}(t)$ and $\tilde{y}(t)$ describes the general dependence between the values taken by each process (equation (A5.4′)). Letting $\grave{q}_k(t)$ and $\grave{y}_k(t)$ be sample functions of $\tilde{q}$ and $\tilde{y}$ respectively, then

$$\Sigma'_{q(t),y(t)}(t_1, t_2) = \Sigma'_{q,y}(t_1, t_2) = \lim_{n\to\infty} \frac{1}{n} \sum_{k=1}^{n} \grave{q}_k(t_1)\grave{y}_k(t_2) \tag{2.34}$$

or also

$$\Sigma'_{q,y}(t_1, t_2) = \int_{-\infty}^{\infty}\int_{-\infty}^{\infty} q\, y\, p(q, t_1, y, t_2)\,\mathrm{d}q\,\mathrm{d}y \tag{2.35}$$

As in Section 2.4.3, the dependence between the values taken by each process can also be described by the autocovariance function

$$\Sigma_{q(t),y(t)}(t_1, t_2) = \Sigma_{q,y}(t_1, t_2) = \Sigma'_{q,y}(t_1, t_2) - \mu_q(t_1)\mu_y(t_2) \tag{2.34′}$$

In general, the correlation structure of two non-stationary r.p. is described by the three functions (cf. equations (2.3) and (2.10))

$$\Sigma'_q(t_1, t_1) = \mathrm{E}[\tilde{q}^2(t_1)] \tag{2.36a}$$

$$\Sigma'_y(t_2, t_2) = \mathrm{E}[\tilde{q}^2(t_2)] \tag{2.36b}$$

$$\Sigma'_{q,y}(t_1, t_2) = \Sigma'_{y,q}(t_2, t_1) \tag{2.36c}$$

which are related by the inequality, analogous to (2.19′) (equation (A3.16))

$$|\Sigma'_{q,y}(t_1, t_2)|^2 \leqslant \Sigma'_q(t_1, t_1)\Sigma'_y(t_2, t_2) \tag{2.37}$$

If $\tilde{q}(t)$ and $\tilde{y}(t)$ are stochastically independent, introduction of (2.32) into (2.35) yields

$$\Sigma'_{q,y}(t_1, t_2) = \mu_q(t_1)\mu_y(t_2) \tag{2.38}$$

where $\mu_q(t_1)$ and $\mu_y(t_2)$ are the relevant *mean values* (2.17a).

If

$$\Sigma_{q,y}(t_1, t_2) = 0 \tag{2.39}$$

$\tilde{q}(t)$ and $\tilde{y}(t)$ are said to be *uncorrelated.*

It is worth noting that, if a vector $\tilde{\mathbf{q}}(t)$ of random processes is considered, the cross-covariance functions can be ordered in a matrix $\boldsymbol{\Sigma}^{\mathbf{q}}(t_1, t_2)$ whose generic entry $\Sigma^{\mathbf{q}}_{ij}(t_1, t_2)$ denotes the cross-covariance function $\Sigma_{q_i, q_j}(t_1, t_2)$ if $i \neq j$, or the autocovariance function $\Sigma_{q_i}(t_1, t_2)$ if $i = j$. Moreover, if a matrix $\tilde{\mathbf{q}}(t)$ of random processes is considered, a four-index matrix $\boldsymbol{\Sigma}^{\mathbf{q}}(t_1, t_2)$ must be considered, with

$$\Sigma^{\mathbf{q}}_{ijkl}(t_1, t_2) = \Sigma_{q_{ij}, q_{kl}}(t_1, t_2) \tag{2.34''}$$

Cross-correlation spectra of two (or more) r.p. can also be defined (equation (A5.55)).

2.5.3 Operations on random processes

Operations on random processes are widely treated in Appendices A3 and A5, to which the reader is referred. However, a brief summary of the main results appears to be suitable for following developments. In particular, if $\tilde{y}(t)$ is a random process that arises as a function of a random process $\tilde{x}(t)$ so that $\tilde{y}(t) = g(\tilde{x}(t))$, the following results hold (equations (A3.1b) and (A3.4)):

$$\mu_y(t) = \int_{-\infty}^{\infty} g(x)\, p(x, t)\, \mathrm{d}x \tag{2.40a}$$

$$s_y^2(t) = \int_{-\infty}^{\infty} g^2(x)\, p(x, t)\, \mathrm{d}x - \mu_y^2(t) \tag{2.40b}$$

and, if $g(x)$ is a monotone function (equation (A3.32))

$$p_y(y, t) = p_x(g^{-1}(y), t) \left| \frac{\mathrm{d}}{\mathrm{d}y}(g^{-1}(y)) \right| \tag{2.41}$$

The derivative in the r.h.s. is replaced by the Jacobian when functions of more random variables are considered.

In the case that these functions are linear combinations of the independent random variables, results of practical interest are allowable on the mean vector and on the covariance matrix of the dependent random variables (equations (A3.26′) and (A3.27′)). Such expressions solve the whole problem of determining the probability law of linear functions of several random variables, when the independent random variables are distributed according to a two parameter *stable distribution*, i.e. their probability law is not changed by a linear transformation.

The normal distribution (Section A6.2) is the most meaningful such distribution for practical applications.

Operations of interest in structural mechanics are also integration (Section A5.2.1) and differentiation (Section A5.2.2). When linear differential equations with constant coefficients and stochastic r.h.s. $\tilde{x}(t)$ are involved by the structural problem model, three different approaches can be introduced in order to study, by the mean and correlation functions, the randomness of the solving random process $\tilde{y}(t)$, as will be seen in Chapter 4 (Section 4.3) in some detail. They make use of the theory of stochastic differential equations, of the impulse response function $h(v)$ (equation (A5.38)) and of the frequency response function $A(\omega)$ (equation (A5.73)) respectively.

2.6 RANDOM VARIABLES

A special case of r.p. is the *random variable*, already introduced in Section 1.2: each sample function is a *constant* $\grave{q}_k$ (i.e. once the initial value of the process has been set, it remains constant). Clearly, the r.v. is a *stationary process* (all properties remain constant in time) but it is *not ergodic* ($m_q(k) = \grave{q}_k$ and $b'_q(v, k) = \grave{q}_k^2$ depend on k).

Examples of r.v. are the yield stress at a point of a structure, or the initial out-of-plumb of a column.

Sometimes (e.g. the wind pressure on a structure during a storm, or the yield stress along a structural steel element [‡]), it is convenient to treat a random process as the sum of a *static* (i.e. constant) part and a *dynamic* (i.e. fluctuating or time-varying) part

$$\tilde{q}(t) = \tilde{q}_1 + \tilde{q}_2(t) \tag{2.42}$$

where $\tilde{q}_1$ is a r.v. and $\tilde{q}_2(t)$ a r.p., usually with zero mean.

The descriptive properties of a random variable can be derived by considering it as a particular stationary non-ergodic random process (see Appendix A2): note however that, if q is a r.v., $B_q(v)$ is a positive constant, $G_q^{\mathrm{v}}(\nu)$ is a Dirac delta function and so is $G_q^{\prime\mathrm{v}}(\nu)$.

2.6.1 Vectors of random variables

Consider n random variables $\tilde{x}_1, \tilde{x}_2, \ldots, \tilde{x}_n$ and denote their respective *mean* (or *expected*) *values* and *variances* by

$$\mu_{x_i} = \int_{-\infty}^{\infty} x_i\, p_{x_i}(x_i)\mathrm{d}x_i = \mathrm{E}[x_i] \tag{2.43a}$$

[‡] In this case, the independent variable is not time but the abscissa along the element axis.

$$s_{x_i}^2 = \int_{-\infty}^{\infty} [x_i - \mu_{x_i}]^2 p_{x_i}(x_i)\,dx_i = E[(x_i - \mu_{x_i})^2] \tag{2.43b}$$

$p_{x_i}(x_i)$ being the PDF of $\tilde{x}_i$.

The *covariance* Σ_{ij} between $\tilde{x}_i$ and $\tilde{x}_j$ is

$$\begin{aligned}\Sigma_{ij} &= \int_{-\infty}^{\infty}\int_{-\infty}^{\infty} (x_i - \mu_{x_i})(x_j - \mu_{x_j})\, p_{x_i,x_j}(x_i, x_j)\,dx_i\,dx_j \\ &= E[(x_i - \mu_{x_i})(x_j - \mu_{x_j})] \\ &= E[x_i x_j] - E[x_i]E[x_j]\end{aligned} \tag{2.44}$$

$p_{x_i,x_j}(x_i, x_j)$ being the joint PDF of $\tilde{x}_i$ and $\tilde{x}_j$.

Note that $\Sigma_{ii} = s_i^2$. Moreover as

$$|\Sigma_{ij}| \leqslant s_{x_i} s_{x_j} \tag{2.44'}$$

the normalized quantity

$$\rho_{ij} = \Sigma_{ij}/s_{x_i} s_{x_j} \tag{2.45}$$

known as the *correlation coefficient*, is between -1 and $+1$. If $\rho_{ij} = 0$, $\tilde{x}_i$ and $\tilde{x}_j$ are said to be *uncorrelated*, while $\tilde{x}_i$ and $\tilde{x}_j$ are said to be perfectly correlated if $\rho_{ij} = \pm 1$.

Two random variables, like two random processes, are *stochastically independent* if

$$p_{x_i,x_j}(x_i, x_j) = p_{x_i}(x_i)\, p_{x_j}(x_j) \tag{2.46}$$

Note that stochastic independence always implies uncorrelation, while, the converse statement is not true in general (Section A2.4).

In matrix notation, given a vector of random variables $\tilde{\mathbf{x}}$, they are characterized by (A2.33)–(A2.37):

(1) the vector $\boldsymbol{\mu}$ of the expected values;
(2) the covariance matrix Σ of the Σ_{ij} defined above; and
(3) the JPDF $p_{x_1,x_2,\ldots,x_n}(x_1, x_2, \ldots, x_n) = p_{\mathbf{x}}(\mathbf{x})$.

2.6.2 Matrices of random variables

In an analogous way, for a matrix $\tilde{\mathbf{x}}$ of random variables $\tilde{x}_{ij}$ $(i = 1, \ldots, n;$ $j = 1, \ldots, r)$, the *mean matrix* is denoted by

$$\boldsymbol{\mu}_{\mathbf{x}} = \{\boldsymbol{\mu}_{\mathbf{x}_1}, \ldots, \boldsymbol{\mu}_{\mathbf{x}_r}\} = [\mu_{ij}^{\mathbf{x}} = \mu_{x_{ij}}] \tag{2.47a}$$

where $\mathbf{x}_j$ means the jth column of $\mathbf{x}$, while the covariance and correlation matrices are

$$\boldsymbol{\Sigma}_{\mathbf{x}} = [\Sigma_{ijkl}^{\mathbf{x}} = \Sigma_{x_{ij},x_{kl}}] \tag{2.47b}$$

$$\boldsymbol{\Sigma}'_{\mathbf{x}} = [\Sigma_{ijkl}'^{\mathbf{x}} = \Sigma'_{x_{ij},x_{kl}}] \tag{2.47c}$$

2.7 POINT PROCESSES; EXTREME VALUES

A random process that describes a random sequence in time of random or deterministic events is said to be a *point process* (p.p.) (Fig. 2.6).

In order to define a p.p., one first needs knowledge of the relevant *counting process* $\tilde{n}(t)$ (Section A6.3) which counts the number of points occurring in an interval $(0, t)$ or, equivalently, of the description of the intervals between each event (*waiting times* or *occurrence intervals*; equation (A6.27)).

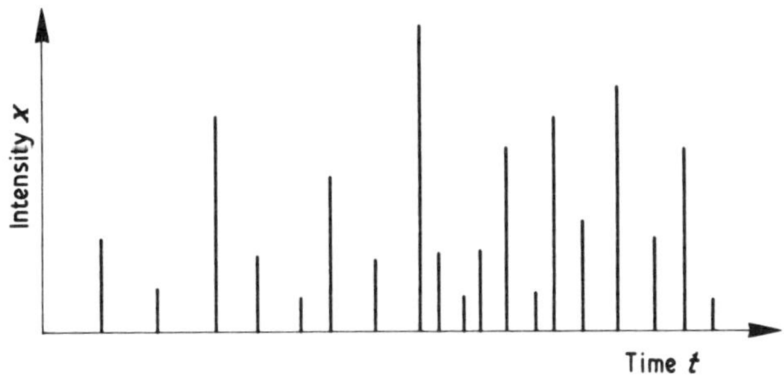

Fig. 2.6 Realization of a point process.

The most usual counting process model is the Poisson process (equation (A6.28))

$$\text{Prob}[\tilde{n}(t_2) - \tilde{n}(t_1) = k] = \frac{[\lambda(t_2 - t_1)]^k}{k!} \exp[-\lambda(t_2 - t_1)] \tag{2.48}$$

defined in Section A6.3.2, where λ is the *mean rate of arrival*. Then the waiting times obey the gamma probability law (Table A2.2) so that the *return period T*, that is the mean occurrence interval, is given by

$$T = \text{E}[t_j - t_{j-1}] = 1/\lambda \tag{2.49}$$

P.p.s are, for example, the violations of any given threshold $\bar{q}$ by a continuous r.p. $\tilde{q}(t)$ (Section A7.4.1): equation (A7.42) (valid for Gaussian processes) states that the mean rate of threshold violations is

$$\lambda_q = c_1 \exp(-\bar{q}^2/c_2) \tag{2.50}$$

showing that the higher the threshold, the larger is the return period T of the overcrossings.

Each *event* corresponding to an *occurrence* in a p.p. may have an *intensity*, which in turn may be a random variable $\tilde{x}$ with CDF $P_x(x)$.

If the p.p. has a deterministic mean rate of arrival λ, the developments of Sections A7.3 and A7.5.1 associate a return period $T(x)$ with the event $\tilde{x} \geqslant x$ (equation (A7.32))

$$T(x) = \frac{1}{1 - \exp[-\lambda(1 - P_x(x))]} \simeq \frac{1}{\lambda}\frac{1}{1 - P_x(x)} \tag{2.51}$$

The latter approximate equality holds if $P_x(x) \ll 1$.

2.7.1 Maxima of random processes

Point processes are also those describing the local maxima of a continuous r.p. $\tilde{q}(t)$ (Section A7.4.2). The maxima of a continuous process $\tilde{q}(t)$ in $(0, t)$ can also be given either as a r.p. $\tilde{y}(t)$ or a r.v. $\tilde{y}_t$ given t. Suppose that the random variable $\tilde{y}_1$ (largest value of $\tilde{q}(t)$ in a unit time interval) is known: in many applications, the process $\tilde{q}(t)$ can be replaced by the point process defined by the sequence of realizations of $\tilde{y}_1$ in successive unit time intervals. Such an idealization is especially suitable when successive occurrences are stochastically independent, so that (equation (A7.25))

$$P_{y_n}(y) = [P_{y_1}(y)]^n \tag{2.52}$$

whence the PDF of $\tilde{y}_n$ is given by

$$p_{y_n}(y) = n[P_{y_1}(y)]^{n-1} p_{y_1}(y) \tag{2.53}$$

For the present problem $\lambda = 1$ and equation (2.51) becomes (equation A7.32)

$$T(y) = \frac{1}{1 - P_{y_1}(y)} \simeq \frac{-n}{\ln[P_{y_n}(y)]} \simeq \frac{n}{1 - P_{y_n}(y)} \tag{2.54}$$

For the accuracy of these approximations see Section A7.3.

Equations (2.52) and (2.53) are also useful for investigating the properties of a p.p. for which the CDF $P_x(x)$ of the intensity of the single event is known. In fact, if $\tilde{y}$ is the maximum value of $\tilde{x}$ in $(0, t)$ and $\tilde{n}$ in $(0, t)$ is a r.v., the CDF of $\tilde{y}$ is given by

$$P_y(x) = \sum_{n=0}^{\infty} [P_x(x)]^n \operatorname{Prob}[\tilde{n} = n] \tag{2.55}$$

When no information is available in addition to the mean value μ_n, $\operatorname{Prob}[\tilde{n} = n]$ is generally assumed to be given by the binomial law $\mathscr{b}(n)$ (Table A8.1) with parameter μ_n.

2.8 BASIC FORMULATIONS OF RELIABILITY CALCULATIONS

It is now possible to put the basis for the calculation of the probability of failure, P_{fail}, on more formal bases than in Chapter 1. In fact, with the introduction of the

JPDF $p(\mathbf{Q})$ of the vector $\mathbf{Q}$ of the relevant random quantities, equation (1.5) becomes

$$P_{\text{fail}} = \int_{\mathscr{U}} p(\mathbf{Q})\,\mathrm{d}\mathbf{Q} \tag{2.56}$$

$$\mathbb{R} = P_{\text{suc}} = \int_{\mathscr{S}} p(\mathbf{Q})\,\mathrm{d}\mathbf{Q} \tag{2.57}$$

where $\mathscr{U}$ and $\mathscr{S}$ are the already defined *unsafe* (or *failure*) and *safe* regions respectively. Because in structural engineering

$$P_{\text{fail}} \ll P_{\text{suc}} \tag{2.58}$$

(cf. Section 1.7), equation (2.56) is much better conditioned from a numerical viewpoint, and therefore used more often than equation (2.57) in applications.

2.8.1 Critical inequality

Extrapolating the presentation of Chapter 1, it can be stated that, for any given structure and mode of failure, the *condition of success* can often be formulated in the form of a *critical inequality*

$$\tilde{S} \leqslant \tilde{R} \tag{2.59}$$

where $\tilde{R}$ and $\tilde{S}$ are two random variables, which measure respectively a generalized *resistance* and the *action* (*load*) or *load effect* homogeneous with $\tilde{R}$.

In other words, the *probability of failure* P_{fail} (equations (1.5) and (2.56)) is given by the probability of violating the critical inequality

$$P_{\text{fail}} = \text{Prob}[\tilde{S} > \tilde{R}] \tag{2.56'}$$

According to the choice of $\tilde{R}$ and $\tilde{S}$ (which in general depend on input, system and output quantities), inequality (2.59) is written in one of several *spaces*, which influences the practical calculation of P_{fail}. The relevant possibilities are the following:

(1) Critical inequality in the *system parameter space*: $\tilde{R}$ is a known system quantity, while $\tilde{S}$ depends on the output quantities, in turn depending on input and system quantities (Fig. 2.7(a)):

$$\tilde{S} = S(\tilde{\mathbf{u}}) = S[\tilde{\mathbf{u}}(\tilde{\mathbf{W}}, \mathscr{K})] \tag{2.60}$$

This is for instance the case of a stress check in a given point in the structure: $\tilde{R}$ coincides with the limit stress (usually the yield stress).

(2) Critical inequality in the *output space*: $\tilde{S}$ is an output quantity, while $\tilde{R}$ must be derived from some system quantities (Fig. 2.7(b)):

$$\begin{aligned} \tilde{S} &= \tilde{u}_i = u_i(\tilde{\mathbf{W}}, \mathscr{K}_1) \\ \tilde{R} &= R(\mathscr{K}_2) \end{aligned} \tag{2.61}$$

where $\mathscr{K}_1$ and $\mathscr{K}_2$ are subsets of $\mathscr{K}$. This is, for example, the case when $\tilde{S}$ is identified with a cross-sectional load effect (bending moment, axial force, shear, torsion) and $\tilde{R}$ with the relevant resisting quantity; another example is given by a displacement-limitation check in which case the second of equations (2.61) becomes simply $R = u_{\text{adm}}$, u_{adm} being a prescribed admissible value.

Sometimes it is useful to distinguish, in the class of the output variables, the ones directly observable in a physical sense, from the *state variables* which are functionally dependent on the basic variables through an appropriate mechanical model, but are not directly measurable.

Note that cases (1) and (2) may coincide under specific system descriptions: for instance, when the randomness of the cross-sectional properties has been independently investigated and is among the data of the problem.

(3) Critical inequality in the *input space* (*load space*): $\tilde{S}$ is a known input quantity, while $\tilde{R}$ depends on the output quantities (Fig. 2.7(c)):

$$\tilde{R} = R[\tilde{\mathbf{u}}) = R[\mathbf{u}(\tilde{\mathbf{W}}, \mathscr{K})] \tag{2.62}$$

Such a situation occurs when the overall carrying capacity of a structure subject to given actions is determined, as is done, for example, in plastic collapse analysis.

The explicit formulation of the critical inequality coincides with the solution of the structural analysis problem in a form suitable for probabilistic treatment: it has therefore been necessary to pay more attention to its alternative forms than is usually done in structural analysis books.

If N different limit states (or modes of failure) are possible, a critical inequality can be written corresponding to each mode:

$$\tilde{S}_k \leqslant \tilde{R}_k \qquad k = 1, 2, \ldots, N \tag{2.63}$$

where the functional forms of l.h.s. and r.h.s., in general, vary with k.

For weakest-link type of behaviour, failure can be identified with violation of at least one of the inequalities (2.63), and the probability of failure is

$$P_{\text{fail}} = \text{Prob}[\tilde{S}_k > \tilde{R}_k \text{ for at least one } k] \tag{2.64}$$

Equation (2.64) can be treated as discussed in Section 1.4.

For failsafe structures, the derivation of the overall probability of failure is not so straightforward. In other cases, the critical inequalities corresponding to different limit states must be considered separately, because of their different consequences.

Sometimes, it may be convenient to consider (2.63) as a *vector inequality*, in the form

$$\tilde{\mathbf{S}} \in \mathscr{S} \tag{2.65}$$

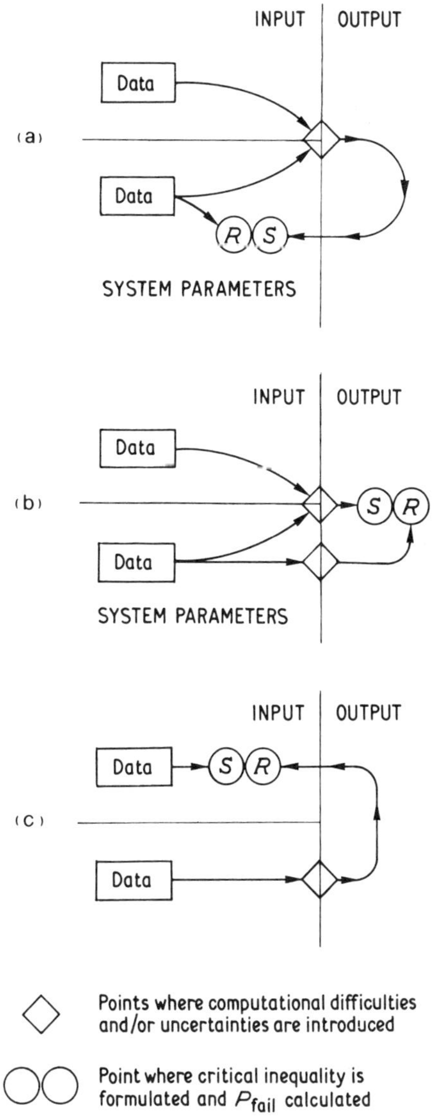

Fig. 2.7 Spaces of random quantities and paths followed to establish critical inequality and computing P_{fail} (diagrammatic).

with $\tilde{\mathbf{S}}$ a vector components $\tilde{S}_k$, and $\mathscr{S}$ a region usually comprising the origin and in general with a fuzzy boundary: this approach will be made clear by the treatment of Section 2.8.2 (equations (2.68), (2.69)) and the examples of applications presented in Chapter 5.

2.8.2 Probability of failure as a convolution integral

Let us return to the expression (2.56) of the probability of failure, and indicate by $p_{S,R}(S, R)$ the JPDF of the random variables $\tilde{S}$ and $\tilde{R}$ (Fig. 2.8). Then, integrating with respect to $\tilde{R}$ first (Fig. 2.9),

$$P_{\text{fail}} = \iint_{[S>R]} p_{S,R}(S,R)\,\mathrm{d}S\,\mathrm{d}R = \int_0^\infty \mathrm{d}S \int_0^S p_{S,R}(S,R)\,\mathrm{d}R \qquad (2.66)$$

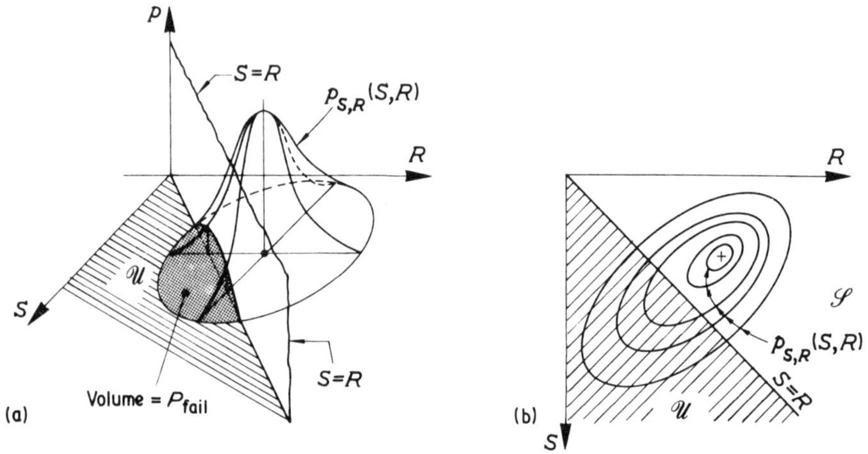

Fig. 2.8 JPDF of load effect S and resistance R, and probability of failure P_{fail}.

If $\tilde{S}$ and $\tilde{R}$ are stochastically independent,

$$p_{S,R}(S,R) = p_S(S)\,p_R(R) \qquad (2.66')$$

and equation (2.66) becomes

$$P_{\text{fail}} = \int_0^\infty p_S(S)\left[\int_0^S p_R(R)\,\mathrm{d}R\right]\mathrm{d}S = \int_0^\infty p_S(S)P_R(S)\,\mathrm{d}S \qquad (2.66'')$$

i.e. the P_{fail} is given by the convolution integral of two functions of S (equation (A3.45)).

Note that $P_R(S)$, the CDF of the r.v. $\tilde{R}$, is by definition

$$P_R(S) = \text{Prob}[R < S] \qquad (2.67)$$

and, therefore, it can be regarded as the *conditional probability of failure* (CPF) of the structure considered when subjected to a well determined load (or load effect) S. The last integral in (2.66″) thus has very clear physical evidence: it is the integral of the probability $p_S(S)\,\mathrm{d}S$ of the load falling in the infinitesimal interval $(S, S + \mathrm{d}S)$ multiplied by the probability (CPF) of the structure failing under that load (load effect).

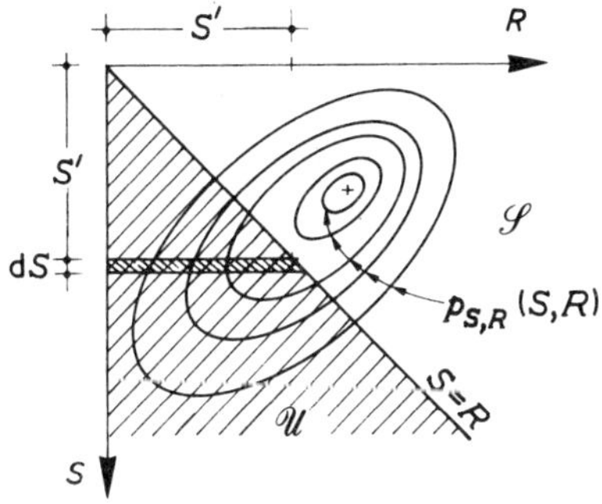

Fig. 2.9 Integration of $p_{S,R}(S, R)$ to obtain P_{fail}.

This latter interpretation allows a generalization of equation (2.66″), particularly useful when the reliability analysis is performed in the *input* (i.e. *load*) space. In fact, provided that the parameters defining the load conditions can be neatly distinguished from the quantities affecting the structural resistance, the probability of failure can be expressed in the form

$$P_{fail} = \int_{\mathscr{D}} p_{\mathbf{W}}(\mathbf{W}) P_R(\mathbf{W}) \, d\mathbf{W} \tag{2.68}$$

where $p_{\mathbf{W}}(\mathbf{W})$ is the JPDF of the random load vector $\tilde{\mathbf{W}}$, the region of integration $\mathscr{D}$ is the (real) space of definition of $\mathbf{W}$ and $P_R(\mathbf{W})$ is an evident extension of the CPF defined above (provided it exists). A simple two-dimensional example of equation (2.68) is shown qualitatively in Fig. 2.10. From Fig. 2.10(a) it is evident that in the case of a *deterministic structure* (i.e. no technological uncertainty), equations (2.68) and (2.56) coincide in the form

$$P_{fail} = \int_{\mathscr{U}} p_{\mathbf{W}}(\mathbf{W}) \, d\mathbf{W} \tag{2.68'}$$

where the *failure* (*unsafe*) *region* $\mathscr{U}$ is defined by

$$\mathscr{U}: \{P_R(\mathbf{W}) = 1 \quad \text{for all} \quad \mathbf{W} \in \mathscr{U}\} \tag{2.69}$$

Conversely, the *safe region* $\mathscr{S}$ is given by

$$\mathscr{S}: \{P_R(\mathbf{W}) = 0 \quad \text{for all} \quad \mathbf{W} \in \mathscr{S}\} \tag{2.69'}$$

In the more general case (*random strength*) the boundary between $\mathscr{S}$ and $\mathscr{U}$ becomes fuzzy (Fig. 2.10(b)), and the function $P_R(\mathbf{W})$ can be regarded as the measure of *membership* of the point $\mathbf{W}$ to the *failure region* [4].

Note that the advantage of equation (2.68) over (2.56) is not only to reduce the dimensions of the integral to those of $\mathbf{W}$, but also to distinguish between two sources of uncertainties, usually of very different nature, which deserve very different treatment. In fact, $p_{\mathbf{W}}(\mathbf{W})$ contains the random uncertainties of the loads

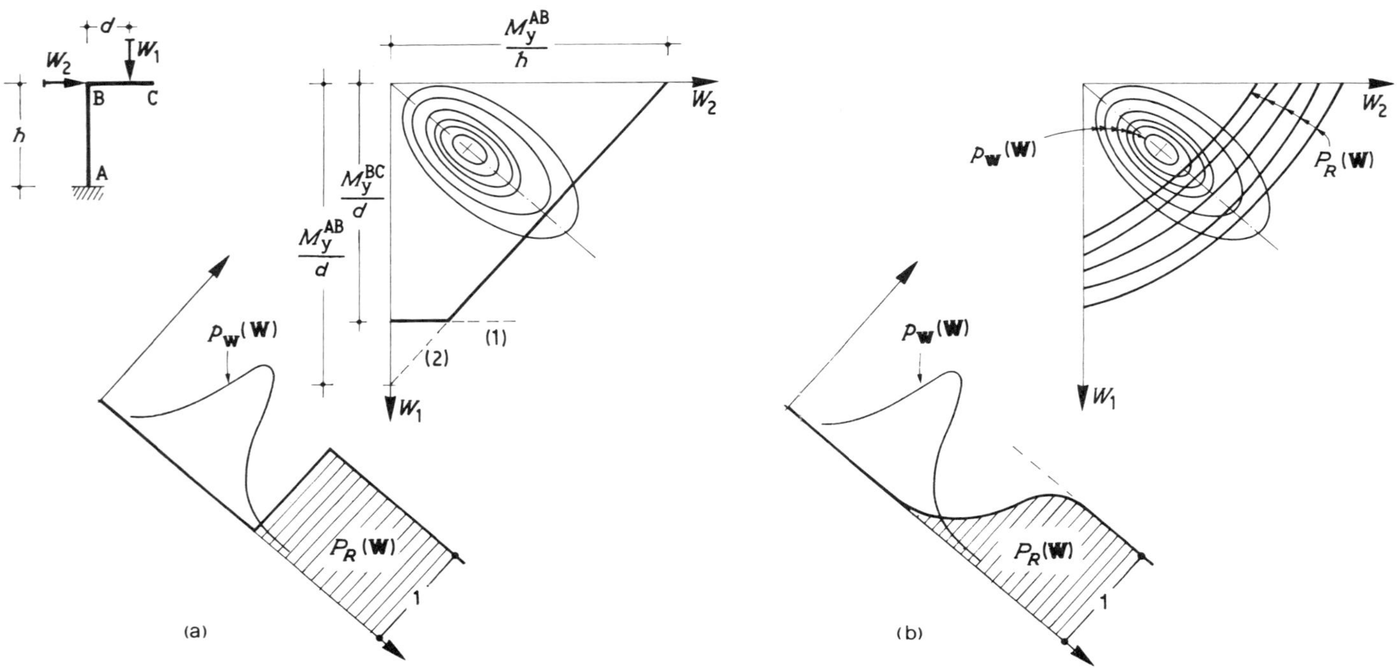

Fig. 2.10 Probability of failure of example structure with two-parameter loading (cf. Fig. 1.6).

which, broadly speaking, should be derived from statistical analyses of the data on the loads (*input* quantities), while $P_R(\mathbf{W})$ essentially contains the *system* uncertainties, and is to be obtained from *probabilistic structural analysis*.

2.8.3 'Level 1' and 'Level 2' approaches

The numerical determination of P_{fail} from any of equations (2.56), (2.64) or (2.68) is seldom easy, especially because of its low values. Also, the statistical data in the relevant low-probability ranges are often incomplete and/or unreliable. Because of these difficulties, simplified procedures are often used in technical problems: these will be treated in detail in Chapter 10, but are briefly introduced here.

Consider (Fig. 2.11) the probability density functions (PDFs) of the two r.v. $\tilde{S}$ and $\tilde{R}$ that appear in the relevant critical inequality (2.59), hence in equation (2.66),

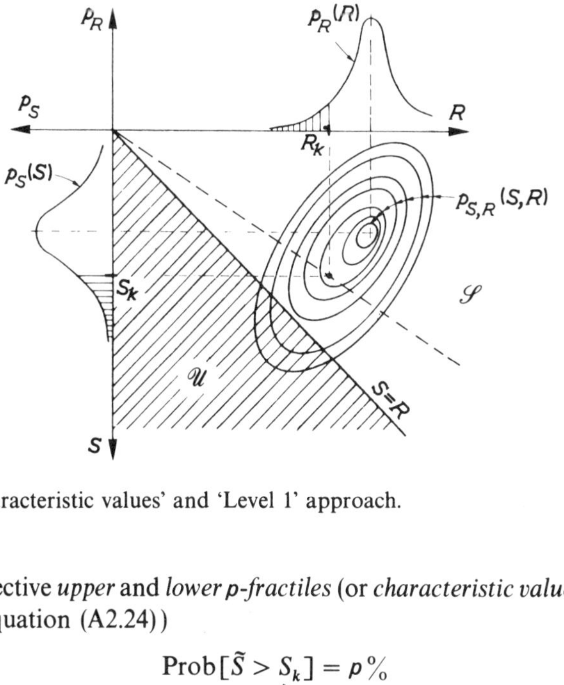

Fig. 2.11 'Characteristic values' and 'Level 1' approach.

and their respective *upper* and *lower p-fractiles* (or *characteristic values*) S_k and R_k, defined by (equation (A2.24))

$$\begin{aligned} \text{Prob}[\tilde{S} > S_k] &= p\,\% \\ \text{Prob}[\tilde{R} < R_k] &= p\,\% \end{aligned} \tag{2.70}$$

p being usually taken equal to a few (2 to 5) per cent. Then at *Level 1* instead of seeking P_{fail}, one simply computes the *characteristic safety factor* (Fig. 2.11)

$$\gamma = R_k/S_k \tag{2.71}$$

It is evident that, in general, the larger γ, the smaller is P_{fail}: a consistent comparison is thus often possible.

Level 2 procedures require determination of only *averages* μ (first moments) and *standard deviations* s or *variances* s^2 (second moments) of the relevant random variables. Then, in the simplest formulation, a *reliability index* β is defined which measures, in units of standard deviations, the distance between the *average* point and the boundary of the *failure* region (see e.g. Fig. 2.12): as a rule, larger values of β imply smaller probabilities P_{fail}. Level 2 procedures assume many alternative forms, which will be presented in Chapter 10. Suffice it here to say that, because of the availability of these procedures, many probabilistic structural calculations are aimed at the determination of first and second moments only.

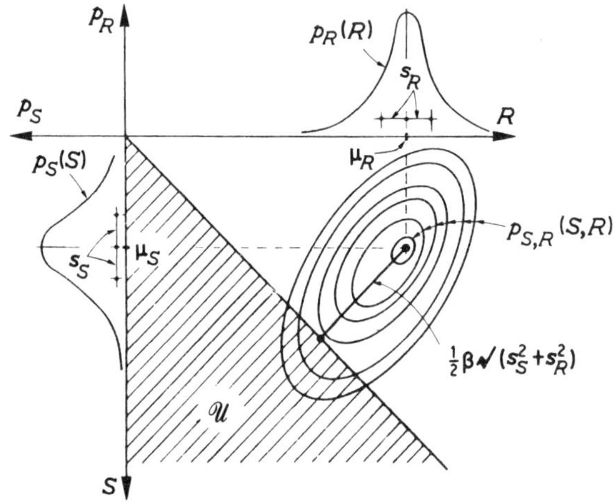

Fig. 2.12 'Level 2' approach.

Finally, a remark appears to be of some interest. For the fundamental problem of two random variables $\tilde{R}$ (resistance) and $\tilde{S}$ (action), typical relations between the probability of failure, the coefficients of variation δ_R and δ_S (2.12′) and the characteristic safety factor γ (2.71) are shown in Fig. 2.13. From inspection of these curves, the probability of failure P_{fail} appears to depend little on δ_S if δ_R is high (curves 10, 11, 12 in Fig. 2.13(b)): this is certainly a good reason to justify the fact that structural engineering has been for a long time devoted mainly to the computation of R rather than to the estimation of S. But in the last two decades structural analysis methods have greatly improved and, at the present time, are very sophisticated and accurate. In other words, in the last two decades δ_R has been reduced more and more so that, as evident from Fig. 2.13(a, c), further improvements in structural design have to be pursued also by accurate definitions of the load properties. The next chapter is wholly devoted to problems of the latter type.

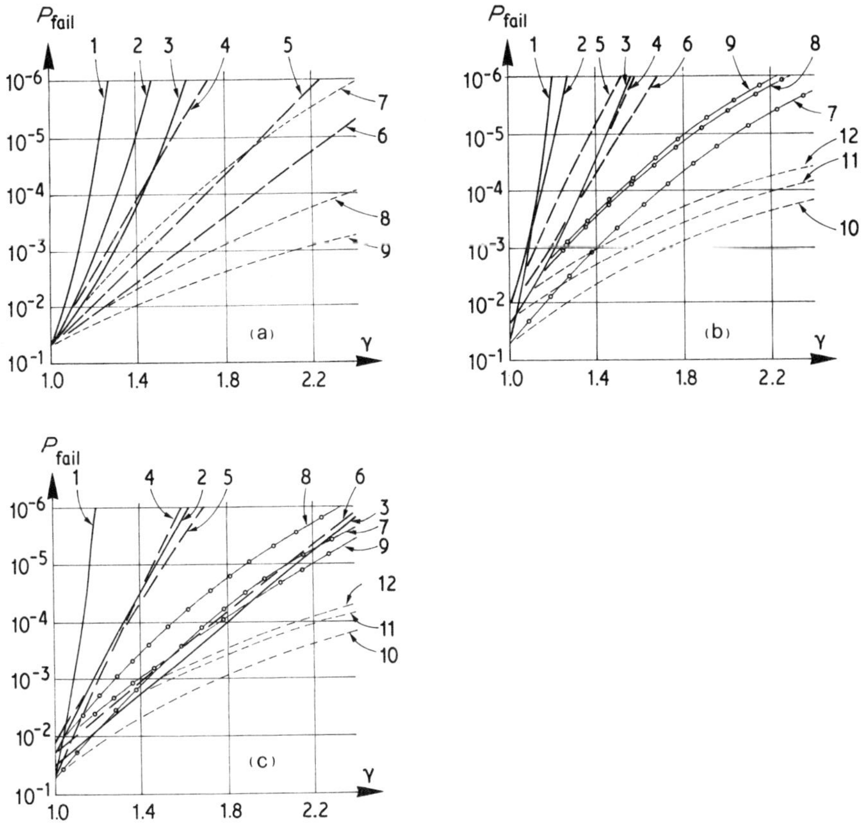

Fig. 2.13 Probability of failure P_{fail} vs characteristic safety factor γ for given distribution and coefficient of variation of resistance (δ_R) and action (δ_S): examples [5].

(a) Deterministic resistance ($\delta_R = 0$)

Load distribution	δ_S 0.1	0.2	0.3
Normal	1	2	3
Extreme type I	4	5	6
Extreme type II	7	8	9

(b) Normally distributed resistance and load

δ_R	δ_S 0	0.1	0.3
0.05	1	2	3
0.10	4	5	6
0.15	7	8	9
0.20	10	11	12

(c) Normally distributed resistance; extreme type I load

δ_R and δ_S as in (b)

REFERENCES[‡]

1. Gnedenko, B. V. (1967) *Theory of Probability*, Chelsea, New York.
2. Bendat, J. S. and Piersol, A. G. (1971) *Random Data: Analysis and Measurement Procedures*, Wiley Interscience, New York.
3. Veneziano, D. (1978) *Random Processes for Engineering Application* (Lecture Notes), Dept. Civil Engineering, M.I.T., Cambridge, MA.
4. Kandel, A. and Byatt, W. J. (1978) Fuzzy sets, fuzzy algebra and fuzzy statistics. *Proc. IEEE*, **66** (12), 1619–30.
5. Ferry-Borges, J. and Castanheta, M. (1971) *Structural Safety*, LNEC, Lisbon.

‡ These lists of references, presented at the end of each chapter, are not intended to be an exhaustive or selected bibliography, but only to acknowledge the sources of specific material.

CHAPTER 3

Models of external actions

After some general considerations on the modelling of random external actions, three common loading conditions are analysed in some detail: live loads, seismic loading and wind loading. Finally, the problem of load combination is dealt with.

3.1 LOAD PROCESSES

The term *external actions* denotes the loads generated by causes physically external to the structure on which they act. The term includes thermal deformations and in general imposed displacements and deformations, like permanent support settlements, and so on.

Such actions have to be modelled paying particular attention to the requirement that the model must carefully reflect only those features of the action that influence the structural response. Therefore, the model depends on the characteristics of both action and structure, and in many cases full adherence to the action in all its complexity is not necessary. For instance, an almost-periodic force $W(t)$ (Fig. 3.1 (a)), with spectral density $G^{\circ}(\omega)$ (Fig. 3.1 (b)), can be satisfactorily described by only its maximum and minimum values W^{+}, W^{-} in the expected lifetime of the structure if (a) fatigue is not a problem, and (b) the first natural frequencies of the structure, corresponding to the real peaks of the structural frequency response function $A(\omega)$ (equation (A5.41 (b)) do not fall in any interval where $G^{\circ}(\omega)$ is significant‡. On the contrary, a more complete description of $W(t)$

‡ In fact, as qualitatively shown by Fig. 3.1 (b), if the peaks of $|A(\omega)|$ fall to the right, and $G^{\circ}(\omega)$ is significant only where $|A(\omega)|$ is approximately constant, the spectral density of the structural response

becomes necessary if significant superposition of the ordinates of the two functions occurs. In seismic engineering, for instance (cf. Fig. 3.12), accurate tests of structural safety require dynamical analysis of the structure subjected to accelerograms fully reflecting the desultory nature of earthquake actions; in some cases, satisfactory results have been found describing the earthquake by a white-noise stochastic process (Table 2.3 and equation (A5.47 (b))), despite the fact that white-noise is physically unrealizable in that it requires that an infinite quantity of energy (Section 2.4.4) be released and despite the fact that earthquakes are not stationary processes.

It is therefore clear that suitable models of actions are the fruits of a combination of observation, engineering insight and experience, and always involve both the environment and the structure model. In fact, the state of the structure at any time depends in general on the whole history of the applied loading environment (which varies in both time and space around the structure) and on all the mechanical characteristics of the structure. The main simplification usually made in load representation is to reduce this general description to an easier, separable form. Thus, transmission of data becomes much easier and the code aims (Chapter 10) can be achieved [1].

In this separable form, the first parameter is a gross, structure-free representation of the loading environment, *per se*. For example, this might be:

(1) the intensity of the (uniformly distributed) occupancy load on the floor of an office building;
(2) the peak ground acceleration of an earthquake motion, a;
(3) the 10 min average wind velocity v.

This environment factor is independent of the mechanical characteristics of the structure and generally involves some temporal and/or spatial 'averaging' (e.g. 10 min average velocity, peak-over-the-event acceleration, average-over-the-floor occupancy load.) Therefore, this first parameter involves primarily *macro-scale time and space dimensions* (of the order of the life of the structure and several orders larger than its size respectively).

The second parameter in the separated description is a factor that represents the interaction of load and structure. For example, this might be:

(1) a live-load reduction factor;
(2) response spectrum for an earthquake load, standardized to unit peak accelerations;
(3) a wind 'gust factor'.

---

is approximately proportional to $G^\circ(\omega)$, hence the relevant response parameters are essentially proportional to the intensity of the force.

On the other hand, if $G^\circ(\omega)$ would fall entirely to the right, where $|A(\omega)| \simeq 0$, the structure would appear not to respond at all. Note however that the frequency response function of a continuous deformable structure has an infinite number of peaks, albeit usually of magnitude decreasing with ω: therefore the condition of negligible response under high-frequency loading must be carefully verified for actual structures.

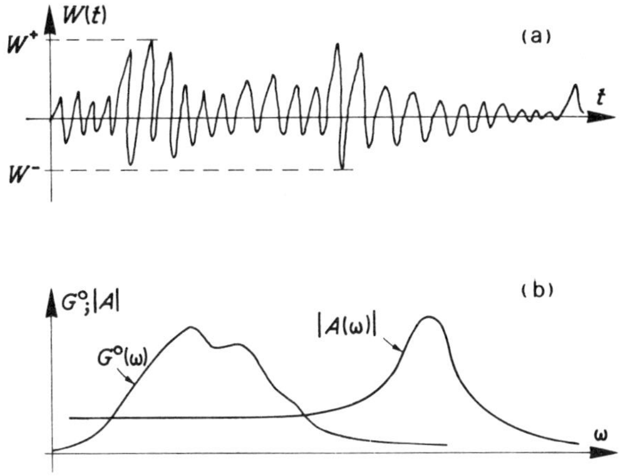

Fig. 3.1 (a) Almost-periodic force $W(t)$; (b) spectral density of $W(t)$ and frequency response function of relevant structure.

This interaction factor normally depends only on gross structural parameters (e.g. the fundamental natural period of the structure or the gross external dimensions of the building) and on only *medium-scale time and space characteristics of the loading environment*; that is, the temporal and spatial loading dimensions involved in this factor are those of the same order as the fundamental period of the structure (Chapter 4) and its gross size.

The final factor in the separated description takes account of the actual structural response under a given imposed load pattern. The method of analysis is thus effectively independent of the load but depends on the detailed mechanical characteristics of the structure (Chapters 4, 5 and 6). The *spatial scale* is of the same order as the dimensions of structural members and cross-sections. The long- and short-term temporal load variations are ignored, unless dynamic effects are accounted for [1].

This separated scheme for the analysis of the state of a structure has grown up in virtually all the types of problems considered by civil engineers: the quantities that cannot easily be classified as *input* or *system* quantities (Section 1.1) are collected into the *load-structure interaction factor*.

Thus the detailed structural analysis can be done without concern for the load causes, while the interaction factor can be specified before the details of the structure are known, avoiding major iterations. The gross loading environment can be studied without regard for the structure involved.

In the following sections of this chapter, attention is focused on the first two factors. Then, Chapters 4 and 5 essentially deal with methodologies for treating the fundamental equations of mechanical systems when the latter are affected by random uncertainty. Finally, in Chapter 6, it will be shown how a consistent theory of structural safety can be set up on the basis of probabilistic methods. This treatment will clearly show the features of practical applications that, in the usual engineering approach, require the introduction of some 'safety coefficients', whose definition is essentially empirical and in little accord with the mathematical rigour of the general principles of the theory of structures and the accurate numerical results allowed by modern computing procedures (Section 2.8.2).

3.2 CLASSIFICATION OF THE ACTIONS

Generally, the description of loads must contain a set of information such as the space distribution and the intensity of forces, the rate of variation in position and/or intensity, their dependence on human activities, the time intervals in which they are active, and so on.

Most often, as already noted in the simple example of Fig. 3.1, such information is not all necessary for a significant test of structural performance. In order to specify the level of detail more easily, a classification of actions can usefully be attempted, specifying for each class the dominant features to be enhanced in the model. Thus, when an action is classified in one or more classes, its significant features are automatically indicated.

A possible classification is the following:

(1) *Static* (or *quasi-static*) *actions*: are applied to the structure without causing significant accelerations.
(2) *Dynamical actions*: cause significant accelerations because of significant variations in time, comparable with the fundamental period of the structure.
(3) *Permanent actions*: are active on the whole lifetime of the structure without changing in intensity, direction and position.
(4) *Time-varying actions:* exhibit different intensities and/or positions at different instants.
(5) *Short-term actions*: act for a time much shorter than the structure's lifetime.
(6) *Long-term actions*: act for a time of an order of magnitude comparable with the structure's lifetime.
(7) *Exceptional actions*: act in dependence on a very unlikely event.
(8) *Fixed actions*: act with a spatial distribution fixed in time.
(9) *Free actions*: act with varying spatial distribution.

Each of the above classes implicitly specifies the information that must be provided in order to obtain the knowledge necessary for reliability analysis.

Wind load acting on a stiff structure, for instance, can be classified as a 1–4–5–9 action (i.e. it belongs to classes 1,4,5 and 9). In contrast the same wind acting on a tall steel building is a 2–4–5–9 action, in that dynamic effects cannot be neglected. In the first case, an intensity of the wind pressure must be specified, possibly through its maximum or mean value in each wind event: the largest of such values in all the events can be assumed in the calculations. The spatial distribution must also be specified (for instance the 'wind profile'). In the second case, the single wind cannot be described simply by the maximum pressure, but its variation with time should be taken into account.

Unfortunately, the parameters specifying the actions cannot be expressed simply by numbers or deterministic functions. The maximum wind pressure in a given time, for instance, cannot be predicted exactly, in that it is a random variable, and, even when nominal values are considered, for instance in Codes of Practice, they should be derived from statistical analysis of a stochastic process: the best knowledge of the loads that will presumably be applied on the structure can only be achieved from a probabilistic point of view.

Thus, let us now consider what kind of information must be supplied in a probabilistic context for each class of actions mentioned above. Consider, first, loads specified by only one parameter $W = W(t)$, as in any of the simple cases shown in Fig. 3.2. Problems deriving from the superposition of different loads will be discussed in Section 3.7.

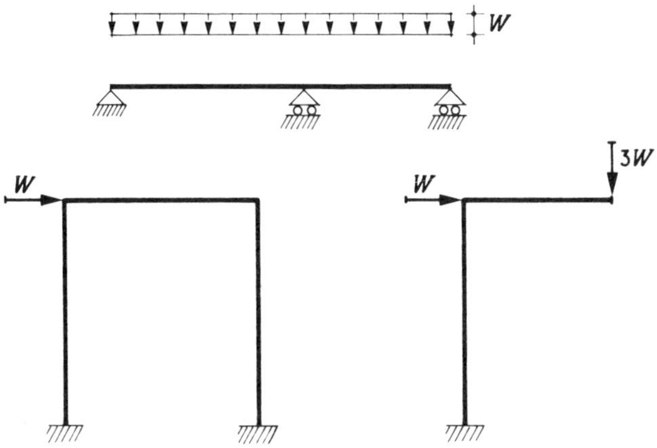

Fig. 3.2 Examples of structural problems.

(1) *Static actions*: $\tilde{W}(t)$ is a stochastic process (Fig. 3.3), but in general only the PDF of its extreme values in the interval $(0, T)$ is necessary (cf. equation (1.4)), T being the *design life* of the structure.
(2) *Dynamical actions*: $\tilde{W}(t)$ is a stochastic process as above, but at least a second-moment description of the process is necessary (i.e. mean value and autocorrelation function) for dynamic analysis of the structure. The second-moment description is generally sufficient when it is possible to assume the load process as stationary with a stable PDF and two degrees of freedom (Section A3.6) and the structural system as linear, as will be shown in Section 4.3. Otherwise, a more complete description must be provided.

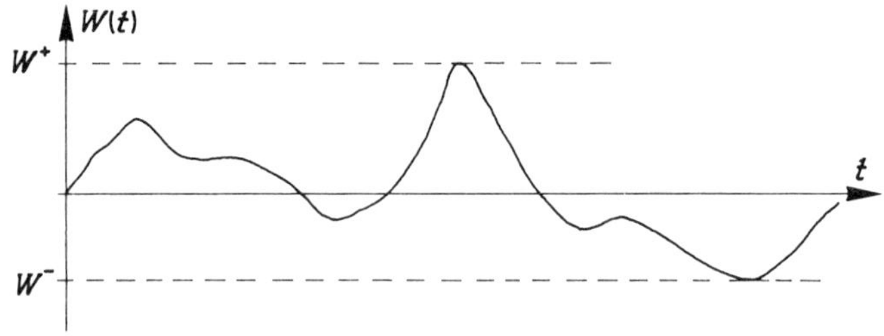

Fig. 3.3 Action with static effect only.

(3) *Permanent actions*: $\tilde{W}(t)$ is a random variable (Section 2.6), fully described by a PDF (Fig. 3.4). This description can be simplified by limiting it to characteristic values or to first and second moments only (cf. Section 2.8.3).

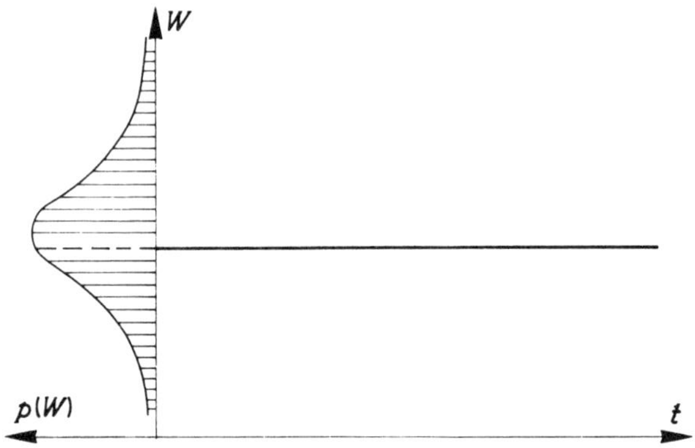

Fig. 3.4 Permanent action.

(4) *Time-varying actions*: Often, even if the action has no dynamical effects, the simple description through extreme values only (as for static actions) is not sufficient. For instance, it may be necessary to take account of the probability of occurrence of loads above each given level, for possible accumulation of damages and/or combination with other loads. Also in this case, second-moment or more complete descriptions can be required, according to the specific problem. In many instances, time-varying actions are essentially discontinuous processes (Figs 3.5 and 3.6). In these cases, they are usually described through the *point process* of the load occurrence (Section 2.7), and the properties of the load when it does act: short- and long-term loads can be distinguished.

(5) *Short-term actions*: $\tilde{W}(t)$ is a discontinuous process, possibly with a large number of repetitions; the description of the occurrence process must be accompanied by the description of some basic load parameters at each occurrence (e.g. intensity, duration, etc.).

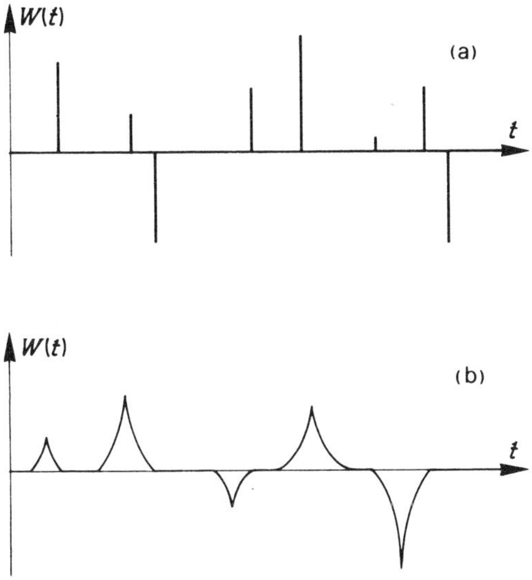

Fig. 3.5 Examples of short-term actions: (a) impulsive process; (b) filtered Poisson process (FPP), equation (A6.43).

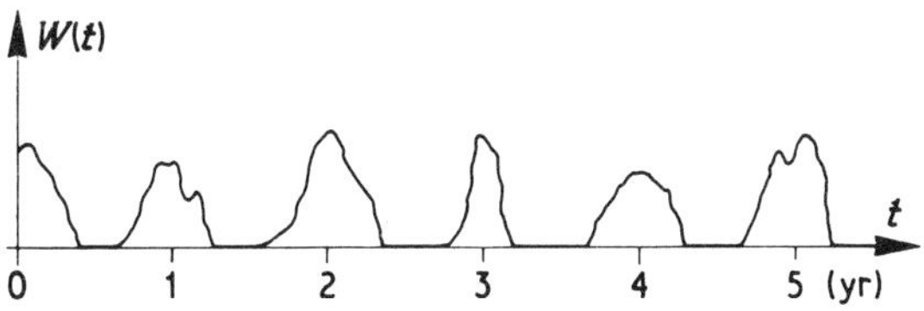

Fig. 3.6 Snow load.

(6) *Long-term actions*: $\tilde{W}(t)$ is a continuous or discontinuous process with comparatively slow time variation once the load has occurred (e.g. snow load in a given region, Fig. 3.6). Besides the basic parameters as for short-term actions, information on the load variation with time must be provided.

(7) *Exceptional* (or *abnormal*) *actions:* the probability of occurrence is necessary for practical applications, while, because of lack of data, the intensity of the load is generally specified by a nominal value only.

(8) *Fixed actions*: besides describing the load intensity, it may be necessary to supply the probability laws of random variables which define the position of the action.

(9) *Free actions*: $\tilde{W}(x,y,t)$ is a stochastic field; at least a second-moment description of the dependence on the space coordinates x and y must be given, besides the appropriate description of the dependence on t.

In the following sections, some common examples of such load descriptions will be considered.

3.3 FLOOR LOADS IN BUILDINGS

Gravity loads are usually categorized into *dead* (or *permanent*) loads and *live* loads: the latter in general include the weight of furniture, people, equipment, etc. Unlike other kinds of loads, such as seismic or wind loads, live load can be taken to be under human control, at least within some limits. In actual fact, live loads are stochastic processes depending upon both spatial and temporal variables.

The gravity floor load process is conveniently idealized as a sum of component load processes (Fig. 3.7). A dead load always acts at a constant random value, while the stochastic variability of live loads is generally modelled by summing a *sustained live load* process and an independent *transient live load* process [2]. The sustained live load process represents live loads which maintain an approximately constant value for the duration of an occupancy, and the transient live load process represents live loads of shorter duration. Combination of sustained and

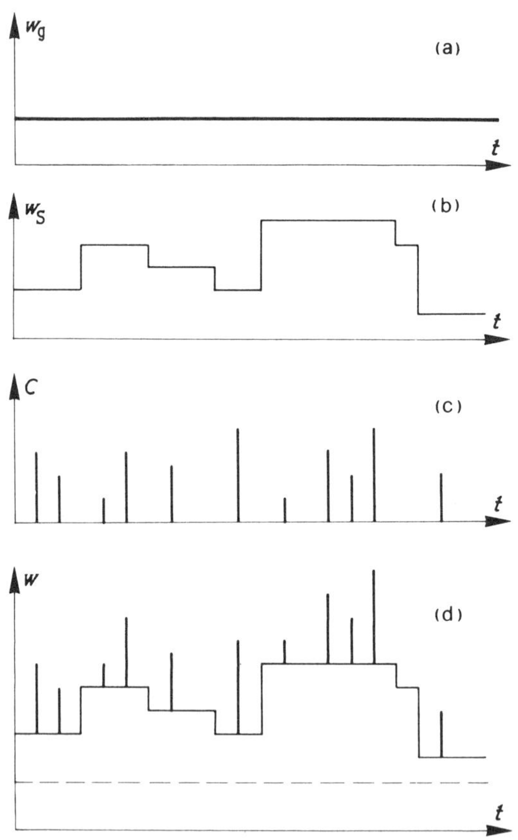

Fig. 3.7 Idealization of floor load processes: (a) permanent load, w_g; (b) sustained live load, w_s; (c) transient live load, C_C; (d) total load, $w = w_g + w_s + C_C$,

transient loads is discussed in general in Section 3.7; here, models for each are presented.

3.3.1 Sustained floor load

Consider first spatial modelling of the sustained load. The instantaneous intensity $\tilde{w}(x,y)$ of the load per unit area at a point of horizontal coordinates (x,y) in a given building can be expressed as the sum of an overall mean value μ_w related to the type of occupancy (offices, housing, and so on) plus a number of zero-mean random variables and/or processes specifying deviations from the mean which account for random differences from building to building, from floor to floor in the same building, and from point to point at the same floor in the same building.

It is thus possible to write [2]

$$\tilde{w}(x,y) = \mu_w + \tilde{Y}_b + \tilde{Y}_f + \tilde{Z}(x,y) \tag{3.1}$$

Here $\tilde{Y}_b$ is the variable specifying the random variation in the mean unit load in a single building with respect to the mean value that would be calculated by averaging loads in all the buildings with the same destination: thus, $\mu_w + \tilde{Y}_b$ is the mean load in the particular building under consideration. The term $\tilde{Y}_f$ specifies the random variations from floor to floor in the same building: hence, $\mu_w + \tilde{Y}_b + \tilde{Y}_f$ is the mean value at the particular floor under consideration in the building. Finally, $\tilde{Z}(x,y)$ is an isotropic stochastic field (i.e. the autocorrelation function of $\tilde{Z}$ depends only on the distance between two points, cf. Section A6.6), independent of $\tilde{Y}_b$ and $\tilde{Y}_f$, representing load intensity variations from point to point on a given floor of a given building.

The model is completely specified after the parameters defining quantities of the stochastic quantities in equation (3.1) have been specified. In particular, the variances s_b^2, s_f^2 and s_Z^2 of $\tilde{Y}_b$, $\tilde{Y}_f$ and $\tilde{Z}(x,y)$ should be estimated depending on the type of building, of occupancy, of the floor level, etc.

In addition, the autocovariance function of $\tilde{Z}(x,y)$ should be given a suitable form. It is to be expected that some continuity exists for sample functions of the instantaneous load intensity. Thus, it can be conjectured that, if at a point P the load is higher than the floor average, the same applies in a more or less large neighbourhood of P. Therefore, a positive correlation can be assumed, presumably decaying with the distance between the points; an autocovariance function which satisfies these conditions is

$$\Sigma_Z(x_1,y_1,x_2,y_2) = \mathrm{Cov}[\tilde{Z}(x_1,y_1), \tilde{Z}(x_2,y_2)] = s_Z^2 \exp(-R^2/D^2) \tag{3.2}$$

with $R^2 = (x_1 - x_2)^2 + (y_1 - y_2)^2$ and D an appropriate constant. In conclusion, the three variances and the autocovariance function parameters should be evaluated in order to make the model operative. Moreover variations in loading from floor to floor can be incorporated in the model by stating that if (x_1, y_1) and (x_2, y_2) are points located on different floors, the covariance of the respective load intensities is expressed by

$$\mathrm{Cov}[\tilde{Z}(x_1, y_1), \tilde{Z}(x_2, y_2)] = c\, s_Z^2 \exp(-R^2/D^2) \tag{3.3}$$

where c is a coefficient relating load intensities at two points on the same vertical: c is usually positive because of the so-called 'stacking effect', i.e. the tendency of occupants to use the spaces in the same way vertically; it should vary with the number of floors separating the two points under consideration, but it is usually kept constant for simplicity's sake [2].

3.3.2 Transient live load

A transient load is applied for a short time, and is commonly modelled in time as a Dirac delta-function, with magnitude equal to the intensity of the maximum load applied during the event. The arrival times of the transient load events are commonly assumed to be Poisson-distributed with assigned mean rate of arrival λ_C (Section A6.3). In space, transient live load events can be characterized by a random number of randomly positioned and randomly sized *cells* of component loads (e.g. persons).

Basic component loads $\tilde{Q}$ are assumed with specified mean value μ_Q and variance s_Q^2. Each load cell contains a random number $\tilde{r}$ of component loads, with mean μ_r and variance s_r^2; it is assumed that $\tilde{Q}$ and $\tilde{r}$ are independent quantities.

The resultant load $\tilde{C}$ of a load cell therefore has mean and variance (equations (A3.14) and (A2.52))

$$\mathrm{E}[\tilde{C}] = \mu_C = \mu_Q \mu_r \tag{3.4}$$

$$\begin{aligned}\mathrm{Var}[\tilde{C}] = s_C^2 &= \mathrm{E}[\mathrm{Var}[Qr|r]] + \mathrm{Var}[\mathrm{E}[Qr|r]] \\ &= \mu_r s_Q^2 + \mu_Q s_r^2 \end{aligned} \tag{3.5}$$

Further, the number of load cells in a given area $\tilde{m}$ is assumed to be Poisson-distributed with parameter λ_m depending on the area. The variables $\tilde{m}$ and $\tilde{C}$ are generally assumed to be independent.

3.3.3 Load-effect models

Structures are often designed and/or analysed element by element. In this case the designer assigns a part of the applied loads to each element according to the *influence surface* of the specific load-effect under consideration. By definition, the ordinate $I(x,y)$ of the influence surface at any point (x,y) provides the magnitude of the load effect which would be induced by application of a unit load at that point. The area over which $I(x, y)$ is non-zero is the influence area, A_I. The load effect under consideration is then given by‡

$$\begin{aligned} H(A_I) &= H_w(A_I) + H_C(A_I) \\ &= \int_{A_I} I(x,y)\, w(x,y)\, \mathrm{d}x\, \mathrm{d}y + \sum_{j=1}^{m(A_I)} C_j I(x_j, y_j) \end{aligned} \tag{3.6}$$

‡ Although the theory of influence surfaces is fully developed for linear elastic structures, their use can be extended to other types of structural behaviour introducing the *tributary areas*.

(a) Deterministic influence surface

Assuming a deterministic influence surface $I(x,y)$, the following results [3] are obtained for the mean value and the variance of the load effect $H(A_I)$ from the theory of integration of random processes (Section A5.5).

Considering the sustained floor loads, one obtains (cf. equations (A5.24) and (A5.25))

$$\mathrm{E}[H_w(A_I)] = \mu_w \int_{A_I} I(x,y)\,\mathrm{d}x\,\mathrm{d}y \tag{3.7}$$

$$\mathrm{Var}[H_w(A_I)] = \int_{A_I}\int_{A_I} I(x_1,y_1)I(x_2,y_2)\,\mathrm{Cov}[\tilde{w}(x_1,y_1),\tilde{w}(x_2,y_2)]\,\mathrm{d}x_1\mathrm{d}y_1\mathrm{d}x_2\mathrm{d}y_2 \tag{3.8}$$

In particular, for A_I on a single floor, equation (A3.26′) yields

$$\begin{aligned}\mathrm{Cov}[\tilde{w}(x_1,y_1),\tilde{w}(x_2,y_2)] &= \mathrm{E}[(\tilde{Y}_b+\tilde{Y}_f+\tilde{Z}(x_1,y_1))(\tilde{Y}_b+\tilde{Y}_f+\tilde{Z}(x_2,y_2))]\\ &= s_b^2+s_f^2+\mathrm{Cov}[\tilde{Z}(x_1,y_1),\tilde{Z}(x_2,y_2)]\end{aligned} \tag{3.8′}$$

analogous to (A3.27′). Thus

$$\begin{aligned}\mathrm{Var}[H_w(A_I)] &= (s_b^2+s_f^2)\left(\int_{A_I} I(x,y)\,\mathrm{d}x\,\mathrm{d}y\right)^2\\ &+\int_{A_I}\int_{A_I} I(x_1,y_1)I(x_2,y_2)\mathrm{Cov}[\tilde{Z}(x_1,y_1),\tilde{Z}(x_2,y_2)]\mathrm{d}x_1\mathrm{d}y_1\mathrm{d}x_2\mathrm{d}y_2\end{aligned} \tag{3.8″}$$

If the further simplifying assumptions are introduced,

$$I(x_2,y_2) = \frac{1}{A_I}\int_{A_I} I(x,y)\,\mathrm{d}x\,\mathrm{d}y + \left(I(x_1,y_1) - \frac{1}{A_I}\int_{A_I} I(x,y)\,\mathrm{d}x\,\mathrm{d}y\right)\mathrm{e}^{-R^2/d^2} \tag{3.8‴}$$

and that

$$\mathrm{E}[\tilde{Z}(x_2,y_2)|\tilde{Z}(x_1,y_1) = Z] = Z\mathrm{e}^{-R^2/D^2} \tag{3.8iv}$$

with the constant $d^2 > D^2$, some algebra and the mean value theorem for integrals lead to

$$\mathrm{Var}[H_w(A_I)] \simeq (s_b^2+s_f^2)\left(\int_{A_I} I(x,y)\mathrm{d}x\mathrm{d}y\right)^2 + \pi D^2 s_Z^2 \int_{A_I} I^2(x,y)\,\mathrm{d}x\,\mathrm{d}y \tag{3.8v}$$

Further, for the transient live load effect with $\lambda_m = \lambda_m(A_I)$ expressing the number of cells per unit area, the following results hold (equations (A6.47) and (A6.48))

$$\mathrm{E}[H_C(A_I)] = \lambda_m \mu_C \int_{A_I} I(x,y)\,\mathrm{d}x\,\mathrm{d}y \tag{3.9}$$

$$\mathrm{Var}[H_C(A_I)] = \lambda_m(s_C^2+\mu_C^2)\int_{A_I} I^2(x,y)\,\mathrm{d}x\,\mathrm{d}y \tag{3.10}$$

(b) Random influence surface

An extension of the foregoing analysis can include cases of static indeterminacy (i.e. cases in which load effects are functions of both loads and material properties) by consideration of a random influence surface, $\tilde{I}(x, y)$, such that

$$\tilde{I}(x,y) = \bar{I}(x,y) + \int_{A_I} g(x,y,x',y')\tilde{\Delta}(x',y')\,dx'\,dy' \tag{3.11}$$

in which $\bar{I}(x, y)$ is a deterministic function of expected material properties; $g(x, y, x', y')$ is the rate of change of $I(x, y)$ from $\bar{I}(x, y)$ from $\bar{I}(x, y)$; $\tilde{\Delta}(x', y')$ is the variation from the expected material properties at the generic point (x', y'), which can be put in the form

$$\tilde{\Delta}(x',y') = \tilde{\bar{\Delta}} + \tilde{\zeta}(x',y') \tag{3.12}$$

In equation (3.12) $\tilde{\bar{\Delta}}$ is a zero-mean random variable which represents the spatial average, over A_I, of variations from expected material properties and $\tilde{\zeta}(x', y')$ is a zero-mean random process (independent of $\tilde{\bar{\Delta}}$) which represents the spatial variation of the deviation from the expected material properties.

With these positions, one obtains for the sustained floor loads

$$\mathrm{E}[H_w(A_I)] = \mu_w \int_{A_I} \bar{I}(x,y)\,dx\,dy \tag{3.13}$$

and after some algebra, if A_I is on a single floor,

$$\begin{aligned}
\mathrm{Var}[H_w(A_I)] \simeq (\mu_w^2 + s_b^2 + s_f^2)&\Bigg[\left(\int_{A_I} \bar{I}(x,y)\,dx\,dy\right)^2 \\
&+ s_{\bar{\Delta}}^2\left(\int_{A_I} K(x,y)\,dx\,dy\right)^2 + \pi D_\zeta^2 s_\zeta^2 \int_{A_I} L^2(x',y')\,dx'\,dy'\Bigg] \\
&+ \pi D_Z^2 s_Z^2\left(\int_{A_I} \bar{I}^2(x,y)\,dx\,dy + s_{\bar{\Delta}}^2 \int_{A_I} K^2(x,y)\,dx\,dy\right. \\
&\left.\qquad + \pi D_\zeta^2 s_\zeta^2 \int_{A_I}\int_{A_I} g^2(x,y,x',y')\,dx\,dy\,dx'\,dy'\right) \\
&- \mu_w^2\left(\int_{A_I} \bar{I}(x,y)\,dx\,dy\right)^2
\end{aligned} \tag{3.14}$$

in which

$$\begin{aligned}
K(x,y) &= \int_{A_I} g(x,y,x',y')\,dx'\,dy' \\
L(x',y') &= \int_{A_I} g(x,y,x',y')\,dx\,dy
\end{aligned} \tag{3.15}$$

and D_ζ and D_Z denote the constant defined in equation (3.2), respectively for $\mathrm{Cov}[\tilde{\zeta}(x'_1,y'_1), \tilde{\zeta}(x'_2,y'_2)]$ and $\mathrm{Cov}[\tilde{Z}(x_1,y_1), \tilde{Z}(x_2,y_2)]$.

Further, for the transient live load effect

$$E[H_C(A_I)] = \lambda_m \mu_C \int_{A_I} \bar{I}(x,y)\,\mathrm{d}x\,\mathrm{d}y \tag{3.16}$$

$$\begin{aligned}\mathrm{Var}[H_C(A_I)] = \lambda_m (s_C^2 + \mu_C^2)\bigg(\int_{A_I} \bar{I}^2(x,y)\,\mathrm{d}x\,\mathrm{d}y + s_\Delta^2 \int_{A_I} K^2(x,y)\,\mathrm{d}x\,\mathrm{d}y \\ + \pi D_\zeta^2 s_\zeta^2 \int_{A_I}\int_{A_I} g^2(x,y,x',y')\,\mathrm{d}x\,\mathrm{d}y\,\mathrm{d}x'\,\mathrm{d}y'\bigg)\end{aligned} \tag{3.17}$$

3.3.4 Equivalent uniformly distributed load (EUDL)

Before discussing the use of the above load model in actual applications, remember that usually the code loads are specified as an *equivalent uniformly distributed load* (*EUDL*). Here equivalence might be intended in more ways, for instance in regard to some particular load effect (i.e. the EUDL may be the uniform load that produces the same stress at a given point as the actual load, Section 3.3.5). Most times the EUDL is given as the average load acting on a given area A

$$\tilde{\bar{w}}(A) = \frac{1}{A}\int_A \tilde{w}(x,y)\,\mathrm{d}A \tag{3.18}$$

Mean value and variance of the r.v. $\tilde{\bar{w}}(A)$ can be obtained from the theory of integration of random processes (Section A5.5).

Consider first the case of an area A, all on one floor, subjected to sustained load. Then, introducing equations (3.1) and (3.2), one gets (equations (A5.16) and (A5.18))

$$\mathrm{E}[\tilde{\bar{w}}(A)] = \frac{1}{A}\int_A \mathrm{E}[\tilde{w}(x,y)]\,\mathrm{d}A = \mu_w \tag{3.19}$$

$$\begin{aligned}\mathrm{Var}[\tilde{\bar{w}}(A)] &= \frac{1}{A^2}\int_A\int_A \{s_b^2 + s_f^2 + \mathrm{Cov}[\tilde{Z}(x_1,y_1), \tilde{Z}(x_2,y_2)]\}\,\mathrm{d}x_1\,\mathrm{d}y_1\,\mathrm{d}x_2\,\mathrm{d}y_2 \\ &= s_b^2 + s_f^2 + s_Z^2 \pi D^2 \frac{1}{A}\left[\mathrm{erf}\left(\sqrt{\frac{A}{D^2}}\right) - (1 - \mathrm{e}^{-A/D^2})\sqrt{\left(\frac{D^2}{\pi A}\right)}\right]^2\end{aligned} \tag{3.20}$$

where erf ( ) denotes the error function

$$\mathrm{erf}(x) = \frac{2}{\sqrt{\pi}}\int_0^x \mathrm{e}^{-y^2}\,\mathrm{d}y = 2\Phi(y\sqrt{2}) - 1 \tag{3.21}$$

and $\Phi(\ )$ is the standardized normal CDF (Table A8.3); the following relation holds

$$\int_0^x \operatorname{erf}(x)\,\mathrm{d}x = x\operatorname{erf}(x) + \frac{1}{\sqrt{\pi}}(\mathrm{e}^{-x^2}-1) \tag{3.21'}$$

With the positions

$$\begin{aligned} k(A) &= \left[\operatorname{erf}\left(\sqrt{\frac{A}{D^2}}\right) - (1-\mathrm{e}^{-A/D^2})\sqrt{\left(\frac{D^2}{A\pi}\right)}\right]^2 \\ s_u^2 &= \pi D^2 s_Z^2 \end{aligned} \tag{3.22}$$

equation (3.20) becomes

$$\operatorname{Var}[\tilde{\bar{w}}(A)] = s_b^2 + s_f^2 + s_u^2\frac{k(A)}{A} \tag{3.23}$$

Consider, now, the case of a reference area A made by the union of a number n of areas $A_1 = \ldots = A_n$ on, respectively, the floors $1, \ldots, n$; the mean load over such areas may be sought, for instance, to calculate the normal force in the column supporting such areas. In this case, equation (3.19) becomes

$$\mathrm{E}[\tilde{\bar{w}}_n(A)] = \mu_w \tag{3.24a}$$

while (3.23) becomes

$$\operatorname{Var}[\tilde{\bar{w}}_n(A)] = s_b^2 + \frac{s_f^2}{n} + s_u^2\frac{k(A)}{nA} + c\frac{n-1}{n}s_Z^2\frac{k(A)}{A} \tag{3.24b}$$

The last equation is easily proved by remembering that

$$\begin{aligned} \tilde{\bar{w}}_n &= \int_A \tilde{w}_n\,\mathrm{d}A = \int_A (\mu_w + \tilde{Y}_b)\,\mathrm{d}A + \sum_{i=1}^n \int_{A_i} (\tilde{Y}_f + \tilde{Z}(x,y))\,\mathrm{d}x\,\mathrm{d}y \\ &= \int_A (\mu_w + \tilde{Y}_b)\mathrm{d}A + \sum_{i=1}^n \int_A (\tilde{Y}_{f_i}/n + \tilde{Z}_i(x,y)/n)\,\mathrm{d}x\,\mathrm{d}y \end{aligned}$$

and, hence, from equations (A5.18), (A5.59) and (A3.27')

$$\begin{aligned} \operatorname{Var}[\tilde{\bar{w}}_n(A)] &= s_b^2 + n\left(\frac{1}{n^2}s_f^2\right) \\ &+ \sum_{i=1}^n \left\{\int_A\int_A \operatorname{Cov}[\tilde{Z}_i(x_1,y_1)/n, \tilde{Z}_i(x_2,y_2)/n]\,\mathrm{d}x_1\,\mathrm{d}y_1\,\mathrm{d}x_2\,\mathrm{d}y_2 \right. \\ &\left. + 2\sum_{j=i+1}^n \left(\int_A\int_A \operatorname{Cov}[\tilde{Z}_i(x_1,y_1)/n, \tilde{Z}_j(x_2,y_2)/n]\,\mathrm{d}x_1\,\mathrm{d}y_1\,\mathrm{d}x_2\,\mathrm{d}y_2\right)\right\} \end{aligned}$$

By introducing equation (3.3), equation (3.24b) is then obtained.

Fitting the data obtained via load surveys in existing buildings, it is possible to estimate the parameters appearing in the preceding equations.

Just as an example, refer to a well known survey on office buildings [4]; rough estimates are:

$$s_b^2 = 0.83\ (\text{kg}\,\text{m}^{-2})^2 \qquad s_f^2 = 7.2\ (\text{kg}\,\text{m}^{-2})^2 \qquad s_Z^2 = 4.76\ (\text{kg}\,\text{m}^{-2})^2$$

$$c = 0.7 \qquad D^2 = 0.835\ \text{m}^2 \qquad \mu_w = 70\ \text{kg}\,\text{m}^{-2}$$

and for an area $A = 20\ \text{m}^2$

$$k(A) = [\text{crf}(4.89) - 0.115]^2 - 0.78$$

The corresponding EUDL over a single floor is described by (equations (3.19) and (3.23))

$$\text{E}[\tilde{\bar{w}}(A)] = 70\ \text{kg}\,\text{m}^{-2} \qquad s_u^2 = 12.49\ (\text{kg}\,\text{m}^{-2})^2 \qquad \text{Var}\,[\tilde{\bar{w}}(A)] = 8.52\ (\text{kg}\,\text{m}^2)^2$$

while the EUDL over $n = 10$ floors is given by

$$\text{E}[\tilde{\bar{w}}_n(A)] = 70\ \text{kg}\,\text{m}^2 \qquad \text{Var}[\tilde{\bar{w}}_n(A)] = 1.61\ (\text{kg}\,\text{m}^{-2})^2$$

These results show that, when more floors are considered, the coefficient of variation of the load is greatly reduced.

Consider now the temporal sequence of changes in loading due to changes of occupancy, rearrangements of rooms, etc. Still making reference, for simplicity's sake, to the EUDL over a given area A, assume that load changes occur according to a Poisson process with mean rate of arrival λ.

Let $P_{\bar{w}}(w)$ be the CDF of the EUDL, for a single load application. The probability that a given value w is not exceeded in the time interval $(0, t)$ can be calculated as follows.

Let $\tilde{k}$ be the number of load changes in $(0, t)$; from the Poisson process formulae (Section A6.3, equation (A6.37))

$$\text{Prob}[\tilde{k} = k \text{ in } (0, t)] = \text{e}^{-\lambda t}\frac{(\lambda t)^k}{k!} \tag{3.25}$$

Hence, assuming that load changes are statistically independent (cf. also equation (A7.48))

$$\begin{aligned}\text{Prob}[\tilde{\bar{w}}(\tau) \leqslant w \text{ for all } \tau\in(0,t)] &= \sum_{k=0}^{\infty} \text{Prob}[\tilde{k}=k \text{ in } (0, t)]\,[P_{\bar{w}}(w)]^k \\ &= \sum_{k=0}^{\infty} \text{e}^{-\lambda t}\frac{[\lambda t P_{\bar{w}}(w)]^k}{k!} \\ &= \exp\{-\lambda t[1 - P_{\bar{w}}(w)]\} \end{aligned} \tag{3.26}$$

Equation (3.26) makes it possible to calculate statistics for extreme values of the EUDL that is definitely the information of interest to the structural engineer.

3.3.5 Equivalent uniformly distributed load from the load effect

As already stated, with respect to a given load distribution and relevant load effect $H(A_I)$, an equivalent uniformly distributed load $\bar{w}_e$ can also be defined as a uniformly distributed load which, in conjunction with an associated deterministic influence function $I(x, y)$, would produce the same load effect $H(A_I)$:

$$\tilde{\bar{w}}_e(A_I) = \frac{\tilde{H}(A_I)}{\int_{A_I} I(x,y)\,dx\,dy} \tag{3.27}$$

with $\tilde{H}(A_I)$ given by equation (3.6).

If the influence surface is random, an EUDL $\bar{w}_e$ can also be defined (cf. (3.11) *et seq.*)

$$\tilde{\bar{w}}_e(A_I) = \frac{\tilde{H}(A_I)}{\int_{A_I} \bar{I}(x,y)\,dx\,dy} \tag{3.28}$$

$$E[\tilde{\bar{w}}_e(A_I)] = \frac{E[H(A_I)]}{\int_{A_I} \bar{I}(x,y)\,dx\,dy} \tag{3.29}$$

$$\mathrm{Var}[\tilde{\bar{w}}_e(A_I)] = \frac{\mathrm{Var}[H(A_I)]}{\left[\int_{A_I} \bar{I}(x,y)\,dx\,dy\right]^2} \tag{3.30}$$

Calculations involving integrals over a rectangular influence area, A_I, can be made more convenient by introducing a reference area A_0.‡

Let $\bar{w}_s$ be the EUDL (A_I) corresponding to sustained loads and $\bar{C}(A_I)$ the one corresponding to transient live loads; the following results can be obtained after some algebra [3].

(a) Deterministic influence surface

Considering the sustained floor loads, one obtains

$$E[\tilde{\bar{w}}_s(A_I)] = \mu_w \tag{3.31a}$$

and under the simplifying hypotheses under which equation (3.8) holds and with A_I on a single floor

$$\mathrm{Var}[\tilde{\bar{w}}_s(A_I)] = s_b^2 + s_f^2 + \frac{\pi D_Z^2 s_Z^2}{A_I}\,\frac{\int_{A_0} I^2(x,y)\,dx\,dy}{\left[\int_{A_0} I(x,y)\,dx\,dy\right]^2} \tag{3.31b}$$

‡ Indicating by $\chi(\ )$ a generic function

$$\int_{A_I} \chi(x,y)\,dx\,dy \simeq A_I \int_{A_0} \chi(x,y)\,dx\,dy = A_I \int_0^1 \int_0^1 \chi(x',y')\,dx'\,dy'$$

Similarly, for transient live loads,

$$\mathrm{E}[\tilde{\tilde{C}}(A_I)] = \lambda_m \mu_C \tag{3.32a}$$

$$\mathrm{Var}[\tilde{\tilde{C}}(A_I)] = \frac{\lambda_m (s_C^2 + \mu_C^2)}{A_I} \frac{\displaystyle\int_{A_0} I^2(x, y)\,\mathrm{d}x\,\mathrm{d}y}{\left[\displaystyle\int_{A_0} I(x, y)\,\mathrm{d}x\,\mathrm{d}y\right]^2} \tag{3.32b}$$

(b) Random influence surface

For sustained floor loads,

$$\mathrm{E}[\tilde{\bar{w}}_\mathrm{s}(A_I)] = \mu_w \tag{3.33a}$$

and, with A_I on a single floor,

$$\begin{aligned}\mathrm{Var}[\tilde{\bar{w}}_\mathrm{s}(A_I)] \simeq s_\mathrm{b}^2 + s_\mathrm{f}^2 + &\frac{(\mu_w^2 + s_\mathrm{b}^2 + s_\mathrm{f}^2}{\left[\displaystyle\int_{A_0} \bar{I}(x, y)\,\mathrm{d}x\,\mathrm{d}y\right]^2}\left[s_{\tilde{\Delta}}^2\left(\int_{A_0} K(x, y)\,\mathrm{d}x\,\mathrm{d}y\right)^2\right.\\ &\left. + \frac{\pi D_\zeta^2 s_\zeta^2}{A_I}\int_{A_0} L^2(x', y')\,\mathrm{d}x'\,\mathrm{d}y'\right]\\ + \frac{\pi D_Z^2 s_Z^2}{A_I\left[\displaystyle\int_{A_0} \bar{I}(x, y)\,\mathrm{d}x\,\mathrm{d}y\right]^2}&\left(\int_{A_0} \bar{I}^2(x, y)\,\mathrm{d}x\,\mathrm{d}y + s_{\tilde{\Delta}}^2\int_{A_0} K^2(x, y)\,\mathrm{d}x\,\mathrm{d}y\right.\\ &\left. + \frac{\pi D_\zeta^2 s_\zeta^2}{A_I}\int_{A_0}\int_{A_0} g^2(x, y, x', y')\,\mathrm{d}x\,\mathrm{d}y\,\mathrm{d}x'\,\mathrm{d}y'\right)\end{aligned} \tag{3.33b}$$

Similarly, for transient live loads,

$$\mathrm{E}[\tilde{\tilde{C}}(A_I)] = \lambda_m \mu_C \tag{3.34a}$$

$$\begin{aligned}\mathrm{Var}[\tilde{\tilde{C}}(A_I)] = \frac{\lambda_m (s_C^2 + \mu_C^2)}{A_I\left[\displaystyle\int_{A_0} \bar{I}(x, y)\,\mathrm{d}x\,\mathrm{d}y\right]^2}&\left(\int_{A_0} \bar{I}^2(x, y)\,\mathrm{d}x\,\mathrm{d}y + s_{\tilde{\Delta}}^2\int_{A_0} L^2(x, y)\,\mathrm{d}x\,\mathrm{d}y\right.\\ &\left. + \frac{\pi D_\zeta^2 s_\zeta^2}{A_I}\int_{A_0}\int_{A_0} g^2(x, y, x', y')\,\mathrm{d}x\,\mathrm{d}y\,\mathrm{d}x'\,\mathrm{d}y'\right)\end{aligned} \tag{3.34b}$$

3.4 SEISMIC LOADING

3.4.1 Earthquakes

Three classes of earthquakes can be recognized: (i) volcanic, (ii) tectonic and (iii) plutonic. Volcanic earthquakes are frequent but weak, and usually are detected only by seismographs near the volcanoes; the released energy may be of chemical origin, or it may be originated by compression of gases or by the motion of the magma. Tectonic and plutonic earthquakes arise in much the same manner; plutonic (or deep-focus) earthquakes originate hundreds of miles beneath the surface, whereas tectonic earthquakes find their originating mechanism near the earth's surface. Tectonic earthquakes constitute the most severe earthquakes, and are the only significant ones as regards effects on buildings. Therefore only tectonic earthquakes will be referred to in the following.

The most reliable explanation for tectonic earthquake production was given by the *elastic-rebound theory* put forward in 1911 by H. F. Reid, who concluded that slow relative displacements of adjacent parts of the earth's crust cause the gradual development of large elastic strains leading finally to fracture. This is followed by very rapid elastic rebound of the masses of rock involved, which come to rest in positions of minimum strain. The net result is the conversion to kinetic energy of the potential strain energy. The sudden release of so much energy is responsible for the havoc caused by earthquakes.

An earthquake gives rise to waves of three kinds:

(1) *Primary* (*PR* or *dilatation waves*), which are longitudinal waves that involve back-and-forth vibrations of particles of matter in the same direction as that in which the waves travel. *PR* waves are essentially pressure waves, like sound waves.
(2) *Secondary* (*SE* or *shear waves*), which are transverse waves that involve vibrations of particles of matter perpendicularly to the direction in which the waves travel. The waves produced by plucking a stretched string are similar to *SE* waves. Both *PR* and *SE* waves are body waves that travel through the earth's interior. *SE* waves, related to shear stress transmission, cannot occur in a liquid, while *PR* waves can.
(3) *Surface* (*SU waves*), which are analogous to sea waves, involve orbital motions of particles of matter, and are limited to the earth's surface. Two kinds of seismic surface waves are known: *Rayleigh waves*, in which displacement of particles is partly vertical and partly in the direction of propagation, and *Love waves*, associated with displacements which are horizontal and at right-angles to the direction of travel. Love waves are propagated only in solids which have a surface of discontinuity of the elastic properties.

The three kinds of seismic waves travel with different velocities, and so arrive at a distant observer at different times after an earthquake; *PR* waves arrive first, *SE* waves next, and finally *SU* waves.

The need for instrumental detection of earthquakes has been felt from very early ages. In the second century AD, the Chinese astronomer, Chang Heng, made the first recorded successful attempt to detect a distant earthquake. This instrument was the first *seismoscope*, i.e. a device to detect small movements of the ground imperceptible to the senses. Seismoscopes, however, are not suitable for seismic engineering purposes, in that they lack the sensitivity and the reliability of *seismographs*, which provide a permanent record, the *seismogram*, of the ground accelerations due to an earthquake. (A typical modern seismograph is illustrated diagrammatically in Fig. 8.1.)

Fig. 3.8 shows features that are commonly present in the seismograms of earthquakes recorded at a site far away from the epicentral region. The record starts with a sudden movement (*P*), when the ground is first disturbed by the arrival of primary waves, followed by a train of waves of small amplitude. After some time, a second sharp movement occurs (*S*), usually of greater amplitude; this marks the first arrival of secondary waves. The primary and secondary waves are initiated at the same instant, but they travel with different velocities. Surface waves travel still more slowly, arriving after the first secondary waves have been recorded; they are responsible for the movements of large amplitude which are seen in Fig. 3.8.‡

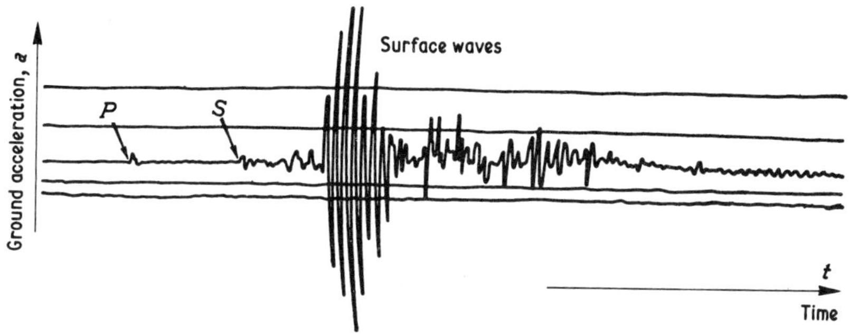

Fig. 3.8 Typical seismogram of a distant earthquake.

Seismograms, evidently, constitute a powerful tool in seismic engineering. In fact they yield the whole time history of the ground acceleration during the earthquake, thus allowing *a posteriori* dynamic analyses of the structure behaviour under the quake. Unfortunately, recorded accelerograms are not sufficient to enable engineers to make *a priori* assessments of seismic safety, as will be discussed later.

The seismogram is also useful to less immediate aims, for instance in determining the position of the *epicentre*, i.e. the part of the earth's surface immediately above the origin, or the *focus* or *hypocentre*, of the earthquake. The time interval, in fact, between the arrival at the recording station of the first primary and the first secondary waves is proportional to the distance of the epicentre from the station. If distances are obtained at three or more stations, the epicentre can be easily located. Keeping sites where seismic activity exists under

‡ The wave velocities depend on the elastic constants of the ground, and are frequency-dependent in viscoelastic media. Indicatively, for a Poisson ratio equal to 0.25, they are related by $v_{PR} = 1.73\, v_{SE}$; $v_{SU} = 0.92\, v_{SE}$.

observation by a suitable *mesh* of seismic observatories is therefore necessary for locating epicentres, hence for describing the state of the earth's crust, especially in regard to *fracture zones*, or *faults*, where earthquakes originate. This helps in forecasting the future localization of earthquakes, and therefore their effects at each site. Also, parameters measuring synthetically the power of earthquakes can be deducted from seismograms. This is fundamental in order to compare earthquakes on an absolute scale: in fact, the amplitude of seismograms at a given site depends not only on the energy released by the earthquake but also on the distance R of the observatory from the hypocentre (or, which is equivalent for practical purposes beyond a few kilometres, from the epicentre).

If, for an earthquake, different records on similar seismographs at various sites are available, a useful comparison can be made by drawing a plot of the maximum amplitude $a_s(R)$ versus epicentral distance of the recording station. A curve of $\ln a_s$ vs R is thus obtained for that earthquake. Inspection shows that the curves plotted for different quakes are all approximately parallel straight lines. This suggests as a possible measure of earthquake power the quantity $\mathcal{M}$ (Richter's *magnitude*), independent of R and defined by

$$\mathcal{M} = \log a_s(R) - \log a_0(R) \tag{3.35}$$

$a_0(R)$ being the $a_s(R)$ of a conventional reference earthquake (log denotes the base 10 logarithm); a_s is measured in micrometres (10^{-6} m) on a standard Woods–Anderson torsion seismometer. The parameter $\mathcal{M}$ defined by equation (3.35) yields a *measure* of the earthquake practically independent of the location of the recording station. Typically, earthquakes are not felt by persons when $\mathcal{M} \leqslant 2$; damage is caused to weak structures near the epicentre when $\mathcal{M} \geqslant 5$; destructive earthquakes correspond to $\mathcal{M} > 6$.

Magnitude, however, should not be confused with earthquake *intensity*, which is by definition a parameter specifying the destructive power (i.e. the severity) of an earthquake at a specified site. Seismologists and seismic engineers are busily engaged in proposing significant definitions for intensity, in that this seems the most suitable parameter for engineering and safety assessment purposes. The well known Mercalli scale and the peak ground acceleration enjoy some trust in this regard, and prove useful in seismic risk analysis. Fig. 3.9 provides, as an example,

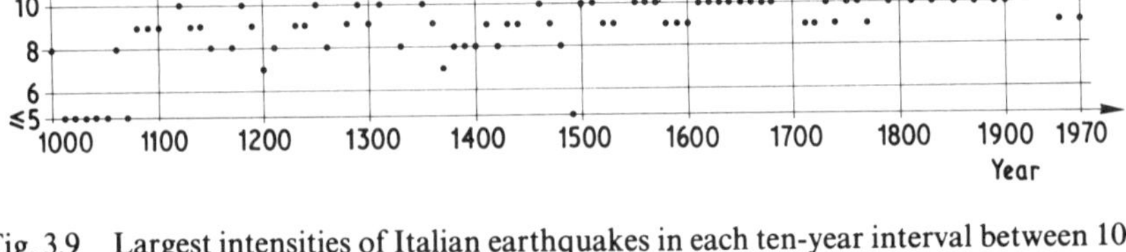

Fig. 3.9 Largest intensities of Italian earthquakes in each ten-year interval between 1000 and 1969 AD (data from CNR, *Progetto Geodinamica*).

the maximum intensities (measured by the Modified Mercalli scale) of the Italian earthquakes over each decade between the years 1000 and 1969. Roughly, it can be said that *magnitude* is an objective measure of the earthquake power, in that it is directly related to the energy released in the quake, while the *intensity* is a subjective measure, related to local conditions: soil stratifications, strength of the buildings, and so on. However, it should not be forgotten that earthquakes are very complex realities, and cannot be completely described by a few parameters. Even the relation between earthquake magnitude and intensity at a given site has a great dispersion. Table 3.1 presents indicative comparisons among some common intensity scales, the peak ground acceleration and the Richter magnitude [5]. But, when it is necessary, more detailed information must be used in order to evaluate seismic structural performance: in fact, magnitude cannot be directly scaled to intensity of ground motion unless factors such as hypocentral distance, soil properties, etc., are taken into account.

3.4.2 Seismic risk analysis

In the following, only sufficiently strong earthquakes at a given site, relevant to engineering decisions, will be considered, i.e. earthquakes with magnitude less than, say, 4, will be neglected. Seismic risk analysis can be roughly set up on a probabilistic basis, provided that some fundamental information is known.

It is quite apparent, from previous considerations on earthquakes and earthquake propagation, that earthquake features are strongly dependent on the site where its effects are to be evaluated. So, we assume that a given site is under examination. The first necessary information is knowledge of the fundamental mechanisms which originate earthquakes felt at the site. For instance, in some cases the presence of an active fracture line (a *fault*) of known length l can be recognized as the origin of the relevant earthquakes. Thus, a first conclusion can be drawn that the earthquake foci will lie on the line AC (Fig. 3.10). If the probability density law $p_X(x)$ of the location $\tilde{x}$ of the earthquake focus B is known (or can be guessed from geology), the combination of these two pieces of information yields the PDF $p_R(R)$ of the focal distance from the site as the result of very simple geometry and transformation of r.v. (Section 2.5.3; also, cf. below the example given by equation (3.49)). The other basic information needed concerns the seismic activity of the fault; precisely, how much energy is released at each earthquake. This information can be given through the distribution of earthquake magnitudes at that fault. Such a distribution follows approximately the exponential law (Table A2.2, with $x = \mathscr{M} - \mathscr{M}_0$.)

$$\text{Prob}[\tilde{\mathscr{M}} \geqslant \mathscr{M}] = 1 - P_{\mathscr{M}}(\mathscr{M}) = \exp[-\zeta(\mathscr{M} - \mathscr{M}_0)] \tag{3.36}$$

where $\mathscr{M}_0$ is the minimum significant magnitude (typically, $\mathscr{M}_0 = 4$), and ζ ranges mostly between 1.5 and 2.3.

Focal distance $\tilde{R}$ and earthquake magnitude $\tilde{\mathscr{M}}$ are the basic data through

Table 3.1 Earthquake intensity scales and Richter magnitude: approximate comparison (after Duff[5]).

Peak ground acceleration ($cm\ s^{-2}$)	*Alternative scales*‡ *JMA*	*MSK*	*MM*	*Modified Mercalli (MM) scale description*	*Richter magnitude* $\mathcal{M}$
1	0	I	I	Not felt. Instrument level	2
	I	II	II	Slight. Felt by sensitive persons	3
	II	III	III	Weak. Felt by many indoors. Rattling. Slight movement	
10	III	IV	IV	Felt by a few outdoors. Rocking. Creaking	4
		V	V	Felt outdoors by many. Furniture disturbed. Hanging objects swing	
	IV	VI	VI	Moderate. Felt by all. Loose objects overturned. Slight damage	5
100		VII	VII	Strong, Considerable damage to weak structures	6
	V	VIII	VIII	Very strong. Overturning of monuments. Partial collapse of structures. Mud ejected	
	VI	IX	IX	Severe. Panic. Considerable damage. Ground cracked. Landslides	7
1000		X	X	Destructive. Most masonry and many wooden houses destroyed or severely damaged. Large ground fissures	
	VII	XI	XI	Ground severely disturbed and water ejected. Tsunami	8
		XII	XII	Total destruction. Large fault movements. Objects thrown into air	9
10000					

‡ MM = modified Mercalli Scale (1931).
JMA = Japanese Meteorological Agency (1949).
MSK = Geofian, USSR (1953).

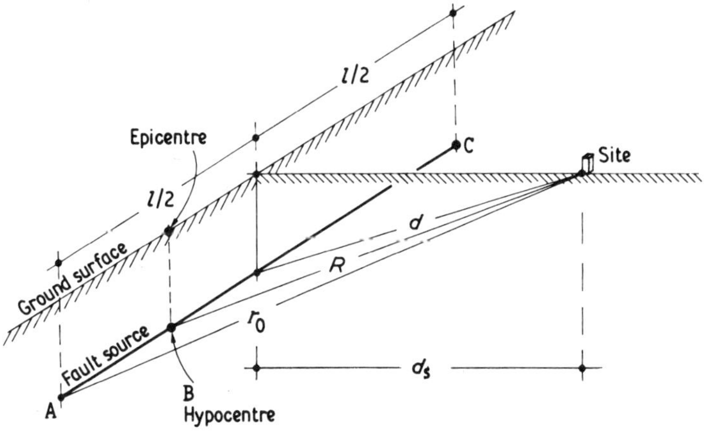

Fig. 3.10 Determination of tha focal distance of the site (after Cornell [6]).

which most significant earthquake features at the site can be expressed, by semi-empirical relations of the type

$$\tilde{q} = \tilde{q}(\tilde{R}, \tilde{\mathcal{M}}) \tag{3.37}$$

For the most widely used earthquake parameters q, appropriate forms of equation (3.37) have been proposed. For instance, for the modified Mercalli (MM) intensity

$$\tilde{\mathcal{I}} = C_1 + C_2\tilde{\mathcal{M}} - C_3 \ln \tilde{R} \tag{3.38}$$

with C_1, C_2, C_3 suitable constants; and for the quake duration $\tilde{d}$, interpolating a table given by Housner [7]

$$\tilde{d}(\mathrm{s}) = -0.33\tilde{\mathcal{M}}^3 + 8.5\tilde{\mathcal{M}}^2 - 56\tilde{\mathcal{M}} + 111 \tag{3.39}$$

Another very significant parameter is the *peak acceleration* $\tilde{a}_p$. In this case, equation (3.37) is often put in the form

$$\tilde{a}_p = e_1[\exp(e_2\tilde{\mathcal{M}})](\tilde{R} + e_4)^{-e_3} \tag{3.40}$$

where e_1, e_2, e_3 and e_4 are empirical constants. Typical laws (3.40) are shown in Table 3.2 [8].

Table 3.2 Attenuation laws, obtained from regression analysis of statistical data; a_p: cm s^{-2}; R: km (rearranged from Rosenblueth [8]).

Western USA	Rock	$a_p = 4\exp(1.26\,\mathcal{M})R^{-1.19}$
	Soil	$a_p = 32\exp(1.22\,\mathcal{M})R^{-1.19}$
Europe		$a_p = 2.88\exp(1.45\,\mathcal{M})R^{-1.1}$
Worldwide		$a_p = 1080\exp(0.5\,\mathcal{M})(R + 25)^{-1.3}$

More generally R may appear in the r.h.s. of (3.40) in the form $g(R)$ where $g(\ )$ is the assumed *attenuation law*.

All parameters q used in practice are decreasing functions of R and increasing functions of $\mathcal{M}$. Therefore, the probability distribution function of any q is easily calculated if the PDFs of R and $\mathcal{M}$ are known. In fact, for a given R,

$$\{\tilde{q} \geqslant q\} \Leftrightarrow \{\tilde{\mathcal{M}} \geqslant \mathcal{M}(R, q)\} \tag{3.41}$$

the function $\mathcal{M}(R, q)$ being the magnitude corresponding to $\tilde{q} = q$ and $\tilde{R} = R$ in (3.37). Hence, the conditional probability of (3.41) is

$$\text{Prob}[\tilde{q} \geqslant q \mid \tilde{R} = R] = \exp\{-\zeta[\mathcal{M}(R, q) - \mathcal{M}_0]\} \tag{3.42}$$

Convolution with the probability density function of the focal distance yields

$$\Pi(q) = \text{Prob}[\tilde{q} \geqslant q] = \int_0^{+\infty} \text{Prob}[\tilde{q} \geqslant q \mid \tilde{R} = R]\, p_R(R)\, \mathrm{d}R \tag{3.43}$$

where

$$\Pi(q) = 1 - P_q(q) \tag{3.44}$$

$P_q(q)$ being in turn the CDF of $\tilde{q}$.

Equation (3.43) yields the probability that when an earthquake occurs the value q of the variable $\tilde{q}$ is exceeded, but does not give any idea of the rate at which risk grows yearly with the structure standing at the site. The information yielded by (3.43) must be completed by consideration of earthquake occurrences. This process is the third, and the last, fundamental datum, and its knowledge can be achieved by means of geological as well as historical research. Poisson processes (Section A6.3) are frequently used as a model for earthquake occurrences: under this assumption, the number $\tilde{n}$ of significant earthquakes originating from the considered source (for instance, a fault) in t years are given by equation (2.48)

$$p_n(n, t) = \text{Prob}[\tilde{n} = n \mid t] = \frac{\mathrm{e}^{-\lambda t}(\lambda t)^n}{n!} \tag{3.45}$$

λ being the mean rate of arrival of significant quakes‡.

The arrival at the site of earthquakes with $\tilde{q} \geqslant q$ is described by selecting in the number $\tilde{n}$ of generic quakes the ones exhibiting the quoted property. It can be proved that the process of occurrence of such 'special' earthquakes is still Poisson, with arrival rate λ_q given by the product of the global rate λ and the probability $\Pi(q)$ that $\tilde{q}$ exceeds the chosen level q (equation (A7.44)), i.e.

$$\lambda_q = \lambda\Pi(q) = \lambda[1 - P_q(q)] \tag{3.46}$$

Thus, the average return period T_q of an earthquake causing exceedance of an

‡ Actually, earthquakes tend to be clustered in sequences, rather than uniformly distributed in time as assumed in the Poisson process. However, a satisfactory model of this process is not available at present.

established value q of $\tilde{q}$ is given, from equation (2.51), by

$$T_q \simeq \frac{1}{\lambda \Pi(q)} \tag{3.47}$$

or, equivalently, the value q of $\tilde{q}$ that is exceeded, on average, every T_q years is given by

$$q \simeq \Pi^{-1}\left(\frac{T_q}{\lambda}\right) \tag{3.48}$$

If T_q is put equal to the *design life* of the structure, equation (3.48) yields the value of the parameter q to be used for structural design, provided the earthquake arrival rate λ and the CDF of q (equation (3.40)) are known.

As an example, one can quote the application of the previous formulae to the peak ground acceleration a_p. Under the assumption that earthquakes originate from a single fault with location of the focus uniformly distributed over the fault length l, the distribution of the annual maximum peak acceleration, $\tilde{a}_{py} = \max(\tilde{a}_p$ in $(0, 1))$, is given, through equations (A7.51), (3.46), (3.42), (3.43) and (3.40) with $e_4 = 0$, by

$$\begin{aligned} P_y(a_p) &= \text{Prob}[\tilde{a}_{py} \leqslant a_p] \\ &= \exp\left\{-\lambda \int_0^{+\infty} \exp\left[-\zeta\left(\frac{1}{e_2}\ln\frac{a_p}{e_1 R^{-e_3}} - \mathscr{M}_0\right)\right] p(R)\,dR\right\} \end{aligned} \tag{3.49}$$

Putting $R^2 = x^2 + d^2$, where d is the distance from the fault and x the abscissa along the fault, and noting that the relation (2.41)

$$p(y)\,dy = p(g^{-1}(y))\,dx \tag{3.50}$$

in this case involves

$$p(R)\,dR = p[(R^2 - d^2)^{\frac{1}{2}}]\,dx = \frac{1}{l}dx \tag{3.50$'$}$$

equation (3.49) becomes

$$\begin{aligned} \text{Prob}[\tilde{a}_{py} \leqslant a_p] &= \exp\left[-\lambda\left(\frac{1}{l}\exp(\zeta\mathscr{M}_0)e_1^{\zeta/e_2}a_p^{-\zeta/e_2}\int_0^{l/2} 2(x^2+d^2)^{-\frac{1}{2}\zeta e_3/e_2}\,dx\right)\right] \\ &\simeq \exp\left(\frac{-\lambda\chi}{l}a_p^{-\zeta/e_2}\int_0^{\infty} 2(x^2+d^2)^{-(\Lambda+1)/2}\,dx\right) \end{aligned} \tag{3.49$'$}$$

where

$$\chi = [\exp(\zeta\mathscr{M}_0)]e_1^{\zeta/e_2} \tag{3.50$''$}$$

$$\Lambda = \zeta\frac{e_3}{e_2} - 1 \tag{3.50$'''$}$$

and the approximation holds if $l \gg d$.

In order to give a closed form to the last integral, remember that denoting by

$$\Gamma(x) = \int_0^\infty e^{-y} y^{x-1}\,dy \tag{3.51}$$

the *gamma function* (see Table A.2), the following relations hold

$$\Gamma(x+1) = x\,\Gamma(x) \tag{3.51$'$}$$

$$\Gamma(nx) = \sqrt{\left(\frac{n^{2nx-1}}{(2\pi)^{n-1}}\right)}\Gamma(x)\Gamma(x+\tfrac{1}{2})\ldots\Gamma(x+(n-1)/n) \qquad n = 2, 3, \ldots. \tag{3.51$''$}$$

$$\int_0^\infty \frac{x^{j-1}}{(1+x)^{j+k}}\,dx = \frac{\Gamma(j)\Gamma(k)}{\Gamma(j+k)} \tag{3.51$'''$}$$

and the last function of j and k is often denoted *beta function*.
Let $n = 2$ and $x = \Lambda/2$ in the above equations and put

$$K = \int_0^\infty 2(x^2+d^2)^{-(\Lambda+1)/2}\,dx = \int_0^\infty 2d^{-(\Lambda+1)}(y+1)^{-(\Lambda+1)/2}\tfrac{1}{2}d\,y^{\frac{1}{2}-1}\,dy$$

$$= d^{-\Lambda}\int_0^\infty y^{-\frac{1}{2}}/(1+y)^{\Lambda/2+\frac{1}{2}}\,dy = d^{-\Lambda}\frac{\Gamma(\Lambda/2)\Gamma(\frac{1}{2})}{\Gamma(\Lambda/2+\frac{1}{2})}$$

$$= \frac{1}{d^\Lambda}\frac{\Gamma(\Lambda/2)\sqrt{\pi}}{\Gamma(\Lambda/2+\frac{1}{2})}\frac{\Gamma(\Lambda/2+\frac{1}{2})}{\Gamma(\Lambda/2+\frac{1}{2})} = \frac{\sqrt{\pi}}{d^\Lambda}\frac{\sqrt{(2\pi)}}{\sqrt{(2^{2\Lambda-1})}}\frac{\Gamma(\Lambda)}{[\Gamma(\Lambda/2+\frac{1}{2})]^2}$$

$$= \frac{2\pi}{(2d)^\Lambda}\frac{\Gamma(\Lambda)}{[\Gamma(\Lambda/2+\frac{1}{2})]^2} \tag{3.51^{iv}}$$

Equation (3.49′) then becomes

$$\text{Prob}[\tilde{a}_{py} \leqslant a_p] = \exp\left[-\frac{\lambda\chi K}{l}a_p^{-\zeta/e_2}\right] \tag{3.49$''$}$$

Consequently, the return period T_a would be approximately

$$T_a \simeq \frac{l}{\lambda\chi K}a_p^{\zeta/e_2} \tag{3.52}$$

and taking logarithms

$$\ln T_a = \ln\left(\frac{l}{\lambda\chi K}\right) + \frac{\zeta}{e_2}\ln a_p \tag{3.53}$$

For a numerical example, consider a particular site in Turkey where the following values of the parameters were calculated [6]

$l = 650$ km	$\mathcal{M}_0 = 5$	$\lambda = 0.0975\ \text{y}l^{-1}$
$\zeta = 1.48$	$d = 44.6$ km	$e_1 = 2000$

$e_2 = 0.8$ $\qquad e_3 = 2$ $\qquad e_4 = 0$

$\Lambda = 2.7$ $\qquad \Gamma(\Lambda) = 1.545$ $\qquad \Gamma(\Lambda/2+\frac{1}{2}) = 0.946$

$K = 5.8795 \times 10^{-5}$ $\qquad \chi = 2.0925 \times 10^9$

In this case, equation (3.53) becomes

$$\ln T_a = \ln 5.42 \times 10^{-2} + 1.85 \ln a_p = -2.91 + 1.85 \ln a_p \qquad (3.53')$$

Equation (3.53′) is plotted in Fig. 3.11: it can be seen that the value $a_p = 0.1\,g \simeq 98$ cm s^{-2} commonly referred to in technical practice, corresponds to a return period, at the quoted site in Turkey, of 273 years.

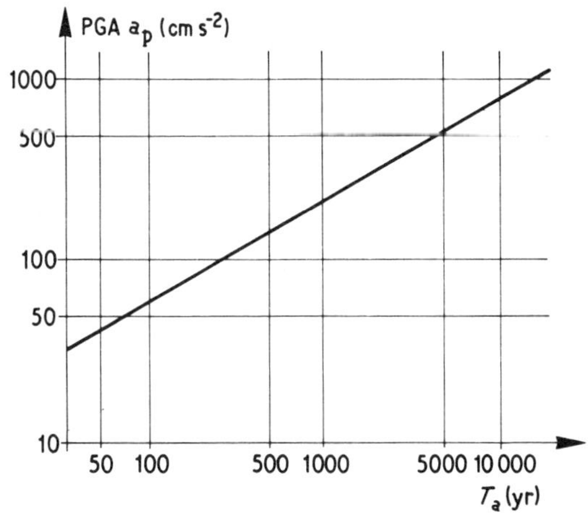

Fig. 3.11 Plot of return period vs peak acceleration (equation (3.53′)).

In conclusion, to investigate the seismic risk at a site of the earth's surface means to describe probabilistically intensity and occurrences of earthquakes at that site. This can be done on the basis of a mere *statistical* study of past earthquakes (e.g. from some *earthquake catalogue*) and of a reliable *attenuation law*. Alternatively, in the *seismogenetic approach* the potential seismic sources for the site under investigation are included in a volume V; this volume is then divided into subvolumes V_i in which the focus spatial distribution can be considered as uniform. On the basis of the geological knowledge of the volume V, then, the CDFs of the magnitude and of the waiting time relevant to a single event with focus in V_i are assigned. Finally, by means of the attenuation law, the CDFs of the same quantities, but referred to the site under investigation, are derived. *Statistical* and *seismogenetic* approaches are sometimes combined, in order to make the best use of all available information.

The occurrence of earthquakes in a given subvolume V_i is often represented by means of the average number $\mu_{n_i}(\mathscr{M})$ of earthquakes with magnitude greater than

$\mathcal{M}$ per year. Equations (3.45), (3.46) and (3.36) give

$$\mu_{n_i}(\mathcal{M}) = \begin{cases} \lambda & \mathcal{M} \leqslant \mathcal{M}_0 \\ \lambda \exp[-(\mathcal{M} - \mathcal{M}_0)] & \mathcal{M} > \mathcal{M}_0 \end{cases} \tag{3.54}$$

However, statistical evidence suggests that a non-linear exponent in the magnitude–frequency law yields a better fitting of historical data. So, quadratic–exponential laws or bi-exponential (exp[exp( )]) laws were proposed, in order to obtain a good fit of the data, and also a low sensitivity to statistical fluctuations.

Some results on the CDF of the waiting time can be obtained according to the developments of Section A6.3.2 (equation (A6.37)), or, more generally, making use of the first-excursion theory illustrated in Section A7.5

3.4.3 Input earthquakes and response spectra

The considerations developed in Section 3.4.2 are suitable for a gross estimate of the danger caused by earthquakes to structures at a given site; they require that the potential effects of an earthquake on buildings be measured in terms of a few parameters or, even better, a single parameter, such as the peak ground acceleration, the MM intensity, etc.

Although this methodology is extremely useful in drawing seismic risk maps and identifying the seismic regions to include in Seismic Codes, recent research shows that it is impossible to describe completely the effects of earthquakes by means of a few parameters, and that the whole time history of the ground motion during the earthquake has its importance. In fact, not only is the very essence of earthquake loading of a dynamical nature, being essentially an imposed random acceleration, but resonance effects may be very significant because the average earthquake power spectrum (equation (2.21)) falls in the range of natural frequencies of most high-rise and normal buildings (Fig. 3.12).

Therefore, appropriate procedures of *structural dynamics* must be used to obtain the *response* of a structure to earthquakes.

It must be recognized that the use of actual earthquake records for step-by-step calculations may be illusory from the viewpoint of structural safety: the scarcity of records, and also the cost of calculations, leads to checking structural response against one or two, at most, recorded earthquakes; on the other hand, calculations have shown examples of structures which would not be damaged by one or two quakes, but would fail under another one of the same intensity.

This is due to randomness of earthquakes, and would make it necessary to check aseismic safety against a statistically significant number of possible earthquakes at the site. The obvious implication is that such earthquakes should be available. Since it is not so, and the accumulation of data for a given site is a very slow process, it is necessary that basic properties of earthquakes be defined, that the influence of local conditions be recognized and that techniques to calculate and verify structural performance consistent with the site seismicity be available.

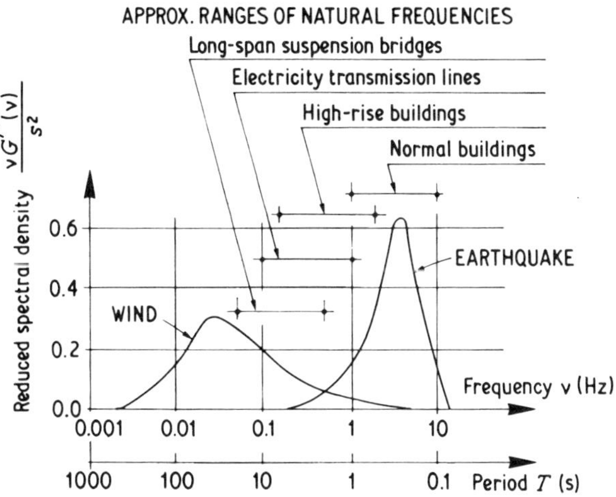

Fig. 3.12 Spectral densities of wind velocity and earthquake acceleration, compared with natural frequencies of main types of buildings (from Ferry-Borges and Castanheta [9]). Note that the ordinate is in dimensionless form (equation (2.25)); the wind spectrum was calculated for $\mu_{v_{10}} = 20$ m s^{-1} and $s^2_{v_{10}} = 2400\ \Upsilon$ m^2 s^{-2} (Υ is a coefficient measuring the soil roughness, usually with a value of between 0.005 and 0.015); the earthquake spectrum corresponds to $\mu_a = 0$ and $s_a^2 = 0.38$ m^2 s^4.

Two procedures are at present used in earthquake engineering, namely the use of *response spectra* and the generation of *artificial earthquakes*.

The response spectrum of an earthquake, first introduced by Housner [10], is a function which associates the natural period T (or the natural pulsation $\omega = 2\pi/T$) of a simple linear oscillator of given damping coefficient ξ (Section 4.3.1) with the maximum value attained by the response when the oscillator is subjected to the considered quake ‡. If the response is identified in the acceleration, velocity or displacement of the mass of the oscillator, one can speak of acceleration, velocity and displacement response spectra. The most widely used is the *velocity spectrum* $S_v(T, \xi)$, which is considered to describe best the structure performance under earthquakes.

Each response spectrum is referred to a single earthquake. By considering the spectra relevant to all earthquakes recorded at a given site and by averaging, one gets *mean response spectra*, such as those plotted in Fig. 3.13.

The most recent earthquake codes introduce a *design spectrum*, obtained as an approximate envelope of all mean response spectra of the relevant geographical region. Such spectra can be directly used to derive the elastic response of one- and multi-degree-of-freedom structures (using modal analysis for the latter) (Section 4.3), as will be seen in Section 10.2.3.

‡ Note that the *response spectrum* thus defined in neither an *input spectrum* nor the *spectrum of the response* (*output spectrum*) of any particular *filter* or oscillator, in the sense of equation (2.1).

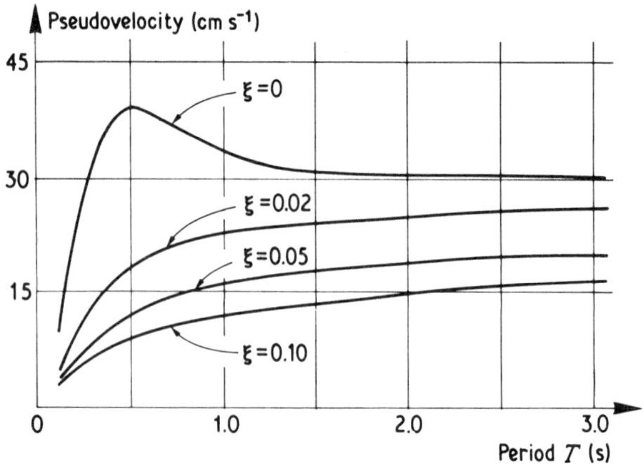

Fig. 3.13. Examples of velocity spectra a given site. ξ damping coefficient (fraction of critical damping) [9].

Response spectra can also be defined and plotted for oscillators exceeding the elastic range, but in this case the superposition law, which is at the basis of modal analysis, does not hold: the extension of the use of inelastic spectra to multi-degree-of freedom systems (as multistorey buildings), although suggested as an approximate procedure by some authors, often gives rise to dangerous errors: this problem will be taken up again in Chapter 10.

Alternatively, appropriate *artificial earthquakes* can be generated and used as input processes in a step-by-step integration of the equations of motion. This procedure is much more time-consuming than spectral analysis, but can be used for elastic and inelastic structures alike.

Response spectra play an important role in the generation of artificial earthquakes: in fact, not only must these retain the main features of past earthquakes recorded at the site or in the relevant area, but they must also be representative of the actual site seismicity and somewhat reflect the originating seismic mechanism. As a matter of fact, notwithstanding their large apparent randomness, earthquakes maintain, at least approximately, some constant properties at a given site: for instance, the average response spectra are very greatly influenced by the local site conditions through soil filtering of seismic waves, and can be taken as a constant for a given site. Many simulation methods are founded on matching spectral properties of actual and artificial earthquakes.

When simulating earthquakes with the purpose of safety analysis of buildings, care must be taken to reproduce closely the characteristics that affect structural behaviour, while poor similarity can be tolerated in the features that do not affect the response: for instance, the frequency content and the autocorrelation function may have rather little influence if quasi-rigid–plastic structures are dealt with. Therefore, engineering earthquake simulation is a compromise, fitting seismic site properties to structural response features.

Though some deterministic approaches have been tried for earthquake

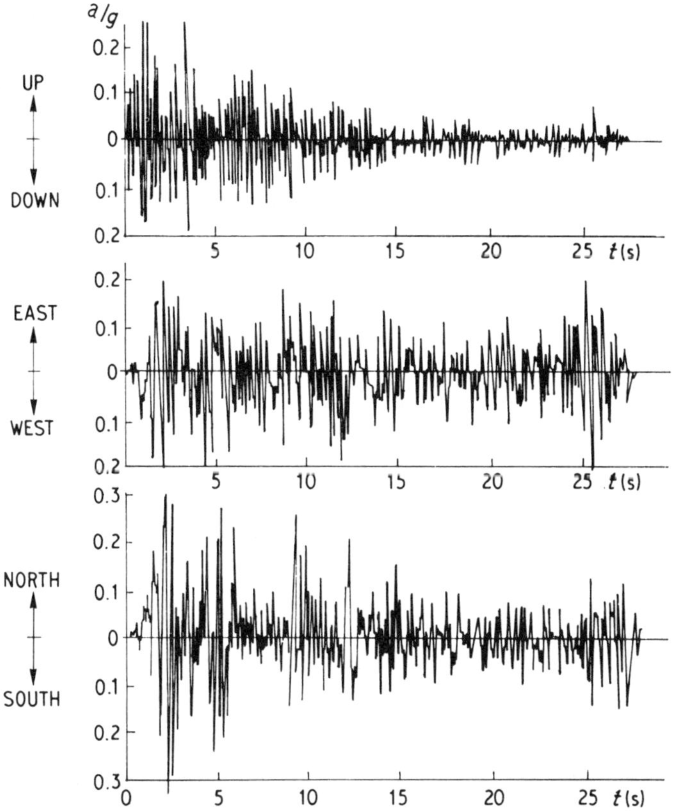

Fig. 3.14 Accelerograms of the Centro, California, earthquake of 19 May 1940, recorded approximately 30 miles from centre of slipped fault. Magnitude $\mathcal{M} = 7$.

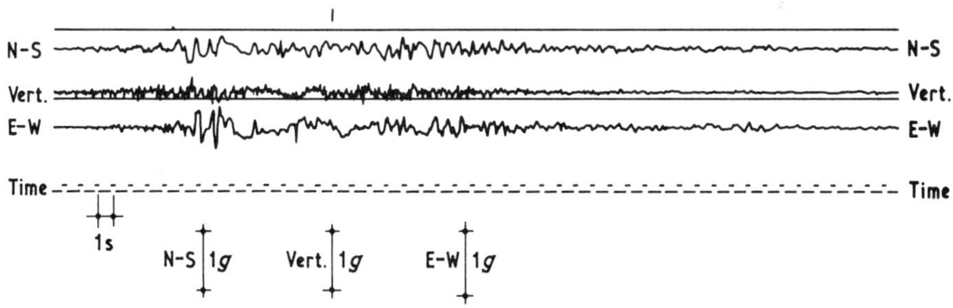

Fig. 3.15 Uncorrected accelerograms of the Southern Italy earthquake, 23 November 1980, recorded at Sturno, i.e. approximately 37 km from the epicentre (from CNEN–ENEL: Terremoto Campano-Lucano del 23 Novembre 1980; prime registrazioni; ENEL, Roma, December 1980).

simulation, mostly by filtering an input movement at a point source through a semi-infinite deformable (possibly non-homogeneous or layered) space, they can hardly take account of the pronounced random character of actual earthquake records, such as those in Figs 3.14 and 3.15.

Grossly, earthquake randomness can be attributed to the circumstance that seismic waves originating at the hypocentre undergo a very large number of reflections and refractions, which are influenced by the unordered locations of geological stratifications, thus superposing each other in a very irregular way. Further, uncertainties arise from the fact that earthquake-recording instruments are very sensitive to the local conditions at the site at which they are placed and to the way in which they have been regulated to start when the quake takes place. (It has been observed that even identical instruments set at a seismic site under similar conditions display different record of the same earthquake.) More errors are added when the record is digitized or manipulated. Thus, earthquakes are random in nature and randomly modified when brought to the analyst's knowledge.

The first attempt to define analytically the random process of earthquake accelerations was made by Housner [10], who proposed to idealize the accelerogram as a series of random pulses, or as a series of sine-wave pulses occurring randomly in time, obeying some specified laws for frequencies and amplitudes.

Later papers suggested the possibility of modelling earthquake accelerograms as segments of white-noise, especially for buildings founded of firm rock. The assumption might appear somewhat misleading, both because white-noise is physically unrealizable (Section 3.1) and because usual accelerograms do not exhibit a stationary character. The first objection is easily overcome by considering white-noise as a mathematical approximation to a limited-band process, intended to evaluate effects on systems with own natural frequencies in the range where the input spectrum is approximately constant. In fact, good agreement between the generated velocity spectra and the spectra calculated directly from recorded earthquakes has been obtained. The second objection is well founded, especially for earthquakes at large epicentral distances (say 200 km or more) in which case there is an effective separation between longitudinal *PR*, transverse *SE* and surface *SU* waves travelling in the earth's crust at rather different velocities (Section 3.4.1).

A recent proposal for numerical simulation of earthquake accelerograms will be presented in Chapter 9 (Section 9.2.4).

3.5 WIND LOADING

3.5.1 Wind phenomenology

Winds are movements of air that arise due to pressure differences in the atmosphere. The greater the difference between the pressures in two regions, the faster the air between them will move. All pressure differences between places on the earth's surface can be attributed directly or indirectly to temperature differences. If one region is warmer than its surroundings, for example, the air above it is heated and expands. The hot air rises, leaving behind a low-pressure zone into which cool air from the high-pressure neighbourhood flows. The horizontal flow towards the heated region at low altitudes is balanced by an outwards flow of the air that has risen, which cools and sinks to replace the air that has moved inwards. In this way, convection cells come into being that convert temperature differences into pressure differences and thus cause winds to occur.

The main variations in wind speed and direction are therefore related to variations of temperature. Periodic trends can thus be expected corresponding to the yearly cycle of seasons, the meteorological cycle (of the duration of about 4 days) and the daily insolation cycle.

Actually, the power spectral density (equation (2.21)) of wind velocity (Fig. 3.16; cf. also Fig. 3.12) do show peaks corresponding to these three cycles, but also another peak in the range of much smaller periods. This fourth peak (often referred to as *micro-meteorological* peak, by contrast with the first three *macro-meteorological* peaks) is related to the turbulence generated in the air flow by friction with the earth's surface.

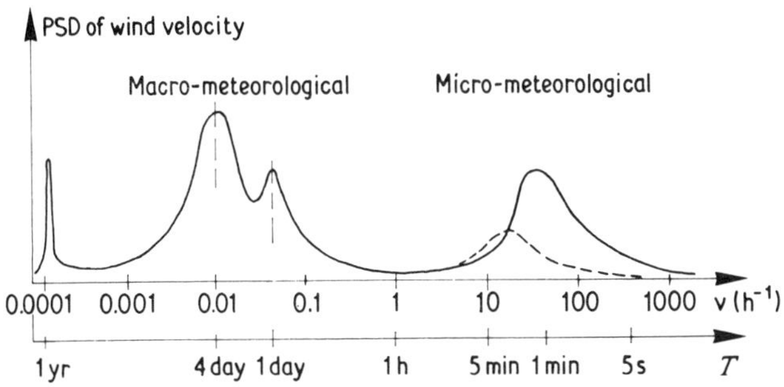

Fig. 3.16 Van der Hoven's wind power spectrum [11]. Note that the PSD is 'reduced' as in Fig. 3.12.

Between micro- and macro-meteorological peaks, the wind spectrum presents a very typical feature, namely a range of very low ordinates (*spectral gap*), approximately between 5 h and 10 min.

On the right of the gap, the spectrum corresponds to the *gusts* due to turbulence of the air flow, and therefore depends very much on the local *roughness* of the earth's surface. In a built-up environment, this roughness reflects the interaction of the buildings with the wind; this interaction modifies the air flow pattern and generates a new flow regime which contributes further small-scale turbulence to the flow. It is this turbulent flow over the surface of a building that determines the wind load on the building, and in particular the local pressure variations (in space and time) on the cladding of the building. The picture can be complicated further by *aeroelastic* phenomena, i.e. by modifications of the air flow induced by motions of the structure itself [12].

In Fig. 3.12, the portion of the wind spectrum on the right of the gap has been compared with the earthquake spectrum and with the natural periods of oscillation of the main types of structural systems. It can be noted that the natural periods of practically all structures are smaller than that associated with the micro-meteorological peak but, while those of the most deformable systems (such as suspension bridges, electrical lines, tall buildings, etc.) fall in a range of significant non-zero ordinates of the wind spectrum, the period of many other structures (e.g. low-rise buildings) is much smaller than even the smaller periods of the winds gusts. The latter structures respond statically to wind actions, while the response of the former ones may involve *dynamic* (and possibly *aeroelastic*) phenomena. A typical example of the PSD of a parameter describing the response to wind action of such a *wind-sensitive structure* is shown in Fig. 3.17. It is proportional to the (input) wind spectrum for periods on the left of the gap, which correspond to *static* components of the response (Section 3.1).

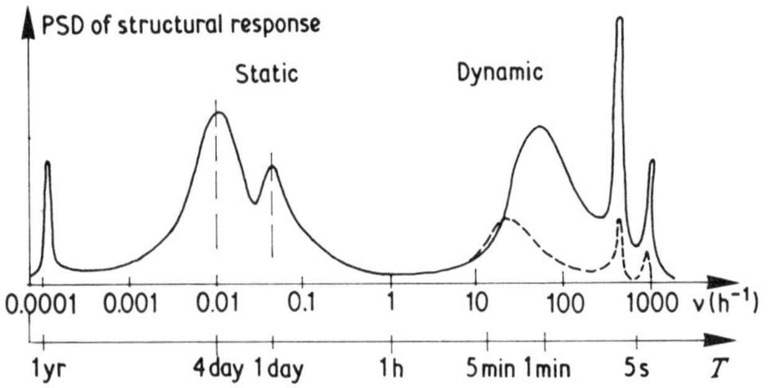

Fig. 3.17 Spectrum of the response of a wind-sensitive structure subject to wind action; spectral density 'reduced' as in Fig. 3.12 [11].

It has already been noted that the *roughness* of the terrain surrounding the site under consideration is the main source of air turbulence. Consequently, the average wind speed in any direction during a storm varies with elevation of the point under consideration (*wind speed gradient*). Typical profiles of wind velocity in the layer affected by the turbulence generated by interaction with the terrain, are plotted in Fig. 3.18: these show average wind speeds in a given direction, recorded simultaneously on the same vertical line at different heights. As shown by such plots, in all cases a level in the atmosphere is reached where the wind is no longer affected by the surface friction. The larger the surface roughness, the higher is this level, commonly named the *gradient level*, which depends also on wind speed.

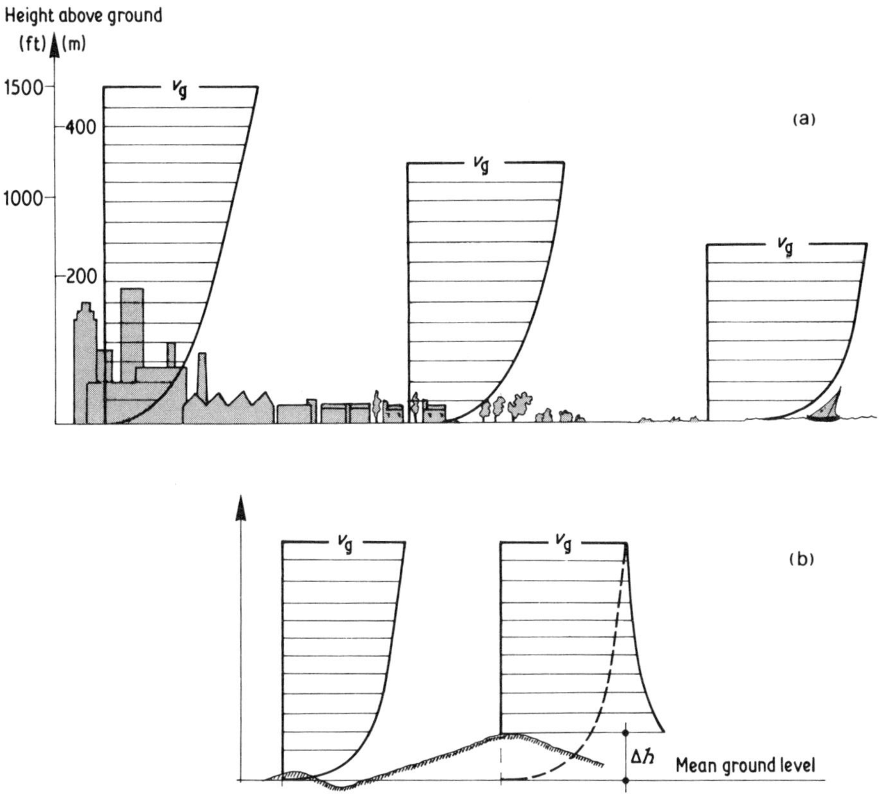

Fig. 3.18 Variation of wind speed with height above ground: (a) over plane terrain of different roughness [13]; (b) over hilly terrain [14].

Typical situations can be classified as:

(1) centres of large cities; very rough, hilly terrain;
(2) suburban areas; towns; city outskirts; wooded areas; rolling terrain; and
(3) flat, open country; open, flat coastal belts; grassland.

The respective gradient levels, usually assumed in the Building Codes, are 300, 400 and 500 m.

Below the gradient level, the wind velocity, averaged over a convenient time interval, can be expressed in the form $v_0^*(t)\chi(z)$, z being the height above ground level and $v_0^*(t)$ the average wind velocity at a reference height z^*. Over horizontal terrain of uniform roughness, the factor $\chi(z)$ can be given an expression of the type

$$\chi(z) = \frac{\ln[(z - z_d)/z_0]}{\ln[(z^* - z_d)/z_0]} \tag{3.55}$$

z_d is a constant named the *zero plane displacement* (typically zero except in centres of large cities where it may be set equal to the minimum between 20 m and $0.75\bar{h}$ ($\bar{h}$ being the average height of buildings in the surrounding area). z_0 is the *roughness length* and ranges from a few centimetres for sand terrain or sea waves to the order of a metre for cities. z^* is the standard level at which measures of wind speed are taken (usually, 10 m above ground).

3.5.2 Wind modelling

It has already been noted that the air flow pattern during a wind storm is characterized by rapid and random variations of both speed and direction in time and in space.

The research effort on wind effects on structures, although rapidly developing in recent years, is by no means comparable with that produced in the field of earthquake engineering, especially from the point of view of results already applicable in structural reliability calculations. This is perhaps mainly due to the circumstance that winds are really a structural problem only in particular regions of the world or for special types of structures, but also to the fact that wind modelling is even more difficult than earthquake modelling, because it involves spatial and temporal variability and therefore multivariate stochastic fields.

A complete model of wind velocity, covering explicitly spatial and temporal variability during a storm, to be used for computer simulation (Chapter 9), can be developed along the following lines.

Taking account of the considerations developed in the previous section, it is convenient to consider the wind velocity $v(\mathbf{x}, t)$ in a given horizontal direction in a point of position vector $\mathbf{x}$ at a time t during a storm as the sum of two terms

$$v(\mathbf{x}, t) = v_0(\mathbf{x}) + v_1(\mathbf{x}, t) \tag{3.56}$$

in which $v_0(\mathbf{x})$ is the *steady* component, equal to the average velocity during the storm, and $v_1(\mathbf{x}, t)$ is a zero-mean process describing the gusts [15].

The steady wind components at different heights z above ground level are usually considered as perfectly correlated to each other and related by deterministic laws of the type illustrated by Fig. 3.18 and equation (3.55).

A more sophisticated model [16], which is presented here for its methodological interest, distinguishes between large and small gusts:

$$v(\mathbf{x}, t) = v_0(\mathbf{x}, t) + v_1(\mathbf{x}, t) + v_2(\mathbf{x}, t) \tag{3.56$'$}$$

The spatial and temporal correlation of the latter two components $\tilde{v}_1(\mathbf{x}, t)$ and $\tilde{v}_2(\mathbf{x}, t)$, related to the air turbulence, varies within very wide limits. Indeed, the size of the eddies in the air flow varies from a few centimetres to kilometres for eddies generated by hills and mountains. As for time variations, the large gusts, $\tilde{v}_1(\mathbf{x}, t)$, have most of their energy in the range of periods larger than about twice the largest natural periods of structural systems, whereas small gusts, $\tilde{v}_2(\mathbf{x}, t)$, have significant energy in a range corresponding to the natural periods of wind-sensitive structures. In conclusion, $\tilde{v}_0$ and $\tilde{v}_1$ are always responsible for static response only, while $\tilde{v}_2$ may excite dynamic response components.

The spatial dependence of the random processes $\tilde{v}_1(\mathbf{x}, t)$ and $\tilde{v}_2(\mathbf{x}, t)$ can be simplified by discretization, i.e. by considering their values only in a finite number n of points, identified by position vectors $\mathbf{x}_1$ $\mathbf{x}_2, \ldots, \mathbf{x}_n$. Large and small gusts are thus described by two vector processes $\mathbf{v}_1(t)$ and $\mathbf{v}_2(t)$ of components $v_{1i}(t)$ and $v_{2i}(t)$ respectively ($i = 1, 2, \ldots, n$).

The gust vectors $\tilde{\mathbf{v}}_1$ and $\tilde{\mathbf{v}}_2$ can be modelled during a storm as vector-valued Gaussian stationary random processes.

The temporal autocorrelation function $B'_{1i}(v)$ (equation (2.3a)) of large gusts at a point $\mathbf{x}_i$, v being the time lag, is assumed to be in the form

$$B'_{1i}(v) = s_1^2 \exp(-v/\bar{v}) \tag{3.57}$$

in which $\bar{v}$ is a parameter giving a measure of the average duration of large gusts. Elaboration of anemometer records (Chapter 8) typically yields standard autocorrelation functions which fit (3.57) well [16]. The spatial cross-correlation functions between simultaneous ($v = 0$) large gusts at points $\mathbf{x}_i$, $\mathbf{x}_j$ (i.e. between the ith and jth components of the vector process $\mathbf{v}_1$) can be assumed in the form

$$B'_{1ij}(R) = s_1^2 \exp(-R_{ij}/\bar{R}) \tag{3.58}$$

where R_{ij} is the distance between $\mathbf{x}_i$ and $\mathbf{x}_j$ and $\bar{R}$ is a parameter related to the spatial dimension of an average gust. In a similar way, the spatial cross-correlation for small gusts $\mathbf{v}_2$ is given by

$$B'_{2ij}(R) = s_2^2 \exp(-R_{ij}/\Lambda) \tag{3.59}$$

where Λ is now the parameter measuring the size of small gusts.

In this way, the variances s_1^2 and s_2^2 of large and small gusts are assumed independent of the point location. All such parameters are related to two basic quantities, namely the steady wind velocity at 10 m above ground, v_{10}, and the surface drag coefficient Υ. Expressions for $\bar{v}$, $\bar{R}$ and Λ, are suggested in the form [16]

$$\bar{v} = \frac{d}{v_{10}} \qquad \bar{R} = \frac{1}{n}\left(\sum v_{0i}\right)\frac{\bar{v}}{2} \qquad \Lambda = \frac{(1/n)\sum v_{0i}}{c v_1} \tag{3.60}$$

with $d = 151$ m, v_{0i} the value of the steady wind velocity v_0 at the point $\mathbf{x}_i$, $c = 7$–8 for the gust component in the direction of the wind and v_1 the fundamental frequency, in hertz, of the structure whose response under wind action has to be evaluated. The total variance of the wind is given by

$$s_v^2 = s_1^2 + s_2^2 = 6.68\Upsilon\, v_{10}^2 \tag{3.61}$$

In order to calculate the characteristics of large gusts, s_1 can be obtained from equation (3.61) neglecting s_2, which is much smaller; then equations (3.57) and (2.21) yield the autocorrelation function and the PSD. With regard to small gusts, the PSD can be obtained as the difference between the assumed PSD of the wind at the site and that of large gusts; finally, the autocorrelation function $B'_{2i}(v)$ can be calculated from equation (2.22).

For more details on computer simulation, the reader is referred to Chapter 9.

3.5.3 Wind velocity modelling (static problems)

The model introduced in the previous section is useful when the structure under investigation behaves dynamically under wind excitations. However, it has already been noted that many common structural engineering problems are concerned with structures in the static response field. In the latter case, the wind action can be idealized (Fig. 3.19) as a sequence of randomly occurring pulses of random duration and random maximum magnitude, i.e. it can be described as a filtered Poisson process (Section A6.3.3). In fact, if only the static behaviour of the structure is involved, the separated form of macro-scale time history introduced in Section 3.11 is of basic interest, while the relatively finer time factor loses any direct meaning and can be taken into account by means of a global gust contribution. In other words, instead of using equation (3.56), the relevant wind velocity is expressed in the form (cf. equation (3.55))

$$v = v_0^* \chi(z)\zeta \tag{3.62}$$

where v_0^* is the steady (mean) velocity at a reference height z^* (usually, $z^* = 10\,\text{m}$ above ground level), $\chi(z)$ takes account of the dependence on height, and ζ is the *gust factor*.

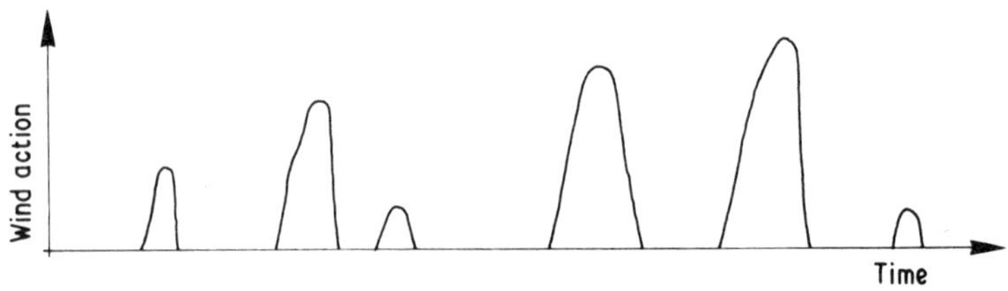

Fig. 3.19 Idealization of the time process of wind action on a structure.

The *wind pressure* p can be derived from the corresponding velocity, by means of the Bernoulli theorem, under the assumption that all the kinetic energy is transformed into pressure

$$p = \tfrac{1}{2}\rho v^2 \tag{3.63}$$

where ρ is the air *density* (mass per unit volume, approximately 1.2 kg m^{-3}). Then, the *wind load* acting on unit area of a solid surface is given by an expression of the type

$$w = c_p p \tag{3.64}$$

in which c_p is the *exposure coefficient*, which depends on the shape of the whole solid body and on the angle between the mean wind direction and the surface, and may be either positive (*pression*) or negative (*suction*).

The *spectral gap* in Van der Hoven's spectrum (Fig. 3.16) indicates that strong winds are events whose duration ranges from a few hours to a few minutes. Therefore, v_0^* should be obtained by averaging over time intervals of this order of magnitude (as short as possible for convenience of measurement).

Table 3.3 shows the ratios between the mean wind speed v_0^* averaged over different intervals, and the speed v_h averaged over an hour, that can be assumed to hold approximately at 10 m above ground level on open terrain. These ratios confirm that the average wind speed is a rather stable measure for averaging intervals between 10 and 60 min; 10 min is actually the most-used interval for wind records.

Table 3.3 Ratio of wind speeds at 10 m above ground averaged over different intervals (rearranged from Sachs [16] and Simiu [18]).

Averaging interval (s)		2	5	10	30	60	100	200	500	1000	3600
v_0^*/v_h	min	1.53	1.47	1.42	1.28	1.24	1.18	1.13	1.07	1.03	1.00
	max.	2.22	2.11	1.95	1.72	1.55	1.44	1.29	1.17	1.09	1.00

An alternative measure of the steady wind component is the so-called *fastest-mile wind speed* v_{fm}, i.e. the largest velocity, in a given time interval, at which air particles pass through a length of one mile. It can be verified that, to good approximation, [18]

$$\frac{v_{fm}^*}{v_{fm}(z)} = \frac{v_0^*}{v_0(z)}\left(1 + \frac{z-10}{10}0.02\right) \tag{3.65}$$

where $v_{fm}(z)$ and v_{fm}^* are the fastest-mile speeds recorded at z and 10 m above ground level respectively, and $v_0(z)$ and $v^*{}_0$ are the corresponding velocities averaged over intervals of 10–60 min.

An example of a continuous record of daily *fastest-mile speed* over a year is shown in Fig. 3.20.

Obviously, data on arrival rate and duration of storms are important in structural reliability calculations (especially with regard to the combination of more short-duration loads, cf. Section 3.7). Their estimation, however, depends on the climatic properties of the site under investigation and, hence, the problem can only be tackled by a micro-zonation founded on statistical measurements (Chapters 7 and 8). In contrast, the effects of the topographic characteristics of the site are included in the factor ζ in equation (3.62), and the wind velocities can be corrected to open country (Fig. 3.20(b)) and analysed over large extensions in order to describe the randomness of the reference velocity v_0^*.

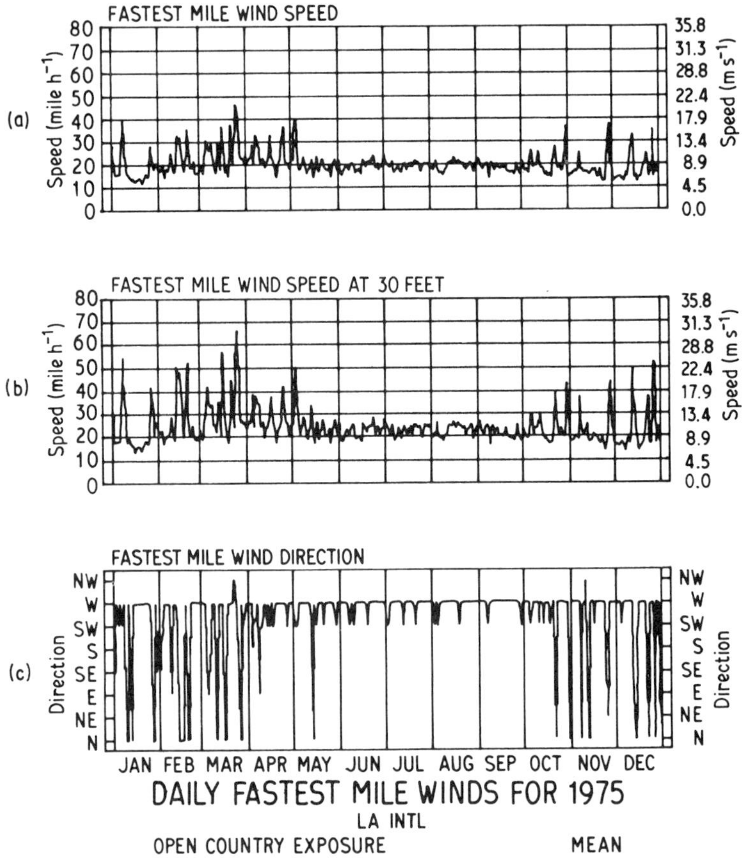

Fig. 3.20 Daily fastest-mile winds during 1975 at Los Angeles International Airport (from Lew and Mart [19]): (a) uncorrected data recorded at 20 ft (6 m); (b) and (c) data corrected to open country at 30 ft (9 m).

3.5.4 Yearly maximum of wind speed; wind direction

In conclusion, wind action is usually specified through a reference velocity v_0^*, which represents an average over 10 min (or a similar interval) of measurements taken 10 m above ground over terrain of uniform standard roughness.

Since, as already said, the average wind speed v_0^* is responsible for a *static* response, only its maximum value v_T^* over an appropriate time interval T is of interest in structural reliability analysis (cf Section 3.2(1)).

For this purpose, the mean arrival rate λ (or the mean occurrence interval $T = 1/\lambda$ (equation (2.49))) of v_0^* must be specified: in other words, it is necessary to associate return periods T to values of wind speed. This can be done on the basis of the CDF, $P_Y(v)$, of the largest wind speed in one year: data and analytical models for $P_Y(v)$ are available to describe the largest wind speeds recorded in each year at an anemometer station (for details see Chapter 7). For any value of the wind speed v, the corresponding return period $T(v)$ can be calculated through the expression (equations (2.54) and (A7.27))

$$T(v) = \frac{1}{1 - P_Y(v)} \tag{3.66}$$

Analysis of the CDF of wind speed $P_Y(v)$ can be made on the basis of extreme statistics. Remembering that the argument v of the distribution P_Y is a speed averaged over a specified time interval (e.g. 10 min), the mechanism originating the random variable v should resemble the one leading to extreme values of an underlying random variable that is the instantaneous wind speed. Therefore, any asymptotic probability distribution of largest values with unlimited upper tail should be an appropriate model for the behaviour of the largest yearly wind speed. These distributions are known as the type I (Fisher–Tippet) and the type II (Frechet) distributions of the largest values, and can be given respectively the forms (Table 2.2 and Section A7.2, equations (A7.5) and (A7.14))

$$P_{\mathrm{I}}(v) = \exp\{-\exp[-c_1(v - v_1')]\} \qquad -\infty < v < \infty \tag{3.67}$$

$$P_{\mathrm{II}}(v) = \exp[-(v_2'/v)^{c_2}] \qquad 0 < v < \infty \tag{3.68}$$

where v_1', c_1, v_2' and c_2 are distribution parameters. There are arguments to prefer the one or the other distribution: recently, the type I distribution tends to be preferred, at least for hurricane-free areas. In fact, it is necessary to distinguish hurricanes and tornadoes from 'current' winds. A hurricane is a large, violent tropical storm, typically 100–200 km in diameter, whose winds spiral inward and upward at velocities of 120 km h^{-1} or more around an *eye* of low pressure; most hurricanes occur on the western side of the Pacific, Indian and North Atlantic Oceans during the late summer and early autumn. Their most violent phases last from a few days to a week; hurricanes usually move at 20–50 km h^{-1} but may move faster or remain in one place for a day or more. A tornado is a small, funnel-shaped rotational storm some hundred metres in diameter that appears to descend from a cumulonimbus (thunderstorm cloud). Velocities in a tornado are apparently several hundred kilometres per hour, though the tornado itself moves at only 30 to 60 km h^{-1}; a tornado usually lasts for less than an hour. Tornadoes most often occur on central continental plains, notably those of the US and Australia.

The parameters of the chosen distribution should be evaluated for the best fit with recorded data. Unfortunately, the method of estimating the moments of the distribution that will be presented in Chapter 7 is not suited for such an analysis. Therefore, more specific approaches are generally employed [20]. A special technique, for instance, has been used for the analysis of anemometer records obtained at different stations [21].

Where hurricanes do not occur, predictions based on Frechet distribution are almost coincident with the ones yielded by type I distributions. The difference in 50 yr winds is of the order of 2% and rises to 8% for 1000 yr winds.

There is reason, therefore, in accepting that in hurricane-free sites yearly extreme wind speeds follow a type I distribution, while extreme speeds associated with hurricanes are better modelled by a type II (Frechet) distribution.

Once the probability distribution of the maximum yearly velocity v_Y^* is known, one obtains the CDF of the maximum velocity in T years by equation (A7.4) or (2.52).

$$P_T(v) = [P_Y(v)]^T \tag{3.69}$$

Equation (3.69) holds under the assumption that the T repetitions are independent. Further, by means of equation (3.66) the return period associated with each wind velocity value is computed.

So far, no reference to the wind direction has been made. However, the wind response of a building is in general sensitive to its direction: therefore, data on the wind speed should always be associated with data on the direction of blowing (e.g. see Fig. 3.20(c)). Joint probability functions of recorded wind velocities and directions have been elaborated. An example is shown in Fig. 3.21; another form of presentation of analogous data is in Fig. 3.22.

This data should lead to the distribution of the extreme wind speed associated with each direction. The calculations could be simplified by assuming that the extreme wind speed is stochastically independent of the direction. In this case, the joint probability density function of speed and direction is simply

$$p_{\theta,T}(\theta, v) = p_T(v)\,p_\theta(\theta) \tag{3.70}$$

where $p_\theta(\theta)$ is the probability density function of the wind direction and $p_T(v)$ the probability density function of the extremum wind speed in T years, easily obtained by derivation of $P_T(v)$ (equation (2.53))

$$p_T(v) = \frac{dP_T}{dv} = T[P_Y(v)]^{(T-1)}\,p_Y(v) \tag{3.71}$$

Under assumption (3.70), if $P_{\text{fail}}(\theta, v)$ is the failure propability of a building under winds blowing from the direction θ with velocity v, the overall probability of failure in T years can be estimated by the integral

$$P_{\text{fail}} \simeq \int_0^\infty \int_0^{2\pi} P_{\text{fail}}(v,\theta)\,p_\theta(\theta)\,p_T(v)\,d\theta\,dv \tag{3.72}$$

which, in rigour, yields only an upper bound to P_{fail}, but, as usual, a close bound in the small probability range.

Sometimes, especially for code purposes, $p_\theta(\theta)$ is assumed to be a uniform distribution. This may be approximate enough at and above gradient level on continental sites far away from the coastline [17], but certainly not for most situations at building level (cf. Figs 3.20(c), 3.21 and 3.22).

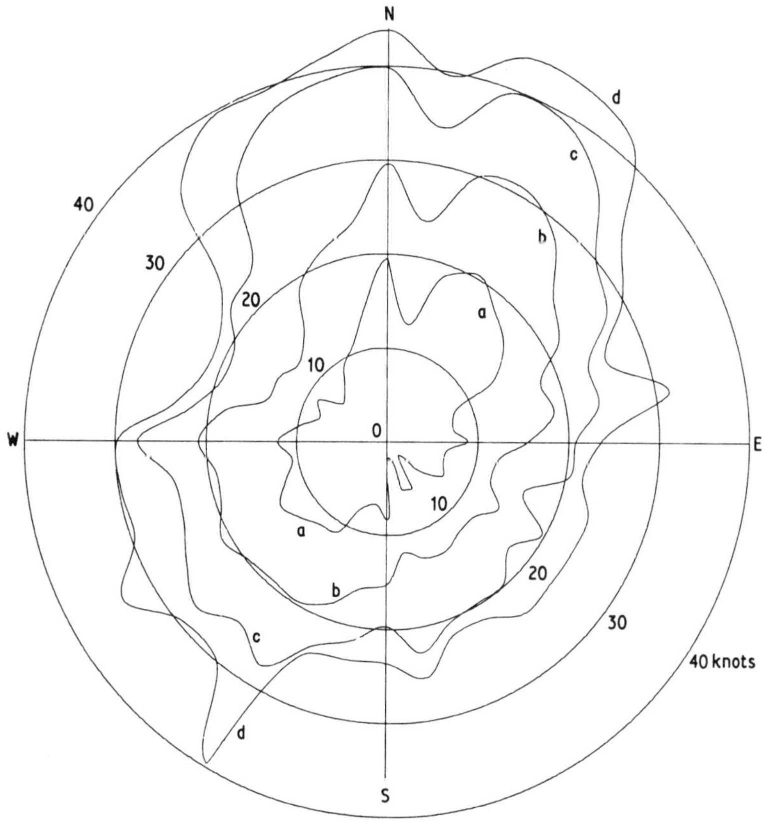

Fig. 3.21 Directional probability function of the 10-min wind velocity recorded over 23 years in Florence (Peretola Airport) at 10 m above ground: level curves of the probability P_θ that the reading falls in each 10° sector and is larger than the corresponding distance from the origin (1 knot = 0.514 m s^{-1}): a, $P_\theta = 10^{-2}$; b, $P_\theta = 10^{-3}$; c, $P_\theta = 10^{-4}$; d, $P_\theta = 10^{-5}$. (Elaboration by A. Borri and P. Spinelli.)

3.6 SNOW LOADING

Note that, in contrast with earthquake and wind loadings, snow has only a static action on structures and, hence, a relatively finer time (and/or space) model of this load is not necessary. Such a circumstance reduces the problem of describing snow action to the analysis of its intensity and its occurrence. However, statistical snow data have been collected from the beginning of the present century, mainly for hydrological purposes, and, therefore, measurements were designed in order to specify the equivalent amount of water in each snow fall rather than to describe the snow accumulation and its weight. The lack of information on the variability of snow intensity and snow accumulation on the ground makes these data very incomplete in view of a correct estimation of snow loading.

In temperate regions, snow occurs, during longer or shorter winter periods, in a sequence that can be idealized to be Poisson distributed, with mean arrival rate that depends strictly on the climatic zone under investigation. Thus, snow load must be classified as short-term action in temperate countries, while it becomes a long-term action for mountain localities or regions near the polar circles. The two extreme cases are characterized by no accumulation of snow from two successive meteorological events (Fig. 3.23(a)) (in the time between the events, the snow melts) or by complete accumulation till the whole cloak of snow is melted in spring (Fig. 3.23(b)). Up to now, intermediate accumulation laws have not been studied in any detail.

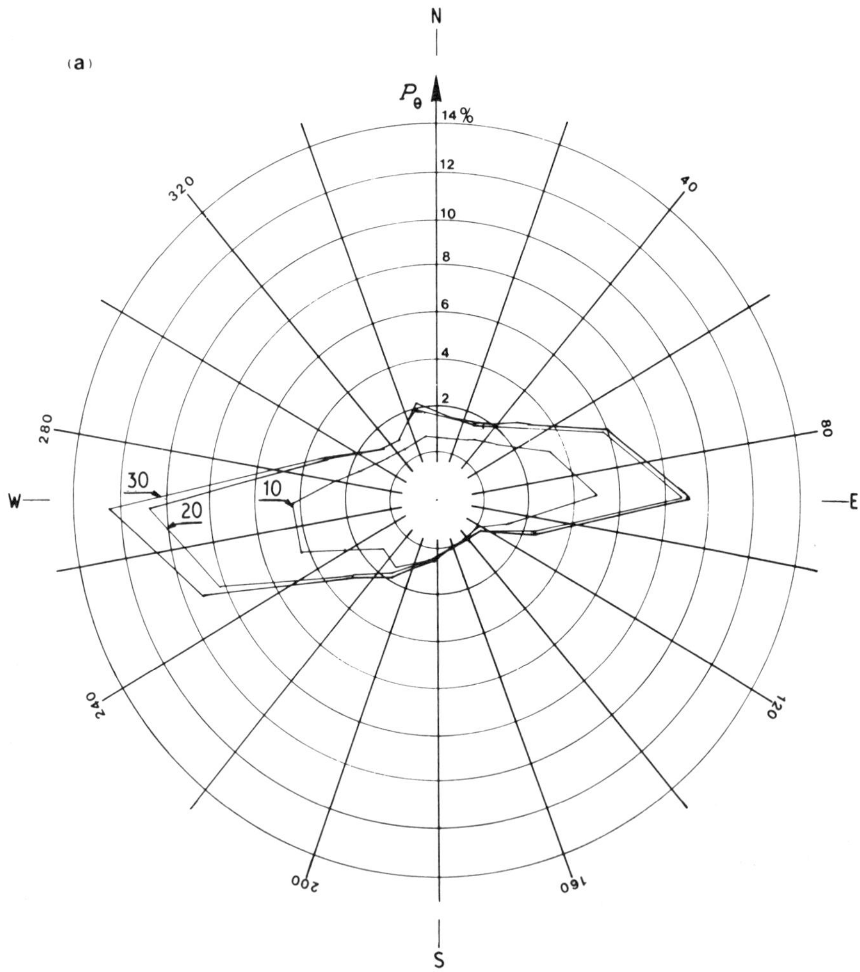

(b)

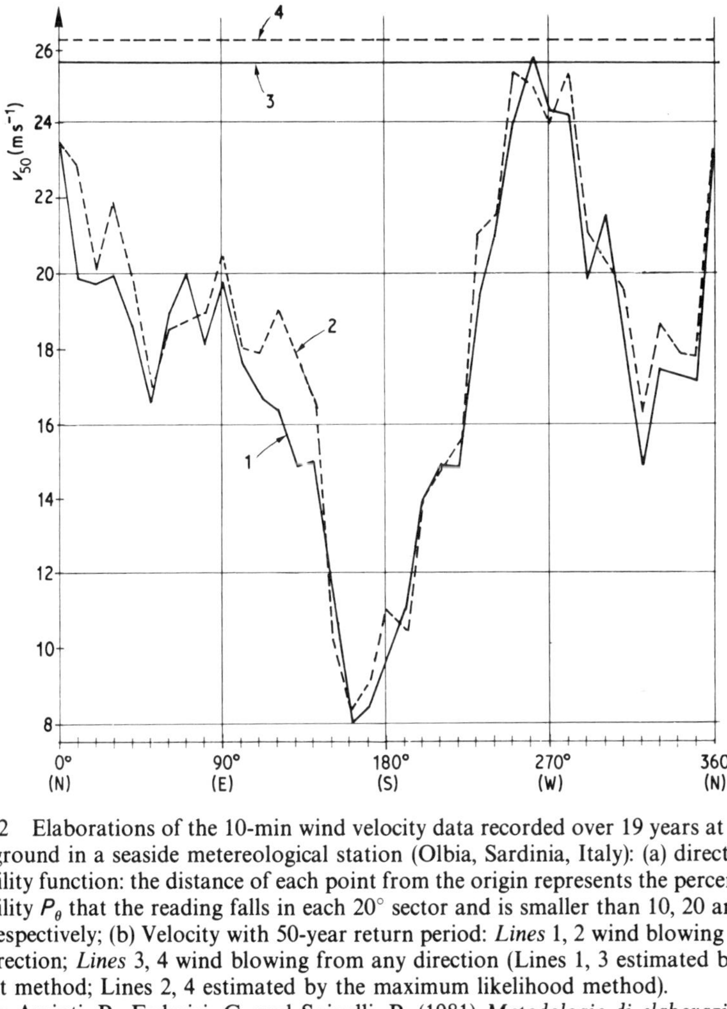

Fig. 3.22 Elaborations of the 10-min wind velocity data recorded over 19 years at 10 m above ground in a seaside metereological station (Olbia, Sardinia, Italy): (a) directional probability function: the distance of each point from the origin represents the percentage probability P_θ that the reading falls in each 20° sector and is smaller than 10, 20 and 30 knots respectively; (b) Velocity with 50-year return period: *Lines* 1, 2 wind blowing from each direction; *Lines* 3, 4 wind blowing from any direction (Lines 1, 3 estimated by the moment method; Lines 2, 4 estimated by the maximum likelihood method).

(From Aminti, P., Federici, G. and Spinelli, P. (1981) *Metodologie di elaborazione e indagine preliminare dei dati di base della ventosita in Italia* (in Italian), Univ. Firenze; Struct. Engrg. Report. UFIST/06/1981.)

To obtain the maximum yearly snow load for unit area on the ground, it is necessary to know not only the stochastic processes $\tilde{h}(t)$ that describe the height of the cloak of snow, but also the snow specific weight $\tilde{\zeta}(t)$ as a (random) function of time.

$$\tilde{w}_{g_y} = \max_t [\tilde{\zeta}(t)\tilde{h}(t)] \qquad 0 \leqslant t \leqslant 1 \text{ yr} \tag{3.73}$$

Equation (3.73) shows that the environmental factors (Section 3.2) must be specified for a structure-free representation of the loading.

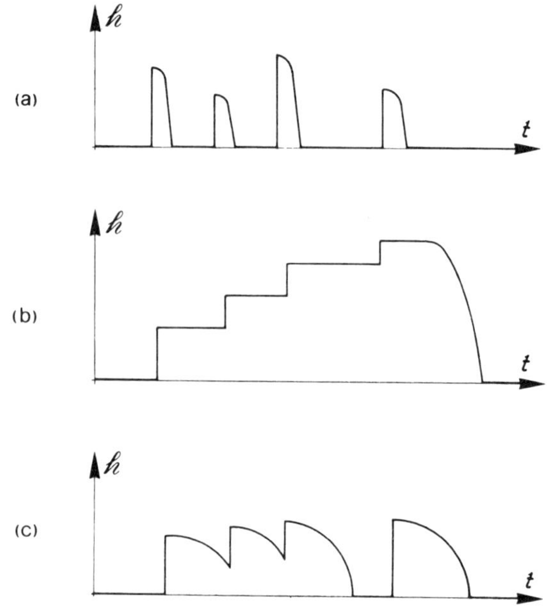

Fig. 3.23 Idealizations of the snow accumulation law.

A simplification can be introduced by the assumption that the snow height h and the relevant specific weight ζ are related by a deterministic monotonic relationship, like that plotted in Fig. 3.24. In this case, the maximum snow load on the ground corresponds to the maximum height of snow. Several investigations developed in the last decade are based on this assumption. From the relevant literature (mostly in German [23,24]), $\tilde{w}_{g_y}$ appears to be distributed according to an extreme type I distribution (equation (A7.5)), truncated at the origin, with coefficient of variation $\delta_{w_{gy}}$ (equation (2.12′)) generally close to 0.45.

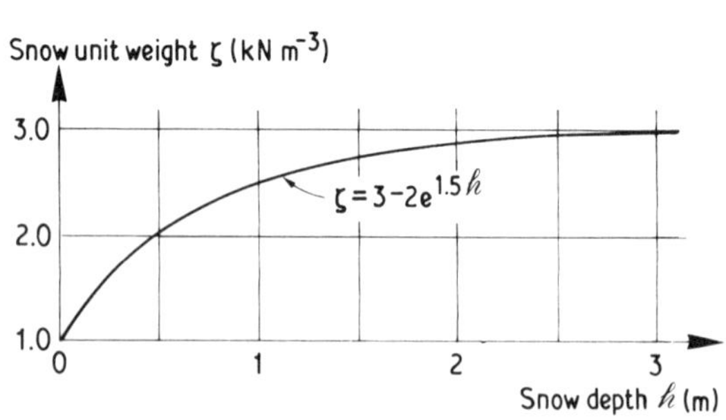

Fig. 3.24 Snow height vs snow specific weight [22].

3.7 THE PROBLEM OF LOAD COMBINATION

Consider a structure subject to one load. If the load environment is characterized by a scalar process $W(t)$, then there is a unique value of this process associated with the first passage into the unsafe domain, provided there is a one-to-one relationship between load effect (as e.g. for a linear elastic structure) and static loading. This value can be found by scanning all the stress locations and/or cross-sections and/or failure modes (Section 1.6) to find the critical one, i.e. the one which becomes unsafe at the lowest load. Under these circumstances, the reliability calculation reduces simply to the search for the extreme (maximum) value of the process $\tilde{W}(t)$ over the design life T of the structure, i.e. to the study of the random variable

$$\tilde{W}_{\max} = \max_T\{\tilde{W}(t)\} \tag{3.74}$$

a variable which is independent of the structure.

Consider next the case when more static loading processes, say $W_1(t)$, $W_2(t), \ldots, W_i(t), \ldots, W_n(t)$, act on the structure. Typical simplifications are generally made in practical reliability analysis and their implications must be outlined. This problem is usually referred to as the *load combination problem.*

The main objectives in this field are to find relatively simple analytical results for cases involving various standard stochastic process models of interest and, if possible, to reduce these results to operations (such as convolution, Section 2.8) involving random variables rather than stochastic processes. A further step consists in relating these results to practical schemes for code checking (i.e. load combination formats) and will be developed in Chapter 10.

3.7.1 Load idealization

In order to obtain operative load combination models, it is generally suitable to adopt the separated form proposed in Section 3.1 and to focus attention on the environment factor allowing for macro-scale time histories. The independence of this factor from the physical characteristics of the structure permits the evaluation of structural reliability by equation (2.68) where only load parameters W_j constant in time are considered.

However, the load process is in general a vector-valued function of time, $\mathbf{W}(t)$. If this variability in time is limited to one element of $\mathbf{W}(t)$, say $W_k(t)$, it is possible to define the most dangerous intensity W_k^s of the load process $W_k(t)$ (generally its maximum or minimum value) during the design life of the structure. Hence reliability analysis can be performed by equation (2.68) on the condition that the quantity W_k^s is considered to describe the kth loading component.

When more than one load component varies in time, there are practical implications upon the problem of representing or modelling the individual load processes. First, of course, at least a temporal model $W_i(t)$ is needed; a random variable model, such as $W_{\max,i}$, is insufficient. Secondly, the model must represent more than simply the extremes of the process; typical durations of (discrete) events must be at least approximately represented for an accurate analysis of the extreme of the combination of such processes. In fact, the extreme of the combination is sensitive to the likelihood that the two processes have high values simultaneously (which in turn depends on how long each process may remain at these levels).

Thus, in addition to load occurrence rate and intensity variation, a correct load combination analysis procedure must also include the random duration of each occurrence and the simultaneous occurrence of different loads. For this purpose, following Section 3.2, the macro-time behaviours of load effects are conveniently categorized into the following three types:

(1) Loads constant or nearly constant during the design life T (Fig. 3.25(a)). The ith constant action is fully described by its random intensity $\tilde{W}_i$ (Section 3.2).
(2) Loads changing at discrete instants of time (Section 2.7), but remaining essentially constant in between (long-duration 'sustained loads'; Fig. 3.7(a) and Fig. 3.25(b)). The jth long-term (or long-duration) action is defined by the mean rate λ_j^{l} of load changes (assumed to be Poisson events) and by the load levels between changes. These levels $\tilde{W}_j^{\mathrm{u}}$ are assumed to be random variables

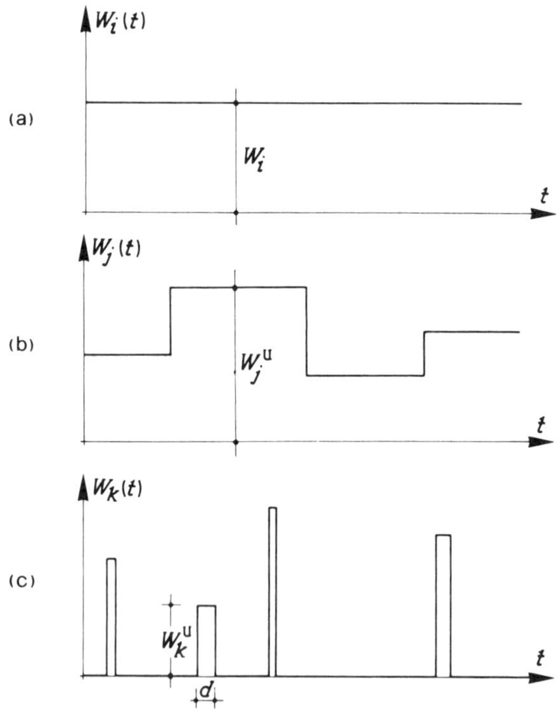

Fig. 3.25 Temporal variability of loads: (a) load constant in time; (b) long-term action; (c) short-term action.

that vary independently from occurrence to occurrence and have the same distribution $p_{W_j^u}(W)$.

(3) Actions occurring infrequently and with short duration (short-term loads; Fig. 3.25(c)). A simple Poisson arrival model of mean rate λ_k is generally introduced to describe the occurrence of the kth short-duration load. Moreover the load intensity $\tilde{W}_k^u$ and duration $\tilde{d}_k$ in each occurrence are assumed to be random variables that vary independently from occurrence to occurrence and have the same distribution $p_{W_k^u}(W)$ and $p_{d_k}(d)$. In such a way the short-duration actions are modelled as filtered Poisson processes (FPP) in the extended sense (Section A6.3.3).

A classification of loads according to the above categories is presented in Table 3.4 for a wide range of structures.

3.7.2 Combination of constant and short-duration loads

Consider a structural problem involving n constant loads and n_s short-duration actions. Three possible combinations of the latter loads can be considered:

(1) the singly acting short-duration load $W_k(t)$, $k = 1, \ldots, n_s$;
(2) the simultaneous occurrence of two short-durations loads $W_k(t)$ and $W_l(t)$; and
(3) the simultaneous occurrence of three short-durations loads $W_k(t)$, $W_l(t)$ and $W_r(t)$.

The simultaneous occurrence of four or more loads is generally very infrequent and hence it may be neglected for risk analysis purposes.

The probabilities of failure P_k, P_{kl} and P_{klr} under the single occurrence of each load combination can be estimated by equation (2.68) in which, in addition to the n constant load parameters $\tilde{\mathbf{W}}$, the kth short-load intensity $\tilde{W}_k^u$ (for P_k), the kth and lth components of the vector $\tilde{\mathbf{W}}^u$ (of length n_s) (for P_{kl}) and the kth, lth and rth components of $\tilde{\mathbf{W}}_l^u$ (for P_{klr}) respectively must be introduced.

The total probability of failure $P_{\text{fail}}(t)$ of the structure in $(0, t)$ can be approximated by the conservative expression [26, 27] (see equations (3.26), (A7.44) and (A7.48))

$$\begin{aligned} P_{\text{fail}}(t) &\leqslant \sum_k [1 - \exp(-\lambda_k t P_k)] + \sum_{k \neq l}\sum [1 - \exp(-\lambda_{kl} t P_{kl})] \\ &\quad + \sum_{k \neq l \neq r}\sum\sum [1 - \exp(-\lambda_{klr} t P_{klr})] \\ &= \sum_k P_{\text{fail}}(t|k) + \sum_{k \neq l}\sum P_{\text{fail}}(t|k,l) + \sum_{k \neq l \neq r}\sum\sum P_{\text{fail}}(t|k,l,r) \end{aligned} \tag{3.75}$$

where $P_{\text{fail}}(t|k)$, $P_{\text{fail}}(t|k,l)$ and $P_{\text{fail}}(t|k,l,r)$ are the probabilites of failure of the structure in $(0, t)$ given the occurrence of each load combination. In (3.75) λ_{kl} and

Table 3.4 Summary of actions‡ on structures (from CIRIA [25]).

	Homes, institutions, offices	Assembly and sports buildings	Warehouse, factory, car park	Storage tanks, silos	Building foundations	Basements, earth-retaining works	Quays, wharves, locks, etc.	Jetties, breakwaters	Permanent offshore structures	Bridge superstructures	Bridge piers and abutments	Transmission towers, masts	Chimneys	Cranes, winding towers	Drilling towers	Formwork and scaffolds	Coffer dams, etc.
Natural actions																	
Self-weight	C	C	C	C	C	C	C	C	C	C	C	C	C	C	C	C	C
Wind loads	S	S	S	S	S	–	S	S	S	S	S	S	S	S	S	–	–
Snow loads	LS	LS	LS	LS	LS	–	–	–	LS	LS	LS	LS	–	–	–	–	–
Ice loads	–	–	–	–	L	–	–	–	L	–	–	L	–	L	L	–	–
Earth pressures	–	–	–	L	C	C	C	–	–	–	C	–	–	–	–	–	L
Water pressures	–	–	–	L	L	L	L	L	L	–	L	–	–	–	–	–	L
Wave forces	–	–	–	–	–	–	–	S	S	–	S	–	–	–	–	–	S
Current forces	–	–	–	–	–	–	L	L	L	–	L	–	–	–	–	–	L
Ground movements	●	●	●	●	●	●	●	●	●	●	●	?	●	●	?	?	?
Temperature and humidity	●	●	●	●	–	–	–	–	●	●	●	–	●	–	–	–	–
Earthquakes	ES	ES	ES	ES	ES	ES	ES	ES	ES	ES	ES	ES	ES	ES	?	?	?

Man-made actions

Imposed dead loads	C	C	C	C	C	C	C	C	C	C	C	–	–	C	–	L	–
Live loads (people)	L	S	L	–	S	–	–	–	L	S	S	–	–	S	–	S	–
Furniture and machinery	L	L	L	L	L	–	L	L	L	–	–	L	L	L	L	–	–
Stored materials	–	–	L	L	L	–	L	L	L	–	–	–	–	–	–	S	–
Vehicles (moving)	–	–	S	–	–	–	S	S	S	S	S	–	–	–	–	S	–
Surge and braking	–	–	S	?	–	–	–	–	–	S	S	–	–	S	–	S	–
Impact and collision	–	–	E	–	E	–	E	E	E	E	E	E	–	–	–	–	E
Explosions	●	●	●	●	–	●	–	–	●	–	–	–	●	–	–	–	–
Mining subsidence	●	●	●	●	●	●	●	●	●	●	●	?	●	●	–	–	?
Wear and tear	●	●	●	●	–	●	●	●	●	●	●	●	●	●	–	–	–

‡ C = Constant or nearly so during design life.
L = Long duration but time-varying.
S = Short duration.
E = Exceptional (abnormal).
● = Not classifiable in general (specific instance must be studied).
– = Not applicable.
? = Not known.

λ_{klr} denote the mean rates of the Poisson events defined by the simultaneous occurrences of the short-duration loads $\tilde{W}_k(t)$ and $\tilde{W}_l(t)$, and $\tilde{W}_k(t)$, $\tilde{W}_l(t)$ and $\tilde{W}_r(t)$. Under the assumption that the expected values μ_{d_k}, μ_{d_l} and μ_{d_r} of the random durations $\tilde{d}_k$, $\tilde{d}_i$ and $\tilde{d}_r$ respectively are such that $\mu_{d_k} \ll \mu_{d_l} \ll \mu_{d_r}$, one obtains

$$\lambda_{kl} \simeq \lambda_k \lambda_l (\mu_{d_k} + \mu_{d_l}) \simeq \lambda_k \lambda_l \mu_{d_l} \tag{3.76}$$

while the mean duration $\mu_{d_{kl}}$ of the simultaneous action of the two loads is

$$\mu_{d_{kl}} \simeq \frac{\mu_{d_k}\mu_{d_l}}{\mu_{d_k} + \mu_{d_l}} \simeq \mu_{d_k} \tag{3.77}$$

In order to explain equations (3.76) and (3.77), consider two independent (0, 1) stochastic processes whose changes from 0 to 1 occur as Poisson events of intensities λ_1 and λ_2 respectively; let the duration of state 1 be deterministic for both processes, and equal to d_1 and d_2 ($d_1 > d_2$) respectively. For both such processes the probability of a zero event at any time t is $(1 - \lambda d)$, while λd is the probability that the complementary event occurs. Because of the assumed independence, the probability that both the processes simultaneously have value 1 is therefore $\lambda_1 d_1 \lambda_2 d_2$ (equation (A1.11)). The mean duration of each simultaneous occurrence can be obtained by considering a time interval $(0, d_2 + d_1)$ with the first process in state 1 from d_2 to $(d_2 + d_1)$, and associating with each point, regarded as the arrival time of the second process, the relevant time interval during which both processes are in the state 1. A duration 0 is associated with time 0; a duration d_2 with all times into the interval from d_2 and d_1; and a duration 0 with time $(d_2 + d_1)$. It is manifest now that the mean duration $\mu_{d_{12}}$ of the simultaneous occurrence is the ratio between the area of the trapezium obtained $((d_1 + d_2 + d_1 - d_2)d_2/2 = d_1 d_2)$ and $d_2 + d_1$, which proves equation (3.77). Furthermore, the mean value of the total time at which both processes are in the state 1 during the interval 0 to t can be obtained as the product with t of the probability that both the processes have value 1 simultaneously $(\lambda_1 \lambda_2 d_1 d_2)$. Therefore the mean number of simultaneous occurrences in 0 to t is

$$\frac{\lambda_1 \lambda_2 d_1 d_2 t}{d_1 d_2/(d_1 + d_2)} = \lambda_1 \lambda_2 (d_1 + d_2)t$$

It follows that the mean arrival rate λ_{12} is given by $\lambda_1 \lambda_2 (d_1 + d_2)$ and also equation (3.76) is proved.

From equations (3.76) and (3.77), then, the mean rate of simultaneous occurrence of $\tilde{W}_k(t)$, $\tilde{W}_l(t)$ and $\tilde{W}_r(t)$ becomes

$$\lambda_{klr} \simeq \lambda_{kl} \lambda_r (\mu_{d_{kl}} + \mu_{d_r}) \simeq \lambda_k \lambda_l \lambda_r \mu_{d_l} \mu_{d_r} \tag{3.78}$$

High reliability levels (i.e. exponents much greater than unity) are generally considered in structural analysis. In this case, the three terms to be summed in equation (3.75) become

$$P_{\text{fail}}(t|k) \simeq \lambda_k t P_k$$
$$P_{\text{fail}}(t|k,l) \simeq \lambda_{kl} t P_{kl} = \Pi_{kl}(t) P_{kl} \tag{3.79}$$
$$P_{\text{fail}}(t|k,l,r) \simeq \lambda_{klr} t P_{klr} = \Pi_{klr}(t) P_{klr}$$

where $\Pi_{kl}(t)$ and $\Pi_{klr}(t)$ coincide, for $\lambda t \ll 1$, with the probabilities that at least one simultaneity occurs in $(0,t)$ (Fig. 3.26; equation (A6.37)). The condition $\lambda t \ll 1$ also denotes that the probabilities that two or more simultaneous occurrences arrive in $(0,t)$ are negligible, and hence Π_{kl} and Π_{klr} approach the probability that one simultaneity occurs in 0 to t. By contrast, if $\lambda t \ll 1$ is not satisfied, Π_{kl} and Π_{klr} lose their probability meaning. But, with reference to the two independent (0, 1) stochastic processes analysed above, the probability that two actions occur simultaneously during the time interval $(0,t)$ can easily be computed, according to equation (A6.37), by equation (A7.48) and hence, remembering equation (3.76)

$$\text{Prob}[\tilde{W}_k \text{ and } \tilde{W}_l \text{ occur simultaneously at least once in } (0,t)]$$
$$= 1 - \exp(-\lambda_{kl} t) = 1(-\exp(-\lambda_k \lambda_l \lambda_k \mathrm{t}) \tag{3.76'}$$

while

$$\text{Prob}[\tilde{W}_k \text{ and } \tilde{W}_l \text{ occur simultaneously only once in } (0,t)]$$
$$= \lambda_{kl} t \exp(-\lambda_{kl} t) \tag{3.76''}$$

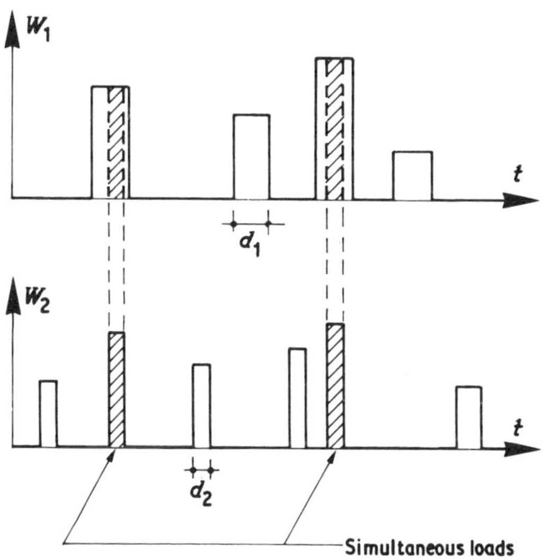

Fig. 3.26 Idealization of stochastic load processes $W_1(t)$ and $W_2(t)$.

3.7.3 Combination with long-duration actions

If one long-duration action, with mean rate λ^l, must also be considered, equation (3.79) is employed to estimate the probability of success $P_{suc}(\bar{t}) = 1 - P_{fail}(\bar{t})$ between two load changes, i.e. in the interval $\bar{t} = 1/\lambda^l$, while the total reliability during the design life T is calculated by

$$P_{suc}(T) = [P_{suc}(\bar{t})]^{T/\bar{t}} \tag{3.80}$$

where $T/\bar{t}$ is the expected number of load changes.

Equation (3.80) can be generalized to load combinations involving more (say n^l) long-duration actions according to the guidelines of Table 3.5. For this purpose, introduce the expected number of changes of the jth long-term load $\tilde{W}_j$

$$r_j = T/\bar{t}_j \tag{3.81}$$

and order the r_j so that

$$r_1 > r_2 > r_3 > \ldots > r_n \tag{3.82}$$

Then the n different verifications, with number of load changes N_{ij} specified by the table rows, must be performed and the largest value of the calculated probabilities of failure must be considered for reliability assessment [9].

Table 3.5 Bases for load combination of long-duration actions.

	Action j =	1	2	3	...	n^l
	Number of time intervals in reference period =	r_1	r_2	r_3	...	r_{n^l}
Combination	Number of load changes					
1	N_{1j}	r_1	r_2/r_1	r_3/r_2	...	r_{n^l}/r_{n^l-1}
2	N_{2j}	1	r_2	r_3/r_2	...	r_{n^l}/r_{n^l-1}
⋮	⋮	⋮	⋮	⋮		⋮
n	$N_{n^l j}$	1	1	1	...	r_n

Such a procedure can be regarded as the one researching the most unfavourable load combination according to the relation

$$\begin{aligned} &\max_{\bar{t}_1}\{W_1(t), W_2(t), \ldots, W_n(t)\} \\ &= \max_j\{W_1(t), W_2(t), \ldots, W_{j-1}(t), \max_{\bar{t}_1} Z_j(t)\} \qquad j = 1, n \end{aligned} \tag{3.83}$$

where

$$Z_j(t) = W_j(t) + \sum_{k=j}^{n-1} W_{k+1}(t, \bar{t}_k) \tag{3.84}$$

$W_{k+1}(t, \bar{t}_k)$ being the maximum value of $W_{k+1}(t)$ in $(0, \bar{t}_k)$ within each interval of length $\bar{t}_k$: these values form the so-called t_k-*duration envelope* of $W_k(t)$.

3.7.4 General approach: some remarks

The approach of Section 3.7.3 can also be applied to problems concerned with short-duration loads by introducing a non-zero discrete probability mass at zero (resulting in occurrence of time intervals with no load), but the computation effort required by equation (3.75) is generally less.

A very simplified combination rule was also proposed [28]: it considers the n combinations

$$y_i = y_i\{W_1(t_0), W_2(t_0), \ldots, W_{i-1}(t_0), W_{i+1}(t_0), \ldots, W_n(t_0), \max_T(W_i(t))\} \qquad i = 1, \ldots, n \quad (3.85)$$

where t_0 is an arbitrary fixed time point. However, in this way one estimates a lower bound of the failure probability. A more accurate approach to the load combination problem should be formulated as an application of the random process crossing theory (Appendix A7), whose results, however, cannot yet be given in formulae of practical use [29].

REFERENCES

1. Cornell, A. C. and Larrabee, R. D. (1977) Representation of loads for code purposes, *Proc. 2nd Int. Conf. on Structural Safety and Reliability* (ICOSSAR'77), Munich, pp. 135–48.
2. Peir, J. C. and Cornell, C. A. (1973) Spatial and temporal variability of live loads. *J. Struct. Div., Proc. ASCE*, **99** (ST5), May.
3. Reid, C. and Turkstra, C. J. (1980) *Serviceability Limit States. Probabilistic Description*, Report ST 80, McGill University, Montreal, Canada.
4. Mitchell, G. R. and Woodgate, R. W. (1971) *Floor Loadings in Office Buildings: the Results of a Survey*, Current Paper CP3/71, Building Research Station, Garston, U.K.
5. Duff, C. G. (1980) Design ground response spectra, *Reference Seismic Ground Motions in Nuclear Safety Assessment*, OECD, Paris.
6. Cornell, C. A. (1968) Engineering seismic risk analysis. *Bull. Seismol. Soc. Am.*, **58**, 1583–606.
7. Housner, G. W. (1972) Earthquke ground motion, *Proc. Int. Conf. on Planning and Design of Tall Buildings*, Lehigh University, Bethlehem, USA.
8. Rosenblueth, E. (1980) Characteristics of earthquakes, in *Design of Earthquake Resistant Structures* (ed. E. Rosenblueth), Pentech Press, London.
9. Ferry-Borges, J. and Castanheta, M. (1971) *Structural Safety*, LNEC, Lisbon.
10. Housner, G. W. (1947) Characteristics of strong motion earthquakes. *Bull. Seismol. Soc. Am.*, **37**, 19–31.
11. Davenport, A. G. (1970) Wind loading on tall buildings, in *Wind Effects on High-Rise Buildings*, Proc. Symp. held at Northwestern University, Evanston, Illinois.
12. Mayne, J. F. and Cook, N. J. (1978) *On Design Procedures for Wind Loading*, Current Paper CP 25/78, Building Research Station, Garston, UK.
13. Cohen, E. and Vellozzi, J. (1970) Proposed American standard building code requirements for minimum design wind loads, in *Wind Effects on High-Rise Buildings*, Proc. Symp. held at Northwestern University, Evanston, Illinois.
14. *Bericht über den Stand der Arbeiten an Regeln zur Erfassung der Windwirkungen auf Bauwerke* (in German), draft, August 1978.

15. Biètry, J., Sacrè, C. and Simiu, E. (1978) Mean wind profiles and change of terrain roughness, *J. Struct. Div., Proc. ASCE,* **104** (ST10), October, 1585–95.
16. Saul, W. E., Tayachandran, P. and Peyrot, A. M. (1976) Response to stochastic wind of *n*-degree tall buildings, *J. Struct. Div., Proc. ASCE,* **102** (ST5), May.
17. Sachs, P. (1972) *Wind Forces in Engineering*, Pergamon Press, Oxford.
18. Simiu, E. and Scanlan, R. H. (1978) *Wind Effects on Structures; An Introduction to Wind Engineering*, Wiley-Interscience, New York, p. 62.
19. Lew, M., and Mart, G. C. (1979) Microzonation in wind engineering, *J. Struct. Div., Proc. ASCE,* **105** (ST6), June.
20. Schuëller, G. I. and Panggabean H. (1975) *Ermittlung der Bemessung-windgeschwindigkeit unter Zugrundelegung eines Zuverlässigkeitskonzeptes*, Beitr. Z. Anwendung der Aerolastik in Bauwesen, Tech. Univ. München, Heft 2.
21. Simiu, E., Bietry, J. and Filliben, J. J. (1978) Sampling error in estimation of extreme winds, *J. Struct. Div., Proc. ASCE,* **104** (ST3), March.
22. Ferry Borges, J. (1976) *Les actions aleatoires* (in French), LNEC, Lisbon, Report No. 476.
23. Luy, H. and Rackwitz, R. (1978) *Darstellung und Auswertung vor Schneehöhenmesungen in der Bundesrepublik Deutschland*, SFB 96, Heft 31/78, TU München.
24. Gränzer, M. and Rieman, H. (1980) *Statistische Auswertung langjähriger Schneemessungen zur Ermittlung der Schneelastverteilung in Bundesgebiet*, Thesis, Instituts für Bautechnik, Berlin.
25. CIRIA (1977) *Rationalisation of Safety and Serviceability Factors in Structural Codes*, Report 63, CIRIA, London.
26. Wen, Y. K. (1977) Statistical combination of extreme loads. *J. Struct. Div., Proc. ASCE,* **103** (ST5), 1079.
27. Casciati, F. and Faravelli, L. (1980) Calibration of the load enhancement factors. *Engineering Structures*, **2**, 113–22.
28. Turkstra, C. J. (1970) Application of Bayesian decision theory, in *Structural Reliability and Codified Design* (ed. N. C. Lind), University of Waterloo Press.
29. Ditlevesen, O. and Madsen, N. O. (1981) Probabilistic modelling of man-made load processes and their individual and combined effects, *Proc. 3rd Int. Conf. on Structural Safety and Reliability* (ICOSSAR '81), Trondheim, pp. 103–34.

CHAPTER 4

Elastic structures

This chapter presents the methodologies for the treatment of static and dynamic problems of elastic structural analysis and reliability under random uncertainties.

4.1 ELASTIC STRUCTURAL ANALYSIS

The simplest assumptions of structural theory, and at the same time those most widely used in technical practice, are elastic behaviour and quasi-static loads. In the framework of reliability theory, linear elastic behaviour is usually introduced in conjunction with a 'weakest-link' type of failure (Section 1.5).

In many important instances, however, the 'quasi-static' assumption is not acceptable, and dynamic analysis is necessary. Therefore, Section 4.3 of this chapter will be devoted to the analysis of linear structures under random dynamic excitation.

As anticipated in Section 1.1, throughout this book only discrete (or discretized) structures will be considered; the fundamental concepts related to the transformation of a structure into a mesh of 'finite elements' will be assumed to be known to the reader. Also, in dynamic problems, the structural mass will be 'lumped' into a finite number of points or rigid elements.

However, as anticipated in Section 1.4, special problems arise in the discretization of *uncertain* structures, in relation to the joint stochastic properties of the finite elements into which the structure has been subdivided. In fact, as will soon become evident to the reader (but could already be expected), *stochastic independence* (or, at the other extreme, *perfect correlation*) of the relevant random quantities greatly simplifies the required calculations. Therefore, it would be preferable to choose finite elements such that their properties are stochastically independent of each other, but this is only possible if an 'extinction interval' d_{ext} (Section 2.4.3) of the relevant r.p. exists and is smaller than the chosen finite element size. (For instance, considering the chain in Fig. 1.3 as a model of a continuous tie and the r.p. $\tilde{\sigma}_{\mathrm{y}}(z)$ representing the yield stress along the tie axis z, the number n of elements should satisfy the inequality $d_{\mathrm{ext}} \leqslant L/n$, where L is the overall length of the tie; cf. also Section 11.4.) Often, however, this condition cannot be met by a finite element model able to reproduce with acceptable approximation the mechanical behaviour of the parent structure.

Systematic procedures for tackling these problems and choosing the *best* compromise between *mechanical* approximation, *probabilistic* and computational difficulties are not yet available, and recourse must be made to technical judgement and experience.

4.2 LINEAR ELASTIC STRUCTURES UNDER STATIC LOADING

For a linear elastic discrete (or discretized structure), the relationship between generalized displacements $\mathbf{u}$ (taken as typical response quantities) and generalized loads $\mathbf{W}$ is [1]

$$\mathbf{K}\mathbf{u} = \mathbf{W} \tag{4.1}$$

where $\mathbf{K}$ is the *stiffness matrix*. The strain vector $\boldsymbol{\varepsilon}$ is then related to the displacement vector by

$$\boldsymbol{\varepsilon} = \mathbf{Q}\mathbf{u} \tag{4.2}$$

the matrix $\mathbf{Q}$ depending essentially on the structural geometry. In turn, the stress vector $\boldsymbol{\sigma}$ is related to the strain vector by the elastic constitutive law‡

$$\boldsymbol{\sigma} = \mathbf{G}\boldsymbol{\varepsilon} \tag{4.3}$$

Therefore, when the displacements have been calculated, strains and stresses can easily be obtained by linear transformations. Loads, in turn, are related to the (generalized) stresses by the equation

$$\mathbf{W} = \mathbf{Q}^{\mathrm{T}}\boldsymbol{\sigma} \tag{4.4}$$

where the superscript T indicates the *transpose* of a matrix. Combining equations (4.2), (4.3) and (4.4) and comparing with equation (4.1), one obtains

$$\mathbf{K} = \mathbf{Q}^{\mathrm{T}}\mathbf{G}\mathbf{Q} \tag{4.5}$$

‡ Note that in this *discretized* model, the *vectors* $\boldsymbol{\sigma}$ and $\boldsymbol{\varepsilon}$ collect the (generalized) stresses and strains in *all significant points* of the structure: they correspond, in a continuous model, to the tensor fields $\sigma_{ij}(\mathbf{x})$ and $\varepsilon_{ij}(\mathbf{x})$. The non-zero elements in the matrix $\mathbf{G}$ are only those which relate elements in $\boldsymbol{\sigma}$ and $\boldsymbol{\varepsilon}$ pertaining to the same structural point.

In (4.1), the elements of **W** and **K** are respectively input and system quantities that can be affected by random uncertainties: consequently, the output (response) will also be random. The cases of random **W** and/or **K** (sometimes denoted respectively *environmental* and *technological* uncertainty, cf Section 1.2) will be treated in detail in the following sections.

In formulating the critical inequality (2.59) for a linear elastic structure, it is usual to express the *generalized load effect* S as a linear function of either *input* or *output* (response) quantities:

$$S = \mathbf{D}^{\mathrm{T}}\mathbf{W} = \mathbf{L}^{\mathrm{T}}\mathbf{u} \tag{4.6}$$

or, in index notation,

$$S = D_j W_j = L_i u_i \tag{4.6$'$}$$

where $\mathbf{D}^{\mathrm{T}} = \{D_1, D_2, \ldots\}$ and $\mathbf{L}^{\mathrm{T}} = \{L_1, L_2, \ldots\}$ are vectors of appropriate constants, related by (equation 4.1)

$$\mathbf{D}^{\mathrm{T}}\mathbf{K} = \mathbf{L}^{\mathrm{T}}; \qquad \mathbf{K}^{\mathrm{T}}\mathbf{D} = \mathbf{L} = \mathbf{K}\mathbf{D} \tag{4.7}$$

due to the symmetry of the stiffness matrix.

The load effect intensity S is a random variable if any of the quantities on the r.h.s. of (4.6) is random.

For instance, S may represent the normal stress in a given point of the structure, to be compared with the yield stress $R = \sigma_{\mathrm{y}}$, or a significant displacement, to be compared with an acceptable value. In some cases it may be necessary to verify more non-independent inequalities, as discussed in general in Section 2.8.1.

4.2.1 Random loads and deterministic structure

(a) Single load application

In case of random loads and deterministic structure, equations (4.1) and (4.6) become

$$\mathbf{K}\tilde{\mathbf{u}} = \tilde{\mathbf{W}} \tag{4.8}$$

$$\tilde{S} = \mathbf{D}^{\mathrm{T}}\tilde{\mathbf{W}} \tag{4.9}$$

The JPDF of the displacements **u** is given by

$$p_{\mathbf{u}} = p_{\mathbf{W}}(\mathbf{K}\mathbf{u}) \cdot \mathrm{Det}\,\mathbf{K} \tag{4.10}$$

which is a special case of equation (A3.61) (see equation (2.41)).

The PDFs of $\boldsymbol{\varepsilon}, \boldsymbol{\sigma}$ and S can be obtained by analogous formulae. In particular, if the loads $\tilde{\mathbf{W}}$ are normally distributed, so are all other quantities‡.

First- and second-order moments of the output quantities are always easily determined from those of the loads. In particular (equations (A3.26′) and

‡ More generally, if $p_{\mathbf{W}}(\mathbf{W})$ is *stable*, the other relevant PDFs are immediately known. In fact, remember that by definition the functional form of a *stable* distribution is not changed by a linear transformation: the *normal* (*Gaussian*) distribution is typically stable (Section A3.6).

(A3.27′)) [2],

$$\boldsymbol{\mu}_{\mathbf{u}} = \mathbf{C}\boldsymbol{\mu}_{\mathbf{W}} \tag{4.11}$$

$$\boldsymbol{\Sigma}_{\mathbf{u}} = \mathbf{C}\boldsymbol{\Sigma}_{\mathbf{W}}\mathbf{C}^{\mathrm{T}} \tag{4.11′}$$

where, to simplify notation,

$$\mathbf{C} = \mathbf{K}^{-1} \tag{4.12}$$

In turn, average and standard deviation of the load effect S are

$$\begin{aligned} \mu_S &= \mathbf{L}^{\mathrm{T}}\boldsymbol{\mu}_{\mathbf{u}} = \mathbf{D}^{\mathrm{T}}\boldsymbol{\mu}_{\mathbf{W}} \\ s_S^2 &= \mathbf{L}^{\mathrm{T}}\boldsymbol{\Sigma}_{\mathbf{u}}\mathbf{L} = \mathbf{D}^{\mathrm{T}}\boldsymbol{\Sigma}_{\mathbf{W}}\mathbf{D} \end{aligned} \tag{4.13}$$

The above formulae (and the analogous ones holding for $\boldsymbol{\varepsilon}$ and $\boldsymbol{\sigma}$) show that, in linear elastic structures, first- and second-order moments of the relevant quantities are linearly related, and in particular:

(1) the mean output quantities are obtained by loading the (deterministic) structure with the average loads; and
(2) the covariance matrices of output quantities are given by linear transformations of the covariance matrix of the loads; the degree of correlation and the coefficients of variation can vary (and usually increase) greatly in these transformations.

(b) Multiple independent load applications

Consider now a deterministic structure subject to a sequence of stochastically independent short-time static loads.

According to equation (4.6), the generalized load effect at the kth load application is

$$S_k = \mathbf{D}^{\mathrm{T}}\mathbf{W}_k \tag{4.14}$$

This value can be regarded as the kth realization of the random variable $\tilde{S}$ (equation (4.9)).

The structure fails if the critical inequality (2.59) is violated *at least in one* load application. Therefore, the problem is similar to that of multiple independent failure modes illustrated in Section 1.4. In particular, if all involved probabilities are very small, to good approximation

$$P_{\text{fail}} = \sum_k P_{\text{f}k} \tag{4.15}$$

$P_{\text{f}k}$ being the probability of failure in the kth load application.

If all loads are taken from the same *population* (i.e. have the same PDF) and the structure does not *deteriorate* in time, then

$$P_{\text{f}k} = P_{\text{f}1} \qquad \text{all } k \tag{4.16}$$

and

$$P_{\text{fail}} = nP_{\text{f}1} \tag{4.17}$$

n being the number of (independent) load applications.

It is evident from (4.17) that the reliability of any structure decreases with the number of load applications. More generally, provided that the strength R does not vary in the successive load applications, this sort of problem can be tackled by introducing the *largest value* of S in the critical inequality (see Section 3.1):

$$\max_k S \leqslant R \tag{4.18}$$

This approach, by which a r.p. is reduced to a r.v., is the subject of the *theory of extreme values* (Appendix A7). However, the derivation of the r.v. '$\max_k S$' from a multivariate r.p. $\tilde{\mathbf{W}}$ can be a very difficult problem, which requires a number of drastic approximations to become tractable (Section A7).

(c) Continuously varying loads

Consider now the case in which the loads are represented by a generic multivariate random process $\tilde{\mathbf{W}}(t)$, but assume that their variations in time are still slow enough for the *static* equations to hold. Assume also that technological uncertainty is negligible. Then, the basic equation (4.1) becomes

$$\mathbf{K}\tilde{\mathbf{u}}(t) = \tilde{\mathbf{W}}(t) \tag{4.19}$$

In (4.19), $\tilde{\mathbf{W}}(t)$, and thence $\tilde{\mathbf{u}}(t)$, are multivariate stochastic processes

$$\begin{aligned} \tilde{u}_i(t) &= C_{ij}\tilde{W}_j(t) \\ \dot{\tilde{u}}_i(t) &= C_{ij}\dot{\tilde{W}}_j(t) \end{aligned} \tag{4.20}$$

Let $\tilde{\mathbf{W}}(t)$ be an n-dimensional Gaussian process; then, indicating the load components and their derivatives respectively by

$$z_i = \begin{cases} W_i & i = 1, 2, \ldots, n \\ \dot{W}_{i-n} & i = n+1, n+2, \ldots, n \end{cases} \tag{4.21}$$

the instantaneous JPDF of $z_i(t)$ is given by (equation (A2.38) and Section A6.2)

$$p_{\mathbf{z}}(z_1, z_2, \ldots, z_{2n}|t)$$

$$= \frac{1}{(2\pi)^n |\text{Det } \Sigma^{\mathbf{z}}_{ij}(t,t)|^{1/2}} \exp\left[-\tfrac{1}{2} \sum_{i,j=1}^{2n} \Xi^{\mathbf{z}}_{ij}(t,t)\cdot(z_i - \mu_{z_i})(z_j - \mu_{z_j}) \right] \tag{4.22}$$

where $\Sigma^{\mathbf{z}}_{ij}(t_1, t_2)$ and $\Xi^{\mathbf{z}}_{ij}(t_1, t_2)$ are elements of the matrix of the covariance functions for $z_i(t)$ and of its inverse respectively.

The processes $\tilde{\mathbf{u}}(t)$ and $\tilde{S}(t)$, which are obtained from $\tilde{\mathbf{W}}(t)$ by linear transformations, are also Gaussian. Introducing the notation

$$y_i = \begin{cases} u_i & i = 1, 2, \ldots, n \\ \dot{u}_{i-n} & i = n+1, n+2, \ldots, 2n \end{cases} \tag{4.23}$$

the instantaneous JPDF of $Y_i(t)$ is given by

$$p_{\mathbf{y}}(y_1, y_2, \ldots, y_{2n}|t) = \frac{1}{(2\pi)^n |\mathrm{Det}\, \Sigma_{ij}^{\mathbf{y}}(t,t)|^{1/2}} \exp\left[-\tfrac{1}{2} \sum_{i,j=1}^{2n} \Xi_{ij}^{\mathbf{y}}(t,t) \cdot (y_i - \mu_{y_i})(y_j - \mu_{y_j})\right] \tag{4.24}$$

where again $\Sigma_{ij}^{\mathbf{y}}(t_1, t_2)$ and $\Xi_{ij}^{\mathbf{y}}(t_1, t_2)$ are respectively elements of the matrix of the covariance functions of $y_i(t)$ and those of its inverse.

Analogously, the JPDF of the load effect S and its time variation $\dot{S}$ is given by

$$p_{\mathrm{S}}(S, \dot{S}|t) = \frac{1}{2\pi |\mathrm{Det}\, \Sigma_{ij}^{\mathbf{S}}|^{1/2}} \times \exp\left(-\frac{\Sigma_{22}^{\mathbf{S}}(S-\mu_S)^2 - 2\Sigma_{12}^{\mathbf{S}}(S-\mu_S)(\dot{S}-\mu_{\dot{S}}) + \Sigma_{11}^{\mathbf{S}}(\dot{S}+\mu_{\dot{S}})^2}{2|\mathrm{Det}\, \Sigma_{ij}^{\mathbf{S}}|}\right) \tag{4.25}$$

Here, $\Sigma_{ij}^{\mathbf{S}} = \Sigma_{ij}^{\mathbf{S}}(t_1, t_1)$ are the elements of the matrix of the covariance functions relevant to the vector $\mathbf{S}(S_1 = S;\ S_2 = \dot{S})$, which can be shown to be given by (equations (A3.27′), (A5.28) and (A5.29)

$$\begin{aligned} \Sigma_{11}^{\mathbf{S}}(t_1, t_2) &= \sum_{i \neq j=1}^{n} D_i D_j\, \Sigma_{ij}^{\mathbf{z}}(t_1, t_2) \\ \Sigma_{12}^{\mathbf{S}}(t_1, t_2) &= \frac{\partial \Sigma_{11}^{\mathbf{S}}(t_1, t_2)}{\partial t_2} \\ \Sigma_{22}^{\mathbf{S}}(t_1, t_2) &= \frac{\partial^2 \Sigma_{11}^{\mathbf{S}}(t_1, t_2)}{\partial t_1 \partial t_2} \end{aligned} \tag{4.26}$$

Assume that violation of the critical inequality is a rare event, so that the probability of more than one violation in the relevant time interval $(0, T)$ is negligible. Then, the probability of unsatisfactory performance is to good approximation given by equations (A7.40) and (A7.49), and remembering that $1 - \exp(-c) \simeq c$ if $c \ll 1$)

$$\begin{aligned} P_{\mathrm{fail}} &= \mathrm{Prob}[S > R \text{ at least once in } 0, T] \\ &\simeq \int_0^T \mathrm{d}t \int_0^\infty \dot{S}\, p_{\mathrm{S}}(R, \dot{S}|t)\, \mathrm{d}\dot{S} \end{aligned} \tag{4.27}$$

and for stationary or almost stationary processes (i.e. processes such that the instantaneous properties change slowly in comparison with the variation of the function (equation (A7.42)) [3]

$$P_{\mathrm{fail}} = \mathrm{Prob}[S > R \text{ at least once in } 0, T] \tag{4.28}$$

$$\simeq \frac{1}{2\pi}\int_0^T \frac{s_{S_2}(t)}{s_{S_1}(t)} \exp\left(\frac{R-\mu_S}{2s_{S_1}(t)}\right) \mathrm{d}t$$

in which (equations (A3.27′) and (A5.28))

$$s_{S_1}^2(t) = \mathrm{Var}[S(t)] = \sum_{i \neq j=1}^{n} D_i D_j \Sigma_{ij}^{\mathbf{z}}(t,t) = \sum_{i \neq j=1}^{n} D_i D_j B_{ij}^{\mathbf{z}}(0)$$

$$s_{S_2}^2(t) - \mathrm{Var}[\dot{S}(t)] = D_i D_j \left.\frac{\partial^2 \Sigma_{ij}^{\mathbf{z}}(t_1,t_2)}{\partial t_1 \partial t_2}\right|_{t_1=t_2=t} \qquad (4.29)$$

(d) An example

Consider the simple structure of Fig. 4.1. The components of the stochastic load vector $\tilde{\mathbf{W}}$ are the vertical force $\tilde{W}_1(t)$ and the horizontal force $\tilde{W}_2(t)$, while d and h are constant, deterministic quantities; dynamic effects are assumed negligible.

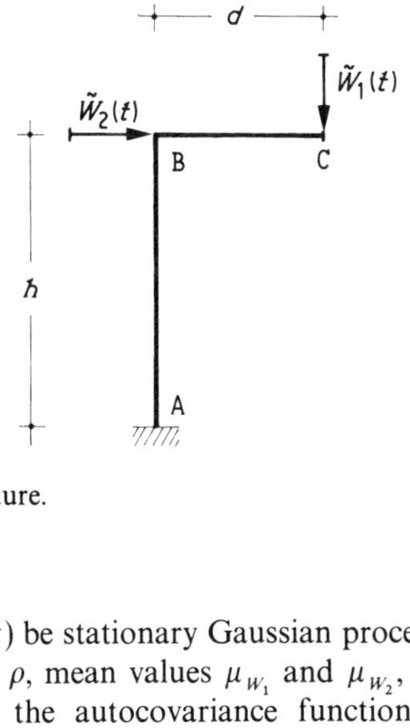

Fig. 4.1 Example structure.

Let $\tilde{W}_1(t)$ and $\tilde{W}_2(t)$ be stationary Gaussian processes with (constant) cross-correlation coefficient ρ, mean values μ_{W_1} and μ_{W_2}, and variances $s_{W_1}^2$ and $s_{W_2}^2$ respectively. Further, the autocovariance functions $B_{W_1}(v)$ and $B_{W_2}(v)$ are assumed to be

$$B_{W_i}(v) = s_{W_i}^2 \exp(-\zeta_i v^2) \qquad i = 1, 2 \qquad (4.30)$$

Hence, the cross-covariance function is (equation (A2.37))

$$\begin{aligned} B_{12}(v) &= B_{21}(v) \\ &= \rho\sqrt{[B_{W_1}(v) B_{W_2}(v)]} \\ &= \rho s_{W_1} s_{W_2} \exp\left(-\frac{\zeta_1+\zeta_2}{2} v^2\right) \end{aligned} \qquad (4.31)$$

By assumption, the significant load effect with regard to the structure's performance is the bending moment M_A in the built-in section A:

$$\tilde{S}(t) = \tilde{M}_A(t) = \tilde{W}_1(t)d + \tilde{W}_2(t)h \tag{4.32}$$

The mean value of $\tilde{M}_A(t)$, which is evidently a stationary process too, is (equations (A3.5a) and (A3.9) or equation (A3.26′))

$$\mu_M = \mu_{W_1} d + \mu_{W_2} h \tag{4.33}$$

while its variance is (equations (A3.5b), (A3.10) or equation (A3.27′))

$$s_M^2 = s_{W_1}^2 d^2 + s_{W_2}^2 h^2 + 2dh\rho s_{W_1} s_{W_2} \tag{4.34}$$

and its autocovariance function is (equation (A3.27′))

$$B_M(v) = d^2 B_{W_1}(v) + 2dhB_{12}(v) + h^2 B_{W_2}(v) \tag{4.35}$$

Consider now the rate of variation of M_A with time, and denote it by $\dot{M}_A$. Because of the commutative property of differentiation and averaging of stationary r.p. (Section A5.2.2),

$$\mu_{\dot{M}} = \dot{\mu}_M = 0 \tag{4.36}$$

while the cross-correlation function between $\tilde{M}_A(t)$ and $\tilde{\dot{M}}_A(t)$ is

$$\begin{aligned} B_{M\dot{M}}(v) &= -\dot{B}_M(v) \\ &= 2\zeta_1 v d^2 B_{W_1}(v) + 2(\zeta_1 + \zeta_2) v dh B_{12}(v) + 2\zeta_2 v h^2 B_{W_2}(v) \end{aligned} \tag{4.37}$$

and the autocovariance of $\tilde{\dot{M}}_A$ is (equation A5.30))

$$\begin{aligned} B_{\dot{M}}(v) &= -\ddot{B}_M(v) \\ &= (2\zeta_1 - 4\zeta_1^2 v^2) d^2 B_{W_1}(v) + 2dh[(\zeta_1 + \zeta_2) - (\zeta_1 + \zeta_2)^2 v^2] B_{12}(v) \\ &\quad + (2\zeta_2 - 4\zeta_2^2 v^2) h^2 B_{W_2}(v) \end{aligned} \tag{4.38}$$

Finally, putting $v = 0$ one gets

$$\begin{aligned} B_M(0) &= s_M^2 = s_{W_1}^2 d^2 + 2dh\rho s_{W_1} s_{W_2} + s_{W_2}^2 h^2 \\ B_{M\dot{M}}(0) &= 0 \\ B_{\dot{M}}(0) &= s_{\dot{M}}^2 = 2\zeta_1 s_{W_1}^2 d^2 + 2(\zeta_1 + \zeta_2) dh\rho s_{W_1} s_{W_2} + 2\zeta_2 s_{W_2}^2 h^2 \end{aligned} \tag{4.39}$$

Hence, the instantaneous JPDF of M_A and $\dot{M}_A$ is given by

$$p(M, \dot{M} \mid t) = \frac{1}{2\pi s_M s_{\dot{M}}} \exp\left[-\left(\frac{(M - \mu_M)^2}{2s_M^2} + \frac{\dot{M}^2}{2s_{\dot{M}}^2} \right) \right] \tag{4.40}$$

The probability that $\tilde{M}_A(t)$ exceeds a given (deterministic) *limit moment* $R = M_y$ in the interval $(0, T)$ can now be obtained from equation (4.27) as

$$P_{\text{fail}} = \text{Prob}[\tilde{M}_A > M_y \text{ in } 0, T] \simeq \int_0^T \int_0^\infty \dot{M} p(M_y, \dot{M} \mid t)\, \mathrm{d}\dot{M}\, \mathrm{d}t \tag{4.41}$$

But from equation (4.40) we have (see equation (4.28))

$$\int_0^\infty \dot{M}p(M_y, \dot{M}|t)\,\mathrm{d}\dot{M} = \frac{1}{2\pi}\frac{s_{\dot{M}}}{s_M}\exp\left(-\frac{(M_y-\mu_M)^2}{2s_M^2}\right) \tag{4.42}$$

which introduced into (4.41) yields

$$P_{\text{fail}} = \text{Prob}[\tilde{M}_A > M_y \text{ in } 0, T] \simeq \frac{T}{2\pi}\frac{s_{\dot{M}}}{s_M}\exp\left(-\frac{(M_y-\mu_M)^2}{2s_M^2}\right) \tag{4.43}$$

That is, under the set assumptions, the probability of failure is to good approximation proportional to the length of the design life T. The analogy of this result with the case of independent load applications (4.17) is remarkable.

(e) Reliability under combined loads: a discussion

Very often, loads such as W_1 and W_2 in Fig. 4.1 have different origin (e.g. gravity, earthquake, wind) and are described by different stochastic processes. In particular, they can act simultaneously or singularly.

Although in principle the problem can be tackled by the study of the vector process $\mathbf{W}(t)$, as done in the preceding sections, simpler procedures must be used in practical applications (Section 3.7). A comparison with the classical treatment of structural safety can also be significant.

In the latter treatment, a *central safety factor* $\gamma^0 > 1$ is introduced, as the ratio between the limit moment (strength) M_y and the mean load effect (4.33)

$$\gamma^0 = M_y/\mu_M \tag{4.44}$$

Introducing γ^0 into equation (4.43), one gets after some algebra

$$P_{\text{fail}} \simeq \frac{T}{2\pi}\frac{s_{\dot{M}}}{s_M}\exp\left(-\frac{(\gamma^0-1)^2}{2\delta_M^2}\right) \tag{4.45}$$

where

$$\delta_M = s_M/\mu_M \tag{4.46}$$

is the coefficient of variation (c.o.v.) of $\tilde{M}_A$.

For a prescribed probability of failure, under the set conditions, the central safety factor must be

$$\gamma^0 = 1 + \delta_M\sqrt{\left[-2\ln\left(\frac{2\pi P_{\text{fail}} s_M}{T s_{\dot{M}}}\right)\right]} \tag{4.47}$$

If the two loads W_1 and W_2 do not act simultaneously, to reach the same probability of failure, different safety factors must be applied, namely

$$\gamma_i^0 = 1 + \delta_i\sqrt{\left[-2\ln\left(\frac{2\pi P_{\text{fail}}}{T\sqrt{(2\zeta_i)}}\right)\right]} \qquad i = 1, 2 \tag{4.48}$$

respectively if either W_1 or W_2 acts (δ_i is the c.o.v. of $\tilde{W}_i$). Note that in general, $\gamma_i^0 \neq \gamma^0$. In (4.48) $s_{\dot{M}}^2/s_M^2 = 2\zeta_i$ is obtained from (4.39), while $\gamma_i^0 = M_y/\mu_{W_1} d$

$= M_{\mathrm{ly}}/\mu_1, \gamma_2^0 = M_{\mathrm{y}}/\mu_{W_2}h = M_{\mathrm{y}}/\mu_2$. μ_i being the expected value of the moment produced in the cross-section A by the single load W_i.

To see briefly what equations (4.47) and (4.48) mean, let us simplify the input data, and put $\zeta_1 = \zeta_2 = \zeta$, $s_{W_1} = s_{W_2} = s_W$, $T\sqrt{(2\zeta)} = r$ and $\delta_1 = \delta_2 = \delta$. It follows from equation (4.39) that

$$s_M^2 = s_W^2(d^2 + 2dh\rho + h^2)$$
$$s_M^2 = s_W^2(2d^2 + 4dh\rho + 2h^2) = 2\zeta s_M^2 \tag{4.49}$$

and

$$\gamma_1^0 = \gamma_2^0 = 1 + \delta\sqrt{\left[-2\ln\left(\frac{2\pi P_{\mathrm{fail}}}{r}\right)\right]} = \gamma'^0$$
$$\gamma^0 = 1 + \delta_M\sqrt{\left[-2\ln\left(\frac{2\pi P_{\mathrm{fail}}}{r}\right)\right]} \tag{4.50}$$

whence

$$\frac{\gamma^0 - 1}{\gamma'^0 - 1} = \frac{\delta_M}{\delta}$$

The maximum value of the above ratio is found to be 1 for cross-correlation coefficient $\rho = 1$, while the minimum occurs when $\rho = -1$ and is

$$\left(\frac{\gamma^0 - 1}{\gamma'^0 - 1}\right)_{\mathrm{min}} = \frac{\mu_1 - \mu_2}{\mu_1 + \mu_2}$$

In order to get an easy comparison of the numerical implication of the above considerations, consider the case $\rho = 0$, $d = h$, $\delta = 0.1$. It is soon noted that

$$\delta_M = \frac{s_M}{\mu_M} = \frac{s_W\sqrt{(d^2 + h^2)}}{\mu_W(d + h)} = \frac{\delta}{\sqrt{2}} = 0.071$$

Note moreover that r is an (approximate) equivalent number of independent load applications. In fact, it is possible to arrive at (4.45) by starting with a scheme of independent tests and regarding the probability

$$P'_{\mathrm{fail}} = \frac{1}{2\pi}\exp\left(-\frac{(\gamma^0 - 1)^2}{2\delta_M^2}\right) \tag{4.45'}$$

as the probability of failure in one test. Then, expanding the binomial in a series and discarding terms of higher order (as $P'_{\mathrm{fail}} \ll 1$), equation (4.45) yields

$$P_{\mathrm{fail}} = 1 - (1 - P'_{\mathrm{fail}})T\sqrt{(2\zeta)} \simeq rP'_{\mathrm{fail}} \tag{4.45''}$$

whence the meaning of measure of independent load repetitions for the quantity r. For the considered example, put $r = 10$ and $P_{\mathrm{fail}} = 10^{-4}$. Then, one gets

$$\gamma'^0 = 1 + 4.40\delta = 1.440$$
$$\gamma^0 = 1 + 3.11\delta = 1.311$$

The introduction of the combined factor γ^0 thus allows one to reduce the design moment at the section A with respect to the sum of the single load effects. Note that the difference as regards the probability of failure is highly significant. In fact, confusing γ^0 with γ'^0 (4.45′) and (4.45″) yields

$$P_{\text{fail}} = \frac{r}{2\pi}\exp\left(-\frac{(\gamma'^0 - 1)^2}{2\delta_M^2}\right) = 6.22 \times 10^{-9}$$

i.e. almost the square of the requested P_{fail}.

4.2.2 Random loads and random structure

(a) Forming the stiffness matrix

Not only the loads but also the stiffness matrix **K** can be affected by random uncertainties. In this case, equation (4.1) is written

$$\tilde{\mathbf{K}}\tilde{\mathbf{u}} = \tilde{\mathbf{W}} \tag{4.51}$$

As shown by equation (4.5), the stiffness matrix is formed by the product of two matrices, namely **Q**, which describes the geometry of the structure, and **G**, which is the matrix of the local elastic coefficients. While the actual process of fabrication and construction of the structure can cause some randomness in the elements of the matrix **Q**, this is often not the main cause of the uncertainty, and can be neglected in comparison with the variability of the elements of **G**. In conclusion, in index notation we write‡

$$\tilde{K}_{ij} = Q_{li}\tilde{G}_{lk}Q_{kj} \tag{4.52}$$

The mean value of a generic element K_{ij} and the covariance between any two elements K_{ij} and K_{rs} are respectively (equations (A3.26′) and (A3.27′))

$$\mu_{ij}^{\mathbf{K}} = Q_{li}\mu_{lk}^{\mathbf{G}}Q_{kj} \tag{4.53}$$

$$\Sigma_{ijre}^{\mathbf{K}} = Q_{li}Q_{kj}Q_{mr}Q_{ne}\Sigma_{lkmn}^{\mathbf{G}} \qquad l, i, k, j, m, r, n, e = 1, 2, 3, \ldots \tag{4.54}$$

Often the elastic coefficients are uncorrelated, i.e.

$$\Sigma_{lkmn}^{\mathbf{G}} = 0 \qquad \text{if} \quad l \neq m \quad \text{or} \quad k \neq n \tag{4.55}$$

In this case, equation (4.54) becomes

$$\Sigma_{ijre}^{\mathbf{K}} = Q_{li}Q_{kj}Q_{lr}Q_{ke}\Sigma_{lklk}^{\mathbf{G}} \tag{4.54′}$$

but in general

$$\Sigma_{ijre}^{\mathbf{K}} \neq 0 \qquad \text{when } i \neq r \quad \text{or} \quad j \neq e \tag{4.55′}$$

i.e. the elements of the stiffness matrix **K** are correlated (even when the elements of **G** are not).

‡ Summation over repeated indices is implied.

(b) Finding the displacements

The solution of equation (4.51) in terms of $\tilde{\mathbf{u}}$ is much more complicated than that of equation (4.8) (whether the load vector $\mathbf{W}$ is random or not) because the inversion of a random matrix is not an easy task.

General procedures become impracticable if too many elements of $\mathbf{K}$ are random. However, approximate solutions exist for matrix inversion and determination of first- and second-order moments of the output quantities.

To this aim, $\tilde{\mathbf{K}}$ is put in the form

$$\tilde{\mathbf{K}} = \overline{\mathbf{K}} + \tilde{\mathbf{H}} \tag{4.56}$$

where $\overline{\mathbf{K}}$ is the deterministic *expected stiffness matrix* (i.e. the matrix with elements $\overline{K}_{ij} = \mu^{\mathbf{K}}_{ij}$, (4.53)) and $\tilde{\mathbf{H}}$ the random matrix of *technological imperfections* (average $\mu^{\mathbf{H}}_{ij} = 0$). Then, taking into account the intrinsic symmetry of $\tilde{\mathbf{K}}$,

$$\tilde{\mathbf{K}}^{-1} = (\overline{\mathbf{K}} + \tilde{\mathbf{H}})^{-1} = (\mathbf{I} + \overline{\mathbf{K}}^{-1}\tilde{\mathbf{H}})^{-1} \cdot \overline{\mathbf{K}}^{-1} = \overline{\mathbf{K}}^{-1} \cdot (\mathbf{I} + \overline{\mathbf{K}}^{-1}\tilde{\mathbf{H}})^{-1} \tag{4.57}$$

where $\mathbf{I}$ is the unit matrix ($I_{ij} = 1, i = j; = 0, i \neq j$).

Assuming, as likely, that the largest eigenvalue of $\overline{\mathbf{K}}^{-1}\tilde{\mathbf{H}}$ is much smaller than unity with sufficient probability, it becomes possible to develop in series the power in the last term of equation (4.57) and to truncate the series at the second-order term, obtaining the required approximate inversion formula [4]

$$\tilde{\mathbf{K}}^{-1} = \overline{\mathbf{K}}^{-1} . [\mathbf{I} - \overline{\mathbf{K}}^{-1}\tilde{\mathbf{H}} + (\overline{\mathbf{K}}^{-1}\tilde{\mathbf{H}})^2] \tag{4.58}$$

Introducing now, for simplicity of further formulae, the notation

$$\mathsf{K} = \overline{\mathbf{K}}^{-1} \qquad \tilde{\mathbf{C}} = \tilde{\mathbf{K}}^{-1} \tag{4.59}$$

(4.58) is written

$$\tilde{\mathbf{C}} = \mathsf{K}[\mathbf{I} - \mathsf{K}\tilde{\mathbf{H}} + (\mathsf{K}\tilde{\mathbf{H}})^2] \tag{4.60}$$

Now let $\Sigma^{\mathbf{H}}_{rslk}$ be the four-dimensional covariance matrix of $\tilde{\mathbf{H}}$, and $C_{ij,\,mn}$ and $C_{ij,\,mnre}$ be other matrices defined by

$$\begin{aligned} C_{ij,\,mn} &= \left.\frac{\partial C_{ij}}{\partial H_{mn}}\right|_{\mathbf{H}=0} \\ C_{ij,\,mnre} &= \left.\frac{\partial^2 C_{ij}}{\partial H_{mn}\partial H_{re}}\right|_{\mathbf{H}=0} \end{aligned} \tag{4.61}$$

Then, average and covariance matrix of $\tilde{\mathbf{C}}$ can be obtained from (4.60)

$$\begin{aligned} \mu^{\mathbf{C}}_{ij} &= \mathsf{K}_{ij} + \tfrac{1}{2} C_{ij,\,mnre}\Sigma^{\mathbf{H}}_{mnre} \\ \Sigma^{\mathbf{C}}_{ijmn} &= C_{ij,\,rs} C_{mn,\,lk}\Sigma^{\mathbf{H}}_{rslk} \qquad i, j, m, n, r, e, l, k = 1, 2, 3, \ldots \end{aligned} \tag{4.62}$$

Equations (4.62) are a particular form of the approximate expressions that can be derived for the mean and the covariance of a vector $\mathbf{y} = \mathbf{y}(\mathbf{x})$ ($\mathbf{x}$ being a vector of random variables) by means of Taylor series expansions of the multivariate function about $\boldsymbol{\mu}_{\mathbf{x}}$. Then, the second-order approximation of the expected value is

$$\boldsymbol{\mu}_{\mathbf{y}} = \mathbf{y}(\boldsymbol{\mu}_{\mathbf{x}}) + \frac{1}{2}\sum_i \sum_j \left.\frac{\partial^2 \mathbf{y}}{\partial x_i \partial x_j}\right|_{\boldsymbol{\mu}_{\mathbf{x}}} \Sigma_{ij}^{\mathbf{x}} \tag{4.62'}$$

while the first-order approximation to $\boldsymbol{\Sigma}_{\mathbf{y}}$ is

$$\boldsymbol{\Sigma}_{\mathbf{y}} = \sum_i \sum_j \left.\frac{\partial \mathbf{y}}{\partial x_i}\right|_{\boldsymbol{\mu}_{\mathbf{x}}} \left.\frac{\partial \mathbf{y}}{\partial x_j}\right|_{\boldsymbol{\mu}_{\mathbf{y}}} \Sigma_{ij}^{\mathbf{x}} \tag{4.62''}$$

With the further assumption of stochastic independence between loads and stiffness (as is reasonable for linear structural models), and introducing in the calculation of the covariance matrix the equality (equation A3.11)

$$\mathrm{E}[xy] = \mu_x \mu_y \tag{4.63}$$

(which holds exactly only if x and y are independent), the formulae for averages and covariance of the output quantities u_l are finally obtained

$$\begin{aligned}
\mu_{u_i} &= \mu_{ij}^{\mathbf{C}} \mu_{W_j} \\
\Sigma_{ij}^{\mathbf{u}} &= \mathrm{E}[(u_i - \mu_{u_i})(u_j - \mu_{u_j})] \\
&= \mathrm{E}[(C_{ir} W_r - \mu_{ir}^{\mathbf{C}} \mu_{W_r})(C_{je} W_e - \mu_{je}^{\mathbf{C}} \mu_{W_e})] \\
&= \mathrm{E}[C_{ir} W_r C_{je} W_e] - \mu_{ir}^{\mathbf{C}} \mu_{W_r} \mu_{je}^{\mathbf{C}} \mu_{W_e} \\
&\simeq \mathrm{E}[C_{ir} C_{je}] \mathrm{E}[W_r W_e] - \mu_{ir}^{\mathbf{C}} \mu_{je}^{\mathbf{C}} \mu_{W_r} \mu_{W_e} \\
&\qquad + \mathrm{E}[C_{ir} C_{je}] \mu_{W_r} \mu_{W_e} - \mathrm{E}[C_{ir} C_{je}] \mu_{W_r} \mu_{W_e} \\
&\qquad + \mu_{ir}^{\mathbf{C}} \mu_{je}^{\mathbf{C}} \mathrm{E}[W_r W_e] - \mu_{ir}^{\mathbf{C}} \mu_{je}^{\mathbf{C}} \mathrm{E}[W_r W_e] \\
&\qquad + \mu_{ir}^{\mathbf{C}} \mu_{je}^{\mathbf{C}} \mu_{W_r} \mu_{W_e} - \mu_{ir}^{\mathbf{C}} \mu_{je}^{\mathbf{C}} \mu_{W_r} \mu_{W_e} \\
&= \Sigma_{irje}^{\mathbf{C}} \mu_{W_r} \mu_{W_e} + \mu_{ir}^{\mathbf{C}} \mu_{je}^{\mathbf{C}} \Sigma_{re}^{\mathbf{W}} + \Sigma_{irje}^{\mathbf{C}} \Sigma_{re}^{\mathbf{W}}
\end{aligned} \tag{4.64}$$

Of course, the approximation of equation (4.64) depends on the approximation of the assumptions introduced in its derivation (truncation of the series in equation (4.58), validity of (4.63), independence of $\tilde{\mathbf{W}}$ and $\tilde{\mathbf{K}}$).

4.3 LINEAR STRUCTURAL DYNAMICS

So far, problems connected with dynamic loading have not been treated in this chapter. However, in both civil and other branches of engineering, there are many instances of structures for which dynamic effects cannot be neglected: such are, for example, the effects of air turbulence on aircraft wings, of wind and/or ground accelerations on tall buildings, of sea waves on offshore structures, etc.

In most dynamic problems, as in all the instances quoted above, the greater uncertainties lie in the loading process rather than in the structural properties. Indeed, whereas for quasi-statically varying loads it may be possible to try and separate uncertainties in the loads and in the structure, which are usually of different origin and nature, this distinction is impossible in dynamic problems, in which the actual forces themselves depend on the structural properties. To reduce complications, therefore, most specific procedures developed for stochastic structural dynamics assume deterministic structures, and this hypothesis will be accepted in the following treatment.

A dynamic structural problem consists, in general, of finding the response process $\mathbf{u}(\mathbf{x}, t)$ in all points of the structure (indicated by the position vector $\mathbf{x}$) and at all times t, when the loading (input) process $\mathbf{W}(\mathbf{x}, t)$ and the structural (system) properties are given. In stochastic dynamics, this problem transforms into finding the properties of a multidimensional random time process $\tilde{\mathbf{u}}(\mathbf{x}, t)$ when the properties of the input r.p. $\tilde{\mathbf{W}}(\mathbf{x}, t)$ are given. Note that, because of the already introduced *discrete structure* assumption (Section 4.1), the number of relevant vectors $\mathbf{x}$, and hence the number of dimensions of $\tilde{\mathbf{u}}$, is finite (although it may be large).

Any time process, and in particular any random process, can be studied in the *time domain* (i.e. as a function of time) or in the *frequency domain* (i.e. through its *spectral* properties, cf. Section 2.4.4). Since either approach can be more or less convenient, depending on the particular problem, both are presented in the following.

4.3.1 The simple oscillator: definition and differential treatment

It is convenient to introduce structural dynamics problems with reference to a *simple elastic oscillator*, i.e. an elastic system whose deformation is fully defined by one scalar parameter (a *one-degree-of-freedom* or simple oscillator). In particular, reference will be made to the shear frame shown in Fig. 4.2, in which the horizontal beam is assumed to be fully rigid, and the axial deformability of the columns is neglected: thus, the deformation of the structure is identified by the horizontal displacement u of the beam.

(a) Governing equations

Let the simple oscillator of Fig. 4.2 be loaded by a horizontal force $W(t)$, which has significant variations in intervals of time comparable with the period of free oscillations of the frame (*natural period* T_0): in this case, dynamic effects cannot be neglected in the derivation of the response process (Section 3.2).

Let M be the mass of the transverse beam, while the mass of the columns is assumed to be negligible. Indicate by K and c the *elastic stiffness* and the (linear) *damping constant* of the oscillator, given respectively by

$$K = \sum_i \frac{12EJ_i}{h^3}$$
$$c = \sum_i \frac{12\zeta J_i}{h^3} \qquad (4.65)$$

where the symbols have the usual meaning of the elementary theory of structures:

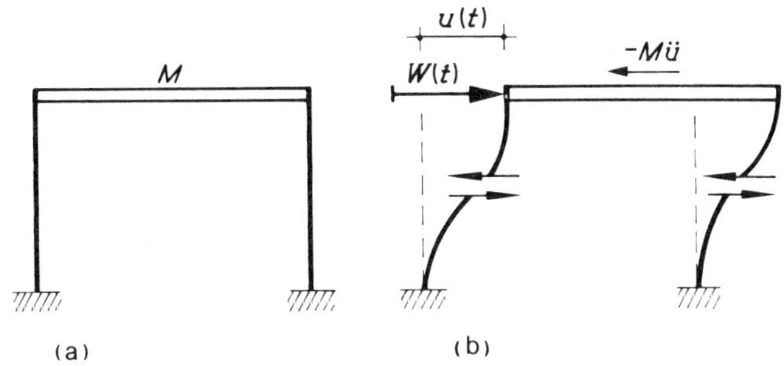

Fig. 4.2 Simple (one-degree-of-freedom) oscillator: (a) undeformed; (b) deformed.

in particular, ζ is the linear damping modulus in simple tension

$$\sigma = \zeta \dot{\varepsilon} \tag{4.66}$$

and the summation index i indicates the columns (only 2 in Fig. 4.2).

As usual, one or two superimposed dots will denote simple and double derivation with respect to time.

The well known equation of motion of the simple oscillator under the forcing function (*excitation*) $W(t)$ is

$$M\ddot{u} + c\dot{u} + Ku = W(t) \tag{4.67}$$

Equation (4.67) is usually standardized by dividing by M and introducing the *undamped natural pulsation* ω_0, the *damping coefficient* ξ and the *active acceleration* $w(t)$, defined respectively by

$$\omega_0 = \sqrt{\left(\frac{K}{M}\right)} \qquad \xi = \frac{c}{2\omega_0 M} \qquad w(t) = \frac{W(t)}{M} \tag{4.68}$$

The *natural pulsation* ω_0 is related to the *natural period* T_0 and *natural frequency* ν_0 of the (undamped) oscillator by

$$\omega_0 = \frac{2\pi}{T_0} = 2\pi\nu_0 \tag{4.68'}$$

Equation (4.67), standardized, becomes

$$\ddot{u} + 2\xi\omega_0\dot{u} + \omega_0^2 u = w(t) \tag{4.69}$$

If $w(t)$ is a deterministic function of time, (4.69) is a simple linear differential equation with constant coefficients, extensively studied in all textbooks of structural dynamics. The real part of its general solution is

$$u(t) = u_p(t) + e^{-\xi\omega_0 t}\{C_1 \sin[\omega_0\sqrt{(1-\xi^2)}t] + C_2 \cos[\omega_0\sqrt{(1-\xi^2)}t]\} \qquad \text{if } 0 \leqslant \xi < 1 \tag{4.69'}$$

$$u(t) = u_p(t) + e^{-\xi\omega_0 t}\{C_1 \exp[\omega_0\sqrt{(\xi^2-1)}t] + C_2 \exp[-\omega_0\sqrt{(\xi^2-1)}t]\} \qquad \text{if } \xi > 1 \tag{4.69''}$$

where $u_p(t)$ is a particular solution of equation (4.69) ($w(t) \neq 0$) and the second term on the r.h.s. is the general solution of the homogeneous equation associated with (4.69) ($w(t) = 0$), with C_1 and C_2 appropriate constants. The particular solution $u_p(t)$ is often the solution corresponding to homogeneous initial conditions, i.e. $u_p(0) = 0$, $\dot{u}_p(0) = 0$: in this case,

$$\left\{u(0) = 0,\ \dot{u}(0) = 0\right\} \Rightarrow \begin{cases} u(t) = u_p(t) \\ C_1 = C_2 = 0 \end{cases} \tag{4.70}$$

$\xi = 1$ identifies *critical damping*, which marks the passage between an oscillatory and a monotonic solution of the homogeneous equation. If $\xi < 1$, a *damped natural pulsation*

$$\omega_d = \omega_0 \sqrt{(1 - \xi^2)} \tag{4.68''}$$

can be defined and introduced in equation (4.69′).

In the case of random $\tilde{w}(t)$, more relevant in the present context, equation (4.69) can be tackled essentially by three alternative procedures, analogous to those used in the deterministic case; namely, (i) the theory of stochastic differential equations, (ii) impulse response analysis, and (iii) frequency response analysis. These three procedures are described in the following.

(b) The equation of motion treated as a stochastic differential equation

The theory of stochastic differential equations [2, 5] is systematically outlined in Appendix A5 (Section A5.2.3). Here, with reference to (4.69), let the input r.p. $\tilde{w}(t)$ be defined through its mean value function $\mu_w(t)$ and autocovariance function $\Sigma_w(t_1, t_2)$: remember that this is a complete description for a Gaussian process (Section A6.2), but is incomplete for a more general process. Let $\mu_u(t)$ and $\Sigma_u(t_1, t_2)$ be the mean value and autocovariance function of the output (response) process $\tilde{u}(t)$. Since differentiation is a linear operator (Section 2.5.3), taking expectations of both members in equation (4.69) one gets (equation (A5.33))

$$\ddot{\mu}_u + 2\xi_0 \dot{\mu}_u + \omega_0^2 \mu_u = \mu_w \tag{4.71}$$

while the initial conditions take the form

$$\mu_u(0) = u_0 \qquad \mu_{\dot{u}}(0) = \dot{u}_0 \tag{4.71'}$$

Consider now the cross-covariance $\Sigma_{w,u}(t_1, t_2)$ between the input and output functions. It can be proved by elementary algebra that the cross-covariance function satisfies the following ordinary differential equation (A5.34)

$$\frac{\partial^2 \Sigma_{w,u}(t_1, t_2)}{\partial t_2^2} + 2\xi\omega_0 \frac{\partial \Sigma_{w,u}(t_1, t_2)}{\partial t_2} + \omega_0^2 \Sigma_{w,u}(t_1, t_2) = \Sigma_w(t_1, t_2) \tag{4.72}$$

with initial conditions

$$\Sigma_{w,u}(t_1, 0) = 0 \qquad \left.\frac{\partial \Sigma_{w,u}(t_1, t_2)}{\partial t_2}\right|_{t_2 = 0} = 0 \tag{4.72'}$$

Similarly the autocovariance function of $\tilde{u}(t)$ is given by (equation (A5.35))

$$\frac{\partial^2 \Sigma_u(t_1, t_2)}{\partial t_1^2} + 2\xi\omega_0 \frac{\partial \Sigma_u(t_1, t_2)}{\partial t_1} + \omega_0^2 \Sigma_u(t_1, t_2) = \Sigma_{w,u}(t_1, t_2) \tag{4.73}$$

with initial conditions

$$\Sigma_u(0, t_2) = 0 \qquad \left.\frac{\partial \Sigma_u(t_1, t_2)}{\partial t_1}\right|_{t_1 = 0} = 0 \tag{4.73'}$$

Note that, as immediately shown by equations (4.71) and (4.71′), the mean value of the response coincides with the displacement of the frame loaded by the *expected excitation*, i.e. by a deterministic excitation equal to the mean value of the actual input process. Similarly, equations (4.72) and (4.72′) show that, looking at t_1 as a parameter and t_2 as the independent variable, $\Sigma_{w,u}(t_1, t_2)$ is the family of responses to the inputs $\Sigma_w(t_1, t_2)$, under homogeneous initial conditions; from equations (4.73) and (4.73′), the same applies to $\Sigma_u(t_1, t_2)$ by exchanging the roles of t_1 and t_2 and by taking $\Sigma_{w,u}(t_1, t_2)$ as the forcing function.

After the autocovariance function of the response has been calculated, the spectral density function can be obtained by double Fourier transform (equation (A5.54))

$$G_u(\omega_1, \omega_2) = \frac{1}{4\pi^2} \int_{-\infty}^{\infty} \int_{-\infty}^{\infty} \Sigma_u(t_1, t_2) \exp[-i(\omega_1 t_1 - \omega_2 t_2)] \, dt_1 \, dt_2 \tag{4.74}$$

If $\xi > 0$, as usually is true in real systems, the motion of the oscillator tends to vanish with elapsing time if the excitation is removed: when this condition is verified, the oscillator is said to be stable.

(c) Steady-state motion

The motion of a stable time-invariant oscillator subjected to a (weakly) stationary random input $\tilde{w}(t)$ tends to a (weakly) stationary r.p. $\tilde{u}^{\mathrm{s}}(t)$ independent of the initial conditions, usually called *steady-state motion*, which can also be regarded as the oscillator response corresponding to any arbitrary initial conditions set at time $t = -\infty$.

Let the mean value and the autocovariance function of $u^{\mathrm{s}}(t)$ be respectively

$$\begin{aligned} \mu_u^{\mathrm{s}}(t) &= \mu_u^{\mathrm{s}} \\ \Sigma_u^{\mathrm{s}}(t_1, t_2) &= B_u^{\mathrm{s}}(t_2 - t_1) = B_u^{\mathrm{s}}(v) \end{aligned} \tag{4.75}$$

where $v = t_2 - t_1$. From equation (4.73) we obtain

$$\frac{d^2 B_u^{\mathrm{s}}(v)}{dv^2} + 2\xi\omega_0 \frac{dB_u^{\mathrm{s}}(v)}{dv} + \omega_0^2 B_u^{\mathrm{s}}(v) = B_{w,u}^{\mathrm{s}}(t_1, t_2) \tag{4.76}$$

Since the l.h.s. of (4.76) depends only on the difference $t_2 - t_1 = v$, the same holds for the r.h.s.

$$B_{w,u}^{\mathrm{s}}(t_1, t_2) = B_{w,u}^{\mathrm{s}}(t_2 - t_1) = B_{w,u}^{\mathrm{s}}(v) \tag{4.77}$$

i.e. the input and the output processes are *jointly stationary*.

Since μ_w and μ_u^s are constant, (4.71) yields immediately

$$\mu_u^s = \mu_w/\omega_0^2 \tag{4.78}$$

while (4.72) becomes, taking account of (4.77),

$$\frac{d^2 B^s_{w,u}}{dv^2} + 2\xi\omega_0 \frac{dB^s_{w,u}}{dv} + \omega_0^2 B^s_{w,u}(v) = B_w(v) \tag{4.79}$$

With regard to the initial conditions, it is only required that

$$\lim_{v\to\pm\infty} B^s_{w,u}(v) = \lim_{v\to\pm\infty} \dot{B}^s_{w,u}(v) = \lim_{v\to\pm\infty} B^s_u(v) = \lim_{v\to\pm\infty} \dot{B}^s_u(v) = 0 \tag{4.80}$$

and that both B^s_u and $B^s_{w,u}$ be symmetric functions of v, whence

$$\dot{B}^s_u(0) = \dot{B}^s_{w,u}(0) = 0 \tag{4.81}$$

The existence of the first derivatives in equation (4.81), or of those involved in equations (4.72) and (4.73), are necessary conditions for the process $\tilde{u}(t)$ (or $\tilde{u}^s(t)$) to be differentiable. Unfortunately, most times neither $\tilde{w}(t)$ nor $\tilde{u}(t)$ can be treated as differentiable processes. The outlined approach thus becomes unserviceable, in that the processes involved do not enjoy the necessary properties of regularity. This remark justifies the need for a weaker formulation of the problem, which will now be discussed.

4.3.2 The simple oscillator: solution by integral approaches

Integral approaches, or *input–output methods* (which include *impulse response analysis* and *frequency response analysis*) [6, 7] can be used successfully to study the response of a linear filter, in both the *deterministic* and *stochastic* realms. They essentially consist of (i) decomposition of the input function $w(t)$ into the linear combination of a (possibly infinite) number of elementary input functions, (ii) solution in terms of elementary output functions, each corresponding to an elementary input, and (iii) linear recomposition of the elementary outputs into the total output functions. Approaches of this sort present several computational advantages in the case of a random input $\tilde{w}(t)$.

(a) Response analysis by the impulse response function

This approach is based on the study of the oscillator response to an *impulsive* input, which can be described as (Fig. 4.3(a))

$$w_\tau(t) = \mathscr{W}_\tau \partial(t-\tau) \tag{4.82}$$

where $\mathscr{W}_\tau$ is a real constant and $\partial(t-\tau)$ is Dirac's delta-function, i.e. the function defined by

$$\begin{aligned} &\partial(t-\tau) = 0 && \text{for all } t \neq \tau \\ &\int_{\tau-\varepsilon}^{\tau+\varepsilon} \partial(t-\tau)\,dt = 1 && \text{for all } \varepsilon > 0 \end{aligned} \tag{4.83}$$

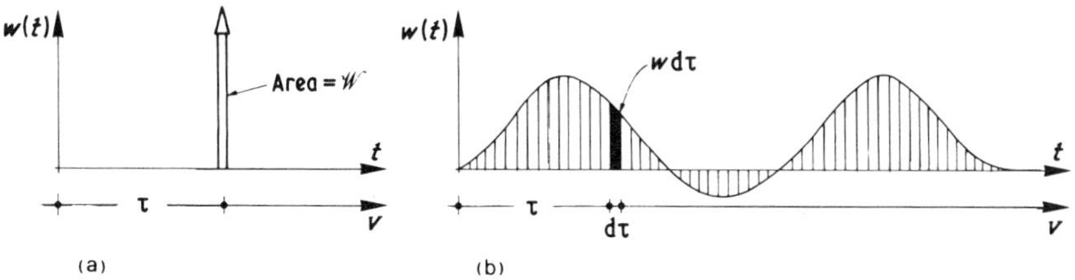

Fig. 4.3 Forcing functions: (a) finite impulse applied at time τ; (b) general input decomposed into a sum of elementary impulses.

We now introduce equation (4.82) into equation (4.69), assume that the oscillator is at rest at $t = 0$, and integrate both members from $t = 0$ to any $t > \tau$:

$$\dot{u}_\tau(t) + 2\xi\omega_0 u_\tau(t) + \omega_0^2 \int_0^t u_\tau(t)\,\mathrm{d}t = \mathscr{W}_\tau \tag{4.84}$$

where $u_\tau(t)$ indicates the sought response to the impulse (4.82).

It is immediately obvious that

$$u_\tau(t) = \dot{u}_\tau(t) = 0 \qquad 0 \leqslant t < \tau \tag{4.85}$$

From equations (4.84) and (4.85), taking account of the regularity of $u_\tau(t)$,

$$\dot{u}_\tau(t) = \mathscr{W}_\tau \qquad u_\tau(t) = 0 \qquad t = \tau + \varepsilon, \quad \varepsilon \to 0 \tag{4.86}$$

For $t > \tau$, the equation of motion is the homogeneous equation

$$\ddot{u}_\tau + 2\xi\omega_0\dot{u}_\tau + \omega_0^2 u_\tau = 0 \tag{4.87}$$

for which (4.86) give the initial conditions. It is thus easy to find

$$u_\tau(t) = \mathscr{W}_\tau h^*(t, \tau) \tag{4.88}$$

where $h^*(t, \tau)$ is the *impulse response function* (or *Green's function*) (equation (A5.38)) of the oscillator, which is given by (cf. equations (4.69′) and (4.69″))

$$h^*(t,\tau) = \begin{cases} 0 & t \leqslant \tau;\ \text{any } \xi \\ (1/\omega_\mathrm{d})\,\mathrm{e}^{-\xi\omega_0(t-\tau)} \sin[\omega_\mathrm{d}(t-\tau)] & t > \tau;\ \xi < 1 \\ (t-\tau)\mathrm{e}^{-\omega_0(t-\tau)} & t > \tau;\ \xi = 1 \\ (1/2\omega_\mathrm{d})\,\mathrm{e}^{-\xi\omega_0(t-\tau)}\,(\mathrm{e}^{\omega_\mathrm{d}(t-\tau)} - \mathrm{e}^{-\omega_\mathrm{d}(t-\tau)}) & t > \tau;\ \xi > 1 \end{cases} \tag{4.89}$$

Fig. 4.4 shows typical functions $h^*(t, \tau)$.

Any time function $w(t)$ can be considered as the superposition of an infinite number of impulses of the type (4.82). Because of the system linearity, the response at any time t can be obtained as the superposition of the responses to all applied impulses: therefore, assuming $w(t) = 0$ for $t \leqslant 0$ and remembering the first of equations (4.89)

$$u_\mathrm{p}(t) = \int_0^t w(\tau)\,h^*(t,\tau)\,\mathrm{d}\tau = \int_{-\infty}^{\infty} w(t-v)\,h(t,v)\,\mathrm{d}v \tag{4.90}$$

where $v = t - \tau$ and $h(t, v) = h^*(t, t - v)$.

Note that, if the system is time-invariant, the impulse response function depends on the delay $v = t - \tau$ but not on the arrival time τ (equation (A5.38)):

$$h^*(t, \tau) = h(v) \tag{4.90'}$$

and (4.90) can be written

$$u_p(t) = \int_{-\infty}^{+\infty} w(t-v)\, h(v)\, dv \tag{4.90''}$$

The r.h.s. of equation (4.90) or (4.90″), known as the Duhamel integral‡, gives the output $u_p(t)$ as a weighted integral of the input process, be it deterministic or random. In the latter case, mean square integration (Appendix A5, Section A5.2) gives the mean value function (equation (A5.63))

$$\mu_{u_p}(t) = \int_0^t \mu_w(\tau)\, h^*(t, \tau)\, d\tau \tag{4.91}$$

and the autocovariance function (equations (A5.64) and (A5.65)

$$\Sigma_{u_p}(t_1, t_2) = \int_0^{t_1} \int_0^{t_2} \Sigma_w(\tau_1, \tau_2)\, h^*(t_1, \tau_1)\, h^*(t_2, \tau_2)\, d\tau_1\, d\tau_2. \tag{4.92}$$

The spectral density function (equations (2.21–2.21″)), i.e. the covariance function of the Fourier transform process of $u_p(t)$, defined by

$$u_F(\omega) = \frac{1}{2\pi} \int_{-\infty}^{\infty} u_p(t)\, e^{-i\omega t}\, dt \tag{4.93}$$

can be obtained from (equation (A5.54)

$$G_{u_p}(\omega_1, \omega_2) = E[(u_F(\omega_1) - \mu_F(\omega_1))\,(u_F(\omega_2) - \mu_F(\omega_2))] \tag{4.94}$$

Hence, after $\Sigma_{u_p}(t_1, t_2)$ has been calculated, $G_{u_p}(\omega_1, \omega_2)$ can be determined from equation (4.74).

Note that (4.90) gives only the particular solution of the equation of motion (4.69) corresponding to homogeneous initial conditions (4.70).

If the initial conditions are not homogeneous ($u_0 \neq 0$ or $\dot{u}_0 \neq 0$), the general integral of (4.69) can be put, for $\xi < 1$, in the form (4.69′)

$$u(t) = C_2 e^{-\xi\omega_0 t} \cos(\omega_d t) + C_1\, e^{-\xi\omega_0 t} \sin(\omega_d t) + \int_0^t w(\tau) h^*(t, \tau)\, d\tau \tag{4.95}$$

where the constants C_1 and C_2 satisfy the initial conditions, i.e.

$$C_2 = u_0 \qquad C_1 \omega_d - \xi\omega_0 C_2 = \dot{u}_0 \tag{4.95'}$$

Note moreover that, in general, for non-homogeneous initial conditions, the response $u(t)$ is not stationary, even if the input function $w(t)$ is stationary. However, if the system is stable (i.e. if $\xi > 0$), the influence of the initial conditions tends to vanish with increasing time t: it is indeed evident from equation (4.95) that for sufficiently large t the first two terms on the r.h.s. become negligible. It is

‡ It is worth noting that the Duhamel integral is a convolution integral according to the definition of equation (A3.45).

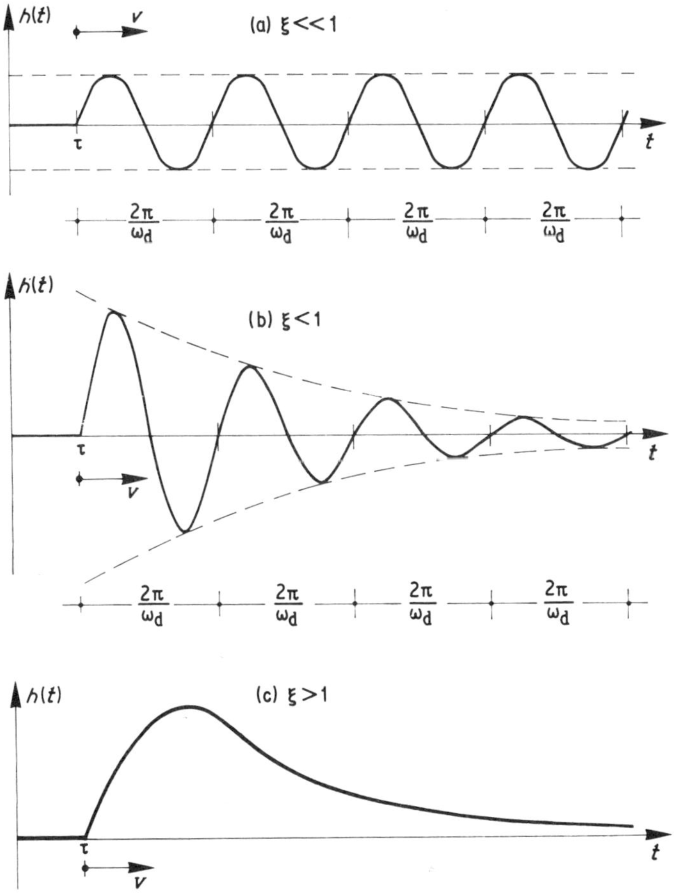

Fig. 4.4 Typical impulse response functions: (a) damping coefficient $\xi \ll 1$; (b) damping coefficient $\xi < 1$; (c) damping coefficient $\xi > 1$.

thus clear why any particular response of a time-invariant system subjected to a stationary input tends to coincide with the already defined steady-state response $u^s(t)$. The properties of the r.p. $u^s(t)$ can be calculated by assuming the starting time of the excitation at $t = -\infty$ instead of $t = 0$; thus the mean value and the autocovariance functions μ_u^s and $B_u^s(v)$ are obtained from equations (4.91) and (4.92), setting the lower limit of the integrals at $-\infty$ (equations (A5.67)–(A5.69)

$$\mu_u^s = \mu_W \int_{-\infty}^{t} h^*(t, \tau)\,d\tau = \mu_W \int_0^{+\infty} h(v)\,dv \tag{4.96}$$

$$B_u^s(v) = \int_0^{+\infty} \int_0^{+\infty} h(v_1)h(v_2)B_W(v - v_2 + v_1)\,dv_1\,dv_2 \tag{4.97}$$

The spectral density of the steady-state response is, again, obtained as the Fourier transform of the autocovariance

$$G_u^s(\omega) = \frac{1}{2\pi}\int_{-\infty}^{\infty} B_u^s(v)\exp(-i\omega v)\,dv \tag{4.98}$$

Since $u^s(t)$ is a stationary process, $G_u^s(\omega)$ is a real-valued function of the pulsation ω; moreover, as already remarked, $B_u^s(v)$ is a symmetric function of v. Hence, equation (4.98) can be given the simpler expression (equation (2.21″))

$$G_u^s(\omega) = \frac{1}{\pi}\int_{0}^{+\infty} B_u^s(v)\cos(\omega v)\,dv \tag{4.98'}$$

(b) Response analysis by the frequency response function

The use of the technique described in the previous section to obtain the spectral density of the output may require cumbersome calculations: in other words, this technique is more suited for analysis in the time domain. An alternative approach can be followed to obtain directly the output properties in the frequency domain: from these, the output properties in the time domain can be obtained by inverse Fourier transforms.

To introduce this technique, let the oscillator be loaded by a harmonic excitation (equation (A5.39))

$$w_\omega(t) = \mathscr{W}_\omega e^{i\omega t} \tag{4.99}$$

where $\mathscr{W}_\omega$ is a real constant.

Introducing (4.99) into the equation of motion (4.69), a particular solution is obtained in the form

$$u_\omega(t) = \mathscr{W}_\omega A(\omega)\, e^{+i\omega t} \tag{4.100}$$

in which

$$A(\omega) = \frac{1}{\omega_0^2 - \omega^2 + 2i\xi\omega_0\omega} \tag{4.101}$$

The complex-valued function $A(\omega)$ is known as the *frequency response function* (equation (A5.41b)) of the oscillator with own natural (undamped) pulsation ω_0 and damping coefficient ξ (equation (4.68)).

Consider now the Fourier transform of a generic excitation $w(t)$ starting at time $t = 0$:

$$w_F(\omega) = \frac{1}{2\pi}\int_{-\infty}^{\infty} w(t)\, e^{-i\omega t}\,dt \tag{4.102}$$

(the lower limit of the integral in (4.102) can be indifferently set at $-\infty$ or 0, because by assumption $w(t) = 0$ for $t < 0$). Obviously, the inverse Fourier transform of $w_F(\omega)$ is again $w(t)$

$$w(t) = \int_{-\infty}^{\infty} w_F(\omega)\, e^{i\omega t}\,d\omega \tag{4.102'}$$

Equation (4.102′) shows that generic excitation $w(t)$ can be regarded as the superposition of an infinite number of elementary harmonic excitations of the type (4.99).

Again (cf. equation (4.90)), because of the system linearity, the response function $u(t)$ is the superposition of the responses $u_\omega(t)$ to each $w_\omega(t)$ (equation (4.100)):

$$u_p(t) = \int_{-\infty}^{\infty} u_\omega(t)\,\mathrm{d}\omega = \int_{-\infty}^{\infty} A(\omega)\, w_F(\omega)\, e^{i\omega t}\,\mathrm{d}\omega \tag{4.103}$$

As evident from (4.103) and (4.102), $u(t)$ is a weighted integral of the Fourier transform $w_F(\omega)$ of the excitation. The mean value function of $w_F(\omega)$ is soon calculated by taking expectations of both members in equation (4.102), i.e.

$$\mu_{w_F}(\omega) = \frac{1}{2\pi}\int_{-\infty}^{\infty} \mu_w(t)\, e^{-i\omega t}\,\mathrm{d}t \tag{4.104}$$

and coincides with the Fourier transform of the mean value function of the excitation. The autocovariance function of $w_F(\omega)$ is the spectral density function $G_w(\omega_1, \omega_2)$ of the input $w(t)$, i.e. is the double Fourier transform of the autocovariance function of $w(t)$ (equation (A5.54))

$$G_w(\omega_1, \omega_2) = \frac{1}{4\pi^2}\int_{-\infty}^{\infty}\int_{-\infty}^{\infty} \Sigma_w(t_1, t_2)\exp[-i(\omega_1 t_1 - \omega_2 t_2)]\,\mathrm{d}t_1\,\mathrm{d}t_2 \tag{4.105}$$

The spectral density of the response is related to the spectral density of the input by the simple formula (equation (A5.73))

$$G_{u_p}(\omega_1, \omega_2) = G_w(\omega_1, \omega_2) A(\omega_1) A^\dagger(\omega_2) \tag{4.106}$$

which is the basic tool in the applications of frequency response analysis. $A^\dagger(\omega)$ indicates the complex conjugate of $A(\omega)$.

The mean value and autocovariance functions of the response can be obtained from equation (4.103) by means of the mean-square theory of integration of stochastic processes (Section A5.2.1, equation (A5.24)).

$$\mu_{u_p}(t) = \int_{-\infty}^{\infty} A(\omega)\mu_{w_F}(\omega) e^{i\omega t}\,\mathrm{d}\omega \tag{4.107}$$

and (equations (A5.72) and (A5.73))

$$\Sigma_{u_p}(t_1, t_2) = \int_{-\infty}^{\infty}\int_{-\infty}^{\infty} A(\omega_1)A^\dagger(\omega_2)\exp(i\omega_1 t_1 - i\omega_2 t_2) G_w(\omega_1, \omega_2)\,\mathrm{d}\omega_1\,\mathrm{d}\omega_2 \tag{4.108}$$

Remember that equation (4.103) yields only a particular integral of equation (4.69): as easily shown, it satisfies the initial conditions

$$\begin{aligned} u_p(0) = u_{p0} &= \int_{-\infty}^{+\infty} A(\omega) w_F(\omega)\,\mathrm{d}\omega \\ \dot{u}_p(0) = \dot{u}_{p0} &= i\int_{-\infty}^{+\infty} \omega A(\omega) w_F(\omega)\,\mathrm{d}\omega \end{aligned} \tag{4.103′}$$

with u_{p0} and $\dot{u}_{p0}$ random variables. Noting that the frequency response function relevant to the linear transformation of differentiation is $i\omega$ (equations (A5.39) and (A5.41a)), and making use of equations (4.108) and (A5.54), it is easy to obtain mean and variance of u_{p0} and $\dot{u}_{p0}$ as follows:

$$\mu_{u_{p0}} = \int_{-\infty}^{\infty} A(\omega)\mu_{w_F}(\omega)\, d\omega$$

$$\mu_{\dot{u}_{p0}} = i\int_{-\infty}^{\infty} \omega A(\omega)\mu_{w_F}(\omega)\, d\omega \tag{4.103''}$$

$$s^2_{u_{p0}} = \int_{-\infty}^{\infty}\int_{-\infty}^{\infty} G_w(\omega_1,\omega_2)A(\omega_1)A^\dagger(\omega_2)\, d\omega_1\, d\omega_2$$

$$s^2_{\dot{u}_{p0}} = \int_{-\infty}^{\infty}\int_{-\infty}^{\infty} G_w(\omega_1,\omega_2)\omega_1 A(\omega_1)\omega_2 A^\dagger(\omega_2)\, d\omega_1\, d\omega_2 \tag{4.103'''}$$

The general integral of equation (4.69), for $\xi < 1$, is given by

$$u_0(t) = C_1 e^{-\xi\omega_0 t}\sin(\omega_d t) + C_2 e^{-\xi\omega_0 t}\cos(\omega_d t) + \int_{-\infty}^{\infty} A(\omega)_F(\omega)e^{i\omega t}\, d\omega \tag{4.109}$$

where C_1 and C_2 are constants which satisfy the relevant initial conditions $u(0) = u_0$ and $\dot{u}(0) = \dot{u}_0$: even if u_0 and $\dot{u}_0$ are deterministic constants, C_1 and C_2 are random variables.

However, $u_p(t)$ (equation (4.103)) and the corresponding $\dot{u}(t)$ vanish deterministically for $t \to -\infty$: therefore, equation (4.103) represents a motion starting at $t = -\infty$, which, if the excitation is stationary, coincides with the already defined steady-state response $u^s(t)$.

This observation leads to some consequences.

First, remembering an analogous observation in the previous subsection (a), the following identity can be established from equations (4.90) and (4.103):

$$u^s(t) = \int_{-\infty}^{+\infty} A(\omega)\, w_F(\omega)\, e^{i\omega t}\, d\omega = \int_{-\infty}^{\infty} h(v)\, w(t-v)\, dv \tag{4.110}$$

whence, taking account of equations (4.102) and (4.102′) one gets

$$h(v) = \frac{1}{2\pi}\int_{-\infty}^{\infty} A(\omega)\, e^{i\omega v}\, d\omega$$

$$A(\omega) = \int_{-\infty}^{\infty} h(v)\, e^{-i\omega t}\, dv \tag{4.111}$$

i.e. impulse and frequency response functions are related by Fourier transforms (equations (A5.41′) and (A5.42)), except that the coefficient $1/(2\pi)$ has been interchanged.

Secondly, again provided $w(t)$ is stationary, equations (4.106)–(4.108) yield for the steady-state motion $u^{s}(t)$, taking account of (4.104),

$$\mu_u^{s} = \frac{\mu_W}{2\pi} \int_{-\infty}^{+\infty} \int_{-\infty}^{+\infty} A(\omega)\,\mathrm{e}^{-\mathrm{i}\omega t}\,\mathrm{d}t\,\mathrm{d}\omega \tag{4.112}$$

$$B_u^{s}(v) = \int_{-\infty}^{\infty} |A(\omega)|^2 G_W(\omega)\mathrm{e}^{\mathrm{i}\omega v}\,\mathrm{d}\omega \tag{4.113}$$

$$G_u^{s}(\omega) = G_W(\omega)|A(\omega)|^2 \tag{4.114}$$

(c) Some remarks

Note that all approaches described in the preceding sections allow one to relate the mean value and autocorrelation functions of input and response but not their full distributions. The information obtained is thus incomplete, unless some specific conditions are satisfied by the input, namely the *stability* property that holds in case of Gaussian processes. In this case, very important in practical applications, the output process is also Gaussian and fully defined by mean and autocovariance functions (as already remarked).

In the differential approach (Section 4.3.1(b)), the processes involved are required to be differentiable; this is not necessary in the integral approaches (via either impulse or frequency response functions). Moreover, the latter approaches lead in general to more tractable algorithms, because, as is well known, numerical integration poses less problems than numerical solution of differential equations.

Finally, the extension of integral approaches to structures with more degrees of freedom is straightforward (at least in case of linearity), while the differential treatment becomes extremely cumbersome.

For all these reasons, integral approaches are almost always preferred in stochastic structural dynamics. Therefore, the following section, devoted to multi-degree-of-freedom elastic structures, will not treat differential approaches explicitly.

4.3.3 Multi-degree-of-freedom structures

(a) Governing equations

As anticipated in Section 4.1, the mass of any deformable structure will be considered lumped into afinite number m of rigid elements: the displacements of these fully describe the deformation pattern of the structure. If each element has χ degrees of freedom ($\chi \leqslant 6$), the whole *discretized* stucture has $n = m\chi$ degrees of freedom. Just to give an example, the mass of a multistorey frame can be lumped (provided some conditions are met) into rigid floor beams: the deformation of the frame is described by the vector $\mathbf{u}$ collecting the (horizontal) displacements u_i of each floor (Fig. 4.5(a)). In another example, the mass of a truss can be lumped into the joints: the deformation of the truss is then described by a vector $\mathbf{u}$, whose elements are the components of the displacement of each point along the three coordinate axes.

The forces acting on the structure are also described by a vector $\mathbf{W}$, defined in such a way that

$$\mathbf{W}^T\mathbf{u} = \sum_i W_i u_i$$

is the work done. Elastic and damping properties of the structure are appropriately discretized into the elements connecting the masses, and are described by a stiffness matrix $\mathbf{K}$ and a damping matrix $\mathbf{c}$: it can be shown that both $\mathbf{K}$ and $\mathbf{c}$ are symmetric and positive definite.

Without going further into the procedure for dynamic modelling of a real structure (which can be found in the appropriate textbooks, e.g. Zienkiewicz [8]) the equation of motion takes the matrix form

$$\mathbf{M}\ddot{\mathbf{u}} + \mathbf{c}\dot{\mathbf{u}} + \mathbf{K}\mathbf{u} = \mathbf{W}(t) \tag{4.115}$$

where $\mathbf{M}$ is another symmetric positive definite matrix, called the mass matrix, which often (as in the examples of Fig. 4.5, but not in general) reduces to the diagonal matrix of the lumped masses.

Since, as stated in the introductory paragraph of Section 4.3, no technological uncertainty will be introduced in the treatment of dynamic problems, $\mathbf{M}$, $\mathbf{c}$ and $\mathbf{K}$ are assumed to be *deterministic* matrices, while $\tilde{\mathbf{W}}(t)$ is a multivariate random process. Reduction of equation (4.115) to standard form requires the introduction of a new displacement vector $\mathbf{u}'$ defined by‡

$$\mathbf{u}' = \mathbf{M}^{1/2}\mathbf{u} \tag{4.116}$$

where $\mathbf{M}^{1/2}$ is the square-root matrix of the (positive definite) matrix $\mathbf{M}$. Then, (4.115) becomes

$$\mathbf{M}^{1/2}\ddot{\mathbf{u}}' + \mathbf{c}\mathbf{M}^{-1/2}\dot{\mathbf{u}}' + \mathbf{K}\mathbf{M}^{-1/2}\mathbf{u}' = \mathbf{W}(t) \tag{4.115$'$}$$

and pre-multiplying by $\mathbf{M}^{-1/2}$, one gets

$$\ddot{\mathbf{u}}' + \mathbf{Z}\dot{\mathbf{u}}' + \mathbf{Y}\mathbf{u}' = \mathbf{M}^{-1/2}\mathbf{W}(t) = \mathbf{w}(t) \tag{4.115$''$}$$

with

$$\mathbf{Z} = \mathbf{M}^{-1/2}\mathbf{c}\mathbf{M}^{-1/2} \qquad \mathbf{Y} = \mathbf{M}^{-1/2}\mathbf{K}\mathbf{M}^{-1/2} \tag{4.117}$$

symmetric, positive definite, matrices. The solution of the problem will then be obtained in terms of the new variables $\mathbf{u}'$ whence the actual displacements $\mathbf{u}$ can be obtained by applying again the transformation (4.116).

In particular, the mean value vector function and the covariance matrix of the actual displacements $\mathbf{u}$ can be calculated by

$$\boldsymbol{\mu}_{\mathbf{u}}(t) = \mathbf{M}^{-1/2}\boldsymbol{\mu}'_{\mathbf{u}}(t) \qquad \boldsymbol{\Sigma}_{\mathbf{u}}(t_1, t_2) = \mathbf{M}^{-1/2}\boldsymbol{\Sigma}_{\mathbf{u}'}(t_1, t_2)\mathbf{M}^{-1/2} \tag{4.116$'$}$$

The general solution of (4.115″) is cumbersome to obtain, unless $\mathbf{Z}$ and $\mathbf{Y}$ have

‡ Equation (4.115) can also be standardized by premultiplying by $\mathbf{M}^{-1}$: in this case, however, neither the symmetry of $\mathbf{Y}$ and $\mathbf{Z}$ (4.115″) nor the orthogonality of the relevant eigenvectors is guaranteed.

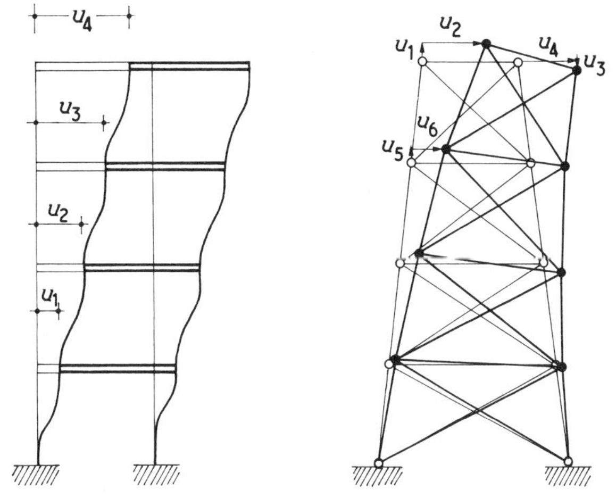

Fig. 4.5 Examples of multi-degree-of-freedom oscillators.

common eigenvectors. In the latter case, and for subcritical damping, the solution can be put in the form

$$\mathbf{u}'(t) = \mathbf{u}'_{\mathrm{p}}(t) + \sum_{k=1}^{n} \mathbf{u}'_k \exp(-\xi_k \omega_{0k} t)\,[C_{1k} \sin(\omega_{dk} t) + C_{2k} \cos(\omega_{dk} t)] \qquad (4.116'')$$

where $\mathbf{u}'_{\mathrm{p}}(t)$ is a particular solution; ω_{0k} (*undamped natural pulsations*) are the positive square roots of the eigenvalues of the matrix $\mathbf{Y}^{\ddagger}$; $\mathbf{u}'_k$ are corresponding n-dimensional eigenvectors (*natural modes of oscillation*); $2\xi_k\omega_{0k}$ are the eigenvalues of the matrix $\mathbf{Z}^{\ddagger}$ $(0 \leqslant \xi_k < 1$ by assumption); $\omega_{dk} = \omega_{0k}\sqrt{(1-\xi_k^2)}$ are the *damped natural pulsations*; and C_{1k}, C_{2k} are appropriate constants.

The solution of equation (4.115″) in the presence of a multivariate random input $\tilde{\mathbf{w}}(t)$ could in principle be pursued through conceptually straightforward generalizations of the procedures illustrated in Sections 4.3.2(*a*) and (*b*), which will be illustrated in subsections (*b*) and (*c*) below [6]. However, it will be evident that such treatments lead to cumbersome calculations, and *modal analysis* (Section 4.3.4) is most often introduced.

(b) Response analysis by the matrix of frequency response functions

Consider the n-degree-of-freedom structure loaded by forces which vary in time according to the harmonic law

$$\mathbf{w}_{\omega}(t) = \mathbf{W}_{\omega} e^{i\omega t} \qquad (4.118)$$

‡ The *eigenvalues* of a $n \times n$ matrix $\Lambda\,(\Lambda_{jk})$ are the solutions of the equation

$$\mathrm{Det}\,|\Lambda_{jk} - \delta_{jk}\lambda| = 0$$

It can be proved that for a symmetric positive definite matrix the n solutions λ_k are real and positive. If the matrix characterizes a system with random properties, the relevant eigenvalues can be obtained according to the procedures discussed in Section 4.4.1.

where $\mathbf{W}_\omega$ is an n-dimensional constant vector. It is easy to show that, when the input (4.118) is introduced, the solution of (4.115″) is (in analogy to (4.100))

$$\mathbf{u}'_\omega(t) = \mathbf{A}(\omega)\mathbf{W}_\omega \mathrm{e}^{\mathrm{i}\omega t} \tag{4.119}$$

where $\mathbf{A}(\omega)$ is an $n \times n$ symmetric matrix (*matrix of frequency response functions*), defined by

$$\mathbf{A}(\omega) = (-\omega^2\mathbf{I} + \mathrm{i}\omega\mathbf{Z} + \mathbf{Y})^{-1} \tag{4.120}$$

Remembering that $\mathbf{u}'_\omega$ and $\mathbf{W}_\omega$ are n-dimensional vectors, equation (4.119) shows that the generic element $A_{jk}(\omega)$ of $\mathbf{A}(\omega)$ is the kth component of the motion caused by the jth force component acting with intensity $\mathscr{W}_{\omega j} = 1$ and pulsation ω.

Take now a generic forcing function $\mathbf{w}(t)$. By developments similar to those in Section 4.3.2(b), the response $\mathbf{u}'_\mathrm{p}(t)$ can be put in the form

$$\mathbf{u}'_\mathrm{p}(t) = \int_{-\infty}^{+\infty} \mathbf{A}(\omega)\mathbf{w}_\mathrm{F}(\omega)\mathrm{e}^{\mathrm{i}\omega t}\,\mathrm{d}\omega \tag{4.121}$$

in which $\mathbf{w}_\mathrm{F}(\omega)$ is the Fourier transform of $\mathbf{w}(t)$,

$$\mathbf{w}_\mathrm{F}(\omega) = \frac{1}{2\pi}\int_{-\infty}^{\infty} \mathbf{w}(t)\mathrm{e}^{-\mathrm{i}\omega t}\,\mathrm{d}t \tag{4.122}$$

If $\tilde{\mathbf{w}}(t)$ is a stochastic process, and $\mathbf{G}_\mathbf{w}(\omega_1, \omega_2)$ its matrix of cross-spectral densities, the spectral density matrix $\mathbf{G}_{\mathbf{u}'_\mathrm{p}}(\omega_1, \omega_2)$ of the vector process $\mathbf{u}'_p(t)$ is, in analogy to equation (4.106),

$$\mathbf{G}_{\mathbf{u}'_\mathrm{p}}(\omega_1, \omega_2) = \mathbf{A}(\omega_1)\mathbf{G}_\mathbf{w}(\omega_1, \omega_2)\mathbf{A}^\dagger(\omega_2) \tag{4.123}$$

The mean value vector function of $\mathbf{u}'_\mathrm{p}(t)$ is

$$\boldsymbol{\mu}_{\mathbf{u}'_\mathrm{p}}(t) = \int_{-\infty}^{\infty} \mathbf{A}(\omega)\boldsymbol{\mu}_{\mathbf{w}_\mathrm{F}}(\omega)\,\mathrm{e}^{\mathrm{i}\omega t}\,\mathrm{d}\omega \tag{4.124}$$

with

$$\boldsymbol{\mu}_{\mathbf{w}_\mathrm{F}}(\omega) = \frac{1}{2\pi}\int_{-\infty}^{\infty} \boldsymbol{\mu}_\mathbf{w}(t)\mathrm{e}^{-\mathrm{i}\omega t}\,\mathrm{d}t \tag{4.124′}$$

If the excitation is stationary and the system time-invariant, the response tends to a steady-state motion $\mathbf{u}'^{\mathrm{S}}(t)$, whose spectral density and mean vector can be obtained from equations (4.123) and (4.124) as

$$\mathbf{G}^\mathrm{S}_{\mathbf{u}'_\mathrm{p}}(\omega) = \mathbf{A}(\omega)\mathbf{G}_\mathbf{w}(\omega)\mathbf{A}^\dagger(\omega) \tag{4.125}$$

$$\boldsymbol{\mu}^\mathrm{S}_{\mathbf{u}'} = \frac{\boldsymbol{\mu}_\mathbf{w}}{2\pi}\int_{-\infty}^{\infty}\int_{-\infty}^{\infty} \mathbf{A}(\omega)\mathrm{e}^{-\mathrm{i}\omega t}\,\mathrm{d}t\,\mathrm{d}\omega \tag{4.126}$$

The autocovariance matrix function of the steady-state response $\mathbf{u}^{\mathrm{S}'}(t)$ is obtained through Fourier transformation of the spectral density:

$$\mathbf{B}^\mathrm{S}_{\mathbf{u}'}(v) = \int_{-\infty}^{\infty} \mathbf{G}^\mathrm{S}_{\mathbf{u}'}(\omega)\mathrm{e}^{\mathrm{i}\omega v}\,\mathrm{d}\omega \tag{4.127}$$

In conclusion, this approach requires the matrix inversion (4.120), and integrations on products of the inverted matrix: the difficulties of its application thus appear self-evident.

(c) *Response analysis by the matrix of impulse response functions*

To build up the matrix of impulse response functions is even more complicated than to build up the matrix of frequency response functions. Several procedures are possible, among which the simplest is perhaps the following one.

With a technique already used in Section 4.2.1(*c*), introduce a $2n$-dimensional *state vector* defined by equation (4.23) or, in vector notation,

$$\mathbf{y} = \begin{pmatrix} \mathbf{u}' \\ \dot{\mathbf{u}}' \end{pmatrix} \tag{4.23$'$}$$

The time derivative of $\mathbf{y}$ is

$$\dot{\mathbf{y}} = \begin{pmatrix} \dot{\mathbf{u}}' \\ \ddot{\mathbf{u}}' \end{pmatrix} \tag{4.23$''$}$$

Introducing further the $2n \times 2n$ matrix

$$\mathbf{H} = \begin{bmatrix} \mathbf{0} & -\mathbf{I} \\ \mathbf{Y} & \mathbf{Z} \end{bmatrix} \tag{4.128}$$

it is easy to show that equation (4.116) can be written as

$$\dot{\mathbf{y}}(t) + \mathbf{H}\mathbf{y} = \mathbf{z}(t) \tag{4.129}$$

where $\mathbf{z}(t)$ is a $2n$-dimensional input vector defined by

$$\mathbf{z} = \begin{pmatrix} \mathbf{0} \\ \mathbf{w} \end{pmatrix} \tag{4.130}$$

In this way, the n-dimensional second-order vector differential equation (4.116) has been formally transformed into a $2n$-dimensional first-order equation.

The solution of equation (4.129) can be written in the form‡

$$\mathbf{y}(t) = \exp[-\mathbf{H}t]\,\mathbf{y}_0 + \int_0^t \exp[-\mathbf{H}(t-\tau)]\,\mathbf{z}(\tau)\,\mathrm{d}\tau \tag{4.131}$$

provided that $\mathbf{w}(t) = 0$ when $t < 0$. In equation (4.131), $\mathbf{y}_0$ is the vector of initial conditions

$$\mathbf{y}_0 = \begin{pmatrix} \mathbf{u}'(0) \\ \dot{\mathbf{u}}'(0) \end{pmatrix} \tag{4.132}$$

‡ An algebraic function $g[\boldsymbol{\Lambda}]$ of a square matrix $\boldsymbol{\Lambda}$ (in particular, the exponential, $\exp[\boldsymbol{\Lambda}]$) is a matrix defined as follows. Let $\mathbf{x}$ be the diagonal matrix of the (complex) eigenvalues x_k of $\boldsymbol{\Lambda}$, $\mathbf{Z}$ a matrix such that

$$\mathbf{x} = \mathbf{Z}^{-1}\boldsymbol{\Lambda}\mathbf{Z} \tag{i}$$

and χ the diagonal matrix defined by

$$\chi_{jk} = \begin{cases} g(x_k) & j = k \\ 0 & j \neq k \end{cases}$$

Then, applying the inverse transformation of (i),

$$g[\boldsymbol{\Lambda}] = \mathbf{Z}\boldsymbol{\chi}\mathbf{Z}^{-1}$$

Noting that, from equation (4.130),

$$\mathbf{z}(\tau) = \begin{bmatrix} \mathbf{0} \\ \mathbf{I} \end{bmatrix} \mathbf{w}(\tau) \tag{4.130$'$}$$

and comparing with equations (4.131) and (4.90), it appears that the $2n \times n$ matrix

$$\exp[-\mathbf{H}(t-\tau)] \begin{bmatrix} \mathbf{0} \\ \mathbf{I} \end{bmatrix} \tag{4.133}$$

plays a role analogous to the impulse response function $h^*(t, \tau)$ for the simple oscillator. In particular, the displacement components $\mathbf{u}(t)$, i.e. the first n components of $\mathbf{y}(t)$, are given by the first n rows of (4.133), which form the sought *matrix of impulse response functions*

$$\mathbf{h}^*(t, \tau) = \begin{cases} \mathbf{0} & t \leqslant \tau \\ [\mathbf{I}, \mathbf{0}] \exp[-\mathbf{H}(t-\tau)] \begin{bmatrix} \mathbf{0} \\ \mathbf{I} \end{bmatrix} & t \geqslant \tau \end{cases} \tag{4.134}$$

In the applications, the matrix $\mathbf{h}^*(t, \tau)$ should be determined from equations (4.128) and (4.134). Again (as in the case of frequency response analysis) this involves very complicated operations on matrices.

Once the matrix $\mathbf{h}^*(t, \tau)$ has been determined, the structural response with homogeneous initial conditions, is expressed by

$$\mathbf{u}'_{\mathrm{p}}(t) = \int_0^t \mathbf{h}^*(t, \tau)\, \mathbf{w}(\tau)\, \mathrm{d}t \tag{4.135}$$

The mean value vector and the matrix of autocovariance functions of $\mathbf{u}'_{\mathrm{p}}(t)$ are given by

$$\mu_{\mathbf{u}'_{\mathrm{p}}}(t) = \int_0^t \mathbf{h}^*(t, \tau) \mu_{\mathbf{w}}(\tau)\, \mathrm{d}\tau \tag{4.136}$$

$$\Sigma_{\mathbf{u}'_{\mathrm{p}}}(t_1, t_2) = \int_0^{t_1} \int_0^{t_2} \mathbf{h}^*(t_1, \tau_1)\, \Sigma_W(\tau_1, \tau_2) \mathbf{h}^{*\mathrm{T}}(t_2, \tau_2)\, \mathrm{d}\tau_1\, \mathrm{d}\tau_2 \tag{4.137}$$

while the matrix of spectral density functions could be calculated by Fourier transformation of the autocovariance function

$$\mathbf{G}_{\mathbf{u}'_{\mathrm{p}}}(\omega_1, \omega_2) = \frac{1}{4\pi^2} \int_{-\infty}^{\infty} \int_{-\infty}^{\infty} \Sigma_{\mathbf{u}'}(t_1, t_2) \exp[-\mathrm{i}(\omega_1 t_1 - \omega_2 t_2)]\, \mathrm{d}t_1\, \mathrm{d}t_2 \tag{4.138}$$

Finally, it can be shown that, very similarly to the time-invariant simple oscillator, $\mathbf{h}(v) = \mathbf{h}^*(t, t-v)$ and $\mathbf{A}(\omega)$ are related by (cf. equation (4.111))

$$\begin{aligned} \mathbf{h}(v) &= \frac{1}{2\pi} \int_{-\infty}^{+\infty} \mathbf{A}(\omega) \mathrm{e}^{\mathrm{i}\omega v}\, \mathrm{d}\omega \\ \mathbf{A}(\omega) &= \int_{-\infty}^{+\infty} \mathbf{h}(v) \mathrm{e}^{-\mathrm{i}\omega v}\, \mathrm{d}t \end{aligned} \tag{4.139}$$

The first of equations (4.139) and equation (4.120) yield an alternative method of obtaining the matrix of impulse response functions for time-invariant multi-degree-of-freedom oscillators.

4.3.4 Modal analysis

As already outlined, great, often overwhelming, difficulties are encountered when the integral methods described in Sections 4.3.3(b) and (c) are directly applied to complicated excitations and *a fortiori* to stochastic inputs, even if the output is sought only in terms of the first two moments.

Therefore, approximate procedures play an important role in the actual solution of stochastic dynamic problems. The most important and widely used of these is *modal analysis* [8] to which this section is devoted: it leads to exact solutions for the motion of undamped structures, and to good approximations for realistic values of damping.

(a) Undamped structures

Putting $\mathbf{Z} = \mathbf{0}$, hence $\xi_k = 0$, the equation of motion (4.116) and its solution (4.116′) specialize into

$$\ddot{\mathbf{u}}' + \mathbf{Y}\mathbf{u}' = \mathbf{w}(t) \tag{4.140}$$

$$\mathbf{u}'(t) = \mathbf{u}'_{\mathrm{p}}(t) + \sum_{k=1}^{n} \mathbf{u}'_k[C_{1k}\sin(\omega_{0k}t) + C_{2k}\cos(\omega_{0k}t)] \tag{4.140′}$$

Equation (4.140′) shows that the free motion (i.e. the general solution of the homogeneous equation associated with (4.140)) is the sum of n harmonic oscillations, each with pulsation ω_{0k} and shape (*mode*) $\mathbf{u}'_k$: with respect to each *mode*, the structure behaves as a single-degree-of-freedom oscillator.

Modal analysis consists of treating each mode independently from each other, and also from the forced oscillation part $\mathbf{u}'_{\mathrm{p}}(t)$. To this aim, the solution of equation (4.140) is sought in the form

$$\mathbf{u}'(t) = \mathbf{U}'\mathbf{f}(t) \tag{4.141}$$

where $\mathbf{U}'$ is the *modal matrix* defined by‡

$$\mathbf{U}' = \{\mathbf{u}'_1, \mathbf{u}'_2, \mathbf{u}'_3, \ldots, \mathbf{u}'_n\} = \begin{bmatrix} u'_{11} & u'_{21} & u'_{31} & \cdots & u'_{n1} \\ u'_{12} & u'_{22} & u'_{32} & \cdots & u_{n2} \\ u'_{13} & u'_{23} & u'_{33} & \cdots & u'_{n3} \\ \vdots & \vdots & \vdots & & \vdots \\ u'_{1n} & u'_{2n} & u'_{3n} & \cdots & u'_{nn} \end{bmatrix} \tag{4.142}$$

‡ The eigenvectors $\mathbf{u}_k$ are defined to within an arbitrary constant. For a symmetric positive definite matrix $\mathbf{Y}$, they are *real* and *orthogonal* and can be *normalized*, i.e. chosen in such a way that

$$\mathbf{u}_h^{\mathrm{T}}\mathbf{u}_k = \begin{cases} 1 & \text{if} \quad h = k \\ 0 & \text{if} \quad h \neq k \end{cases}$$

It is easy to show that, by the very definition of $\mathbf{u}_k$,

$$\mathbf{U}'^{\mathrm{T}}\mathbf{Y}\mathbf{U}' = \mathbf{\Omega} \tag{4.143}$$

where $\mathbf{\Omega}$ is the diagonal matrix of the eigenvalues of $\mathbf{Y}$

$$\mathbf{\Omega} = \begin{bmatrix} \omega_{01}^2 & 0 & \cdots & 0 \\ 0 & \omega_{02}^2 & \cdots & 0 \\ \vdots & \vdots & & \vdots \\ 0 & 0 & \cdots & \omega_{0n}^2 \end{bmatrix} \tag{4.144}$$

Then, introducing (4.141) into (4.140), noting that $\mathbf{U}'$ (equation (4.142)) does not change in time, and pre-multiplying by $\mathbf{U}'^{\mathrm{T}}$, a set of n independent differential equations in each component of $\mathbf{f}(t)$ is obtained

$$\ddot{f}_k(t) + \omega_{0k}^2 f_k(t) = u'_{kj} w_j(t) = w'_k(t) = \mathbf{u}'^{\mathrm{T}}\mathbf{w} \tag{4.145}$$

$f_k(t)$ and $w'_k(t)$ are the *modal* (or *normal*) *components* of the response (*motion*) and the input (*load*) respectively; $f_k(t)$ are also called simply *modal* or *normal coordinates*.

(b) Damped structures

The uncoupling of the set of equations (4.140) into the n independent equations (4.145) derives from the fact that $\mathbf{\Omega}$ (equation (4.143)) is a diagonal matrix.

If the position (4.141) is introduced into the damped equation of motion (4.115″), n uncoupled equations are obtained only if the matrix

$$\mathbf{U}'^{\mathrm{T}}\mathbf{Z}\mathbf{U}' = \mathbf{\Psi} \tag{4.146}$$

is also a diagonal matrix, but this is rigorously true only under very special conditions‡. However, in most practical cases, the system is lightly damped, and the cross-terms in the matrix $\mathbf{\Psi}$ are negligible, so that the equations of motion can be uncoupled without significant errors into

$$\ddot{f}_k(t) + 2\xi_k\omega_{0k}\dot{f}_k + \omega_{0k}^2 f_k = u_{kj} w_j(t) = w'_k(t) \tag{4.147}$$

ξ_k and ω_{0k}^2 have already been defined (cf. equation (4.116′)) as the eigenvalues of $\mathbf{Z}$ and $\mathbf{Y}$, and $2\xi_k\omega_{0k}$ can be shown to be the elements of the (approximately) diagonal matrix $\mathbf{\Psi}$.

It is worth repeating that, notwithstanding the error made in uncoupling equations (4.147), this approximation is generally accepted, and is indeed included in most computer programs for dynamic analysis of multi-degree-of-freedom structures.

‡ Namely, it can be proved that $\mathbf{\Psi}$ is diagonal if, and only if, n real ζ_k ($k = 0, 1, \ldots, n-1$) exist such that [9]

$$\mathbf{Z} = \sum_k \zeta_k \mathbf{Y}^k$$

By the Cayley-Hamilton theorem this is equivalent to stating that $\mathbf{Z}$ is an analytical matrix function of $\mathbf{Y}$: in turn, this implies that $\mathbf{Z}$ and $\mathbf{Y}$ have common eigenvectors.

Note that $\mathbf{U}'$ is the matrix of the eigenvectors of the matrix $\mathbf{Y}$ defined by equation (4.143), i.e. of the mode shapes of the solutions of the *undamped* equation (4.140)

(c) General solutions in terms of modal coordinates

The modal coordinates $f_k(t)$, appearing in equation (4.147) (or (4.145)) can be obtained by applying any of the methods holding for one-degree-of-freedom oscillators. For instance, applying the impulse response function procedure (the most used in the present context), one gets (for homogeneous initial conditions)

$$f_k(t) = \int_0^t w'_k(t)\, h^*_k(t, \tau)\, \mathrm{d}\tau \tag{4.148}$$

where $h^*_k(t, \tau)$ is the impulse response function pertaining to the kth natural mode of vibration.

The matrix of impulse response functions, analogous to that introduced in Section 4.3.3(c), is in the present case the diagonal matrix

$$\mathbf{h}^*_\mathrm{d}(t, \tau) = \begin{bmatrix} h^*_1(t, \tau) & 0 & \dots & 0 \\ 0 & h^*_2(t, \tau) & \dots & 0 \\ \vdots & \vdots & & \vdots \\ 0 & 0 & \dots & h^*_n(t, \tau) \end{bmatrix} \tag{4.149}$$

and the particular solution of equation (4.115″), in vector form, can finally be written

$$\mathbf{u}'_\mathrm{p}(t) = \mathbf{U}' \int_0^t \mathbf{h}^*_\mathrm{d}(t, \tau)\, \mathbf{U}'^\mathrm{T} \mathbf{w}(\tau)\, \mathrm{d}\tau \tag{4.150}$$

Comparison of equations (4.150) and (4.135) yields the equality

$$\mathbf{h}^*(t, \tau) = \mathbf{U}' \mathbf{h}^*_\mathrm{d}(t, \tau)\, \mathbf{U}'^\mathrm{T} \tag{4.151}$$

valid under the same approximations which allowed uncoupling of the equations of motion, and much more manageable than equations (4.128) and (4.134). Equations (4.136)–(4.138) can then be applied to obtain mean value, autocovariance and spectral density of $\mathbf{u}'_\mathrm{p}$ with the same approximation.

In a perfectly analogous way, the frequency response approach can be followed. Introduce the Fourier transforms of the *modal* components of response and input

$$f_{\mathrm{F}k}(\omega) = \frac{1}{2\pi} \int_{-\infty}^{+\infty} f_k(t)\, \mathrm{e}^{-\mathrm{i}\omega t}\, \mathrm{d}t \tag{4.152}$$

$$w'_{\mathrm{F}k}(\omega) = \frac{1}{2\pi} \int_{-\infty}^{+\infty} w'_k(t)\, \mathrm{e}^{-\mathrm{i}\omega t}\, \mathrm{d}t \tag{4.153}$$

and define the matrix

$$\mathbf{A}_\mathrm{d}(\omega) = \left[\frac{I_{ji}}{\omega_{0j}^2 - \omega^2 + 2\mathrm{i}\,\xi_j \omega_{oj}^2 \omega} \right] \tag{4.154}$$

i.e. the diagonal matrix of the frequency response functions of the n normal modes of vibration. Again, equation (4.154) is much more manageable than its counterpart, equation (4.120).

Then, in analogy to equation (4.103)

$$\mathbf{f}(t) = \int_{-\infty}^{+\infty} \mathbf{A}_{\mathrm{d}}(\omega)\mathbf{w}'_{\mathrm{F}}(\omega)\mathrm{e}^{\mathrm{i}\omega t}\,\mathrm{d}\omega \tag{4.155}$$

whence $\mathbf{u}'_{\mathrm{p}}(t)$ is given again by equation (4.121), if the approximate equality

$$\mathbf{A}(\omega) = \mathbf{U}'\mathbf{A}_{\mathrm{d}}(\omega)\mathbf{U}'^{\mathrm{T}} \tag{4.156}$$

is introduced. Equations (4.123)–(4.127) can then be used, with the same approximation under which equation (4.156) holds.

(d) An important practical result of modal analysis

If the modal components of loads, $w'_k(t)$, are stochastically independent processes, and the system is lightly damped, the modal components of the motion, $f_k(t)$, are also stochastically independent. Then, consider any component of the response vector

$$u'_{\mathrm{p}i}(t) = U'_{ik} f_k(t) \tag{4.157}$$

(or any other quantity linearly related to the modal component f_k). Each term of the sum on the r.h.s. of equation (4.157) is stochastically independent of each other, so that the autocovariance function of $u_{\mathrm{p}i}$ is (equation (A5.59))

$$\Sigma_{u'_{\mathrm{p}i}}(t_1, t_2) = U'^2_{ik}\Sigma_{f_k}(t_1, t_2) \tag{4.158}$$

where on the r.h.s. Σ_{f_k} is the autocovariance function of $f_k(t)$ and the sum over the dummy index k is implied. Putting $t_1 = t_2 = t$ in equation (4.158),

$$s^2_{u_{\mathrm{p}i}}(t) = U'_{ik} s^2_{f_k}(t) \tag{4.159}$$

where s^2 indicate the relevant variances, and again summation over k is implied on the r.h.s.

Equation (4.159), although rigorously valid only under the assumptions set at the beginning of this paragraph, is much used in estimating the maximum value of a stationary response from the maxima of the component processes. In fact, maxima of stationary processes with zero mean value are roughly proportional to the variance: for example for the Rayleigh distribution, $\mathrm{E}[\tilde{x}_{\max}] \simeq 1.2533 s_x$ (equation (A7.46)).

4.3.5 A numerical example of modal analysis

In order to illustrate previous considerations, consider a simple structural problem, namely a two-storey shear frame acted on by an horizontal acceleration of the foundation, $\tilde{a}(t)$ (Fig. 4.6).

It is assumed that masses M_1, M_2 are lumped on rigid beams, while columns are made of linearly elastic material. The elastic stiffness matrix $\mathbf{K}$ is made by simply assembling floor stiffnesses k_j $(j = 1, 2)$ expressed by

$$k_j = \sum_{i=1}^{n_j} \frac{12EJ_{ij}}{h_j^3}$$

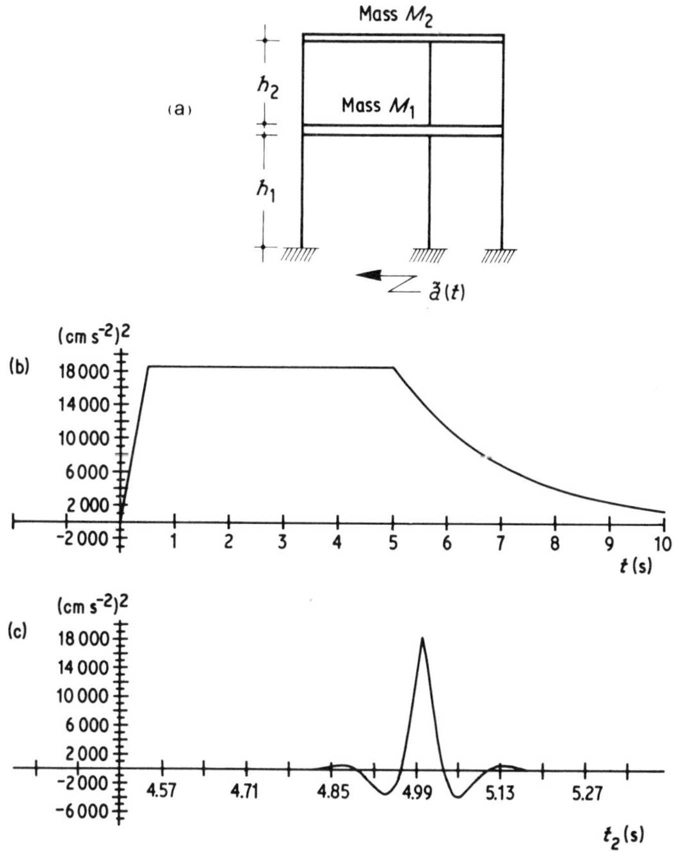

Fig. 4.6 (a) Example frame; (b) Variance of base acceleration $\tilde{a}(t)$ vs time t; (c) Covariance function $\Sigma_a(t_1\, t_2)$ of $\tilde{a}(t)$ vs t_2, for $t_1 = 5$ s.

where E is the elastic modulus of the material, J_{ij} is the cross-sectional inertia of the ith column at the jth floor, n_j is the number of columns at the jth floor, and h_j is the height between two subsequent floors. Assuming that at all floors,

$$k_j = 2 \times 10^5 \text{ kg cm}^{-1}$$

the stiffness matrix is soon formed

$$\mathbf{K} = \begin{bmatrix} k_1 + k_2 & -k_2 \\ -k_2 & k_2 \end{bmatrix} = \begin{bmatrix} 4 & -2 \\ -2 & 2 \end{bmatrix} \times 10^5 \text{ kg cm}^{-1}$$

The mass matrix $\mathbf{M}$ in the present case is a diagonal matrix. Assuming that

$$M_1 = M_2 = 1.5 \times 10^3 \text{ kg s}^2 \text{ cm}^{-1}$$

one gets

$$\mathbf{M} = \begin{bmatrix} M_1 & 0 \\ 0 & M_2 \end{bmatrix} = \begin{bmatrix} 1.5 & 0 \\ 0 & 1.5 \end{bmatrix} \times 10^3 \text{ kg s}^2 \text{ cm}^{-1}$$

To build up a consistent damping matrix **c** is a difficult task. Most times it is formed by a linear combination of the mass and stiffness matrices

$$\mathbf{c} = \alpha \mathbf{M} + \beta \mathbf{K}$$

with

$$\alpha = 2\xi \frac{\omega_{01}\omega_{02}(\omega_{01} - \omega_{02})}{\omega_{01}^2 - \omega_{02}^2}$$

$$\beta = 2\xi \frac{\omega_{01} - \omega_{02}}{\omega_{01}^2 - \omega_{02}^2}$$

ξ being the damping factor expected for the structure at hand. Note that in the above expressions dependence of damping on the velocity of deformation has been neglected and/or summarized by attributing a suitable value to ξ.

Usually the following values can be attributed to ξ for reinforced concrete structures:

(1) $\xi = 0.01$ when the structure behaves elastically;
(2) $\xi = 0.04$ when cracking of concrete takes place during the motion; and
(3) $\xi = 0.08$ when plastic hinges are activated.

In the following, the first case will be considered ($\xi = 1\%$).

Let us now calculate the standard stiffness matrix **Y** (equation (4.117)). One gets

$$\mathbf{M}^{-1/2} = \begin{bmatrix} 2.582 & 0 \\ 0 & 2.582 \end{bmatrix} \times 10^{-2} \text{ s kg}^{1/2} \text{ cm}^{-1/2}$$

and (equation (4.117))

$$\mathbf{Y} = \begin{bmatrix} 266.67 & -133.33 \\ -133.33 & 133.33 \end{bmatrix} \text{s}^{-2}$$

The eigenvalues ω_{0k}^2 of the above matrix (i.e. the square values of the undamped natural pulsations) are collected in the matrix

$$\Omega = \begin{bmatrix} 50.93 & 0 \\ 0 & 349.07 \end{bmatrix} \text{s}^{-2}$$

and the matrix of normal modes is given by

$$\mathbf{U}' = \begin{bmatrix} 0.5257 & -0.8507 \\ 0.8507 & 0.5257 \end{bmatrix} \text{s kg}^{1/2} \text{ cm}^{1/2}$$

with each mode ordered on columns of $\mathbf{U}'$.

It is now possible to compute coefficients α and β

$$\alpha = 2 \times 0.01 \times \frac{7.14 \times 18.68(7.14 - 18.68)}{50.93 - 349.07} = 0.103\,\mathrm{s}^{-1}$$

$$\beta = 2 \times 0.01 \times \frac{7.14 - 18.68}{50.93 - 349.07} = 0.774 \times 10^{-3}\,\mathrm{s}$$

whence the following damping matrix is obtained

$$\mathbf{C} = \begin{bmatrix} 464.75 & -154.92 \\ 154.92 & 309.84 \end{bmatrix} \mathrm{kg\,s\,cm}^{-1}$$

The corresponding standard matrix $\mathbf{Z}$ (equation (4.117)) is

$$\mathbf{Z} = \begin{bmatrix} 0.310 & -0.103 \\ -0.103 & 0.207 \end{bmatrix} \mathrm{s}^{-1}$$

whose eigenvalues are given by

$$2\xi_1\omega_{01} = 0.143\,\mathrm{s}^{-1} \qquad 2\xi_2\omega_{02} = 0.374\,\mathrm{s}^{-1}$$

It can be checked that the product $\mathbf{U}'^{\mathrm{T}}\mathbf{Z}\mathbf{U}'$ yields a diagonal matrix, thus allowing completely rigorous decoupling of the motion of modal coordinates.

The base acceleration $\tilde{a}(t)$ has been modelled as a stochastic process simulating the occurrence of rare earthquakes according to the procedure by Ruiz and Penzien (see Chapter 9, Section 9.2.4, neglecting the second filtering). Such base acceleration has been conceived as a quake of duration 10 s, with expected maximum acceleration equal to $0.4g$ and mean value equal to zero.

The filter simulating the action of surface ground layers is here assumed for numerical simplicity to be characterized by the following parameters

$$\omega_0 = 50\,\mathrm{s}^{-1} \qquad \xi_0 = 0.5$$

while the shaping function $\chi(t)$ is characterized by equation (9.40) with

$$t_1 = 0 \qquad t_2 = 5\,\mathrm{s} \qquad c = 0.5\,\mathrm{s}^{-1}$$

The variance of such acceleration is given in Fig. 4.6(b) while a typical example of covariance function $\Sigma_a(t_1, t_2)$ is given in Fig. 4.6(c).

The loading process $\tilde{\mathbf{W}}(t)$ is expressed by

$$\tilde{\mathbf{W}}(t) = -\mathbf{M}\mathbf{1}\tilde{a}(t)$$

where $\mathbf{1}$ denotes the column vector with all components equal to 1. The standardized load process $\tilde{\mathbf{w}}(t)$ is then given by

$$\tilde{\mathbf{w}}(t) = -\mathbf{M}^{1/2}\mathbf{1}\tilde{a}(t)$$

whence the covariance matrix function of the standard loading

$$\Sigma_{\mathbf{w}}(t_1, t_2) = \mathbf{M}^{1/2}\begin{bmatrix} 1 & 1 \\ 1 & 1 \end{bmatrix}\mathbf{M}^{1/2}\Sigma_a(t_1, t_2) = \mathbf{M}_0\Sigma_a(t_1, t_2)$$

Since the input acceleration has zero mean value, by the linearity of the structure all output quantities also have zero mean value. It is mainly interesting then to compute second moments of output functions. Most important response parameters are the floor displacements and shear forces. The covariance matrix functions of both shear and displacements can be put in the form

$$\Sigma_{\zeta}(t_1, t_2) = \int_0^{t_1} \int_0^{t_2} \Lambda_{\zeta}\, \mathbf{h}_{\mathrm{d}}^{*}(t_1, \tau_1) \mathbf{B} \mathbf{h}_{\mathrm{d}}^{*\mathrm{T}}(t_2, \tau_2) \mathbf{C}_{\zeta} \Sigma_{a}(\tau_1, \tau_2)\, \mathrm{d}\tau_1\, \mathrm{d}\tau_2$$

valid for $\zeta = u$ (displacements) and $\zeta = S$ (shear forces).

The matrix $\mathbf{B}$ is given by

$$\mathbf{B} = \mathbf{U}'^{\mathrm{T}} \mathbf{M}_0 \mathbf{U}'$$

while the matrices Λ_{ζ}, $\mathbf{C}_{\xi}$ depend on the ζ index as follows:

$$\Lambda_u = \mathbf{M}^{-1/2} \mathbf{U}', \qquad \mathbf{C}_u = \mathbf{U}'^{\mathrm{T}} \mathbf{M}^{-1/2}$$

$$\Lambda_S = \Xi \mathbf{K} \mathbf{M}^{-1/2} \mathbf{U}' \qquad \mathbf{B}_S = \mathbf{U}'^{\mathrm{T}} \mathbf{M}^{-1/2} \mathbf{K} \Xi^{\mathrm{T}}$$

Ξ being a matrix whose generic element Ξ_{ij} is given by

$$\Xi_{ij} = \begin{cases} 1 & j \geqslant i \\ 0 & j < i \end{cases}$$

Example plots of the variance and covariance functions of the shear forces $\Sigma_S(t, \tau)$ are given in Fig. 4.7(a), (b) and (c).

4.4 ELASTIC COLUMNS

As is well known from the theory of structures, there exist many types of structures in which the variations of geometry connected with deformations under load cause the response to be non-linear, even if the material is linear elastic.

One such class of structures are those prone to buckling, of which the compressed column is the simplest, and at the same time the most important, example. It is therefore worth while to take a brief look at this problem in a probabilistic context.

Let us recall first that there are two alternative ways of studying a column: the first is to search the points of bifurcation of equilibrium (which is an eigenvalue problem); the second is to consider the unavoidable imperfections present in the column‡ and to obtain the load corresponding to some specific limit condition. These two approaches will be examined in the following sections.

‡ Indeed, from the probabilistic point of view, to find a perfectly straight column loaded by exactly centred forces is an event realized with zero probability.

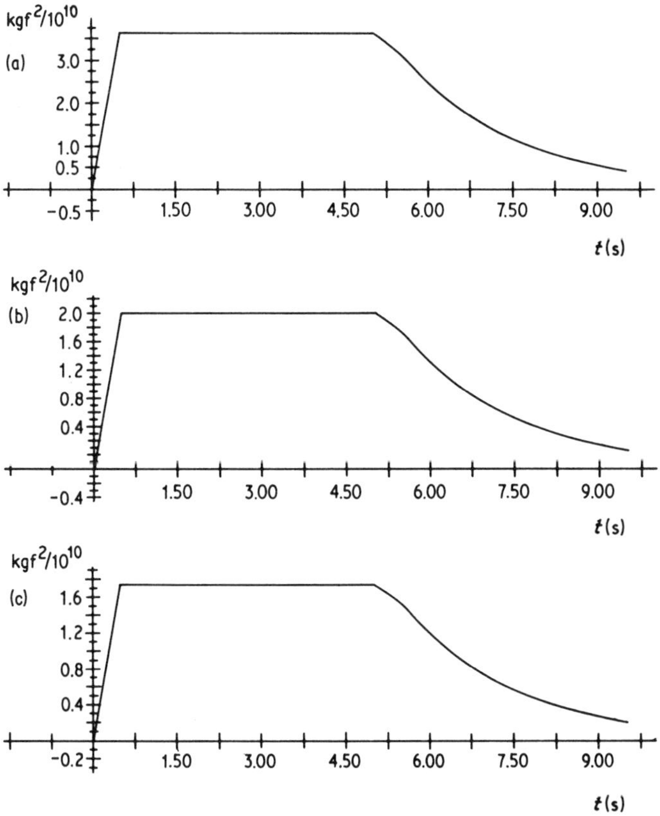

Fig. 4.7 Results of example of Fig. 4.6 (a) Variance of the total shear force at the first floor vs time; (b) covariance of simultaneous total shear forces at first and second floors vs time; (c) variance of the total shear force at the second floor.

4.4.1 The column problem as a random eingenvalue problem

Under the hypotheses of the classical theory (small strains, initially straight column, perfectly centred loads, homogeneous elastic material), the mathematical model that describes the column problem is the differential equation:

$$\frac{\partial^2}{\partial x^2} EJ(x) \frac{\partial^2 u(x)}{\partial x^2} + W \frac{\partial^2 u(x)}{\partial x^2} = 0 \tag{4.160}$$

with the initial conditions

$$\begin{aligned} EJ(0)\frac{\partial^2 u(0)}{\partial x^2} - \zeta_1 \frac{\partial u(0)}{\partial x} &= 0 \qquad & u(0) = 0 \\ EJ(l)\frac{\partial^2 u(l)}{\partial x^2} + \zeta_2 \frac{\partial u(l)}{\partial x} &= 0 \qquad & u(l) = 0 \end{aligned} \tag{4.161}$$

In equations (4.160) and (4.161), $u(x)$ denotes the lateral displacement at distance x from an end, l is the length of the column, EJ is the flexural stiffness and W is the applied axial compressive load. ζ_1 and ζ_2 are the stiffnesses of the end springs ($\zeta_1 = \zeta_2 = 0$ for simple supports, and $\zeta_1 = \zeta_2 = \infty$ for fixed supports) (Fig. 4.8).

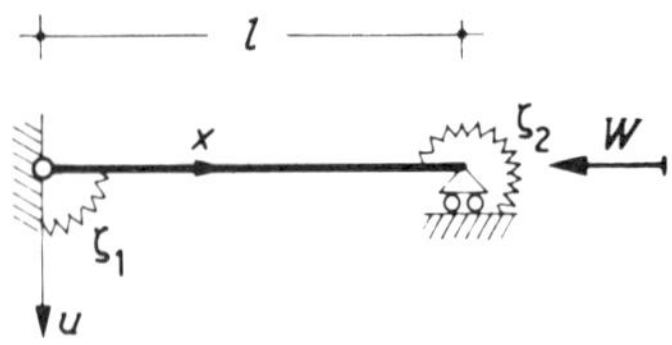

Fig. 4.8 'Perfect' column.

By introducing the usual substitutions

$$x = zl \qquad 0 \leqslant z \leqslant 1$$
$$EJ(z) = E_0 J_0 [1 + D(z)]$$
$$\zeta_1 = \zeta_{10}(1 + c_{\zeta_1})$$
$$\zeta_2 = \zeta_{20}(1 + c_{\zeta_2})$$
$$\chi = \Lambda^2 = \frac{Wl^2}{E_0 J_0}$$

Equations (4.160) and (4.161) reduce to

$$[1 + D(z)]u''(z)\}'' + \chi u''(z) = 0 \tag{4.160'}$$

$$\begin{aligned} &[1 + D(0)]u''(0) - (\zeta_{10} l/E_0 J_0)(1 + c_{\zeta_1})u'(0) = 0 \qquad u(0) = 0 \\ &[1 + D(l)]u''(l) + (\zeta_{20} l/E_0 J_0)(1 + c_{\zeta_2})u'(l) = 0 \qquad u(l) = 0 \end{aligned} \tag{4.161'}$$

where a prime denotes differentiation with respect to z. In the classic buckling case χ is unknown and equations (4.160′) and (4.161′) describe an eigenvalue problem. It has non-trivial solutions ($u(z) = u_j(z)$, where u_j is the eigenfunction relevant to the jth buckling mode shape) corresponding to the eigenvalues $\chi = \chi_j$, $j = 1,2, \ldots,$ and the jth buckling load is given by

$$W_j = \chi_j E_0 J_0 / l^2 \tag{4.162}$$

Now let the spring support stiffnesses ζ_1 and ζ_2 be random variables, while the flexural stiffness EJ is assumed to be a space-stationary (homogeneous) random function. All random functions and random variables are considered to be stochastically independent. The determination of the buckling loads W_j becomes a *random eigenvalue problem.*

For use in equations (4.160′) and (4.161′), uncertainties in beam–column properties can be restated as follows: c_{ζ_1} and c_{ζ_2} are random variables with zero mean; $D(z)$ is a homogeneous random function with zero mean. These three quantities can be considered to be small perturbations about an unperturbed case ($c_{\zeta_1} = c_{\zeta_2} = D = 0$), when their magnitudes are small compared with unity. Then, the result of the standard perturbation analysis‡ for χ_j is

$$\chi_j = \chi_j^0 + \frac{1}{C_j}\left(\int_0^1 D(z)L_j(z)\,\mathrm{d}z + \frac{\zeta_{20}l}{E_0J_0}K_j(0)\,c_{\zeta_1} + \frac{\zeta_{20}l}{E_0J_0}K_l(l)\,c_{\zeta_2}\right) \tag{4.163}$$

where

$$L_j(z) = [u_j^{0\prime\prime}(z)]^2 \qquad K_j(z) = [u_j^{0\prime}(z)]^2$$
$$C_j = \int_0^1 [u_j^{0\prime}(z)]^2\,\mathrm{d}z \tag{4.164}$$

and χ_j^0 and $u_j^0(z)$ refer to the unperturbed case.

If (4.163) is employed under random uncertainty, the small perturbation assumption is reduced to requiring the variances of the perturbations to be small. In this case, taking the expected value of (4.163) (equations (A5.16) and (A3.26′))

$$\mathrm{E}[\chi_j] = \chi_j^0 \tag{4.165}$$

i.e. the mean value of the jth eigenvalue is independent of any random perturbations and equal to the eigenvalue for the unperturbed case. With regard to the variance, the following expression is obtained (equations (A5.18) and (A3.27′))

$$\mathrm{Var}[\chi_j] = \frac{1}{C_j^2}\left[\int_0^1\int_0^1 \mathrm{E}[D(z_1)D(z_2)]L_j(z_1)L_j(z_2)\,\mathrm{d}z_1\,\mathrm{d}z_2 + \left(\frac{\zeta_{10}l}{E_0J_0}K_j(0)\right)^2 s_{c_{\zeta_1}}^2 + \left(\frac{\zeta_{20}l}{E_0J_0}K_j(l)\right)^2 s_{c_{\zeta_2}}^2\right] \tag{4.166}$$

$\mathrm{E}[D(z_1)D(z_2)]$ in equation (4.166) can be derived from equations (A3.13) and (A3. 14), taking account of the assumption of stochastic independence of the

‡ Consider the linear differential equation

$$\ell(x) + \chi x = 0$$

where $\ell(\;)$ is a differential operator, with prescribed boundary conditions and known eigenvalues χ_j and orthonormal eigenfunctions x_j. The eigenvalues and eigenfunctions of an eigenvalue problem corresponding to a *perturbed* differential equation

$$\ell(y) - \varepsilon Z y + \chi y = 0$$

where Z denotes a given function which is continuous in the fundamental domain and ε a perturbation parameter, can be calculated by an approximate method that is known as the *perturbation analysis* [10]. It is founded on the statement that the new eigenvalues $\bar{\chi}_j$ as well as the new eigenfunctions y_{ij} may be expanded in powers of the perturbation parameter ε, e.g.

$$\bar{\chi}_j = \chi_j + \varepsilon\chi_{j1} + \varepsilon^2\chi_{j2} + \cdots$$

Equation (4.163) only contains the first approximation for which

$$\chi_{j1} = \int Z x_j^2\,\mathrm{d}z$$

relevant random functions,

$$E[D(z_1)D(z_2)] = B'_{c_E}(z_1 - z_2) + B'_{c_J}(z_1 - z_2) + B'_{c_E}(z_1 - z_2)B'_{c_J}(z_1 - z_2) \quad (4.167)$$

In order to analyse the accuracy of the results of equations (4.165) and (4.166), a simulation method (Chapter 9) can be adopted. It consists of generating on a digital computer a random sample of columns and computing the eigenvalue of each column in the sample. The corresponding sample mean and variance (equations (A3.63) and (A3.66)) approximate the exact solution, the approximation improving with increasing sample size.

The comparison of the results of the perturbation and simulation approaches to the estimation of the variance on the first four eigenvalues, showed [11] that the perturbation method yields results of sufficient accuracy provided the coefficient of variation of the spring stiffnesses is kept within reasonable limits. Further, if the spring stiffnesses are deterministic, it provides a correct solution for non-uniform beam columns over a much wider range of statistical variation of E and J than would be found in reality.

Other examples of *random eigenvalues* are the natural pulsations of systems with random uncertainties (excluded in Section 4.3). The treatment of this case is very similar to that of buckling: in particular, both simulation and perturbation approaches can be employed.

4.4.2 The 'imperfect column' problem

There are many alternative, but essentially equivalent, ways of introducing unavoidable imperfections in the column problem. One way is to consider that the load W is applied with random eccentricities $\tilde{e}_1$, $\tilde{e}_2$ at ends $x = 0$ and $x = l$ respectively. Referring, for simplicity's sake, to the case of a supported beam with constant EJ, the boundary conditions of equation (4.160) become (Fig. 4.9 (a))

$$\begin{aligned} \tilde{u}(0) &= \tilde{u}(l) = 0 \\ \frac{d^2}{dx^2}\tilde{u}(l) &= -\frac{W\tilde{e}_2}{EJ} \\ \frac{d^2}{dx^2}\tilde{u}(0) &= -\frac{W\tilde{e}_1}{EJ} \end{aligned} \quad (4.168)$$

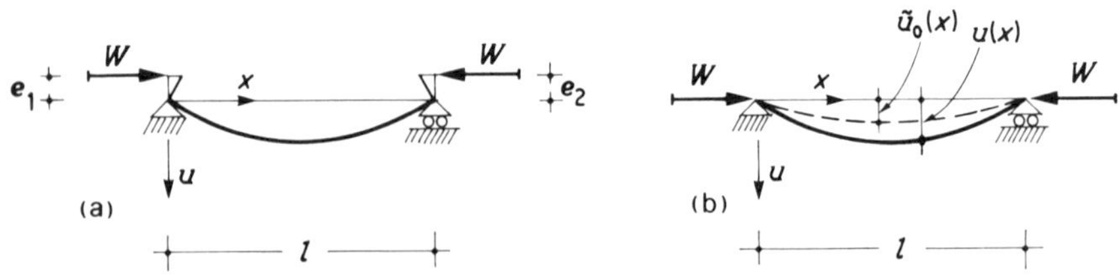

Fig. 4.9 Eccentrically compressed and imperfect columns.

The solution of equation (4.160) with the random boundary conditions (4.168) can be written in the form

$$\tilde{u}(x) = \tilde{e}_1\left(\frac{\sin[\Lambda(l-x)/l]}{\sin\Lambda} - \frac{l-x}{l}\right) + \tilde{e}_2\left(\frac{\sin[\Lambda x/l]}{\sin\Lambda} - \frac{x}{l}\right) \tag{4.169}$$

where $\Lambda = \sqrt{(Wl2/EJ)}$ The PDF of $\tilde{u}(x)$ can be determined from the JPDF of the eccentricities e_1 and e_2 by suitable transformation of variables [5] (Section A3.7).

If, on the contrary, the load is assumed to be perfectly centred at both ends, but small random variations of the column centreline are possible (Fig. 4.9 (b)), equation (4.160) takes the form (remember that EJ is now assumed constant)

$$\frac{d^4\tilde{u}}{dx^4} + \frac{W}{EJ}\frac{d^2\tilde{u}}{dx^2} = -\frac{W}{EJ}\frac{d^2\tilde{u}_0}{dx^2} \tag{4.170}$$

where $\tilde{u}_0(x)$ denotes the initial out-of-straightness, while the boundary conditions (4.168) reduce to

$$\begin{aligned} \tilde{u}(0) = \tilde{u}(l) = 0 \\ \frac{d^2}{dx^2}\tilde{u}(0) = \frac{d^2}{dx^2}\tilde{u}(l) = 0 \end{aligned} \tag{4.168'}$$

Equation (4.170) can be integrated twice, leading to the equation

$$\frac{d^2\tilde{u}}{dx^2} + \frac{W}{EJ}\tilde{u} = -\frac{W}{EJ}\tilde{u}_0 \tag{4.171}$$

which is an equation of the same type that appeared in Section 4.3.1 as the governing equation of the undamped linear oscillator. Following the same approaches, the first two moments of the stochastic process $\tilde{u}(x)$ can be derived from those of the random function $\tilde{u}_0(x)$.

However, the strength of the element rather than the lateral displacement $u(x)$ of the centreline is the most meaningful parameter of the structural response. An investigation of this question, which for small slenderness values cannot ignore the inelastic behaviour of the column material, will be presented in Chapter 6.

REFERENCES

1. Martin, H. C. (1966) *Introduction to Matrix Methods of Structural Analysis*, McGraw-Hill, New York.
2. Veneziano, D. (1978) *Random Processes for Engineering Applications* (Lecture Notes), Dept. Civil Engrg., M. I. T.
3. Bolotin, V. V. (1969) *Statistical Methods in Structural Mechanics*, Holden Day, San Francisco.
4. Esteva, L. (1975) Second-moment probabilistic analysis of statically loaded non-linear structures. *Proc. 2nd Int. Conf. on Application of Statistics and Probability in Soil and Structural Engineering* (ICASP 2), Aachen, pp. 115–30.

5. Srinivasan, S. K. and Vasudevan, R. (1971) *Introduction to Random Differential Equations and Their Applications*, Elsevier, Amsterdam.
6. Lin, Y. K. (1976) *Probabilistic Theory of Structural Dynamics*, Krieger, Huntington, New York (Reprint).
7. Robson, J. D. (1963) *An Introduction to Random Vibrations*, Edinburgh University Press.
8. Zienkiewicz, O. C. (1971) *The Finite Element Method in Engineering Science*, McGraw-Hill, New York.
9. Caughey, T. K. and O'Kelly, M. E. J. (1965) *J. Appl. Mech. ASME*, **32**, 583.
10. Courant, J. and Hilbert, D. (1961) *Methods of Mathematical Physics*, Interscience, New York.
11. Shinozuka, M. and Astill, C. J. (1972) Random eigenvalue problems in structural analysis, *AIAA J*, **16**, 456–62.

CHAPTER 5

Plastic structures

The theory of plastic limit analysis under both static (monotonic) and variable repeated loads is extended in this chapter to structures with random strengths.

5.1 STRUCTURAL ANALYSIS BEYOND THE ELASTIC RANGE

The limitations of elastic structural theory with respect to the assessment of actual resistance, hence of safety, have long been recognized. Indeed, to accept as valid the elastic behaviour up to the critical point of the structure is equivalent to identifying the crisis with the first attainment of a critical stress condition in any point of the structure, i.e. to assuming a *weakest-link* type of behaviour (cf. Section 1.5). This is in practice true in some cases, such as statically determinate structures or structures made of *brittle* material, but in most structures a stress *redistribution* allows the loads to grow further beyond the level corresponding to the first critical condition.

The *ultimate* load condition, corresponding to complete *collapse* of the structure, can be obtained, under a number of assumptions, by means of *plastic limit analysis*, whose bases were set up some 30–40 years ago, and which has since developed into a complete, logically consistent and experimentally confirmed theory (see e.g. Baker *et al.* [1], Neal [2] and Hodge [3]). Plastic limit analysis, in its original form, assumes an ideally ductile behaviour of the structural material, neglects the effects of changes of geometry on equilibrium, and tackles directly the search for the *limit* (i.e. *collapse*) *condition*, avoiding step-by-step calculations of the structural response during stress redistribution phases.

The advantages of such an approach from the point of view of simplicity are evident, and all the more so in the *probabilistic* context of this book. Indeed, stress redistribution is beneficial for structural safety, because it also reduces the effects of random local weaknesses: for instance, it is easy to show that if the yield forces N_y in the n elements of the bundle of Fig. 1.4(a) are stochastically independent, with common mean μ_N and coefficient of variation $\delta_N = s_N/\mu_N$, the average and the coefficient of variation of the limit load W_p are, respectively (equations (A3.9) and (A3.15))

$$\mu_W = n\mu_N \qquad \delta_W = \frac{\delta_N}{\sqrt{n}} \tag{5.1}$$

i.e. the effect of the random strength variations tend to vanish with increasing n.

However, to find the stress distribution at an intermediate stage in a statically redundant structure with random material and/or geometrical properties would be extremely complicated, because of the many added uncertainties and computational difficulties in the inelastic response range.

Therefore, the present Chapter 5 is devoted to the extension of *plastic limit analysis* to random-strength structures. The basic hypotheses of plastic limit analysis will be assumed to hold, but the limit strength of the structural material will be defined by random quantities. It is usual in limit analysis to find a *limit load condition* to be compared with the applied loads, i.e. to formulate the *critical inequality* in load space (Section 2.8.1); therefore, *probabilistic limit analysis*, for statically applied loads, consists of the determination of the *conditional probability of failure* (CPF) under given loads $P_R(\mathbf{W})$ to be introduced in equation (2.68) to find the probability of collapse P_{fail}; the relevant PDF of the applied loads $p_W(\mathbf{W})$ will be assumed, in this chapter, to be given.

In the second part of the chapter the case of variable repeated loads will be examined as an extension of the known problem of *incremental collapse* in deterministic limit analysis. In the present context, an appropriate probabilistic definition of the acting loads must be formulated.

Finally, the problem of accumulation of plastic strains will be formulated briefly, which leads to a fading out of the distinction between incremental collapse and shakedown, and to speak of a (non-stationary) stochastic process of *plastic adaptation*, for which appropriate bounds can be found.

In actual structures, a number of *limit states* can be defined, which go from *serviceability limit states* (which, under some provisions, could be studied by elastic theory) to *ultimate limit states*: alternative limit states will be examined in Chapter 6, while this chapter will be confined to that particular ultimate state (*plastic collapse*) to which limit analysis refers.

According to the already stated general rule of this book, the structural systems are considered as *discrete*. In particular, from the viewpoint of limit analysis, this means that yielding can only take place in well defined locations related to the finite elements into which the structure is subdivided, or, for structures made up by one-dimensional elements, in a finite number of cross-sections.

5.2 PLASTIC LIMIT ANALYSIS WITH RANDOM STRENGTHS (PROBABILISTIC LIMIT ANALYSIS)

Consider a structure made of ductile material and loaded by actions represented by a vector $\mathbf{W}$. According to *limit analysis* theory, *collapse* is identified with the development of *plastic hinges* (i.e. zones of material with stresses in the yield limit condition) in such a number and location to allow a movement of the whole or of a part of the structure without requiring deformation of the zones with stresses below the yield limit, i.e. such to transform the structure into a *plastic collapse mechanism.*

The load conditions at the limit of collapse $\mathbf{W}_p$ are determined on the basis of two theorems that will now be briefly recalled, and later extended to random-strength structures.

5.2.1 General formulation and theorems of limit analysis

Let $\mathbf{W}$ be the applied load vector and $\boldsymbol{\sigma}$ the vector of (generalized) stresses in the relevant *points* (sections, nodes or elements) of the structure (cf. Section 4.2). The condition of equilibrium between $\boldsymbol{\sigma}$ and $\mathbf{W}$ is (equation (4.4))

$$\mathbf{W} = \mathbf{Q}^{\mathrm{T}}\boldsymbol{\sigma} \tag{5.1'}$$

The condition of *admissibility* of the stress vector is usually expressed in the form

$$\boldsymbol{\phi}(\boldsymbol{\sigma}) = \mathbf{H}\boldsymbol{\sigma} - \boldsymbol{\chi} \leqslant \mathbf{0} \tag{5.2}$$

with $\mathbf{H}$ and $\boldsymbol{\chi}$ an appropriate *matrix* and an appropriate *vector* respectively. Inequality (5.2) corresponds to a piecewise linear *yield condition* in each structural *point.*

The condition of *compatibility* between the vectors of the (generalized) strains and displacements, $\boldsymbol{\varepsilon}$ and $\mathbf{u}$ respectively, is (equation (4.2))

$$\boldsymbol{\varepsilon} = \mathbf{Q}\mathbf{u} \tag{5.3}$$

In limit analysis, elastic strains are assumed to be zero, and the vector $\boldsymbol{\varepsilon}$ represents *plastic strains.*

Plastic strains are only possible where one component of (5.2) is satisfied with the equality sign; moreover, they must satisfy the *normality rule*, i.e.

$$\boldsymbol{\varepsilon} = \mathbf{H}^{\mathrm{T}}\boldsymbol{\lambda}_p \tag{5.4}$$

where $\boldsymbol{\lambda}_p$ is a vector of positive components called *plastic multipliers.*

The active hyperplanes of the yield domain (5.2) are selected by the condition

$$\boldsymbol{\phi}^{\mathrm{T}} \boldsymbol{\lambda}_{\mathrm{p}} = 0 \tag{5.4'}$$

The *kinematic theorem* of limit analysis proves that collapse occurs if a vector $\mathbf{u}$ (*collapse mechanism*) exists such that the work done by the applied loads $\mathbf{W}$

$$L_{\mathrm{w}} = \mathbf{W}^{\mathrm{T}} \mathbf{u} \tag{5.5}$$

is larger than the corresponding internal plastic work

$$L_{\mathrm{d}} = \boldsymbol{\sigma}^{\mathrm{T}} \boldsymbol{\varepsilon} \tag{5.6}$$

Combining this condition with the previous equations (5.1)–(5.4), it can be stated that collapse occurs if, and only if, at least one vector $\mathbf{u}$ exists such that

$$\mathbf{W}^{\mathrm{T}} \mathbf{u} > \boldsymbol{\chi}^{\mathrm{T}} \boldsymbol{\lambda}_{\mathrm{p}} \tag{5.7}$$

with

$$\mathbf{Q}\mathbf{u} = \mathbf{H}^{\mathrm{T}} \boldsymbol{\lambda}_{\mathrm{p}} \qquad \boldsymbol{\lambda}_{\mathrm{p}} \geqslant 0 \tag{5.7'}$$

A mechanism that satisfies inequality (5.7) is called a *kinematically sufficient mechanism*; a load condition $\mathbf{W}_{\mathrm{K}}$ for which at least one kinematically sufficient mechanism exists is defined as a *kinematically sufficient load condition.*

Conversely, it can be proved (*static theorem*) that collapse does *not* occur if there exists a *statically admissible stress field*, i.e. a stress vector in equilibrium with the applied loads and not violating the strength inequality at any point of the structure (in other words, satisfying both (5.1′) and (5.2)).

A load condition $\mathbf{W}_{\mathrm{S}}$ for which at least one statically admissible stress field exists is called a *statically admissible load condition.*

The purpose of limit analysis is to identify load conditions $\mathbf{W}_{\mathrm{R}}$ which are at the same time *kinematically sufficient* and *statically admissible*, i.e. that correspond to the threshold of collapse‡. Use of the two theorems yields approximate procedures that allow one to bound $\mathbf{W}_{\mathrm{R}}$ from above and from below.

If, as is often done in limit analysis, the load condition is given to within a (positive) *load factor* α, i.e. in the form

$$\mathbf{W} = \mathbf{W}_0 + \alpha \mathbf{W}_{\mathrm{a}} \tag{5.8}$$

with $\mathbf{W}_0$ and $\mathbf{W}_{\mathrm{a}}$ fully defined, it is possible to define *kinematically sufficient* load factors α_{K}, *statically admissible* load factors α_{S}, and the *collapse* load factor α_{p} (corresponding to kinematically sufficient load conditions $\mathbf{W}_{\mathrm{K}} = \mathbf{W}_0 + \alpha_{\mathrm{K}} \mathbf{W}_{\mathrm{a}}$, statically admissible load conditions $\mathbf{W}_{\mathrm{S}} = \mathbf{W}_0 + \alpha_{\mathrm{S}} \mathbf{W}_{\mathrm{a}}$, and the load condition at the threshold of collapse $\mathbf{W}_{\mathrm{p}} = \mathbf{W}_0 + \alpha_{\mathrm{p}} \mathbf{W}_{\mathrm{a}}$, respectively), and to prove that α_{p} is unique and all α_{K} and α_{S} satisfy the inequalities

$$\alpha_{\mathrm{K}} \geqslant \alpha_{\mathrm{p}} \geqslant \alpha_{\mathrm{S}} \tag{5.9}$$

‡ Remember that *plastic collapse* is the only limit state considered in plastic limit analysis, hence in this chapter $\mathbf{W}_{\mathrm{p}}$ and $\mathbf{W}_{\mathrm{R}}$ coincide.

5.2.2 The probability of plastic collapse under given loads

If the vector χ in the strength inequalities (5.2) is a vector of random components, the probability $P_R(\mathbf{W})$ of collapse under given loads $\mathbf{W}$ (CPF) coincides with the probability that $\mathbf{W}$ is *kinematically sufficient*, and with the probability that $\mathbf{W}$ is *not* statically admissible (equal to one minus the probability that $\mathbf{W}$ is statically admissible)‡: i.e. formally

$$P_R(\mathbf{W}) = \text{Prob}[\exists \mathbf{u} \in \mathscr{V} : \mathbf{W}^T\mathbf{u} > \tilde{\chi}^T \boldsymbol{\lambda}_p] \tag{5.10}$$

$$P_R(\mathbf{W}) = 1 - \text{Prob}[\exists \boldsymbol{\sigma} \in \mathscr{T} : \mathbf{H}\boldsymbol{\sigma} \leqslant \tilde{\chi}] \tag{5.11}$$

In equations (5.10) and (5.11), $\mathscr{V}$ and $\mathscr{T}$ are respectively the set of all possible *mechanisms* (vectors of compatible displacements) and *stress vectors* in equilibrium with $\mathbf{W}$ (i.e. satisfying (5.7′) and (5.1) respectively).

For a discrete structure, $\mathscr{V}$ is usually a set of discrete points, containing a finite number of mechanisms. Let $\mathbf{u}_i$ be one of these mechanisms, and $\boldsymbol{\lambda}_{pi}$ the corresponding vector of plastic multipliers.

The associated plastic dissipated work

$$\tilde{L}_{di} = \tilde{\chi}^T \boldsymbol{\lambda}_{pi} \tag{5.12}$$

is the linear combination of the components of a random vector $\tilde{\chi}^T$, hence a random variable, whose PDF can be calculated with comparative ease.

Let now E_i be the event '$\mathbf{u}_i$ is kinematically sufficient', i.e.

$$[E_i] = [\mathbf{W}^T\mathbf{u}_i > \tilde{L}_{di}] \tag{5.13}$$

Let also P_{Ki} be the probability that event E_i occurs (equation (1.7))

$$P_{Ki} = P_{fi} = \text{Prob}[E_i] \tag{5.14}$$

According to the kinematic theorem of limit analysis, the considered structure *collapses* if *at least one* of the events (5.13) occurs: in other words, the mechanisms $\mathbf{u}_i$ can be considered as alternative modes of failure and the formulae of Section 1.4 can be applied. In particular (equations (1.5)–(1.8))

$$\begin{aligned} P_R(\mathbf{W}) &= \text{Prob}[E_1 \cup E_2 \cup E_3 \cup \ldots] \\ &= P_{f1} + P_{f2} + P_{f3} + \ldots - \text{Prob}[E_1 \cap E_2] - \text{Prob}[E_1 \cap E_3] - \ldots \\ &\qquad + \text{Prob}[E_1 \cap E_2 \cap E_3] + \ldots \end{aligned} \tag{5.15}$$

and, if all events E_i are stochastically independent

$$\begin{aligned} P_R(\mathbf{W}) = {} & P_{f1} + P_{f2} + P_{f3} + \ldots - P_{f1}P_{f2} - P_{f1}P_{f3} - \ldots \\ & + P_{f1}P_{f2}P_{f3} + P_{f1}P_{f2}P_{f4} + \ldots \end{aligned} \tag{5.16}$$

However, the possible mechanisms usually share one or more plastic hinges, and the events E_i are correlated: therefore, the calculation of the cross-terms of

‡ If χ is a *deterministic* vector, these probabilities are indeterminate if $\mathbf{W} = \mathbf{W}_R$, and otherwise either 0 or 1.

equation (5.15) becomes rather cumbersome, and many procedures have been developed to calculate $P_R(\mathbf{W})$ approximately, and in particular to bound it on both sides. The most simple such bounds are given by (equations (1.6)–(1.8))

$$\max_i P_{fi} \leqslant P_R(\mathbf{W}) \leqslant \sum_i P_{fi} \tag{5.17}$$

where the sum must be extended over all elements of $\mathscr{V}$.

Remember (Section 1.4) that, if

$$P_R(\mathbf{W}) \ll 1 \qquad \text{and} \qquad P_{fi} \ll 1, \qquad \text{all } i \tag{5.18}$$

i.e. in the range of the small probabilities acceptable in structural engineering, the upper bound in inequality (5.17) is a very good approximation of $P_R(\mathbf{W})$.

It has thus been shown that the *kinematic* theorem of limit analysis yields directly a procedure for evaluating $P_R(\mathbf{W})$.

Consider now the definition of $P_R(\mathbf{W})$ given by equation (5.11), corresponding to the *static* theorem of limit analysis. In this case the definition set $\mathscr{T}$ is *not* a discrete set, and therefore bounds like those found via the kinematic approach cannot be established directly, because the individual probabilities are also large, and their sum would soon exceed unity.

However, both static and kinematic approaches can be used quite effectively and rigorously in probabilistic limit analysis owing to the two theorems that will be proved in the following section.

5.2.3 Bounding theorems of probabilistic limit analysis

It is to be underlined that the r.h.s. inequality in (5.17) is proved only if the sum is extended over all possible mechanisms. Therefore, if not all these mechanisms are known, or if their number is very large, inequality (5.17) becomes inapplicable. This is not a speculative possibility, especially in the range of very small probabilities (i.e. just when the r.h.s. of (5.17) appears most approximate) as will be discussed at the end of Section 5.2.7(b). However, the simple example of Fig. 5.1 is sufficient to show that, if the limit moment $M_y(x)$ of a beam is a random process along the beam axis x, many alternative locations are possible for the plastic hinges (i.e. many alternative mechanisms are significant), unless the bending moment diagram $M(x)$ presents sharp and well localized vertices. In consequence of this, to calculate in full rigour the upper bound in (5.17) may become lengthy and cumbersome (and even impossible for continuous structures).

Note moreover that the bound is less approximate, the more the failure modes are correlated, as they certainly become when close alternate locations of plastic hinges are considered.

Conversely, a *lower* bound to $P_R(\mathbf{W})$ can be obtained if a limited number n of mechanisms (i.e. a finite subset $\mathscr{V}_n \subset \mathscr{V}$) is considered. Let E be the event *collapse of the structure under the given loads* $\mathbf{W}$, so that

$$P_R(\mathbf{W}) = \text{Prob}[E] \tag{5.19}$$

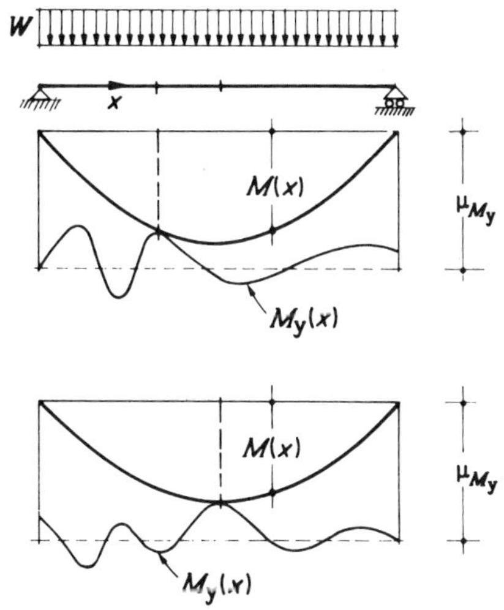

Fig. 5.1 Simple supported beam under uniform load: two possible samples of stochastic limit moment and corresponding locations of plastic hinge.

and E_K the event 'one of the mechanisms of the set $\mathscr{V}_n$ is kinematically sufficient', i.e.

$$[E_K] = [\exists \mathbf{u} \in \mathscr{V}_n\colon\ \mathbf{W}^T\mathbf{u} \geqslant \tilde{\chi}^T \lambda_p] \tag{5.20}$$

If the event E_K is realized, the load condition $\mathbf{W}$ is *kinematically sufficient*, and therefore, by the first theorem of limit analysis, collapse occurs; in other words, E_K *implies* E‡

$$[E_K] \Rightarrow [E] \tag{5.21}$$

Consider now m stress vectors $\boldsymbol{\sigma}_j$, forming a *discrete* set $\mathscr{T}_m \subset \mathscr{T}$, and denote by E_S the event that *none* among the stress vectors $\boldsymbol{\sigma}_j \in \mathscr{T}_m$ in equilibrium with $\mathbf{W}$ is statically admissible. By the second theorem of limit analysis, if collapse occurs. no statically admissible stress field exists; thence

$$[E] \Rightarrow [E_S] \tag{5.22}$$

In conclusion [4], from implications (5.21) and (5.22), taking account of equation (A1.9′)

$$P_K(\mathbf{W}) \leqslant P_R(\mathbf{W}) \leqslant P_S(\mathbf{W}) \tag{5.23}$$

where $P_R(\mathbf{W})$ is defined by equation (5.19), and

$$\begin{aligned} P_K(\mathbf{W}) &= \mathrm{Prob}\,[E_K] = \mathrm{Prob}\,[\exists \mathbf{u} \in \mathscr{V}_n\colon\ \mathbf{W}^T\mathbf{u} > \tilde{\chi}^T\lambda_p] \\ P_S(\mathbf{W}) &= \mathrm{Prob}\,[E_S] = 1 - \mathrm{Prob}\,[\exists \boldsymbol{\sigma} \in \mathscr{T}_m\colon\ \mathbf{H}\boldsymbol{\sigma} \leqslant \tilde{\chi}] \end{aligned} \tag{5.24}$$

‡ *Implication* between two events A and B, indicated by the symbol $\Rightarrow$, can be defined from the conditional probability of B given A (Appendix A1):

$$[A] \Rightarrow [B] \quad \text{only if} \quad \mathrm{Prob}\,[B|A] = 1$$

$\mathscr{V}_n$ and $\mathscr{T}_m$ being discrete subsets of the set $\mathscr{V}$ of all possible mechanisms and of the set $\mathscr{T}$ of all equilibrated stress fields.

Thus, it is proved that an upper (*safe*) and a lower (*unsafe*) bound to the actual CPF $P_R(\mathbf{W})$ can be obtained through consideration of a limited number of stress vectors and mechanisms: the upper bound $P_S(\mathbf{W})$ is the probability that *none* of the investigated stress vectors (in equilibrium with **W** by assumption) is *statically admissible*, while the lower bound $P_K(\mathbf{W})$ is the probability that *at least one* of the investigated mechanisms is *kinematically* sufficient under **W**.

Of course, if the loads applied on the structure are described by a JPDF $p_W(\mathbf{W})$ and the probability of collapse is given, according to equation (2.68), by

$$P_{\text{fail}} = \int_{\mathscr{D}} p_W(\mathbf{W})\, P_R(\mathbf{W})\, \mathrm{d}\mathbf{W} \tag{5.25}$$

convolution of $p_W(\mathbf{W})$ with $P_S(\mathbf{W})$ and $P_K(\mathbf{W})$ will yield upper and lower bounds to P_{fail};

$$P_{\text{K fail}} \leqslant P_{\text{fail}} \leqslant P_{\text{S fail}} \tag{5.25$'$}$$

with

$$\begin{aligned} P_{\text{S fail}} &= \int_{\mathscr{D}} p_W(\mathbf{W}) P_S(\mathbf{W})\, \mathrm{d}\mathbf{W} \\ P_{\text{K fail}} &= \int_{\mathscr{D}} p_W(\mathbf{W}) P_K(\mathbf{W})\, \mathrm{d}\mathbf{W} \end{aligned} \tag{5.25$''$}$$

It has been noted [5] that equation (5.21) and the corresponding l.h.s. inequality in (5.23) remain valid if other failure modes (e.g. buckling modes) are included besides the plastic collapse mechanisms.

In contrast, the validity of equation (5.22) is strongly dependent on the assumptions of limit analysis. However, it can be interpreted as a *discretization* theorem for the infinite set of equilibrated stress vectors, in that it allows one to obtain significant results working on a finite number of stress vectors. It can also be proved [6] that under realistic *regularity* assumptions, discretization can be such as to ensure convergence of $P_S(\mathbf{W})$ to the actual CPF $P_R(\mathbf{W})$ with increasing number m of stress vectors introduced in $\mathscr{T}_m$.

5.2.4 Applications of the bounding theorems

(a) Kinematic theorem

If only one mechanism is considered, the probability P_{Ki} that it is kinematically sufficient under loads **W** obviously yields a lower bound to the probability that **W** is kinematically sufficient. Thus, if n mechanisms are considered, a value of $P_K(\mathbf{W})$ is given by

$$P_K(\mathbf{W}) = \max_i (n) P_{Ki} \leqslant \max_i P_{Ki} \tag{5.26}$$

where $\max_i(n) P_{Ki}$ is the largest P_{Ki} among the n considered mechanisms, and $\max_i P_{Ki}$, as in (5.17), is the largest P_{Ki} among all possible mechanisms.

Also, from elementary logic (cf. Section A1.1 and Section 1.6), the probability of the event E_K, defined by equation (5.20), is not larger than the sum of the probabilities that each of the considered mechanisms is kinematically sufficient, i.e. making use of definition (5.14),

$$P_K(\mathbf{W}) \leqslant \sum_{i=1}^{n} P_{Ki} \qquad (5.17')$$

In other words, the sum of the probabilities of collapse according to a limited number of mechanisms is rigorously an upper bound to $P_K(\mathbf{W})$, not to $P_R(\mathbf{W})$. However, in the range of very small probabilities, the r.h.s. of (5.17′) yields a close estimate of the l.h.s., so that it can be assumed without significant errors that

$$P_K(\mathbf{W}) = \sum_{i=1}^{n} P_{Ki} \qquad (5.17'')$$

If the events E_i, associated with each of the considered mechanisms, were stochastically independent, a better value of $P_K(\mathbf{W})$ (i.e. a value larger than equation (5.17″)) but still satisfying inequality (5.23) in full rigour) would be given by a formula like (5.16). However, the most probable collapse mechanisms (by which the set $\mathscr{V}_n$ should be formed for a good approximation) often share one or more plastic hinges, and therefore the events E_i are correlated. Hence, reliable evaluations of the lower bound probability $P_K(\mathbf{W})$ are given only by formulae like (5.15), of which (5.16) yields only very rough estimates.

It is evident how it is possible to construct an increasing sequence of values of $P_K(\mathbf{W})$, by considering first one mechanism, then two, then three, etc.

$$P_K^{(1)}(\mathbf{W}) = P_{K1} \qquad (5.27)$$

$$\begin{aligned} P_K^{(2)}(\mathbf{W}) &= P_{K1} + P_{K2} - \mathrm{Prob}[E_1 \cap E_2] \\ &= P_K^{(1)}(\mathbf{W}) + P_{K2} - \mathrm{Prob}[E_1 \cap E_2] \end{aligned} \qquad (5.27')$$

$$\begin{aligned} P_K^{(3)}(\mathbf{W}) &= P_{K1} + P_{K2} + P_{K3} - \mathrm{Prob}[E_1 \cap E_2] - \mathrm{Prob}[E_1 \cap E_3] \\ &\qquad - \mathrm{Prob}[E_2 \cap E_3] + \mathrm{Prob}[E_1 \cap E_2 \cap E_3] \\ &= P_K^{(2)}(\mathbf{W}) + P_{K3} - \mathrm{Prob}[E_1 \cap E_3] - \mathrm{Prob}[E_2 \cap E_3] \\ &\qquad + \mathrm{Prob}[E_1 \cap E_2 \cap E_3] \end{aligned} \qquad (5.27'')$$

Therefore, it is not necessary to choose the full set $\mathscr{V}_n$ *a priori*: further mechanisms can be introduced if and when a better approximation of $P_R(\mathbf{W})$ is required.

The joint probabilities in equation (5.27″) can be calculated according to the following procedure, if the strength vector $\tilde{\chi}$ is Gaussian (or in general, has a stable distribution). Let E_i be the event defined by equation (5.13), and put

$$\begin{aligned} L_i &= \mathbf{W}^{\mathrm{T}}\mathbf{u}_i \\ \tilde{x}_i &= \tilde{L}_{\mathrm{d}i} - L_i = \tilde{\chi}^{\mathrm{T}}\boldsymbol{\lambda}_{\mathrm{p}i} - L_i \\ \boldsymbol{\Lambda}_{\mathrm{p}} &= \begin{bmatrix} \lambda_{\mathrm{p}11} & \cdots & \lambda_{\mathrm{p}r} \\ \vdots & & \vdots \\ \lambda_{\mathrm{p}l1} & \cdots & \lambda_{\mathrm{p}lr} \end{bmatrix} \end{aligned} \tag{5.28}$$

so that the $\tilde{x}_i$ are expressed in vector form by

$$\tilde{\mathbf{x}} = \boldsymbol{\Lambda}_{\mathrm{p}}\tilde{\chi} - \mathbf{L} \tag{5.29}$$

and the mean vector and covariance matrix of the random vector $\tilde{\mathbf{x}}$ are expressed by (equations (A3.26′) and (A3.27′))

$$\begin{aligned} \boldsymbol{\mu}_{\mathbf{x}} &= \boldsymbol{\Lambda}_{\mathrm{p}}\boldsymbol{\mu}_{\chi} - \mathbf{L} \\ \Sigma_{\mathbf{x}} &= \boldsymbol{\Lambda}_{\mathrm{p}}\Sigma_{\chi}\boldsymbol{\Lambda}_{\mathrm{p}}^{\mathrm{T}} \end{aligned} \tag{5.29′}$$

Since the JPDF of the strength parameters $\tilde{\chi}_i$ is stable by assumption, the joint distribution of the $\tilde{x}_i$ is of the same type as $\tilde{\chi}_i$. In particular, if $\tilde{\chi}$ is a Gaussian vector, $\tilde{\mathbf{x}}$ is Gaussian too, and the JPDF of its components is given by equation (A2.38).

After this stage, the most practicable procedure to calculate the joint probabilities in equation (5.27″) is to integrate on the above density function.

So the joint probability of l events E_i can be calculated remembering that joint occurrence of such events means that the associated mechanisms $\mathbf{u}_i$ are all kinematically sufficient, i.e. $\mathbf{W}^{\mathrm{T}}\mathbf{u}_i > \tilde{L}_{\mathrm{d}i}$ and $\tilde{x}_i > 0$ for $i = 1, \ldots, l$. Therefore, in equation (5.27″) it is possible to introduce

$$\mathrm{Prob}\left[\bigcap_{i=1}^{l} E_i\right] = \int_0^{\infty} \int_0^{\infty} p(x_1, \ldots, x_l)\,\mathrm{d}x_1 \ldots \mathrm{d}x_l \tag{5.30}$$

In practice, simultaneous consideration of many mechanisms, unless equation (5.17″) is accepted, leads to rather cumbersome calculations, while the circumstance that the calculated probabilities are always on the unsafe side would suggest the introduction of more and more mechanisms in the procedure.

(b) Static theorem

When $P_{\mathrm{R}}(\mathbf{W})$ is to be approximated by the *static approach*, $P_{\mathrm{S}}(\mathbf{W})$ (equation (5.24)) is usually obtained from the complementary probability

$$\Pi_{\mathrm{S}}(\mathbf{W}) = 1 - P_{\mathrm{S}}(\mathbf{W}) = \mathrm{Prob}[\,\exists\boldsymbol{\sigma} \in \mathcal{T}_m\!: \mathbf{H}\sigma \leqslant \tilde{\chi}] \tag{5.31}$$

i.e. the probability that at least one of the considered stress vectors is statically admissible.

For an appropriate choice of the set $\mathcal{T}_m$ of the stress vectors to be examined, the meaningful vectors in the complete set $\mathcal{T}$ must be recognized first. To this aim, we define a stress vector $\boldsymbol{\sigma} \in \mathscr{T}$ to be *reducible* if another equilibrated vector $\boldsymbol{\sigma}' \in \mathscr{T}$ exists such that

$$\mathbf{H}\boldsymbol{\sigma}' <: \mathbf{H}\boldsymbol{\sigma} \tag{5.32}$$

the symbol $<:$ mean that the inequality $<$ holds only for the components of $\mathbf{H}\boldsymbol{\sigma}$ that are larger than zero.

It can be proved [6] that, irrespective of the strength vector χ, no collapse mechanism exists associated with a reducible stress vector; hence, it would be useless to investigate reducible stress vectors for statical admissibility, either in classical limit analysis or in probabilistic limit analysis. This point is well exemplified by the elementary case of Fig. 5.2. It is quite obvious that only the *non-reducible* bending moment diagram in the dashed area can be associated with some collapse mechanism. In conclusion, in the static approach to limit analysis, $\mathscr{T}$ is conveniently replaced by $\mathscr{T}^0$, the set of *non-reducible* equilibrated stress vectors.

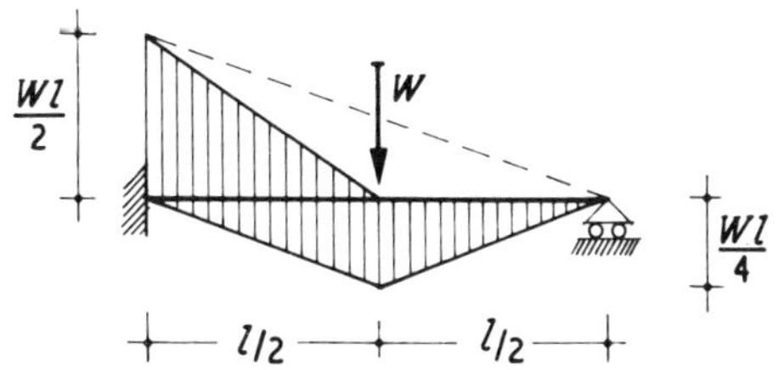

Fig. 5.2 Range of non-reducible bending moment diagrams in a simple example.

Consider two non-reducible equilibrated stress fields, like those shown in Fig. 5.3(a) and (b); let P_1, P_2 and $P_{1,2}$ be the probability that respectively the first stress field, the second stress field and their outer envelope (Fig. 5.3(c)) be statically admissible. Obviously, the probability that at least one of the two fields is statically admissible is given by

$$\Pi_S(\mathbf{W}) = P_1 + P_2 - P_{1,2} \tag{5.33}$$

Equation (5.33) can be generalized in the following way. Take n equilibrated stress vectors $\boldsymbol{\sigma}_j$, and define $\chi^{\max,n}$ as the vector collecting the largest components of the n vectors $\mathbf{H}\boldsymbol{\sigma}_j$

$$\chi_l^{\max,n} = \max_j (H_{lk}\sigma_{kj}) \qquad j = 1, 2, \ldots, n \tag{5.34}$$

where $\chi_l^{\max,n}$ is the lth component of the vector $\chi^{\max,n}$, and σ_{kj} the kth component of the vector σ_j. Summation on the repeated index k is implied within the brackets. Obviously‡

$$[\chi^{\max,n} \leqslant \chi] \Leftrightarrow [\mathbf{H}\boldsymbol{\sigma}_j \leqslant \chi \quad \text{for all } j \in (1, 2, \ldots, n)] \tag{5.35}$$

‡ $[A] \Leftrightarrow [B]$ means that A implies B and B implies A.

Then let

$$P_{1,2,\ldots,n} = \text{Prob}\,[\chi^{\max,n} \leqslant \tilde{\chi}] \tag{5.36}$$

be the probability that the outer envelope, and therefore the n stress vectors $\boldsymbol{\sigma}_j$ ($j = 1, 2, \ldots, n$), be admissible.

Making use of definition (5.36), the probability $\Pi_S(\mathbf{W})$ that at least one among m stress vectors is statically admissible is given by

$$\begin{aligned}\Pi_S(\mathbf{W}) &= P_1 + P_2 + \ldots + P_m - P_{1,2} - P_{1,3} - \ldots - P_{2,3} - P_{2,4} - \ldots \\ &\quad + P_{1,2,3} + P_{1,2,4} + \ldots + P_{2,3,4} + \ldots\ldots \\ &= \sum_{l=1}^{m} (-1)^{l+1} \sum_{k_1,k_2,\ldots,k_l} P_{k_1,k_2,\ldots,k_l}\end{aligned} \tag{5.37}$$

Putting $m = 1$, then $m = 2$, $m = 3$, etc., in equation (5.37), an increasing sequence of values of $\Pi_S(\mathbf{W})$ can be established (in analogy to (5.30) for P_K). To obtain this sequence systematically, construct a sequence of matrices $\mathbf{Y}_k$ with N_k columns ($k = 0, 1, 2, \ldots$) and as many rows as the elements of χ. The first matrix of the sequence ($k = 0$) is a column matrix of zero elements:

$$\mathbf{Y}_0 = \mathbf{0} \qquad N_0 = 1 \tag{5.38}$$

The following matrices ($k \geqslant 1$) are built up by the rule

$$\begin{aligned} N_k &= 2N_{k-1} \\ \mathbf{Y}_k^h &= \mathbf{Y}_{k-1}^h \\ \mathbf{Y}_k^{h+N_{k-1}} &= \max\,[\mathbf{Y}_{k-1}^h; \mathbf{H}\boldsymbol{\sigma}_{k-1}] \\ h &= 1, 2, \ldots, N_{k-1} \end{aligned} \tag{5.38'}$$

where $\mathbf{Y}_k^h$ denotes the hth column of the matrix $\mathbf{Y}_k$, and $\max\,[\mathbf{Y}_k^h; \mathbf{H}\boldsymbol{\sigma}_k]$ the column constructed by taking in each row the largest of the components of the two vectors $\mathbf{Y}_k^h$ and $\mathbf{H}\boldsymbol{\sigma}_k$.

To each matrix, associate an N_k-dimensional vector $\boldsymbol{P}_k$ by defining its hth component as

$$P_k^h = \text{Prob}\,[\mathbf{Y}_k^h \leqslant \tilde{\chi}] \tag{5.39}$$

In parallel to the matrix sequence, consider the sequence of vectors $\mathbf{q}_k$, each with N_k components, defined by the position

$$\mathbf{q}_0 = \mathbf{1} \qquad N_0 = 1 \tag{5.40}$$

and the rule

$$\begin{aligned} N_k &= 2N_{k-1} \\ q_k^h &= q_{k-1}^h \\ q_k^{h+N_{k-1}} &= -q_{k-1}^h \\ h &= 1, \ldots, N_{k-1} \end{aligned} \tag{5.40'}$$

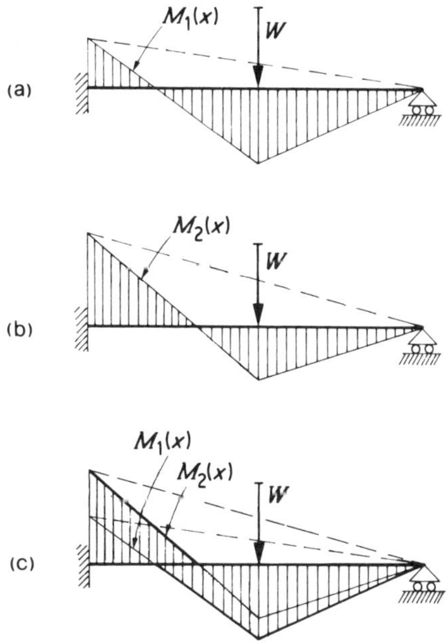

Fig. 5.3 Two non-reducible stress fields for the example of Fig. 5.2 and their outer envelope: (a) $P_1 = \text{Prob}\,[|M_1| \leqslant \tilde{M}_y \text{ for all } x]$; (b) $P_2 = \text{Prob}\,[|M_2| \leqslant \tilde{M}_y \text{ for all } x]$; (c) $P_{1,2} = \text{Prob}\,[\max(|M_1|, |M_2|) \leqslant \tilde{M}_y \text{ for all } x]$

It is easily recognized that equation (5.37) for any k is equivalent to

$$P_S(\mathbf{W}) = 1 - \Pi_S(\mathbf{W}) = \boldsymbol{P}_k^T \mathbf{q}_k = P_k^h q_k^h \text{ (sum over } h) \qquad (k = 1, 2, \ldots, m) \tag{5.41}$$

as P_k^0 is equal to unity. Thus, the r.h.s. depends on the number k of stress fields and on the particular stress fields that have been considered.

The above treatment (which will be illustrated by an example in Section 5.2.7(a)) leads to a conceptually simple way of calculating $\Pi_S(\mathbf{W})$, provided the probabilities (5.39) can be obtained. However, this procedure is affected by evident limitations as to the maximum number m of stress vectors that can be introduced in the calculations. In fact, adding one stress vector requires that the time of calculation is doubled: therefore, the time necessary for calculations increases as 2^m, and the procedure soon becomes very time-consuming. (It seems that in most cases the maximum practicable value of m is of the order 10–15, depending on the characteristics of the available computer.) The value of m affects only the accuracy of the static approximation: thus, the practical impossibility of increasing m beyond certain limits does not allow one to obtain close approximations of $P_R(\mathbf{W})$ via the static approach. On the other hand, it is to be noted that the time required for a probabilistic *static* analysis of structures of increasing complexity grows almost linearly with the dimensions of the relevant vectors $(\boldsymbol{\sigma}, \mathbf{u})$ while the time required for the kinematic analysis grows much more rapidly.

Remembering also that the static approximation is always on the *safe* side, it can be tentatively concluded that the kinematic approach is more suited for the analysis of simple structures, where close evaluations of $P_R(\mathbf{W})$ are possible by considering practically all relevant mechanisms, but the static approach is to be preferred when dealing with complex structures.

5.2.5 Domain with constrained failure probability in the load space

In the applications to multiparameter load conditions, the functions $P_R(\mathbf{W})$, $P_K(\mathbf{W})$ and $P_S(\mathbf{W})$ are usually represented by their *level surfaces*. For any given constant $p \in (0, 1)$, these surfaces are the boundaries of the three domains in load space defined by

$$\begin{aligned} \mathscr{S}_p: &\quad \{[\mathbf{W} \in \mathscr{S}_p] \Leftrightarrow [P_R(\mathbf{W}) \leqslant p]\} \\ \mathscr{S}_{Sp}: &\quad \{[\mathbf{W} \in \mathscr{S}_{Sp}] \Leftrightarrow [P_S(\mathbf{W}) \leqslant p]\} \\ \mathscr{S}_{Kp}: &\quad \{[\mathbf{W} \in \mathscr{S}_{Kp}] \Leftrightarrow [P_K(\mathbf{W}) \leqslant p]\} \end{aligned} \tag{5.42}$$

It can be seen that for any p the domains $\mathscr{S}_p$, $\mathscr{S}_{Sp}$ and $\mathscr{S}_{Kp}$ comprise the origin of the load space ($\mathbf{W} = 0$), and that by definition

$$[p_1 \leqslant p_2] \Leftrightarrow [\mathscr{S}_{p_1} \subseteq \mathscr{S}_{p_2};\ \mathscr{S}_{Sp_1} \subseteq \mathscr{S}_{Sp_2};\ \mathscr{S}_{Kp_1} \subseteq \mathscr{S}_{Kp_2}] \tag{5.43}$$

Inequalities (5.23) then lead to

$$\mathscr{S}_{Sp} \subseteq \mathscr{S}_p \subseteq \mathscr{S}_{Kp} \quad \text{for all} \quad p \in (0, 1) \tag{5.44}$$

In deterministic limit analysis, as already noted (Section 5.2.2), $P_R(\mathbf{W})$ is a 0–1 function. Use of the approximating techniques based on the static and kinematic theorems leads to two other 0–1 functions, $P_S(\mathbf{W})$ and $P_K(\mathbf{W})$. Denoting by $\mathscr{S}$, $\mathscr{S}_S$ and $\mathscr{S}_K$ the domains bounded by the surfaces of discontinuity of $P_R(\mathbf{W})$, $P_S(\mathbf{W})$ and $P_K(\mathbf{W})$ respectively, definitions (5.42) become

$$\begin{aligned} \mathscr{S}_p &= \mathscr{S} \\ \mathscr{S}_{Sp} &= \mathscr{S}_S \quad \text{for all } p\text{: } (0 \leqslant p < 1) \\ \mathscr{S}_{Kp} &= \mathscr{S}_K \end{aligned} \tag{5.42'}$$

(With respect to plastic collapse, $\mathscr{S}$ coincides with the *safe region* defined in Chapters 1 and 2.) Then, (5.44) yields the well known relationship

$$\mathscr{S}_S \subseteq \mathscr{S} \subseteq \mathscr{S}_K \tag{5.44'}$$

In view of applications, it appears important to note that, as easily proved [8]

$$\mathscr{S}_{0.5p}^{OC} \subseteq \mathscr{S}_p \subseteq \mathscr{S}_{2p}^{IC} \tag{5.45}$$

where $\mathscr{S}_{0.5p}^{OC}$ is the smallest convex domain containing $\mathscr{S}_{0.5p}$, and $\mathscr{S}_{2p}^{IC}$ the largest convex domain contained in $\mathscr{S}_{2p}$. (This property has been called the *semi-*

convexity of $\mathscr{S}_p$, and is weaker than the well known convexity of $\mathscr{S}$ in classical limit analysis.)

It is also possible to prove that $\mathscr{S}_p$ is *starlike* with respect to the origin, i.e. if a load vector $\mathbf{W} \in \mathscr{S}_p$, then for any scalar $\alpha \in (0, 1)$ also $\alpha\mathbf{W} \in \mathscr{S}_p$.

(a) Determination of external approximations $\mathscr{S}_{\mathrm{K}p}$

The simplest procedure to find domains $\mathscr{S}_{\mathrm{K}p}$ is to consider a few collapse mechanisms $\mathbf{u}_i$, each independent of the others. In other words, a mechanism $\mathbf{u}_i$ is chosen and the lower bound $P_{\mathrm{K}}(\mathbf{W})$ is identified with the probability of collapse associated with that mechanism $P_{\mathrm{K}i}$ (equation (5.14))

$$P_{\mathrm{K}}(\mathbf{W}) = P_{\mathrm{K}i} = \mathrm{Prob}\,[\mathbf{W}^{\mathrm{T}}\mathbf{u}_i > \tilde{L}_{\mathrm{d}i}] \tag{5.46}$$

Let

$$P_{Li}(x) = \mathrm{Prob}\,[\tilde{L}_{\mathrm{d}i} \leqslant x] \qquad 0 \leqslant x \leqslant \infty \tag{5.47}$$

be the CDF of the r.v. $\tilde{L}_{\mathrm{d}i}$ (equation (5.12)) that can easily be obtained from that of $\tilde{\chi}$. Since any distribution law is a non-decreasing function of the argument it is easy to obtain the following implication from definitions (5.46) and (5.47):

$$[\mathbf{W}^{\mathrm{T}}\mathbf{u}_i \leqslant x] \Leftrightarrow [P_{\mathrm{K}}(\mathbf{W}) \leqslant P_{Li}(x)] \tag{5.48}$$

Hence, for any value of p ($0 \leqslant p \leqslant 1$), the probability (5.46) does not exceed p in the half-space defined by

$$\mathbf{W}^{\mathrm{T}}\mathbf{u}_i \leqslant x(p) \tag{5.49}$$

$x(p)$ being the inverse function of (5.47). According to the third definition (5.42), the half-space (5.49) is the $\mathscr{S}_{\mathrm{K}p}$ domain relative to the mechanism $\mathbf{u}_i$ and the probability p.

Varying the mechanism $\mathbf{u}_i$, the half-space (5.49) changes because in general both $\mathbf{u}_i$ and $\lambda_{\mathrm{p}i}$, hence the CDF $P_{Li}(x)$, vary. Since the domain $\mathscr{S}_p$ must be contained in all the $\mathscr{S}_{\mathrm{K}p}$ relative to the same probability p, a convex piecewise linear outer bound to $\mathscr{S}_p$ is obtained as the intersection of several half-spaces (5.49)‡. An example of the application of this procedure will be presented in Section 5.2.7(b); as will be evident, the calculations are greatly simplified if the JPDF of the components of the strength vector χ are stable, because the changes of mechanism correspond to changes in the parameters of the linear combination (5.12) which, in this case, do not alter the analytical expression of the probability distributions.

If n mechanisms are considered simultaneously, the probability $P_{\mathrm{K}}(\mathbf{W})$ corresponds to the *joint* probability of the events E_i ($i = 1, \ldots, n$). Since

$$P_{\mathrm{K}}(\mathbf{W}) \geqslant P_{\mathrm{K}i}(\mathbf{W}) \tag{5.50}$$

a better approximating domain is in general obtained. However, as already noted in Section 5.2.4(a), the probabilities $P_{\mathrm{K}i}$ do not refer to stochastically independent

‡ This *intersection* procedure is equivalent to the *maximization* indicated by equation (5.26). Note in particular that the intersection of the two half-spaces relative to a mechanism $\mathbf{u}_i$ and its opposite $-\mathbf{u}_i$ is a *strip* bounded by two parallel hyperplanes.

events. (Stochastic independence is possible only between mechanisms which do not share any *plastic hinge*, but to limit the choice of the mechanisms in this way cannot lead to any acceptable approximation of $\mathscr{S}_p$.) Therefore, equation (5.16) cannot be used to calculate $P_K(\mathbf{W})$, which requires taking account of correlations between the mechanisms considered.

(b) Determination of internal (safe) approximations $\mathscr{S}_{Sp}$

Similar procedures can be set up for obtaining domains $\mathscr{S}_{Sp}$ approximating $\mathscr{S}_p$ from the inside. Consider an equilibrated stress vector $\boldsymbol{\sigma}_j$ and let P_j be the probability that it is statically admissible. Then a value of the upper bound probability $P_S(\mathbf{W})$ is

$$P_S(\mathbf{W}) = 1 - P_j = p \tag{5.51}$$

For any given p and stress vector $\boldsymbol{\sigma}_j$, equation (5.51) and the second of equations (5.42) yield an $\mathscr{S}_{Sp}$ domain. $\mathscr{S}_p$ is external to all $\mathscr{S}_{Sp}$ relative to a given p, and therefore an internal approximation to $\mathscr{S}_p$ can be obtained as the union of the $\mathscr{S}_{Sp}$ corresponding to several equilibrated stress vectors.

A better approximation of $\mathscr{S}_p$ is obtained by identifying $P_S(\mathbf{W})$ with the probability that all the considered stress vectors are statically admissible, i.e.

$$P_S(\mathbf{W}) = 1 - \Pi_S(\mathbf{W}) = p \tag{5.52}$$

with $\Pi_S(\mathbf{W})$ given by equation (5.37).

5.2.6 The case of loads defined by one parameter

Now let the loads acting on the random-strength structure be defined in the form (5.8). The CPF under loads $\mathbf{W}_0 + \alpha\mathbf{W}_a$ coincides with the CDF of the *collapse load factor* $\tilde{\alpha}_R$: in fact, by definition of α_R,

$$\text{Prob}[\exists \mathbf{u}\colon (\mathbf{W}_0 + \alpha\mathbf{W}_a)^T\mathbf{u} > \tilde{L}_{di}] = \text{Prob}[\tilde{\alpha}_R < \alpha] = P_R(\alpha) \tag{5.53}$$

For any given value of α, the procedures illustrated in Sections 5.2.3 and 5.2.4 furnish two bounds to $P_R(\alpha)$, analogous to (5.23) (Fig. 5.4)

$$P_K(\alpha) \leqslant P_R(\alpha) \leqslant P_S(\alpha) \tag{5.54}$$

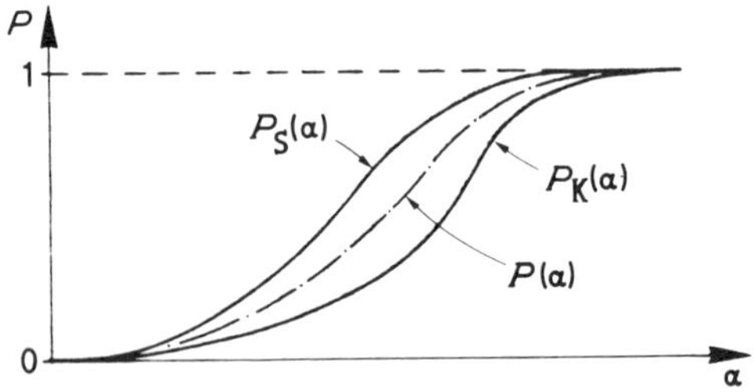

Fig. 5.4 Bounds to the CDF of α_R.

The bounding curves $P_K(\alpha)$ and $P_S(\alpha)$, usually much easier to calculate than the *exact* CPF $P_R(\alpha)$, can be viewed as the CDF of two random variables, respectively a *kinematically sufficient load factor* α_K and a *statically admissible load factor* α_S

$$\begin{aligned} P_K(\alpha) &= \mathrm{Prob}[\alpha_K < \alpha] \\ P_S(\alpha) &= \mathrm{Prob}[\alpha_S < \alpha] \end{aligned} \tag{5.55}$$

Note that, for any value of probability p, $\tilde{\alpha}_K$, $\tilde{\alpha}_R$ and $\tilde{\alpha}_S$ respect inequalities (5.9) whereas inequalities (5.54) become equalities for $\alpha - 0$ and $\alpha \to \infty$.

Any pair of curves $P_K(\alpha)$, $P_S(\alpha)$ allows one also to bound the average value μ_α of $\tilde{\alpha}_R$, and its variance s_α^2. In fact, given a $P_K(\alpha)$ and a $P_S(\alpha)$, any CDF $P(\alpha)$ comprised between $P_K(\alpha)$ and $P_S(\alpha)$ must satisfy inequalities (5.54), and also

$$\begin{aligned} &P(0) = 0 \qquad P(\infty) = 1 \\ &p(\alpha) = \frac{\mathrm{d}P}{\mathrm{d}\alpha} \geqslant 0 \qquad \text{for all } \alpha \in (0, \infty) \end{aligned} \tag{5.56}$$

Now let $\partial p(\alpha)$ be an *admissible variation* of the function $P(\alpha)$, i.e. an infinitesimal function such that $P(\alpha) + \partial P(\alpha)$ satisfies conditions (5.54) and (5.56): obviously

$$\partial P(0) = \partial P(\infty) = 0 \tag{5.57}$$

It is also easy to show that [4]

$$\lim_{\alpha \to \infty} [\alpha\, \partial P(\alpha)] = \lim_{\alpha \to \infty} [\alpha^2\, \partial P(\alpha)] = 0 \tag{5.58}$$

Then, from the definition of the average μ_α associated with $P(\alpha)$ (equation (2.17a)),

$$\mu_\alpha = \int_0^\infty \alpha \frac{\mathrm{d}P}{d\alpha}\,\mathrm{d}\alpha \tag{5.59}$$

the variation of μ_α due to ∂P is

$$\partial \mu_\alpha = \int_0^\infty \alpha \frac{\mathrm{d}(\partial P)}{\mathrm{d}\alpha}\,\mathrm{d}\alpha = [\alpha\, \partial P]_0^\infty - \int_0^\infty \partial P\,\mathrm{d}\alpha = -\int_0^\infty \partial P\,\mathrm{d}\alpha \tag{5.60}$$

Now, from $P_K(\alpha)$ only positive variations ($\partial P \geqslant 0$) are possible, while from $P_S(\alpha)$ only negative variations ($\partial P \leqslant 0$) are possible; therefore $\mu_{\alpha K}$ and $\mu_{\alpha S}$ are respectively the largest and the smallest of the possible μ_α:

$$\mu_{\alpha K} \geqslant \mu_\alpha \geqslant \mu_{\alpha S} \tag{5.61}$$

in which

$$\mu_{\alpha K} = \int_0^\infty \alpha \frac{\mathrm{d}P_K}{\mathrm{d}\alpha}\,\mathrm{d}\alpha \tag{5.62a}$$

$$\mu_{\alpha S} = \int_0^\infty \alpha \frac{\mathrm{d}P_S}{\mathrm{d}\alpha}\,\mathrm{d}\alpha \tag{5.62b}$$

Since inequalities (5.61) hold for any possible $P(\alpha)$, they also hold for the *true* (but unknown) CPF $P_R(\alpha)$: in other words, (5.61) yields the required bounds to the average of α_R.

Note now that a curve $P_K(\alpha)$ can be obtained from the collapse mechanism of the *expected structure* (i.e. of the deterministic structures with strength vector $\chi = \mu_\chi$, μ_χ being the *average* strength vector): because of the linearity between α_K and χ for a given mechanism (equations (5.46) and (5.12)), the average $\mu_{\alpha K}$ corresponding to this curve $P_K(\alpha)$ coincides with the collapse load factor $\alpha_{\bar{R}}$ of the (deterministic) expected structure. Then, the first inequality (5.61) proves that the average collapse load of the random structure cannot be larger than the collapse load of the expected structure: in other words, *on average, randomness decreases the resistance of a plastic structure.*

In order to bound the variance of the collapse load factor α_R, consider again a possible CDF curve $P(\alpha)$, and take the second moment of the PDF $dP/d\alpha$ with respect to the axis $\alpha = \mu_{\alpha R}$, $\mu_{\alpha R}$ being the average of α_R (equation (2.17c)):

$$\Upsilon = \int_0^\infty (\alpha - \mu_{\alpha R})^2 \frac{dP}{d\alpha} d\alpha \tag{5.63}$$

Taking account of (5.58), the variation of T for an *admissible variation* ∂P of $P(\alpha)$ is

$$\partial\Upsilon = \int_0^\infty (\alpha - \mu_{\alpha R})^2 \frac{d(\partial P)}{d\alpha} d\alpha = -2\int_0^\infty (\alpha - \mu_{\alpha R})\,\partial P\, d\alpha \tag{5.64}$$

Consider now the possible curve $P_1(\alpha)$ defined by (Fig. 5.5(a))

$$\begin{aligned} P_1(\alpha) &= P_K(\alpha) \qquad && 0 \leqslant \alpha < \mu_{\alpha R} \\ P_1(\alpha) &= P_S(\alpha) && \alpha > \mu_{\alpha R} \end{aligned} \tag{5.65}$$

discontinuous at $\alpha = \mu_{\alpha R}$. Admissible variations ∂P from $P_1(\alpha)$ must verify the following condition

$$\begin{aligned} \partial P_1(\alpha) &\geqslant 0 \qquad && 0 \leqslant \alpha < \mu_{\alpha R} \\ \partial P_1(\alpha) &\leqslant 0 && \alpha > \mu_{\alpha R} \end{aligned} \tag{5.66}$$

whence, for $P(\alpha) = P_1(\alpha)$

$$\partial\Upsilon \geqslant 0 \tag{5.66$'$}$$

Hence,

$$\Upsilon_1 = \int_0^\infty (\alpha - \mu_{\alpha R})^2 \frac{dP_1}{d\alpha} d\alpha \tag{5.67}$$

is the minimum value of Υ in the set of possible CDF $P(\alpha)$.

In order to find the maximum value of Υ, consider the family of CDF curves defined by (Fig. 5.5(b))

$$P_2(\alpha, C) = \begin{cases} P_S(\alpha) & \alpha\colon P_S(\alpha) < C \\ C & \alpha\colon P_S(\alpha) \geqslant C \geqslant P_K(\alpha) \\ P_K(\alpha) & \alpha\colon P_K(\alpha) \geqslant C \end{cases} \tag{5.68}$$

where C is a parameter ($0 \leqslant C \leqslant 1$), and note that for any possible CDF $P(\alpha)$

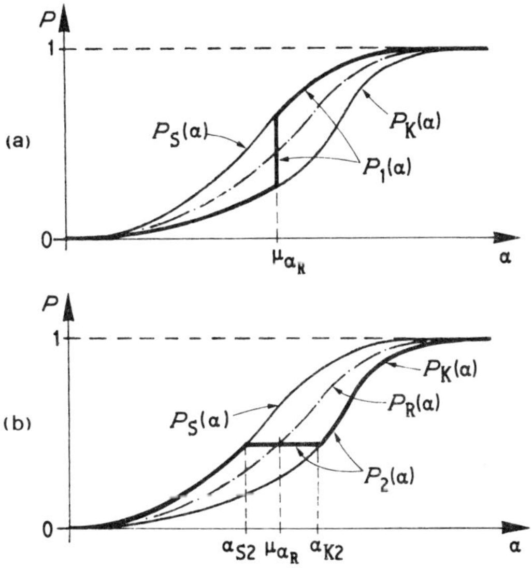

Fig. 5.5 Curves $P(\alpha)$ to bound the variance of α_R.

(Fig. 5.5(b)) it is possible to find a CDF of the type (5.68) with a larger value of Υ: in fact, it is sufficient to take $C = P(\mu_{\alpha R})$, note that

$$\begin{cases} P_2(\alpha, C) - P(\alpha) \geqslant 0 & \alpha < \mu_{\alpha R} \\ P_2(\alpha, C) - P(\alpha) \leqslant 0 & \alpha > \mu_{\alpha R} \end{cases} \tag{5.69}$$

and remember equation (5.64). The maximum Υ must therefore be sought in the family of functions P_2 (5.68). Admissible variations in this set are

$$\partial P_2 = \begin{cases} \text{const.} = \partial C & \alpha_{S2} \leqslant \alpha \leqslant \alpha_{K2} \\ 0 & \alpha < \alpha_{S2} \quad \text{or} \quad \alpha > \alpha_{K2} \end{cases} \tag{5.70}$$

where α_{S2} and α_{K2} are defined by

$$P_S(\alpha_{S2}) = P_K(\alpha_{K2}) = C \tag{5.71}$$

The stationarity condition for Υ in the family of curves P_2 is

$$\partial\Upsilon = -2\int_{\alpha_{S2}}^{\alpha_{K2}} (\alpha - \mu_{\alpha R})\,\partial C\,\mathrm{d}\alpha = 0 \tag{5.72}$$

whence

$$(\alpha_{K2} - \mu_{\alpha R})^2 - (\alpha_{S2} - \mu_{\alpha R})^2 = 0 \tag{5.73}$$

or

$$\mu_{\alpha R} = (\alpha_{K2} + \alpha_{S2})/2 \tag{5.74}$$

Therefore, the largest value of Υ corresponds to the curve P_2 with C given by equations (5.71) and (5.74).

The variance of α_R is the particular value of Υ obtained by introducing in equation (5.63) the true CPF $P_R(\alpha)$: therefore, it is bounded between the maximum and minimum values of Υ just found.

Note that the above values of Υ depend on $\mu_{\alpha R}$: if this is known only approximately (e.g. is bounded in the interval (5.61)), the same uncertainty is carried over to the determination of $P_1(\alpha)$ and $P_2(\alpha)$, hence of the bounds to the variance of α_R.

5.2.7 Numerical examples and discussion

(a) Three-bar truss

As a first example, consider the simple truss in Fig. 5.6 made up of three elastic–perfectly plastic bars whose limit normal forces at the yield threshold are three independent Gaussian random variables $\tilde{N}_{y1}$, $\tilde{N}_{y2}$, $\tilde{N}_{y3}$, with the same expected value $\mu_{N_i} = \mu_N$ and the same standard deviation $s_{N_i} = s_N = 0.1\mu_N$. For any value N_y of the limit normal force, let $P_r(N_y)$ be the probability that $\tilde{N}_{yr} \leqslant N_y$ in the rth rod.

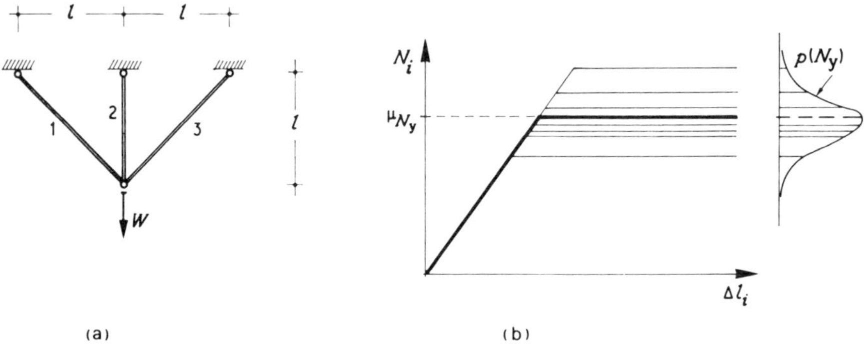

Fig. 5.6 Example three-bar truss [7].

In general, it is useful to execute a preliminary study of the deterministic *expected structure.*

With reference to any of the two mechanisms in Fig. 5.7, the collapse load for the expected structure is found to be

$$W_{\bar{R}} = (1 + \sqrt{2})\mu_N \tag{5.75}$$

Let us now apply the static procedure to find an upper bound on the CDF of the limit load $\tilde{W}_R$. To this aim, establish a number of equilibrated stress vectors, say 10, allowing one to explore a neighbourhood of the collapse stress vector in the expected structure by taking the stress vectors as in Table 5.1. Note that the vector in column 6 is the stress vector at the collapse threshold in the expected structure.

Each of the vectors is in equilibrium with the force $W_{\bar{R}}$. Hence, by multiplying

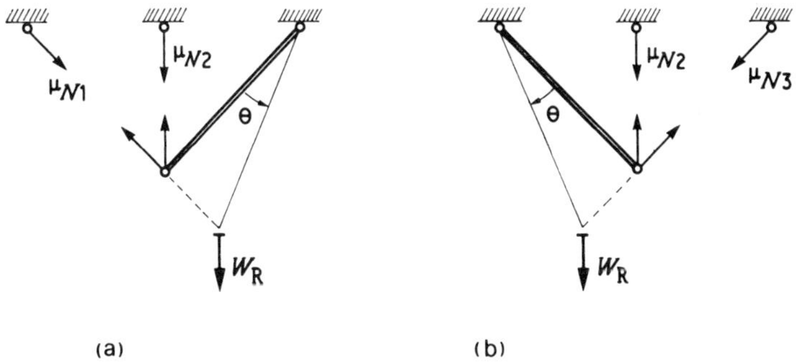

Fig. 5.7 Collapse mechanisms of three-bar truss.

Table 5.1 Three-bar truss (Fig. 5.6). Stress vectors in equilibrium with load $W_{\bar{R}}$

$j =$	1	2	3	4	5	6	7	8	9	10
N_1/μ_N	0.984	0.937	1.030	0.958	1.010	1.000	0.971	0.945	0.992	0.903
N_2/μ_N	1.020	1.090	0.956	1.060	0.981	1.000	1.040	1.080	1.010	1.140
N_3/μ_N	0.984	0.937	1.030	0.958	1.010	1.000	0.971	0.945	0.992	0.903

the columns in Table 5.1 by $\alpha = W/W_{\bar{R}}$ one gets 10 stress vectors equilibrating any value W of the applied force.

To fix ideas, let $W = W_{\bar{R}}$. The sequence of matrices $\mathbf{Y}_k$ and vectors $\mathbf{q}_k$ (equations (5.38′) and (5.40)) would be as follows ($k = 0, 1, 2, \ldots, 10$)

$$\mathbf{Y}_0 = \begin{bmatrix} 0 \\ 0 \\ 0 \end{bmatrix} \qquad \mathbf{q}_0 = [1] \tag{5.76a}$$

$$\mathbf{Y}_1 = \begin{bmatrix} 0 & 0.984 \\ 0 & 1.020 \\ 0 & 0.984 \end{bmatrix} \qquad \mathbf{q}_1 = [1 \quad -1] \tag{5.76b}$$

$$\mathbf{Y}_2 = \begin{bmatrix} 0 & 0.984 & 0.937 & 0.984 \\ 0 & 1.020 & 1.090 & 1.090 \\ 0 & 0.984 & 0.937 & 0.984 \end{bmatrix} \qquad \mathbf{q}_2 = [1 \quad -1 \quad -1 \quad 1] \tag{5.76c}$$

$$\mathbf{Y}_3 = \begin{bmatrix} 0 & 0.984 & 0.937 & 0.984 & 1.030 & 1.030 & 1.030 & 1.030 \\ 0 & 1.020 & 1.090 & 1.090 & 0.956 & 1.020 & 1.090 & 1.090 \\ 0 & 0.984 & 0.937 & 0.984 & 1.030 & 1.030 & 1.030 & 1.030 \end{bmatrix} \tag{5.76d}$$

$$\mathbf{q}_3 = [1 \quad -1 \quad -1 \quad 1 \quad -1 \quad 1 \quad 1 \quad -1]$$

$$\vdots \qquad \vdots$$

Introducing all 10 stress vectors collected in Table 5.1, one gets ($m = 10$) $N_{10} = 2^m = 1024$. By virtue of the assumed statistical independence of the limit normal forces in the three bars, one gets (equation (5.39)) for $W = W_{\bar{R}}$

$$P_k^h = \text{Prob}[\mathbf{Y}_k^h \leqslant \tilde{\chi}] = \prod_{r=1}^{3} [1 - P_r(Y_k^{hr})] \tag{5.77}$$

Y_k^{hr} being the element in the hth column and in the rth row of the matrix $\mathbf{Y}_k$. For any other load $W = \alpha W_{\bar{R}}$ one gets similarly

$$P_k^h(\alpha) = \prod_{r=1}^{3} [1 - P_r(\alpha Y_k^{hr})] \tag{5.77'}$$

Hence, equation (5.41) soon yields the static approximation

$$P_S(\alpha) = \sum_{h=0}^{1024} q_{10}^h P_{10}^h(\alpha) = \text{Prob}[\tilde{\alpha}_S \leqslant \alpha] \tag{5.78}$$

where it should be understood that $P_S(\alpha)$ is now the CDF of a statically admissible factor $\tilde{\alpha}_S$.

It can be noted that a less sharp approximation $P'_S(\alpha)$ might be stopped at only the first stress vector in Table 5.1. In this case in fact, taking $k = 1$ in equation (5.77′)

$$P'_S(\alpha) = P_1^0(\alpha) - P_1^1(\alpha) = 1 - P_1^1(\alpha) \tag{5.79}$$

Another remark is that also the probability $P_E(\alpha)$ that the structure shows elastic behaviour under the force $W = \alpha W_{\bar{R}}$ is at the same time a static approximation. Elastic normal forces in the structure can be readily computed

$$N_{1E} = \frac{1}{\sqrt{2(1+\sqrt{2})}} \alpha W_{\bar{R}} = 0.293 \alpha W_R$$

$$N_{2E} = \frac{\sqrt{2}}{1+\sqrt{2}} \alpha W_{\bar{R}} = 0.586 \alpha W_R$$

$$N_{3E} = \frac{1}{\sqrt{2(1+\sqrt{2})}} \alpha W_{\bar{R}} = 0.293 \alpha W_R$$

and $P_E(\alpha)$ is soon calculated by a formula similar to equation (5.77)

$$P_E(\alpha) = 1 - \prod_{r=1}^{3} [1 - P_r(N_{rE})] \tag{5.80}$$

Let us now consider the kinematic procedure, and let us, for simplicity, consider only the mechanism in Fig. 5.7(a), where $\tilde{N}_{y1}$ and $\tilde{N}_{y2}$ should be put in place of μ_{N_1} and μ_{N_2}. The limit value of the kinematic load $\tilde{W}_K$ associated with the considered mechanism is a random variable, since the forces N_{y1} and N_{y2} and random too.

The virtual work relation between the external force and the internal limit forces is

$$-\tilde{N}_{y1}\theta_1\sqrt{2}-\tilde{N}_{y2}\theta_1+\tilde{W}_K\theta_1=0$$

whence

$$\tilde{W}_K=\tilde{N}_{y1}\sqrt{2}+\tilde{N}_{y2}=\tilde{\alpha}_K W_0 \tag{5.81}$$

if W_0 is a reference load (for instance the already used $W_{\bar{R}}$).

Equation (5.81) shows that $\tilde{W}_K$ is a linear combination of the basic random variables $\tilde{N}_{y1}$ and $\tilde{N}_{y2}$. Therefore, the expected value is

$$\mu_{W_K}=(1+\sqrt{2})\mu_N \tag{5.82a}$$

and the variance is

$$s^2_{W_K}=3s^2_N \tag{582b}$$

Moreover, by the particular assumption that the $\tilde{N}_{yr}$ are Gaussian random variables, as the Gaussian distribution is a *stable* distribution, $\tilde{W}_K$ is also Gaussian, and its probability distribution function is now fully specified. A kinematic approximation is then given by

$$P_K(W)=\text{Prob}[\tilde{W}_K\leqslant W] \tag{5.83}$$

or, taking the reference load $W_0=\mu_{W_K}=W_{\bar{R}}$ (equation (5.75))

$$P_K(\alpha)=\text{Prob}[\tilde{W}_K/\mu_{W_K}\leqslant\alpha_K]=\text{Prob}[\tilde{\alpha}_K\leqslant\alpha] \tag{5.84}$$

where it should be understood that $P_K(\alpha)$ is the CPF of a kinematically admissible factor α_K.

The four probability functions $P_E(\alpha)$, $P'_S(\alpha)$, $P_S(\alpha)$ and $P_K(\alpha)$ are plotted together in Fig. 5.8. It can be deduced from overall inspection of the figure that

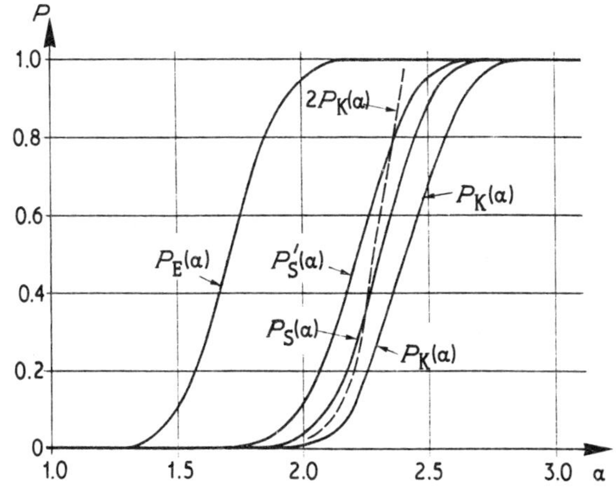

Fig. 5.8 Three-bar truss: CDF of elastic limit load factor α_E and bounds to CDF of plastic collapse factor α_R (after Baratta (7)).

$P_S(\alpha)$, and even $P'_S(\alpha)$, successfully evidence most of the total benefit allowed by ductility in regard to the collapse load factor, in that the relevant CDFs are shifted to the right towards $P_E(\alpha)$ considerably in comparison of the difference between $P_E(\alpha)$ and $P_K(\alpha)$.

The result is, unfortunately, quite different if small collapse probabilities are looked at.

In particular, in this example, only two symmetrical collapse mechanisms are possible, so that the upper bound in inequality (5.17) is obtained simply by doubling the kinematic lower bound given by equation (5.83), and, as shown by the dashed line in Fig. 5.8, it approximates the true collapse probability better than $P_S(\alpha)$ in the small probability range.

Still with reference to the same simple structure, the dependence of the static approximation on the number k of stress vectors introduced in equation (5.41) is highlighted by Fig. 5.9, where several calculated values of $P_S(\alpha)$ are shown by black dots: it appears that, for a given α, these values decrease in a rather irregular way with increasing k (depending on the choice of the sequence of stress vectors). For comparison, some values of the other upper bound $2P_K(\alpha)$ are also shown in Fig. 5.9: it can be seen that there is a difference of the order $5 \div 10\,\%$ between the

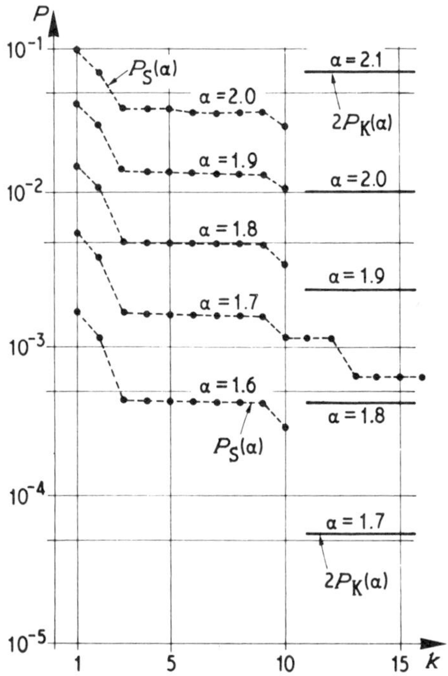

Fig. 5.9 Three bar-truss: upper bound evaluations of collapse probability versus the number k of stress vectors considered (after Baratta [7]).

values of the load factor α corresponding to the same values of the probability of collapse in the range $10^{-2} \div 10^{-5}$, evaluated by means of the two approximate procedures just discussed.

(b) Portal frame

As a second example, consider the simple portal frame shown in Fig. 5.10 under two-parameter loading [8]. Possible plastic deformations are limited to rotations concentrated in the 11 indicated cross-sections.

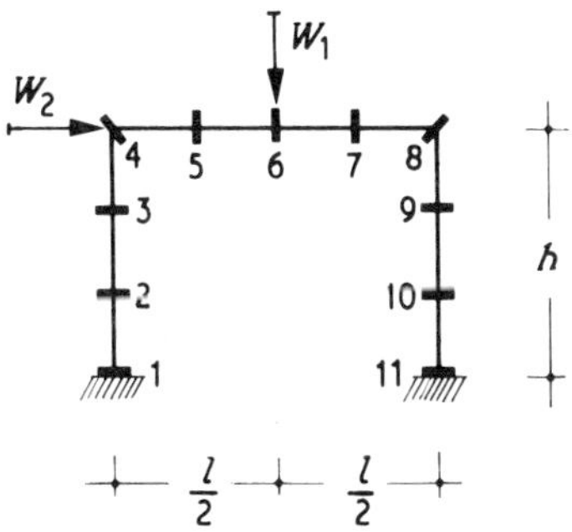

Fig. 5.10 Example portal frame and possible hinge sections; $l = 4.00$ m, $h = 3.00$ m (from Augusti and Baratta [8]).

Under the further hypothesis that positive and negative limit moments $\tilde{M}_{yr}$ are equal in each section, the yield condition shall express the condition that the absolute value of the bending moment M_r in the rth plastic cell is not larger than $\tilde{M}_{yr}$, irrespective of the sign of M_r, i.e.

$$|M_r| \leqslant \tilde{M}_{yr} \tag{5.85}$$

Similar to the previous example, the limit moments in the 11 cells are assumed to be stochastically independent and Gaussian-distributed with the same expected limit moment $\mu_{M_y} = 10$ t m and variance $s^2_{M_y} = (0.1\,\mu_{M_y})^2$. For any value M_y of the limit bending moment, let $P_r(M_y)$ be the probability that $\tilde{M}_{yr} \leqslant M_y$.

Because of the previous assumptions, the structure behaves symmetrically with respect to the sign of both loads, namely

$$P_R(W_1, W_2) = P_R(-W_1, W_2) = P_R(-W_1, -W_2) = P_R(W_1, -W_2) \tag{5.86}$$

Calculations can therefore be limited to the first quadrant in the load space ($W_1 \geqslant 0$; $W_2 \geqslant 0$).

Let us now develop first the kinematic procedure. There are only three possible collapse mechanisms for the expected structure, that are shown in Fig. 5.11 and labelled by $i = 1, 2, 3$ respectively. The dissipated plastic power appearing in equation (5.12) is, respectively,

$$\begin{aligned} \tilde{L}_{d1} &= (\tilde{M}_{y4} + 2\tilde{M}_{y6} + \tilde{M}_{y8})\theta \\ \tilde{L}_{d2} &= (\tilde{M}_{y1} + 2\tilde{M}_{y6} + 2\tilde{M}_{y8} + \tilde{M}_{y11})\theta \\ \tilde{L}_{d3} &= (\tilde{M}_{y1} + \tilde{M}_{y4} + \tilde{M}_{y8} + \tilde{M}_{y11})\theta \end{aligned} \tag{5.87}$$

The value of θ is arbitrary and it will be put at $\theta = 1$ in the following.

Since the $\tilde{M}_{yr}$ have been assumed to be normally distributed, the $\tilde{L}_{di}$ are also normally distributed, with means and variances given by (in t m and $t^2 m^2$)

$$\begin{aligned} \mu_{L_1} &= 40 \qquad & s^2_{L_1} &= 6 \\ \mu_{L_2} &= 60 \qquad & s^2_{L_2} &= 10 \\ \mu_{L_3} &= 40 \qquad & s^2_{L_3} &= 4 \end{aligned} \tag{5.88}$$

whence it is possible to infer readily the individual kinematic probabilities P_{Ki}. It is now easy to obtain the domains $\mathscr{S}_{K_p}$ for any value of p, for instance $p = 0.1$ and $p = 0.01$. First, from the normal distribution tables, one obtains

$$\frac{L_d - \mu_{L_i}}{s_{L_i}} = \begin{cases} -1.28 \\ -2.23 \end{cases} \quad \text{when} \quad P_{K_i}(L_d) = p = \begin{cases} 0.1 \\ 0.01 \end{cases} \tag{5.89}$$

Equation (5.49) then provides for each considered mechanism

$$p = 0.1 \qquad \begin{aligned} 2W_1 \qquad &\leqslant 36.85 \qquad & i &= 1 \\ 2W_1 + 3W_2 &\leqslant 55.95 & i &= 2 \\ 3W_2 &\leqslant 37.44 & i &= 3 \end{aligned} \tag{5.90a}$$

$$p = 0.01 \qquad \begin{aligned} 2W_1 \qquad &\leqslant 34.30 \qquad & i &= 1 \\ 2W_1 + 3W_2 &\leqslant 52.64 & i &= 2 \\ 3W_2 &\leqslant 35.34 & i &= 3 \end{aligned} \tag{5.90b}$$

The required domains $\mathscr{S}_{Kp}$ are obtained by the *intersection* of the half-spaces (5.90a) and (5.90b) respectively; their polygonal boundaries $\mathscr{C}_{Kp}$ are drawn in Fig. 5.12(a) and (b).

In order to construct an equilibrated moment diagram $\mathbf{M}_j$ for use in the *static procedure*, consider one of the three collapse mechanisms in Fig. 5.11(a). For a given ratio

$$c = W_2/W_1 \tag{5.91}$$

the equality

$$\mathbf{W}^{\mathrm{T}}\mathbf{u}_i = \sum_r \mu_{M_{yr}} \theta_r = \mu_{M_{yr}} \sum_r \theta_r$$

defines the kinematically sufficient loads for the expected structure and that mechanism

$$W_{1Ki,\bar{\mathbf{R}}}(c) = W^{c}_{1i} \qquad W_{2Ki,\bar{\mathbf{R}}}(c) = cW^{c}_{1i} \tag{5.91$'$}$$

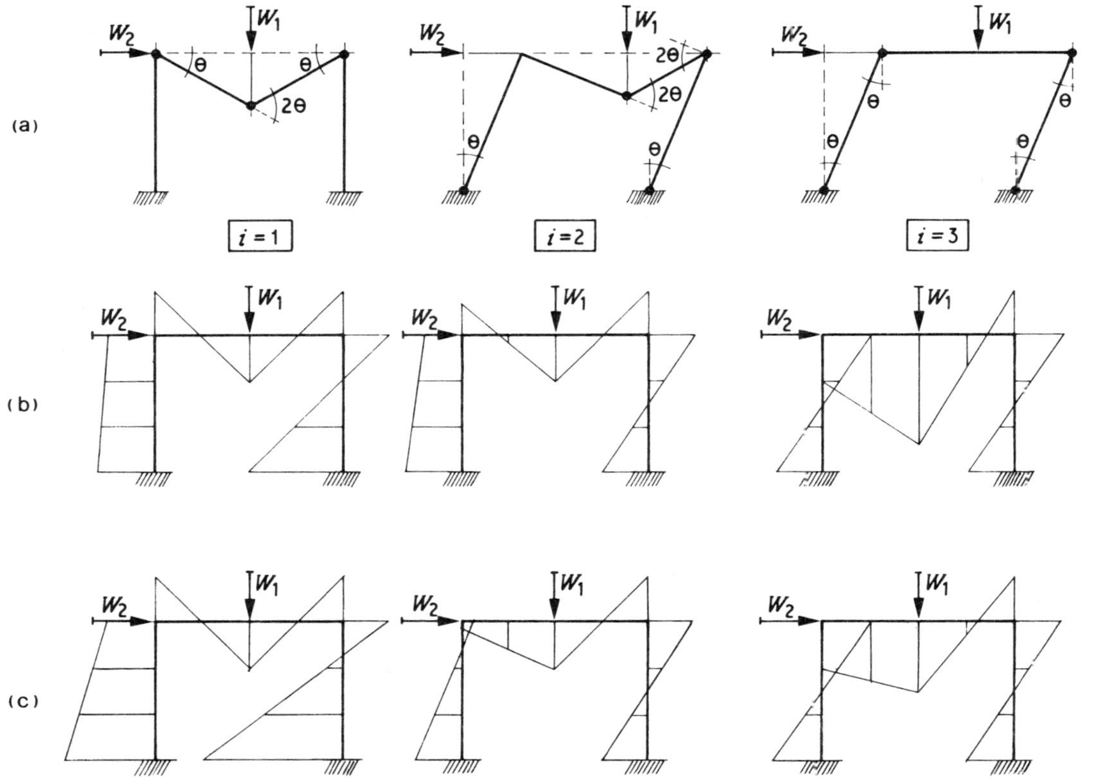

Fig. 5.11 Portal frame: (a) collapse mechanisms of expected structure; (b) equilibrated moment diagrams $\mathbf{M}_i$; $c = W_2/W_1 = 0.5$; (c) equilibrated moment diagrams $\mathbf{M}_i$; $c = W_2/W_1 = 0.8$ (from Augusti and Baratta [8]).

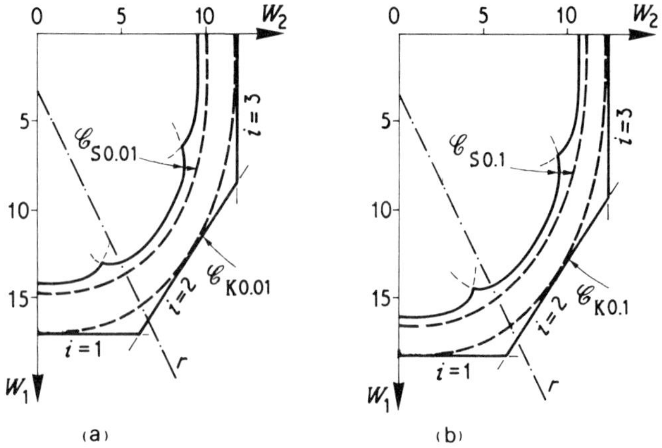

Fig. 5.12 Portal frame: examples of approximate collapse domains; (a) $p = 0.01$; (b) $p = 0.1$ [8].

Following the usual procedures of deterministic limit analysis, it is easy to set up the moment diagram $\mathbf{M}_i^K(c)$ which is in equilibrium with the loads (5.91′), with values $\pm \mu_{M_y}$ at the active plastic hinges. This diagram is unique for mechanisms $i = 2$ and $i = 3$. An infinity of diagrams that satisfy the above condition exist for $i = 1$ (which is a partial collapse mechanism): the diagram satisfying the further condition

$$\sum_{r=1}^{11} (M_{1r}^K(c))^2 = \text{minimum} \tag{5.91″}$$

has been chosen in the present example. Note however that it is not necessary that any correspondence exists between the failure mechanisms and the equilibrated stress fields: the above considerations only reflect a *criterion* that can be followed for choosing three stress fields among many others.

Anyway, three equilibrated diagrams can now be associated with any point (W_1, W_2) in the load space by homotetical reduction

$$\mathbf{M}_i = \mathbf{M}_i(W_1, cW_1) = \frac{W_1}{W_{1i}^c} \mathbf{M}_i^K(c) \tag{5.92}$$

By way of illustration, the diagrams $\mathbf{M}_i$ for two values of $c = W_2/W_1$ are shown in Fig. 5.11(b) and (c).

The probability P_{si} that the single $\mathbf{M}_i$ is not statically admissible is given by (remembering that the limit moments are independent)

$$P_{Si}(W_1, W_2) = 1 - \prod_{r=1}^{11} [1 - P_r(M_{ir})] \tag{5.93}$$

Using equation (5.93) the domains $\mathscr{S}_{Sp_i}$ have been calculated for $p = 0.1$ and $p = 0.01$. Fig. 5.12(a) and (b) show the boundaries $\mathscr{C}_{Sp}$ of the *union* of these domains, which gives the required inner bound for $\mathcal{S}_p$ according to the remark of Section 5.2.5(b).

As already pointed out, both the inner and the outer bound domains can be improved by considering jointly the three moment diagrams and the three collapse mechanisms according to the rules (5.41) and (5.27″) respectively. Since only three diagrams and mechanisms are here taken into consideration, the relevant calculations are comparatively simple. The boundaries of the resulting domains are plotted by heavy dashed lines in Fig. 5.12, showing an improvement more evident on the static (inner) domain than on the kinematic (outer) domain. In the latter, the improvement seems confined in the neighbourhood of the intersections of the straight lines, where two mechanisms have comparable probabilities.

Consider now the loads acting on the portal frame defined in Fig. 5.10 as depending on a single parameter α,

$$W_1 = W_{10} + \alpha W_{1\alpha}$$
$$W_2 = \alpha W_{2\alpha} = c\alpha W_{1\alpha}$$

and put $W_{10} = 3\,\mathrm{t}$, $W_{1\alpha} = 7\,\mathrm{t}$, $c = 0.5$ (Fig. 5.13(a)). The load condition lies on the dash-dotted line r in Fig. 5.12. The collapse mechanism of the expected structure is $i = 2$ in Fig. 5.11; the kinematically sufficient factor is given by

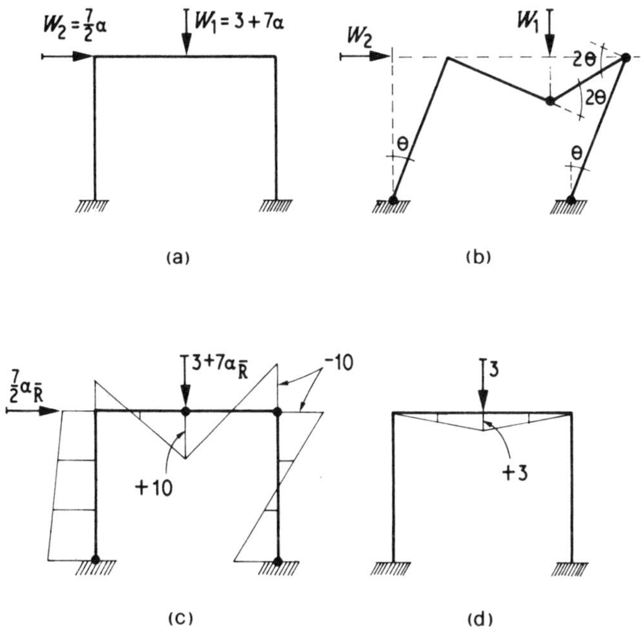

Fig. 5.13 Portal frame under mono-parametric load (forces in tonnes, moments in t m) (from Augusti and Baratta [4]).

(Fig. 5.13(b))

$$\tilde{\alpha}_{K2} = \frac{\tilde{M}_{y1} + 2\tilde{M}_{y6} + 2\tilde{M}_{y8} + \tilde{M}_{y11} - W_{10}\,l/2}{W_{1\alpha}(l/2 + 0.5h)} \tag{5.94}$$

whence the average and variance of $\tilde{\alpha}_{K2}$ are given by

$$\mu_{\alpha_{K2}} = \frac{6\mu_{M_y} - 0.5\,W_{10}l}{3.5\,W_{1\alpha}} = 2.204$$
$$s^2_{\alpha_{K2}} = (0.129)^2 \tag{5.95}$$

The (Gaussian) distribution curve of $\tilde{\alpha}_{K2}$, $P_{K2}(\alpha)$, is plotted in Fig. 5.14(a) and for the low-probability range in semilogarithmic scale in Fig. 5.14(b).

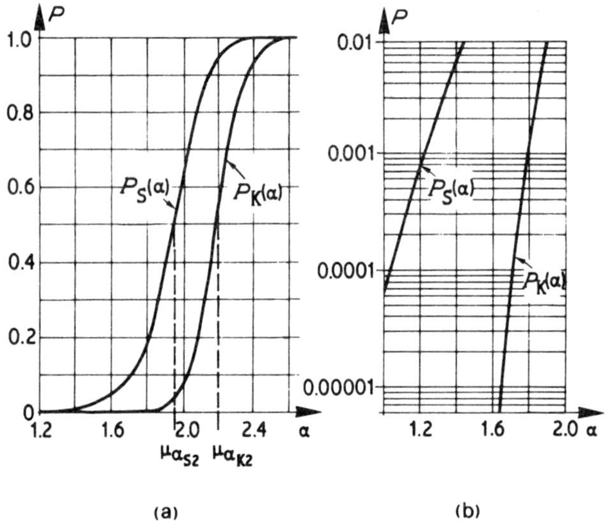

Fig. 5.14 Bounds to CPF of portal frame: mono-parametric load condition [4].

As for the static procedure, reference can be made to the moment diagram in Fig. 5.13(c), which is the stress field at the collapse threshold of the expected structure and would coincide with the $\mathbf{M}_{2\bar{R}}$ in Fig. 5.11(b). Note however that, because of the permanent load W_{10}, moments equilibrating W_{10} would be considered independent of α.

Since the structure is redundant, any diagram $\mathbf{M}_0$ in equilibrium with W_{10} can be taken as well. In the present application the diagram of Fig. 5.13(d) has been used. Equation (5.93) should then be applied introducing the moments (for $i = 2$)

$$M_{2\bar{R},r} = M_{0r} + \alpha(M^{K}_{2r} - M_{0r})/\mu_{\alpha_{K2}} \tag{5.96}$$

in place of M_{ir}.

The curve $P_{S2}(\alpha)$ plotted in Fig. 5.14(a) and (b) has been obtained; the corresponding mean and variance are

$$\begin{aligned} \mu_{\alpha_{S2}} &= 1.944 \\ s^2_{\alpha_{S2}} &= (0.178)^2 \end{aligned} \tag{5.97}$$

The results allow a good evaluation of the expected collapse factor μ_{α_R}. In fact, from inequalities (5.61)

$$\mu_{\alpha_R} = 2.074 \pm 0.130 \tag{5.98}$$

which approximates μ_{α_R} with an error of less than $\pm 6.3\%$.

Much less close are the bounds that have been calculated for the variance, namely

$$(0.06)^2 \leqslant s^2_{\alpha R} \leqslant (0.28)^2 \tag{5.99}$$

These results could be improved by increasing the number of investigated stress vectors and collapse mechanisms, according to the procedure outlined in Section 5.2.4 (a) and (b), as made in the previous example.

Varying the value of $c = W_{2\alpha}/W_{1\alpha}$, it is possible to repeat the above calculations to obtain upper and lower bounds for the CDF of α_R for any value of c. Let $P'_S(\alpha)$ be the static upper bound calculated working on the ultimate bending moment diagram of the expected structure, and let $P_S(\alpha)$ be the improvement of this upper bound that can be achieved by introducing up to 10 moment diagrams in the calculations according to the procedure of equation (5.41). It is found that in the present example, for almost every value of c, the kinematic bound calculated on the basis of the actual collapse mechanism for the expected structure is a good (although unsafe) approximation of the true collapse probability out of the range of very small values of the collapse probability. So $P_{Ki}(\alpha)$, for proper value of i, can be considered as a reliable reference for investigating the efficiency of the static approximation.

Such efficiency can be judged, as in the truss example, by comparison with the CDF of the elastic limit factor $\tilde{\alpha}_E$, i.e. by checking what fraction of the total advantage allowed by ductility can be evidenced by the static procedure. To this aim an *effectiveness index* of the CDF $P_S(\alpha)$ can be introduced by the following position

$$I[P_S(\alpha)] = \frac{\displaystyle\int_0^\infty [P_E(\alpha) - P_S(\alpha)]\mathrm{d}\alpha}{\displaystyle\int_0^\infty [P_E(\alpha) - P_{Ki}(\alpha)]\mathrm{d}\alpha} = \frac{\mu_S - \mu_E}{\mu_{Ki} - \mu_E} \tag{5.100}$$

where $P_E(\alpha)$ is the CDF of the elastic limit factor.

The index I yields an *overall evaluation* of the improvement in safety that can be evidenced by the approximation $P_S(\alpha)$ in comparison with the actual improvement that is made available by the structural ductility. Obviously, the index cannot point out significant differences in the range of small probabilities.

Fig. 5.15 shows a plot of the expected values $\mu_E(c)$ (the elastic limit factor), $\mu_S(c)$ (the statically admissible factor on the basis of 10 moment diagrams) and $\mu_K(c)$

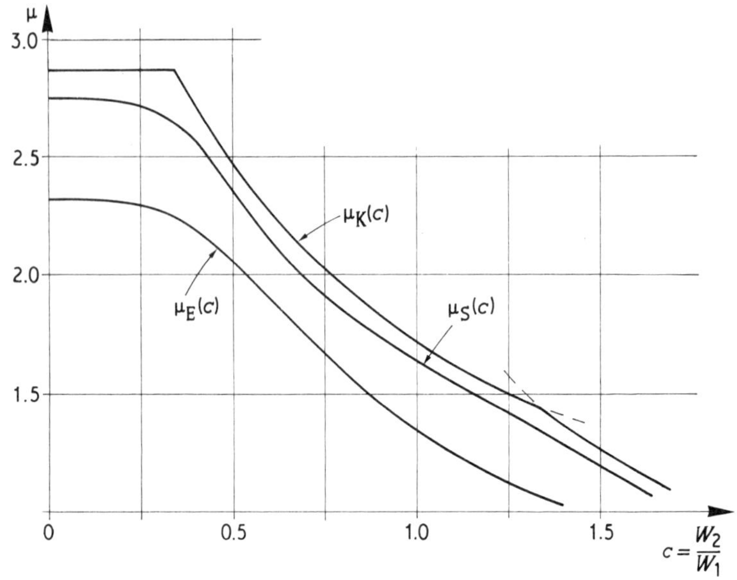

Fig. 5.15 Portal frame: mean values of elastic limit factor, statically admissible factor and kinematically sufficient factor vs force ratio c [7].

(the kinematically admissible approximations corresponding to the three possible failure mechanisms of the expected structure) as functions of the ratio c. As can be argued from the diagram, the estimates of the expected value of the actual collapse factor $\tilde{\alpha}_R$ are rather close, and show the acceptable effectiveness of the static approximation, especially taking into account that the upper curve could also be improved.

Fig. 5.16 shows in semilogarithmic plots the curves $P_E(\alpha)$, $P'_S(\alpha)$, $P_S(\alpha)$ and $P_{K1}(\alpha)$ for $c = 0.05$, 0.50 and 1.35. It can be noted that the effectiveness index (equation (5.100)) takes in all the cases values above 70% proving that static approximations are able to show most of the advantage allowed by plastic redistribution.

The examples referred to above prompt us to remark that *accuracy* is the crucial point for the static method, as already pointed at the end of Section 5.2.4 (b), while *uncertainty* is the weakness of the kinematic method.

In fact, as regards the kinematic approximation, it should be kept in mind that in the range of small probabilities the calculations can be simplified simply by summing the individual probabilities pertaining to every collapse mechanism considered in the calculations, i.e. putting

$$P_R(\alpha) = \sum_{i=1}^{n} P_{Ki}(\alpha) \qquad (5.101)$$

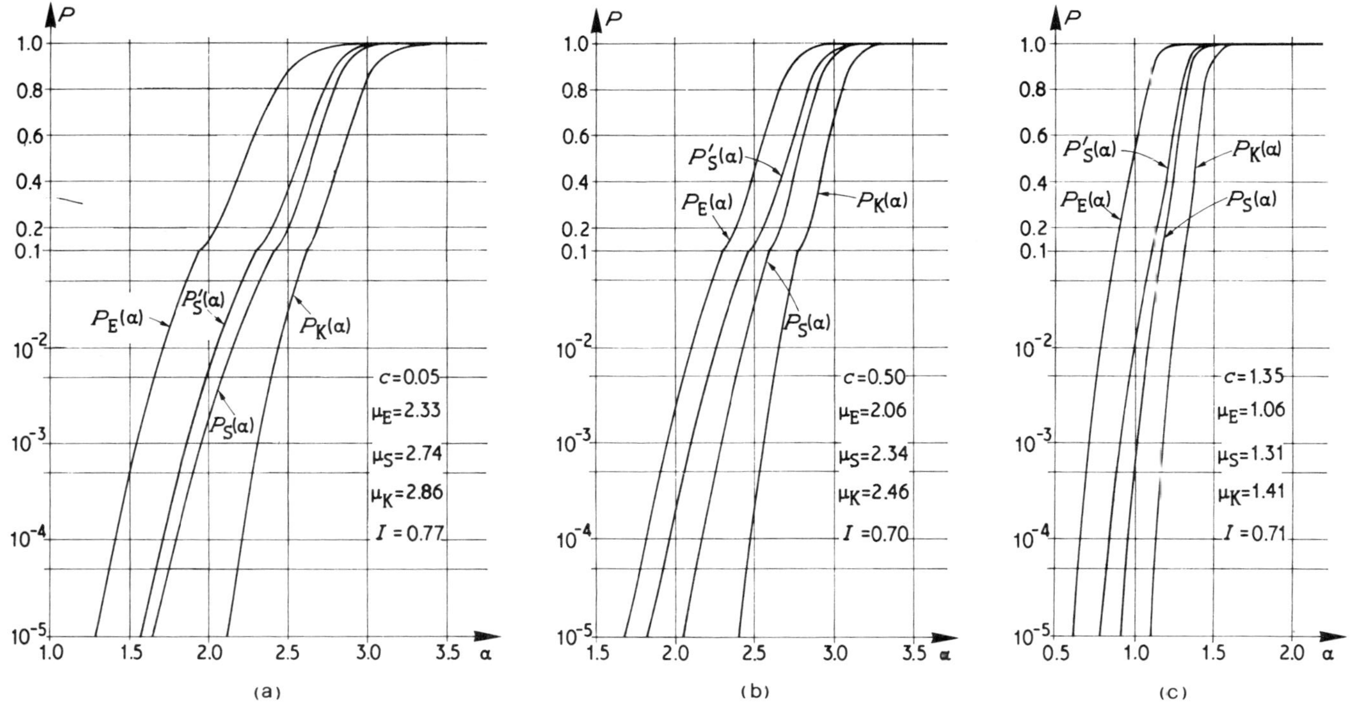

Fig. 5.16 Portal frame: CDF of elastic limit, statically admissible and kinematically sufficient factors for three different values of force ratio c [7].

Equation (5.101), as well as its rigorous version (5.27″), is rather difficult to use, despite its apparent ease. Numerical experimentation, in fact, has proved that in the range of small probabilities there is a frequent interchange of the *dominant mechanism* (i.e. the mechanism that has the largest individual probability of collapse), which many times does not coincide with any of the apparently most likely mechanisms (for instance, those of the expected structure). Thus, it is very difficult, in this range of probabilities, to recognize all significant mechanisms to be included in equation (5.101). In the case of the frame in Fig. 5.10, more than 200 possible mechanisms were found, and many of these became significant in the range of small probabilities. It is therefore to be expected that failure may occur in a different mode from the one given by deterministic limit analysis, if failure is caused by abnormal weakness of the resisting members. For example, if $c = 2.1$ and if loads corresponding to a factor $\alpha < 0.7$ are applied to the frame, failure is expected to occur according to mechanism 3 (Fig. 5.11), although this is the most dangerous failure mechanism for the expected structure only if $c \geqslant 3.2$. It appears, therefore, that deterministic analyses may not be sufficient to detect the real reasons for structural malfunctions, and a probabilistic point of view may be necessary for a realistic discussion of structural failures.

5.3 PLASTIC STRUCTURES UNDER TIME-VARYING LOADS

It is well known [1–3] that classical limit analysis can be extended to the case of loads $\mathbf{W}'(t)$ varying arbitrarily (and quasi-statically) in time, but remaining limited in some way, for instance by inequalities of the type

$$0 \leqslant \mathbf{W}'(t) \leqslant \mathbf{W} \tag{5.102}$$

where, as usual, a vector inequality is intended as an inequality between all corresponding components of the vectors‡. In this case, in fact, even if the loads $\mathbf{W}$ are not able to cause collapse of the structure, a process of *incremental collapse* may be set up, in which plastic work is dissipated in each loading cycle, leading either to an unbounded increase of permanent displacements (which form an *incremental collapse mechanism*) or to rupture by alternating plasticity (*low-cycle fatigue*) in one or more plastic hinges. In contrast, if the total plastic work (and the displacements) remain limited, even under infinite cycles of applications of loads respecting inequalities (5.102), the structure is said to *shake down*. Kinematic and static theorems have been formulated for this problem too, and will be briefly recalled in Section 5.3.1; the following sections will deal with the extension of the incremental-collapse/shakedown analysis to a probabilistic context.

‡ It is in general convenient to introduce loads that may change sign by increasing the number of positive definite components of $\mathbf{W}$.

5.3.1 The theorems of shakedown and incremental collapse

Let $\boldsymbol{\sigma}_E(t)$ be the *elastic* stress vector connected with the time-varying loads $\mathbf{W}'(t)$, and $\boldsymbol{\sigma}_0$ a self-equilibrated stress field (i.e. a stress vector in equilibrium with nil applied loads). Using the same symbols as in Section 5.2.1, the *static theorem* (also called *shakedown theorem*) states that the structure *shakes down* (i.e. incremental collapse does *not* take place) if (cf. equations (5.1) and (5.1′))

$$\begin{aligned} \boldsymbol{\chi}_E + \mathbf{H}\boldsymbol{\sigma}_0 - \boldsymbol{\chi} &\leqslant 0 \\ \mathbf{Q}^T\boldsymbol{\sigma}_0 &= \mathbf{0} \end{aligned} \tag{5.103}$$

where

$$\boldsymbol{\chi}_E = \max_t \{\mathbf{H}\boldsymbol{\sigma}_E(t)\} \tag{5.104}$$

In other words, the structure cannot be lead to incremental collapse if a self-equilibrated stress vector $\boldsymbol{\sigma}_0$ exists such that, added to vector $\boldsymbol{\chi}_E$ of maximum elastic stress effects, it does not violate the strength inequality.

On the other hand, the *kinematic theorem of incremental collapse* states that incremental collapse takes place if a plastic mechanism $\mathbf{u}$ exists such that the internal work calculated by maximum elastic stresses exceeds the internal work dissipated by plastic stresses activated in the mechanism. In symbols (cf. equations (5.3) and (5.4))

$$\begin{aligned} \boldsymbol{\chi}_E^T \boldsymbol{\lambda}_p &= 1 \\ \mathbf{Q}\mathbf{u} &= \mathbf{H}^T\boldsymbol{\lambda}_p \\ \boldsymbol{\lambda}_p &\geqslant \mathbf{0} \\ \boldsymbol{\chi}_E^T \boldsymbol{\lambda}_p &\geqslant \boldsymbol{\chi}^T\boldsymbol{\lambda}_p \end{aligned} \tag{5.105}$$

A load condition $\mathbf{W}$ such that, when the time-varying loads $\mathbf{W}'(t)$ verify inequality (5.102), also (5.103) is verified (and therefore the structure shakes down) will be named in the following a *statically admissible load condition.* Conversely, a load condition $\mathbf{W}$ such that inequality (5.105) holds is named in turn a *kinematically sufficient load condition.* It can be proved that, if the maximum load vector is assigned in the form $\alpha\mathbf{W}$ (i.e. to within a scalar factor α), the statically admissible load conditions $\alpha_{SV}\mathbf{W}$ and the kinematically sufficient conditions $\alpha_{KV}\mathbf{W}$ define two classes of coefficients α_{SV} and α_{KV} which respect the inequalities

$$\alpha_{SV} \leqslant \alpha_{RV} \leqslant \alpha_{KV} \tag{5.106}$$

with α_{RV} unique. In other words, only one load condition of the family $\alpha\mathbf{W}$ exists (and is denoted by $\alpha_{RV}\mathbf{W}$) such that both (5.103) and (5.105) hold, i.e. is at the same time statically admissible and kinematically sufficient (under variable loads).

5.3.2 Probabilistic analysis; random structural strength

Let $\mathbf{W}$, the range in which the load vector can vary (5.102), be assigned deterministically and assume that the strength vector $\tilde{\boldsymbol{\chi}}$ is random, with known

JPDF. Denote by $P_{RV}(\mathbf{W})$ the probability that the structure does *not* shake down when the loads vary in $\mathbf{W}$, and by $\mathscr{V}_V$ the set of all possible incremental collapse mechanisms of the structure‡: $P_{RV}(\mathbf{W})$ is given, according to (5.105), by

$$P_{RV}(\mathbf{W}) = \text{Prob}[\exists\ \mathbf{u} \in \mathscr{V}_V\text{:}\quad \tilde{\chi}_E^T \boldsymbol{\lambda}_p \geqslant \tilde{\chi}^T \boldsymbol{\lambda}_p] \tag{5.107}$$

Conversely, the static formulation of the shakedown problem leads to a different expression for the same probability. If $\mathscr{T}_0$ is the set of self-equilibrated stress vectors $\boldsymbol{\sigma}_0$, inequality (5.103) yields

$$P_{RV}(\mathbf{W}) = 1 - \text{Prob}[\exists \boldsymbol{\sigma}_0 \in \mathscr{T}_0 \text{:}\ \chi_E + \mathbf{H}\boldsymbol{\sigma}_0 - \tilde{\chi} \leqslant 0] \tag{5.108}$$

Equations (5.107) and (5.108) are equivalent to (5.10) and (5.11) in static limit analysis: also, in the present case, both are very difficult to calculate. However, extension of the theorems of shakedown and incremental collapse to random-strength structures provides a useful tool for practical calculations, in that the extended theorems allow one to work on limited subsets of $\mathscr{V}_V$ and $\mathscr{T}_0$, as will be seen below.

Let E_V be the event 'incremental collapse of the structure' (coincident with the event in the square brackets of equation (5.107)). Take a set $\mathscr{V}_{Vn} \subseteq \mathscr{V}_V$ of n mechanisms among the possible incremental collapse mechanisms of the structure, and denote by E_{KV} the event that one among these n mechanisms is kinematically sufficient under variable loads respecting (5.102), i.e. with already known symbols,

$$[E_{KV}] = [\exists\ \mathbf{u} \in \mathscr{V}_{V_n}\text{:}\ \chi_E^T \boldsymbol{\lambda}_p \geqslant \tilde{\chi}^T \boldsymbol{\lambda}_p] \tag{5.109}$$

According to the kinematic theorem of incremental collapse

$$[E_{KV}] \Rightarrow [E] \tag{5.110}$$

Now let

$$P_{KV}(\mathbf{W}) = \text{Prob}[E_{KV}] \tag{5.111}$$

while by definition

$$P_{RV}(\mathbf{W}) = \text{Prob}[E_V] \tag{5.112}$$

Then (in analogy to inequality (5.23))

$$P_{KV}(\mathbf{W}) \leqslant P_{RV}(\mathbf{W}) \tag{5.113}$$

i.e. the probability of incremental collapse according to a mechanism of the limited set $\mathscr{V}_{V_n}$ is not larger than the actual probability of incremental collapse.

Let now $\mathscr{T}_{0m} \subseteq \mathscr{T}_0$ be a set of m self-equilibrated stress fields $\boldsymbol{\sigma}_{0j}$ ($j = 1, 2, \ldots, m$), denote by $[E_{SV}]$ the event that *none* of these stress fields is statically admissible under loads varying in the range (5.102), and indicate by $P_{SV}(\mathbf{W})$ the probability of this event

$$\begin{aligned} P_{SV}(\mathbf{W}) &= \text{Prob}[E_{SV}] \\ &= 1 - \text{Prob}[\exists \boldsymbol{\sigma}_0 \in \mathscr{T}_{0m}\text{:}\ \chi_E + \mathbf{H}\boldsymbol{\sigma}_0 - \tilde{\chi} \leqslant 0] \end{aligned} \tag{5.114}$$

‡ Note that, in general, $\mathscr{V}_V \supseteq \mathscr{V}$.

(Note that the event in the square brackets in the last term of (5.114) is the event that one of the fields of $\mathscr{T}_{0m}$ *is* statically admissible.) Then, according to the shakedown theorem

$$[E_{\mathrm{RV}}] \Rightarrow [E_{\mathrm{SV}}] \tag{5.115}$$

whence

$$P_{\mathrm{RV}}(\mathbf{W}) \leqslant P_{\mathrm{SV}}(\mathbf{W}) \tag{5.116}$$

i.e. the probability of not finding an admissible field in a discrete set of m self-equilibrated fields is not smaller than the actual probability of the structure not shaking down.

Inequalities (5.113) and (5.116), which are perfectly analogous to (5.23), represent the kinematic and static theorems of probabilistic collapse analysis under variable repeated loads (probabilistic incremental collapse analysis) [9, 10].

Note that, in incremental collapse, the significant components of the load vector may include thermal effects and other applied deformations (which have no effect on static plastic collapse). Note also that, if the applied loads also include dynamic effects, the formulation of the static theorem does not vary, but $\mathscr{T}_0$ must include all stress vectors consequent to elastic motions depending on all possible initial conditions (i.e. initial self-stresses, initial displacements, initial velocities) and calculated in the absence of the forcing cause; while χ_{E} must take into account all stress vectors associated with the elastic motions consequent to the applied loading. Unfortunately, an equivalent kinematic theorem has not been found for the dynamic case, and therefore the kinematic theorem of probabilistic incremental collapse cannot include dynamic loading.

Finally, if the load range $\mathbf{W}$ is defined to within a factor α, the probabilities $P_{\mathrm{SV}}(\alpha\mathbf{W})$ and $P_{\mathrm{KV}}(\alpha\mathbf{W})$ become the CDF of the already defined statically admissible factor α_{SV} and kinematically sufficient factor α_{KV} respectively, while $P_{\mathrm{RV}}(\alpha\mathbf{W})$ is the CDF of the actual incremental collapse factor α_{RV}. Instead of (5.106), the following inequalities hold between the above-defined CDF's

$$P_{\mathrm{KV}}(\alpha) \leqslant P_{\mathrm{RV}}(\alpha) \leqslant P_{\mathrm{SV}}(\alpha) \tag{5.117}$$

Relations between the average values and the variances of α_{KV}, α_{RV} and α_{SV} perfectly analogous to those found in Section 5.2.6 can be established by quite similar arguments. In particular, it turns out also that the incremental collapse factor of the *expected structure* (i.e. the deterministic structure with average strength values) is larger than the average of the incremental collapse factors of the random-strength structures.

5.3.3 An example of probabilistic shakedown analysis

Consider again the portal frame described and studied, under static loads, in Section 5.2.7(b) (Fig. 5.10). Let it now be loaded by two (slowly) varying forces

$W'_1(t)$ and $W'_2(t)$, comprised in the ranges

$$0 \leqslant W'_1(t) \leqslant W_1 \qquad 0 \leqslant W'_2(t) \leqslant W_2 \tag{5.118}$$

For a deterministic structure, four regions can be recognized in the **W**-space, as plotted qualitatively in Fig. 5.17, namely the elastic region $\mathscr{S}_E$, the shakedown region $\mathscr{S}_V$, the incremental collapse region $\mathscr{U}_{IC}$ and the region $\mathscr{U}$ of plastic collapse under fixed loads (instantaneous or 'static' collapse).

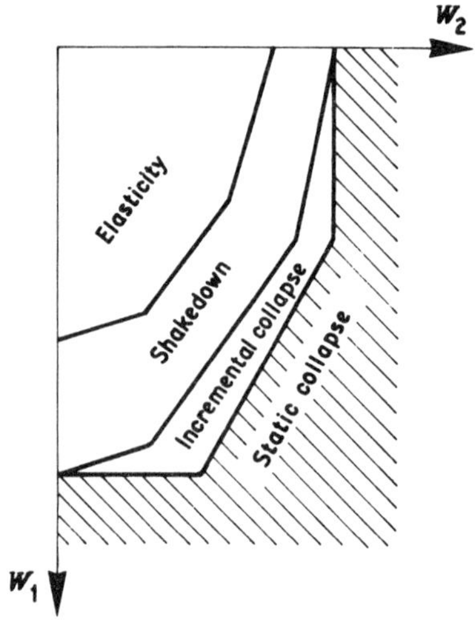

Fig. 5.17 Shakedown and collapse regions of the frame in Fig 5.10, when the strengths are deterministic.

Denoting by $\mathscr{S}$ the safe region with respect to the instantaneous (static) collapse

$$\mathscr{S} = \mathscr{S}_V + \mathscr{U}_{IC} = \mathscr{D} - \mathscr{U} \tag{5.119}$$

where $\mathscr{D}$ is the space of definition of the loads **W**. (Note that in many cases, $\mathscr{U}_{IC}$ vanishes.)

When the structural strength vector is random, the boundaries between the four above-defined regions become fuzzy.

As an example, the shakedown analysis of the portal frame will now be presented, with the same definitions of the random strengths as in Section 5.2.7(b), i.e. in each of the 11 critical sections $|M_{yr}| = \tilde{M}_{yr}$ $(r = 1, \ldots, 11)$, $\tilde{M}_{yr}$ being a set of independent random variables distributed according to the Gaussian probability law $P_r(M_y)$ with mean value $\mu_{M_y} = 10\,\text{t m}$ and standard deviation $s_{M_y} = 1\,\text{t m}$ in every section [9, 10].

The inner and outer approximate domains which result from the analysis coincide with the level curves of the bounding incremental collapse probabilities P_{SV} and P_{KV}, drawn in the plane of the load parameters W_1 and W_2. To obtain the curves $P_{SV}(W_1, W_2) = p$, consider the three bending moment diagrams (5.92);

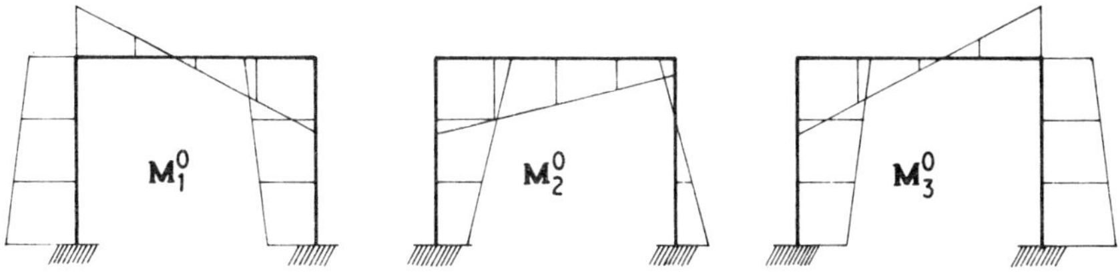

Fig. 5.18 Self-equilibrated bending moment diagrams [9].

then three self-equilibrated diagrams (Fig. 5.18) are associated with each pair (W_1, cW_1) by setting

$$\mathbf{M}_j^0(W_1, W_2) = \mathbf{M}_j(W_1, W_2) - \mathbf{M}^E(W_1, W_2) \qquad c = W_2/W_1, j = 1, 2, 3 \tag{5.120}$$

where $\mathbf{M}^E(W_1, W_2)$ denotes the elastic bending moment diagram relevant to the loads W_1 and W_2. Now, introduce the vectors $\mathbf{M}^E_{\max}$ and $\mathbf{M}^E_{\min}$ denoting the maximum and minimum bending moment diagrams and for each pair $(W_1, c\,W_1)$ put

$$M_{rj} = \max\{|M^E_{r\max} + M^0_{jr}|, |M^E_{r\min} + M^0_{jr}|\} \tag{5.104'}$$

$$P_j = \prod_{r=1}^{11} [1 - P_r(M_{rj})]$$

$$P_{js} = \prod_{r=1}^{11} [1 - P_r(\max\{M_{rj}, M_{rs}\})] \qquad j, s = 1, 2, 3; s > j \tag{5.121}$$

$$P_{123} = \prod_{r=1}^{11} [1 - P_r(\max\{M_{r1}, M_{r2}, M_{r3}\})]$$

From (5.114) and (5.104′) it is possible to obtain the expression for P_{SV}:

$$1 - P_{SV}(W_1, W_2) = P_1 + P_2 + P_3 - P_{12} - P_{23} - P_{13} + P_{123} \tag{5.122}$$

Function $P_{SV}(W_1, cW_1)$ is plotted in Fig. 5.19 in a semilogarithmic diagram versus the load parameter W_1 for a number of values of the ratio c, in the range $P_{SV} \leqslant 0.1$. Intersecting these curves by horizontal lines it is possible to obtain points of the corresponding level curves. Examples of these curves are shown in Fig. 5.20.

Also for the kinematic curves, the three collapse mechanisms considered when obtaining the level curves, $P_{SV}(W_1, W_2) = p$, are analysed. Furthermore, it is convenient, in order to keep the calculations simple, to consider each failure mechanism separately, and to calculate only the individual failure probabilities. By this simplification, it turns out that the level curves are composed of a number of straight lines. In order to illustrate the procedure, consider, for instance, mechanism $\mathbf{u}_2$ (Fig. 5.11). The equation of virtual work for this mechanism is

$$3W_2 + 2.1816\,W_1 = M_{y1} + 2M_{y6} + 2M_{y8} + M_{y11} \tag{5.123}$$

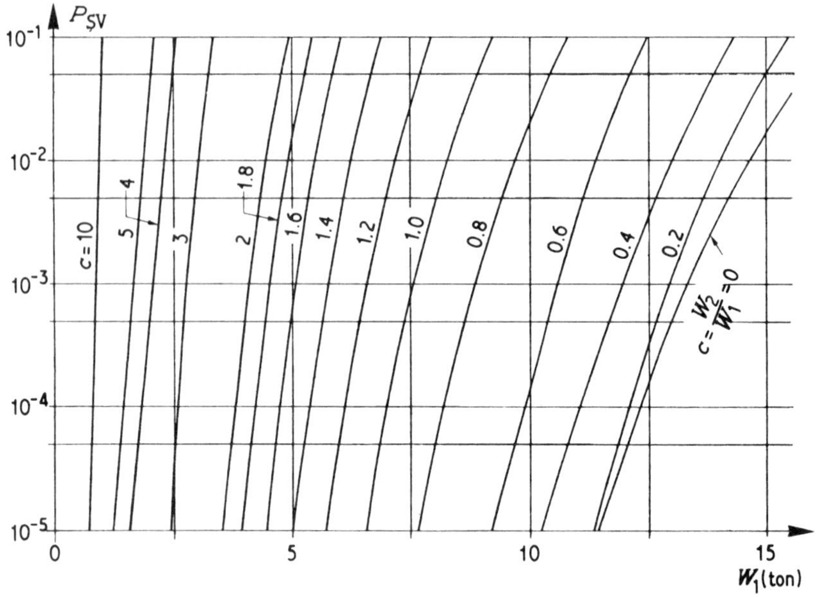

Fig. 5.19 Upper bound probability P_{SV} vs load parameter W_1 [10].

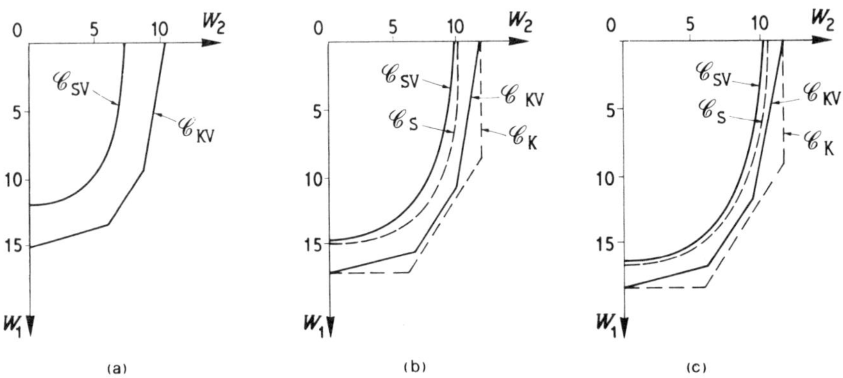

Fig. 5.20 Examples of approximate collapse domains of portal frame: solid lines, variable repeated loads; broken lines, static (monotonic loads): (a) $p = 0.00001$; (b) $p = 0.01$; (c) $p = 0.1$ [10].

which differs from equation (5.87) and (5.90) because of the different meaning of the external work, as outlined by the comparison of equation (5.105) with equation (5.5).

The second member of (5.123) is a Gaussian random variable $\tilde{L}_{dV2}$ with mean value and variance given by equation (5.88). In order to keep the failure probability below p, it is necessary that the first member of equation (5.123) remains smaller than $\mu_{L_2} + u_P s_{L_2}$, u_P being the value of the standard normal variable u which is not exceeded with probability p. For $p = 0.1$, the equation of the first side of the boundary is given by (cf. equation (5.90a))

$$3W_2 + 2.1816W_1 = 6\mu_{My} - 1.28\sqrt{10}s_{My} \tag{5.123'}$$

Similarly, the equation for the alternating plasticity mechanism relevant, for instance, to section 1 is:

$$M^E_{1max} - M^E_{1min} = 0.1818W_1 + 0.8864W_2 = 2\tilde{M}_{y1} \tag{5.124}$$

For $p = 0.1$, the equation of another side of the boundary is thus given by

$$0.0909W_1 + 0.4432W_2 = \mu_{My} - 1.28\,s_{My} \tag{5.124'}$$

Note now that the same equation (5.124) is obtained when section 11 is considered as critical, and rupture occurs if equation (5.124) holds in section 1 or in section 11. Therefore, if p_1 is the probability that equation (5.124) holds in only one section, the compound probability will be

$$(1 - p_1)^2 = 1 - p \tag{5.125}$$

whence

$$p_1 = 1 - \sqrt{(1 - p)} \tag{5.125'}$$

Replacing u_p by u_{p1}, it is possible to obtain directly the overall probability of rupture relevant to both mechanisms: for $p = 0.1$, equation (5.125') yields $p^1 = 0.05$, $u_{p1} = 1.65$, and equation (5.124') gives

$$0.0909W_1 + 0.4432W_2 = \mu_{My} - 1.65s_{My} \tag{5.126}$$

Examples of curves P_{KV} are shown in Fig. 5.20. It is soon noted that only the three mechanisms of Fig. 5.11(a) yield significant bounds on the probability of incremental collapse. This is clearly due to the definition (5.118) assigned to the set of admissible loading; however, even under these conditions, the shakedown domains are significantly reduced with respect to the corresponding domains relevant to monotonic leading.

5.3.4 Significance of 'incremental collapse' and 'shakedown' in a probabilistic context

As evident, the classical formulation of the theory of limit analysis under variable repeated loading implies that *shakedown* and *incremental collapse* are looked at from two different viewpoints. In fact, if the static (*shakedown*) theorem is satisfied, dissipated plastic work and diplacements of the structure do remain limited regardless of the number and specific succession of the load applications, provided only that at any time the loads are comprised within the given finite region in the load space: the shakedown theorem is in other words a *sufficient* condition for survival of the structure (under the assumptions of limit analysis theory).

On the other hand, if the kinematic (*incremental collapse*) theorem is satisfied, unbounded plastic work is dissipated and unlimited deformations develop if some appropriate form of cycling load process is set up, such that the loads still respect at all times bounds of the type (5.102), but the number of load cycles increases indefinitely with time. However, at any finite time, dissipated plastic work and permanent deformations remain finite unless the load vector falls at least once in the instantaneous (or 'static') collapse region $\mathscr{U}$ (cf. Fig. 5.17).

The first conclusion that can be drawn from the above considerations is that, assuming perfectly plastic structural behaviour and identifying collapse with unbounded displacements, fully reliable bounds to the probability of collapse, P_{fail}, can be obtained by applying the *kinematic approach* with respect to the collapse under the largest loads (instantaneous collapse), and the *static approach* when the effects of possible load repetitions must be taken into account (incremental collapse). In other words, the reliability bounds (5.25″) should more safely be written

$$
\begin{aligned}
P_{\text{S fail}} &= \int_{\mathscr{D}} p_{\text{m}}(\mathbf{W})\, P_{\text{SV}}(\mathbf{W})\, \text{d}\mathbf{W} \\
P_{\text{K fail}} &= \int_{\mathscr{D}} p_{\text{m}}(\mathbf{W}) P_{\text{K}}(\mathbf{W})\, \text{d}\mathbf{W}
\end{aligned}
\tag{5.127}
$$

where $P_{\text{SV}}(\mathbf{W})$ and $P_{\text{K}}(\mathbf{W})$ have the now well defined meaning, and $p_{\text{m}}(\mathbf{W})$ is the JPDF of the extreme loads (*maxima*) during the assigned structural lifetime.

Looking more closely at the significance of the incremental collapse/shakedown concepts, they can be regarded as the possible asymptotic outcomes (at infinite time) of the process of plastic *adaptation* [11]. In fact, all structures, subjected to any loading history, try to adapt themselves to resist the applied loads in the best possible way: in ductile structures, this is achieved by means of plastic deformations and stress redistributions which move the region of elastic response, trying to fit it to the domain of the applied loads. If the latter domain can be contained in the former, the process of plastic adaptation will eventually stop and the structure shake down; if not, the 'adaptation' of the structure is never complete, and plastic deformations may continue indefinitely (as depicted diagrammatically in Fig. 5.21(a) and (b) respectively). In either case, adaptation tends to increase the mean time between subsequent plastic deformations.

However, it is conceptually contradictory to restrict a stochastic load process for an infinite time within a deterministically bounded domain. One must instead consider probabilistically defined *reliable load domains*, i.e. domains in which the loads fall with sufficiently large probability. But such domains grow larger in time, as can easily be demonstrated by considering the example of a load process consisting of a sequence of stochastically independent load applications from the same population [11]. Assume that at each load application, the probability of the loads falling in a domain $\mathscr{X}_1$ (Fig. 5.22) is p_{w1}: to have the same probability for 2, 3, . . . , n independent load applications, one should take increasing domains

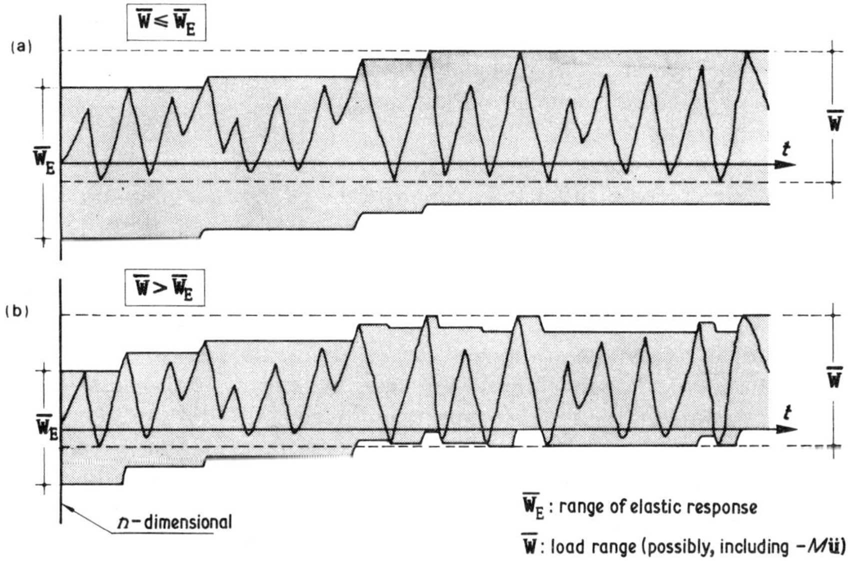

Fig. 5.21 Basic cases of elastic–plastic response to random loads: (a) shakedown; (b) partial adaptation.

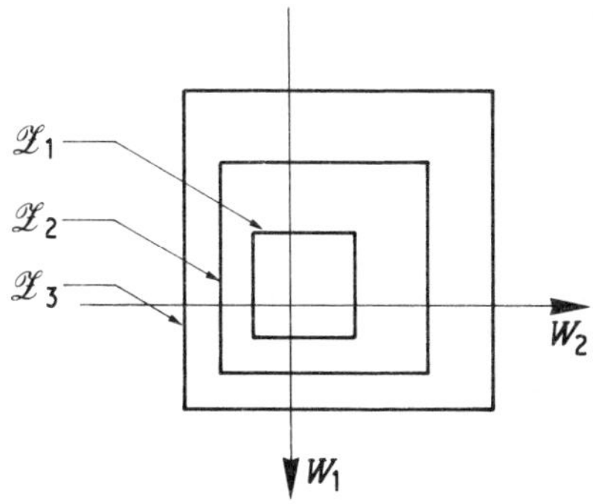

Fig. 5.22 Domains with a given probability of containing the loads in 1, 2, 3, . . . applications (diagrammatic).

$\mathscr{L}_2, \mathscr{L}_3, \ldots, \mathscr{L}_n$ which correspond to single-load probabilities $p_{w2}, p_{w3}, \ldots,$ p_{w3} defined by

$$p_{w1} = p_{w2}^2 = p_{w3}^3 = \ldots = p_{wn}^n \tag{5.128}$$

In conclusion, one should define a finite lifetime and obtain the JPDF of the extreme loads $p_m(\mathbf{W})$ already introduced in (5.127). But it would be meaningless to use this JPDF to obtain a *probability of incremental collapse*, because the number of violations of the shakedown domain during the lifetime is certainly finite regardless of whether the shakedown domain is violated or not, and therefore the permanent deformations of the structure also remain finite unless the instantaneous collapse domain, which remains the only one with a definite meaning, is violated too. Therefore, this probability of incremental collapse can only be regarded as a surely safe estimate of the actual probability of failure, closer than the upper bound of equations (5.127) [12].

However, a more significant result than the bounds given by equations (5.127) can only be obtained if the time evolution of the plastic adaptation process is considered. In short, one should start from probabilistic data on the loading process (in particular, the probability of excursion out of any limited domain in a given time interval), and then try to derive from these data the probability of reaching given values of permanent deformation in the same time interval. These results would allow one to obtain the reliability of the structure under consideration with respect both to limit states defined by specific values of a permanent deformation (or displacement) and to complete collapse (i.e. infinite displacements).

Such information, evidently much more complete than those yielded by limit and shakedown analysis, is however very difficult to obtain, and usually recourse is made to numerical simulation techniques as will be seen in Chapter 9.

It is also possible to make use of the techniques which allow one to bound, on the one hand, any displacement of plastic structures under loads varying within the shakedown domain and, on the other hand, the probability that a random process of known properties violates any given threshold in a given time interval, and obtain the bounding procedure that will be illustrated in the following sections.

5.3.5 Plastic adaptation as a stochastic process

As indicated in the preceding section, the correct way to tackle the problem of plastic adaptation under time-varying loads is to try to establish relations on the time variation of the average interval between two plastic excursions, the rate of accumulation of plastic displacements (or the accumulated displacements), and the internal plastic dissipation (or the plastic deformation). To this aim, bounding procedures can be set up, as will be now illustrated [13].

Assuming that the structure has been discretized into constant strain/constant stress finite elements, let $\boldsymbol{\varepsilon}_{\mathrm{E}}$ and $\boldsymbol{\varepsilon}_{\mathrm{P}}$ be the elastic and the plastic strain vectors, so that

$$\boldsymbol{\varepsilon} = \boldsymbol{\varepsilon}_{\mathrm{E}} + \boldsymbol{\varepsilon}_{\mathrm{p}} = \mathbf{Q}\mathbf{u} \tag{5.129}$$

is the total strain vector, $\mathbf{u}$ being the displacement vector, and let

$$\boldsymbol{\sigma} = \mathbf{G}\boldsymbol{\varepsilon}_{\mathrm{E}} \tag{5.130}$$

be the stress vector, $\mathbf{G}$ being the matrix of the elastic coefficients.

Let the yield condition in the stress space be piecewise linear (or linearized) so that plastic deformation is regulated as follows [14]:

$$\begin{aligned} \boldsymbol{\varepsilon}_p &= \mathbf{H}^T \boldsymbol{\lambda}_p \\ \dot{\boldsymbol{\lambda}}_p &> \mathbf{0} \\ \boldsymbol{\phi} &= \mathbf{H}\boldsymbol{\sigma} - \boldsymbol{\Upsilon}\boldsymbol{\lambda}_p - \boldsymbol{\chi} \leqslant \mathbf{0} \\ \boldsymbol{\phi}^T \boldsymbol{\lambda}_p &= 0 \\ \boldsymbol{\phi}^T \dot{\boldsymbol{\lambda}}_p &= 0 \end{aligned} \tag{5.131}$$

where the symbols have the same meaning as before, with the introduction of the work-hardening matrix $\boldsymbol{\Upsilon}$ governing the translation of the faces of the plastic boundary after yielding.

If the structure is excited by a (vector) load $\mathbf{W}(t)$, equation (4.115) of the motion, taking into account the instantaneous plastic redistribution of stresses, becomes

$$\begin{aligned} \mathbf{M}\ddot{\mathbf{u}} + \mathbf{c}\dot{\mathbf{u}} + \mathbf{K}\mathbf{u} + \mathbf{e}\mathbf{u} &= \mathbf{W}(t) + \mathbf{Q}^T\mathbf{G}\boldsymbol{\varepsilon}_p \\ \mathbf{u}(0) = \mathbf{u}_0 \qquad & \dot{\mathbf{u}}(0) = \dot{\mathbf{u}}_0 \end{aligned} \tag{5.132}$$

$\mathbf{e}$ being a matrix depending on the fundamental configuration around which the structure vibrates and accounting for small changes in the structure geometry. In the following, $\mathbf{e}$ will be incorporated in the stiffness matrix by putting $\mathbf{K}' = \mathbf{K} + \mathbf{e}$.

Consider then in the interval $(0, \tau)$ the fictitious response $\mathbf{u}_e(t)$, calculated under the assumption that the structure behaves indefinitely elastically and with arbitrary (i.e. not necessarily the same as the actual process) initial conditions:

$$\begin{aligned} \mathbf{M}\ddot{\mathbf{u}}_e + \mathbf{c}\dot{\mathbf{u}}_e + \mathbf{K}'\mathbf{u}_e &= \mathbf{W}(t) \\ \mathbf{u}_e(0) = \mathbf{u}'_0 \qquad & \dot{\mathbf{u}}_e(0) = \dot{\mathbf{u}}'_0 \end{aligned} \tag{5.133}$$

The residual plastic deformation developed from $t = 0$ to τ can be evaluated (in most cases bounded from above) on the basis of knowledge of the fictitious response. Let $\mathbf{u}_p(\tau)$ be the residual displacement vector to be evaluated. It can be recognized that

$$\mathbf{u}_p = (\mathbf{K}')^{-1}\mathbf{Q}^T\mathbf{G}\mathbf{H}^T\boldsymbol{\lambda}_p \tag{5.134}$$

and that such a vector can be described by specifying the value of its scalar product with an appropriate test load vector $\mathbf{W}_t$[14].

It can be proved that for any test load $\mathbf{W}_t$

$$\mathbf{W}_t^T\mathbf{u}_p(\tau) \leqslant Z_0 + \tfrac{1}{2}\boldsymbol{\lambda}_S^T(\tau)\mathbf{Y}\boldsymbol{\lambda}_S(\tau) = \zeta(\tau) \tag{5.135}$$

provided that

$$\boldsymbol{\phi}_e(t) = \mathbf{H}\boldsymbol{\sigma}_e(t) + \mathbf{H}\boldsymbol{\sigma}_t - \mathbf{Y}\boldsymbol{\lambda}_S - \boldsymbol{\chi} \leqslant 0 \qquad \text{for all } t \in (0, \tau) \tag{5.136}$$

where $\boldsymbol{\sigma}_e(t)$ are the instantaneous stresses in the fictitious response, and the matrix $\mathbf{Y}$ is given by

$$\mathbf{Y} = \boldsymbol{\Upsilon} - \mathbf{H}[\mathbf{G}\mathbf{Q}(\mathbf{K}')^{-1}\mathbf{Q}^T\mathbf{G} - \mathbf{G}]\mathbf{H}^T \tag{5.137}$$

In (5.135) Z_0 is the sum of the kinetic and potential energies corresponding to the difference between the actual $(\mathbf{u}_0, \dot{\mathbf{u}}_0)$ and fictitious $(\mathbf{u}'_0, \dot{\mathbf{u}}'_0)$ initial conditions‡;

‡ Note that if $\mathbf{u}_0 = \mathbf{u}'_0$, $\dot{\mathbf{u}}_0 = \dot{\mathbf{u}}'_0$, $Z_0 = 0$.

$\boldsymbol{\sigma}_t$ is the elastic stress field in the structure under the test load $\mathbf{W}_t$ and $\boldsymbol{\lambda}_S$ is any vector of constant plastic multipliers, that is the quantity to act in order that the response $\mathbf{u}_e(t)$ verifies inequality (5.135). The $\boldsymbol{\lambda}_S$ are kept constant in the interval $(0, \tau)$, but can be changed when bounds at different instants τ are wanted. Note that the larger $\boldsymbol{\lambda}_S$ (i.e. the more the original elastic domain must modify in order to encounter adaptation), the larger is the upper bound, r.h.s. in inequality (5.135)[§].

The above procedure can be applied to investigate approximately the plastic adaptation process of any elastic–plastic structure acted on by a stochastic load $\tilde{\mathbf{W}}(t)$. In this case, in fact, both $\tilde{\mathbf{u}}_e(t)$ and $\tilde{\boldsymbol{\sigma}}_e(t)$, and then $\tilde{\boldsymbol{\Phi}}_c(t)$, are random processes linearly related to the input loading $\tilde{\mathbf{W}}(t)$, and can be calculated by the methods explained in Chapter 4.

The stress vector $\boldsymbol{\sigma}_t$ is a constant vector, uniquely defined after the test load has been set up, and the vector of plastic multipliers can be chosen quite arbitrarily, as well as the fictitious initial conditions. After this has been done, $\zeta(\tau)$ in (5.135) is a definite quantity, and the probability

$$P_u(\zeta|\tau) = \text{Prob}\,[\mathbf{W}_t^T\tilde{\mathbf{u}}_p(\tau) \leqslant \zeta(\tau)] \tag{5.138}$$

is larger than

$$\text{Prob}\,[\tilde{\boldsymbol{\Phi}}_e(t) \leqslant \mathbf{0} \quad \text{for all } t \in (0, \tau)] = P(\tau, \boldsymbol{\lambda}_S, \mathbf{u}'_0, \dot{\mathbf{u}}'_0) \tag{5.139}$$

Equation (5.139) thus yields a lower bound on the CDF of $\mathbf{W}_t^T\mathbf{u}_p(\tau)$, i.e. the problem of evaluating the CDF of the residual displacement is reduced to the calculation of the first excursion probability of the random process $\tilde{\boldsymbol{\Phi}}_e(t)$. Different values of $\zeta(\tau)$ in (5.138) can be obtained suitably varying $\boldsymbol{\lambda}_S$, $\mathbf{u}'_0$ and $\dot{\mathbf{u}}'_0$.

The largest effort of the structure to adapt itself to the input loading corresponds to the largest value of the probability (5.139). There is, in general, more than one vector of plastic multipliers $\boldsymbol{\lambda}_S$ that maximize (5.139) in connection with varying $\mathbf{u}_0$ and $\dot{\mathbf{u}}'_0$. If the condition

$$\boldsymbol{\lambda}_S^T \mathbf{Y} \boldsymbol{\lambda}_S = \text{minimum} \tag{5.140}$$

is also applied, one picks up the best possible upper bound

$$1 - P_0(\tau) = P[\tau|(\boldsymbol{\lambda}_S, \mathbf{u}'_0, \dot{\mathbf{u}}'_0)_{\text{opt}}] \tag{5.139'}$$

and the maximum possible adaptation effort of the structure for the given loads. In this case one can say that *full adaptation* has occurred at the cost of allowing a large value of the residual displacement.

In order to calculate the probabilities (5.139), reference can be made to Section A7.4.1, equation (A7.49). In particular, it can be proved that

$$P_0(\tau) \leqslant \sum_{i=1}^{\infty} \int_0^{\tau} \nu_{0i}^+(t)\,\mathrm{d}t = P_u^+(\tau) \tag{5.141}$$

[§] Note that $\mathbf{W}_t$ can be chosen in such a way that the l.h.s. of equation (5.135) is just a definite component of the vector $\mathbf{u}_p$.

with ν_{0i}^+ the expected rate of upcrossing of the ith component of the stochastic vector process $\tilde{\mathbf{\Phi}}_e(t)$ through the level $\mathbf{0}$. As well, the same probability can also be bounded from below by

$$P_0(\tau) \geqslant \max_t \max_i \text{Prob}[\phi_{ei}(t) > 0 \quad \text{for } t \in (0, \tau)] = P_u^-(\tau) \tag{5.142}$$

It can be concluded that $P_u^+(\tau)$ is an upper bound on the probability $[1 - P_u(\zeta|\tau)]$ that $\mathbf{W}^T\mathbf{u}_p$ exceeds any given value ζ. By contrast, $P_u^-(\tau)$ is not a lower bound for $[1 - P_u(\zeta|\tau)]$, in that the inequality signs in (5.142) and (5.138)–(5.139) are not in accord; therefore $P_u^-(\tau)$ can only be viewed at as on *approximation* of $[1 - P_u(\zeta|\tau)]$.

5.3.6 An example of plastic adaptation

Consider the simple one-bay shear frame in Fig. 5.23 consisting of a rigid transverse beam of mass M, supported by n deformable columns [13]. The constitutive law of the columns is assumed given by the simple anholonomic shear force vs horizontal beam displacement diagram in Fig. 5.24. In this way, the structure is reduced to a one-degree-of-freedom system, and all the vector and matrix quantities in equation (5.132) become scalars. The parameter governing the frame deformation is the horizontal beam displacement u.

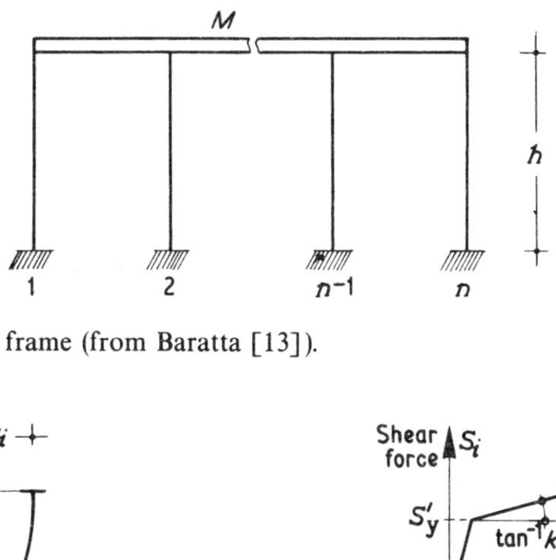

Fig. 5.23 Example frame (from Baratta [13]).

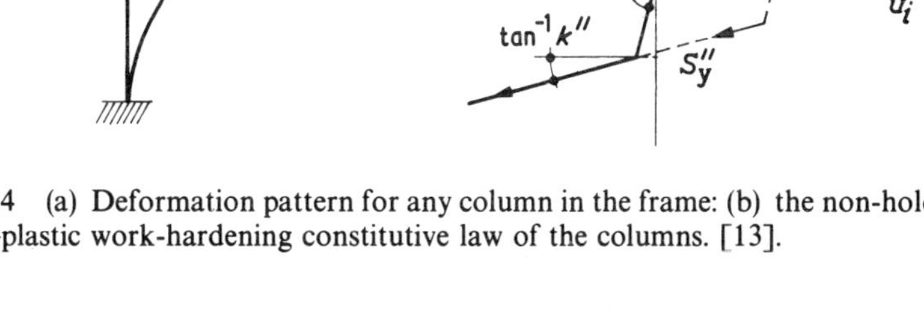

Fig. 5.24 (a) Deformation pattern for any column in the frame: (b) the non-holonomic elastic–plastic work-hardening constitutive law of the columns. [13].

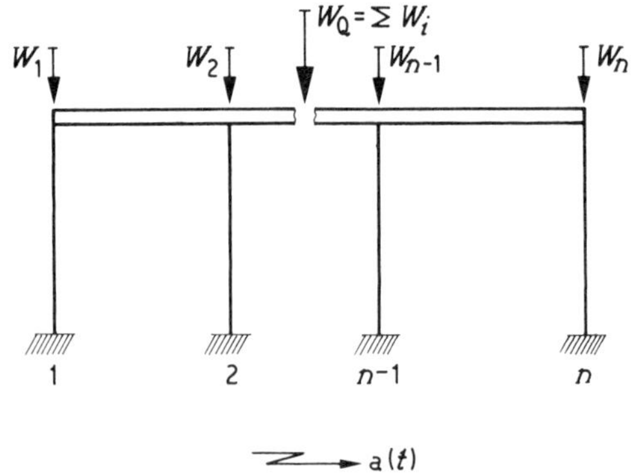

Fig. 5.25 The spatial distribution of loads on the frame: W_i, permanent gravitational loads; $a(t)$, earthquake load (ground acceleration) [13]

The loading pattern consists of permanent vertical loads on the beam, whose resultant is W_Q (Fig. 5.25), and of a horizontal variable force W_F corresponding to short ground oscillations $\tilde{a}(t)$ (simulating earthquakes) arriving according to a Poisson process with intensity λ (Fig. 5.26). Each oscillation (Fig. 5.27) is assumed to be described by a separable random process (cf. equation (2.9))

$$\tilde{a}(t) = \tilde{a}_0(t - t_j)e^{-\Lambda(t - t_j)} \tag{5.143}$$

with $\tilde{a}_0(t - t_j)$ a Gaussian stationary process with zero mean value and spectral density as in Fig. 5.28.

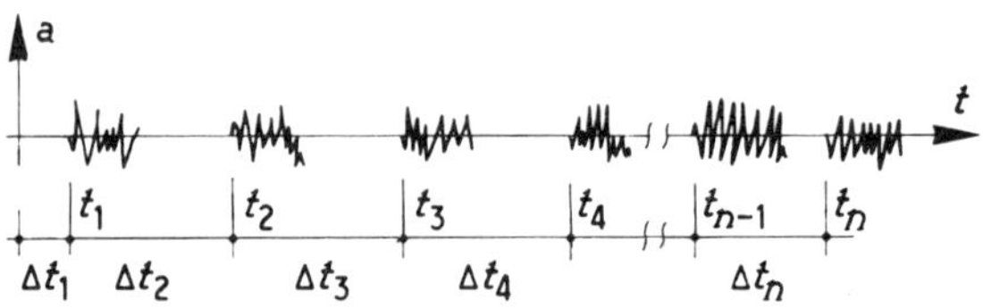

Fig. 5.26 The accidental load (ground shocks) occurrence process.

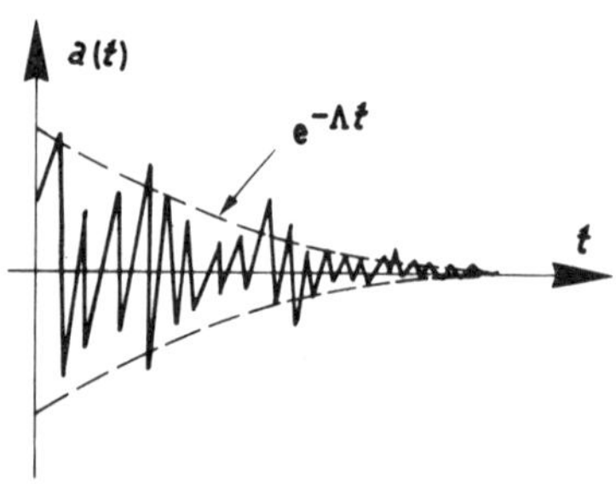

Fig. 5.27 Sample function for a single ground shock.

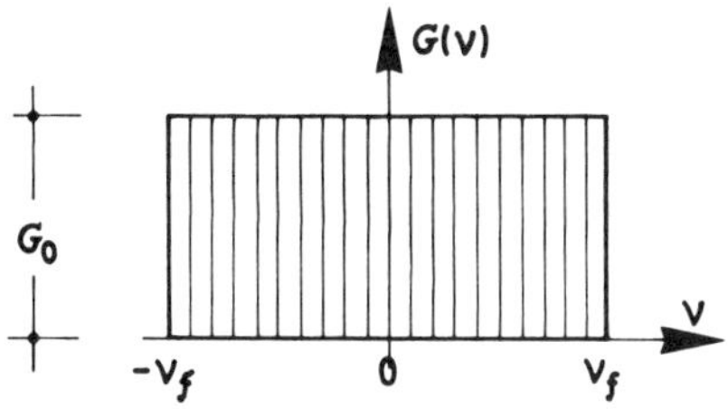

Fig. 5.28 Assumed spectral density of single shocks.

The coefficient e accounting for second-order geometrical effects in (5.132) can be calculated by a simple application of the principle of virtual work. According to the deformation pattern assumed for the column, it turns out to be

$$e = \frac{6}{5}\frac{W_Q}{h} \tag{5.144}$$

The yield condition of the structure can be set up as follows

$$\begin{aligned} \mathbf{S} - \Upsilon\boldsymbol{\lambda}_p - \mathbf{S}'_y &\leqslant \mathbf{0} \\ -\mathbf{S} + \Upsilon\boldsymbol{\lambda}_p - \mathbf{S}'_y &\leqslant \mathbf{0} \end{aligned} \tag{5.145}$$

where $\mathbf{S}$ is the n-vector of shear forces in the columns, $\mathbf{S}'_y$ and $\mathbf{S}''_y$ are the limit forces, and $\boldsymbol{\lambda}_p$ is the n-vector of plastic components of displacement in the n column playing the role of the plastic multipliers. Note that despite the fact that the frame has only one degree of freedom on the whole, different plastic deformations must be considered in every column. The strain-hardening $n \times n$ matrix $\boldsymbol{\Upsilon}$ is defined by

$$\boldsymbol{\Upsilon}_{ij} = I_{ij}\frac{k''_i k'_i}{k'_i - k''_i} \tag{5.146}$$

k'_i and k''_i are defined in Fig. 5.24.

The quantity of interest to be evaluated being $u_p(\tau)$, the residual horizontal displacement at time τ, the test load can be conveniently chosen as a horizontal unit force applied at the beam. Initial conditions of the elastic fictitious motion are set equal to zero, coincident with the actual ones. In this way, fixing any constant value of the plastic multipliers $\boldsymbol{\lambda}_S$, the probability that

$$\tilde{u}_p(\tau) \geqslant \tfrac{1}{2}\boldsymbol{\lambda}_S^T\mathbf{Y}\boldsymbol{\lambda}_S \tag{5.147}$$

is bounded from above by the probability that the inequalities

$$\begin{aligned} \tilde{\mathbf{S}}_e(t) + \mathbf{S}_0 - \boldsymbol{\Upsilon}\boldsymbol{\lambda}_{PS} - \mathbf{S}'_y &\leqslant \mathbf{0} \\ -\tilde{\mathbf{S}}_e(t) - \mathbf{S}_0 + \boldsymbol{\Upsilon}\boldsymbol{\lambda}_{PS} - \mathbf{S}''_y &\leqslant \mathbf{0} \end{aligned} \tag{5.148}$$

are not fulfilled at least once in $(0, \tau)$. In (5.148) $\tilde{\mathbf{S}}_e$ is the random process of elastic shear forces in the column, while $\mathbf{S}_0$ is a reference shear force.

Without loss of generality, the unit time can be taken such that $\lambda = 1$ (i.e., on average, τ earthquakes occur in the interval $(0, \tau)$. Leaving out the details of the calculations, Fig. 5.29 shows the upper bound $P_u^+(\tau)$ and the *approximation* $P_u^-(\tau)$ of the probability that a given residual displacement u_p is exceeded in the interval

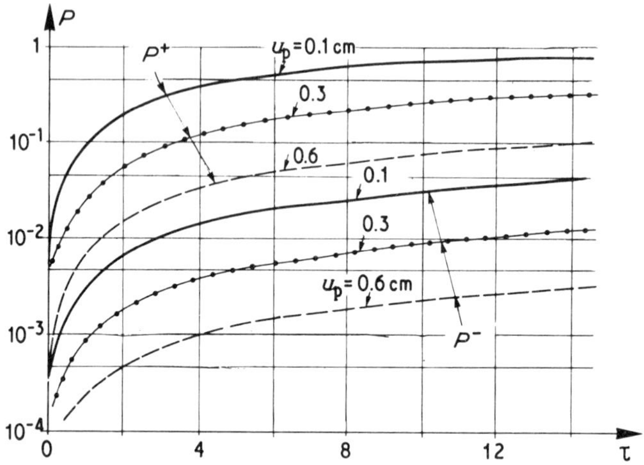

Fig. 5.29 Upper bounds (P^+) and estimates (P^-) of the probability that the residual beam displacement does not exceed u_p in the time $(0, \tau)$; the unit time is taken such that $\lambda = 1$ (from Baratta [13]).

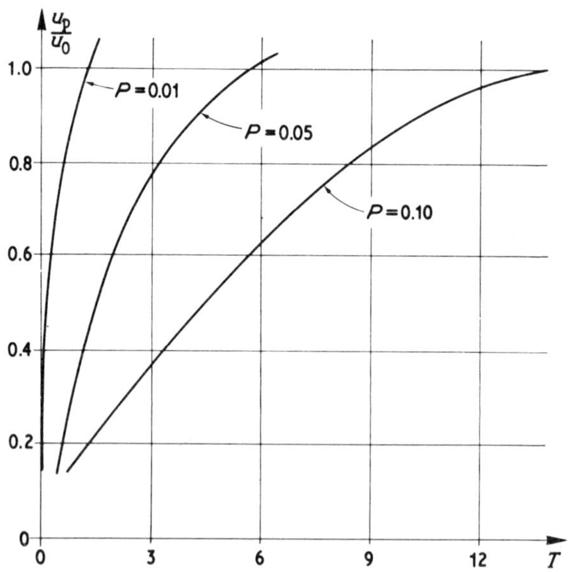

Fig. 5.30 Adaptation parameter u_p/u_0 vs time between plastic excursions T: P is the probability that such time is overestimated; $\lambda = 1$. [13].

$(0, \tau)$. The curve marked by $u_p = 0.6$ cm corresponds to *full adaptation* of the frame, and is the maximum value that can be calculated by inequality (5.147).

Let $u_0 = 0.6$ cm be the residual displacement for full plastic adaptation. Then, for $u_p \leqslant u_0$, the ratio u_p/u_0 is a measure of the maximum strength reserve (in the sense of adaptation) that the structure has displayed to resist the external load up to time τ. This can be applied to estimate the time T between two subsequent plastic excursions. A plot of T versus the values of the adaptation parameter u_p/u_0 is plotted in Fig. 5.30 for some values of the probability that this time is overestimated. It can be observed that as the adaptation process proceeds (increasing u_p/u_0), the frequency of plastic excursion decreases.

REFERENCES

1. Baker, J. F., Horne, M. R. and Heyman J. (1956) *The Steel Skeleton*, vol. 2, Cambridge University Press.
2. Neal, B. G. (1977) *The Plastic Methods of Structural Analysis*, 3rd edn, Chapman and Hall, London.
3. Hodge, P. G. (1959) *Plastic Analysis of Structures*, McGraw-Hill, New York.
4. Augusti, G. and Baratta, A. (1972) Limit analysis of structures with stochastic strength variations. *J. Struct. Mech.*, **1** (1), 43–62.
5. Augusti, G. (1975) Some observations on the calculation of structural failure probability. *Meccanica, AIMETA*, **10**, 61–3.
6. Augusti, G. and Baratta, A. (1975) Probabilistic limit analysis and design of structures: the static approach. *Proc. 2nd Int. Conf. on Applications of Statistics and Probability in Soil and Structural Engineering* (ICASP 2), Aachen, vol. 1, pp. 7–26.
7. Baratta, A. (1978) Trattazione numerica del metodo statico nell'analisi limite stocastica (in Italian), *Costruzioni Metalliche*, **30**(1), 1–7.
8. Augusti, G. and Baratta, A. (1973) Theory of probability and limit analysis of structures under multi-parameter loading; in *Foundation of Plasticity* (ed. A. Sawczuk), Noordhoff, Leyden, pp. 347–64.
9. Augusti, G. and Baratta, A. (1973) Limit and shakedown analysis of structures with stochastic strengths. *2nd Int. Conf. on Structural Mechanics in Reactor Technology*, Berlin, vol. 5, Paper M7/8.
10. Augusti, G. and Baratta, A. (1973) Plastic shakedown of structures with stochastic local strengths. *AIPC/IABSE Symp.*, Lisboa, Preliminary Publ., pp. 287–92.
11. Baratta, A. (1978) On shakedown analysis philosophy in engineering design. *Seminar 'Matériaux and Structures sous Chargement Cyclique'*, Palaiseau.
12. Casciati, F. and Sacchi, G. (1975) On the longevity of structures having random resistance. *Proc. 2nd Int. Conf. on Applications of Statistics and Probability in Soil and Structural Engineering* (ICASP 2), Aachen, vol. 1, pp. 101–14.
13. Baratta, A. (1977) Plastic adaptation of structures under stochastic excitation. *J. Struct. Mech.*, **5**(4), 421–50.
14. Maier, G. (1973) Upper bounds on deformations of elastic workhardening structures in the presence of dynamic and second-order geometric effects. *J. Struct. Mech.*, **2** (4), 265–80.

CHAPTER 6

Structural reliability and structural design

Probabilistic treatment of ultimate and serviceability limit states is developed: to this aim some more sophisticated models of structural behaviour than those presented in Chapters 4 and 5 are necessary. The second part of the chapter deals with design procedures (in particular, optimization procedures) which take random uncertainties into account.

6.1 LIMIT STATES

While in Chapter 3 models of external actions on structures were presented, Chapters 4 and 5 put in a probabilistic context the two models of structural behaviour, namely the linear elastic and the perfectly plastic, which are generally known to today's engineers in the deterministic formulation. It is now possible to tackle the analysis and design of actual structures, at the same time taking into account uncertainties in the load processes and in the structural response. This will be done, as usual in modern approaches in civil engineering, with respect to several possible *limit states*, of which those examined in Chapters 4 and 5 are only two limit (albeit basic) cases. Therefore, more models of structural behaviour are necessary; the most important will be identified in the following discussion of limit-state design, and presented in successive sections.

6.1.1 Definition

A structure, or part of a structure, is considered unfit for use when it exceeds a particular state, called a limit state, beyond which it infringes one of the criteria governing its performance or use [1].

Among the *limit states* thus defined, one distinguishes:

(1) *ultimate limit states*, which correspond to ruinous collapse or at least to large deformations which make it necessary to replace the structure;
(2) *serviceability limit states*, which are related to normal use and durability of the structure.

Limit-state design is a *decision process* in which 'various uncertainties have to be accounted for in assessing the design variables in order to obtain an acceptable failure probability' [1]. However, different safety margins can be accepted for different limit states because their consequences differ considerably. These consequences can be placed in two categories:

(1) risk of death and/or injuries to people, concern for public reaction to possible failure (*intangibles*);
(2) economic damages due to loss of use of structure and/or need for replacement or repair (*tangibles*).

Therefore, the consequences of failures must be evaluated from the viewpoint of the risk to *life and limb* and/or the entity of economical losses: it is worth noting that an ultimate limit state does not necessarily imply very serious or disastrous consequences, while, on the other hand, a serviceability limit state may have very heavy economical consequences (although, usually, few *intangible* losses).

6.1.2 Classification of limit states

A typical list of limit states is as follows [1]:

(*a*) *Ultimate limit states*

LS.U1 Loss of static equilibrium of a part or the whole of the structure considered as a rigid body (as may happen for a supporting wall).

LS.U2 Localized rupture in a point or section of the structure (e.g. reinforced-concrete structures).

LS.U2′ Excessive deformation (as may happen if the structure is comparatively slender and the analysis has been performed without taking into account second-order geometrical effects).

LS.U3 Attainment of the elastic limit at some points in the structure (except in the case of *brittle* material behaviour, this is a *conventional* limit state, sometimes introduced in order to use simple linear elastic structural analysis).

LS.U4 Transformation of the structure into a mechanism (cf. Chapter 5).

LS.U5 General or local instability.

LS.U6 Fatigue fracture due to repeated loads (usually, service loads; e.g. bridges, mechanical elements).

(*b*) *Serviceability limit states*

These can be distinguished in three categories:

LS.S1 Deformation that may adversely affect the appearance or efficiency of the structure.

LS.S2 Local damage which might require excessive maintenance or lead to corrosion.

LS.S3 Vibration (from causes such as wind gusts or moving machinery) which can cause discomfort and/or alarm.

For a sound design, in addition to verifying reliability with respect to each listed limit state, the structure must be protected against unforeseen circumstances. Therefore, it must be such that 'it should not be damaged to an extent disproportionate to the original accidental cause' [1]. Also, fire resistance and durability should be considered.

6.1.3 Limit-state models

For limit-state design, a suitable model must be chosen for each specific limit state and this 'should incorporate appropriate variables allowing for the uncertainties with respect to actions, the response of the structure as a whole and the behaviour of individual elements of the structure' [1].

The models presented in Chapters 4 and 5 represent well the ultimate limit states LS.U3 and LS.U4 respectively. Other models are necessary for the study of other limit states: some will be presented in Section 6.2. Then, Section 6.3 deals with probabilistic structural analysis with respect to the ultimate limit states listed in subsection 6.1.2(a). In all these cases, two alternative designs can be compared through the probability of attaining the limit state under consideration (probability of structural failure according to that mode, P_{fail}), or equivalently through the probability of survival (*reliability*, $\mathbb{R} = P_{\text{suc}} = 1 - P_{\text{fail}}$) with respect to that limit state (cf. Section 2.8).

In many cases, the intensity of the loads acting on the structure is assumed to depend on only one scalar (and positive) parameter α, so that stochastic diagrams like those of Fig. 6.1 could in principle be plotted. Assuming further that dynamic effects are negligible and that the structural properties do not vary in time (as they might, because of *deterioration* due to load repetitions), the value of α that corresponds to the attainment of the ith limit state is a random variable $\tilde{\alpha}_i$, representing the *load-carrying capacity of the structure* (Fig. 6.1). Then, equations (2.66) and (2.68) (Section 2.8) become, for deterministic loads of intensity $\alpha = \alpha^*$:

$$P_{\text{fail}} = \int_0^{\alpha^*} p_i(\alpha)\,\mathrm{d}\alpha = P_i(\alpha^*) \tag{6.1}$$

and for random loads of PDF $p_\alpha(\alpha)$:

$$\begin{aligned} P_{\text{fail}} &= \int_0^{\infty} p_\alpha(\alpha)\mathrm{d}x \int_0^{\alpha} p_i(\alpha')\,\mathrm{d}\alpha' \\ &= \int_0^{\infty} p_\alpha(\alpha) P_i(\alpha)\,\mathrm{d}\alpha \end{aligned} \tag{6.2}$$

where $p_i(\alpha)$ and $P_i(\alpha)$ are respectively the PDF and the CDF of the relevant load-carrying capacity factor‡.

It is thus evident that probabilistic analysis requires first a deterministic relation (the *structural model*) between the structural parameters (defining material properties and geometrical quantities) and the load-carrying capacity of the structure. The problem is thus, in principle, reduced to the description of a random variable which is a known function of random design quantities whose joint properties must be known or assigned (Appendix A3 and Section A5.5).

Cases in which the time variability of the loads and/or the response cannot be neglected, and in which therefore this simple presentation is somewhat complicated, have already been encountered both in Chapter 4 (e.g. Sections 4.2.1 (c) and 4.3) and in Chapter 5 (Section 5.3, and in particular 5.3.5). However, it can be assumed that structural behaviour with respect to ultimate limit states has in all cases the *binary* (0–1) character that has been assumed so far: i.e. the structure *fails* if the limit state is reached, is *successful* if not, and no other alternative exists.

In contrast, serviceability failure conditions generally indicate only the beginning of a continuous sequence of undesired states, which are associated with an increasing amount of *damage*. Therefore, in order to compare different designs with regard to serviceability limit states, more detailed information is required than the simple probability of attainment (that may even be practically 1 during the design life of the structure). This problem will be neglected for the moment and taken up again in Section 6.5.

6.2 NON-LINEAR STATIC AND DYNAMIC RESPONSE

It is well known that non-linearities complicate structural analysis, and in particular probabilistic structural analysis, to a very great extent (cf. Section 5.1), so that established systematic methods of treatment are still lacking (except numerical procedures, which will be treated in Chapter 9).

Nevertheless, non-linear response enters into many limit states (LS.U2, LS.U5, etc.) and therefore it is worth while to give some mention of the way in which non-linear analysis can be treated under uncertainty.

6.2.1 Non-linear response under static loads

(a) General solution

Consider a structure whose response equation can be put in the form

$$\mathscr{L}(\mathbf{u}, \boldsymbol{\zeta}) = \mathbf{W} \tag{6.3}$$

‡ Remember that, since $\alpha \geqslant 0$ by assumption, the cumulative density function (CDF) $P(\alpha)$ coincides with the *conditional probability of failure* (CPF) under loads of intensity α; this concept allows an immediate generalization to multiparameter loading (Section 2.8).

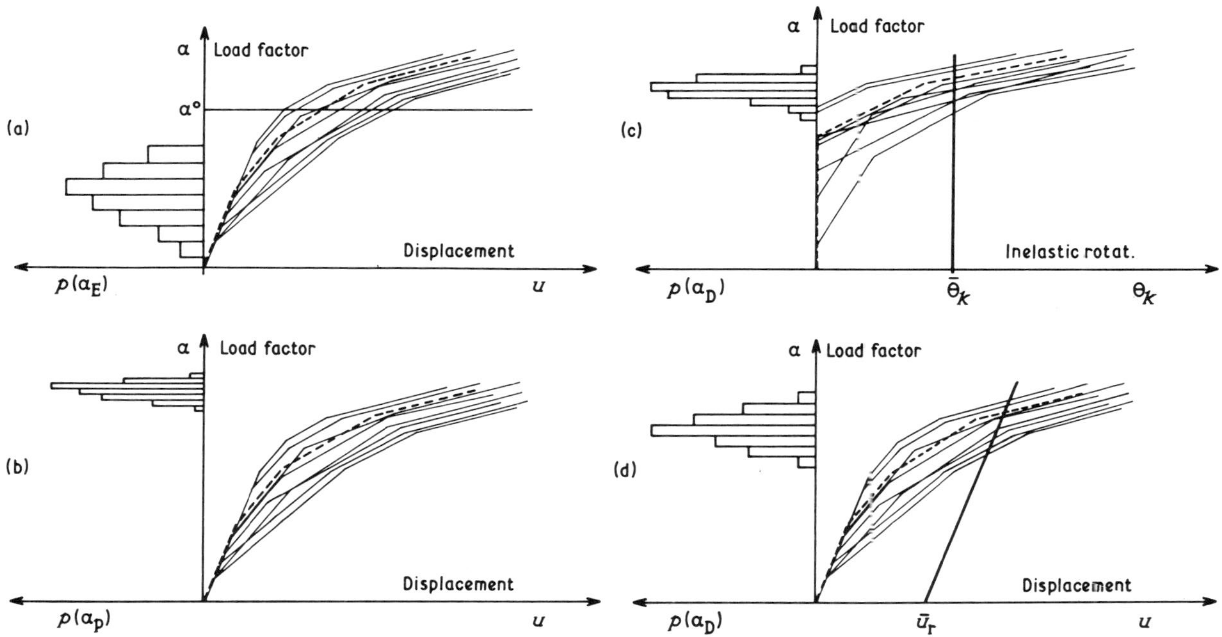

Fig. 6.1 Stochastic plots of load factor α vs significant response parameter and corresponding PDF of load-carrying capacity factor for different limit states (qualitative examples): (a) elastic limit (α_E); (b) plastic limit (α_p); (c) threshold inelastic rotation (α_D); (d) threshold residual displacement (α_D).

where $\mathscr{L}$ is a non-linear algebraic operator on the (generalized) displacements $\mathbf{u}$ and the structural parameters $\boldsymbol{\zeta}$. If the system (6.3) can be solved for $\mathbf{u}$, no difficulty arises, in principle, to find the JPDF of the response parameters u_i, if the loads W_i are random variables with known JPDF. In fact, equation (A3.61) yields immediately

$$p_{\mathbf{u}}(\mathbf{u}) = p_{\mathbf{W}}(\mathbf{u}, \boldsymbol{\zeta}) \cdot J \tag{6.4}$$

where J is the absolute value of the Jacobian determinant whose elements are defined by

$$J_{ij} = \frac{\partial W_i}{\partial u_j} = \frac{\partial \mathscr{L}^i}{\partial u_j} \tag{6.5}$$

However, first and second moments of the response vector are not given by simple formulae, except when non-linearities are small, in which case equation (6.3) can be approximated by a linear system analogous to (4.1), and equation (4.11) used with satisfactory accuracy.

More complicated is the case in which some or all the structural parameters $\boldsymbol{\zeta}$ are random. The mathematical problem can be formulated by writing the response equation (6.3) in the form

$$\mathscr{L}(\mathbf{u}, \tilde{\boldsymbol{\zeta}}) = \mathscr{L}(\mathbf{u}, \boldsymbol{\mu}_\zeta + \tilde{\mathbf{H}}) = \tilde{\mathbf{W}} \tag{6.6}$$

where $\mathscr{L}$ is a deterministic algebraic operator, $\boldsymbol{\mu}_\zeta$ the (deterministic) vector of the average structural properties, and $\tilde{\mathbf{H}}$ the vector of the random imperfections analogous to $\tilde{H}_{ij}$ (equation (4.56)) in the linear case.

Let m be the number of random imperfections. If the JPDF of $(W_1, W_2, \ldots, W_n, H_1, \ldots, H_m)$ is known, the JPDF of $(u_1, u_2, \ldots, u_n, H_1, \ldots, H_m)$ is given by the already known formula

$$p_{\mathbf{u},\mathbf{H}}(\mathbf{u}, \mathbf{H}) = p_{\mathbf{W},\mathbf{H}}[\mathscr{L}(\mathbf{u}, \boldsymbol{\mu}_\zeta + \mathbf{H}), \mathbf{H}] . J' \tag{6.7}$$

where J' is the absolute value of the determinant of the matrix defined by

$$\left.\begin{aligned} J'_{ij} &= \frac{\partial W_i}{\partial u_j} = \frac{\partial \mathscr{L}^i}{\partial u_j} && \text{for} \quad 1 \leqslant i \leqslant n, \qquad 1 \leqslant j \leqslant n \\ J'_{ij} &= \frac{\partial H_{i-n}}{\partial u_j} = 0 && \text{for } n+1 \leqslant i \leqslant n+m, \quad 1 \leqslant j \leqslant n \end{aligned}\right\} \tag{6.8}$$

$$\left.\begin{aligned} J'_{ij} &= \frac{\partial W_i}{\partial H_{j-n}} = \frac{\partial \mathscr{L}^i}{\partial H_{j-n}} && \text{for} \quad 1 \leqslant i \leqslant n, \quad n+1 \leqslant j \leqslant n+m \\ J'_{ij} &= \frac{\partial H_{i-n}}{\partial H_{j-n}} = I_{ii} && \text{for } n+1 \leqslant i \leqslant n+m, \; n+1 \leqslant j \leqslant n+m \end{aligned}\right\} \tag{6.8'}$$

Therefore

$$J' = J \tag{6.8''}$$

Finally, the required PDF $p_{\mathbf{u}}(\mathbf{u})$ is obtained by an m-fold integration

$$p_{\mathbf{u}}(\mathbf{u}) = \int_{-\infty}^{+\infty} \cdots \int_{-\infty}^{+\infty} p_{\mathbf{u},\mathbf{H}}(\mathbf{u}, \mathbf{H}) \, d\mathbf{H} \tag{6.9}$$

It is evident that the computations connected with the above procedure become overwhelming if the number m of random structural parameters is larger than a few units.

(b) Second-moment analysis

For non-linear response problems, as in many others of probabilistic structural theory, first and second moments of the response quantities may be the most important (if not the only) information that can be obtained with sufficient accuracy and/or confidence. An available procedure [2] is summarized below.

Consider a structure whose response is described by equation (6.3); assume that the loads are proportional to a factor α, i.e.

$$\tilde{\mathbf{W}} = \alpha \tilde{\mathbf{W}}_0 \tag{6.10}$$

and let α increase in steps ($\alpha_1 < \alpha_2 < \ldots < \alpha_m < \ldots$), while the stochastic parameters $\tilde{\mathbf{W}}$ and $\tilde{\zeta}$ remain constant.

Denoting by $\tilde{\mathbf{u}}^{(m)}$ the response vector at the mth step, by $\tilde{\mathbf{K}}^{(m)}$ the *tangent stiffness* matrix at the same stage

$$\tilde{K}_{ij}^{(m)} = \left.\frac{\partial \mathscr{L}^i}{\partial u_j}\right|_{\mathbf{u} = \mathbf{u}^{(m)}} \tag{6.11}$$

and $\tilde{\mathbf{C}}^{(m)}$ its inverse

$$\tilde{\mathbf{C}}^{(m)} = (\tilde{\mathbf{K}}^{(m)})^{-1} \tag{6.12}$$

the recursive formula for the response vector is

$$\tilde{\mathbf{u}}^{(m+1)} = \tilde{\mathbf{u}}^{(m)} + \tilde{\mathbf{C}}^{(m)}[\alpha_{m+1}\tilde{\mathbf{W}}_0 - \mathscr{L}(\mathbf{u}^{(m)}, \zeta)] \tag{6.13}$$

The main source of difficulties lies in the inversion (6.12) of the stochastic matrix (6.11). At each step m, the approximate formula (4.60) can be adopted subject to the limitations pointed out in its derivation, and with the matrices $\bar{\mathbf{K}}$ and $\tilde{\mathbf{H}}$ now defined by

$$\bar{K}_{ij} = \left.\frac{\partial \mathscr{L}^i}{\partial u_j}\right|_{\mathbf{u} = \boldsymbol{\mu}_{\mathbf{u}};\ \zeta = \boldsymbol{\mu}_{\zeta}} \tag{6.14}$$

$$\tilde{H}_{ij} = \frac{\partial \mathscr{L}^i}{\partial u_j} - \bar{K}_{ij} \tag{6.15}$$

where the suffixes (m) and the indication $\mathbf{u} = \mathbf{u}^{(m)}$ have been omitted for simplicity.

Then, the expression for averages and covariance matrices at each step of the incremental procedure can be obtained along the lines illustrated in Section 4.2.2. These expressions can be applied recursively, as in a deterministic step-by-step calculation, but the relevant averages and covariance matrices must be calculated and stored at each step [2]. Improvements of the procedure to reduce the computational effort seem possible and are the object of current researches.

(c) Simplified use of extremum principles

It is well known that extremum principles are very useful in structural mechanics and find a wide range of applications. In the framework of probabilistic structural analysis, extremum principles furnish procedures for calculating the probability that the response vector $\mathbf{u}$ falls in some special subset Λ_p of the set of Ξ possible responses.

For the sake of simplicity and brevity, the following treatment will be limited to structures with some (unspecified) random property, subjected to given loads. A more general treatment can be found in the literature [3].

Assume that the actual response $\mathbf{u}_m$ be characterized by a *minimum* relationship of the type

$$\phi(\mathbf{u}_m) = \min_{\mathbf{u} \in \Lambda} \phi(\mathbf{u}) \tag{6.16}$$

Λ being a *random subset* of Ξ characterized by a functional $\mathscr{F}$ ‡ yielding the Prob$[\mathbf{u} \in \Lambda]$ for every $\mathbf{u} \in \Xi$:

$$\mathscr{F} : \mathbf{u} \in \Xi \rightarrow \text{Prob}[\mathbf{u} \in \Lambda] \tag{6.17}$$

and $\phi(\mathbf{u})$ a deterministic functional defined on Ξ specifying the physical conditions imposed on the solution vector, described by its *contour* (level line, level surface)

$$\phi(\mathbf{u}) = \text{const.} \tag{6.18}$$

This obviously would not be true if ϕ were also a random functional. Coming back to equation (6.16), it is obvious that for any numerical value ϕ_0

$$\{\phi(\mathbf{u}_m) \geqslant \phi_0\} \Rightarrow \{\phi(\mathbf{u}) \geqslant \phi_0 \text{ for all } \mathbf{u} \in \Lambda\} \tag{6.19}$$

and therefore

$$\text{Prob}[\phi(\mathbf{u}_m) \geqslant \phi_0] \leqslant \text{Prob}[\mathbf{u} \in \Lambda] \text{ for all } \mathbf{u} \in \Xi : \phi(\mathbf{u}) \geqslant \phi_0 \tag{6.20}$$

Similarly, the complementary inequality can be demonstrated

$$\text{Prob}[\phi(\mathbf{u}_m) \leqslant \phi_0] \geqslant \text{Prob}[\mathbf{u} \in \Lambda] \text{ for all } \mathbf{u} \in \Xi : \phi(\mathbf{u}) \leqslant \phi_0 \tag{6.21}$$

Vice versa, if $\mathbf{u}_M$ is characterized by a *maximum* relationship

$$\phi_M(\mathbf{u}_M) = \max_{\mathbf{u} \in \Lambda_M} \phi_M(\mathbf{u}) \tag{6.22}$$

where ϕ_M is another suitable functional and Λ_M an appropriate random subset of Ξ, the following inequalities can be defined

$$\text{Prob}[\phi_M(\mathbf{u}_M) \geqslant \phi_{0M}] \geqslant \text{Prob}[\mathbf{u} \in \Lambda_M] \text{ for all } \mathbf{u} \in \Xi : \phi_M(\mathbf{u}) \geqslant \phi_{0M} \tag{6.23}$$

$$\text{Prob}[\phi_M(\mathbf{u}_M) \leqslant \phi_{0M}] \leqslant \text{Prob}[\mathbf{u} \in \Lambda_M] \text{ for all } \mathbf{u} \in \Xi : \phi_M(\mathbf{u}) \leqslant \phi_{0M} \tag{6.24}$$

With reference to problems in which equation (6.16) holds, take a discrete subset $(\mathbf{u}_1, \ldots, \mathbf{u}_n)$ of Ξ such that

$$\phi(\mathbf{u}_j) \leqslant \phi_0 \qquad j = 1, 2, \ldots, n \tag{6.25}$$

‡ $\mathbf{u} \in \Lambda$ means that $\mathbf{u}$ enjoys the property that characterizes Λ. If Λ is a deterministic subset, the functional is the index function of Λ. Otherwise, the functional $\mathscr{F}$ specifies the probability that $\mathbf{u}$ enjoys said property.

One obtains immediately

$$\mathrm{Prob}[\phi(\mathbf{u}_\mathrm{m}) \leqslant \phi_0] \geqslant \mathrm{Prob}[(\mathbf{u}_1 \in \Lambda) \cup (\mathbf{u}_2 \in \Lambda) \cup \ldots \cup (\mathbf{u}_n \in \Lambda)] \quad (6.26)$$

which gives a more accurate bound than inequality (6.21).

An analogous inequality can be proved if equation (6.22) holds. The above inequalities can be used for response prediction: assume, in fact, that the following bound has been obtained (equation (6.20))

$$\mathrm{Prob}[\phi(\mathbf{u}_\mathrm{m}) \geqslant \phi_0] \leqslant P \quad (6.27)$$

where the event under probability means that the solution falls outside the domain Λ_p defined by

$$\phi(\mathbf{u}) \leqslant \phi_0 \quad (6.28)$$

Then

$$\mathrm{Prob}[\mathbf{u}_\mathrm{m} \notin \Lambda_\mathrm{p}] \leqslant P \quad (6.29)$$

Hence the response $\mathbf{u}_\mathrm{m}$ can be forecast with a certain *degree of approximation* (depending on the extension of the region Λ_p) and with a certain *degree of confidence* (depending on the probability P).

As a simple illustration, consider the case that the possible responses $\mathbf{u}$ are fully characterized by two parameters, and that the set of possible responses covers the whole real plane, where level lines (6.18) can be drawn (Fig. 6.2).

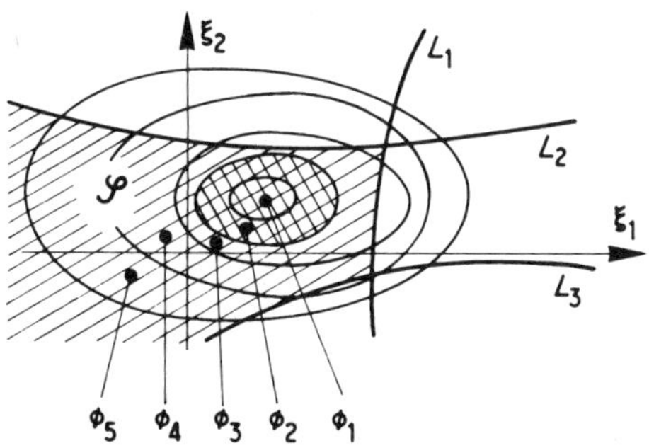

Fig. 6.2 Regions Φ_j: $\phi(\mathbf{w}) \leqslant \phi_j$; safe region $\mathscr{S}$.

Take an ordered sequence of constants $\phi_1 < \phi_2 < \phi_3 < \ldots$ and denote by Φ_j the regions defined by

$$\{\mathbf{u} \in \Phi_j\} \Leftrightarrow \{\phi(\mathbf{u}) \leqslant \phi_j\} \qquad j = 1, 2, 3, \ldots \quad (6.30)$$

Thus, by (6.26), a lower bound can be established for the probability that $\mathbf{u}_\mathrm{m}$ falls in each region Φ_j. Let

$$\mathrm{Prob}[\mathbf{u}_\mathrm{m} \in \Phi_j] \geqslant P_j \quad (6.31)$$

Now combine the above treatment with critical inequalities (2.63). Take for instance three inequalities ($k = 1, 2, 3$), and let the boundaries of the respective safe regions be the lines indicated by L_1, L_2 and L_3 in Fig. 6.2. The overall safe region $\mathscr{S}$ (hatched in the figure) is defined by

$$\{\mathbf{u} \in \mathscr{S}\} \Leftrightarrow \{\phi_k(\mathbf{u}) < L_k \text{ for all } k = 1, 2, 3\} \tag{6.32}$$

Only responses $\mathbf{u} \in \mathscr{S}$ are acceptable. Taking the largest j such that

$$j_{\mathrm{m}} = j \tag{6.33}$$

$$\Phi_j \subseteq \mathscr{S} \tag{6.34}$$

($j_{\mathrm{m}} = 2$ in Fig. 6.2) and taking account of (6.31)

$$\mathrm{Prob}[\mathbf{u}_{\mathrm{m}} \in \mathscr{S}] \geqslant \mathrm{Prob}[\mathbf{u}_{\mathrm{m}} \in \Phi_j] \geqslant P_j \qquad j = j_{\mathrm{m}} \tag{6.35}$$

which finally gives a (lower) bound to the probability of successful performance.

Some applications of the considerations developed in this subsection have been formulated with reference to the evaluation of plastic deformations of structures with random yield strength. It can also be shown that the bounds of the collapse load of rigid–plastic random strength structures, which have been presented in Chapter 5, are particular cases of the bounds that can be obtained by means of extremum principles.

(d) Limitations of non-linear analysis and alternative approach

The approaches so far illustrated in this section present significant limitations with regard to practical applications. The main one is perhaps the implicit assumption that the functional relationship between input and response quantities is known in closed form: in fact, the procedures require differentiations of the non-linear operator $\mathscr{L}(\ )$ (or, equivalently, knowledge of the *contours* (level surfaces) of $\phi\ (\ )$). Moreover, the procedures of subsections (a) and (c) require cumbersome calculations, because of the manifold numerical integrations (often ill conditioned) in the calculation of the relevant probabilities.

It must also be noted that the stress–strain relation has been assumed to be holonomic. Such a model of course does not represent *inelastic* material behaviour, unless no significant unloading occurs: but it is well known that such an assumption cannot in general be made (even for monotonically increasing loads) when the stress–strain law is strongly non-linear, and therefore a major limitation on the significance of the mechanical model is thus introduced.

The only approach which allows one in general to circumvent these difficulties (but often at the cost of long computing times and sometimes of the lack of a synthetic clear view of the structural situation) is *numerical simulation*, i.e. the performance of a large number of (numerical) *simulated experiments*. Each *experiment* is the solution of a deterministic problem, in which a structure with artificially chosen properties is followed throughout a single artificially generated loading history. Structural properties and loading history are possible *realizations* of random variables and/or stochastic processes. Since the problem analysed is deterministic, modern computing methods allow the introduction of very complicated (and non-holonomic) laws and therefore an accurate modelling of the mechanical problem. The set of responses obtained from the simulated experiments (each a possible realization of the actual situation) is finally investigated by statistical techniques (Chapter 7) to obtain the required answers in probabilistic terms. *Numerical simulation* methods and procedures will be the sole object of the whole of Chapter 9, where examples will be presented.

6.2.2 Non-linear response under dynamic loads

The preceding considerations are all the more valid when dynamic effects are not negligible. In fact, the only way of tackling in general a *deterministic* non-linear dynamic structural problem is to perform a numerical step-by-step integration of the equations of motion. The response history is divided into short time increments and the response is calculated during each increment for a linear system whose properties were determined at the beginning of the interval. Thus the non-linear analysis is approximated as a sequence of analyses of successively changing linear systems.

For this purpose, putting at any instant of time t

$$\begin{aligned} \ddot{\mathbf{u}}(t+\Delta t) &= \ddot{\mathbf{u}}(t) + \Delta\ddot{\mathbf{u}}(t) \\ \dot{\mathbf{u}}(t+\Delta t) &= \dot{\mathbf{u}}(t) + \Delta\dot{\mathbf{u}}(t) \\ \mathbf{u}(t+\Delta t) &= \mathbf{u}(t) + \Delta\mathbf{u}(t) \end{aligned} \tag{6.36}$$

the equation of motion (equation (4.115)) is written in the incremental form

$$\mathbf{M}\,\Delta\ddot{\mathbf{u}}(t) + \mathbf{c}(t)\Delta\dot{\mathbf{u}}(t) + \mathbf{K}(t)\Delta\mathbf{u}(t) = \Delta\mathbf{W}(t) \tag{6.37}$$

and usually formally simplified into

$$\mathbf{K}_0(t)\Delta\mathbf{u}(t) = \Delta\mathbf{W}_0(t) \tag{6.38}$$

fully equivalent to a static incremental-equilibrium relationship.

In equation (6.37) $\mathbf{K}(t)$ is the stiffness matrix of the structure considered at time t and $\Delta\mathbf{u}(t)$ is the vector of the increments of the displacements representing the deformed shape of the structure. $\mathbf{c}(t)$ is the damping matrix at time t and $\Delta\dot{\mathbf{u}}(t)$ the incremental velocity vector; $\mathbf{M}$ is the mass matrix and $\Delta\ddot{\mathbf{u}}(t)$ is the incremental acceleration vector, both defined for the specified set of displacement coordinates. Note that it has been assumed that the mass matrix $\mathbf{M}$ does not change with time. Furthermore $\Delta\mathbf{W}(t)$ is the vector of the increments of the externally applied loads $\mathbf{W}(t)$ and all the variations are referred to the considered time step. If the action is a motion of the foundation ground, $\mathbf{W}(t)$ becomes $-\mathbf{MI}a(t)$, where $a(t)$ is the ground acceleration.

In equation (6.38) the dynamic behaviour is accounted for by the inclusion of inertial and damping effects in the effective load ($\Delta\mathbf{W}_0$) and stiffness ($\mathbf{K}_0$) terms. With reference to a three-point recurrence scheme of integration for second-order equations (Newmark algorithm), the last terms can be written, for $\mathbf{K}(t)$ and $\mathbf{c}(t)$

constant during the step, as

$$\mathbf{K}_0(t) = \mathbf{K}(t) + \mathbf{M}(\beta\Delta t^2) + \gamma\mathbf{c}(t)/(\beta\Delta t) \tag{6.39}$$

$$\begin{aligned}\boldsymbol{\Delta}\mathbf{W}_0(t) = {} & \boldsymbol{\Delta}\mathbf{W}(t) + \mathbf{M}[\dot{\mathbf{u}}(t)/(\beta\Delta t) + \tfrac{1}{2}\ddot{\mathbf{u}}(t)/\beta] \\ & + \mathbf{c}(t)[\gamma\dot{\mathbf{u}}(t)/\beta + (\tfrac{1}{2}\gamma/\beta - 1)\Delta t\ddot{\mathbf{u}}(t)]\end{aligned} \tag{6.40}$$

where $\mathbf{u}(t)$, $\dot{\mathbf{u}}(t)$ and $\ddot{\mathbf{u}}(t)$ denote the displacements, velocities and accelerations at time t, Δt is the time interval defining the integration step, and β and γ are appropriate constants.

The most common scheme puts $\beta = \frac{1}{6}$, $\gamma = \frac{1}{2}$ and corresponds to a linear acceleration assumption which leads to a quadratic variation of the velocity vector and a cubic variation of the displacement vector during each time step Δt.‡

$$\mathbf{u}(\tau) = \mathbf{u}(t) + \dot{\mathbf{u}}(t)\tau + \ddot{\mathbf{u}}(t)\tau^2/2 + \boldsymbol{\Delta}\ddot{\mathbf{u}}\tau^3/(6\Delta t) \qquad \text{for } 0 \leqslant \tau \leqslant \Delta t \tag{6.41}$$

Obviously, such a procedure is founded on the solution of equation (6.38), i.e. on the performance of an incremental elasto–plastic static analysis when elasto–plastic structures are considered.

Consider now in particular an r-degree-of-freedom elasto–perfectly plastic frame subjected to a set of dynamic loads $\mathbf{W}(t)$. For the steps during which no plastic hinge forms and no local unloading occurs, the motion is governed by equation (6.37) whose solution may be expressed, from equation (6.38), by

$$\boldsymbol{\Delta}\mathbf{u}(t) = \mathbf{K}_0^{-1}(t)\boldsymbol{\Delta}\mathbf{W}_0(t) \tag{6.42}$$

These displacements induce in the structure a bending moment diagram equivalent to the diagram due to the statical loads $\boldsymbol{\alpha}(\Delta t, t)$

$$\boldsymbol{\alpha}(\Delta t, t) = \mathbf{K}(t)\boldsymbol{\Delta}\mathbf{u}(t) = \mathbf{K}(t)\mathbf{K}_0^{-1}(t)\boldsymbol{\Delta}\mathbf{W}_0(t) \tag{6.43}$$

By substituting equation (6.43) into equation (9.112), the dynamic behaviour, governed by the incremental equation of motion (6.37), can be investigated by means of linear programming while the whole elasto–plastic dynamic analysis is

‡ Under this hypothesis, for any given time increment, the analysis procedure consists of the following operations:

(1) $\mathbf{u}(t)$ and $\dot{\mathbf{u}}(t)$ are known either from values at the end of the preceding step or as initial conditions; $\mathbf{c}(t)$ and $\mathbf{K}(t)$ for the interval Δt are evaluated and the initial acceleration is given by

$$\ddot{\mathbf{u}}(t) = \mathbf{M}^{-1}[\mathbf{W}(t) - \mathbf{c}(t)\dot{\mathbf{u}}(t) - \mathbf{K}(t)\mathbf{u}(t)]$$

which is a rearrangement of the equation of equilibrium for time t.

(2) $\boldsymbol{\Delta}\mathbf{W}_0(t)$ and $\mathbf{K}_0(t)$ are computed from equations (6.39) and (6.40).

(3) The displacement increment vector $\boldsymbol{\Delta}\mathbf{u}(t)$ is given by equation (6.38) and with it

$$\boldsymbol{\Delta}\dot{\mathbf{u}}(t) = 3\boldsymbol{\Delta}\mathbf{u}(t)/\Delta t - 3\dot{\mathbf{u}}(t) - \Delta t\ddot{\mathbf{u}}(t)/2$$

$$\boldsymbol{\Delta}\ddot{\mathbf{u}}(t) = 6\boldsymbol{\Delta}\mathbf{u}(t)/\Delta t^2 - 6\dot{\mathbf{u}}(t)/\Delta t - 3\ddot{\mathbf{u}}(t)$$

(4) The displacements and velocities at the end of the increment are given by equation (6.37′):

$$\mathbf{u}(t+\Delta t) = \mathbf{u}(t) + \boldsymbol{\Delta}\mathbf{u}(t) \qquad \dot{\mathbf{u}}(t+\Delta t) = \dot{\mathbf{u}}(t) + \boldsymbol{\Delta}\dot{\mathbf{u}}(t)$$

The analysis for the considered step is finished and the process is repeated for the next time interval.

performed by means of a Newmark integration procedure, [4]. In particular, one can solve the equivalent static analysis and obtain the bending moment distribution. Then the sought displacements are evaluated by any elastic method and hence velocities and accelerations are computed. Finally equations (6.39) and (6.40) provide the effective load and stiffness terms for the next step.

Such a linear programming approach, however without considering explicitly inertia forces, is illustrated in detail in Section 9.6.2 (equation (9.112)). The main point is the evaluation of the critical value Δt_c of the time increment at which new plastic hinges form or local unloadings occur during the time step under investigation. Further details of numerical procedures for deterministic non-linear structural problems can be found in the specialized literature, but it is clear that, in the case of random uncertainties, it becomes natural (and usually necessary ‡) to look for information on the random properties of the output by inserting the numerical solution of the equations of motion into a *simulation* scheme.

Sometimes, also in non-linear dynamic problems, the simulation procedure is made quicker and more effective by the use of the *response spectrum* to the input process (introduced in Section 3.4.3). The practical use of the response spectra, which are currently available for earthquake loading processes only, will be discussed in Chapter 10.

6.2.3 Imperfection-sensitive structures

As already noted in Section 4.4, many instances are known of structures which are very sensitive to the presence of imperfections. The conceptually most correct way of tackling this problem is indeed to consider imperfections as *random perturbations* $Y_1, Y_2, \ldots, Y_m$ of the ideally perfect structural model, and to search for the JPDF of relevant structural response parameters $u_1, u_2, \ldots, u_n$, by means of a formula of the type (6.4), from the JPDF of imperfections and acting loads $p(Y_1, Y_2, \ldots, Y_m, W_1, W_2, \ldots, W_r)$ and the functional relationships

$$u_i = u_i(Y_1, Y_2, \ldots, Y_m, W_1, W_2, \ldots, W_r) \tag{6.44}$$

However, very few problems of imperfection-sensitive structures have been tackled, and the general formulae are hardly ever used.

Indeed, in most problems of structural stability, load conditions depending on only one scalar parameter are considered, and the problem is reduced to the search for the value of the parameter (*load-carrying capacity* or *critical load*) that corresponds to loss of stability of the structure. This type of problem will be treated in the following, assuming that imperfections (and possibly other structural properties) are random: thus, the load-carrying capacity is also a random variable. A more basic discussion on structural stability in a probabilistic context will be found in Chapter 11; remember also that the buckling of *elastic* columns has already been treated in Section 4.4. In many instances, the load

‡ A noteworthy exception is the bounding procedure presented in Sections 5.3.5 and 5.3.6.

parameter is expressed as a *stress* σ: typically, in columns, σ is the axial force divided by the cross-sectional area. The symbol σ will therefore be used in the following to indicate a generic load parameter.

(a) Effects of imperfections on load-carrying capacity

The load-carrying capacity of a structure is in general decreased by the presence of an *imperfection*, but the amount of the reduction varies greatly from one structural type to another, and depends also on the material behaviour, as appears from the qualitative plots in Fig. 6.3. It can be easily seen that in the case of Fig. 6.3(a) (typically, a very slender column) the load-carrying capacity may be very close to the ideal buckling (bifurcation) load σ_{nom} while even small imperfections cause

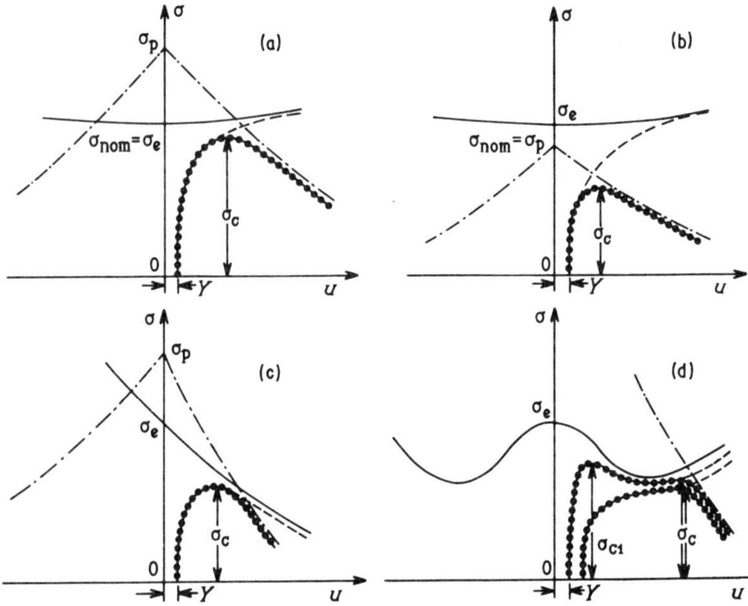

——— elastic behaviour (ideally *perfect* structure);
- - - - elastic behaviour (*imperfect* structure);
—·— plastic behaviour (*perfect* structure);
•••••• actual elastic–plastic behaviour.

Fig. 6.3 Typical examples of response of imperfection–sensitive-structures:
(a) symmetric buckling and stable elastic post-buckling, $\sigma_e < \sigma_p$;
(b) symmetric buckling; $\sigma_p < \sigma_e$;
(c) asymmetric buckling, $Y > 0$, $\sigma_e < \sigma_p$;
(d) symmetric buckling, unstable post-buckling and elastic recovery; $\sigma_e \ll \sigma_p$.
Definitions: u, response parameter; Y, significant imperfection parameter; σ, load parameter; σ_e, elastic bifurcation load (Euler buckling load); σ_p, plastic yield load (squash load); σ_{nom}, buckling load of perfect structure; σ_c, actual load-carrying capacity.

significant decreases of σ_c in the cases of Fig. 6.3(b) and 6.3(c). However, apart from the variation in the scale of the abscissae, the relation between carrying capacity σ_c and significant imperfection Y is always of the type shown in Fig. 6.4, i.e. has a pointed maximum for $Y = 0$, and is concave upwards on both sides.

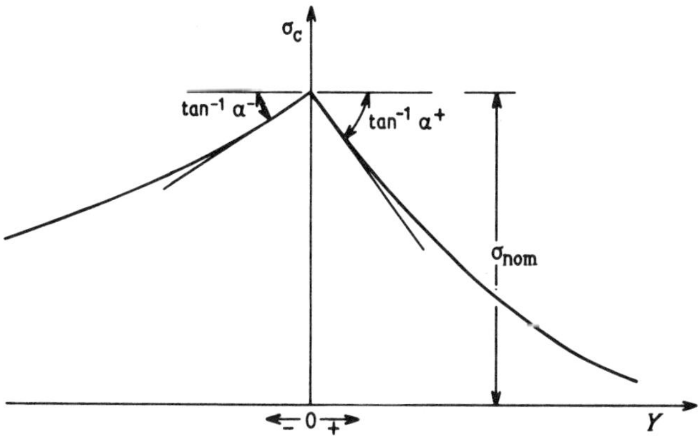

Fig. 6.4 Load-carrying capacity σ_c vs imperfection parameter Y (typical plot).

Because of the non-holonomy of this relationship, equation (6.4) must be generalized to obtain the PDF of the load capacity, $p_c(\sigma)$. In particular, under the assumption of one relevant imperfection parameter (implicitly accepted in Figs 6.3 and 6.4), and no other random variable,

$$p_c(\sigma) = \frac{p_Y(Y^+)}{|\sigma_c/\mathrm{d}Y|_{Y=Y^+}} + \frac{p_Y(Y^-)}{|\mathrm{d}\sigma_{c'}/\mathrm{d}Y|_{Y=Y^-}} \tag{6.45}$$

where $p_Y(Y)$ is the PDF of the imperfection (assumed known), Y^+ and Y^- are the two values (respectively positive and negative) of Y that correspond to $\sigma_c = \sigma$, and $|\mathrm{d}\sigma_c/\mathrm{d}Y|$ the absolute value of the derivative of the σ_c–Y relationship. The use of equation (6.45) is illustrated in Fig. 6.5: note the peculiar shape assumed by $p_c(\sigma)$ which cannot go beyond $\sigma_c = \sigma_{nom}$, the buckling load of the perfect structure. If the initial slopes α^+ and α^- are finite (as they are if the buckling load falls in the inelastic range), the upper limit corresponds to a sharp *cut-off*, i.e. to a *discontinuity* of $p_c(\sigma)$ (Fig. 6.5(c)).

A very simple example can be useful to make things clearer. Consider the rigid rod in Fig. 6.6(a), hinged at the lower end with an elastic–perfectly plastic spring (Fig. 6.6(b)) and compressed by a force W applied with an eccentricity e at the top end.

The load capacity is given by

$$W_c(e + \theta_y l) = M_y \tag{6.46}$$

M_y, θ_y and l being defined in Fig. 6.6. It is convenient to introduce the imperfection

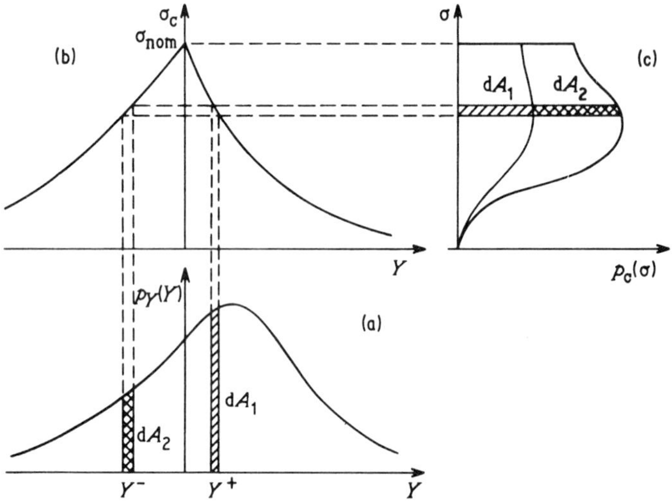

Fig. 6.5 Construction of the PDF of load-carrying capacity $p_c(\sigma)$.

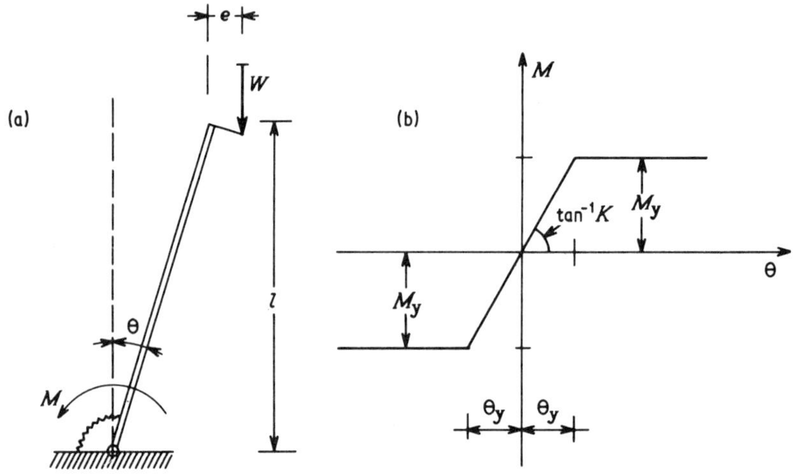

Fig. 6.6 Elastic–plastic single-degree-of-freedom column model.

and load parameters defined by

$$Y = \frac{e}{\theta_y l}; \qquad \sigma = \frac{Wl}{K} \tag{6.47}$$

so that equation (6.46) can be rearranged into

$$\sigma_c = \frac{1}{1+|Y|} \tag{6.46'}$$

where $|Y|$ is the absolute value of Y. Equations (6.45) and (6.46′) then yield

$$p_c(\sigma) = (1+Y)^2[p_Y(Y) + p_Y(-Y)] \qquad Y \geqslant 0 \qquad (6.45')$$

Fig. 6.7(b) shows some examples of PDFs given by (6.45′) for each of the (normal) PDFs $p_Y(Y)$ in Fig. 6.7(a). The corresponding average values and standard deviations are tabulated in Fig. 6.7(c).

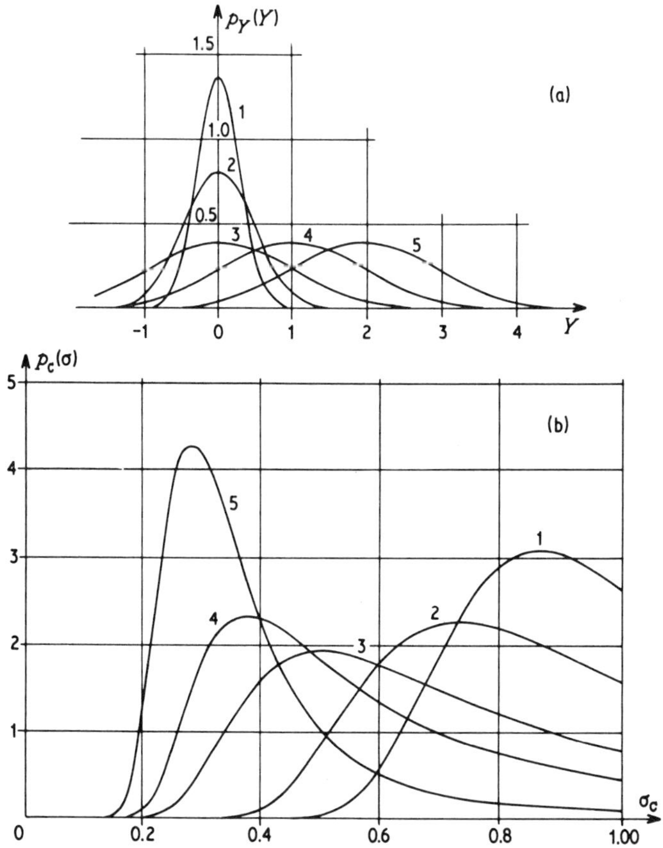

Curve no.	Imperfection Y		Load capacity σ_c		
	Average μ_Y	Standard deviation s_Y	Average μ_Y	Standard deviation s_Y	
1	0	0.3	0.86	0.111	
2	0	0.5	0.75	0.146	(c)
3	0	1	0.62	0.187	
4	1	1	0.53	0.195	
5	2	1	0.37	0.151	
	assumed		calculated		

Fig. 6.7 PDF of imperfection Y and corresponding PDFs of load-carrying capacity of single-degree-of-freedom elastic–plastic column model ($\sigma_{nom} = 1.00$).

(b) Imperfect inelastic columns

Only few complete solutions of technically significant problems of the above type are available. Among these, a treatment of hinged elastic–plastic columns [5] is now illustrated, while a numerical simulation approach to the same problem will be presented in Chapter 9 (Section 9.3.1).

The imperfection was assumed as an initial out-of-straightness in the form of a half sine wave (Fig. 6.8(a))‡

$$\tilde{u}_0(x) = \tilde{u}_{0m} \sin\left(\frac{\pi x}{l}\right) \tag{6.48}$$

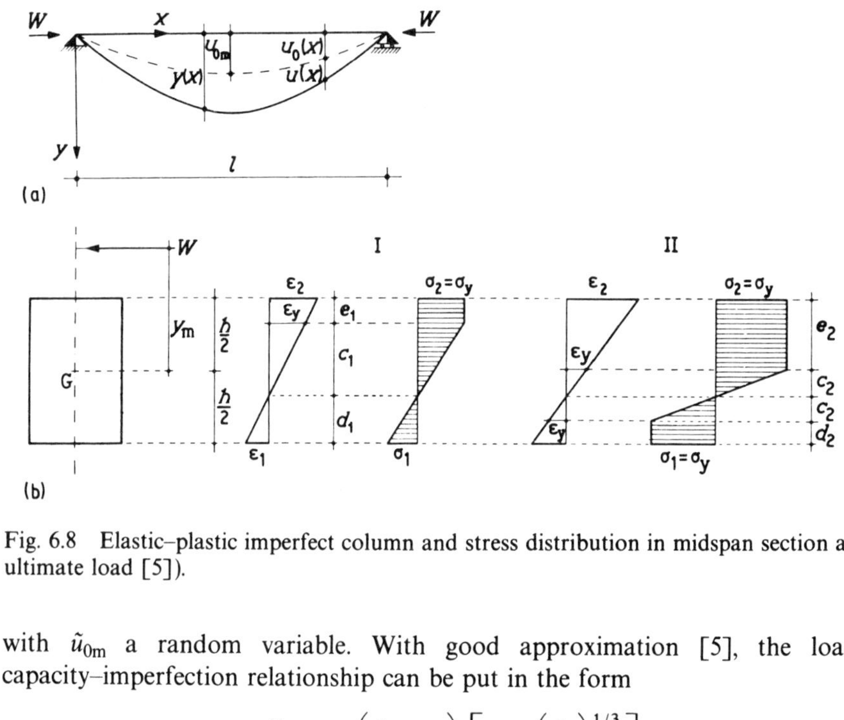

Fig. 6.8 Elastic–plastic imperfect column and stress distribution in midspan section at ultimate load [5]).

with $\tilde{u}_{0m}$ a random variable. With good approximation [5], the load capacity–imperfection relationship can be put in the form

$$Y = \frac{u_{0m}}{\imath} = 3\left(\frac{\sigma_y}{\sigma_c} - 1\right)\left[1 - \left(\frac{\sigma_c}{\sigma_e}\right)^{1/3}\right] \tag{6.49}$$

where $\imath$ is the core radius of the column cross section§, σ_y the yield stress, σ_e the Euler buckling stress and σ_c the *critical stress* (load capacity). Assuming $\tilde{Y}$ to be normally distributed with zero mean and standard deviation s, the yield stress to be $\sigma_y = 2800$ kg cm^{-2}, and the slenderness ratio $\lambda_B = l/r$ to vary between 20 and

‡ The imperfection can be more completely described by a stochastic process, but then numerical simulation becomes compulsory to get any result (see e.g. [6]).

§ Note that, for a rectangular section of depth h, the radius of gyration $r = h/\sqrt{12}$ and the core radius $\imath = h/6 = r/\sqrt{3}$.

200, several PDF curves of the critical stress σ_c have been calculated (Fig. 6.9). All these curves show the expected cut-off at the buckling stress of the perfect strut

$$\sigma_{nom} = \min(\sigma_y; \sigma_e) \tag{6.50}$$

For a more realistic treatment, one must remember that the *yield stress* of a given material is also a random quantity, for instance normally distributed with a standard deviation s_{σ_y} equal to 10% of its mean value μ_{σ_y}. Moreover, it seemed appropriate also to introduce random variability of the *slenderness ratio* λ_B since this quantity may be different from the design value because of defective restraints and/or approximations introduced in the calculations. Also λ_B has been assumed to be normally distributed, and its standard deviation s_λ has been taken equal to 10% of the average value μ_λ. With these assumptions, the probability density curves of the critical stress σ_c take the form shown in Fig. 6.10(a): the upper cut-off disappears because large values of σ_{nom} are now possible, albeit with little probability. However, the curves are still very asymmetric, and the average collapse stress μ_{σ_c} remains significantly smaller than the *nominal collapse stress*, i.e. the collapse stress of what can be defined *expected column* (straight column with

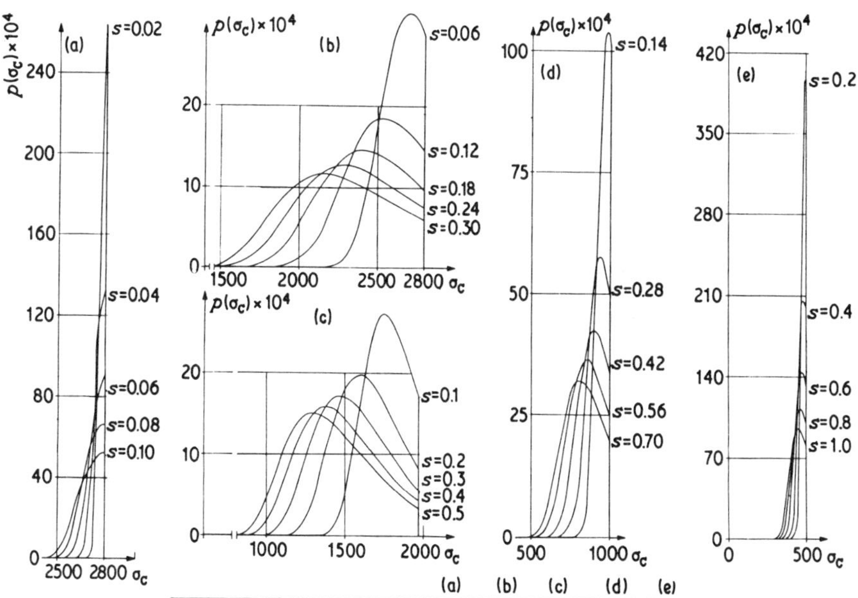

	(a)	(b)	(c)	(d)	(e)
Slenderness ratio λ_B	20	60	100	140	200
Euler buckling stress σ_e (kg cm^{-2})	49 348	5483	1974	1007	493

Yield stress $\sigma_y = 2800$ kg cm^{-2}

Fig. 6.9 PDFs of column critical stress (load-carrying capacity) for deterministic yield stress σ_y and slenderness ratio λ_B, and normally distributed imperfection $\tilde{Y}$, with average $\mu_Y = 0$ and standard deviation s indicated [5].

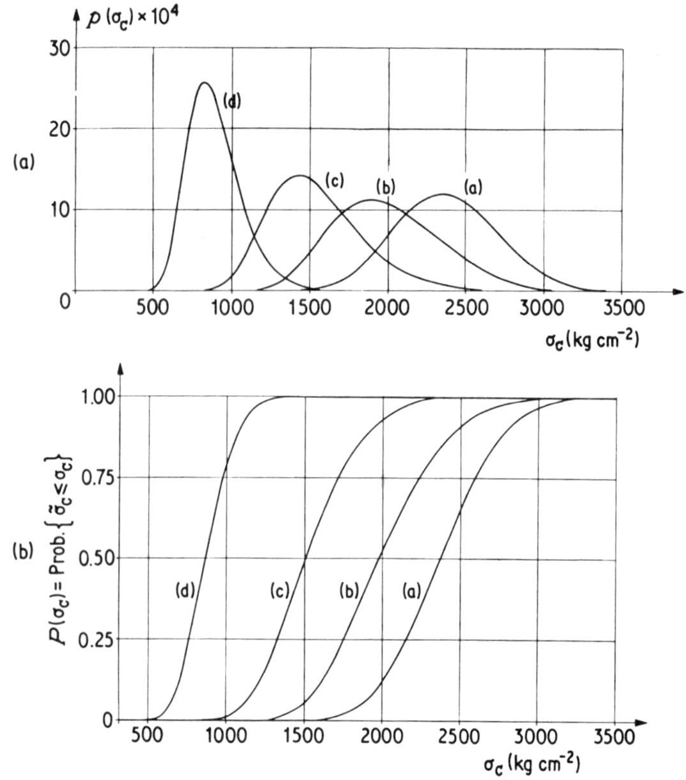

Fig. 6.10 PDFs and CDFs of critical stress of elastic–plastic column; random, normally distributed imperfection, yield stress and slenderness ratio [5]:

	Average	*Standard deviation*
Yield stress σ_y (kg cm^{-2})	$\mu_{\sigma_y} = 2800$	$s_{\sigma_y} = 280$
Imperfection parameter Y	$\mu_Y = 0$	$s = 0.003\,\mu_\lambda$
Slenderness ratio λ_B	$\mu_\lambda =$ 60 (a), 80 (b), 100 (c), 140 (d)	$s_\lambda = 0.1\,\mu_\lambda$

average properties) (Fig. 6.11)

$$\sigma_{\text{nom}} = \min(\mu_{\sigma_y};\ \mu_{\sigma_e}) \tag{6.50'}$$

(equation (6.50′) reduces to (6.50) if σ_y and λ_B are deterministic),

The lower tails of the PDF curves $p_c(\sigma)$ are of the utmost importance in actual design. In particular, to any given probability of collapse P_B of the column under a compression $\sigma = W/A$, there corresponds an *admissible* stress σ_{cam} in the design of

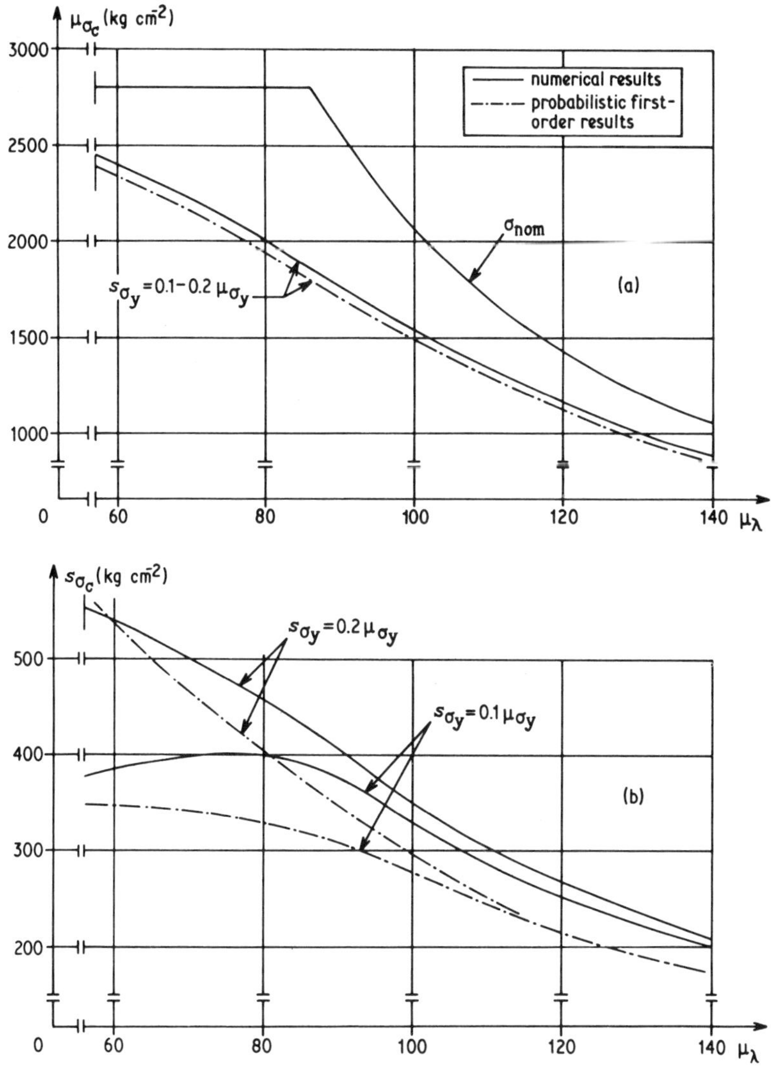

Fig. 6.11 Nominal collapse stress σ_{nom}, average collapse stress μ_{σ_c} and standard deviation s_{σ_c} of collapse stress of imperfect column (standard deviation of yield stress as indicated; other data as in Fig. 6.10) (after Augusti and Baratta [5a].)

the column

$$P_{\text{B}} = \text{Prob}[\tilde{\sigma}_{\text{c}} \leqslant \sigma_{\text{cam}}] = \int_0^{\sigma_{\text{cam}}} p_{\text{c}}(\sigma)\text{d}\sigma = P_{\text{B}}(\sigma_{\text{cam}}) \tag{6.51}$$

A nominal *safety factor* γ^0 can also be defined as the ratio between the nominal collapse stress and the admissible stress

$$\gamma^0 = \frac{\sigma_{\text{nom}}}{\sigma_{\text{cam}}} \tag{6.52}$$

The variation of γ^0 with the other material and geometrical properties can be investigated by several types of plots, as exemplified in Fig. 6.12 [5]. Note in particular that γ^0 is approximately linear in $\log(P_B(\sigma_{cam}))$, at least in the range of very small probabilities (Fig. 6.12(b)), and is largest for intermediate slenderness ratios such that $\sigma_y \simeq \sigma_e$ (Fig. 6.12(a)).

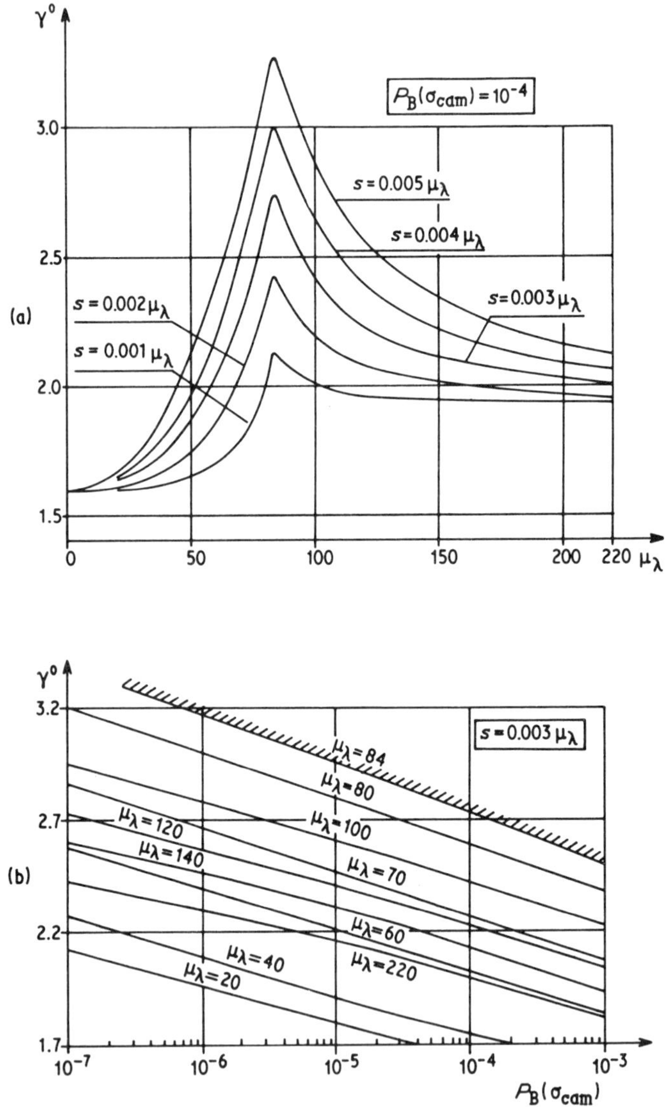

Fig. 6.12 Elastic–plastic column: examples of plots of nominal safety factor γ^0 (6.52) vs average slenderness μ_λ(a) and vs probability of failure (b):
$\mu_{\sigma_y} = 2800 \text{ kg cm}^{-2}$; $s_{\sigma_y} = 280 \text{ kg cm}^{-2}$; $s_{\lambda_B} = 0.1\,\mu_{\lambda_B}$; other data as indicated [5].

Other studies, and in particular numerical simulation studies including the effects of residual stresses, have essentially confirmed, for pin-ended columns, the results summarized above.

It is worth noting that pin-ended elements can usually be assumed in bracings and trusses. In other structural types (e.g. rectangular frames), rigid connections, transmitting bending moment as well as shear, are essential to the structural continuity. In this case the probabilistic evaluation of the column strength would require an accurate definition of the boundary conditions, which generally depend on the geometrical and mechanical random parameters of the whole structure and, in case of elastic–plastic behaviour, can also change as the applied loads increase: also, probabilistic modelling of imperfections should be quite different. Therefore, the results obtained for isolated pin-ended columns are not necessarily valid in other cases: however, they are often used in practice for want of more appropriate data, but great care should be recommended in this operation.

6.3 RELIABILITY WITH RESPECT TO ULTIMATE LIMIT STATES

6.3.1 Loss of static equilibrium (LS.U1)

This limit state is comparitively simple to deal with when the resisting system and the acting loads are well defined in probabilistic terms. This is usually the case, for instance, of a support wall, where the overall stability check reduces to the basic case (R, S) discussed in Chapter 1: some difficulties may only lie in the probabilistic description of S from the soil thrust. In contrast, the problem of slope stability (Fig. 6.13) is very cumbersome also in a deterministic context since

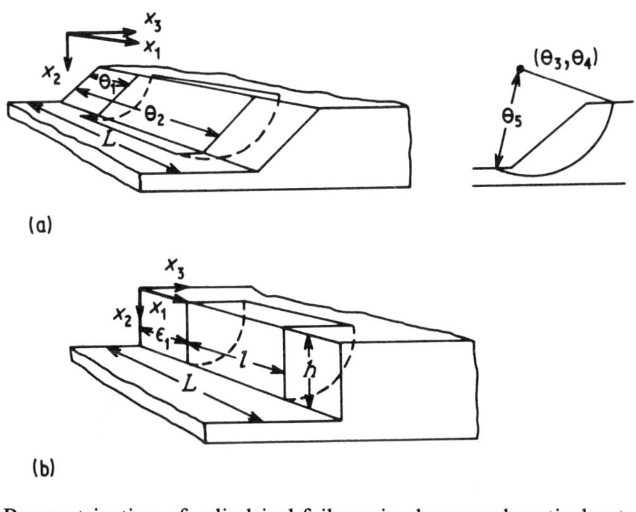

Fig. 6.13 Parametrization of cylindrical failures in slopes and vertical cuts [7].

the shape of the failure mechanism and hence the resisting geometry has to be determined. Obviously, the problem becomes formidable to handle when the basic parameters vary randomly in all three spatial directions.

Nevertheless, under the simplifying assumption of time independence (or of deterministic time dependence), some investigations that consider explicitly the uncertainty of the variation of soil properties have been developed recently. The proposed models are increasingly sophisticated from the probabilistic viewpoint: indeed, in the most recent studies (e.g. [7]), shear resistance is modelled as a three-dimensional random field, which may be isotropic, homogeneous, non-isotropic or non-homogeneous. In contrast, in order to obtain tractable analytical models, the mechanical aspect is greatly simplified: in particular, the failure mode is usually assumed to be known, and the material is assumed to obey an associated plastic flow law. As a consequence of these approximations, the proposed models have led to calculated slope reliabilities much lower than those that can be derived from practical experience [7a]. Therefore, the cited investigations can be regarded as first contributions to a rational treatment of slope stability problems, but cannot as yet be regarded to yield valid quantitative indications.

Given the random variability of the geometry of the problem, this is another case in which, at the present time, the most accurate *probabilistic* information can be obtained from the use of numerical simulation procedures (cf. Section 6.2.1(d)), which do not require excessive simplifications of the mechanical model.

6.3.2 Localized rupture (LS.U2) or excessive deformation (LS.U2′)

In the case of *brittle* structural behaviour, the ultimate limit state practically coincides with the first attainment of the elastic limit in any *point* of the structural solid. On the other hand, the rupture of a *section* (that requires some spreading of plastic deformation) is the ultimate limit state currently considered in non-linear analysis of reinforced-concrete structures.

Excessive deformation, elastic or permanent, may also coincide with an ultimate limit state, either because of geometrical second-order effects that have been neglected in the calculations, or because the structural elements do not have sufficient *ductility* to absorb the deformation without damage, or because the permanent change of geometry is sufficient to put the structure out of service.

All these cases, except that of brittle behaviour which will be treated in Section 6.3.3, require the use of procedures capable of dealing with non-linear response, such as those presented in Section 6.2.

6.3.3 Attainment of the elastic limit (LS.U3)

(a) General statements

This limit state is often introduced in order to treat (deterministic or probabilistic) structural problems by means of linear models, which neglect material non-linearities as well as the effects of changes of geometry. These models have been presented in detail, for both cases of static and dynamic response, in Chapter 4.

However, redistribution of stresses makes this limit state a conventional one in most instances: between the attainment of the elastic limit in a point of the structure and actual collapse there is usually a more or less wide interval. A notable exception is the case of a brittle material, which causes a *weakest-link* behaviour with regard to overall failure, i.e. the limit strength of the structure is reached as soon as one element reaches its limit strength (Section 1.5).

In the case in which this is only a conventional limit state the calculation of reliability can be performed assuming the same *critical inequality* (in terms of stresses) in all points of the structure and thus individuating the *most stressed* point as the point in which the failure condition is first reached by a proportional increase of stresses.

However, the randomness of the strength of the single elements must be taken into account in the actually brittle case: as a matter of fact, *brittle fracture* depends on the local stress in the points where the most dangerous structural defects occur. The origin of such defects is not important to the development of the theory, and therefore a general approach can be developed starting from the *weakest-link* assumption, as summarized below.

(b) Probabilistic approach to brittle rupture

Consider a body of volume V under uniform stress conditions (i.e. subjected to the same stress σ in every point $\mathbf{x} \in V$)‡ and define the *local strength* $\tilde{R}$ such that its CDF $P_R(\sigma)$ is the probability of (local) rupture when the stress is equal to σ. Then, assuming that rupture can be initiated at any defective point in the body, and that n is the average number of such defects per unit volume, equations (1.4) and (A7.4(a)) yield the CDF $P_{RV}(\sigma)$ of the strength of the whole solid of volume V

$$P_{RV}(\sigma) = 1 - [1 - P_R(\sigma)]^{nV} \tag{6.53}$$

nV being the total number of defects. Since this number is very large, it is convenient to introduce a minimum asymptotic representation. The distributions must be limited in the tail of interest by the minimum strength σ_0 of the defective element; hence, type III extreme distributions (equation (A7.21)) can be adopted:

$$P_{RV}(\sigma) = 1 - \exp\left[-\left(\frac{\sigma - \sigma_0}{\sigma_1 - \sigma_0}\right)^c\right] \tag{6.54}$$

where the parameter σ_1 depends on the volume considered, and is generally given in the form

$$\sigma_1 - \sigma_0 = \sigma_c (V_0/V)^{1/c} \tag{6.54'}$$

V_0 being a reference volume (*standard volume*). Thus, equation (6.54) takes the

‡ In case of a bi- or tri-dimensional stress state, an *equivalent* (scalar) *stress* σ, deduced in accord to any appropriate strength criterion, must be introduced in the following treatment.

form, consistent with (6.53),

$$P_{RV}(\sigma) = 1 - \exp\left[-\frac{V}{V_0}\left(\frac{\sigma - \sigma_0}{\sigma_c}\right)^c \right] \tag{6.54''}$$

Depending on the magnitude of the exponent c, the relevant PDF (equation (A7.22)) takes on the various shapes of Fig. 6.14. Furthermore, the mean value μ_R and the standard deviation s_R of the defect resistance are given by equations (A7.23) and (A7.24):

$$\mu_R = \sigma_0 + \sigma_c (V_0/V)^{1/c}\Gamma(1+1/c) \tag{6.55}$$

$$s_R = \{\sigma_c^2 (V_0/V)^{2/c}[\Gamma(1+2/c) - \Gamma^2(1+1/c)]\}^{\frac{1}{2}} \tag{6.56}$$

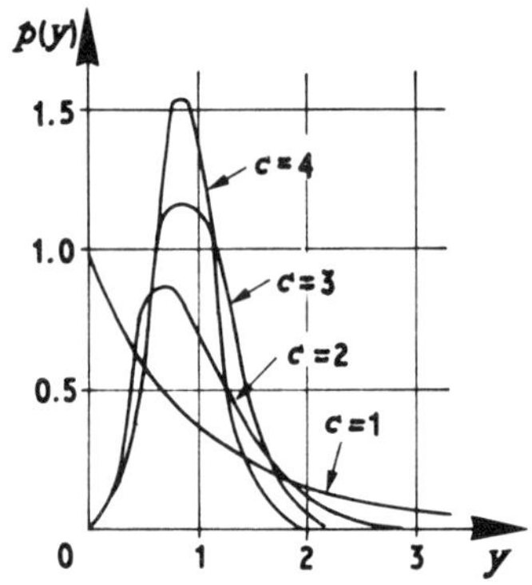

Fig. 6.14 PDFs of brittle strength for different values of c corresponding to the CDFs given by equation (6.54''). $y = \dfrac{\sigma - \sigma_0}{\sigma_c}\left(\dfrac{V}{V_0}\right)^{1/c}$ (after [9]).

In order to extend the treatment to non-uniform stress states, note that, if the volume V is made up by elementary volumes $\mathrm{d}V$, equation (6.54'') can be formally written

$$P_{RV}(\sigma) = 1 - \exp\left[-\frac{\int_V \mathrm{d}V}{V_0}\left(\frac{\sigma - \sigma_0}{\sigma_c}\right)^c \right] \tag{6.57}$$

Thus, if the stress σ is not uniform the CDF of the overall strength can be postulated, consistently with equation (6.57), in the form

$$P_V(\sigma) = 1 - \exp\left[-\frac{1}{V_0}\int_{\sigma > \sigma_0} \mathrm{d}V \left(\frac{\sigma(\mathbf{x}) - \sigma_0}{\sigma_c}\right)^c \right] \tag{6.58}$$

where σ on the l.h.s. is now a parameter which measures the external load and $\sigma(\mathbf{x})$ under the integral is the local stress.

The integrand on the r.h.s. of equation (6.58) can be called *specific hazard* (or risk of failure per unit volume) and is equivalent to the hazard rate introduced in Section 6.4.1.

Experimental results have confirmed the good accuracy of formula (6.54), but have also pointed out some discrepancies when the theory was employed to estimate the scale effect, i.e. to infer the properties of a finite solid from those of small specimens. In fact, experimental coefficients of variation of overall resistance are generally much lower than the values calculated from the tests of elements: such results appear to indicate that perfectly brittle behaviour is never achieved in nature.

6.3.4 Transformation of the structure into a mechanism (LS.U4)

The ultimate limit state defined by plastic collapse of the structure has been treated in detail in Chapter 5.

The simplicity of the mechanical model (comparable with the linear elastic model) has made it possible to include the case of time-varying loads.

6.3.5 Instability (LS.U5)

This type of failure, already discussed in Section 6.2.3, is of essentially random character: as a matter of fact, the corresponding load-carrying capacity depends directly on the value taken by random quantities, the *imperfections*, which do not exist in the idealized mechanical model. Therefore, as long since well known to engineers, instability collapse loads exhibit a very marked dispersion, generally much larger than the loads corresponding to other types of collapse: indeed, instability failure, more than any other failure, should be approached from a *probabilistic* viewpoint, as is increasingly being recognized in structural engineering.

(a) Compressed columns

In Section 6.2.3 a probabilistic treatment of pin-ended elastic–plastic columns was presented. In particular, it was shown how, under some simplifying assumptions, plots of the nominal *safety factor* γ^0 versus the column slenderness ratio λ_B (or its average μ_λ) can be obtained for any given value of the probability of collapse and of the random properties of the imperfections (Fig. 6.12).

In order to use these results for actual column design (or reliability evaluation), they can be reported as plots of admissible stress

$$\sigma_{\text{cam}} = \sigma_{\text{nom}}/\gamma^0 \tag{6.59}$$

This is done in the example of Fig. 6.15(a), where the average slenderness ratio is used as abscissa, and the standard deviation of the imperfection as parameter, while other quantities are assumed to be given. Then, in Fig. 6.15(b), one of these curves is shown compared with a commonly used design curve: their substantial agreement is remarkable.

As already pointed out at the end of Section 6.2.3, only results relevant to pin-ended columns are available in any detail at the present time, and these are more or less empirically extended to take care of other instability cases.

(b) Overall stability: plane frames

Besides the stability check of each imperfection-sensitive structural element, an analysis that takes into account the second-order geometrical effects (P–Δ effects) must be developed for all structures in which the displacements from the initial geometry are not negligible.

Unfortunately, no synthetic structural analysis procedure is available (not even in a deterministic context) which can take care at the same time of local stability of single elements and global structural stability. Therefore, for each particular structural type, a specific study must be performed: elastic–plastic frames are of particular interest, as they constitute a class of instability-prone structures very frequently encountered in civil engineering.

The first-order analysis of these structures can be performed step-by-step by specifically developed procedures, examples of which will be presented in Chapter 9; it is usually assumed that the distributed bending deformations obey an indefinitely elastic law, while inelastic deformations concentrate in *plastic hinges* (cf. Chapter 5).

To tackle the instability problem, the assumption of small displacements must be removed, so that changes of geometry influence the equilibrium equations. Then the state of the structure must be investigated taking into account both the elastic and the geometrical stiffness matrices. Because of the non-negligible contribution of the latter, any solving procedure must be iterated until the axial forces calculated for the structural elements coincide with the values adopted in the determination of the geometrical stiffness.

In order to reduce the computational efforts of the iterations, approximate methods have been suggested in which the stresses are calculated in the initial geometry, while non-linearities are taken into account by a suitable set of fictitious shears.

In any approach, failure of the frame is indicated by appropriate instability conditions‡. In particular, in the more exact approach the stability criterion concerns the sign of the diagonal elements of the solving matrix, on which a pivotal step is performed each time a plastic hinge forms. In the second case, the eigenvalue of the matrix must be evaluated to estimate the critical load value.

As numerical procedures are necessary for both structural analysis and stability criterion, reliability analysis of such structures can be generally performed only by simulation techniques (Chapter 9). However, no conceptual difficulty arises if the external forces are defined by only one stochastic process $\alpha(t)$. In particular, the structural design lifetime T should be discretized into a finite number of intervals

‡ Failure, in fact, does not necessarily involve the formation of a plastic collapse mechanism as for the rigid-plastic frames of Chapter 5.

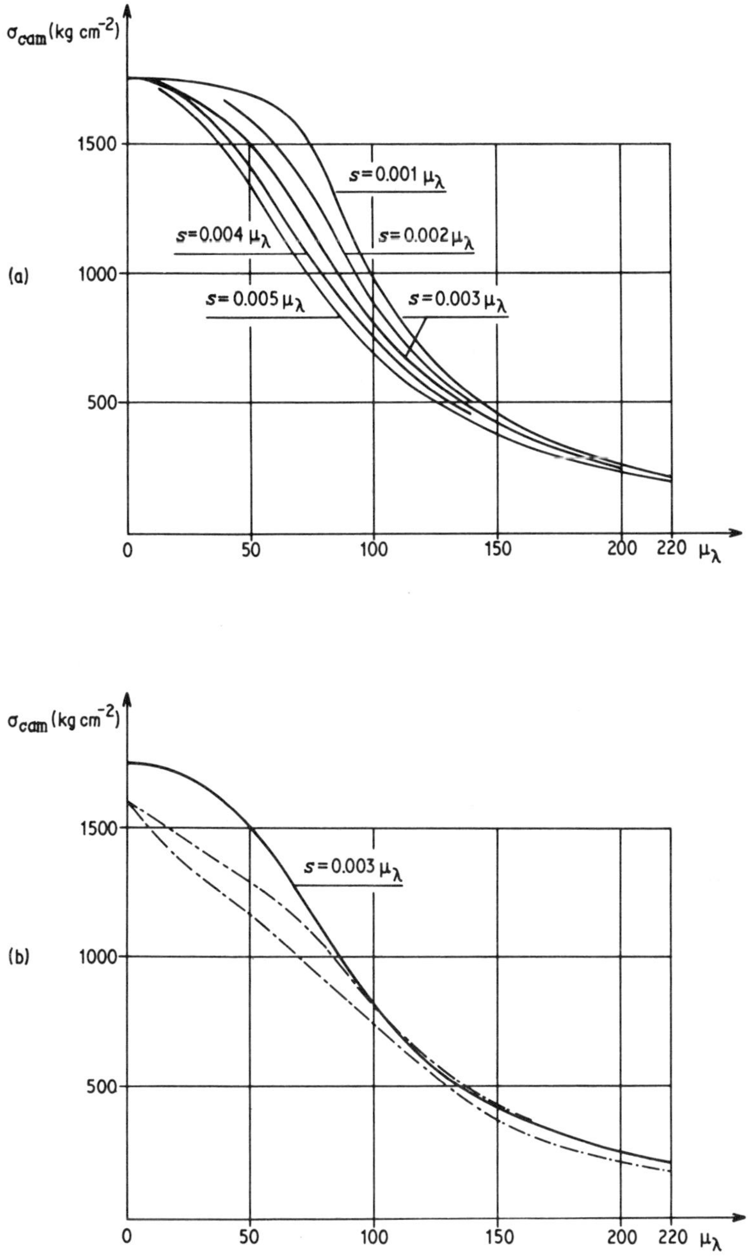

Fig. 6.15 (a) Admissible stress σ_{cam} of pin-ended column vs average slenderness ratio for several standard deviations s of imperfection parameter; (b) comparison with (dash-dotted) column design curves (—·—·—) of the current Italian Recommendations for steel structure design [5].

such that the structural responses in two successive intervals are independent of each other (Section 3.6).

By contrast, if the forces acting on the structure are described by a vector stochastic process $\boldsymbol{\alpha}(t)$ and the structural behaviour is geometrically non-linear, the inelastic deformations that may develop along structural members before any time t can influence the structural resistance to the forces occurring at time t. It follows that:

(1) The structural lifetime T cannot be discretized in intervals of independent structural behaviours; therefore equation (3.80) is no longer valid and the reliability of the structure can be evaluated by equation (3.75) only for $\tau = T$.
(2) The probability of failure at any time under given loads $P_R(\boldsymbol{\alpha}, t)$ depends on the whole loading path. Hence the probabilities of failure on the r.h.s. of equation (3.75) cannot be calculated as in equations (6.1) or (6.2).

Therefore a correct safety analysis of the structure would require the convolution, at any time t $(0 < t \leqslant T)$, of the PDF of the multivariate stochastic process of the acting forces with the CPF $P_R(\boldsymbol{\alpha}, t)$.

From a computational point of view, a rough estimate of the structural reliability can be obtained by performing a sufficiently large number of simulated time-history analyses in $(0, T)$.

A different approach to safety analysis consists of determining lower and upper bounds to the probability of failure P_{fail}. This approach is possible if the *order* in which the simultaneous loads reach their final value is known and is always the same [8].

6.3.6 Fatigue (LS.U6)

Fatigue failure is another case where use of probabilistic concepts appears logically essential. Basically, fatigue failure depends on the accumulation of a large number of small damages caused by successive load cycles. In fact, fluctuating stresses can cause a process of localized progressive structural changes in the material, which generally results in lowering the resistance for subsequent stresses. In particular, submicroscopic cracks may progress to microscopic cracks, whose development is terminated by macroscopic fracture. Failure by fatigue is the result of this cumulative damage.

The amount of damage caused by every cycle depends mainly on the maximum load-effect intensity; therefore, the resultant damage at the end of the loading history is known only if the number of cycles and the maximum load effect of each cycle are known. As fatigue failure usually requires, in practical cases, some millions of load cycles, it is clear that full knowledge of the input data can rarely be achieved, and probabilistic treatment is the only rationally possible one.

Current practice in fatigue-resistant design is based on the knowledge of the relation between (uniaxial) stress range σ and number n of cycles of constant amplitude σ leading to fracture. The number n is often denoted as the *fatigue life* of the given material or the considered structural joint or detail.

In actual fact, the σ–n relation has a marked random character even under the most accurately controlled test conditions (Fig. 6.16). In particular, a large statistical variation of the fatigue life corresponds to small standard deviations of the stress range amplitude.

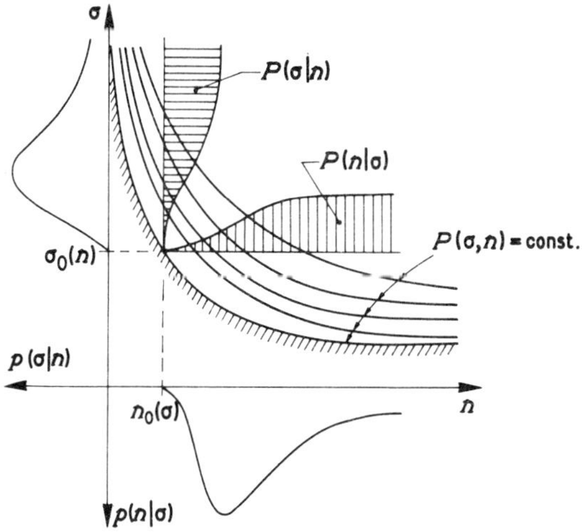

Fig. 6.16 Typical random relationship between stress range and number of cycles (after [9]).

For the CDF $P(n|\sigma)$, many authors propose a type III (Weibull) extreme distribution of minimum values (equation (A7.21)):

$$P(n|\sigma) = 1 - \exp\left[-\left(\frac{n - n_0}{n_c - n_0} \right)^c \right] \tag{6.60}$$

so that the *failure rate function* (or hazard rate, cf. Section 6.4.1), defined by the ratio

$$\Pi(n|\sigma) = -\frac{p(n|\sigma)}{1 - P(n|\sigma)} \tag{6.61}$$

takes the form

$$\Pi(n|\sigma) = \frac{1}{n_c - n_0} \left(\frac{n - n_0}{n_c - n_0} \right)^{c-1} \tag{6.62}$$

Of the three parameters appearing in the preceding equations, n_0 is the minimum life below which no failure is expected for stress amplitude σ. The *shape parameter* c is an indicator of fatigue sensitivity: $c > 1$ corresponds to an increasing $\Pi(n|\sigma)$ and, hence $c = 1$ is a lower limit of fatigue sensitivity. Finally, the parameter n_c, often called *characteristic life*, is a central measure of the fatigue life.

The exponent on the r.h.s. of equation (6.60) is a measure of fatigue *damage* and is often indicated as D, so that equation (6.60) is more simply written

$$P(n|\sigma) = 1 - e^{-D} \tag{6.60'}$$

Note in particular that $D = 1$ corresponds to $n = n_c$.

Consider now a structural element subjected to n_t cycles of loading with random stress range amplitudes. A cumulative damage rule can be derived [9, 9a] under the assumptions that (i) the probability of fatigue failure $P(n_t)$ has an expression of the same type (6.60)

$$P(n_t) = 1 - \exp\left(-\frac{n_t - n_{0t}}{n_{ct} - n_{0t}} \right)^{c_t} \tag{6.63}$$

and (ii) the damages produced at each stress level are independent of each other; then, assuming a finite number N of stress ranges σ_i, each attained n_i times, and applying the law of compound probabilities (equation (1.10)),

$$P(n_t) = \prod_{i=1}^{N} P(n_i|\sigma_i) \tag{6.64}$$

Introducing (6.63) and (6.60), equation (6.64) yields

$$\exp\left(-\frac{n_t - n_{0t}}{n_{ct} - n_{0t}} \right)^{c_t} = \exp\left[-\sum_{i=1}^{N} \left(\frac{n_i - n_{0i}}{n_{ci} - n_{0i}} \right)^{c_i} \right] \tag{6.65}$$

and hence

$$\left(\frac{n_t - n_{0t}}{n_{ct} - n_{0t}} \right)^{c_t} = \sum_{i=1}^{N} \left(\frac{n_i - n_{0i}}{n_{ci} - n_{0i}} \right)^{c_i} \tag{6.66}$$

In equations (6.65) and (6.66), n_i, n_{ci}, n_{0i} and c_i are respectively the fatigue life, the characteristic life, the minimum life and the shape parameter for the stress level σ_i.

The l.h.s. of equation (6.66) can be regarded as the *total damage D* in the n_t random cycles, and each term on the r.h.s. the damage D_{ni} caused by the n_i cycles of amplitude σ_i‡

$$D = \sum_{i=1}^{N} D_{ni} \tag{6.67}$$

When the minimum lives are all zero ($n_{0i} = 0$ for all i), the following non-linear damage rule is obtained.

$$D = \left(\frac{n_1}{n_{c1}} \right)^{c_1} + \left(\frac{n_2}{n_{c2}} \right)^{c_2} + \ldots + \left(\frac{n_N}{n_{cN}} \right)^{c_N} \tag{6.67'}$$

which, when all shape parameters c_i are equal to unity, becomes the well-known Miner's linear damage rule.

If Miner's rule is applied, most experimental failures occur with $D < 1$; it is

‡ Another cumulative damage rule, commonly employed when successive sequences with the same random amplitude have to be considered, consists of summing the probabilities of failure relevant to each sequence.

usually stated that the expected damage at failure has the value

$$\mu_D = 0.83 \tag{6.67''}$$

The average number of stress cycles leading to failure (*expected fatigue life* μ_n) can be obtained from

$$\mu_n = \mu_D \frac{\chi_1}{E[\tilde{\sigma}^{\chi_2}]} \tag{6.68}$$

where μ_D takes the value given by equation (6.67″) (or another appropriate value suggested by experiments); χ_1 and χ_2 are parameters obtained by fitting the experimental σ–n curve with a relation of the type

$$\ln n = \ln \chi_1 - \chi_2 \ln \sigma \tag{6.68'}$$

The coefficient of variation of n may be expressed as

$$\delta_n = \sqrt{(\delta_{1n}^2 + \delta_{2n}^2)} \tag{6.69}$$

where δ_{1n} is the uncertainty associated with the scatter of fatigue data and δ_{2n} is the uncertainty associated with the inaccuracy in the determination of the mean life. Reliability-based criteria for fatigue resistant design have also been developed from equations (6.68) and (6.69), assuming a Weibull distribution for the random variable $\tilde{n}$.

The above fatigue model is subjected to many criticisms, because the notion of fractional damage does not appear to describe satisfactorily the actual mechanical behaviour. An alternative theory of metal fatigue is based on the assumption that the propagation of a dominant crack is the primary cause of fatigue failure, and that failure occurs when the crack reaches a *critical length* l_c, related to the applied stress σ by a relation of the type (*Griffith's criterion*)

$$l_c = C/\sigma^2 \tag{6.70}$$

Therefore, if the material presents an initial crack of known length l_0, the analysis of the fatigue process becomes the study of the growth of a single crack from l_0 to the critical length. Assume that when a sequence of stresses $\sigma_1, \ldots, \sigma_k$ extends the crack from l_0 to l, the following formula holds:

$$\frac{2}{\sqrt{l_0}} - \frac{2}{\sqrt{l}} = Z(\sigma_1) + Z(\sigma_2) + \ldots + Z(\sigma_k) \tag{6.71}$$

where $Z(\sigma)$ is an appropriate function of the applied stress. When the structural element is subjected to random stresses $\tilde{\sigma}_1, \tilde{\sigma}_2, \ldots, \tilde{\sigma}_k$ at random instants of time $\tilde{t}_1, \tilde{t}_2, \ldots, \tilde{t}_k$, equation (6.71) becomes the stochastic equation

$$\sqrt{(l_0/l)} = 1 - \tfrac{1}{2}\sqrt{l_0} \int_0^t Z(\tilde{\sigma}(\tau))\, d\tilde{N}(\tau) \tag{6.72}$$

where $\tilde{N}(\tau)$ denotes the random number of instants in (0, t) in which the stresses are applied and the integral is a Stieltjes integral (Section A2.2). From equation (6.70) the stress $\tilde{\sigma}_k(t)$ needed to cause failure at time t can be shown to be given by

$$\tilde{\sigma}_k(t)/\sigma_v = 1 - \tfrac{1}{2}\sqrt{l_0} \int_0^t Z(\tilde{\sigma}(\tau))\,d\tilde{N}(\tau) \tag{6.73}$$

where σ_v is the stress for which the critical crack length is l_0.

In order to handle (6.73), the random variable $d\tilde{N}(\tau)$ must be explained. For this purpose, assume that

$$\begin{aligned} \text{Prob}\,[d\tilde{N}(\tau) = 1] &= p_1(\tau)\,d\tau + o(d\tau) \\ \text{Prob}\,[d\tilde{N}(\tau) = m] &= o(d\tau) \qquad m > 1 \\ \text{Prob}\,[d\tilde{N}(\tau) = 0] &= 1 - p_1(\tau)\,d\tau + o(d\tau) \end{aligned} \tag{6.74}$$

where $p_1(\tau)$ is the so-called *product density of degree 1*. Equations (6.74) imply that every one of the moments of the random variable $d\tilde{N}(\tau)$ is equal to $p_1(\tau)\,d\tau + o(d\tau)$ and hence

$$\mu_{N(\tau)} = \int_0^\tau p_1(\tau)\,d\tau \tag{6.74$'$}$$

By this approach, one obtains for the mean value of the stress $\tilde{\sigma}_k$ needed to cause failure at time t the formal expression

$$E[\tilde{\sigma}_k(t)] = \sigma_v\left(1 - \tfrac{1}{2}\sqrt{l_0} \int_0^t E[Z(\sigma(\tau))]p_1(\tau)\,d\tau\right) \tag{6.75}$$

The preceding summary treatments of fatigue should make it clear that, in this as well as in many other cases where probabilistic approaches would be most appropriate (slope stability, structural buckling, etc.), two types of treatment are generally available at present. The first aim at practical applications and, hence, are often theoretically objectionable and/or badly approximated; the second, more accurate, lead to incomplete results which need further developments to meet technical requirements. In all these cases, the structural engineer must be very cautious in adopting, for his own calculations, the results of other researchers or engineers whenever the limitations relevant to those results are not explicitly specified: in the latter case, a rough numerical simulation (Chapter 9) is sometimes better than unreasonable use of analytical results.

6.4 GENERAL APPROACH TO RELIABILITY ANALYSIS; MULTIPLE MODES OF FAILURE

6.4.1 Time variation of reliability

(a) General formulation

In Section 1.3, the safe region $\mathcal{S}$, the unsafe region $\mathcal{U}$ and their common boundary $\mathcal{C}$ were defined. Reliability analysis consists essentially in relating the probabilities of success and failure and the positive (outward) crossings of the surface $\mathcal{C}$ by the multivariate response process $\mathbf{u}(t)$ (Section A7.5). Let

$$\mathbb{R}(0) = P_{\text{suc}}(0) = 1 \tag{6.76}$$

i.e. assume that at the initial time the system is in the safe region $\mathscr{S}$ with probability one. Further, let the crossings of the surface $\mathscr{C}$ be a regular stream of events, and denote by $P_{\text{fail}}(t)$ the probability that, during the time interval $(0, t)$, the system leaves the safe region *at least once*, while the probabilities that during this interval *one*, *two*, etc., positive crossings occur are denoted by $P_1(t)$, $P_2(t)$, etc., respectively. Evidently, if $\mathbb{R}(t)$ is the probability of success (*reliability*) in $(0, t)$,

$$\mathbb{R}(t) = 1 - P_{\text{fail}}(t) = 1 - \sum_{n=1}^{\infty} P_n(t) \tag{6.77}$$

The probabilities $P_n(t)$ are related to the moments of the number $\tilde{n}(t)$ of positive crossings during time interval $(0, t)$ by [10]

$$\mathrm{E}[\tilde{n}^r(t)] = \sum_{n=1}^{\infty} n^r P_n(t) \qquad r = 1, 2, \ldots \tag{6.78}$$

$\tilde{n}(t)$ is a random function taking integer values. From equations (6.77) and (6.78) one obtains lower estimates for the reliability function

$$\mathbb{R}(t) \geqslant 1 - \mathrm{E}[\tilde{n}(t)] \tag{6.79}$$

$$\mathbb{R}(t) \geqslant 1 - \tfrac{11}{6}\mathrm{E}[\tilde{n}(t)] + \mathrm{E}[\tilde{n}^2(t)] - \tfrac{1}{6}\mathrm{E}[\tilde{n}^3(t)] \tag{6.79'}$$

$$\vdots$$

Analogous upper estimates are

$$\mathbb{R}(t) \leqslant 1 - \tfrac{3}{2}\mathrm{E}[\tilde{n}(t)] + \tfrac{1}{2}\mathrm{E}[\tilde{n}^2(t)] \tag{6.80}$$

$$\mathbb{R}(t) \leqslant 1 - \tfrac{25}{12}\mathrm{E}[\tilde{n}(t)] + \tfrac{35}{24}\mathrm{E}[\tilde{n}^2(t)] - \tfrac{5}{12}\mathrm{E}[\tilde{n}^3(t)] + \tfrac{1}{24}\mathrm{E}[\tilde{n}^4(t)] \tag{6.80'}$$

$$\vdots$$

The first estimate (equation (6.79)) is the most interesting for applications. In fact, the mathematical expectation $\mathrm{E}[\tilde{n}(t)]$ of the number of positive crossings can be calculated by comparatively simple methods (Section A7.4.1).

Moreover, if the level of the reliability is sufficiently high (as generally for ultimate limit states), so that $\mathrm{E}[\tilde{n}(t)] \ll 1$ for all considered values t, equation (6.79) yields a conservative, but accurate, estimate of the reliability function

$$\mathbb{R}(t) \simeq 1 - \mathrm{E}[\tilde{n}(t)] \tag{6.81}$$

The value yielded by equation (6.81) is also close to the estimate based on the hypothesis that the positive crossings of the surface $\mathscr{C}$ form a Poisson process (equation (A7.49)),

$$\mathbb{R}(t) = \exp\{-\mathrm{E}[\tilde{n}(t)]\} \tag{6.82}$$

When $\mathbb{R}(0) \neq 1$, the estimates (6.81) and (6.82) are to be replaced by the following (equations (A7.47) and (A7.55))

$$\mathbb{R}(t) \simeq \mathbb{R}(0)\{1 - \mathrm{E}[\tilde{n}(t|\mathscr{S})]\} \tag{6.83}$$

$$\mathbb{R}(t) \simeq \mathbb{R}(0)\exp\{-\mathrm{E}[\tilde{n}(t|\mathscr{S})]\} \tag{6.84}$$

Here $\tilde{n}(t|\mathscr{S})$ is the number of positive crossings of the surface $\mathscr{C}$ during the time interval $[0, t]$ for a sample of trajectories beginning from the safe region $\mathscr{S}$. Note that $\mathrm{E}[\tilde{n}(t|\mathscr{S})] \neq \mathrm{E}[\tilde{n}(t)]$, and that the calculation of this conditional mathematical expectation $\mathrm{E}[\tilde{n}(t|\mathscr{S})]$ is associated with very serious difficulties. For example if $\mathbf{u}(t)$ is a stationary Gaussian process, the conditional process $\mathbf{u}(t|\mathscr{S})$ is non-stationary as well as non-Gaussian.

An alternative formulation of the reliability function is obtained by introducing the *hazard rate* (or *force of mortality*) $\Pi(t)$

$$\Pi(t) = \frac{\mathrm{d}\mathbb{R}(t)/\mathrm{d}t}{\mathbb{R}(t)} \tag{6.85}$$

Thus, assuming $\mathbb{R}(0) = 1$ (equation (6.76)), $\mathbb{R}(t)$ can be written

$$\mathbb{R}(t) = \exp\left(-\int_0^t \Pi(\tau)\,\mathrm{d}\tau\right) \tag{6.86}$$

It is usually assumed that $\mathbb{R}(\infty) = 0$; which is equivalent to stating that $\int_0^\infty \Pi(t)\mathrm{d}t$ diverges.

The moments of the time to failure are given, in terms of the reliability function, by

$$\mathrm{E}[\tilde{t}^n] = \int_0^\infty t^n \frac{\mathrm{d}\mathbb{R}(t)}{\mathrm{d}t}\,\mathrm{d}t = n\int_0^\infty \tilde{t}^{(n-1)}\mathbb{R}(t)\,\mathrm{d}t \tag{6.87}$$

(b) Analytical forms of reliability functions

Some functions which have been used in reliability studies are the following [11].

(i) *Exponential distribution* (Table A.2)

$$\mathbb{R}(t) = \exp(-\lambda t) \qquad \lambda > 0 \tag{6.88a}$$

$$\Pi(t) = \lambda \tag{6.88b}$$

$$\mathrm{E}[\tilde{t}^n] = n!/\lambda^n \tag{6.88c}$$

(ii) *Gamma distribution* (Table A.2) .

$$\mathbb{R}(t) = \frac{1}{\Gamma(m+1)}\int_{vt}^\infty y^m \mathrm{e}^{-y}\,\mathrm{d}y \qquad m > -1,\ v > 0 \tag{6.89a}$$

$$\Pi(t) = \frac{v^{m+1}t^m}{\Gamma(m+1)}\,\frac{\mathrm{e}^{-vt}}{\mathbb{R}(t)} \tag{6.89b}$$

$$\mathrm{E}[\tilde{t}^n] = \frac{\Gamma(m+n+1)}{\Gamma(m+1)}\,\frac{1}{v^n} \tag{6.89c}$$

(iii) *Weibull distribution* (Section A7.2.3)

$$\mathbb{R}(t) = \exp(-t^c/t_c) \qquad c,\ t_c > 0 \tag{6.90a}$$

$$\Pi(t) = c\,t^{c-1}/t_c \tag{6.90b}$$

$$\mathrm{E}[\tilde{t}^n] = \Gamma(n/c+1)\,t_c^{n/c} \tag{6.90c}$$

(*iv*) *Truncated normal distribution* (Table A.2 and Section A2.3)

$$\frac{d\mathbb{R}(t)}{dt} = \frac{c\phi(c(t-t_0))}{\Phi(ct_0)} \qquad c>0,\ -\infty<t_0<\infty \tag{6.91a}$$

where $\phi(x) = \dfrac{1}{\sqrt{(2\pi)}}\,\mathrm{e}^{-x^2/2} \qquad \Phi(x) = \dfrac{1}{\sqrt{(2\pi)}}\displaystyle\int_{-\infty}^{x}\mathrm{e}^{-y^2/2}\,dy$

$$\Pi(t) = \frac{c\phi(c(t-t_0))}{\Phi(-c(t-t_0))} \tag{6.91b}$$

$$\mathrm{E}[\tilde{t}] = t_0 + \frac{1}{c}\frac{\phi(ct_0)}{\Phi(ct_0)} \tag{6.91c}$$

$$\mathrm{E}[\tilde{t}^2] = t_0^2 + \frac{1}{c^2} + \frac{t_0}{c}\frac{\phi(ct_0)}{\Phi(ct_0)} \tag{6.91d}$$

(*v*) *General lifetime distribution*

$$\frac{d\mathbb{R}(t)}{dt} = \frac{\mathrm{e}^{-t}t^c}{\Gamma(c+1)}\sum_{n=0}^{\infty} C_n L_n^c(t) \qquad c>-1,\ C_0=1 \tag{6.92a}$$

$$\mathbb{R}(t) = \sum_{n=0}^{\infty}\binom{n+c}{c}C_n\sum_{j=0}^{n}(-1)^j\binom{n}{j}Y_j(t) \tag{6.92b}$$

$$\Pi(t) = \Gamma(c+1)\left[\sum_{n=0}^{\infty}C_n\sum_{j=0}^{n}\binom{n+c}{n-j}\frac{(-1)^j}{j!}Y_j(t)\right]^{-1}\left(\frac{d\mathbb{R}(t)}{dt}\right) \tag{6.92c}$$

$$\mathrm{E}[\tilde{t}^n] = \frac{\Gamma(n+c+1)}{\Gamma(c+1)}\sum_{j=0}^{n}(-1)^j\binom{n}{j}C_j \tag{6.92d}$$

where the $L_n^c(t)$ are generalized Laguerre polynomials

$$L_n^c(t) = \sum_{j=0}^{n}\binom{n+c}{n-j}\frac{(-t)^j}{j!}$$

the C_n are suitable parameters and the $Y_j(t)$ are incomplete gamma functions defined by

$$Y_j(t) = \frac{1}{\Gamma(c+j+1)}\int_t^{\infty} y^{c+j}\mathrm{e}^{-y}\,dy$$

Many other reliability functions have been proposed on the basis of empirical studies, mathematical convenience, and as a result of more fundamental theoretical models for failure phenomena.

The exponential distribution is the most widely used, also when there is little, if any, direct evidence for its validity in the given situation: in fact, just as a central limit theory of some generality (Section A4.4) exists and appears to lend some theoretical basis to the use of the normal law in many statistical procedures, a limit theorem for the distribution of positive random variables suggests the use of an exponential lifetime distribution.‡

6.4.2 Multiple modes of failure

(a) Alternative modes

Assume first that m *alternative* modes of failure exist for the structure under investigation. The results of Section 1.4 can be obtained noting that in this case the safe region $\mathscr{S}$ is the intersection of the m regions $\mathscr{S}_k$ corresponding to each failure mode, and consequently the surface $\mathscr{C}$ consists of m pieces $\mathscr{C}_1, \mathscr{C}_2, \ldots, \mathscr{C}_m$. Let the analytical prolongation of $\mathscr{C}_k$ be denoted by $\mathscr{C}'_k$. The probabilities of crossing each surface $\mathscr{C}_k$ at least once during the time interval $(0, t)$ are denoted by $P_{\mathscr{C}_k}(t)$; the analogous probabilities corresponding to the surfaces $\mathscr{C}'_k$ are denoted by $P_{\mathscr{C}'_k}$. Evidently $P_{\mathscr{C}_k}(t) \leqslant P_{\mathscr{C}'_k}(t)$. Hence (equation (1.8))

$$P_{\text{fail}}(t) \leqslant \sum_{k=1}^{m} P_{\mathscr{C}'_k}(t) \tag{6.93}$$

For sufficiently rare crossings the inequality sign in (6.93) can be replaced by the sign of equality with acceptable approximation (cf. equation (1.12)). Thus, for the reliability function, the following expression is obtained:

$$\mathbb{R}(t) \simeq 1 - \sum_{k=1}^{m} \mathrm{E}[\tilde{n}_k(t)] \tag{6.93$'$}$$

where $\mathrm{E}[\tilde{n}_k(t)]$ is the mathematical expectation of the number of positive crossings through the surface $\mathscr{C}'_k$ in $(0, t)$.

(b) Modes 'in cascade'

Suppose now that two limit states are *in cascade*, i.e. the second limit state can only occur after the first one has been attained. Then, in this case, the global structural reliability is equal to the reliability for the first limit state. However, different *costs* can be associated to each limit state (cf. Section 6.6.2 below).

Combinations of alternative and cascading limit states are possible, as in the following example, taken from [12].

(c) An example

Consider the single-storey two-bay mild-steel frame shown in Fig. 6.17. It supports a uniform dead load of $0.35\,\text{t m}^{-1}$, plus a mass of 20 t, with centroid at a significant height above the frame beam. This structure is also subjected to a (random) horizontal action, proportional to a single coefficient α, which results in a horizontal force 3.5α at the beam level, and in two vertical forces 20α acting

‡ Hazard rates depending on more than one parameter $\Pi(t)$ would be very useful in some engineering applications: at the present time, some models are under investigation.

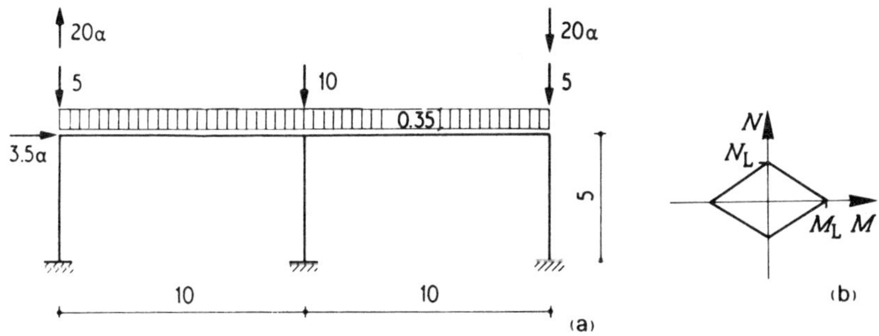

Fig. 6.17 Example frame (loads and lengths in tonnes and metres respectively) and yield interaction profile [12].

(respectively downwards and upwards) on the external columns. All loads are assumed random: the dead loads are normally distributed around the indicated values (with coefficient of variation 0.01), while the parameter α has an extreme distribution of type II (maxima) with average $\mu_\alpha = 1$ and coefficient of variation δ_α equal to several values in the range 0.1–0.7.

The live load factors corresponding to the three (ultimate) limit states described below have been evaluated by a simulation procedure (Chapter 9), and then used to investigate the probabilities of failure. The yield stresses have been assumed as random, statistically independent in the relevant sections (top and bottom column sections, and beam sections at joints), and normally distributed with average 2400 kg cm^{-2} and coefficient of variation 0.05. (Since the yield stress is the only *strength* variable considered, the chosen c.o.v. should include random variations of section geometry.)

(*i*) *Overall plastic collapse* (p)

The load factor α'_p corresponding to the development of a plastic collapse mechanism has been sought, following standard procedures, by means of a linear programming approach.

In the following, the symbol α_p will indicate the load factor, analogous to α'_p but taking into account the P–Δ effect at incipient collapse, calculated by introducing a suitable fictitious shear force.

(*ii*) *Column instability* (B)

The load factor which causes buckling of the most stressed column of the example frame (indicated by α_B in the following) is evaluated in the crude way described below.

Indicating by N the axial force in the most stressed column, the true relationship $N = N(\alpha)$ can be approximated by a bilinear relationship ABC (Fig. 6.18(a)). In this way, if the critical (buckling) axial force N_B is known, the corresponding load factor α_B is obtained immediately. In turn, under the

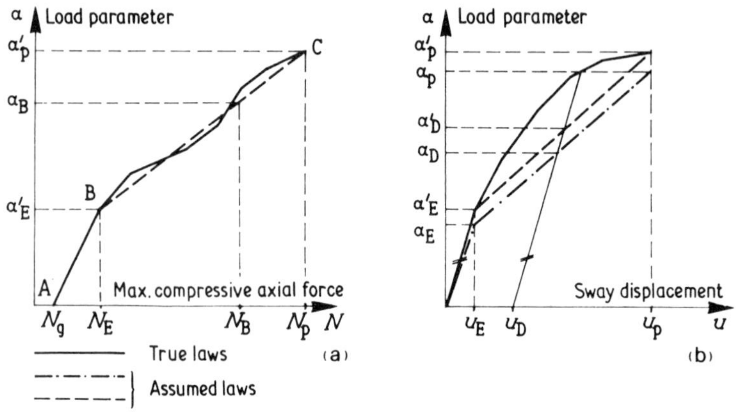

Fig. 6.18 Assumption introduced for taking into account the buckling and the permanent deformation limit states for the frame in Fig. 6.17. The suffixes indicate: g, dead load only; E, first yield; B, buckling (instability); p, plastic mechanism; D, limit deformation (The 'true laws' including P–Δ effect are not shown) [12].

assumption that column instability can only occur when plastic hinges have developed at both end sections, the buckling force N_B can be obtained from the following relationship [12]

$$\frac{N_B}{N_c} \pm 0.4\frac{1-N_B/N_y}{1-N_B/N_e} = 1 \tag{6.94}$$

where N_c is the critical load of the compressed column, related to the imperfection parameter Y (equation (6.49)) assumed normally distributed with $\mu_Y = 0.001\,h/i$ and $\delta_Y = 0.24$; N_y denotes the yielding axial force and N_e the Eulerian buckling load of the column being considered.

(*iii*) *Permanent sway displacement* (D)
The corresponding (approximate) load factor, calculated by the approach of Section 9.5.2(b) (Figs 6.18(b) and 9.19), will be indicated by α'_D, while α_D will denote the analogous factor which takes into account the P–Δ effect.

Some results of such computations are presented in Fig. 6.19 and Table 6.1.

Then, assume that (a) overall plastic collapse can only occur after the permanent sway limit state has been attained, and (b) the two limit states above are independent of column buckling. Thus, we have both typical situations: two limit states in cascade, and a third, independent one. The total probability of failure is equal to the probability of failure in either the D or the B limit state, i.e.

$$P_{\text{fail}} = P_D + P_B - \text{Prob}[D \cap B] \tag{6.95}$$

Neglecting the third term on the r.h.s., which is usually small (and exactly nil if the

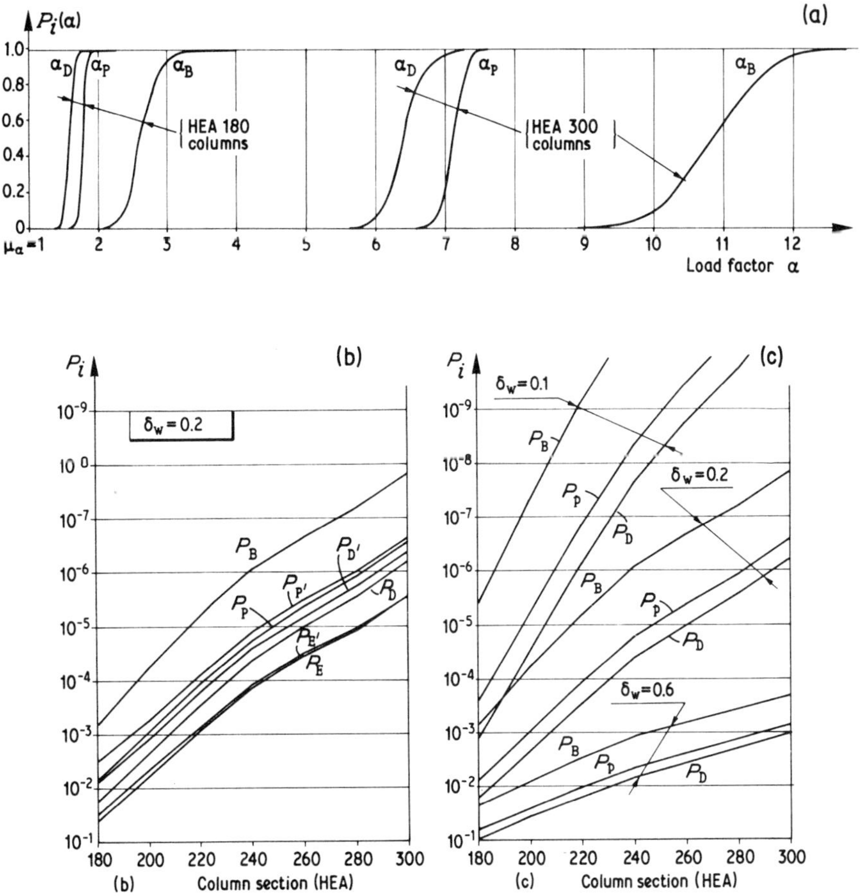

Fig. 6.19 Example of Fig. 6.17: (a) Typical examples of conditional probabilities of failure for the considered limit states. (b), (c) Probabilities P_i of attaining each limit state (typical plots) [12].

conditions for attaining the two limit states are mutually exclusive in the considered time interval) equation (6.95) reduces to

$$P_{\text{fail}} = P_{\text{D}} + P_{\text{B}} \tag{6.95$'$}$$

and the corresponding reliability is written

$$\mathbb{R} = 1 - P_{\text{fail}} = 1 - (P_{\text{D}} + P_{\text{B}}) \tag{6.95$''$}$$

The probability of failure in the p mode, P_{p}, will only appear in a cost-benefit analysis (Sections 6.7.1).

Table 6.1 Calculated probabilities of failure of model frame (Fig. 6.17) with respect to each limit state (from Augusti and Casciati [12])

Column section	Probability of failure	Coefficient of variation of live load 0.1	0.2	0.3	0.4	0.5	0.6	0.7
HEA 180	P_P	2.6 E−04	7.3 E−03	2.4 E−02	4.0 E−02	5.3 E−02	6.3 E−02	6.9 E−02
	P_B	3.8 E−06	6.6 E−03	4.3 E−03	1.0 E−02	1.6 E−02	2.2 E−02	2.6 E−02
	P_D	1.3 E−04	1.7 E−02	4.3 E−02	6.4 E−02	7.9 E−02	9.0 E−02	9.6 E−02
HEA 200	P_P	5.8 E−06	9.1 E−04	5.5 E−03	1.2 E−02	1.9 E−02	2.6 E−02	3.0 E−02
	P_B	4.6 E−08	5.8 E−05	7.8 E−04	2.5 E−03	4.9 E−03	7.6 E−03	9.8 E−03
	P_D	2.9 E−05	2.2 E−03	1.0 E−02	2.0 E−02	2.9 E−02	3.7 E−02	4.2 E−02
HEA 220	P_P	1.5 E−07	1.2 E−04	1.3 E−03	3.9 E−03	7.2 E−03	1.1 E−02	1.3 E−02
	P_B	7.8 E−10	6.2 E−06	1.6 E−04	7.1 E−04	1.6 E−03	2.9 E−03	4.0 E−03
	P_D	7.4 E−07	2.9 E−04	2.4 E−03	6.4 E−03	1.1 E−02	1.6 E−02	1.9 E−02
HEA 240	P_P	4.7 E−09	1.8 E−05	3.5 E−04	1.3 E−03	2.8 E−03	4.7 E−03	6.2 E−03
	P_B	2.0 E−11	8.3 E−07	3.9 E−05	2.3 E−04	6.1 E−04	1.2 E−03	1.8 E−03
	P_D	2.3 E−09	4.3 E−05	6.4 E−04	2.2 E−03	4.3 E−03	6.8 E−03	8.8 E−03
HEA 260	P_P	3.5 E−10	4.3 E−06	1.3 E−04	5.9 E−04	1.4 E−03	2.5 E−03	3.5 E−03
	P_B	1.6 E−12	2.1 E−07	1.5 E−05	1.0 E−04	3.1 E−04	6.5 E−04	1.0 E−03
	P_D	1.7 E−10	1.0 E−05	2.3 E−04	9.5 E−04	2.1 E−03	3.6 E−03	4.9 E−03
HEA 280	P_P	3.6 E−11	1.2 E−06	5.3 E−05	2.9 E−04	7.6 E−04	1.5 E−03	2.1 E−03
	P_B	1.8 E−13	6.2 E−08	6.3 E−06	5.2 E−05	1.7 E−04	3.9 E−04	6.2 E−04
	P_D	1.8 E−10	2.9 E−06	9.5 E−05	4.6 E−04	1.1 E−03	2.1 E−03	2.9 E−03
HEA 300	P_P	2.5 E−12	2.8 E−07	1.8 E−05	1.2 E−04	3.6 E−04	7.6 E−04	1.2 E−03
	P_B	1.5 E−14	1.5 E−08	2.4 E−06	2.4 E−05	8.7 E−05	2.1 E−04	3.5 E−04
	P_D	1.1 E−11	6.2 E−07	3.2 E−05	1.9 E−04	5.4 E−04	1.1 E−03	1.6 E−03

6.5 SERVICEABILITY LIMIT STATES

The probabilistic treatment of serviceability limit states, up to the present time, has received far less attention than ultimate limit states. This is mainly due to the fact that serviceability limit states usually involve only economical consequences (and therefore cause less concern), and moreover their *gradual* effects, already discussed in Section 6.1.3, make a complete classification and a full treatment very difficult if not impossible: many different problems arise, and general models lose value.

In some instances, the limit state can be identified with the overcrossing of a (more or less conventional) *discomfort threshold*; and very high probabilities can be accepted of attaining this threshold during the design life of the structure. In this case, other quantities become significant, such as the intervals between successive threshold crossings, the maximum value of a response parameter after each crossing, and the total time in which the threshold is violated (Fig. 6.20). Such an approach can be (and often is) accepted when the structure behaves elastically,

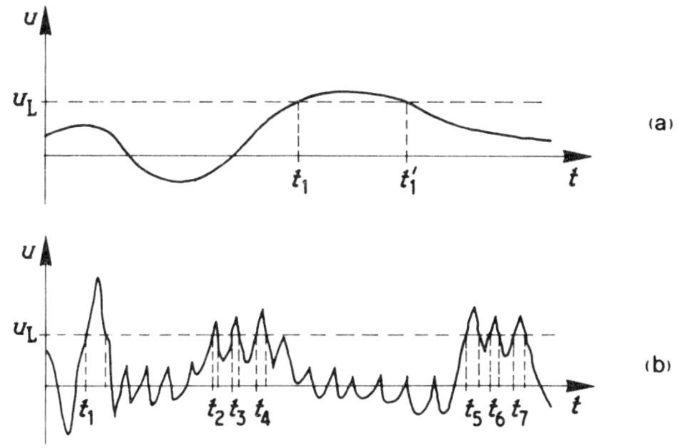

Fig. 6.20 Different realizations of a stochastic process: (a) slow variations, with few threshold crossings; (b) rapid variations, with frequent threshold crossings.

so that, once the external action has decreased, the structure returns below the threshold without any permanent structural damage, but only with the economical and/or psychological consequences involved by having been temporarily out of service (or uncomfortable).

In other cases (threshold attained in the inelastic field; creep; . . .), the main problem becomes damage accumulation (Section 5.2.5) and hence the whole time history of the problem should be considered.

Such a classification of serviceability limit-states founded on the required methodological approach, appears better for applications, while the one presented in Section 6.1.2(b) referred more to the physical aspects. Both types of limit state would require, for a rational design approach, the estimation of the JPDF $p(\mathbf{u}, \tau)$ of the maximum values of the response parameters $\tilde{\mathbf{u}}$ since the most recent time of full serviceability and of the time $\tilde{\tau}$ at which the serviceability state associated with $\mathbf{u}$ was entered. The critical inequality (equation (1.17)) for these limit states is in fact

$$\mathbf{u} \leqslant \mathbf{u}_L \tag{6.96}$$

where $\mathbf{u}_L$ is an assigned vector that defines the undesired event.

Multidimensional problems generally require very sophisticated analytical tools. Therefore, the rest of this section considers inequality (6.96) in the special, simple case in which only one response parameter (e.g. a deflection at a given point) is significant.

6.5.1 Serviceability limit states attained in the elastic range

(a) General discussion

When the structure behaves elastically up to $\mathbf{u} = \mathbf{u}_L$, it may go out of service at many independent intervals. These occurrences are related to excessive values of the external actions (and possibly of the dynamic effects): when these overloads are removed, inequality (6.96) is again satisfied.

Static elastic deflection at a point in time is a function of the load at that time, according to the models developed in Chapter 4 (Section 4.2). By contrast, dynamical analysis requires knowledge of the previous load time history (Section 4.3). In any case, the methods of Chapter 4 permit one to describe the general features of the response stochastic process $\tilde{\mathbf{u}}(t)$, once the stochastic process of loading is known.

In the case of static vertical loads, the expression for a monodimensional displacement process $\tilde{u}(t)$ can be written making use of the influence coefficient $I(x, y)$ (Section 3.2.3) for instantaneous deflection, so that

$$\tilde{u}(t) = \int_{A_I} I(x, y)\,\tilde{w}(x, y, t)\,\mathrm{d}x\,\mathrm{d}y \tag{6.97}$$

where $\tilde{w}(x, y, t)$ is the load per unit area as a function of position and time and A_I is the influence area (possibly random). Equation (6.97) permits one to adopt the expressions of Section 3.3 to investigate the process $\tilde{u}(t)$.

If the random function $\tilde{u}(t)$ is fully known, the results of Sections A7.4 and A7.5 can be employed to study the point processes of crossing times and local maxima, and the distribution of local maxima.

Moreover, for narrow-band processes, where the upcrossings are not independent of each other, but tend to cluster, the envelope process (A7.52) can be introduced instead of $\tilde{u}(t)$ to obtain results of practical interest. It is worth noting, however, that these results are very scarce as they reduce to the mean upcrossing rate (equation (A7.43)) and, in a very special case ($\tilde{u}(t)$ stationary and Gaussian) to the PDF of the magnitude of the peak of $\tilde{u}(t)$ (equation (A7.45)). Further developments are the object of many current researches, but the whole subject is a complicated one, especially with regard to multidimensional responses. Researches are especially devoted to extend the results of Section A7.4, and define the random properties of the duration of each violation. In this regard, a simple result can be obtained when $\tilde{u}(t)$ (or its envelope) is a stationary random process (Fig. 6.21); in fact the total duration L of the threshold violation during the interval T can be written in the form

$$\begin{aligned} L &= L_1 + L_2 + L_3 + \ldots \\ &= \int_0^T \phi(u(t) - u_\mathrm{L})\,\mathrm{d}t \end{aligned} \tag{6.98}$$

where $\phi(\ \ )$ is a binary function of the type shown in Fig. 6.21(b). Then the mean value of this total duration is given by

$$\mu_L = \int_0^T \mathrm{E}[\phi(u(t) - u_\mathrm{L})]\,\mathrm{d}t = [1 - P_u(u_\mathrm{L})]\,T \tag{6.99}$$

as $\phi(u(t) - u_\mathrm{L})$ is equal to 0 with probability $P_u(u_\mathrm{L})$ and 1 with probability $1 - P_u(u_\mathrm{L})$, $P_u(u)$ being the CDF of $\tilde{u}(t)$.

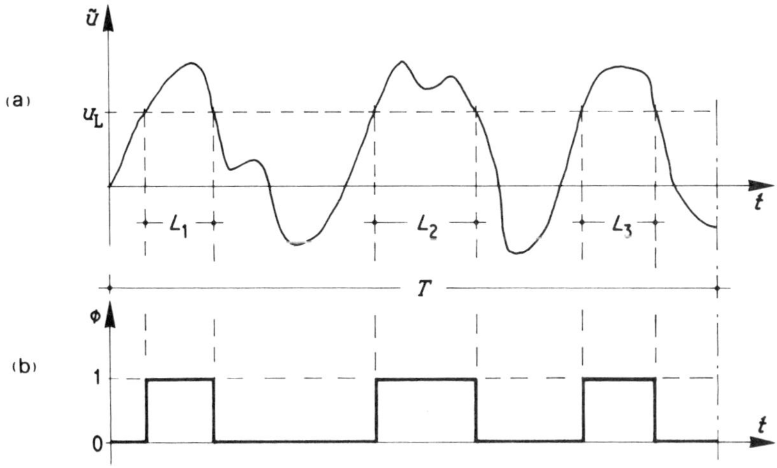

Fig. 6.21 Duration of threshold violations of a stochastic process.

(*b*) *A result for buildings under vertical loads*

No general result is available on the JPDF $p(u, \tau)$ that was indicated to be of primary interest for design purposes. A particular result can be obtained under some specific assumptions: the structure is subject only to sustained and transient live load (Section 3.3)‡ and the engineer is only interested in the first upcrossing of a displacement threshold during each tenancy as the non-structural repair is undertaken in conjuction with normal redecoration at the conclusion of the tenancy [13].

Under these assumptions, the maximum value of the response parameter results from the combination of the arbitrary sustained live load effect and the maximum transient live load effect that occurs in the duration of the sustained load. The distribution of the maximum transient load effect during the holding time of a sustained load may be satisfactorily approximated by the distribution of the maximum transient load effect during the mean holding time of sustained loads. Thus

$$u_{\mathrm{m}} = \max\{u(t)|d_i\} \simeq u_{s_i}(t) + \max\{u_C(t)|\mu_d\} \qquad (6.100)$$

where $u_{s_i}(t)$ is the sustained live load effect during tenancy i, $u_C(t)$ is the transient live load effect, μ_d is the mean duration of sustained loads, and d_i is the duration of tenancy i.

For estimated load models in conjunction with deterministic influence surfaces, one can prove [13] that the distribution of $\max\{u_C(t)|\mu_d\}$ has the mean corresponding approximately to its 0.55 fractile and a variance that is small by comparison with the variance of $u_{s_i}(t)$. Elastic-deflection influence coefficients are essentially deterministic, so that

$$\max\{u(t)|d_i\} = u_{s_i}(t) + u^0 \qquad (6.101)$$

‡ Note that for serviceability limit states, the effects of permanent loads can often be taken as nil (initial values).

where

$$u^0 = E[\max\{u_C(t)|\mu_d\}] \\ = P_{u_C}^{-1}[1 + (\lambda_s/\lambda_C) \ln 0.55] \tag{6.101'}$$

under the assumption of Poisson arrivals of transient load events for which $P_{uC}(u)$ is the CDF of the transient load effect, $\lambda_s = 1/\mu_d$ is the mean rate of occurrence of sustained load events and λ_C is the mean rate of occurrence of transient load events (Section 3.3).

It follows that

$$P_{u_m}(u) = P_{u_{s_i}}(u - u^0) \tag{6.102}$$

Now, the joint probability distribution function of the maximum load effect during a tenancy, $\tilde{u}_m$, and the time of completion of the tenancy $\tilde{\tau}$, is

$$p_{u_m,\tilde{\tau}}(u,t) = p_{\tilde{\tau}}(t)p_{u_m}(u) \\ = (1 - e^{-\lambda_s})p_{u_{s_i}}(u - u^0) \tag{6.103}$$

as $\tilde{u}_m$ and $\tilde{\tau}$ are independent.

This joint probability distribution function of deflection and time may be expressed in terms of the distribution functions of the equivalent uniformly distributed sustained and transient live loads.

In fact let $\bar{C}_{u^0}$ and $\bar{w}_{eu}$ be the equivalent uniformly distributed transient live load corresponding to u^0 and the EUDL corresponding to $u_m = u$ respectively (Section 3.3.5). Moreover, let $p_{\bar{w}_{s_u}}(w)$ be the PDF of an EUD sustained live load and $P_C(C)$ denote the CDF of an EUD transient live load; then

$$p_{u_m,\tilde{\tau}}(u,t) = (1 - e^{-\lambda_s})p_{w_{su}}(\bar{w}_{eu} - \bar{C}_{u^0}) \tag{6.103'}$$

where

$$\bar{C}_{u^0} = P_C^{-1}[1 + (\lambda_s/\lambda_C) \ln 0.55] \tag{6.103''}$$

6.5.2 Serviceability limit states involving inelastic behaviour

(a) Random creep

It is well known that tests on creep properties of most structural materials yield a large scatter in response quantities: an obvious conclusion is that creep properties are inherently random and that this randomness must be introduced explicitly in the treatment of the experimental data. However, this aspect has not yet been extensively studied, and only a few papers have been published: yet, it appears a very interesting and promising area for further investigations. As an example, a brief probabilistic treatment of creep effects in redundant structures with randomly inhomogeneous creep properties is presented in this section [14]. The treatment is limited to truss-type structures (uniaxial stress state in all points), subjected to time-invariant loads. Also, geometrical and elastic structural properties will be assumed to be deterministic and independent of displacements.

A different problem is illustrated in the following section (6.5.2(b)), namely, the growth of permanent displacements under a stochastically evolving live load.

It must also be stressed that both treatments are based on the assumption of a memoryless material, i.e. the actual creep strain rate is not dependent on the load history but depends only on the present stress.

Assume that the creep strain rate is given by the modified Norton equation

$$\dot{\varepsilon}_c(x) = \dot{\varepsilon}_0 [1 + \zeta\tilde{\phi}(x)]\left(\frac{\sigma}{\sigma_r}\right)^r \tag{6.104}$$

where x is the axial coordinate in each monodimensional bar, σ_r, $\dot{\varepsilon}_0$ and r are deterministic material constants, and ζ is a random parameter, defining the creep properties of each bar. Random variations of material properties in different points of the bar are introduced by the random function $\tilde{\phi}(x)$, assumed to be a normal ergodic Markovian stochastic process (Section A6.4) such that

$$|\zeta\tilde{\phi}(x)| \ll 1 \tag{6.105}$$

with sufficiently large probability. The expected value, variance and autocorrelation function are respectively

$$\mathrm{E}[\zeta\tilde{\phi}] = \zeta \tag{6.106}$$

$$\mathrm{Var}[\zeta\tilde{\phi}] = 2\zeta + o(\zeta^2) \tag{6.107}$$

$$\Sigma(x_1, x_2) = B(x_0) = 2\zeta \exp(-c|x_0|) + o(\zeta^2) \tag{6.108}$$

where c is a material constant and

$$x_0 = x_2 - x_1 \tag{6.109}$$

With the position

$$\tilde{\phi}_L = \frac{1}{L}\int_0^L \tilde{\phi}(x)\,\mathrm{d}x \tag{6.110}$$

L being the length of the bar, the average strain rate in the specimen is given by

$$\dot{\varepsilon}_L = \frac{1}{L}\int_0^L \dot{\varepsilon}_c(x)\,\mathrm{d}x = \dot{\varepsilon}_0(1 + \zeta\tilde{\phi}_L)\left(\frac{\sigma}{\sigma_r}\right)^r \tag{6.111}$$

The statistics of $\dot{\varepsilon}_L$ are readily derived from the statistics of $\zeta\tilde{\phi}_L$ (Section A5.2.1):

$$\mathrm{E}[\zeta\tilde{\phi}_L] = \frac{1}{L}\int_0^L \mathrm{E}[\zeta\tilde{\phi}]\,\mathrm{d}x = \zeta \tag{6.112}$$

$$\begin{aligned}\mathrm{Var}[\zeta\tilde{\phi}_L] &= \frac{1}{L^2}\int_0^L\int_0^L \Sigma(x_1, x_2)\,\mathrm{d}x_1\,\mathrm{d}x_2 \\ &= \frac{4\zeta}{(cL)^2}(cL + \mathrm{e}^{-cL} - 1)\end{aligned} \tag{6.113}$$

Now let **W** and **u** be the vectors of nodal forces and displacements in the truss, **N** and **Z** the normal forces and elongations in the bars, **Q** the geometrical compatibility matrix. The equilibrium and compatibility equations can be put respectively in the form

$$\mathbf{W} = \mathbf{Q}^{\mathrm{T}}\mathbf{N} \tag{6.114}$$

$$\mathbf{Z} = \mathbf{Q}\mathbf{u} \tag{6.115}$$

Now write the constitutive equation (6.111) for every bar‡

$$\dot{Z}_{ci} = \dot{\varepsilon}_{Li} L_i = \dot{\varepsilon}_{0i}[1 + \zeta \tilde{\phi}_{Li}]\,(N_i/N_{ri})^{r_i} \tag{6.116}$$

where

$$N_{ri} = \sigma_{ri} A_i \tag{6.117}$$

$L_i, A_i, \sigma_{ri}, r_i$ being the length, cross-sectional area, and creep characteristics of the ith bar (all deterministic quantities). For simplicity of further developments, ζ is assumed constant throughout the structure: thus, the vector of creep elongation rates (equation (6.104))

$$\dot{\mathbf{Z}}_{\mathrm{c}} = \dot{\mathbf{Z}}_{\mathrm{c}}(\mathbf{N}, \zeta \tilde{\boldsymbol{\phi}}_L) \tag{6.118}$$

is a known function of the axial forces N_i and the random parameters $\tilde{\phi}_{Li}$.

At the initial time, put $\mathbf{Z}_{\mathrm{c}} = 0$, so that displacements and axial forces are given by

$$\left.\begin{aligned} \mathbf{Q}^{\mathrm{T}}\mathbf{G}\mathbf{Q}\mathbf{u} &= \mathbf{K}\mathbf{u} = \mathbf{W} \\ \mathbf{N} &= \mathbf{G}\mathbf{Z}_{\mathrm{e}} = \mathbf{G}\mathbf{Z} \end{aligned}\right\} \quad t = 0 \tag{6.119}$$

where $\mathbf{K}$ is the elastic stiffness matrix, $\mathbf{G}$ the Hooke's diagonal matrix and $\mathbf{Z}$ the total elongation

$$\mathbf{Z} = \mathbf{Z}_{\mathrm{c}} + \mathbf{Z}_{\mathrm{e}} \tag{6.120}$$

For $t \geqslant 0$, rates of stresses and displacements are given by

$$\left.\begin{aligned} \dot{\mathbf{N}} &= \mathbf{G}\dot{\mathbf{Z}}_{\mathrm{e}} = \mathbf{G}[\mathbf{Q}\dot{\mathbf{u}} - \dot{\mathbf{Z}}_{\mathrm{c}}(\mathbf{N}, \zeta \tilde{\boldsymbol{\phi}}_L)] \\ \mathbf{K}\dot{\mathbf{u}} &= \mathbf{Q}^{\mathrm{T}}\mathbf{G}\dot{\mathbf{Z}}_{\mathrm{c}}(\mathbf{N}, \zeta \tilde{\boldsymbol{\phi}}_L) \end{aligned}\right\} \quad t \geqslant 0 \tag{6.121}$$

remembering that by assumption

$$\dot{\mathbf{W}} = \mathbf{0} \tag{6.122}$$

Equation (6.121) governs the transit creep redistribution of stresses in a structural Hooke–Norton system.

Since both $\dot{\mathbf{u}}$ and $\dot{\mathbf{N}}$ are random functions of time, because of the randomness in the constitutive equations introduced by the parameters $\tilde{\boldsymbol{\phi}}_L$, to solve these differential equations exactly appears a difficult task: perhaps, techniques similar to the ones summarized in Appendix A5 could be suggested.

However, the results most significant for engineering purposes are contained in the asymptotic solution for $t \to \infty$. Assuming that the stress rates tend to zero as time tends to infinity, the governing equations can be written

$$\left.\begin{aligned} \mathbf{Q}^{\mathrm{T}}\mathbf{N} &= \mathbf{W} \\ \mathbf{Q}\dot{\mathbf{u}} &= \dot{\mathbf{Z}}_{\mathrm{c}}(\mathbf{N}, \zeta \tilde{\boldsymbol{\phi}}_L) \end{aligned}\right\} \quad t \to \infty \tag{6.123}$$

‡ Note that no summation over repeated indices is implied in this section.

and an approximate solution is possible for some properties of the asymptotic stress field. For this purpose, perturbation analysis, already introduced in Section 4.3.1, can be useful: in this scheme, the asymptotic stress and rates of displacement are written as power series [14]:

$$\mathbf{N} = \mathbf{N}_0 + \zeta\mathbf{N}_1 + \zeta^2\mathbf{N}_2 + \mathbf{o}(\zeta^3) \tag{6.124}$$

$$\dot{\mathbf{u}} = \dot{\mathbf{u}}_0 + \zeta\dot{\mathbf{u}}_1 + \zeta^2\dot{\mathbf{u}}_2 + \mathbf{o}(\zeta^3) \tag{6.125}$$

where ζ is the already used perturbation parameter defining the intensity of random fluctuations of creep properties in the bars. All quantities refer to $t = \infty$, and from now throughout this section tildes ($\tilde{}$) will be omitted for simplicity's sake from response quantities.

Since the applied loads do not depend on ζ, the equilibrium equations become

$$\mathbf{Q}^{\mathrm{T}}\mathbf{N}_0 = \mathbf{W} \tag{6.126a}$$

$$\mathbf{Q}^{\mathrm{T}}\mathbf{N}_1 = \mathbf{0} \tag{6.126b}$$

$$\mathbf{Q}^{\mathrm{T}}\mathbf{N}_2 = \mathbf{0} \tag{6.126c}$$

and the compatibility conditions yield immediately

$$\mathbf{Q}\dot{\mathbf{u}}|^{(j)} = \dot{\mathbf{Z}}_{\mathrm{c}}(\mathbf{N}, \zeta\tilde{\boldsymbol{\phi}}_L)\,|^{(j)} \qquad j = 0, 1, 2 \tag{6.127}$$

where the symbol $|^{(j)}$ denotes the jth derivative with respect to ζ at the point $\zeta = 0$.

From equation (6.127), taking into account equations (6.124) and (6.125), some algebra leads to

$$\mathbf{Q}\dot{\mathbf{u}}_0 = \dot{\mathbf{Z}}_{\mathrm{c}}(\mathbf{N}_0, \mathbf{0}) \tag{6.128a}$$

$$\mathbf{Q}\dot{\mathbf{u}}_1 = \dot{\mathbf{Z}}_{\mathrm{c}}(\mathbf{N}_1, \mathbf{0})|^{(1)} \tag{6.128b}$$

$$2\mathbf{Q}\dot{\mathbf{u}}_2 = \dot{\mathbf{Z}}_{\mathrm{c}}(\mathbf{N}_2, \mathbf{0})|^{(2)} \tag{6.128c}$$

Inspection of equations (6.126) and (6.127) shows that $\dot{\mathbf{u}}_0$ and $\mathbf{N}_0$ are the steady-state solution for deterministic creep properties ($\zeta = 0$).

The ith components of the r.h.s.s of (6.128) are respectively given by equation (6.116) and

$$\dot{Z}_{ci}(\mathbf{N}, \mathbf{0})|^{(1)} = \frac{\dot{\varepsilon}_{0i} L_i}{(N_{ri})^{r_i}}\,[\tilde{\phi}_{Li}(N_{0i})^{r_i} + r_i(N_{0i})^{r_i - 1} N_{1i}] \tag{6.129a}$$

$$\dot{Z}_{ci}(\mathbf{N}, \mathbf{0})|^{(2)} = \frac{2\dot{\varepsilon}_{0i} L_i r_i (N_{0i})^{r_i}}{(N_{ri})^{r_i}}\left[\tilde{\phi}_{Li}\frac{N_{1i}}{N_{0i}} + \frac{r_i - 1}{2}\left(\frac{N_{1i}}{N_{0i}}\right)^2 + \frac{N_{2i}}{N_{0i}}\right] \tag{6.129b}$$

The complete solution of equations (6.126)–(6.128) can be obtained by first solving equations (6.126a) and (6.128a) for $\mathbf{N}_0$ and $\dot{\mathbf{u}}_0$ as an ordinary creep problem unaffected by uncertainty. Once $\mathbf{N}_0$ and $\dot{\mathbf{u}}_0$ are available, $\mathbf{N}_1$ and $\dot{\mathbf{u}}_1$ can be calculated by coupling equations (6.126b) and (6.128b). It is seen from equation (6.129a) that the r.h.s. of (6.128b) is linear in $\mathbf{N}_1$: therefore $\mathbf{N}_1$ and $\dot{\mathbf{u}}_1$ are obtained by solving the following linear system

$$\mathbf{Q}^T\mathbf{N}_i = \mathbf{0}$$
$$\mathbf{Q}\dot{\mathbf{u}}_1 + \mathbf{C}_1\mathbf{N}_1 = \mathbf{C}_2\tilde{\boldsymbol{\phi}}_L \tag{6.130}$$

$\mathbf{C}_1$ and $\mathbf{C}_2$ being suitable diagonal matrices depending on material parameters and on the deterministic creep solution ($\mathbf{N}_0$, $\dot{\mathbf{u}}_0$).

Expected values and variances of $\dot{\mathbf{u}}_1$ and $\mathbf{N}_1$ can easily be deduced by means of the techniques described for a linear response of structures under random constant load (Chapter 4 and Appendix A3).

Similar considerations apply to the calculation of $\mathbf{N}_2$ and $\dot{\mathbf{u}}_2$, which are again determined as the solution of a linear system of equations, whose coefficients and slack terms depend on the same parameters as equation (6.130) and on the probabilistic properties of the products $\tilde{\phi}_{Li}N_{1i}$ and N_{1i}^2, that can be calculated after equation (6.130) has been solved.

As an example, for the very simple bar structure shown in Fig. 6.22 it can be shown that

$$\begin{aligned} \mathrm{E}[\dot{u}] &\geqslant \dot{u}_0 \\ \delta_{\dot{u}} &\leqslant \zeta\sqrt{2} \end{aligned} \tag{6.131}$$

i.e. the expected asymptotic displacement rate is larger than the rate calculated in the absence of inhomogeneity, and its coefficient of variation $\delta_{\dot{u}}$ is bounded by

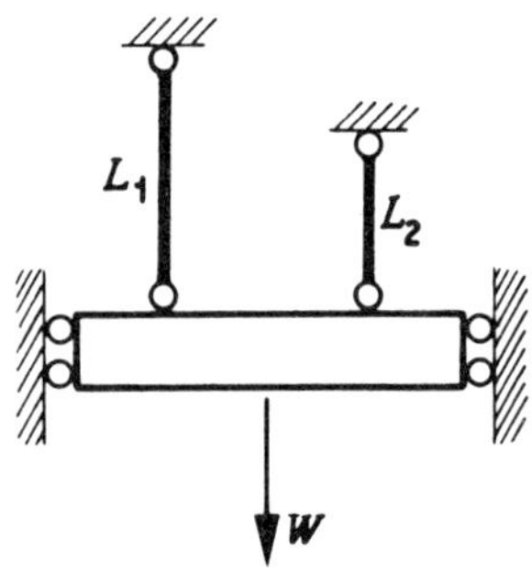

Fig. 6.22 Simple bar structure for study of random creep response (after Broberg and Westlund [14]).

$\zeta\sqrt{2}$. Further, if cross-sectional areas and creep properties of the two bars are equal, the expected value and the variance of the asymptotic displacement rate of the structure are bounded from above by the same properties calculated for a single bar with length L_2, loaded by the force

$$W_1 = \frac{W}{1 + (L_2/L_1)^{1/r}} \tag{6.132}$$

(b) Creep under stochastic live loads

The magnitude of a permanent deflection which changes over time in response to sustained loading depends on the whole previous load history. A possible model of such a deflection, again with special reference to residential building floors (Section 3.3; cf. also Section 6.5.1(b)), is

$$\tilde{u}_c(T) = \int_0^T \int_{A_I} I(x, y, T, t)\tilde{w}_t(x, y, t)\,dx\,dy\,dt \tag{6.133}$$

where $\tilde{u}_c(T)$ is the deflection at time T, $I(x, y, T, t)$ is the influence coefficient relating deflection at time T to load at time t, and $\tilde{w}_t(x, y, t)$ is the load as a function of position and time t.

The following approximations simplify the application of the model [13].

(1) Creep deflections are sensitive to the holding time of a load, and creep deflections due to transient live load events are negligible:

$$\tilde{u}_c(T) \simeq \int_0^T \int_{A_I} I(x, y, T, t)[\tilde{w}(x, y, t) + \tilde{w}_g(x, y)]\,dx\,dy\,dt \tag{6.134}$$

where $\tilde{w}$ denotes the sustained live load and $\tilde{w}_g(x, y)$ is the dead load as a function of position.

(2) Creep deflections are primarily a function of early loading:

$$\tilde{u}_c(T) \simeq \int_0^T \int_{A_I} I(x, y, T, t)[\tilde{w}(x, y, t) + \tilde{w}_g(x, y)]\,dx\,dy\,dt \qquad T \leqslant T^0$$

$$\tilde{u}_c(T) \simeq \int_0^{T^0} \int_{A_I} I(x, y, T, t)[\tilde{w}(x, y, t) + \tilde{w}_g(x, y)]\,dx\,dy\,dt \qquad T > T^0$$

where T^0 is the time after which creep effect increments are negligible.

(3) The mean holding time of sustained live loads, μ_d, is greater than the critical time for creep effects, T^0, so that:

$$\begin{aligned}\tilde{u}_c(t) &\simeq \int_0^T \int_{A_I} I(x, y, T, t)[\tilde{w}(x, y, 0) + \tilde{w}_g(x, y)]\,dx\,dy\,dt \\ &\simeq \int_{A_I} \left[\int_0^T I(x, y, T, t)\,dt\right][\tilde{w}(x, y, 0) + \tilde{w}_g(x, y)]\,dx\,dy \\ &\simeq \int_{A_I} I_T(x, y)[\tilde{w}(x, y, 0) + \tilde{w}_g(x, y)]\,dx\,dy \qquad T \leqslant T^0\end{aligned} \tag{6.135a}$$

$$\tilde{u}_c(T) \simeq \int_{A_I} I_{T^0}(x, y)[\tilde{w}(x, y, 0) + \tilde{w}_g(x, y)]\,dx\,dy \qquad T > T^0 \tag{6.135b}$$

where

$$I_T(x, y) = \int_0^T I(x, y, T, t)\,dt$$

is the influence coefficient for deflection under initial sustained loading to time T and $\tilde{w}(x, y, 0)$ is the sustained live load during first occupancy.

(4) The influence coefficient for deflection under sustained loading is a function of time-dependent moduli of elasticity.

It follows from the stochastic serviceability model that, if a creep failure occurs, it occurs during the first tenancy and the relevant serviceability parameter is the maximum total deflection prior to repair during the tenancy.

The maximum deflection during the first tenancy is the resultant of total (elastic plus creep) dead load and sustained live load deflections and a transient live load deflection component (equation (6.100)):

$$\tilde{u}_{\mathrm{m}} = \max\{\tilde{u}(t)\,|\,d_0\} = \tilde{u}_{w_g}(T^0) + \tilde{u}_{s_0}(T^0) + \max\{u_C(t)\,|\,\Delta t\} \qquad d_0 > T^0 \tag{6.136}$$

where $\tilde{u}_{w_g}(T)$ is the total deflection, at time T, due to dead load; $\tilde{u}_{s_0}(T)$ is the total deflection, at time T, due to sustained live load of first tenancy; $\tilde{u}_C(t)$ is the elastic deflection due to transient live loads; $\tilde{u}(t)$ is the total deflection; Δt is the interval of time between creep failure and repair; T^0 is the characteristic time of attaining the threshold creep deflection; and d_0 is the duration of first tenancy.

Approximate the distribution of the maximum transient live load effect during Δt by the distribution of the maximum transient live load effect during the expected value of Δt:

$$\mathrm{E}[\Delta t] \simeq \mathrm{E}[d_0 - T^0] \simeq \mu_d - T^0 \tag{6.137}$$

This distribution (Section 6.4.1) has a mean approximately corresponding to its 0.55 fractile and a variance that is small by comparison with the variance of complementary deflection components (equation (6.101)). Thus,

$$\max\{\tilde{u}(t)\,|\,d_0\} = \tilde{u}_{w_g}(T^0) + \tilde{u}_{s_0}(T^0) + u' \tag{6.138}$$

where

$$u' = \mathrm{E}[\max\{\tilde{u}_C(t)\,|\,\Delta t] = P_{u_C}^{-1}\left[1 + \frac{\ln 0.55}{\mathrm{E}[\Delta t]\lambda_C}\right] \tag{6.139}$$

It follows (equation (6.102)) that in this case

$$P_{u_{\mathrm{m}}}(u) = P_{(u_{w_g} + u_{s_0})}(u - u') \tag{6.140}$$

The joint probability density function of maximum load effect, $\tilde{u}_{\mathrm{m}}$, and time of associated repair, $\tilde{\tau}$, is (equation (6.103′)):

$$p_{u_{\mathrm{m}},\tau}(u, t) = p_{u_{\mathrm{m}}}(u)\,p_{\tilde{\tau}}(t) \tag{6.141}$$

if the PDF on the r.h.s. are assumed independent of each other.

The creep-dependent load effect $u_{w_g}(T^0) + u_{s_0}(T^0)$, is the sum of correlated dead load effect and sustained live load effect components. The distribution of the resultant sustained load effect may be derived from a comprehensive model of sustained floor loads. The dead load process and the sustained live load process are independent and the distribution of dead load intensity has a variance that is generally small by comparison with that of sustained live loads, so that dead load intensity is well approximated by its expected value. A realistic model of the total sustained load process, compatible with the previously described sustained live load model, is given by equation (3.1), while an equivalent uniformly distributed sustained load $\tilde{w}$ is analysed in Section 3.3.5(b).

In the last case, one also obtains (equation (6.103″))

$$p_{u_m,\tau}(u,t) = p_\tau(t)\, p_{\tilde{w}}(\bar{w}_{u-u'}) \tag{6.142}$$

where $\bar{w}_{u-u'}$, is the equivalent uniformly distributed total sustained load value which, in conjunction with the mean influence surface for sustained load deflection, produces a deflection $(u-u')$.

If changes of occupancy are Poisson events, the time to the first change of occupancy is exponentially distributed with mean μ_d. At this time $\bar{t}$, one can assume $p_\tau(\bar{t}) = 1$ and so equations (6.141) and (6.142) are made easier to calculate.

(c) Inelastic deformations

Limitations on the deformative parameters can also be made necessary by the accumulation of plastic deformations under repeated loads. Some examples of such a behaviour have already been encountered in Chapter 5 (Sections 5.3.5 and 5.3.6) and others will be found in Chapter 9: a very simple and evident one is shown in Fig. 6.23.

In such cases, the non-linearity and the non-holonomy of the consititutive law of the material predominate and drastic simplifying assumptions are necessary to

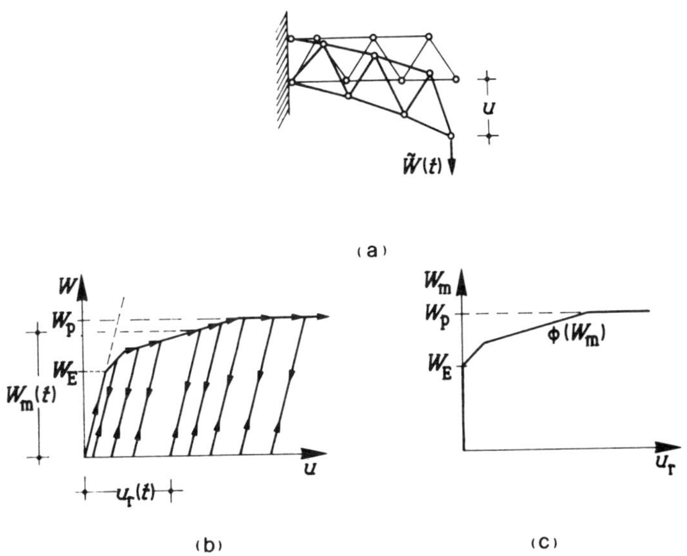

Fig. 6.23 Accumulation of inelastic deformations (partly after [9]).

tackle the problem, such as (i) deterministic filter, (ii) elasto–plastic input–output relationship, (iii) absence of self-stresses in the initial conditions, (iv) stochastic processes of load that cannot change sign during the structural lifetime.

Under these assumptions the output process $\tilde{u}(t)$ can be regarded as the sum of the residual deformation $\tilde{u}_r(t)$ and the elastic one $\tilde{u}_e(t) = \tilde{W}(t)/K$ where K is a constant. Moreover, assume that the residual deformation at time t depends only on $\tilde{W}_m(t)$, the maximum value of the load in $(0, t)$, and express this relationship in the form (Fig. 6.23(c))

$$\tilde{u}_r(t) = \phi[\tilde{W}_m(t)] \tag{6.143}$$

Assume further that the PDF of $\tilde{W}_m(t)$ is known. From equation (A3.32) it follows that

$$p_{u_r}(u_r, t) = p_{W_m}(\phi^{-1}(u_r), t)\frac{d\phi^{-1}(u_r)}{du_r} \tag{6.143'}$$

Also, from equation (A3.61), the JPDF of the elastic deformation and its derivative is given by

$$p_{u_e, \dot{u}_e}(u_e, \dot{u}_e, t) = p_{W, \dot{W}}(u_e K, \dot{u}_e K, t)K^2 \tag{6.144}$$

If the elastic limit W_y is not a frequent value, the function $u_r(t)$ becomes a step function, with derivative equal to zero everywhere except at a finite number of points where it is not defined. Therefore $\dot{u} \simeq \dot{u}_e$ and, from equation (6.144), one obtains [15]

$$p_{u_e, \dot{u}}(u_e, \dot{u}, t) = p_{W, \dot{W}}(u_e K, \dot{u} K, t)K^2 \tag{6.145}$$

Further, as equation (A3.42) involves that

$$p_{u, \dot{u}}(u, \dot{u}, t) = \int_{-\infty}^{\infty} p_{u_e, \dot{u}}(u_e, \dot{u}, t)\, p_{u_r}(u - u_e, t)\, du_e \tag{6.146}$$

the mean number of upcrossings of the level u_L of the non-stationary process $\tilde{u}(t)$ is given by equations (A7.38) and (A7.40)

$$E[\tilde{\nu}^+_{u_L}(t)] = \int_0^t dt \int_0^{\infty} \dot{u}(t) p_{u, \dot{u}}(u_L, \dot{u}, t)\, d\dot{u} \tag{6.147}$$

If the system is not deterministic, the function $\phi(W_m)$ is not deterministic and the foregoing treatment fails. Simulation procedures must therefore be employed to investigate the randomness of the response due to the uncertainties in the system parameters (Section 9.6 and, for the dynamic case, Section 6.2.2).

However, the basic problem is the description of the variability in time of the response properties and the approaches coincide with the methods developed for the ultimate limit state related to gross deformations (Section 6.3). They can be summarized as follows:

(i) Structures whose carrying capacity does not vary with the deformations
In this case, the mechanical behaviour is independent of time and its randomness can be investigated by simulating the response independently of the initial conditions. Thus, if for instance the load is described by a filtered Poisson process (FPP, Section A6.3.3) one determines by simulation the characteristics of the

random variables relevant to the 'response function' of the output filtered Poisson process. More generally, the input and output functions must be idealized by suitable analytical models in order to avoid cumbersome time-history simulations.

(*ii*) *Structures whose carrying capacity does vary with the deformations*
In this case, the analytical model that idealizes the output function at time t must allow for the inelastic deformation accumulated before and hence Markov processes (Section A6.4) must be introduced. Here, the transition probabilities from a state to another must generally be obtained by simulation.

The usefulness of hybrid approaches (simulation in addition to suitable stochastic models) for the solution of such problems will be discussed in Section 9.4; some examples of damage accumulation analysis will be presented in Section 9.5.

6.6 STRUCTURAL DESIGN AS A DECISION PROCESS

The design of structures, like most engineering design, is generally a trial-and-error procedure: on the basis of past experience, the engineer conceives a first tentative design; then he verifies the resulting structure, points out defects and/or overestimations, and corrects the tentative design accordingly; the whole process is iterated until both safety and economic requirements are met‡. Practical design is thus founded on a succession of structural analyses, essentially aimed at choosing the *best* (i.e. at *optimizing*) geometric and mechanical properties among the possible solutions of the structural problem under investigation.

With the development of computers, interactive programs are being proposed to perform this iterative procedure. However, fully automatic optimization algorithms have so far been formulated only under very restrictive conditions; namely, one limit state is taken into account and optimum design corresponds to the solution that minimizes (or maximizes) an appropriate simple function of the design parameters (e.g. the cost, the weight, etc.).

Therefore, even within a fully *deterministic* framework, the computer is not yet (and never will be) able to replace, but only to aid, the designer: this must be kept in mind, otherwise inappropriate meaning can be attributed to analytical and numerical design procedures.

Moreover, the consequences of the design choices (a set of *decisions*) depends on a series of factors and events that (as the reader of this book should well understand by now) are not known with certainty. Therefore, rational criteria for selecting the *best* design (under the stated limitation!) should take such uncertainties into account explicitly.

‡ This discussion is not intended to include problems of *architectural* design process but only the choice of the supporting structure.

The most immediate step in this direction is to substitute the mere *initial cost* H_I of the structure with the *total anticipated expenditure* in the form

$$H = H_I + P_{fail} H_{fail} \tag{6.148}$$

where H_{fail} is the expenditure consequent upon a failure of probability P_{fail}.

Recently calculated curves of total anticipated expenditure as functions of P_{fail} are shown in Fig. 6.24: each of them has a minimum corresponding to a well determined value of P_{fail}. (Other peculiarities of the shape of these curves, which will be seen in the following to be essentially reproduced in all available results, will be discussed later.)

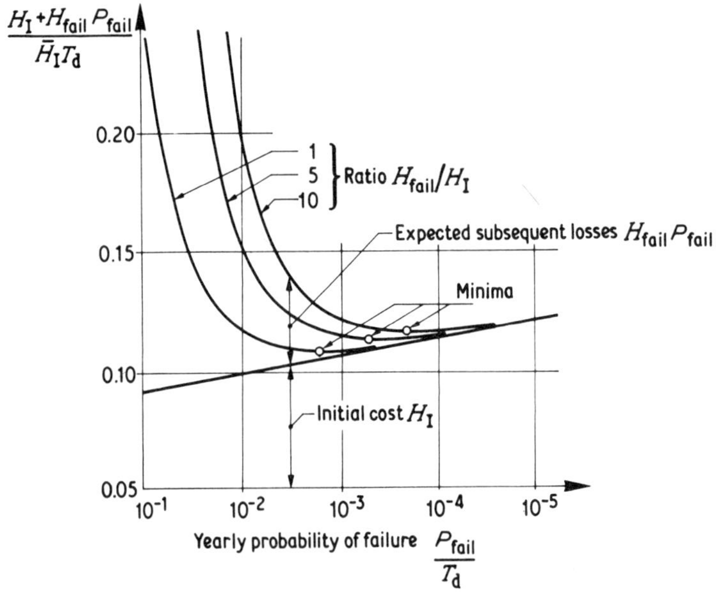

Fig. 6.24 Initial cost H_I and total anticipated expenditure H (equation 6.148) vs probability of failure P_{fail} ($\bar{H}_I$, initial cost for a given reference reliability; T_d, design life in years) (after [1]).

More generally, recourse can be made (and more and more frequently is explicitly made) to *decision theory*, which will be briefly illustrated in the following section.

6.6.1 Elements of decision theory

(a) Decision with given information

Suppose that the designer can take a decision (e.g. select a design x) among a number n of available alternative solutions $x_1, x_2, \ldots, x_n$ (that are the potential *courses of action*). Let m be the number of possible states $y_j (j = 1, \ldots, m)$, each of which can occur at random as the outcome of the action taken. As a result of having taken action x_i and having found state y_j, the engineer will receive a value $\mathbb{U}(x_i, y_j)$, a numerical measure (not necessarily monetary) of the consequences of the pair x_i, y_j. A *decision tree* (Fig. 6.25) displays graphically the formulation in the case of discrete and finite state and action spaces. (Both the state and action spaces might logically be treated as infinite and/or continuous.) Now, it is necessary to define a suitable function $\mathbb{U}$, the *utility function*, so that its expected values corresponding to each action i

$$E[\mathbb{U}_i] = \mathbb{U}(x_i, y_1)\text{Prob}[y_1] + \ldots \mathbb{U}(x_i, y_m)\text{Prob}[y_m] \qquad (6.149)$$

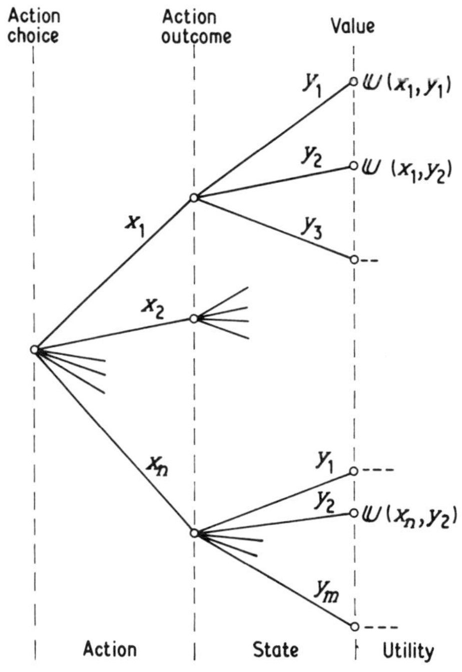

Fig. 6.25 Decision tree for discrete state and action spaces.

or, in case of continuous y-space,

$$E[\mathbb{U}_i] = \int_0^\infty \mathbb{U}(x_i, y)p(y)\,dy \qquad (6.149')$$

can be compared with each other, in order to take a rational decision among the n possible courses of action. (For more details, the reader is referred to specialistic texts [17, 18].)

It is worth noting that $\mathbb{U}$ is a *potential* function, i.e. a reference value and the scale can be chosen arbitrarily.

In practice, therefore, one must specify all branches and all possible paths along the decision tree, and assign appropriate probabilities to each state. Then the expected utilities are computed for the paths consequent to each action: the action which corresponds to the maximum expected utility is usually identified with the best possible design decision (*design for maximum expected utility*).

(b) Decision while acquiring information

The probabilities of each state can be reassessed whenever new information is obtained. Let $\text{Prob}^{\text{I}}[y_i]$ be the *prior* probability of the state y_i. As a result of an experiment, an outcome $\dot{z}_k$ is obtained. From the equality

$$\text{Prob}[y_i|z_k]\text{Prob}[z_k] = \text{Prob}[z_k|y_i]\text{Prob}^{\text{I}}[y_i] \tag{6.150}$$

one obtains the *posterior* probability $\text{Prob}^{\text{II}}[y_i]$ (equation (A1.19))

$$\text{Prob}^{\text{II}}[y_i] \equiv \text{Prob}[y_i|\dot{z}_k] = \frac{\text{Prob}[\dot{z}_k|y_i]\text{Prob}^{\text{I}}[y_i]}{\sum_j \text{Prob}[\dot{z}_k|y_j]\text{Prob}^{\text{I}}[y_j]} \tag{6.151}$$

Equation (6.151), often denoted *Bayes' rule*, states that the posterior probability of a state is the product of the *sample likelihood* given y_i, $\text{Prob}[\dot{z}_k|y_i]$, and the prior probability of y_i, divided by a normalizing constant. The problem is therefore to establish the sample likelihoods for the problem under investigation: then (6.151) allows one to derive the posterior probability from the prior one and the outcome of the experiment.

Often the designer has the option of getting (at some cost) the outcome of an experiment before making its choice of action. Sometimes, several types or levels of experimentation are possible, and the decision-maker must choose among them in a suitable manner. The decision tree is thus completed as shown in Fig. 6.26.

Common structural engineering decision problems are related to the selection (action choice) of the structural type among a discrete set and of the structural element sizes among either a continuous or a discrete range of possibilities. The state is generally assumed discrete (with *failure* or *success* the only alternatives) when ultimate limit states are considered, and continuous for serviceability limit states.

Experiments may concern the load description and the material properties when the engineer is designing a new building, or the determination of the actual structural properties when the behaviour of existing structures is investigated.

6.6.2 Definition of the utility function

If only one specific failure mode is considered, structural utility $\mathbb{U}$ is commonly expressed in the form

$$\mathbb{U}(y) = \mathbb{B}^\circ - H_{\text{I}} - H_{\text{fail}}(y) \tag{6.152}$$

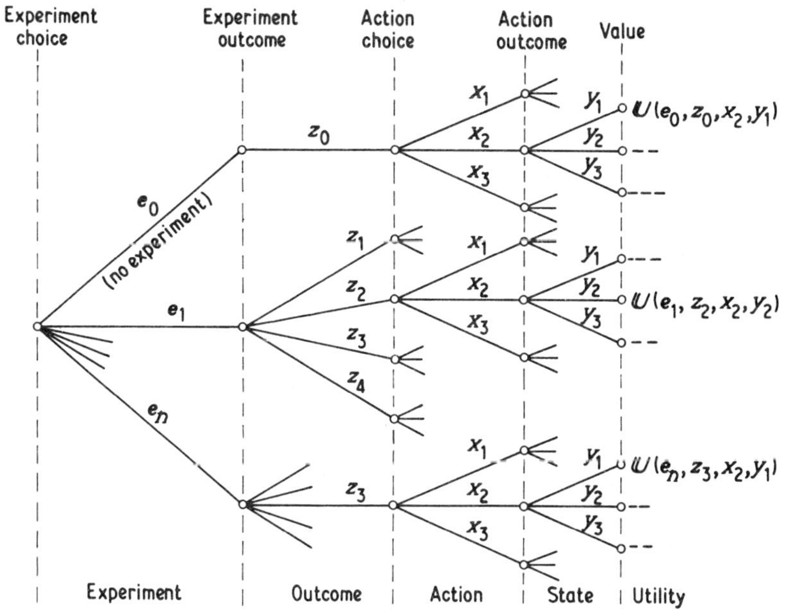

Fig. 6.26 Complete decision tree.

where $\mathbb{B}^\circ$ is the benefit that would be derived from a fully successful structure and H_I the initial structural cost. The quantity y is the state variable (cf. also Section 2.8.1) for the considered failure mode and $H_{\text{fail}}(y)$ denotes the corresponding failure cost function. Noting that y is a random variable, a generalized measure of structural utility with regard to the considered failure mode k is then the expected utility given by equation (6.149) or (6.149′).

Thus, in order to evaluate expected utility, one must define utility as a function of a state variable, determine the probability mass or density function of that state variable with reference to appropriate load and structural response models and relevant design constraints, and finally integrate the derived utility density function over the range of the state variable. However, (6.149) and (6.149′) can be simplified when the failure cost is a binary function, i.e. is nil for a certain range of the state variable, and takes the constant value H_{fail} beyond a *failure point*. In this case, that generally corresponds to ultimate limit states, (6.149) and (6.149′) can be written in the form:

$$\mathrm{E}[\mathbb{U}(y)|T] = \mathbb{B}^\circ - H_\mathrm{I} - H_{\text{fail}} P_{\text{fail}} \qquad (6.152')$$

where P_{fail} is the probability of failure in the considered mode during the structural lifetime. Note that the second and third term on the r.h.s. of (6.152′) coincide with the total anticipated expenditure (6.148).

Suppose now that more failure modes are possible, and denote the (binary) cost of each by $H_{\text{fail}}(k, j)$ and the corresponding probability by $P_{\text{fail}}(k, j)$; more precisely, $(k, 0)$ will indicate each independent mode ($k = 1, 2, 3, \ldots$) and (k, j) ($j = 1, 2, \ldots$) each mode *in cascade* after the kth mode (Section 6.4.2).

By definition

$$P_{\text{fail}}(k, 0) \geqslant P_{\text{fail}}(k, 1) \geqslant P_{\text{fail}}(k, 2) \geqslant \ldots \tag{6.153}$$

while it is customary to assume

$$H_{\text{fail}}(k, 0) \leqslant H_{\text{fail}}(k, 1) \leqslant H_{\text{fail}}(k, 2) \leqslant \ldots \tag{6.153$'$}$$

Then, the expected utility takes the form

$$E[\mathbb{U}|T] = \mathbb{B}^{\circ} - H_{\text{I}} - L(T) \tag{6.154}$$

with the *expected losses* $L(T)$ given by

$$L(T) = \textstyle\sum_k \{H_{\text{fail}}(k, 0) P_{\text{fail}}(k, 0) - \sum_j [H_{\text{fail}}(k, j) P_{\text{fail}}(k, j) - H_{\text{fail}}(k, j-1) P_{\text{fail}}(k, j-1)]\} \tag{6.154$'$}$$

Note that all probabilities appearing in the preceding equations refer to a specific time interval, namely the design life of the structure T, which was not explicitly indicated in the formulae for simplicity of notation.

Often, the expected losses $L(T)$ include only the costs directly consequent to failure, while the loss of a possible benefit is taken into account by a decrease in the expected benefit with respect to $\mathbb{B}^{\circ}$. Thus, equation (6.154) takes the final form

$$\text{E}[\mathbb{U}|T] = \mathbb{B}(T) - H_{\text{I}} - L(T) \tag{6.155}$$

where $\mathbb{B}(T)$ is the benefit that the structure is expected to yield during its *design life* T, taking into account the possibility that it may fail at a time $t_{\text{fail}} < T$.

For significant economic comparisons, gains and losses appearing in the expression of utility must be reported to a given time, usually the present time. For this purpose, the simple assumption of constant (yearly) *interest rate* r' will be accepted in the following, so that the present value H_0 of a sum of money to be earned (or lost) at a future time t, H_t, is given by the well-known formula

$$H_0 = H_t(1 + r')^{-t} = H_t \mathrm{e}^{-rt} \tag{6.156}$$

The quantity

$$r = \ln(1 + r') \tag{6.156$'$}$$

is called the *force of interest*.

Moreover, it will be assumed that the time to failure t_{fail} is exponentially distributed, in accord with equations (6.88′) and (A7.47):

$$\begin{aligned} p_{t_{\text{fail}}}(t) &= p_1 \exp(-p_1 t) \\ P_{t_{\text{fail}}}(t) &= 1 - \exp(-p_1 (t)) \end{aligned} \tag{6.157}$$

where p_1 is the probability of failure per year (*failure rate*).

(a) Benefit

The benefit derived from a successful structure may be very difficult to assess, but the associated utility can be mapped into a monetary scale at that value which the relevant decision-maker identifies as the '*true*' value of the structure. The benefit derived from a structure may be a fixed value or it may accrue at a time rate.

An estimate of the benefit of a fully serviceable structure is obtained by estimating the value of the original investment in terms of the interest in returns.

Applying equation (6.156), the present value of the benefits $\mathbb{B}^\circ(t)$ derived for the structure in the case of success for an interval $(0, t)$ is

$$\mathbb{B}^\circ(t) = \begin{cases} \int_0^t b(1+r')^\tau \,\mathrm{d}\tau = \int_0^t b' \mathrm{e}^{-r\tau} \,\mathrm{d}\tau & \text{for} \quad t \leqslant T \\ \int_0^T b\,\mathrm{e}^{-r\tau} \,\mathrm{d}\tau & \text{for} \quad t > T \end{cases} \tag{6.158}$$

where b is the yearly benefit of a successful structure (*rate of benefit*) which becomes zero after failure. Note that this last assumption implies that, even if the structure fails shortly after the beginning of its useful life, it is not renewed during the rest of the design life T.

Then the expectation of the present value of the benefit, to be introduced in equation (6.155), is

$$\mathbb{B}(T) = \mathrm{E}[\mathbb{B}^\circ(t)|\mathrm{T}] = \int_0^\infty \mathbb{B}^\circ(t) p_{t_{\text{fail}}}(t)\,\mathrm{d}t \tag{6.159}$$

Assuming b constant, and introducing (6.157),

$$\mathbb{B}(T) = \frac{\mathrm{b}}{r}\left(1 + \frac{\rho_1}{\rho_1 + r}\right)\{1 - \exp[-(\rho_1 + r)T]\} \tag{6.159'}$$

(b) Cost of construction

The cost of construction is the cost of site preparation plus the cost of building. These costs are relatively well defined for individual cases. In general, it has been suggested [13] that the expected cost of site preparation can be taken approximately equal to the cost of the structure, and the expected cost of the building approximately three times the cost of the structure;

$$H_\mathrm{I} \simeq H_\mathrm{I}^\mathrm{s} + 3H_\mathrm{I}^\mathrm{s} = 4H_\mathrm{I}^\mathrm{s} \tag{6.160}$$

where H_I^s is the cost of the structure.

The cost of a structure is relatively well defined as a function of the quantities of materials used in the structure. Approximate unit costs for some in-place materials and unit total costs for some types of buildings are shown in Table 6.2 as functions of two reference values.

The cost of construction also contains the cost of possible experiments that the designer may decide to introduce according to Section 6.6.1(b).

(c) Cost of failure

The cost of a failure is the sum of the loss of reputation, plus the material damage involved, plus either the loss of benefit associated with permanent termination of

Table 6.2 Approximate relative construction costs (elaborated from Reid and Turkstra [13]).

Concrete, ready mixed (material) (*reference value*)	1.00
Concrete placing (incl. labour and equipment)	
beams	0.53
elevated slabs	0.35
Reinforced concrete in place (incl. forms and reinforcing)	
beams	7.67
elevated slabs	4.07
Total building costs (excl. architect's fees and land cost)	
apartments (*reference value*)	1.00
factories	0.72
hospitals	2.88
offices	1.64
schools	1.48

rental income subsequent to failure or the cost of effecting repairs to render the building serviceable again.

The first policy (no repair) is generally assumed when ultimate limit states are analysed. In this case, the loss of benefit is often accounted for in the benefit term (equation (6.158)), but the main cost of failure is that related to the damage that the falling-down structure causes to stored materials and machinery, to the productive activity inside the building and to living people.

The first two contributions can be expressed generally in monetary terms, even if their evaluation depends strictly on the structure type and on the social context in which the event takes place. Suppose that a loss L_t has been calculated and that failure occurs at a time t; applying again equation (6.156) one gets the present value of the loss

$$L^0(t) = \begin{cases} L_t e^{-rt} & \text{for } t < T \\ 0 & \text{for } t > T \end{cases} \tag{6.161}$$

T again being the design life of the structure. By analogy with equation (6.159), the expectation of the present value of the loss to be introduced in equation (6.155) is

$$L(T) = \mathrm{E}[\tilde{L}^0(t)|T] = \int_0^T L^0(t) p_{t_{\text{fail}}}(t)\,dt \tag{6.162}$$

and making use of (6.157)

$$L(T) = \frac{p_1}{p_1 + r}\{1 - \exp[-(p_1 + r)T]\} \tag{6.163}$$

On the contrary, only a cynical estimation of the material losses due to dead or injured people is possible (for instance, in terms of insurance premiums): this problem will be discussed in some detail in Section 6.7.4.

The second policy (repair) is generally considered when the achievement of a serviceability limit state determines the actual design of the structure.

Table 6.3 Serviceability failure characteristics [13].

Observed u/L	*Type of building*	*Failure characteristics*
0.0025	Office	Cracks in block partitions at doors, crushing of blocks at corners, cracks in masonry walls
0.0039	School	Gaps below partitions
0.0064	Office	Dishing of floor noticeable to staff, filing cabinets and desks needed shimming
0.0067	Office	Filing cabinets, etc., tilted
0.0074	Office	Walls out of plumb, doors out of square, furniture rolled, tenants moved
0.0083	Dormitories	Lifting at corners and distortion of walls
0.0088	Undetermined	Partitions and door frames racked; damage to engineer's reputation
0.0133	Office	Obvious shimming of counters; 'cheap office building'

The reduction of serviceability of systems associated with static structural response is often assumed to be a function of the ratio of maximum deflection to span length (or storey height), u/L, a parameter that is widely incorporated in design codes. A sufficiently general and precise relationship between serviceability and u/L has not been established, but some information is summarized in Table 6.3. On the basis of such information, the following model can be suggested in connection with serviceability limit states:

(1) $H_{\text{fail}}(u/L) = 0$ if $u/L \leqslant 0.002$.
(2) $H_{\text{fail}}(u/L) = H_{\text{fail}}$ if $u/L \geqslant 0.008$, where H_{fail} is the cost of complete unserviceability. (Complete unserviceability is that state in which the building is not marketable for normal use in a competitive market situation.)
(3) The interpolation function for cost of failure is linear in u/L between $u/L = 0.002$ and $u/L = 0.008$:

$$H_{\text{fail}}(u/L) = \left(\frac{u/L - 0.002}{0.006}\right) H_{\text{fail}} \qquad 0.002 < u/L < 0.008$$

Obviously, as the costs are expected present values and generally are the sum of structural and non-structural repair costs, they must be expressed as a function of time to failure since completion of the building.

Note finally that, when the threshold value 0.002 is attained in the elastic range, the actual cost of failure can also depend on the duration of the threshold violation.

6.7 PROCEDURES FOR OPTIMAL DESIGN UNDER UNCERTAINTIES

Following the discussion of Section 6.6, it appears clearly that a *rational analytical* design procedure would have the aim of selecting some structural parameters (*design parameters*) in such a way that some *objective function* is maximized. The most widely used objective function is the *expected utility*.

These procedures are still at the research stage, but it is worth while to illustrate them here (with the aid of simple, but complete examples) because of their foreseeable development in the near future.

It is however to be repeated that the present aim of these procedures is limited to the rational choice of some scalar parameters, and not to wider problems of *design*, such as the choice of the structural topology.

6.7.1 Ultimate limit states

Consider first the problem of design for ultimate limit states, so that a binary utility function has to be considered and equations (6.159) and (6.163) can be employed.

The expected utility (equation (6.155)) depends on the design parameters x_j through the probabilities p_{1i} of occurrence of each relevant limit state i. In turn, such p_{1i} are related to the design parameters by means of equation (2.68), with the chosen x_j determining the CPF relative to the ith limit state (mode of failure). In the case of deterministic structural properties‡ equation (2.68) is replaced by equation (2.68′), and the x_j determine the relevant unsafe region. In both cases, the dependence of the expected utility $\mathbb{U}$ on the design parameters x_j in general is not an analytical function.

For a rational design, the design parameters x_j should be chosen so as to maximize the expected utility $\mathbb{U}$ (equation (6.155)). However, as already noted, $\mathbb{U} = \mathbb{U}(x_j)$ is not an analytical function; the simplest way to evaluate its maximum would therefore be to compare its values at a number of suitably chosen points. This is comparatively easy when there is only one design parameter to choose, but becomes prohibitively long in the case of two or more parameters.

(a) An example of optimal design with one parameter

Before presenting a procedure for more parameters, consider again the example introduced in Section 6.4.2 (Fig. 6.17) [12]. The only design parameter is the common cross-section of the columns, which must be chosen among the standard European wide-flange HEA series (individuated by the section width in millimetres, which is approximately 5% larger than the depth).

Assume now that failure can occur according to one of more alternative limit states, the consequences of which correspond to two possible costs, H_C (cost of

‡ Remember that, while the chosen x_j are deterministic, the actual value taken by the quantity measured by x_j may be random.

complete collapse) and H_D (cost of partial collapse), $H_D \leqslant H_C$; then (equation (6.163))

$$\mathcal{L} = \frac{p_1}{p_1 + r}\{1 - \exp[-(p_1 + r)T]\}\,(p_C H_C + p_D H_D) \qquad (6.164)$$

where p_C and p_D are the probabilities that the loss in case of failure is H_C or H_D respectively. As already illustrated (Section 6.4.2), three different limit states have been assumed to be possible: they are indicated by D (excessive permanent deformation), p (full plastic collapse) and B (buckling). Limit state p can occur only if D occurs; limit state B is independent of D and p. The structure fails (i.e. does not yield any more benefit) if any of the three limit states occur: the cost of D is H_D, the cost of p or B is H_C. Under these assumptions, in equations (6.159′) and (6.164),

$$p_1 = p_{1D} + p_{1B} \qquad (6.165)$$

$$p_C = \frac{p_{1B} + p_{1p}}{p_1} \qquad p_D = \frac{p_{1D} - p_{1p}}{p_1} \qquad (6.166)$$

where p_{1i} (i = D, B, p) is the rate of occurrence of each limit state.

From the previous relationships, and indicating by H_I the initial cost of the structure, the expected utility (equation (6.155)) is

$$\begin{aligned} \mathrm{E}[\mathbb{U}|T] &= \mathbb{B} - \mathcal{L} - H_I \\ &= \{1 - \exp[-(p_{1D} + p_{1B} + r)T]\}\left[\frac{b}{r}\right. \\ &\quad \left. - \frac{p_{1D} - p_{1p}}{p_{1D} + p_{1B} + r}\left(\frac{b}{r} + \frac{H_D(p_{1D} - p_{1p})}{p_{1D} + p_{1B}} + \frac{H_C(p_{1B} + p_{1p})}{p_{1D} + p_{1B}}\right)\right] - H_I \end{aligned} \qquad (6.167)$$

The above expression can easily be modified if different limit states are considered.

Some significant results are given in Fig. 6.27 for different values of the interest rate r', of the coefficient of variation of the load, and of the ratios of the failure costs H_C and H_D to the initial cost H_I. Comparing the values calculated for each HEA section, the design which maximizes the expected utility (*optimum*) can be recognized immediately.

(b) A procedure and an example with three design parameters

To generalize this design procedure to problems with more than one design parameter, it has been suggested to introduce an approximate analytical expression for $\mathbb{U}(x_i)$ and its derivatives and to apply a known algorithm to search for the unconstrained extremum of a generic function [19]. This procedure requires also the introduction of an auxiliary scalar (state) variable R_i, which can be viewed as the measure of a *conventional resistance*, and is defined in the case of a deterministic structure as the factor that brings the average load vector on the boundary of the unsafe region $\mathcal{U}_i$ (Fig. 6.28): the suffix i indicates the limit state considered. In the case of a structure with random strength, $\mathcal{U}_i$ is the unsafe region of the fictitious deterministic structure obtained by taking the characteristic value (lower 5% fractiles) of all random quantities influencing the structural strength. The variable R_i coincides in this case with the state parameter y introduced in Section 6.5.1.

Obviously R_i is a function of the design parameters x_j. This functional dependence can be approximated by calculating R_i exactly at a number N of points x_{jk} $(k = 1, \ldots, N)$ in the range of likely design parameters x_j. Let R_{ik} be

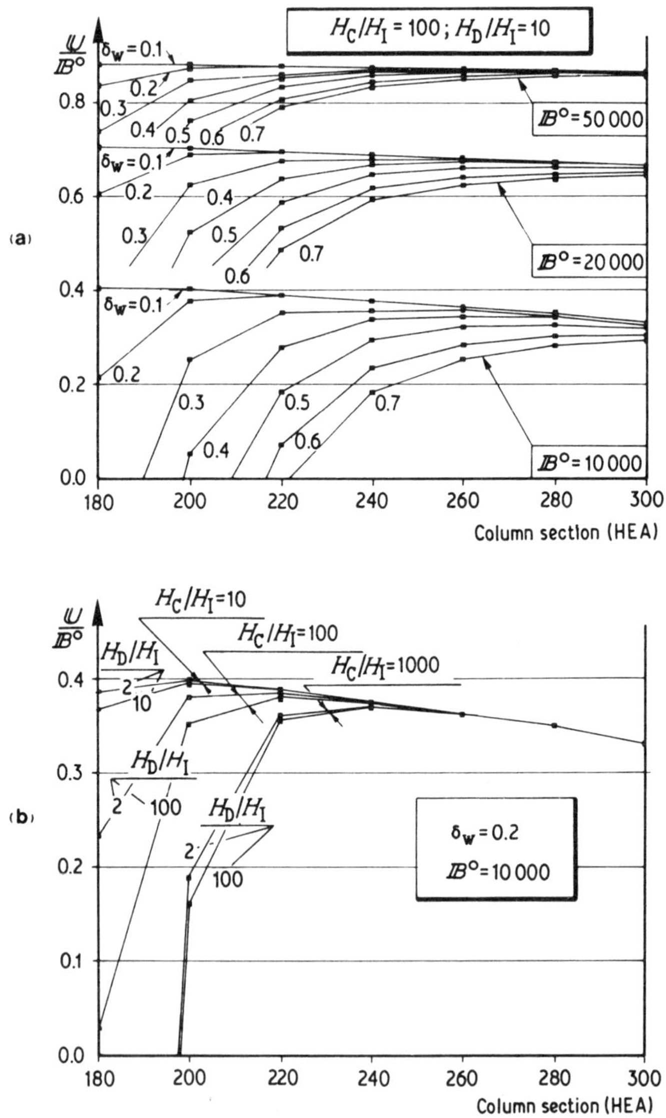

(C)

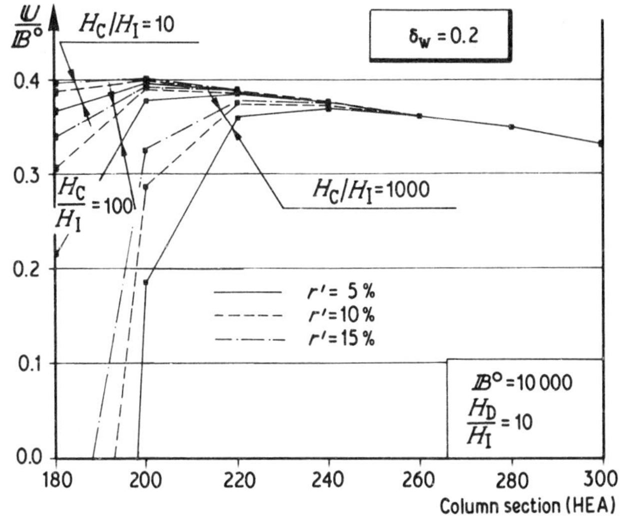

Fig. 6.27 Design for maximum utility for the frame of Fig. 6.17 with one design parameter: (a) expected utilities for different present benefits $\mathbb{B}^0$ (yearly interest rate $r' = 5\%$) (b) influence of expected failure losses on utility ($r' = 5\%$); (c) influence of interest rate r' on expected utility. The monetary unit is equivalent to 1 kg of steel [12].

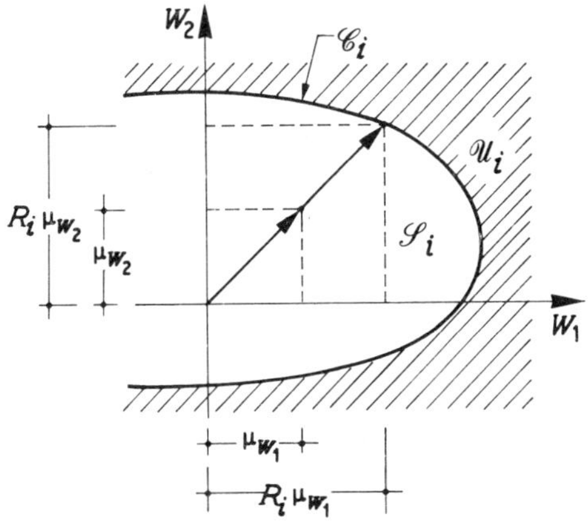

Fig. 6.28 Definition of conventional resistance R_i for multi-parameter load and ith limit state.

the calculated values and $g_k(x_j)$ be appropriate shape functions (i.e. functions such that $g_k(x_{jl}) = I_{lk}$; $k, l = 1, \ldots, N$); then

$$R_i(x_j) = \sum_{k=1}^{N} R_{ik} g_k(x_j) \tag{6.168}$$

An analogous treatment can be developed for the initial cost H_I.

Clearly, for given random loads, p_{1i} and R_i are strictly related. The further assumption has been introduced that p_{1i} depends on x_j only through R_i: this does not seem likely to cause too gross errors. It is also reasonable, in the range of small failure probabilities, to assume an exponential dependence of p_{1i} on R_i:

$$p_{1i} \simeq c_{1i} \exp(-c_{2i} R_i) \tag{6.169}$$

The two constants in each expression (6.169) are best chosen to fit the value p_{1i} calculated numerically from equation (2.68) or (2.68′), at the two points x_{jk} which are presumed to correspond to the largest and smallest R_{jk}.

It is worth noting that in this way the probabilistic analysis of the structure, that involves the greatest computational effort, must be performed at only two points of the design parameter space.

From equations (6.167)–(6.169) the approximate expressions for $\mathbb{U}(x_j)$ and $\partial\mathbb{U}/\partial x_j$ can be obtained explicitly. In particular, after some algebra, differentiation of equation (6.167) yields

$$\begin{aligned}
\frac{\partial \mathbb{U}}{\partial x_j} = &-\frac{\partial H_{\mathrm{I}}}{\partial x_j} + \frac{\partial p_{1\mathrm{D}}}{\partial x_j}\left[-\frac{b + H_{\mathrm{D}}(p_{1\mathrm{B}} + p_{1\mathrm{p}} + r) - H_{\mathrm{C}}(p_{1\mathrm{p}} + p_{1\mathrm{B}})}{(p_{1\mathrm{D}} + p_{1\mathrm{B}} + r)^2} + \exp[-(p_{1\mathrm{D}} + p_{1\mathrm{B}} + r)T]\right.\\
&\times \left.\frac{b - H_{\mathrm{D}}(p_{1\mathrm{D}} - p_{1\mathrm{p}}) - H_{\mathrm{C}}(p_{1\mathrm{p}} + p_{1\mathrm{B}})](p_{1\mathrm{D}} + p_{1\mathrm{B}} + r)T + b + H_{\mathrm{D}}(p_{1\mathrm{B}} + p_{1\mathrm{p}} + r) - H_{\mathrm{C}}(p_{1\mathrm{p}} + p_{1\mathrm{B}})}{(p_{1\mathrm{D}} + p_{1\mathrm{B}} + r)^2}\right]\\
&+ \frac{\partial p_{1\mathrm{B}}}{\partial x_j}\left[-\frac{b + H_{\mathrm{C}}(p_{1\mathrm{D}} + r) - H_{\mathrm{D}}(p_{1\mathrm{D}} - p_{1\mathrm{p}}) - H_{\mathrm{C}}\, p_{1\mathrm{p}}}{(p_{1\mathrm{D}} + p_{1\mathrm{B}} + r)^2} + \exp[-(p_{1\mathrm{D}} + p_{1\mathrm{B}} + r)T]\right.\\
&\times \left.\frac{b - H_{\mathrm{D}}(p_{1\mathrm{D}} - p_{1\mathrm{p}}) - H_{\mathrm{C}}(p_{1\mathrm{p}} + p_{1\mathrm{B}})](p_{1\mathrm{D}} + p_{1\mathrm{B}} + r)T + b + H_{\mathrm{C}}(p_{1\mathrm{D}} + r)H_{\mathrm{D}}(p_{1\mathrm{D}} - p_{1\mathrm{p}}) - H_{\mathrm{C}} p_{1\mathrm{p}}}{(p_{1\mathrm{D}} + p_{1\mathrm{B}} + r)^2}\right]\\
&- \frac{\partial p_{1\mathrm{p}}}{\partial x_j}\frac{(H_{\mathrm{C}} - H_{\mathrm{D}})}{p_{1\mathrm{D}} + p_{1\mathrm{B}} + r}\{1 - \exp[-(p_{1\mathrm{D}} + p_{1\mathrm{B}} + r)T]\}
\end{aligned} \tag{6.170}$$

Near the optimum design it is likely that

$$p_{1i} \ll r \tag{6.171}$$

and therefore equation (6.170) can be approximated to

$$\frac{\partial \mathbb{U}}{\partial x_j} = \frac{\partial H_{\mathrm{I}}}{\partial x_j} + \frac{\partial p_{1\mathrm{D}}}{\partial x_j} C_{\mathrm{D}} + \frac{\partial p_{1\mathrm{B}}}{\partial x_j} C_{\mathrm{B}} + \frac{\partial p_{1\mathrm{p}}}{\partial x_j} C_{\mathrm{p}} \tag{6.172}$$

with

$$\begin{aligned}
C_{\mathrm{D}} = \frac{1}{r^2}\{&-[b + H_{\mathrm{D}} r - H_{\mathrm{C}}(p_{1\mathrm{B}} + p_{1\mathrm{p}})] + \mathrm{e}^{-rT}[b(1 + rT)\\
&+ H_{\mathrm{D}} r - H_{\mathrm{C}}(p_{1\mathrm{B}} + p_{1\mathrm{p}})(1 + rT)]\}
\end{aligned} \tag{6.173a}$$

$$C_{\mathrm{B}} = \frac{1}{r^2}\{-(b + H_{\mathrm{C}} r) + \mathrm{e}^{-rT}[b(1 + rT) + H_{\mathrm{C}} r]\} \tag{6.173b}$$

$$C_{\mathrm{p}} = -\frac{H_{\mathrm{C}} - H_{\mathrm{D}}}{r}(1 - \mathrm{e}^{-rT}) \tag{6.173c}$$

Further it can be noted that all terms of the coefficients C_i (i = D, B, p) are constant with respect to x_i except $H_C(p_{1B}+p_{1p})$ in C_D. However, this term influences the value of the coefficient very little and therefore to consider C_D also constant does not introduce a significant error. Then, equations (6.172), (6.168) and (6.169) allow one to express the gradient vector of $\mathbb{U}(x_i)$ in the following simple way:

$$\frac{\partial \mathbb{U}}{\partial x_j} = -\sum_k H_{1k}\frac{\partial g_k(x_j)}{\partial x_j} - \sum_i c_{1i}c_{2i}c_i \exp(c_{2i}R_i)\left(\sum_k R_{ik}\frac{\partial g_k(x_j)}{\partial x_j}\right) \quad (6.174)$$

The expressions obtained so far make it possible to use a known algorithm for maximizing $\mathbb{U}$: a steepest-ascent procedure, which makes use of the conjugate gradient method, has been chosen in [19]. Once a first approximation of the maximum utility point x_{jm} has been found, the procedure is repeated, re-evaluating the expressions of R_i and p_{1i} (equations (6.168) and (6.169)) from a new set of points x_{jk} chosen in the neighbourhood of x_{jm} in a way such to converge to the true maximum. The iterative procedure for the choice of the points x_{jk} is diagrammatically illustrated, in the case of a single design parameter ($j = 1$), in Fig. 6.29.

In [19], the above-described optimization procedure has been applied, with different sets of data, on the portal frame shown diagrammatically in Fig. 6.30; the main results are summarized in Table 6.4. The design parameters are the cross-

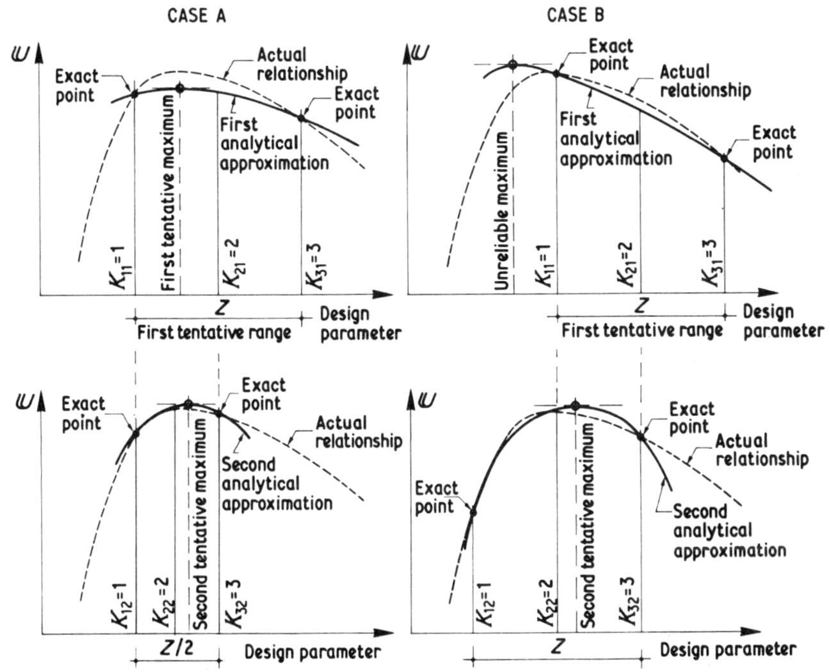

Fig. 6.29 Scheme of the iterative generalized procedure for maximizing expected utility.

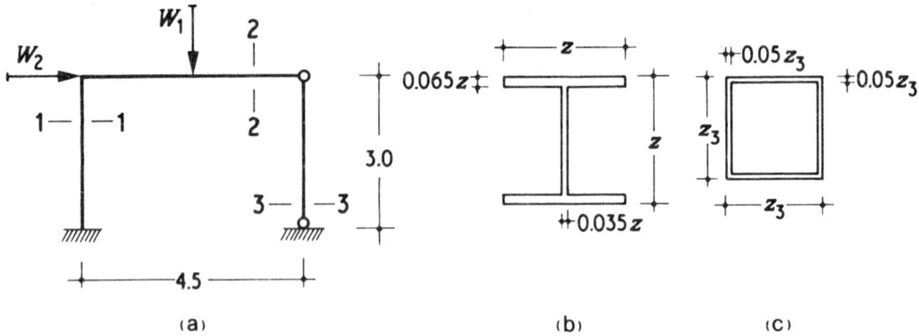

Fig. 6.30 Example frame and sections to be designed [19]:
(a) Structural diagram (lengths in metres); (b) sections 1–1 ($z = z_1$) and 2–2 ($z = z_2$); (c) section 3–3.

sections of columns and beam, each fully determined by one linear measure, indicated by z_1, z_2 and z_3 respectively, as shown in Fig 6.30 (b) and (c). The ratios chosen for the section measures correspond approximately to standard rolled sections.

The assumed 'economic' parameters r', H_C and H_D are indicated explicitly in Table 6.4; $\mathbb{B}^\circ = 3000$ is used throughout; b is calculated from equation (6.158) with $T = 50$ years (yielding $b = 160$ for $r' = 0.05$ and $b = 420$ for $r' = 0.15$); the monetary unit is equivalent to 1 kg of steel: thus, H_I is easily calculated from the structural weight. Other non-structural costs of construction are not included in the data, as they are essentially constant with respect to the chosen 'design'; they should be deducted from the calculated utility, without otherwise modifying results. It is worth underlining that, although the data have been chosen in apparently realistic ranges, the scope of the example was mainly to test the optimization procedure, and secondarily to give first gross indications on parameter sensitivity and general trends, but certainly not to yield quantitative design suggestions.

C_D (equation (6.170a)) had been calculated introducing $p_{1B} + p_{1p} = 10^{-4}$ and 10^{-5}: the final results differed very little, and only those corresponding to 10^{-4} are shown in the table.

The frame can be subjected to two forces W_1 and/or W_2 as indicated, each distributed according to an extreme law of type II (maxima). Three loading conditions (W_1 alone; W_2 alone; W_1 and W_2 together but statistically independent) have been considered, with the averages μ_{W_1} and μ_{W_2} (in tonnes) and coefficients of variation δ_{W_1} and δ_{W_2} indicated in columns (c)–(f) of Table 6.4.

Mechanical and geometrical properties of the frame are assumed to be deterministic, except for some cases (indicated in columns (g) of Table 6.4) in which

(1) the yield stress σ_y is normally distributed around the average $\mu_{\dot{\sigma}_y} = 2400$ kg cm^{-2} with coefficient of variation 0.05;

Table 6.4 Example of optimization (re. Fig. 6.30) [19].

Case	H_C	H_D	*Loading condition*				σ_y and	End of first cycle					Final design							
	$\times 10^{-3}$	$\times 10^{-3}$	μ_{W_1}	δ_{W_1}	μ_{W_2}	δ_{W_2}	Y_0	z_1	z_2	z_3	$E[U_1]$	$E[U_1^0]$	z_1	z_2	z_3	$E[U]$	p_D	p_B	p_P
no.	(*a*)	(*b*)	(*c*)	(*d*)	(*e*)	(*f*)	(*g*)	(*j*)	(*k*)	(*l*)	(*m*)	(*n*)	(*o*)	(*p*)	(*q*)	(*r*)	(*s*)	(*t*)	(*u*)
$r' = 0.05$																			
1	1500	3	4.0	0.2	–	–	Determ.	*152*	*264*	*107*	2335	2401	*65*	*259*	*79*	2530	1.5 E – 06	8.4 E – 08	1.5 E – 06
2	1500	3	4.0	0.1	–	–	Random	*162*	*224*	*127*	2490	2544	*55*	*290*	*63*	2729	5.3 E – 07	2.6 E – 08	5.1 E – 07
3	1500	3	4.0	0.1	–	–	Determ.	*164*	*226*	*129*	2477	2535	*48*	*295*	*61*	2723	5.0 E – 07	2.6 E – 08	4.9 E – 07
4	1500	3	4.0	0.1	–	–	Determ.	*152*	*244*	*117*	2460	2514	*48*	*295*	*61*	2722	5.0 E – 07	2.6 E – 08	4.9 E – 07
5	1500	3	–	–	1.0	0.1	Determ.	*233*	*0*	*0*	2787	2795	*183*	*0*	*0*	2868	2.1 E – 07	0	2.1 E – 07
6	1500	3	–	–	1.0	0.1	Random	*280*	*0*	*0*	2680	2702	*231*	*0*	*0*	2781	5.9 E – 07	0	5.9 E – 07
7	1500	3	–	–	1.0	0.2	Determ.	*285*	*0*	*0*	2669	2692	*236*	*0*	*0*	2772	6.4 E – 07	0	6.4 E – 07
8	1500	3	2.8	0.1	0.7	0.2	Random	*229*	*209*	*97*	2422	2503	*162*	*228*	*64*	2558	1.1 E – 06	5.7 E – 08	1.1 E – 06
9	1500	3	2.8	0.1	0.7	0.2	Determ.	*231*	*211*	*99*	2411	2489	*171*	*228*	*67*	2570	2.2 E – 07	1.2 E – 07	2.2 E – 07
10	1500	3	2.8	0.1	0.7	0.2	Determ.	*321*	*323*	*147*	2588	1923	*171*	*226*	*67*	2570	4.5 E – 07	1.1 E – 07	4.5 E – 07
11	1500	15	2.8	0.1	0.7	0.2	Determ.	*231*	*211*	*99*	2411	2489	*169*	*235*	*66*	2556	1.2 E – 07	9.4 E – 08	1.2 E – 07
12	750	3	2.8	0.1	0.7	0.2	Determ.	*227*	*207*	*95*	2432	2512	*160*	*225*	*64*	2566	2.1 E – 06	1.1 E – 07	2.1 E – 06
13	750	15	2.8	0.1	0.7	0.2	Determ.	*227*	*207*	*95*	2432	2512	*165*	*228*	*64*	2582	2.5 E – 07	1.7 E – 07	2.5 E – 07
$r = 0.15$																			
14	1500	3	2.8	0.1	0.7	0.2	Determ.	*226*	*205*	*94*	2441	2520	*157*	*221*	*63*	2584	2.5 E – 07	1.5 E – 07	2.5 E – 07
15	1500	15	2.8	0.1	0.7	0.2	Determ.	*226*	*205*	*94*	2441	2520	*158*	*222*	*63*	2578	2.6 E – 06	1.4 E – 07	2.6 E – 06
16	750	3	2.8	0.1	0.7	0.2	Determ.	*221*	*200*	*91*	2462	2539	*154*	*216*	*62*	2602	4.9 E – 06	2.7 E – 07	4.9 E – 06
17	750	15	2.8	0.1	0.7	0.2	Determ.	*221*	*200*	*91*	2462	2539	*154*	*216*	*62*	2602	4.9 E – 06	2.7 E – 07	4.9 E – 06

(2) the out-of-straightness Y_0 of the pin-ended column is distributed according to a truncated extreme CDF of type I (minima):

$$p(Y_0) = \begin{cases} 1 - \exp\{-\exp[\zeta(Y_0 - \chi)]\} & Y_0 > 0 \\ 0 & Y_0 < 0 \end{cases} \qquad (6.175)$$

with

$$\zeta = \frac{5.9}{l/1000} \qquad \chi = \frac{l}{1280}$$

l being the column length ($l = 300$ cm). In the deterministic cases, the values of σ_y and Y_0 are taken equal to the characteristic values (lower and upper 5% fractiles respectively) of the above distributions: this should make comparison of results more significant.

The three limit states D, B and p, described earlier in the first example have been taken into account. D corresponds to either a permanent horizontal displacement 0.6 cm = 3 m/500 of the point of application of W_2 or a permanent vertical displacement 0.9 cm = 4.5 m/500 of the point of application of W_1 (Fig. 6.30): the calculation of these displacements is analogous to that described in Section 9.6.1. With regard to B, this is due to the axial load in the right-hand (pin-ended) column reaching the buckling value, related to Y_0 by the functional relationship (6.49). This can happen either in the elastic range or after one plastic hinge has developed somewhere in the frame: the lower load value in the two conditions is taken as that corresponding to the limit state B. Finally, the collapse mechanism p implies, in general, the development of two plastic hinges.

It is worth anticipating that numerical results show that in all optimal designs, the limit states D and p coincide (cf. columns (s) and (u), Table 6.4), so that in equation (6.166)

$$p_D = 0 \qquad p_C = 1 \qquad (6.176)$$

It is not yet clear whether this coincidence reflects a general trend of optimal designs or is related to the considered limit states and/or the low redundancy of the example structure.

For each application of the optimization procedure, the points x_{jk} are fixed in the vertices and the centre of a cube in the z_1, z_2, z_3 space (Fig. 6.31). The shape

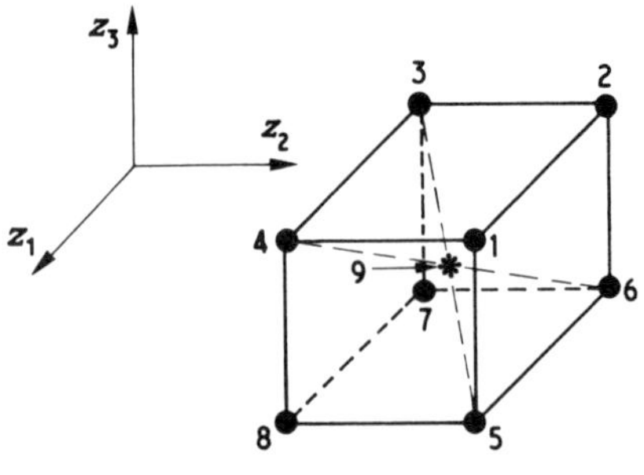

Fig. 6.31 Points x_{jk} for numerical calculations.

functions g_k of common use in finite element methods are adopted. After a first optimum has been determined, this point is automatically chosen as the centre of a new cube, with halved sides, and all calculations start again. The value $\mathrm{E}[\mathbb{U}_{n-1}^{\circ}]$ is the expected utility calculated at the previous optimum point with updated analytical approximations. The cycles are iterated until the cube sides becomes smaller than a prescribed value (10 mm in the example) or the difference $\mathrm{E}[\mathbb{U}_n] - \mathrm{E}[\mathbb{U}_{n-1}]$ becomes appropriately small.

For the determination of the constants in equation (6.169), p_{1i} are calculated at the vertices 1 and 7 of the cube (Fig. 6.31): they have been obtained numerically from equations (2.68) or (2.68′). In the cases of a random structure, $P_{Ri}(\mathbf{W})$ has been obtained by a Monte Carlo simulation procedure with 100 trials for each point. When W_1 and W_2 act together, the unsafe domains $\mathscr{U}_i$ (or the domains $P_{Ri}(\mathbf{W}) \geqslant$ constant) have been attributed the shape shown in Fig. 6.32, and the structural strength analysis is performed with the ratios W_2/W_1 corresponding to the three indicated lines, obtaining the marked points. Note that P_{1i}, by definition, are yearly probabilities: therefore the assumed loads are implicitly yearly maxima (supposed simultaneous in the combined load conditions).

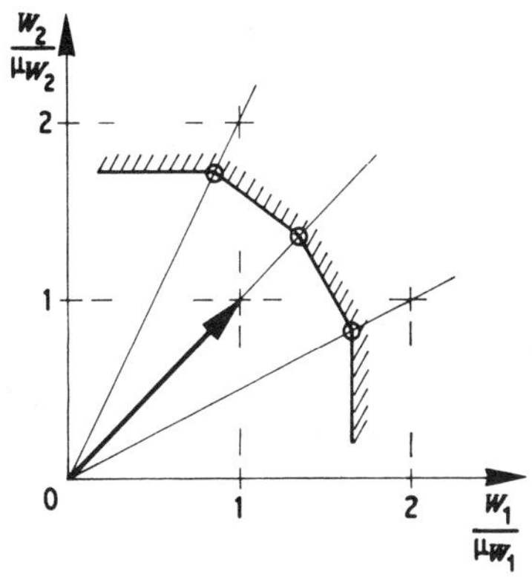

Fig. 6.32 Assumed shape of failure regions [19].

6.7.2 Serviceability limit states

With regard to serviceability limit states, it is worth quoting the following model [13], which can be applied under the assumption that the cost of failure does not depend on the duration of the relevant threshold violation. No more general model is known to the authors.

Reconsider first definition (6.152) in a deterministic context. Note that, if H_{fail} is binary (i.e. is 0 if failure does not occur and H_{fail}^* if it occurs), the utility is also binary, i.e. either $\mathbb{U} = \mathbb{U}_{\mathrm{suc}}$ or $\mathbb{U} = \mathbb{U}_{\mathrm{fail}}$. On the other hand, in the case of serviceability limit states, H_{fail} is either strictly increasing or strictly decreasing over a finite interval of y_k, and is a continuous and continuously differentiable function. It follows that, again in a deterministic context, the utility function

(1) has finite bounds;
(2) is monotonic and continuous between these bounds.

According to Section 6.6.2, the expected utility function with respect to a particular mode of failure is completely defined by the relevant utility function and the probability distribution function of the utility parameter. But, the characteristics of such a defined serviceability function can be modelled by a linear function of a cumulative probability distribution function such that (Section 6.6.2(c))

$$\mathbb{U}(y_k) = \mathbb{B}^\circ - H_{\text{I}} - H_{\text{fail}} P_{Z_k}(y_k) \tag{6.177}$$

where $P_{Z_k}(y_k)$ is the cumulative probability distribution function of a random variable, Z_k.

It follows that, for H_{fail} = constant,

$$\begin{aligned} \text{E}[\mathbb{U}_k] &= \int_{-\infty}^{\infty} [\mathbb{B}^\circ - H_{\text{I}} - H_{\text{fail}} P_{Z_k}(y_k)] p_{y_k}(y_k)\,\text{d}y_k \\ &= \mathbb{B}^\circ - H_{\text{I}} - H_{\text{fail}} \int_{-\infty}^{\infty} P_{Z_k}(y_k) p_{y_k}(y_k)\,\text{d}y_k \qquad (6.178) \\ &= \mathbb{B}^\circ - H_{\text{I}} - H_{\text{fail}}\, \text{Prob}[y_k > Z_k] \end{aligned}$$

That is, the expected utility of a system with a continuous serviceability function is the same as the expected utility of a system with a binary utility function (equation (6.153)) discontinuous at a random failure point, Z_k, such that a linear function of the cumulative probability distribution function of Z_k models the characteristics of the serviceability function, and in particular the cumulative probability distribution function of Z_k strictly increases from 0 to 1 over a finite interval of Z_k. These two systems are therefore equivalent in a mathematical sense, but the reader is cautioned against assuming conceptual equivalence. Note that a binary utility function is generally not descriptive of reality and that a utility-equivalent cumulative probability distribution function is not a classical statistical distribution function because the underlying process is not a classical statistical process but a utility assessment process.

Further, as outlined in Section 6.6.2(c) (where one assumed that the optimal repair strategy is to achieve 'full serviceability' by 'full instantaneous repair' while the utility parameter y_k is u/L), the failure costs are present-value costs. Therefore the knowledge of the JPDF $p(u, \tau)$ introduced in Section 6.5.1 of the maximum value of the response parameter $\tilde{u}$ since the most recent time of full serviceability and the time $\tilde{\tau}$ at which the serviceability state associated with u was entered, is required in order to compute numerically the expected utility by equation (6.178).

The dependence of $H_{\text{fail}}(u/L)$ on the utility parameter u/L is in fact accounted for in $P_{Z_k}(y_k)$, that in the case of Section 6.5.2(c) is defined by

$$P_{Z_k}(y_k) = \begin{cases} 0 & u/L \leqslant 0.002 \\ \dfrac{u/L - 0.002}{0.006} & 0.002 < u/L < 0.008 \\ 1 & u/L \geqslant 0.008 \end{cases} \qquad (6.179)$$

Given serviceability as a function of u/L and τ, if deflection is a gravity floor load effect, the joint probability distribution function of u/L and τ may be derived from stochastic models of gravity floor loads and structural response (cf. Section 6.5).

6.7.3 Some discussion and open problems

(a) Shape of utility function

In the examples presented in the previous sections (and in all cases known to the authors), the expected utility is a well behaved function of the design parameters, presenting one maximum in correspondence of a well defined point in the parameter space.

It is also very important to note that (again in all known cases) the expected utility decreases dramatically for underdesigned structures, but decreases very slowly for overdesigned structures (as evident from Fig. 6.27: cf. also Fig. 6.24)‡; on the other hand, the probability of failure P_{fail} decreases very rapidly with increasing member size. It may be possible therefore to overdesign a structure at a very low cost in (monetary) utility, increasing the reliability of the structure at the same time by two or three orders of magnitude (Fig. 6.33). This point will be taken up again in Section 6.7.4(b).

(b) Possible improvements in the objective function

Besides improvements in its definition and calculation, the very choice of the '*expected utility*' as the objective function in structural optimization can be questioned on several grounds. First, each interested party (owner, contractor, prospective tenant, society at large) may have different views of what is the '*benefit*' to be expected or hoped for from a construction, and may evaluate differently the costs and the losses. Also, each party has a maximum amount of damage (monetary or other) that they are willing to risk or capable of affording: therefore a '*minimax*' design rule should in some way be integrated into the '*maximum utility*' concept [20].

Perhaps the objective function should not be the 'expected utility', but some sort of '*characteristic utility*' corresponding to a predetermined probability of

‡ This property may appear at first sight strange, especially to practising engineers, who tend to forget that in the usual (short-sighted!) economical evaluations the probability of failure and the associated expected cost $H_{\text{fail}}\,P_{\text{fail}}$ are neglected.

(a)

(b)

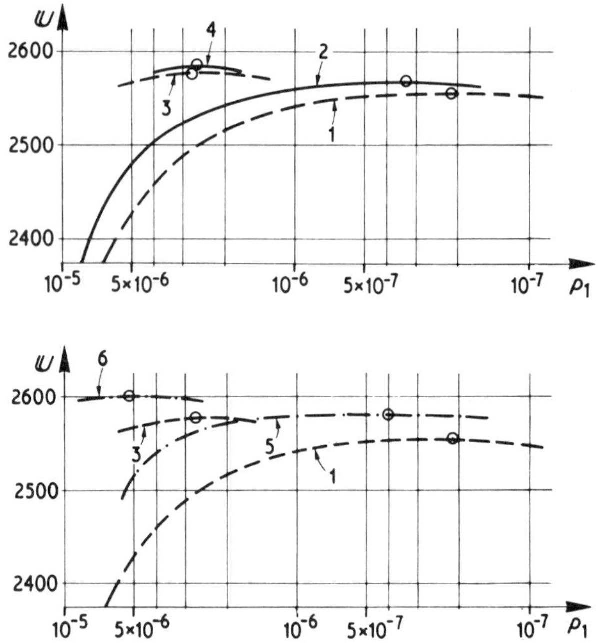

Fig. 6.33 Example frame (Fig. 6.30); deterministic structure, random loads. Expected utility 𝕌 vs probability of failure per year p_1.

(a) $H_C = 1\,500\,000$;

$$\left.\begin{aligned} &\text{line 1: } H_D = 15\,000 \\ &\text{line 2: } H_D = 3\,000 \end{aligned}\right\} r' = 5\%$$

$$\left.\begin{aligned} &\text{line 3: } H_D = 15\,000 \\ &\text{line 4: } H_D = 3\,000 \end{aligned}\right\} r' = 15\%$$

(b) $H_D = 15\,000$;

$$\left.\begin{aligned} &\text{line 1: } H_C = 1\,500\,000 \\ &\text{line 5: } H_C = 750\,000 \end{aligned}\right\} r' = 5\%$$

$$\left.\begin{aligned} &\text{line 3: } H_C = 1\,500\,000 \\ &\text{line 6: } H_C = 750\,000 \end{aligned}\right\} r' = 15\%$$

being attained. Furthermore, the interests of all parties should be taken into account, with appropriate weights.

The answers to these questions is essential in order to increase the significance and applicability of the design procedures of the type discussed.

(c) Dependence on economics

In decision theory the utility approach is regarded as an axiomatic method. One states a set of axioms on the effects of his 'strategies' and on the behaviour of the environment, so that some decisional rules can be derived. However, the above utility approach to the structural optimization problem contains implicitly a dependence of the technical problem on the economic trends at the time of design. So the maximum utility design depends on the present interest rate and on the different elements (material, labour, personal property involved in a failure, . . .) that define the problem. Figs 6.27 and 6.33 thus show that different optimal designs are obtained that generally correspond to different degrees of safety.

Also, comparison of the curves obtained for different values of the interest rate r' illustrates clearly the dependence of the maximum utility design on the economic trends at the time of design. For instance, if r' is assumed to be equal to 15% instead of to 5%, Fig. 6.33 shows that for both the considered cases $H_D = 3000$ and $H_D = 15\,000$ the initial steel weight of the structure decreases by about 10%, the maximum expected utility increases by 0.5%, but the probability of failure per year increases from 3×10^{-7} to 3×10^{-6} approximately. This result was obtained under the assumption that, in both cases, a successful structure yields the same total benefit $\mathbb{B}^\circ$. However, if the same yearly benefit is assumed, the only consequence is a higher total benefit for the structure characterized by a lower interest rate, and Fig. 6.27 pointed out a very low dependence of the optimal design on the variable $\mathbb{B}^\circ$.

If the optimum design is regarded as the most suitable distribution of the available resources capable of providing safety to the analysed structure, the utility approach discussed must be completed by a constraint on the failure probability relevant to the maximum utility design. Without this constraint, in fact, the solution of structural optimization might be an economical optimum that defines a design which is unsatisfactory (unsafe) from a social view point.

(d) Sensitivity to perturbations

Finally, it may be of interest to indicate a possible handicap, so far not examined to the authors' knowledge, of structures designed to the *maximum expected utility* rule. It is known in deterministic structural theory that an apparent *optimal* design can be very sensitive to structural *imperfections* or other forms of *perturbations*. Perhaps, a *probabilistically optimal* design might result very sensitive to gross human errors and other abnormal events, which are usually neglected in the calculations.

In fact, in the design parameter space, some of the different descent paths from the optimal design point (in a neighbourhood such as the analysed cube of minimum side) involve very little decreases of the expected utility. But, in the same neighbourhood, there may be also some other descent paths that lead to very small (sometimes negative) values of the objective function. In other words, the structural problem may be very sensitive to some sort of perturbation, and a high *risk* be associated with the optimal design. To avoid this danger, one can search for the maximum expected utility point in the design parameter space in order to define the region of satisfactory designs, but, once the optimum is determined, the stability of the solution should be investigated and, if necessary, improved.

6.7.4 Taking account of intangibles

Some of the contradictions between *expected utility* and *maximum acceptable damage* can be removed if it is understood that some types of damage cannot be assigned a *price* in monetary terms: human life is the foremost example, as indeed should be obvious. On the contrary, many researchers have tried to include it in the formation of an objective function, obtaining absurd results. This occurs also when a *price* of human life is simply thought of as an additive term implicitly included in the losses. This point can be illustrated with reference to Fig. 6.33 (b), which shows, together with curves 1 and 3, the analogous curves corresponding to half the loss H_C associated with structural failure ($H_C = 750\,000$ instead of $H_C = 1\,500\,000$). However, in this way, very little modification of the maximum utility design and of the associated probability of failure per year is obtained. In other words, the maximum utility design is not sensitive to human life loss, when this is accounted for by a conventional price, unless such high prices are associated with it that the economical aspects of the problem are certainly misrepresented.

(a) Grandori's model

The approach followed by Grandori and co-workers in the study of seismic risk [21] highlights and avoids such absurd results, as illustrated below.

Let $H(y_k)$ be the expected monetary cost (cost of construction plus expected losses), expressed in dollars per person per year, and $\upsilon(y_k)$ the expected number of victims per affected person per year. The use of these functions is supposed to be made by a large community where the economic problem is seen in terms of the entire community: thus y_k can be regarded as a (set of) very general design parameter(s) (e.g. safety factors prescribed by a building code).

The expected number of victims per person per year can be assumed proportional to the number of failures $N(y_k)$:

$$\upsilon(y_k) = \zeta N(y_k) \tag{6.180}$$

with 0.3 as a reasonable value of the coefficient ζ for extensively built-up regions.

The curves $H(y_k)$ and $\upsilon(y_k)$ look qualitatively as shown in Fig. 6.34.

The apparently most obvious approach to balance the expected cost against the expected number of victims can be expressed as follows: assign a monetary value H_v to one life lost and find the optimum value of y_k at the considered site minimizing the *total cost* $H(y_k) + H_v\,\upsilon(y_k)$; i.e. derive the design value y_{k0} from the condition:

$$y_k = y_{k0} \qquad H_{\text{tot}}(y_k) = H(y_k) + H_v\,\upsilon(y_k) = \text{minimum} \tag{6.181}$$

This kind of purely economic approach deserves strong objections, and may lead to absurd consequences. For example, H_v could be defined as the difference between what an individual could produce and what he might consume. In that case, the cynical application of this criterion to the design of an old people's home

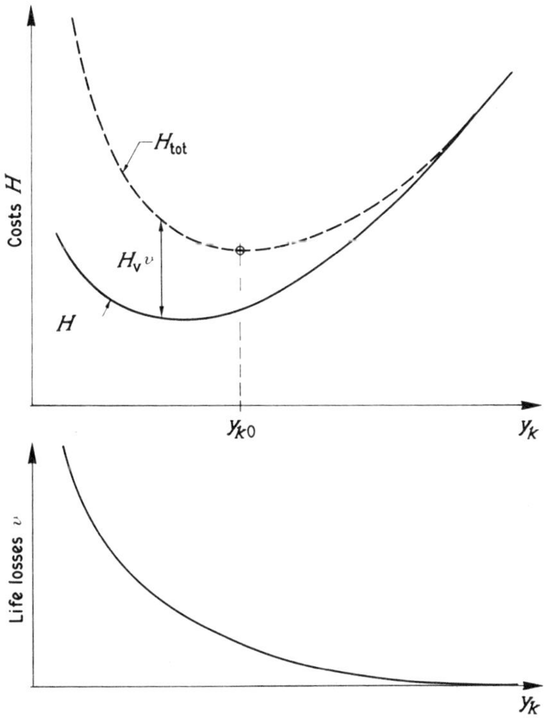

Fig. 6.34 Expected costs H and expected number of victims v vs design parameter y_k.

would assume a negative value for H_v. That is, the loss of a human life would be seen as a positive gain.

But even if a satisfactory value of H_v had been defined as the total amount of the purely economic consequences to the community of the loss of a human life, the application of equation (6.181) would face a fundamental criticism that can be pointed out as follows. Consider the curves $H_{tot}(y_k)$ and $v(y_k)$ (Fig. 6.35). Starting from a generic value of y_k, an increase Δy_k gives rise to a variation ΔH_{tot} in the *total cost* and an increase $-\Delta v$ in lives saved. The ratio between these two variations defines the *total cost of one saved life*

$$\hbar(y_k) = \Delta H_{tot}/-\Delta v \tag{6.182}$$

For example, for a seismic code, y_k may be the lateral force coefficient: then $\hbar(y_k)$ is the amount of money necessary to save one more human life by increasing y_k from a given value y_{k1}. In other words, if for y_{k1} the cost (6.182) is, for instance, one million dollars, this means that a small increase of y_k over y_{k1} leads to an increase of cost and to a decrease of victims such that the cost of each new life saved is one million dollars.

The value of y_k obtained from equation (6.181) obviously corresponds to

$$\hbar(y_{k0}) = 0 \tag{6.182'}$$

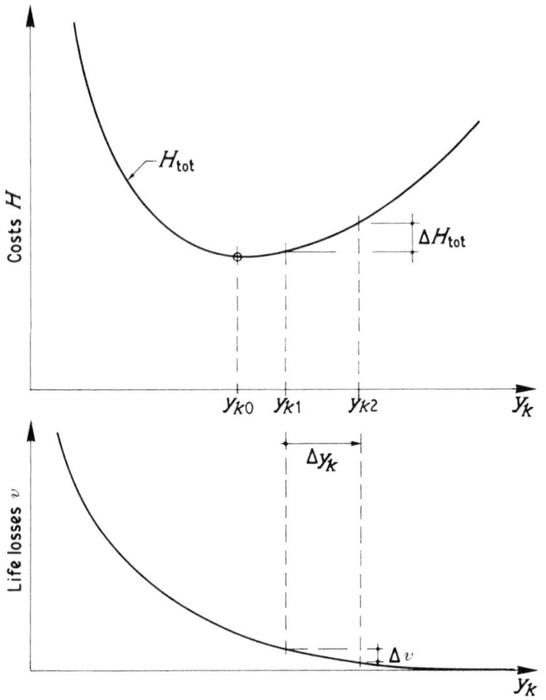

Fig. 6.35 Variation of total expected cost H_{tot} and number of victims v with the design parameter y_k.

So, bluntly speaking, the criterion expressed by equation (6.181), as (6.182) shows, can be summed up like this: it is not worth while for the community to pay a single dollar to save one further human life.

It must be accepted, as a first conclusion, that an *optimum* solution in terms of total cost H_{tot} is not practicable: the choice of a risk involving human life can hardly be reduced to an optimization problem. This does not mean that it is impossible to indicate clear guidelines to the decision-makers, as shown by the following discussion, developed with special reference to seismic codes: in this case let y_k be the seismic coefficient, and $H^*(y_k)$ the *additional* cost of antiseismic design (with respect to a design that neglects the danger of earthquakes).

Suppose that, for each site in the country, the curves $H^*(y_k)$ and $v(y_k)$ for a sample building are available. From these it is easy to derive the curve (Fig. 6.36)

$$\hbar^*(y_k) = \Delta H^*/-\Delta v \tag{6.183}$$

The ratio (6.183) can be defined as the *additional* or *marginal cost of one saved life.*

If the cost of design is shared among all the members of the community, the following position seems reasonable: for a given amount C of the total cost of design for all the sample buildings in the country, the distribution must be such

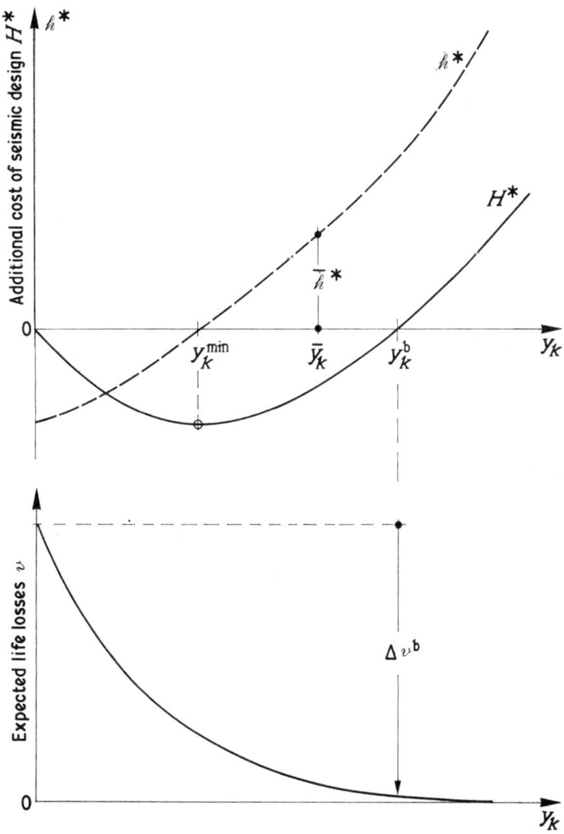

Fig. 6.36 Seismic design: the marginal cost of a saved life and the choice of a lateral force coefficient.

that the expected overall number of victims due to the collapse of these buildings be minimum. It is easy to show [21] that this condition is satisfied when and only when the marginal cost of a saved life $h^*(y_k)$ is the same at all sites. In conclusion, the condition

$$h^* (y_k) = -\frac{\Delta H^*}{\Delta v} = \text{constant} \tag{6.184}$$

optimizes the inter-site distribution of resources set aside for preventing the structural risk.

If the distribution criterion expressed by (6.184) is accepted, the problem of the absolute value of the safety level can be discussed in terms of a single site, considered as a reference site.

Now suppose that Fig. 6.36 is relative to the reference site. The function $H^*(y_k)$ generally has a minimum: let $y_k^{\min}$ be the corresponding value of y_k. Then

$y_k = y_k^{min}$ obviously leads to $\hbar^* = \Delta H^*/-\Delta v = 0$, as already observed with reference to total cost (equation (6.182′)).

It seems obvious that the design coefficient must certainly be greater than y_k^{min}.

Another significant value of y_k is that indicated by y_k^b in Fig. 6.36. It is given by $H^*(y_k^b) = H^*(0)$. Designing with $y_k = y_k^b$ means having the same monetary costs as if there were no seismic code, but it would obviously greatly reduce the number of expected victims.

Another significant value of y_k, which will be called here $\bar{y}_k$ (Fig. 6.36), is defined as the value of y_k which corresponds to a marginal cost $\bar{\hbar}^*$ such that clearly the same sum invested in other risk prevention schemes would save many more lives than just one. For instance, if we establish in a first attempt 1 000 000 \$ $< \bar{\hbar} <$ 2 000 000 \$, we get for California $0.06 < \bar{y}_k < 0.1$ and for Japan $0.15 < \bar{y}_k < 0.20$: quite realistic figures [21].

A thorough consideration of the values y_k^{min}, y_k^b and $\bar{y}_k$, together with the corresponding values of v, can provide the necessary support for decision-making. These values have precise meanings that can be illustrated without any possibility of misunderstanding to those who have to take the decisions.

It must be pointed out that giving a quantitative definition to $\bar{\hbar}^*$ is, to say the least, delicate and difficult. But this type of approach does at least have the advantage of not clashing with logical or moral problems. Fixing a finite limit to the cost of preventing a given risk is in fact inevitable, since this cost cannot be other than a fraction of the gross national product, which is also finite.

(b) An alternative model

An alternative, less sophisticated way of formulating the maximum utility design problem, avoiding the contradictions already emphasized in this section, has been recently suggested [22]. According to this approach, one first searches for the *economically optimal* design, as the one that shows comparatively the largest expected utility, including among the *expected losses* only the purely monetary consequences of the possible failures; one then varies this design, *worsening* it from a purely *utilitarian* point of view, but at the same time increasing its reliability (i.e. decreasing its *probability of failure*). Non-technical considerations (essentially, how much one is willing or forced to *spend*, in the form of decreased economical return, in order to reduce the expected *intangible* losses) will tell which of these *worsened* designs to choose.

The main idea is to start from the design corresponding to the maximum expected utility $\mathbb{U}_{max}$ (economical optimum) and vary its parameters to increase the reliability. The best path in the design parameter space is such that every point corresponds to the smallest reduction of utility (with respect to $\mathbb{U}_{max}$) for any given yearly reliability $1 - p_1$ (equation (6.156)). This path can be denoted as the *slowest utility decrease path*, and coincides with the steepest descent path (from

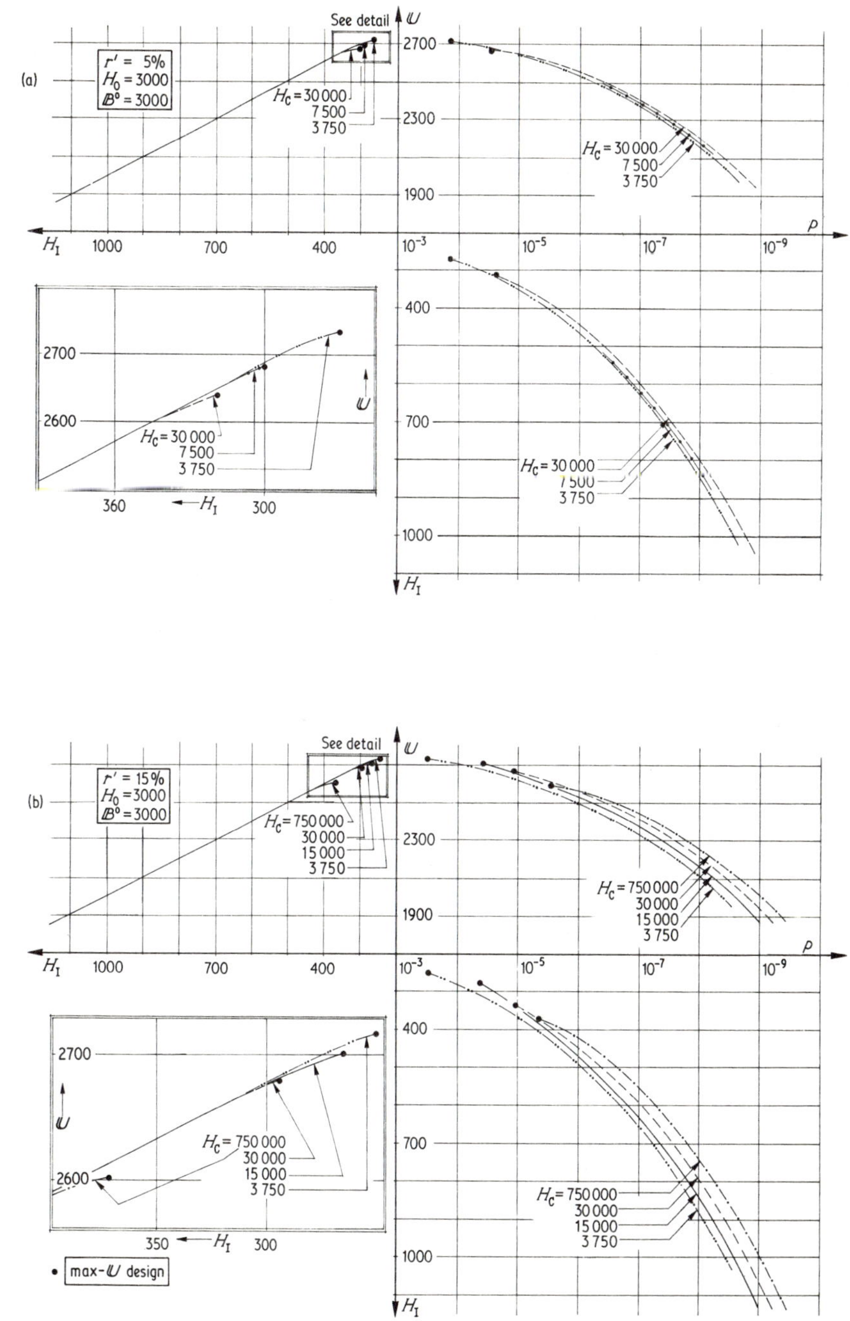

Fig. 6.37 Expected utility $\mathbb{U}$, yearly probability of collapse p, initial cost H_I in $\mathbb{U}_{max}$ point and along *slowest utility decrease* path ($\mathbb{B}^0$ fixed, H_C varying) [22].

the $\mathbb{U}_{\max}$ point) of the function

$$\phi(x_j) = \frac{\mathbb{U}_{\max} - \mathbb{U}}{-p_1} \tag{6.185}$$

To write explicit expressions for the gradient components, necessary for the solving algorithm, it has proved expedient (cf. Section 6.7.1) to define a *conventional resistance* R_i as the factor that brings the average load point on the boundary of the unsafe region (Fig. 6.28). The relation between R_i and the design parameters has been approximated by equation (6.168) and p_{1i} by equation (6.169).

With these positions, the components of the gradient of $\phi(x_j)$ are

$$\begin{aligned}\frac{\partial \phi}{\partial x_j} &= \frac{1}{p_1^2}\left[\frac{\partial \mathbb{U}}{\partial x_j} p_1 + (\mathbb{U}_{\max} - \mathbb{U})\frac{\partial p_1}{\partial x_j}\right] \\ &= \frac{1}{p_1^2}\left[\frac{\partial \mathbb{U}}{\partial x_j} p_1 - (\mathbb{U}_{\max} - \mathbb{U})\sum_i a_i b_i \exp(-b_i R_i)\left(\sum_k R_{ik}\frac{\partial g_k}{\partial x_j}\right)\right]\end{aligned} \tag{6.186}$$

where $\partial \mathbb{U}/\partial x_i$ is given by equation (6.172).

The procedure summarized above to find the slowest utility decrease path has been applied [22] to the same steel frame already shown in Fig. 6.30, with the economical data indicated in Figs 6.37 and 6.38 and, as in Section 6.7.1 (b), a

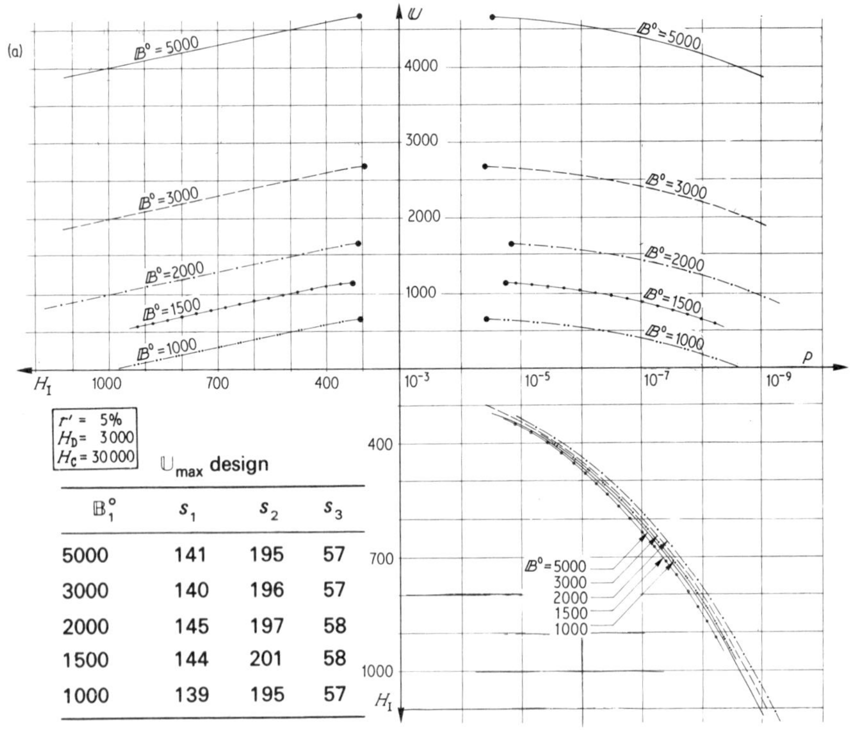

$\mathbb{U}_{max}$ design

$\mathbb{B}_1^\circ$	s_1	s_2	s_3
5000	141	195	57
3000	140	196	57
2000	145	197	58
1500	144	201	58
1000	139	195	57

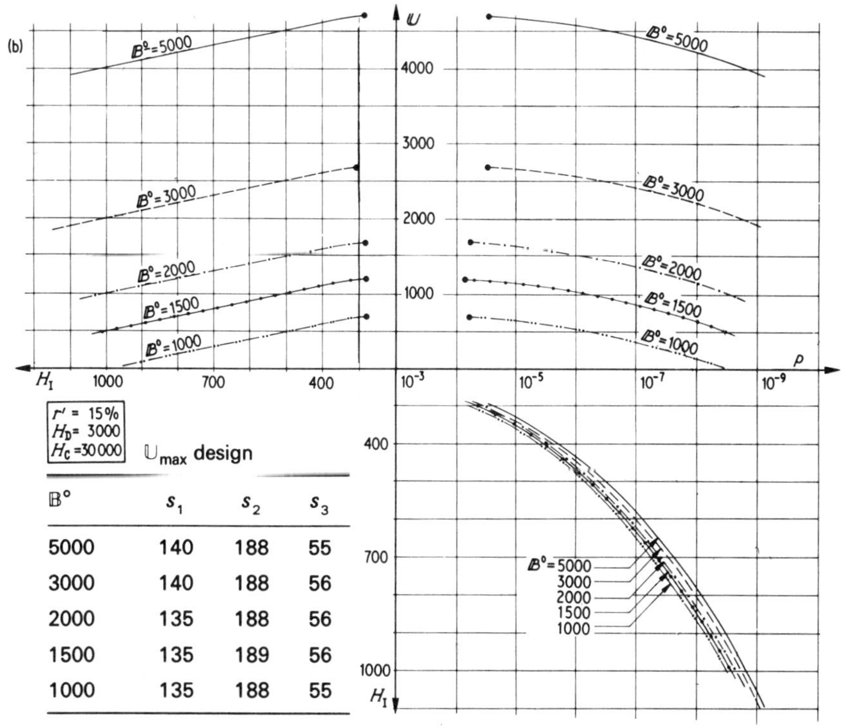

$\mathbb{U}_{max}$ design

$\mathbb{B}^o$	s_1	s_2	s_3
5000	140	188	55
3000	140	188	56
2000	135	188	56
1500	135	189	56
1000	135	188	55

Fig. 6.38 As Fig. 6.37 (but H_C fixed, $\mathbb{B}^0$ varying) [22].

design life $T = 50$ years. All geometrical and mechanical properties of the frame have been assumed to be deterministic.

Figs 6.37 and 6.38 show the plots of the main results obtained. Inspection of these results allows some considerations, which, of course, cannot pretend to have any generality nor quantitative significance, but can give some qualitative indications on the trends of the phenomena investigated.

First of all, the plots of $\mathbb{U}$ vs p_1 show that large increases of reliability can be obtained with comparatively small losses in expected utility. Taking for instance the curve $H_C = 30\,000$ in Fig. 6.37 (a), decreases of $\mathbb{U}$ from 2670 to 2500 or to 2270 (approx. 6% or 15% respectively, but neglecting non-structural costs, which should be deducted from the calculated $\mathbb{U}$) correspond to 100- or 1000-fold decreases in the probability of failure (from 0.3×10^{-4} to 0.4×10^{-6} or to 0.2×10^{-7} per year approximately).

Secondly, for any given set of parameters, there exists a value of p_1 under which the decrease in utility is equal to the increase in initial cost: this is evident from the $\mathbb{U}$–H_I plots. Beyond this point, there would be no convenience whatsoever, for a private entrepreneur, to increase the reliability of the structure. (However, this argument is strictly related to the assumed economical model: for instance, it would be invalidated if H_I contained an insurance premium decreasing with p_1.)

The inserts in Fig. 6.38 show that, at least within the range investigated, the $\mathbb{U}_{max}$ design is insensitive to the value of the benefit $\mathbb{B}^{\circ}$, whose variations essentially result in additive terms of the expected utility. On the other hand, the cost of collapse H_C and the interest rate r' do affect the $\mathbb{U}_{max}$ design; however, all slowest utility decrease paths tend to become very close to each other with decreasing p_1. Therefore, when the design is essentially governed by the reliability, the value of most *economic* data appear to be of little importance on its choice.

Note however that, as already stated several times, these indications hold only in relation to the very narrow part of the actual design process investigated, namely the choice of a few geometrical quantities in a fixed structural layout. With realistic data, results like those presented here may also tell whether the proposed structural design could be economically (and/or socially) beneficial, or perhaps other structural forms and/or materials should be tried for comparison. But these considerations would lead into a very broad range of problems, outside the scope of the present book.

In conclusion, it has been shown that it is actually possible to start from the maximum expected utility design (economical optimum) and follow a continuous path (*slowest utility decrease path*) of alternative designs that correspond to increases of reliability with the smallest possible loss of utility: the results obtained would indicate that increases of 2–3 orders of magnitude in the reliability are possible with reasonable decreases of utility.

Of course, the numerical procedure would become much more cumbersome for a more realistic problem than the example investigated, but its feasibility appears to be proved in principle.

It is to be noted that in this approach, differently from previously known procedures of conditional optimization under prescribed reliability [23], the acceptable reliability must not be chosen *a priori*, but only after inspection of the path of alternative designs shows the economical consequences of its increases.

REFERENCES

1. Joint Committee on Structural Safety (1976) Common unified rules for different types of construction and material. *CEB Bull. d'Information*, **116E**.
2. Esteva, L. (1975) Second moment probabilistic analysis of statically loaded non-linear structures. *Proc. 2nd Int. Conf. on Applications of Statistics and Probability in Soil and Structural Engineering* (ICASP 2), Aachen, vol. 1, pp. 115–29.
3. Baratta, A. (1975) Reliability and extremum principles in structural analysis. *Meccanica*, **10**(4), 246–53.
4. Casciati, F., Faravelli, L. and Gobetti, A. (1980) Elasto-plastic response analysis of seismic-resistant frames via mathematical programming. *Proc. 7th World Conf. Earthquake Engineering*, Istanbul, vol. 5, pp. 521–8.

5. Augusti, G. and Baratta, A. (1971) Teoria probabilistica della resistenza delle aste compresse. *Costruzioni Metalliche*, **23** (1), 44–58 (in Italian). *Construction Metallique*, **8**(2), 5–20 (in French).

5a. Augusti, G. and Baratta, A. (1976) Reliability of slender columns: comparison of different approximations; *Buckling of Structures, IUTAM Symp.*, Cambridge, MA, 1974 (ed. B. Budianski), Springer-Verlag, Berlin and New York, pp. 183–98.

6. Elishakoff, I. (1980) Hoff's problem in a probabilistic setting. *J. Applied Mech., ASME*, **47**(2), 403–8.

7. Veneziano, D. and Antoniano, J. (1979) Reliability analysis of slopes: frequency domain method. *J. Geotech. Eng. Div., Proc. ASCE*, **105** (GT2), 165–82.

7a. Schultze, E. (1979) Slope stability and the bearing capacity of shallow foundations on slopes. *Proc. 3rd Int. Conf. on Applications of Statistics and Probability in Soil and Structural Engineering* (ICASP 3), Sydney, vol. 3, pp. 198–229.

8. Casciati, F. and Faravelli, L. (1980) Safety analysis for random elastic–plastic frames in the presence of second order geometrical effects. *Applied Math. Modelling*, **4** (1), 16–22.

9. Bolotin, V. V. (1969) *Statistical Methods in Structural Mechanics*, Holden-Day, San Francisco.

9a. Augusti, G., Baratta, A. and Casciati, F. (1979) Structural response under random uncertainties. *Proc. 3rd Int. Conf. on Applications of Statistics and Probability in Soil and Structural Engineering* (ICASP 3), Sydney, vol. 3, pp. 220–328.

10. Bolotin, V. V. (1972) Reliability theory and stochastic stability; in *Stability* (ed. H. H. Leipholz), University of Waterloo Press, Ontario, pp. 385–422.

11. Weiss, G. H. (1964) A survey of some mathematical models in the theory of reliability; in *Statistical Theory of Reliability* (ed. M. Zelen), University of Wisconsin Press, Madison.

12. Augusti, G. and Casciati, F. (1977) Multiple modes of structural failure: probabilities and expected utilities. *Proc. Int. Conf. on. Structural Safety and Reliability* (ICOSSAR '77), München; Werner-Verlag, Düsseldorf, pp. 39–57.

13. Reid, S. G. and Turkstra, C. J. (1980) *Serviceability Limit States—Probabilistic Description*, Report ST 80–1, McGill University, Montreal.

14. Broberg, H. and Westlund, R. (1978) Creep in structures with random material properties. *Int. J. Solids Struct.*, **14**, 365–74.

15. Casciati, F. (1976) La sicurezza strutturale nei confronti degli stati limite di esercizio (in Italian). *Proc. 3rd Congr. Naz. AIMETA*, Cagliari, vol. 2, Paper No. 3.

16. CIRIA (1976) *Rationalisation of Safety and Serviceability Factors in Structural Codes*, Report 63, CIRIA, London.

17. Benjamin, J. R. and Cornell, A. C. (1970) *Probability, Statistics and Decision for Civil Engineers*, McGraw-Hill, New York.

18. Raiffa, M. (1968) *Decision Analysis: Introductory Lectures on Choices under Uncertainty*. Addison-Wesley, Reading, MA.

19. Augusti, G. and Casciati, F. (1979) Further studies on structural design for maximum expected utility. *Proc 3rd Int. Conf. on Applications of Statistics and Probability in Soil and Structural Engineering* (ICASP 3), Sydney, vol. 2, pp. 686–700.

20. Baumol, W. J. (1971) *Economic and Operation Analysis*. Prentice Hall, Englewood Cliffs, NJ.

21. Grandori, G. (1979) Earthquake risk. *Proc. Seminar on Earthquakes, Soils and Structures*, Saint Rémy les Chevreuse.
22. Augusti, G., Borri, A. and Casciati, F. (1981) Structural design under random uncertainties : economical return and 'intangible' quantities. *Proc. 3rd Int. Conf. on Structural Safety and Reliability* (ICOSSAR '81), Trondheim; Elsevier, Amsterdam, pp. 483–94.
23. Frangopol, D. and Rondal, J. (1978) Le dimensionnement probabiliste optimal des structures (in French). *Annales ITBTP*, **363**, pp. 21–30.

PART TWO

Operative Methods

CHAPTER 7

Data statistics

In this chapter the possible manipulations on the available observations of a random variable are discussed. The available data collection is first regarded as a discrete random variable approaching the actual underlying variable: so it is possible to define a *sample probability distribution*, a *sample mean*, a *sample variance*, ... Next, the relations between sample and actual properties are investigated, and suitable *estimators* are introduced. Finally, the significance of results from estimators is discussed and judged (*error analysis*; *hypothesis testing*).

7.1 INTRODUCTORY REMARKS

In previous chapters, the possibility of filtering random input data through the equations of the 'theory of structures' has been investigated, in order to turn the description of the structure, of the environment and of the functionality demand into probabilistic forecasting of the performance of the facility concerned.

Of course, the underlying philosophy is that data (such as actual loads, material strength, and so on) cannot be foreseen in detail in the design procedure, but their realization and/or their evolution in time are subjected to uncertainty. Such uncertainties must be quantified and, in the previous chapters, it was understood that the input quantities and the system parameters were described by suitable mathematical models.

Obviously, these mathematical models have to reflect the actual character of the data, as revealed by *experience*. In this case experience coincides with *observations* of data, so that *knowledge* of data should be intended as the availability of a collection of past realizations in situations similar to the one under examination. Such knowledge of data in many cases is not exhaustive for operational purposes for at least two reasons. It may be in fact (and often is) a very partial description of data, because it includes only a number of elements from a larger set on which an unknown measure of probability is thought to be defined. Secondly, such knowledge of data, generally, is not suitable for mathematical manipulations aimed at deriving directly similar 'knowledge' of the parameters describing the performance of the structure. For this purpose, in fact, the structure should be solved under each available set of data: but thus a very partial description of the output quantities of interest would be obtained.

Extrapolation of observations in order to derive probability statements regarding also non-observed data realizations is the object of *statistics*. The question is usually approached by investigating *probabilistic* models which best fit the observed data; such a procedure, besides lending itself to data extrapolation, offers also the description of data in the form assumed in the previous chapters, thus also answering the second point.

7.2 PRELIMINARY OPERATIONS ON SAMPLES

7.2.1 Sample

Let $\tilde{x}$ be a random variable: a set of n repeated measurements of $\tilde{x}$ is called a *sample of size* n (see Section A3.8). It is usually postulated that the n measurements in each sample are independent of each other: hence (see Appendix A1) $\tilde{x}_1, \tilde{x}_2, \ldots, \tilde{x}_n$ are stochastically independent random variables with identical probability distribution function. A sample defined in this manner is called a *random sample of size* n and constitutes an n-dimensional random variable.

It is worth noting that the elements of the sample must be measurements of the same random variable. It follows that the collection of data must not be affected by changes of the technological process or of the environmental conditions, nor by replacement of the measuring instruments. To clarify this point, consider for instance the random variable $\tilde{v} = \tilde{v}^*_{10y}$ defined as the yearly maximum of the wind speed averaged over 10 min recorded at a given site, say Malpensa Airport, Milan. The recorded data are the values averaged over the 10 min preceding every third hour (00.00–03.00–06.00–09.00–12.00–15.00–18.00–21.00, GMT) in the period 1951–1973: the yearly maxima are shown in Table 7.1. These values can be

Table 7.1 Yearly maxima of the wind speed averaged over 10 min at Malpensa Airport, Milan [1].

Year	$v (m\,s^{-1})$	*Year*	$v (m\,s^{-1})$	*Year*	$v (m\,s^{-1})$
1951	18.468	1959	13.851	1967	19.494
1952	16.416	1960	13.338	1968	15.390
1953	14.364	1961	21.546	1969	14.364
1954	12.825	1962	16.416	1970	15.390
1955	11.799	1863	11.286	1971	18.468
1956	14.364	1964	12.825	1972	15.390
1957	18.468	1965	12.825	1973	17.955
1958	15.390	1966	15.390		

considered a sample (of size $n = 23$) of the random variable $\tilde{v}$ only if the climatic conditions did not change and the airport anemometer was not replaced from 1951 to 1973.

7.2.2 Frequency

Let E be the event which occurs if a measurement of $\tilde{x}$ yields a value in a given measurable set. For every random sample of size n, let n_E be the number of sample values x_i that imply the occurrence of the event E. Then the *statistical relative frequency* of E obtained from the given random sample is

$$\text{freq}[E] = n_E/n \tag{7.1}$$

Note that the relative frequencies of mutually exclusive events add; the frequency of a certain event is 1; and the frequency of the impossible event is 0‡.

If a random sample can have only two possible (mutually exclusive) outcomes, say E and E^C, with

$$\text{Prob}[E] = p, \qquad \text{Prob}[E^C] = 1 - p = q \tag{7.1'}$$

(Bernoulli trials), then the random variable $\tilde{n}_E$ has a binomial distribution (equation (A1.14)) and hence (Table A.1 and equation (A3.5))

$$\text{E}[\text{freq}[E]] = \frac{\text{E}[\tilde{n}_E]}{n} = p = \text{Prob}[E] \tag{7.2}$$

$$\text{Var}[\text{freq}[E]] = \frac{\text{Var}[\tilde{n}_E]}{n} = \frac{pq}{n} = \frac{\text{Prob}[E](1 - \text{Prob}[E])}{n} \tag{7.3}$$

The *cumulative frequency distribution of a sample* is defined by

$$F(x) = \text{freq}[E\colon \tilde{x} \leqslant x] = \text{freq}[\tilde{x}_j \leqslant x] = n_E/n \tag{7.4}$$

where n_E is the number of measurements not larger than x. $F(x)$ is a non-decreasing step function, with $F(-\infty) = 0$ and $F(\infty) = 1$.

7.2.3 Bernoulli's theorem

Consider equation (A4.17) and note that there the symbol $\tilde{x}/n$ denotes the frequency freq$[E]$ here defined by equation (7.1). Thus equation (A4.17) expresses Bernoulli's theorem: the relative frequency freq$[E] = \tilde{n}_E/n$ of realizing the event E in n independent trials is a random variable (see Appendix B) which converges to Prob$[E]$ in probability as $n \to \infty$.

A stronger statement of Bernoulli's theorem is provided by equation A4.18, for which freq$[E]$ converges to Prob$[E]$ with probability 1.

Further, the De Moivre–Laplace theorem (equation (A4.31)) states that

‡ The defining properties of the probability theory (see Appendix A1) can be regarded as extrapolated from these corresponding properties of the statistical relative frequency.

freq[E] is asymptotically normal with mean Prob[E] and variance {Prob[E] (1 – Prob[E])}/n.

7.2.4 Class frequency

A sample is a set of data that through appropriate manipulation can give indications of the properties of the underlying random variable.

The preliminary operation usually consists of putting the data in increasing order. For instance, the sample of Table 7.1 can be ordered as shown in Table 7.2.

Table 7.2 Ordered yearly largest wind speed at Malpensa Airport, Milan (1951–1973).

Year	*j*	$v(m\,s^{-1})$	*Year*	*j*	$v(m\,s^{-1})$	*Year*	*j*	$v(m\,s^{-1})$
1963	1	11.286	1956	9	14.364	1962	17	16.416
1955	2	11.799	1969	10	14.364	1973	18	17.955
1954	3	12.825	1958	11	15.390	1951	19	18.468
1964	4	12.825	1966	12	15.390	1957	20	18.468
1965	5	12.825	1968	13	15.390	1971	21	18.468
1960	6	13.338	1970	14	15.390	1967	22	19.494
1959	7	13.851	1972	15	15.390	1961	23	21.546
1953	8	14.364	1952	16	16.416			

The *sample range* D_x for a random sample $(\grave{x}_1, \grave{x}_2, \ldots, \grave{x}_n)$ is the difference between the largest $(\grave{x}_{max})$ and the smallest $(\grave{x}_{min})$ sample value. For the example considered, such *rank statistics* are:

$$\grave{v}_{min} = 11.286\ \mathrm{m\,s^{-1}},\ \grave{v}_{max} = 21.546\ \mathrm{m\,s^{-1}},\ D_v = 10.260\ \mathrm{m\,s^{-1}}$$

The sample range can be partitioned into a finite (or infinite) number H of conveniently chosen *class intervals* $(x^i - \frac{1}{2}\Delta x^i) < x \leqslant (x^i + \frac{1}{2}\Delta x^i)$ $(i = 1, 2, \ldots, H)$ of length $\Delta x^1, \Delta x^2, \ldots, \Delta x^H$ respectively, centred at $x = x^i$ $(x^1 < x^2 < \ldots < x^H)$. Then, for the random sample $(\grave{x}_1, \grave{x}_2, \ldots, \grave{x}_n)$, the *class frequency* n_i is the number of times that $\grave{x}_j$ $(j = 1, \ldots, n)$ falls into the ith class interval and the *relative class frequency* f_i^0 is

$$f_i^0 = \mathrm{freq}[E{:}\,(x^i - \tfrac{1}{2}\Delta x^i) < \grave{x}_j \leqslant x^i + \tfrac{1}{2}\Delta x^i)] = n_i/n \tag{7.5}$$

Further, the *cumulative class frequencies* N_i and the *cumulative relative class frequencies* F_i are defined by

$$N_i = \sum_{k=1}^{i} n_k \tag{7.6}$$

$$F_i = \mathrm{freq}\left[E{:}\,\grave{x}_j \leqslant x^i + \tfrac{1}{2}\Delta x^i\right] = \sum_{k=1}^{i} f_1^0 = \frac{N_i}{n} = F(x^i + \tfrac{1}{2}\Delta x^i) \tag{7.7}$$

where $F(x)$ is the cumulative frequency distribution defined by equation (7.4).

Finally the *class frequency density of the sample* f_i is defined by

$$f_i = \frac{f_i^0}{\Delta x^i} = \frac{F_i - F_{i-1}}{\Delta x^i} \tag{7.8}$$

For the Malpensa Airport example, five classes of the same length $\Delta v = 2.5\,\mathrm{m\,s^{-1}}$ have been considered in the range 10–25 m s^{-1}. The corresponding class frequencies are presented in Table 7.3.

Table 7.3 Class frequencies of yearly largest wind speeds recorded at Malpensa Airport (1951–1973).

v^j (m s^{-1})	Class j	Class range (m s^{-1})	Elements belonging to the jth class	n_i	N_i	f_i^0	f_i (s m^{-1})	F_i
11.25	1	10.0–12.5	v_1 and v_2	2	2	2/23	8/230	3/23
13.75	2	12.5–15.0	v_3 to v_{10}	8	10	8/23	32/230	10/23
16.25	3	15.0–17.5	v_{11} to v_{17}	7	17	7/23	28/230	17/23
18.75	4	17.5–20.0	v_{18} to v_{22}	5	22	5/23	20/230	22/23
21.25	5	20.0–22.5	v_{23}	1	23	1/23	4/230	1

The statistics $F(x), n_i, N_i, f_i^0, F_i$ and f_i allow several graphical representations of the sample distribution. The most meaningful representations are defined and drawn in Fig. 7.1 with reference to the Malpensa Airport example.

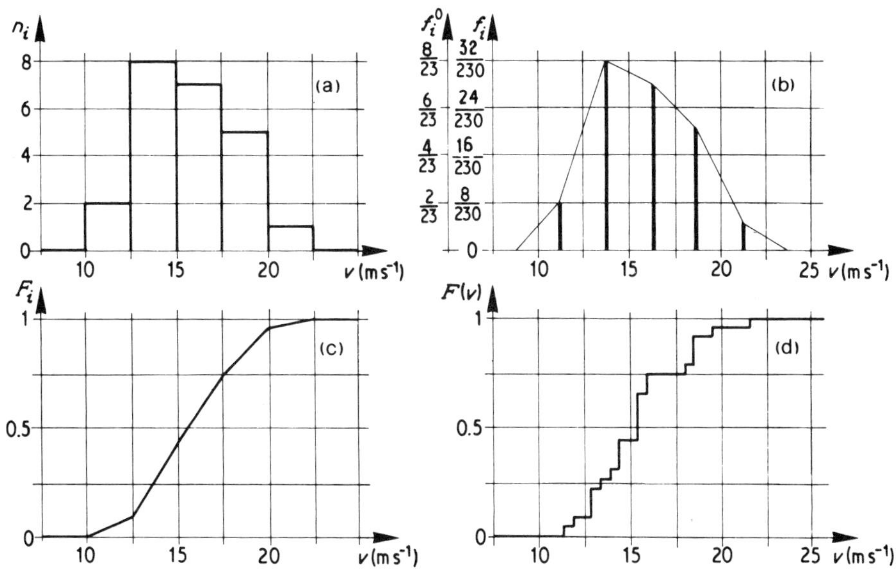

Fig. 7.1 (a) Histogram of class frequency. (b) Bar histogram of class frequency (or class frequency density) and connecting polygon. (c) Cumulative class frequency. (d) Cumulative frequency distribution ($\Delta v_i = 2.5\,\mathrm{m\,s^{-1}}$).

Finally, one defines the *sample p–fractile* (or *fractile of order p*) x^{s}_{kp} by the relation

$$\text{freq}[E: x < x^{s}_{kp}] = F(x^{s}_{kp}) = p \tag{7.9}$$

7.2.5 Measures of location

The sample 0.5-fractile is called the *sample median.* From Fig. 7.1(d) one obtains for the example considered, $v^{s}_{k0.5} = 15.390\ \text{m s}^{-1}$.

The median is the simplest *measure of location* for a given random sample $(\dot{x}_1, \dot{x}_2, \ldots, \dot{x}_n)$, while the most meaningful such measure is the *sample average* (or *mean*) defined (cf. equation (A3.63)) by

$$m_n = \frac{1}{n} \sum_{j=1}^{n} \dot{x}_j \tag{7.10}$$

The quantity m_n can be regarded as a random variable for which equations (A3.65) hold. For the example of Table 7.1 one obtains $m_{23} = 15.4792\ \text{m s}^{-1}$.

In addition to the sample average and median, a further measure of location is the *sample mode,* defined as the value $\dot{x}_j$ of the sample associated with the largest frequency. In the example considered, the sample mode is $15.390\ \text{m s}^{-1}$.

7.2.6 Measures of dispersion

The simplest *measure of dispersion* is the sample range D_x defined in Section 7.2.4. Further, given a random sample $(\dot{x}_1, \dot{x}_2, \ldots, \dot{x}_n)$, the *sample variance* is (equation (A3.66))

$$s_n^2 = \frac{1}{n} \sum_{j=1}^{n} (x_j - m_n)^2 = \frac{1}{n}\left(\sum_{j=1}^{n} x_j^2 \right) - m_n^2 \tag{7.11}$$

The quantity s_n^2 can be regarded as a random variable

$$\tilde{s}_n^2 = \tilde{\chi} = \frac{1}{n} \sum_{j=1}^{n} (\tilde{x}_j - \tilde{m}_n)^2 \tag{7.11$'$}$$

Introducing $\mu_{\tilde{x}} = \mathrm{E}[\tilde{x}]$ (equation (A2.11))

$$\tilde{\chi} = \frac{1}{n} \sum_{j=1}^{n} [(\tilde{x}_j - \mu_x) - (\tilde{m}_n - \mu_x)]^2$$

$$= \frac{1}{n} \sum_{j=1}^{n} [(\tilde{x}_j - \mu_x)^2 - 2(\tilde{x}_j - \mu_x)(\tilde{m}_n - \mu_x) + (\tilde{m}_n - \mu_x)^2]$$

$$= \frac{1}{n} \sum_{j=1}^{n} [(\tilde{x}_j - \mu_x)^2] - 2(\tilde{m}_n - \mu_x)^2 + (\tilde{m}_n - \mu_x)^2$$

$$= \frac{1}{n} \sum_{j=1}^{n} [(\tilde{x}_j - \mu_x)^2] - (\tilde{m}_n - \mu_x)^2 \tag{7.11$''$}$$

Then, defining two further random variables $\tilde{y}_j = (x_j - \mu_x)^2$ $(j = 1, \ldots, n)$ and $y = (\tilde{m}_n - \mu_x)^2$, equation (7.11″) becomes

$$\tilde{\chi} = \frac{1}{n}\tilde{y}_1 + \frac{1}{n}\tilde{y}_2 + \ldots + \frac{1}{n}\tilde{y}_n - \tilde{y} \tag{7.11‴}$$

The mean of $\tilde{\chi}$ can be obtained from equations (A3.26′) and (A3.65).

$$\begin{aligned} \mu_\chi &= \frac{1}{n}\mu_{y_1} + \frac{1}{n}\mu_{y_2} + \ldots + \frac{1}{n}\mu_{y_n} - \tilde{y} \\ &= \frac{1}{n}(s_x^2 + \ldots + s_x^2) - \frac{s_n^2}{n} \\ &= \frac{n-1}{n}s_x^2 = \mathrm{E}[\tilde{s}_n^2] \end{aligned} \tag{7.12}$$

Then, one can also introduce the *estimating sample variance*

$$(s_n^0)^2 = \frac{1}{n-1}\sum_{j=1}^{n}(x_j - m_n)^2 \tag{7.13}$$

characterized by the property

$$\mathrm{E}[(s_n^0)^2] = s_x^2$$

For the wind speed example of Table 7.1 one obtains

$$s_{23}^2 = 6.5598\ \mathrm{m^2\,s^{-2}} \qquad (s_{23}^0)^2 = 6.8580\ \mathrm{m^2\,s^{-2}}$$

The *sample moments* a'_{rn} and the *sample central moments* a_{rn} of order r are defined by

$$a'_{rn} = \frac{1}{n}\sum_{j=1}^{n}\dot{x}_j^r \tag{7.14a}$$

$$a_{rn} = \frac{1}{n}\sum_{j=1}^{n}(\dot{x}_j - m_n)^r \tag{7.14b}$$

For $r = 2$, $a_{2n} = s_n^2$.

7.3 PROBABILITY MODEL AND SAMPLE VALUES

Consider a hypothetical infinite sample of the r.v. $\tilde{x}$, which will be called a *population*. The relevant sample cumulative frequency distribution (*population distribution*, equations (7.4) and (7.7)) does not necessarily coincide with the underlying probability distribution function (equation (A2.2)). In fact, Bernoulli's theorem (Section 7.2.3 and equation (A4.18)) does not state that freq $[E]$ tends asymptotically in a mathematical sense to Prob $[E]$ but only that the event

$|\text{freq}[E] - \text{Prob}[E]| < \varepsilon$, for any positive number ε, tends to be certain as the sample size n increases.

Nevertheless, the following statement is usually accepted as an axiom: *the population distribution is identical to the theoretical probability distribution function of the underlying random variable.* Thus a broad, axiomatic relation between frequency and probability is introduced. This axiom is a direct consequence of the *empirical law* [2] that was already introduced at the beginning of Chapter 2. According to this law, one postulates that if the probability of an event can be shown to be smaller than any positive real number, the occurrence of the event is practically negligible: on the other hand, convergence theorems prove this is the case of the event that the frequency does not tend asymptotically to the probability.

For a more precise formulation, consider the deviate $\tilde{\Delta}_n$

$$\tilde{\Delta}_n = \max_x |\tilde{F}_n(x) - P_x(x)| \tag{7.15}$$

where $\tilde{F}_n(x)$ is the cumulative frequency distribution of a sample of size n (equation (7.4)) and is regarded as a random variable. In fact any sample $(\grave{x}_1, \grave{x}_2, \ldots, \grave{x}_n)$ is only a particular realization of the many collections of n realizations of $\tilde{x}$, and hence $F_n(x)$ is a discrete random variable whose possible values are $0, 1/n, 2/n, \ldots, 1$. Cantelli's theorem (equation (A4.18)) states that, as n tends to ∞, $\tilde{\Delta}_n$ tends to 0 with probability 1. Another theorem (Kolmogorov's theorem) is more significant in that it offers the possibility to estimate the distribution function of $\tilde{\Delta}_n$. In particular if the CDF of $\tilde{x}$, $P_x(x)$, is continuous, Kolmogorov's theorem states that

$$\lim_{n\to\infty} \text{Prob}[\sqrt{n}\tilde{\Delta}_n < y] = \Lambda(y) = \begin{cases} 0 & y \leqslant 0 \\ \sum_{k=-\infty}^{\infty} (-1)^k e^{-2k^2y^2} & y > 0 \end{cases} \tag{7.16}$$

Statistical methods use empirical data to infer the characteristics of the cumulative frequency distribution of the population and hence the properties the CDF $P_x(x)$ associated with the random variable $\tilde{x}$ of interest. In other words, statistical methods aim at deriving probability models from the available sample values.

In some applications, the statistics defined in Sections 7.24–7.26 can be used as direct, rough estimates of the characteristics of the population distribution. Alternatively, one can postulate (on the basis of the physical background) the form of the underlying distribution (Appendix A2):

$$P_x(x) = P_x(x; \eta_1, \eta_2, \ldots) \tag{7.17}$$

and estimate the unknown *population parameters* $\eta_1, \eta_2, \ldots$ from the given random sample. Parameter estimation represents a first basic topic of elementary statistics.

This procedure yields more accurate results if the postulated distribution law is a good representation of reality. Otherwise it can lead to large errors and seriously wrong conclusions. Therefore, any assumptions on the theoretical distribution properties must be checked by statistical tests: these, however, when successful, do not prove a hypothesis, but only show that it has not been proved wrong (and therefore is likely to be right).

Further details will be presented in Section 7.6. Here, note that Kolmogorov's theorem is useful for testing the effectiveness of hypotheses regarding the type of distribution of the random variable under examination. Assume, in fact, that a given probability function $\ddot{P}_x(x)$ has been guessed for $\tilde{x}$ and that $F_n(x)$ has been built up from an observed sample. Then compute

$$\ddot{\Delta}_n = \max_x |F_n(x) - \ddot{P}_x(x)| \tag{7.18}$$

i.e. the realization of the random variable (7.15) corresponding to the observed sample. If the sample size n is sufficiently large, it is possible to calculate from equation (7.16)

$$\text{Prob}[\tilde{\Delta}_n \geqslant \ddot{\Delta}_n] \simeq 1 - \Lambda(\sqrt{n}\,\ddot{\Delta}_n) = \varepsilon \tag{7.19}$$

Let ε be so small that the event $\tilde{\Delta}_n \geqslant \ddot{\Delta}_n$ can be considered practically unrealizable: it should be concluded that the assumption $P_x(x) = \ddot{P}_x(x)$ is wrong and the difference $\ddot{\Delta}_n$ cannot be ascribed to randomness of the sample $(\dot{x}_1, \dot{x}_2, \ldots, \dot{x}_n)$, but to the intrinsic character of the random variable $\tilde{x}$. Thence $\ddot{P}_x(x)$ cannot be used to describe the random variable $\tilde{x}$.

However if ε is sufficiently large, the calculated deviate $\ddot{\Delta}_n$ is then a value that is likely to be found in a single test. In this case, one can conclude that the sample test is not in conflict with the assumed CDF $\ddot{P}_x(x)$. However, this does not prove that $\ddot{P}_x(x)$ is the actual CDF of $\tilde{x}$.

7.4 CLASSICAL PARAMETER ESTIMATION

The probability distributions described in Appendixes A2 and A7 (and in particular normal, log-normal, Poisson and extreme distributions) are generally adopted as theoretical population distributions in statistical models. Normal distributions are particularly convenient as they are defined completely by only two parameters, i.e. the relevant first- and second-order moments. Moreover, their use is often justified by the central limit theorem (equations (A4.29)–(A4.31)).

Once the type or analytical form of the probability model is assumed, the problem is the estimation of the values of the parameters of this function (equation (7.17)). Looking at $(\dot{x}_1, \dot{x}_2, \ldots, \dot{x}_n)$ as the realization of a random event, it is clear that any quantity that can be calculated by manipulation of the elements of $(\dot{x}_1, \dot{x}_2, \ldots, \dot{x}_n)$ is the realization of some particular random variable and this must be kept well in mind when dealing with any parameter of the random variable that is estimated from a random sample.

Let η be a parameter of the random variable $\tilde{x}$ that must be estimated starting from the sample. Any sample statistic $\eta_n(\dot{x}_1, \dot{x}_1, \ldots, \dot{x}_n)$, defined as a function of the given random sample, whose result is interpreted as the actual value of η, is called an *estimator*. The goodness of the estimator should be evaluated by investigating its properties and by ensuring that the general criteria on the testing of statistical hypotheses (Section 7.6) are satisfied.

7.4.1 Errors and properties of estimates

Let $\ddot{\eta}(\dot{x}_1, \dot{x}_2, \ldots, \dot{x}_n)$ be an estimator of a given parameter η, based on the sample $(\dot{x}_1, \dot{x}_2, \ldots, \dot{x}_n)$ of size n. This estimator has a numerical value which should be viewed as a realization of the random variable $\tilde{\eta}_n$, because of the randomness of the sample itself. The error $\mathscr{E}_n$ on the estimate can be synthetically expressed by the mean square difference of the random variable $\tilde{\eta}_n$ with respect to the actual value η:

$$\mathscr{E}_n = \mathrm{E}[(\tilde{\eta}_n - \eta)^2] \tag{7.20}$$

It can be readily checked that

$$\mathscr{E}_n = \mathrm{Var}[\tilde{\eta}_n] + (\mu_{\eta_n} - \eta)^2 \tag{7.21}$$

Thus, it can be recognized that the error $\mathscr{E}_n$ is the sum of two terms: the first one, the variance of the estimator, is due to purely random variations; the second one, of systematic nature and originating from the difference between the expected value of the estimator and the actual parameter value, is equal to the square of the quantity

$$B(\tilde{\eta}_n) = \mu_{\eta_n} - \eta \tag{7.22}$$

which is usually called the *bias* of the estimator. If the considered estimator has zero bias, it is said to be *unbiased.*

The most common property that is required of an estimator is that $\tilde{\eta}_n$ converges to η in probability (equation (A4.4)); if this condition is encountered the estimator is said to be *consistent*. As shown in Section A4.1, a sufficient (but not necessary) condition to meet this requirement is given by the convergence in the mean square sense of $\mathscr{E}_n$ to 0 and hence, for an unbiased estimator, by

$$\lim_{n \to \infty} \mathrm{Var}[\tilde{\eta}_n] = 0 \tag{7.20$'$}$$

It is usually more expressive to assess the error as a function of the actual value of the parameter η. Use is often made of the *coefficient of variation of the estimator:*

$$e_\eta = \frac{\sqrt{\mathscr{E}_n}}{\eta} \tag{7.23}$$

It is soon realized that, for an unbiased estimator, e_η coincides with the coefficient of variation δ_{η_n} of $\tilde{\eta}_n$.

Finally, as estimates with small variance are generally desirable, an estimator is said to be *efficient* when it minimizes the error among all possible unbiased estimators.

The statistical relative frequency freq $[E]$ is an unbiased, consistent estimate of the corresponding probability Prob $[E]$, and the sample mean m_n, the estimating sample variance $(s_n^0)^2$ and the sample moment a'_{rn} are unbiased, consistent estimates of the mean μ_x, of the variance s_x^2 and of the moment $\mathrm{E}[\tilde{x}^r]$ respectively‡.

7.4.2 Derivation of estimates

If the population distribution is described by a given function $p_x(x, \eta_1, \eta_2, \ldots)$, with $\eta_1, \eta_2, \ldots$ parameters to be determined (see equation (7.17)), each population characteristic (like $\mathrm{E}[\tilde{x}]$, $\mathrm{Var}[\tilde{x}]$, $\mathrm{E}[\tilde{x}^r], \ldots$) is a function of the parameters $\eta_1, \eta_2, \ldots$

The *method of moments* considers jointly the estimates $\ddot{\eta}_{1n} = \eta_1(\dot{x}_1, \dot{x}_2, \ldots, \dot{x}_n)$, $\ddot{\eta}_{2n} = \eta_2(\dot{x}_1, \dot{x}_2, \ldots, \dot{x}_n), \ldots, \ddot{\eta}_{mn} = \eta_m(\dot{x}_1, \dot{x}_2, \ldots, \dot{x}_n)$ of m corresponding population parameters $\eta_1, \eta_2, \ldots, \eta_m$) of m corresponding population parameters $\eta_1, \eta_2, \ldots, \eta_m$ and introduces the relations

$$\mathrm{E}[\tilde{x}^r] = q_r(\ddot{\eta}_{1n}, \ddot{\eta}_{2n}, \ldots, \ddot{\eta}_{mn}) = a_{rn}(\dot{x}_1, \dot{x}_2, \ldots, \dot{x}_n) \qquad r = 1, 2, \ldots, m \tag{7.24}$$

Equation (7.24) are obtained by equating the first m sample moments a_{rn} to the corresponding population moments: hence the resulting η_{mn} are function of the sample moments.

An alternative approach is the so-called *method of maximum likelihood* [3]. Given a random sample of size n, the *likelihood function* $L(\dot{x}_1, \dot{x}_2, \ldots, \dot{x}_n)$ is the probability density function defined, in the n-dimensional sample space of sample points $(\dot{x}_1, \dot{x}_2, \ldots, \dot{x}_n)$, by

$$L(x_1, x_2, \ldots, x_n) = p_x(x_1)p_x(x_2)\ldots p_x(x_n) \tag{7.25}$$

where $p_x(x)$ is the PDF of the random variable $\tilde{x}$. For a given sample the value of the likelihood function is a function of the unknown parameters $\eta_1, \eta_2, \ldots$. The values of the estimators of these parameters, $\ddot{\eta}_{1n}, \ddot{\eta}_{2n}, \ldots, \ddot{\eta}_{mn}$, are chosen such that the likelihood function is as large as possible for each available sample $(\dot{x}_1, \dot{x}_2, \ldots, \dot{x}_n)$; they can be obtained by solving the following system of equations:

$$\frac{\partial}{\partial \ddot{\eta}_{kn}} \ln L(x_1, x_2, \ldots, x_n; \ddot{\eta}_{1n}, \ddot{\eta}_{2n}, \ldots, \ddot{\eta}_{mn}) = 0 \qquad k = 1, 2, \ldots, m \tag{7.26}$$

‡ This does not mean that $(s_n^0)^2$ is a better estimator of s_x^2 than s_n^2. The more frequent use of the first in practice is only for the sake of promoting uniformity among reports of experimental data.

Equation (7.26) establishes that the chosen set of estimators is the most probable one. It is worth noting that such a method applies also to multidimensional populations and applies even in the case of non-random samples.

7.5 OPERATIVE PROCEDURES OF INFERENCE

Consider the sample of Table 7.1 and suppose that the CDF $P_v(v)$ of the random variable $\tilde{v}$ must be identified. As already pointed out in Section 7.3, the best approach to such a problem usually starts from the choice of an analytical form for the sought function. In common applications this choice can be greatly simplified by knowledge of the definition range involved by the physical or geometrical meaning of the variable (e.g., in the specific example being considered v is non-negative and hence $P_v(v)$ must only be defined over $0 \leqslant v < \infty$) and by past experience (e.g. a sample of the same random variable in a similar, although different, situation proved to be well fitted by some type of distribution). However, in general, a trial-and-error procedure must be followed.

Suppose first that $\tilde{v}$ is normally distributed‡. The parameters that define the normal distribution (Table 2.2) are the mean μ_v and the standard deviation s_v. From the values already given (Section 7.2), applying equation (7.24),

$$\overset{\prime\prime}{\mu}_v = m_{23} = 15.4792\ \mathrm{m\,s^{-1}} \qquad \overset{\prime\prime}{s}{}_v^2 = s_{23}^2 = 6.56\ \mathrm{m^2\,s^{-2}}.$$

Thus the actual $p_v(v)$ is estimated by

$$\ddot{p}_v(v) = \frac{1}{\sqrt{(2\pi \times 6.5598)}} \int_{-\infty}^{v} \exp\left(-\frac{1}{2}\frac{(v-15.4792)^2}{6.56}\right)$$

$$= 0.156 \int_{-\infty}^{v} \exp\left[0.076(v-15.479)^2\right] \tag{7.27}$$

This function has been plotted, by using Table A.5, together with the cumulative frequency distribution in Fig. 7.2: the comparison of the two lines allows a first evaluation of the goodness of fit of the estimated CDF. In the case that the result is not satisfactory, one considers other distribution types by an analogous procedure until a function that fits the experimental data well is found.

To compare the cumulative frequency distribution graphically with the selected analytical expression for the PDF of the considered random variable $\tilde{x}$ is a practical (though inaccurate) first step. Unfortunately, in most cases plots of PDFs, very different from each other from the conceptual point of view, do not show comparable discrepancies in the significant numerical values. In fact, the intrinsic differences appear mostly in the *tails* of the distributions, where the functions have such small values that it is difficult to evaluate whether these

‡ Note that this choice is certainly wrong because of the non-negative nature of the variable $\tilde{v}$. However, it may be the better inference from the available sample in the neighbourhood of the mean.

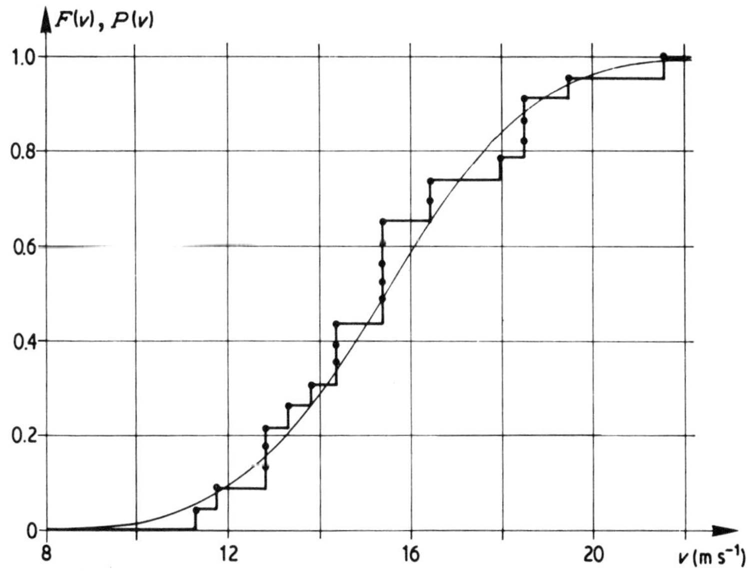

Fig. 7.2 Comparison between the cumulative frequency distribution of the considered sample (cf. Fig. 7.1 (d)) and the normal interpolation (thin solid curve).

differences are significant or not for practical purposes. Moreover, when the comparison is used to judge the goodness of fit of a sample distribution to a mathematical model, often no information on the tails is provided by the cumulative frequency distributions. A well defined criterion is therefore needed to judge whether the form of the cumulative frequency distribution agrees with the assumed model.

Further, even if the distribution type is known, the parameter estimates determined by means of equation (7.24) or (7.26) may be unsatisfactory, so that an improved estimation procedure is necessary to obtain the actual PDF of $\tilde{x}$. For the most common types of distribution, such problems can be tackled by using *probability papers.*

7.5.1 Probability papers

For any given theoretical distribution $P_x(x)$, the probability paper has coordinate scales such that the plots of all CDFs of that type are *straight lines.*

To construct the simplest probability papers, it is first necessary to define a new random variable $\tilde{y}$ which follows the considered distribution with known values $P_y(y)$, and is related to the variable $\tilde{x}$ by a linear relationship: in a plane with x as abscissae and y as ordinates (both on a linear scale), this relationship will obviously appear as a straight line. Then, the ordinates are altered by substituting for each value of y the corresponding $P_y(y)$: any function $P_x(x)$ of the considered type, plotted on this paper, is a straight line, because $P_x(x) = P_x[x(y)] = P_y(y)$.

Consider for instance the probability paper for the Gaussian distribution (Fig. 7.3). In this case the variable $\tilde{y}$ coincides with the standardized variable $\tilde{x}^*$ (equation (A4.27))

$$\tilde{y} = \tilde{x}^* = \frac{(\tilde{x} - \mu_x)}{s_x} \tag{7.28}$$

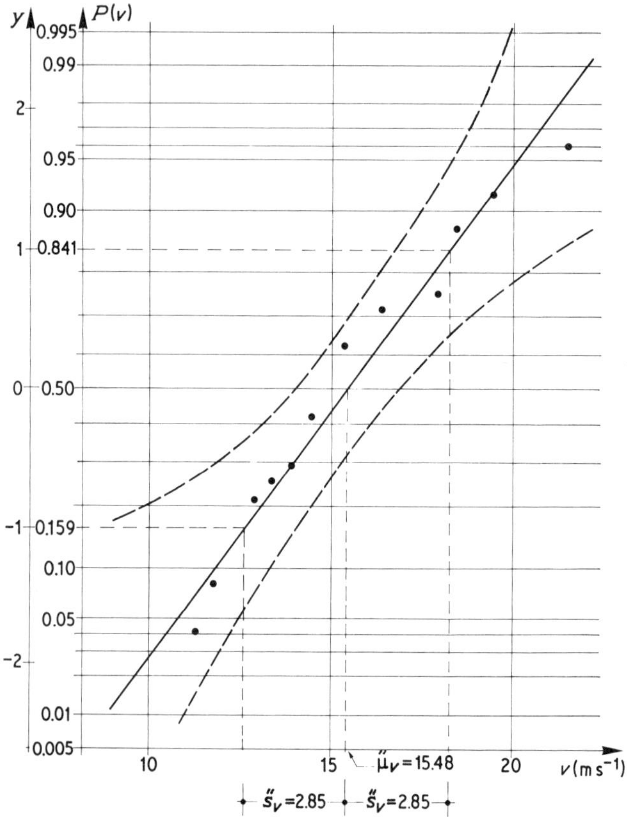

Fig. 7.3 Probability paper for the Gaussian distribution and plot of the given sample.

and $P_y(y)$ is given by Table A.5. For the extreme type I (maxima) distribution, one can assume (Fig. 7.4)

$$\tilde{y} = c(\tilde{x} - x') \tag{7.29}$$

and (Table A.6)

$$P_y(y) = \exp[-\exp(-y)] \tag{7.29'}$$

For the log-normal distribution, it is convenient to plot linearly on the abscissae $\ln \tilde{x}$ instead of $\tilde{x}$. Then, choosing

$$\tilde{y} = \ln\left(\frac{\tilde{x} - \mu_x)}{s_x}\right) \tag{7.30}$$

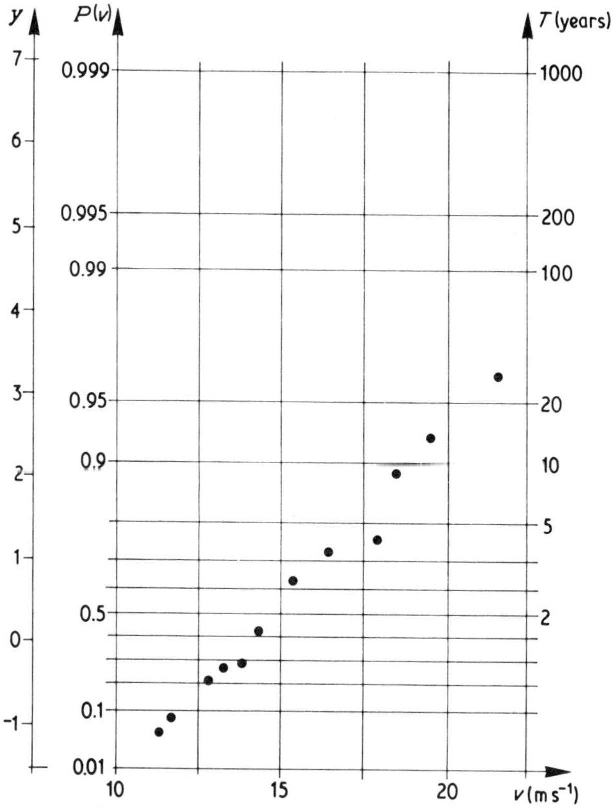

Fig. 7.4 Probability paper for the extreme type I (maxima) distribution and plot of the given sample.

with $P_y(y)$ given again by Table A.5, the required linear relationship is obtained (Fig. 7.5).

7.5.2 Representation of observed data on probability papers

In the described probability papers, $P_y(y) = 0$ if $y = -\infty$ and $P_y(y) = 1$ if $y = +\infty$. It follows that the largest observed value, for which $F(\ \ ) = 1$, cannot be represented. However, if $P_x(x)$ is normal, it assumes the value 1 if and only if $x = +\infty$: therefore the largest observed value (that is finite) is associated with a value of $P_x(\ \ ) < 1$.

Therefore, the definition of the cumulative frequency given by equation (7.4) must be modified in order to meet conditions for correct plotting. The most common modified definition was given by Gumbel [4]

$$F^*(x) = \text{freq}[E\colon \tilde{x} \leqslant x \mid n+1] = n_E/(n+1) \tag{7.31}$$

i.e. the size of the sample is fictitiously increased by one unit.

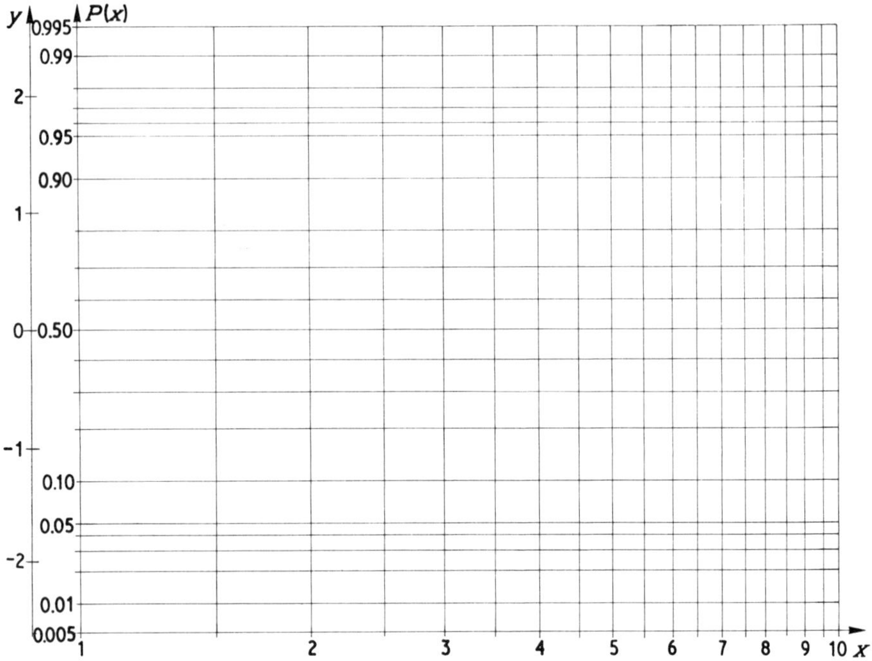

Fig. 7.5 Probability paper for the log-normal distribution.

With reference again to the example of Table 7.1, equation (7.31) yields the values of $F^*(v)$ shown in Table 7.4, and plotted in the probability papers of Figs 7.3 and 7.4.

Using a probability paper, the test for acceptability of the assumed distribution is reduced to the judgement of straightness of the plot, while the estimation of the parameters of the assumed CDF coincides with the determination of the best straight-line interpolation of this plot.

It is immediately evident from Figs 7.3 and 7.4 that the data of the example are much better represented by an extreme type I distribution.

Table 7.4 Modified cumulative frequency $F^*(x)$ of yearly largest with speeds at Malpensa Airport (1951–1973).

$v(m\,s^{-1})$	$F^*(v)$	$v(m\,s^{-1})$	$F^*(v)$	$v(m\,s^{-1})$	$F^*(v)$
11.286	0.0417	14.364 }		16.416 }	0.7083
11.799	0.0833	14.364 }	0.4167	16.416 }	
12.825 }		14.364 }		17.955	0.7500
12.825 }	0.2083	15.390 }		18.468 }	
12.825 }		15.390 }		18.468 }	0.8750
13.338	0.2500	15.390 }	0.6250	18.468 }	
13.851	0.2917	15.390 }		19.494	0.9167
		15.390 }		21.546	0.9583

7.5.3 Least-squares method for plot interpolation

The most usual procedure followed in straight-line interpolation of points plotted on a probability paper‡ is the well known *method of least squares.* Note that, by using this procedure, one also estimates the parameters of the distribution with more accuracy than the methods of Section 7.4 allow.

Let $\dot{x}_1, \dot{x}_2, \ldots, \dot{x}_n$ be a sample of size n and let $\dot{y}_1, \dot{y}_2, \ldots, \dot{y}_n$ be the corresponding values of the auxiliary $\tilde{y}$. By the definition of probability paper, the relationship between the variables $\tilde{y}$ and $\tilde{x}$ must be linear:

$$x = c_1 y + c_2 \qquad y = \frac{x - c_2}{c_1} \tag{7.32}$$

Thus, any straight line, with c_1 and c_2 unknown, defines a set of values $x'_1, x'_2, \ldots, x'_n$ that correspond to the sample values of the auxiliary variable. It follows that, for any straight line, it is possible to calculate the sum ϕ of the squares of the differences $(\dot{x}_i - x'_i)$

$$\phi = \sum_{i=1}^{n} (\dot{x}_i - x'_1)^2 = \sum_{i=1}^{n} [\dot{x}_i - (c_1 y_i + c_2)]^2 \tag{7.33}$$

The method of least squares consists of selecting the straight line for which ϕ is minimum, i.e. the straight line defined by the values c_{10} and c_{20} for which

$$\partial\phi/\partial c_1 = 0, \partial\phi/\partial c_2 = 0. \tag{7.33'}$$

From equation (7.33)

$$\begin{aligned} \frac{\partial\phi}{\partial c_1} &= 2c_1 \sum_{i=1}^{n} y_i^2 + 2\left(c_2 \sum_{i=1}^{n} y_i - \sum_{i=1}^{n} \dot{x}_i y_i\right) \\ \frac{\partial\phi}{\partial c_2} &= 2nc_2 + 2\left(c_1 \sum_{i=1}^{n} y_i - \sum_{i=1}^{n} \dot{x}_i\right) \end{aligned} \tag{7.33''}$$

and the solving system is

$$\begin{aligned} c_{10} \sum_{i=1}^{n} y_i^2 + c_{20} \sum_{i=1}^{n} y_i &= \sum_{i=1}^{n} \dot{x}_i y_i \\ c_{10} \sum_{i=1}^{n} y_i + nc_{20} &= \sum_{i=1}^{n} \dot{x}_i \end{aligned} \tag{7.33'''}$$

whence

$$c_{10} = \frac{\left[\sum_{i=1}^{n} \dot{x}_i y_i - \left(\sum_{i=1}^{n} \dot{x}_i\right)\left(\sum_{i=1}^{n} y_i\right)\right]}{\left[\sum_{i=1}^{n} y_i^2 - \left(\sum_{i=1}^{n} y_i\right)\left(\sum_{i=1}^{n} y_i\right)\right]} \tag{7.34}$$

$$c_{20} = \sum_{i=1}^{n} \dot{x}_i - c_{10} \sum_{i=1}^{n} y_i \tag{7.35}$$

‡ Or, indeed, in the interpolation of a curve of any chosen shape on any type of plotting paper.

On Gaussian probability paper, the relationship between y and x is given by equation (7.28): therefore, c_1 and c_2 are the estimates of the standard deviation s_x and of the mean μ_x respectively. With reference to the example of Table 7.4, equations (7.34) and (7.35) give (Fig. 7.3)

$$c_{10} = \ddot{s}_v = \sum_{i=1}^{n} \dot{x}_i y_i \Big/ \sum_{i=1}^{n} y_i^2 = 50.8665/17.843 = 2.8508 \text{ m s}^{-1}$$

$$c_{20} = \ddot{\mu}_v = \sum_{i=1}^{n} x_i = m_{23} = 15.4792 \text{ m s}^{-1}$$

Note that, in the normal distribution case, the estimate μ_v presents the same value that was obtained by equation (7.24): in fact, because of the symmetry of the assumed distribution,

$$\sum_{i=1}^{n} y_i = 0$$

As shown by the above example, the estimation of the *best* parameters of an assumed distribution can be pursued independently from any judgement on the goodness of fit of the distribution.

7.6 HYPOTHESIS TESTING

7.6.1 Statistical hypotheses

Natural sciences, including engineering science, are essentially founded on observations. Observations are organized by the scientist, who finally discovers common elements in the realizations of the physical experiment that he is carrying on. On this basis he formulates some hypotheses and then builds up a set of logical implications, mostly by mathematical tools, whose final result should agree with experimental evidence.

In statistics, a *hypothesis* is a conjecture about the distribution of a random variable $\tilde{x}$. In particular, a hypothesis concerning some parameter of a distribution (e.g. mean or variance) is called a *parametric hypothesis*: more precisely, a *simple hypothesis* if it states that the parameter has a specific value (e.g. $\mu_x = 5$) or a *composite hypothesis* if it states that the parameter lies in some given interval (e.g. $4 \leqslant \mu_x \leqslant 6$). If the assumption regards the form of the distribution rather than a parameter, it is called a *non-parametric hypothesis* (e.g. $\tilde{x}$ is normally distributed).

7.6.2 Philosophy for testing statistical hypotheses

A *test of the statistical hypothesis* H consists of obtaining a random sample ($\dot{x}_1, \dot{x}_2, \ldots, \dot{x}_n$) and then examining this sample in order to decide whether the hypothesis H is acceptable or not.

Each test specifies an n-dimensional *rejection region* Ξ_n such that H is to be rejected if the test sample ($\dot{x}_1, \dot{x}_2, \ldots, \dot{x}_n$) belongs to Ξ_n. Note that an experiment can suggest the rejection of a hypothesis but is seldom sufficient to convince one that the hypothesis is true. Nevertheless, in practice, if one does not reject a hypothesis after a statistical test, one proceeds as if the hypothesis were true, even though one may suspect that it is not. Therefore, four possible events arise [3] [5]:

(1) H is true and, as $(\dot{x}_1, \dot{x}_2, \ldots, \dot{x}_n) \notin \Xi_n$, it is accepted by the test;
(2) H is false and, as $(\dot{x}_1, \dot{x}_2, \ldots, \dot{x}_n) \in \Xi_n$, it is rejected by the test;
(3) H is true but, as $(\dot{x}_1, \dot{x}_2, \ldots, \dot{x}_n) \in \Xi_n$, it is rejected by the test (*error of the first type*); and
(4) H is false but, as $(\dot{x}_1, \dot{x}_2, \ldots, \dot{x}_n) \notin \Xi_n$, it is accepted by the test (*error of the second type*).

In order to present a formal example of the procedure of testing a hypothesis, let η be a parameter of an assumed distribution and consider the simple parametric hypothesis H: $\eta = \zeta_0$, with ζ_0 a given constant; assume that, if the hypothesis is true, the PDF $p_{\zeta|\zeta_0,n}(\zeta)$ is known, ζ being a sample estimate. Let $\ddot{\eta} = \zeta_1$ be the estimate of the parameter η obtained on the basis of a sample of size n. Then, given a sample $(\dot{x}_1, \dot{x}_2, \ldots, \dot{x}_n)$, the problem reduces to stating a rule for deciding whether or not the relevant $\ddot{\eta} = \zeta_1$ is consistent with the distribution for ζ determined by H ($\eta = \zeta_0$). If ζ_1 is too small or too large with respect to ζ_0, it will be rather unlikely that $\tilde{\zeta}$ does have the distribution involved by H, and H is rejected.

The procedure requires first the selection of a small probability $p_1 = 2p'$ (called the *level of significance* of the test), then the finding of the lower and upper p'-fractiles $\zeta_{p'}$ and $\zeta_{1-p'}$ of $\tilde{\zeta}$ (Fig. 7.6): the hypothesis H is to be rejected if ζ_1 falls outside the interval $(\zeta_{p'}, \zeta_{1-p'})$, i.e. if either $\zeta_1 < \zeta_{p'}$ or $\zeta_1 > \zeta_{1-p'}$.

The hypothesis test outlined above is called a *two-sided test*: in fact, if H is not true, the actual value of η could be either greater or less than ζ_0. In other cases a *one-sided test* may be sufficient: for instance, the composite hypothesis $\eta \geqslant \zeta_0$ would be false only if $\zeta < \zeta_0$.

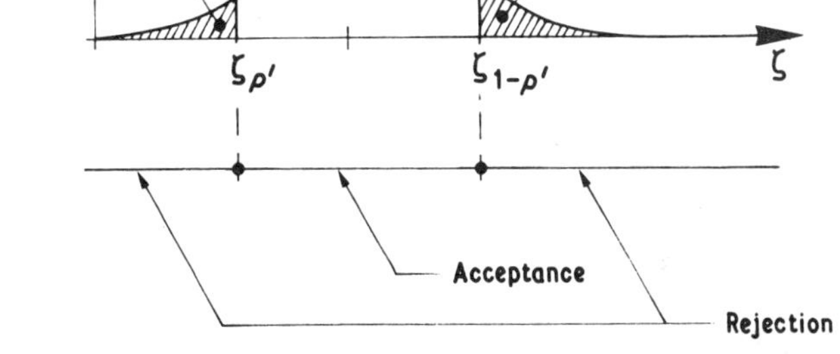

Fig. 7.6 Acceptance and rejection region for the test of statistical hypothesis H.

The theory provides no rules for selecting the value of p_1; usual values are in the range 1 % to 5 %. In fact the probability of a type I error is reduced by reducing the level of significance p_1, but in this way the probability p_2 of a type II error (i.e. the hypothesis is false but it is accepted by the test) is increased. The probability $1-p_2$ is called the *power of the test* and a diagram $(1-p_2)$ vs p_1 is called the *operating characteristic* of the test: an example of operating characteristic is plotted in Fig. 7.7.

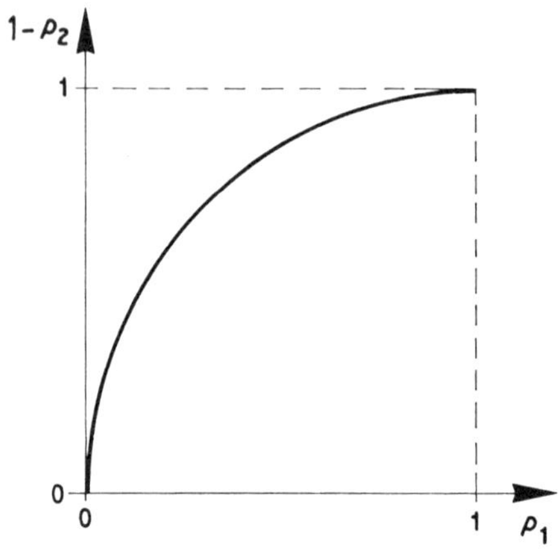

Fig. 7.7 Typical operating characteristic of the significance test of a statistical hypothesis.

However the only way to reduce both p_1 and p_2 is to increase the sample size n for the estimate η. This forms the basis for selecting the necessary sample sizes for statistical experiments (Section 7.6.6).

In conclusion, the problem of analysing the 'reliability' of a statistical hypothesis consists of checking whether the observed discrepancies from the assumed theoretical model depend either on the randomness of the result or on the inadequacy of the theory. The methods employed for this purpose are related to modern decision theory and to the analysis of variance, and are beyond the scope of this book. Only the most common procedures are summarized in Sections 7.6.3–7.6.5.

7.6.3 Tests for mean and variance of a normal distribution

Consider a normally distributed random variable $\tilde{x}$ (this assumption can often be justified by the central limit theorem (Section A4.4)). All sample values are normal variables and the distributions $p_{\zeta|\zeta_0,n}(\zeta)$ of the significant parameters can be calculated explicitly, as shown in the following. In particular, for any sample of size n drawn from a normal population with mean $\mu_{\tilde{x}}$ and variance $s^2_{\tilde{x}}$, the following results are available.

Recalling equation (7.10), the variable

$$u = \frac{m_n - \mu_x}{s_x/\sqrt{n}} \tag{7.36}$$

has a standardized normal distribution ($\mu_u = 0$; $s_u = 1$) (Table A.5).

Recalling equations (7.11) and (7.13), the variable

$$t_{n-1} = \frac{m_n - \mu_x}{s_n/\sqrt{(n-1)}} = \frac{m_n - \mu_x}{s_n^0/\sqrt{n}} \tag{7.37}$$

called the *Student's ratio*, has a t distribution with $n-1$ degrees of freedom: by definition this means that t^2 has an F distribution with 1 and $n-1$ degrees of freedom (see Table A.2 for the definition and Table A.8 for the values). Finally, the variable

$$\chi^2_{n-1} = \frac{(n-1)(s_n^0)^2}{s_x^2} = \frac{ns_n^2}{s_x^2} = \frac{1}{s_x^2}\sum_{j=1}^{n}(\hat{x}_j - m_n)^2 \tag{7.38}$$

has a χ^2 distribution with $n-1$ degrees of freedom (Table A.2 for the definition and Table A.7 for the values). Note that if $\tilde{z}$ is χ_n^2 random variable with n degrees of freedom, $(\tilde{z}/n)$ is asymptotically normal with mean 1 and variance $2/n$.

7.6.4 χ^2 goodness-of-fit test

The test of a non-parametric hypothesis (i.e. a hypothesis on the probability distribution model) is based on a comparison between the shape of either the PDF or the CDF and the histogram or cumulative frequency distribution respectively‡. Analytical tests are also available to compare data and model. The most usual is the so-called χ^2 *test* that is illustrated in the following.

Let $p_x(x)$ be the theoretical density function: the corresponding probability that $\tilde{x}$ lies in the class interval $x^i - \frac{1}{2}\Delta x^i < x < x^i + \frac{1}{2}\Delta x^i$ (expected class frequency) is

$$Y_i = \int_{x^i - \frac{1}{2}\Delta x^i}^{x^i + \frac{1}{2}\Delta x^i} p_{\tilde{x}}(x)\, dx \tag{7.39}$$

Let f_i be the frequency density of the considered sample (equation (7.8)). The discrepancy between the observed frequency f_i and the expected frequency Y_i within each class interval is given by $f_i - Y_i$. Thus, the quantity

$$X^2 = \sum_{i=1}^{k} \frac{(f_i - Y_i)^2}{Y_i} \tag{7.39'}$$

where k is the number of class intervals, can be introduced to measure the total discrepancy for all class intervals.

Under a suitable hypothesis, it is possible to show that the CDF for X^2 is

‡ In the case of histograms, the comparison is complicated by the dependence of the shape on the number of classes used.

approximately the same as for χ^2_{n-1} defined by equation (7.38). Now the number of degrees of freedom $n-1$ is equal to the number of class intervals k minus the number of different independent linear restrictions imposed on the observation. One such restriction is given by

$$\sum_{i=1}^{k} f_i = 1$$

hence

$$n = k - r \tag{7.40}$$

where r is the number of independent parameters of the assumed PDF, e.g. $r = 2$ (mean and variance) if the PDF is normal.

The usual way to apply the χ^2 goodness-of-fit test is to select class intervals so as to have equal expected frequencies Y_i within each interval. Table 7.5 presents the relevant computations for the Malpensa Airport example under the assumption that $\tilde{v}$ is normally distributed with the parameters given by equation (7.35). (Otherwise one could select class intervals of equal width, but in this case it is desirable to have at least five observations in each class interval.)

Table 7.5 χ^2 computation for the sample of yearly maxima of wind speeds at Malpensa Airport.

i	Y_i	$x^i - \frac{1}{2}\Delta x^i$ $(m\,s^{-1})$	$x^i + \frac{1}{2}\Delta x^i$ $(m\,s^{-1})$	n_i	$f_i = \dfrac{n^i}{n}$	$\dfrac{(f_i - Y_i)^2}{Y_i}$	$\sum_{j=1}^{i} n \dfrac{(f_j - Y_j)^2}{Y_j}$
1	0.20	$-\infty$	13.08	5	0.2174	0.0015	0.0348
2	0.20	13.08	14.75	5	0.2174	0.0015	0.0696
3	0.20	14.75	16.21	5	0.2174	0.0015	0.1045
4	0.20	16.21	17.87	2	0.0870	0.0638	1.5729
5	0.20	17.87	$+\infty$	6	0.2609	0.0185	1.9994

From Table 7.5, one derives that $\chi^2 = 1.9994$ for the assumed distribution of the wind speeds at Malpensa Airport. Then at the 5% significance level ($p_1 = 5\%$), the critical value $\chi^2_{p_1}$ coincides with the 95% fractile of the χ^2 distribution with $5-3=2$ degrees of freedom. From Table A.7, $\chi^2_{p_1} = 5.99$, a value much greater than the observed value of 1.9994. Therefore this test does not show a significant contradiction between the distribution of the data and the normal model: since we now form the literature that the distribution of $\tilde{v}$ is not normal, this is an example of a type-II error (but see Section 7.6.6 below).

7.6.5 Confidence region for cumulative frequency

Consider now a random variable $\tilde{x}$ with known CDF $P_x(x)$. For any x', it is possible to choose two positive values $F_L(x')$ and $F_U(x')$ such that

$$F_L(x') \leqslant P_x(x'); \qquad F_U(x') > P_x(x');$$
$$\text{Prob}[F_L(x') \leqslant F(x') < F_U(x')] = 1 - p_1 \tag{7.41}$$

where $F(x')$ is the cumulative frequency for $x = x'$ corresponding to an arbitrary sample and $(1 - p_1)$ is called the *confidence coefficient.* The values $F_L(x')$ and $F_U(x')$ are the lower and upper *confidence limits*, bounding a *confidence interval* for the frequency $F(x')$. The region between the curves $F_L(x)$ and $F_U(x)$ in the (x, F) plane is called the *confidence region* and contains with probability $1 - p_1$ each sample point $(F(\dot{x}_j), \dot{x}_j)$, $(j = 1, \ldots, n)$. Conversely, the confidence region can also be defined by associating, with any value F, the confidence limits x_L and x_U. Such procedures are usually employed, in addition to the procedure of Section 7.5, to perform an objective model verification.

In order to define the confidence region, introduce the relation

$$F = P_x(x_F) = P_u(u_F) \tag{7.42}$$

where $u_F = (x_F - \mu_x)/s_x$; and its inverse

$$u_F = \phi(F) \tag{7.43}$$

Therefore, as the jth sample involves that $F_j(x')$ corresponds to any values x' of x, so it also involves that $u'_{F_j} = \phi(F_j(x'))$ corresponds to x'. Thus u'_F takes its value at random and hence it can be regarded as a random variable $\tilde{u}'_F$ that under usual hypotheses can be assumed normally distributed with

$$\mu_{u_F} = u' \tag{7.44a}$$

$$s_{u_F} = \frac{1}{p_u(u')} \sqrt{\left(\frac{P_u(u')[1 - P_u(u')]}{n}\right)} \tag{7.44b}$$

where $u' = (x' - \mu_x)/s_x$ and $p_u(u) = \mathrm{d}P_{\tilde{u}}(u)/\mathrm{d}u$.

It follows that, for any confidence coefficient $(1 - p_1)$, it is possible to define a real and positive number c in such a way that u_F belongs, with probability $(1 - p_1)$, to the interval u'_{FL} to u'_{FU} where

$$\begin{aligned} u'_{FL} &= \mu_{u'_F} - c s_{u'_F} \\ u'_{FU} &= \mu_{u'_F} + c s_{u'_F} \end{aligned} \tag{7.45}$$

For instance, if $p_1 = 5\%$ (95% confidence) Table A.5 gives $c = 1.96$.

The r.h.s. of equation (7.44b), within the factor $\sqrt{n}$, is plotted in Fig. 7.8 for the normal distribution and the extreme type I (maxima) distribution. The use of such diagrams is very useful in the calculation of u'_{FL} and u'_{FU} by equation (7.45) when $\mu_{u'_F} = u'$ and k are assigned. Then $F_L(x')$ and $F_U(x')$ are derived by equation (7.42) and the points $(x', F_L(x'))$, $(x', F_U(x'))$ are plotted in the plane (x, F) to determine the confidence region. However, the approximations introduced in determining equation (7.44) make equation (7.45) reliable only if $0.15 < P_x(x') < 0.85$.

The 95% confidence region obtained for the Malpensa Airport example, when the wind speed is supposed to be normally distributed, is plotted in Fig. 7.3. Also this test does not show that the available data are in significant contradiction to the normal model.

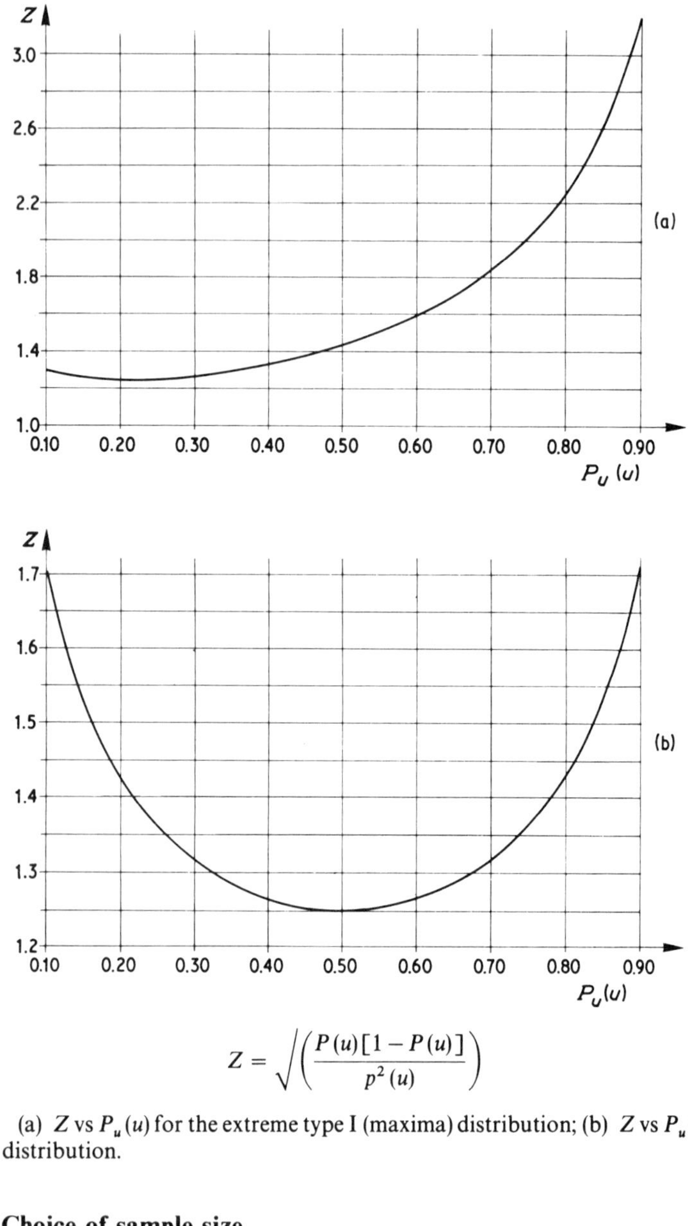

$$Z = \sqrt{\left(\frac{P(u)[1-P(u)]}{p^2(u)}\right)}$$

Fig. 7.8 (a) Z vs $P_u(u)$ for the extreme type I (maxima) distribution; (b) Z vs $P_u(u)$ for the normal distribution.

7.6.6 Choice of sample size

The previous discussion assumed that the sample size n was given in advance. The question now arises of the sample size necessary to make a test significant. The solution of this problem is not easy: in the following it is solved in a very particular case [6], once more with reference to the Malpensa Airport example.

There is reason to believe that the mean value μ_v of the random variable $\tilde{v}$ is (equation (7.35)) 15.4792 m s^{-1}. Further, assume that the standard deviation is known to be $s_{\tilde{v}} = 2.85$ m s^{-1}. What is the proper sample size to test the hypothesis that $\mu_v = 15.4792$ m s^{-1} at the 5% level of significance ($p_1 = 2p' = 5\%$), with a probability $p_2 = 5\%$ of not detecting a difference of 10% between estimated and actual value (type II error)?

An appropriate estimate for μ_v is m_n (equation (7.10)), whose sampling distribution is given by equation (7.36). Therefore the upper and lower limits of the acceptance region for the test are

$$\begin{aligned} \zeta_{p'} &= \frac{s_v}{\sqrt{n}} u_{p'} + \mu_v \\ \zeta_{(1-p')} &= \frac{s_v}{\sqrt{n}} u_{(1-p')} + \mu_v \end{aligned} \tag{7.46}$$

If now the true mean value is in the interval $\mu'_v = \mu_v \pm \Delta$, type II error would occur with probability p_2 if $m_n < \zeta_{(1-p)}$ or $m_n > \zeta_{p'}$, and hence

$$\begin{aligned} \zeta_{p'} = \zeta_{1-p_2} &= \frac{s_v}{\sqrt{n}} u_{(1-p_2)} + \mu_v - \Delta \\ \zeta_{(1-p')} = \zeta_{p_2} &= \frac{s_v}{\sqrt{n}} u_{p_2} + \mu_v + \Delta \end{aligned} \tag{7.47}$$

From equation (7.46) and (7.47), one derives

$$-u_{p_2/2} = u_{(1-p')} = u_{p_2} + \frac{\sqrt{n}}{s_v}\Delta = -u_{(1-p_2)} + \frac{\sqrt{n}}{s_v}\Delta \tag{7.47'}$$

and finally

$$n = \left(\frac{s_v(u_{(1-p')} + u_{(1-p_2)})}{\Delta} \right)^2 \tag{7.48}$$

In the analysed example, $\Delta = 10\% \times 15.4792 = 1.5479$ m s^{-1}, and hence

$$n = \left(\frac{2.85}{1.5479} \right)^2 (1.96 + 1.654)^2 = 44.06$$

is the required sample size. Therefore the available sample of size 23 is quite unsatisfactory to meet the fixed requirements, and this result alone makes the successes of the previous tests irrelevant.

7.7 REGRESSION AND CORRELATION ANALYSIS

If paired samples of two random variables $\tilde{x}$ and $\tilde{y}$, related to each other in some way, are available, two problems can be of interest [7]:

(1) to determine a functional relationship $y = y(x)$, in order to predict which value $\tilde{y}$ will take if $\tilde{x} = x$; and
(2) to investigate the correlation (see Section A2.4) between $\tilde{y}$ and $\tilde{x}$: then the correlation coefficient gives a numerical measure of how useful a (formally deterministic) law between $\tilde{y}$ and $\tilde{x}$ is‡.

7.7.1 Data observed in pairs

The plot of observed pairs of values of two random variables $\tilde{y}$ and $\tilde{x}$ is called a *scattergram*. The scattergram of the yearly maxima of the wind speeds averaged over 10 min at the two Milan airports (Malpensa and Linate) [1] for each year from 1951 to 1973 (Table 7.6) is shown in Fig. 7.9.

Table 7.6 Yearly maxima of the wind speeds averaged over 10 min at the Milan Airports (Malpensa and Linate).

Year	*Malpensa* $v\,(m\,s^{-1})$	*Linate* $v_L\,(m\,s^{-1})$	*Year*	*Malpensa* $v\,(m\,s^{-1})$	*Linate* $v_L\,(m\,s^{-1})$
1951	18.468	16.416	1963	11.286	10.260
1952	16.416	10.260	1964	12.825	13.851
1953	14.364	29.241	1965	12.825	11.286
1954	12.825	11.799	1966	15.390	13.851
1955	11.799	27.189	1967	19.494	13.338
1956	14.364	10.260	1968	15.390	13.851
1957	18.468	9.747	1969	14.364	12.312
1958	15.390	11.286	1970	15.390	12.825
1959	13.851	27.702	1971	18.468	15.390
1960	13.338	14.364	1972	15.390	13.338
1961	21.546	19.494	1973	17.955	23.598
1962	16.416	14.877			

Given a random sample of size $n(\grave{x}_j, \grave{y}_j)$ the following statistics can be defined:

(1) the *sample averages*

$$m_n^x = \frac{1}{n}\sum_{j=1}^{n} \grave{x}_j \qquad m_n^y = \frac{1}{n}\sum_{j=1}^{n} \grave{y}_j \tag{7.49}$$

(2) the *sample variances*

$$(s_n^x)^2 = \frac{1}{n}\sum_{j=1}^{n} (\grave{x}_j - m_n^x)^2 \qquad (s_n^y)^2 = \frac{1}{n}\sum_{j=1}^{n} (\grave{y}_i - m_n^y)^2 \tag{7.50}$$

(3) the *sample covariance*

$$s_n^{x,y} = \frac{1}{n}\sum_{j=1}^{n} (\grave{x}_i - m_n^x)(\grave{y}_j - m_n^y) \tag{7.51}$$

‡ It is worth noting that the correlation coefficient has meaning only if there is a sound basis for believing that there is a relationship between $\tilde{y}$ and $\tilde{x}$: in fact, the correlation coefficient between arbitrary samples of two variables may appear to be significantly different from zero also when it is clear that no relationship exists between the underlying random variables.

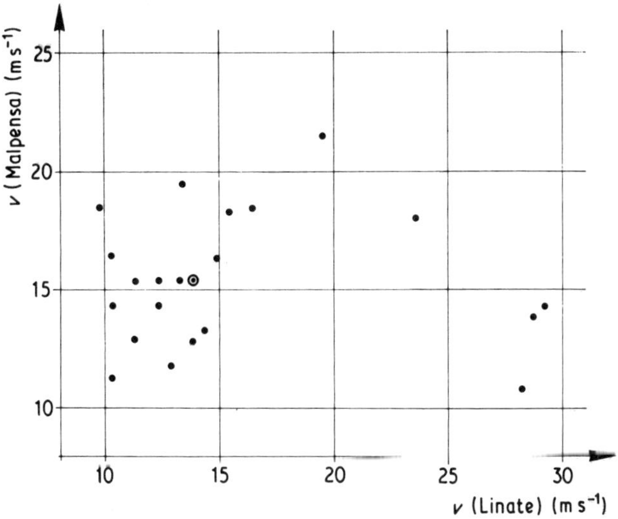

Fig. 7.9 Scattergram of the wind speed data at Malpensa and Linate Airports.

(4) the *sample linear correlation coefficient*

$$r_n^{x,y} = \frac{1}{n}\sum_{j=1}^{n} \frac{\grave{x}_j - m_n^x}{s_n^x}\,\frac{\grave{y}_j - m_n^y}{s_n^y} \tag{7.52}$$

that can be shown to be limited in value to $-1 \leqslant r_n^{x,y} \leqslant 1$.

With reference to the example of Table 7.6, these statistics assume the values:

$$m_{23}^{v} = 15.4792\,\mathrm{m\,s^{-1}};\quad m_{23}^{v_L} = 15.5015\,\mathrm{m\,s^{-1}};\quad s_{23}^{v} = 2.5612\,\mathrm{m\,s^{-1}};$$
$$s_{23}^{v_L} = 5.7404\,\mathrm{m\,s^{-1}};\quad s_{23}^{v,v_L} = -2.5974\,\mathrm{m^2\,s^{-2}};\quad r_{23}^{v,v_L} = -0.1767$$

7.7.2 Regression and correlation

It is natural to believe that, if the points $(\grave{x}_j, \grave{y}_j)$ lie on a straight line, a definite linear functional relationship exists between $\tilde{x}$ and $\tilde{y}$. This would be a very favourable circumstance that would immediately assess dependence between the two variables. Most often the sample is scattered in the plane (x, y), and the degree to which functional dependence influences the association of values of y to x must be investigated.

For this purpose, consider the following relation:

$$Y(x) = C_0 + C_1 x \tag{7.53}$$

If the coefficients C_0 and C_1 are calculated by minimizing the quadratic deviation

$$\phi_y = \sum_{j=1}^{n} [\grave{y}_j - Y(\grave{x}_j)]^2 \tag{7.54}$$

(see Section 7.5.3), the straight line (7.53) is called the *linear regression line* of y on x. However, if linear dependence exists between y and x, x should also depend linearly on y. Hence

$$X(y) = C_2 + C_3 y \tag{7.55}$$

is the regression line of x on y. The coefficients C_2 and C_3 are again calculated by minimizing

$$\phi_x = \sum_{j=1}^{n} [\grave{x}_j - X(\grave{y}_j)]^2 \tag{7.56}$$

It is soon found that, introducing the estimated means (7.49), the straight lines (7.53) and (7.55) are written as

$$Y(x) = m_n^y + C_1(x - m_n^x) \qquad X(y) = m_n^x + C_3(y - m_n^y) \tag{7.57}$$

with

$$C_1 = \frac{s_n^{x,y}}{(s_n^x)^2} \qquad C_3 = \frac{s_n^{x,y}}{(s_n^y)^2} \tag{7.58}$$

The two lines intersect at the point (m_n^x, m_n^y); if $C_1 C_3 = 1$ they coincide, and linear functional dependence between x and y can be assumed.

In real cases the product $C_1 C_3$ will differ somewhat from unity: the closer the product to unity, the sharper the functional relationship. Hence, it is natural to measure the degree of linear dependence by the square root of $C_1 C_3$, which coincides in absolute value with the sample coefficient of linear correlation defined by equation (7.52):

$$r_n^{x,y} = \pm\sqrt{(C_1 C_3)} \tag{7.59}$$

In order to get an idea of what this coefficient means, note that the two regression lines, Z_1 and Z_2 (equations (7.53) and (7.55)), coincide when $C_1 C_3 = 1$ ($r_n^{x,y} = \pm 1$) and are orthogonal when $C_1 = C_3 = 0$ ($r_n^{x,y} = 0$). In Fig. 7.10 the position of the two lines is drawn corresponding to some values of $r_n^{x,y}$.

It is perhaps necessary to note explicitly that the above defined coefficient yields information only on *linear correlation*, and not on general statistical or functional dependence of the two variables. To demonstrate this, consider for instance two random variables related by a circular relationship

$$\begin{aligned} \tilde{x} &= R\cos\tilde{\theta} \\ \tilde{y} &= R\sin\tilde{\theta} \end{aligned} \tag{7.60}$$

with R a numerical constant and $\tilde{\theta}$ a random variable. In this case, if $\grave{\theta}_j$ are n observed values of $\tilde{\theta}$ uniformly scattered on the interval $(0, 2\pi)$, one gets

$$\begin{aligned} m_n^x &= \frac{1}{n} R \sum_{j=1}^{n} \cos\grave{\theta}_j = 0 \\ m_n^y &= \frac{1}{n} R \sum_{j=1}^{n} \sin\grave{\theta}_j = 0 \\ s_n^{x,y} &= R^2 \sum_{j=1}^{n} \cos\grave{\theta}_j \sin\grave{\theta}_j = 0 \end{aligned} \tag{7.61}$$

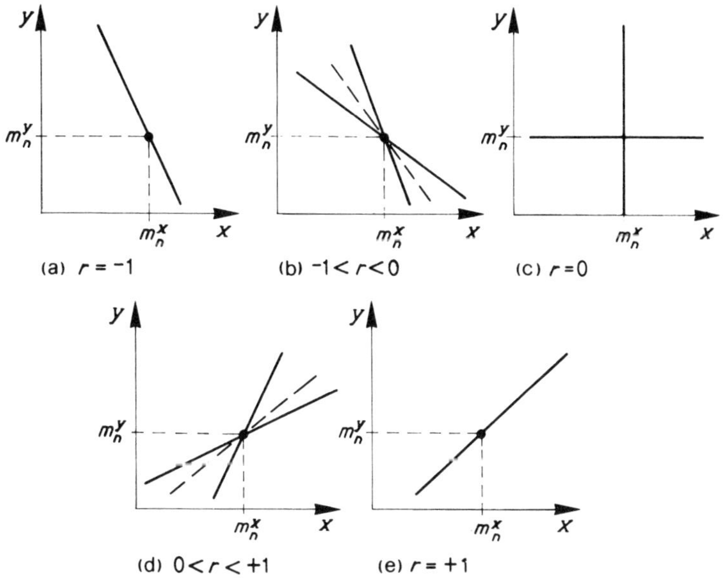

Fig. 7.10 Relative position of the two regression lines for some values of the coefficient of linear correlation ($r = r_n^{x,y}$, equation (7.52)).

and hence

$$r_n^{x,y} = 0$$

This indicates only that no linear dependence exists, despite the fact that $\tilde{x}$ and $\tilde{y}$ are functionally dependent.

7.7.3 Estimation of the linear correlation coefficient

The sample linear correlation coefficient $r_n^{x,y}$ defined by equation (7.52) is an estimator of the correlation coefficient $\rho_{x,y}$ (equation (A2.37)) of the random variables $\tilde{x}$ and $\tilde{y}$, whose joint probability density function is $p_{x,y}(x, y)$.

The confidence in such an estimator can be investigated by referring to *Fisher's function*

$$\omega(r) = \tfrac{1}{2} \ln\left(\frac{1+r}{1-r}\right) \tag{7.62}$$

Since $r_n^{x,y}$, the *a priori* value of the estimator is a random variable because of the randomness of the observations in the sample; also $\tilde{\omega} = \omega(\tilde{r}_n^{x,y})$ is therefore a random variable, which can be proved to be approximately normal with mean and variance given by

$$\begin{aligned} \mu_\omega &= \tfrac{1}{2} \ln\left(\frac{1+\rho_{x,y}}{1-\rho_{x,y}}\right) \\ s_\omega^2 &= \frac{1}{n-3} \end{aligned} \tag{7.63}$$

$\rho_{x,y}$ being the actual unknown value of the coefficient of correlation.

The confidence interval for the correlation coefficient $(\rho_{x,y}-\varepsilon, \rho_{x,y}+\varepsilon)$ with level of significance p_1, can be determined by solving the system

$$p_a = \text{Prob}\left[\tilde{u} \leqslant \frac{\omega(r_n^{x,y}) - \omega(r_n^{x,y}+\varepsilon)}{s_\omega}\right]$$

$$p_b = \text{Prob}\left[\tilde{u} \leqslant \frac{\omega(r_n^{x,y}) - \omega(r_n^{x,y}-\varepsilon)}{s_\omega}\right] \tag{7.64}$$

$$p_1 = p_a + p_b$$

where $\tilde{u}$ is the standardized normal variable.

REFERENCES

1. ENEL (Ente Nazionale Energia Elettrica) (1967) *Dati Anemometrici Risultanti dalle Registrazioni Effettuate dal Servizio Meteorologico dell' Aeronautica* (in Italian), ENEL, Milan.
2. Gnedenko, B. V. (1968) *Theory of Probability*, Chelsea, New York.
3. Benjamin, J. R. and Cornell, C. A. (1970) *Probability, Statistic and Decision for Civil Engineers*, McGraw-Hill, New York.
4. Gumbel, E. J. (1967) *Statistics of Extremes*, Columbia University Press, New York.
5. Hald, A. (1952) *Statistical Theory with Engineering Applications*, Wiley, New York.
6. Hadley, G. (1969) *Elementary Statistics*, Holden-Day, San Francisco.
7. Veneziano, D. (1978) *Random Processes for Engineering Application* (lecture notes), Department of Civil Engineering, M.I.T.

CHAPTER 8

Statistical analysis of random functions

The theoretical framework of stochastic processes is dealt with in Appendices A5 and A6 and summarized in Chapter 2. Here techniques to choose the most suitable models from observed data and to calculate the basic model parameters are dealt with leaving details and more sophisticated arguments to specialist books.

8.1 PROPERTIES OF STRUCTURAL PARAMETERS

Environmental actions on structures and their own properties evolve in time. Hence, they can be described by simple random variables only in a first approximation (sometimes appropriate and useful, sometimes quite misleading).

That many relevant parameters are stochastic processes is easily inferred from very common examples. Loading processes have already been discussed in Chapter 3, but uncertainty is also present in the evolution of material properties. The curing of concrete that influences the material constitutive properties; the relaxation of steel and the creep of concrete in prestressed concrete structures; the setting of the mortar in masonry structures; the corrosion of iron bars in reinforced concrete: all these phenomena are, more or less, affected by uncertainty. Nevertheless, each of them exhibits some well defined characteristics that must be reproduced in the mathematical description.

In some cases (e.g. a single quasi-static load acting on a non-aging structure), the knowledge of the distribution of the maximum value of some definite quantity (e.g. the load intensity) during a given period is sufficient for practical purposes. Then, if the value in a previous period does not influence the subsequent one, such models are conveniently treated by the results of the so-called *statistics of extremes*

(Sections A7.1–A7.3). In these cases the required sample consists of a set of maxima and the techniques developed in the previous chapter apply, as shown by the Malpensa Airport example. In other cases (e.g. several time-varying loads acting quasi-statically on a structure, or cases in which the dynamic response is of importance in the design process), the whole stochastic process must be modelled on the basis of the observed data. This topic is the object of this chapter.

8.2 REMARKS ON DATA ACQUISITION

Many questions arise in stochastic data acquisition, i.e. the set of operations aimed at making available the sample function $\dot{x}(t)$, $t \in (0,T)$, of the underlying process $\tilde{x}(t)$. Several sources of error can, in fact, be recognized in the many phases leading to the final sample (often required in digitized form). Collection, recording and preparation are the phases that can be recognized in data acquisition.

Errors in data collection are strictly related to the instrument needed to convert the physical quantity to be measured to an analogue quantity whose intensity can be quantified and recorded at a sufficient speed. When seismic acceleration is to be recorded, for instance, the most suitable technique is to convert seismic acceleration into an electrical difference of potential that can be measured directly by a voltmeter and recorded on magnetic tape: the instrument used is the seismometer (Fig. 8.1), and the very existence of a wide literature on the subject demonstrates the existence of a potential source of errors, connected with the frequency response function of the mechanical system.

Recording of data also poses delicate problems. First, the tape speed must be kept constant as much as possible; random variations of the speed would cause distortions in the recorded input. Moreover, the tape magnetization process is basically non-linear and discrepancies are produced from recording to reproducing.

Special problems arise when a record is digitized. Essentially, a discretization effect is introduced, in that a continuous phenomenon is now represented by a set of pairs $(t_i, \dot{x}_i(t_i))$ $(i = 1, \ldots, N)$, $\dot{x}(t_i)$ being the sample value at time t_i. For instance, let $v = t_i - t_{i-1}$ be the time interval chosen for the digitization (and kept the same for all t_i). In order that changes in the sign of $\dot{x}(t)$ be shown by the couples $(t_i, \dot{x}(t_i))$, it is necessary that no more than one zero-crossing occurs in the time v. This leads one to assess that the highest frequency ν_k that can be recognized in the digitized sample (Nyquist frequency) is not larger than $1/(2v)$. The power associated with higher frequencies in the sample is distributed over the lower frequencies (Fig. 8.2): this phenomenon is called *aliasing* [1]. Frequencies in the original data above $1/(2v)$ are *folded* back into the frequency range from 0 to $1/(2v)$ and confused with data in this lower range. Two practical methods allow one to tackle this problem:

(i) choose v sufficiently small so that it is physically unreasonable that $\nu > \nu_k$ in the data;

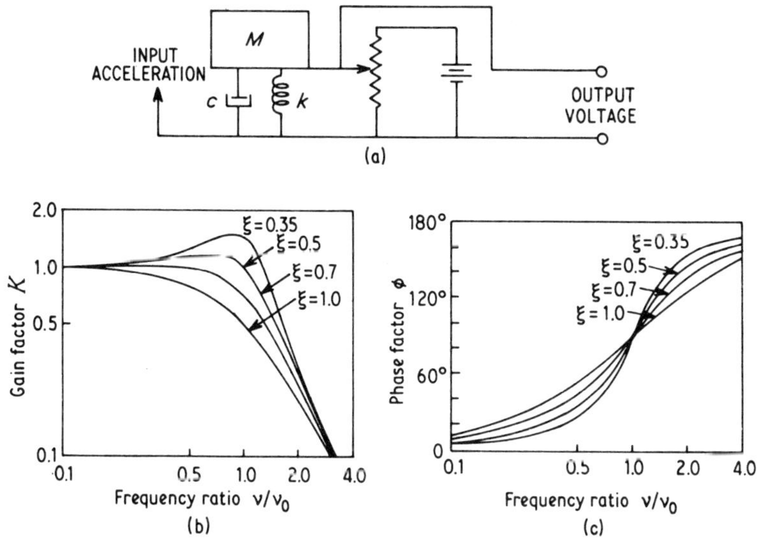

Fig. 8.1 Diagram and typical frequency response characteristics of open loop seismograph. (a) Diagram of idealized accelerometer.

(b) Accelerometer gain factor $K = K|A(\nu)| = \left(\left[1-\left(\frac{\nu}{\nu_0}\right)^2\right]^2 + \left[2\xi\frac{\nu}{\nu_0}\right]^2\right)^{-1/2}$

(c) Accelerometer phase factor $\phi = \tan^{-1}\left(2\xi\frac{\nu}{\nu_0}\Big/\left[1-\left(\frac{\nu}{\nu_0}\right)^2\right]\right)$

ν_0 = undamped natural frequency; ξ = damping coefficient (from Bendat and Piersol [1]).

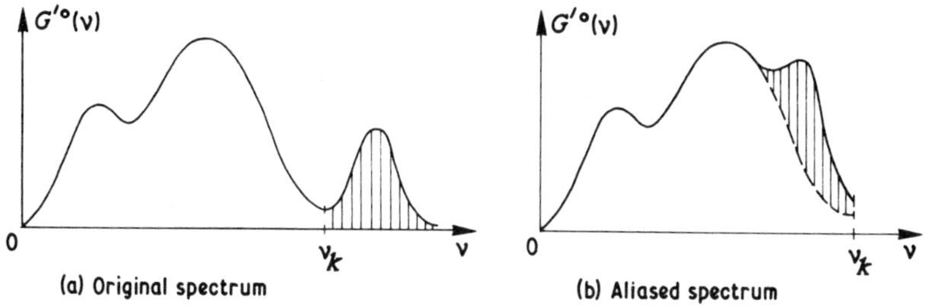

Fig. 8.2 Power spectrum of a record modified by *folding:* (a) true spectrum; (b) aliased spectrum.

(ii) filter the data before sampling so that $\nu > \nu_k$ is no longer contained in the filtered data (see Section 8.3).

On the other hand, the interval v cannot be reduced indefinitely because in that case small errors in the recorded values $\dot{x}(t_i)$ would introduce large errors in the frequency analysis. Moreover, values at very close instants would be highly correlated and would introduce useless increases in cost and length of calculations. Finally, obvious limitations in the number of digits that can be used to represent a numerical value make it practically impossible to reduce the digitization interval beyond a certain limit.

Details on the above problems must be left to specialist books [1]: however, the above summary shows that this is a difficult matter that must be faced with much care.

8.3 TREND REMOVAL AND FILTERING

Often, some particular properties of the random process $\tilde{x}(t)$ are to be investigated in detail.

Consider first the so-called *trend* in a sample function. Specifically, a trend is defined as any component in the data which is non-periodic, or at least has a period longer than the length of the sample record: often the trends indicate the *deterministic* part of the process. The least-squares method (see Section 7.5.3) is widely applied for detection and, possibly, elimination of linear quadratic, cubic, or general polynomial trends (see for example Fig. 8.3).

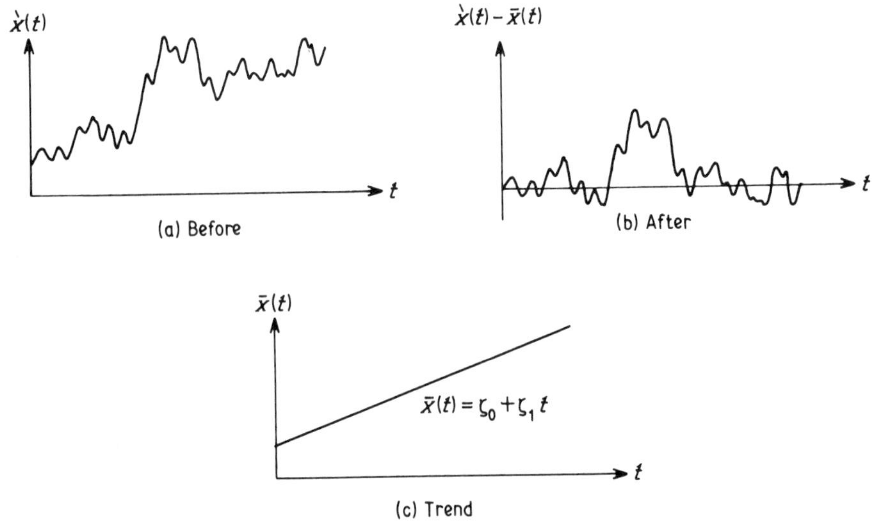

Fig. 8.3 Linear trend removal (typical example): (a) original record; (b) record after elimination of linear trend; (c) linear tend [1].

Trend removal allows isolation of random aspects of the sample data. In particular, after polynomial components have been removed, the significant periodic components can be recognized in the sample function.

Filtering helps in recognizing such periodic components (see for example Fig. 8.4). For instance, let $\grave{x}(t)$ be the sample record and let $\grave{y}(v, t)$ be the output

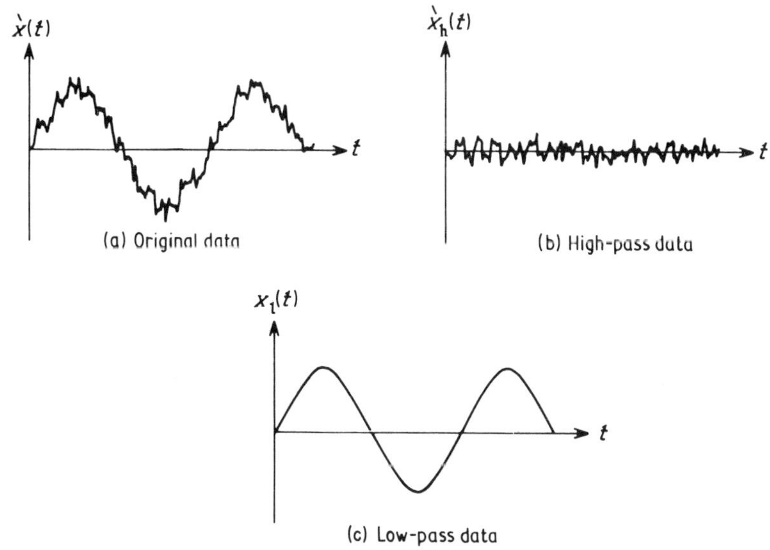

Fig. 8.4 High-pass and low-pass filtering (after [1]).

obtained by passing $\dot{x}(t)$ through a narrow-band filter (Section A6.1), whose bandwidth b is centred on the frequency v where the presence of a periodic component is suspected. The spectral density of the output $\dot{y}(v, t)$ exhibits a peak corresponding to the frequency v and the area of this peak tends to be invariant as the bandwidth b approaches zero.

Thus, filtering is useful to transform a sample function with the purpose of analysing the properties of random processes in connection with some specific frequency intervals or of focussing attention on some special features of the sample function.

Filters can be classified from the point of view of their action on power spectra: Fig. 8.5 (see also Table 2.3) shows both ideal filters and models employed for numerical purposes. *Low-pass filters* are those that eliminate, from the input data, the power displayed at frequencies above a given limit v_2. *High-pass filters* have the opposite effect: they cut off all frequency components under a given threshold value v_1 (cf. also Fig. 8.4). An intermediate category is the *band-pass filter*, which only leaves power in the output at frequencies contained in a given interval (v_1, v_2).

For instance, when $\tilde{x}(t)$, the random process to be analysed, is the action of some dynamic excitation on structures, it may be convenient to eliminate from data the higher-frequency components, which only act as a disturbance in data analysis and have no effect on the structure; on the other hand, low-frequency components can be treated as static actions. Thus, dynamic effects are attributed to some central part of the original spectrum: filtering through a suitable band-pass filter, besides assessing statistical input properties, may therefore be convenient in order to isolate the characteristics that effect the dynamic response of the structure.

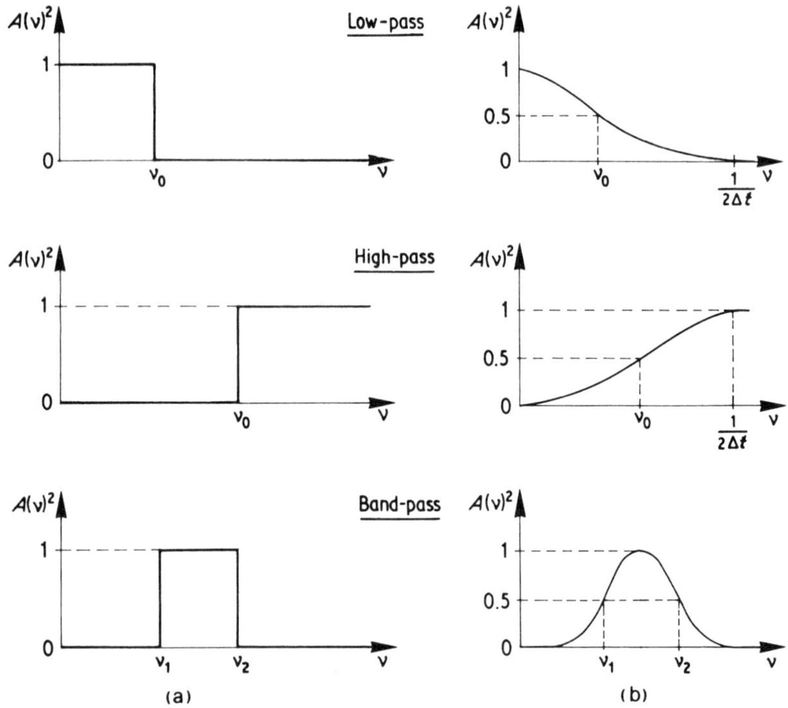

Fig. 8.5 (a) Characteristics of ideal filters. (b) Characteristics of digital filters. $A(\nu)$ is the frequency response function of the filter (equation (A5.41b)).

8.4 TEST FOR STATIONARITY

One of the most important things to verify on a sample function $\dot{x}(t)$ is the assumption of stationarity of the underlying stochastic process $\tilde{x}(t)$ (Appendix A5 and Section 2.3.1).

Most times the hypothesis of stationarity is introduced only to simplify treatment and, possibly, simulation: for example, many stationary models of earthquake acceleration have been developed and successfully used in seismic engineering, despite the fact that seismic accelerograms have long since been recognized to exhibit strong non-stationary character.

In a number of cases, the stationarity of a stochastic process is suggested by the mechanism causing the phenomenon. If it can be recognized that the basic source does not evolve in time, i.e. the general environmental conditions are constant in time, then it appears logical to believe that stationarity holds. For instance, if the bumps of the asphalt covering of a road are being recorded, there is reason to accept that the record should exhibit the same character for the whole length of road built by the same technique. In contrast, the process will not be stationary if, for instance, the average composition of the covering varies along the road.

The usual case, however, is that one must decide about stationarity on the basis of observation of a single sample function.

This can be accomplished under two conditions. First, the length of the sample should be sufficiently large to show up any possible non-stationary characters; in particular, it must be pointed out that if the non-stationary trend needs a very long time to reveal itself compared to the time for which the random excitation can act on the structure, the non-stationarity can be neglected. The process of occurrence of earthquakes can be viewed as an example of this latter category: since geological transformations require a very long time in comparison to the lifetime of man-made facilities, the process can be considered stationary for seismic engineering purposes, despite the fact that, on the geological timescale, significant trends may be present. The second condition is that the non-stationary character of the process is reflected by the available sample record.

It is usually difficult to assess full stationarity; on the other hand, weak stationarity (equation (A5.5)) only requires that the expected value and the autocorrelation function are time-independent.

A usual technique for testing stationarity using a single sample record of length T is summarized in the following. Let

$$\Delta t = \frac{T}{N} \ll T \tag{8.1}$$

be a time interval such that no statistical dependence exists between any two segments of length Δt of the sample. For a stationary process, if v_{ext} is the extinction time of the autocorrelation function (Section 2.4.3), it is sufficient to take $\Delta t \gg v_{\text{ext}}$. It is thus possible to divide the sample into N subsequent samples, statistically independent from each other. The mean square value (Section 2.4.4)

$$b_i' = \int_{t_i}^{t_i+\Delta t} \dot{x}^2(t)\mathrm{d}t \tag{8.2}$$

can then be computed for each segment.

The problem is now to decide if these sample values are all realizations of the same random variable. If it is so, the process can be considered stationary. The question can be investigated, for instance, by using the so-called *run test.*

Since the sequence is discrete, it is possible to define a numerical value of the mean square value, say b_0', that divides the sequence into two sets of equal size, the first set defined by the condition $b_i' > b_0'$ and the second set defined by $b_i' < b_0'$ (b_0' is the *median* value). Then, it is possible to associate to the sequence b_i' a binary sequence by taking the value 0 if $b_i' > b_0'$ and the value 1 if $b_i' < b_0'$. For instance, for $N = 14$, a sequence of the type shown in Table 8.1 is obtained.

Then define the *run number* r of the sequence as the number of sequential states in the binary variable. In the example of Table 8.1, $r = 9$. Note however that the run number is the realization of a random variable $\tilde{r}$, because of the randomness of the process underlying the sample function.

It can be proved that, if the calculated values of the mean square value are all

Table 8.1 Example of runs about median.

$i =$	1	2	3	4	5	6	7	8	9	10	11	12	13	14
Binary sequence	0	1	1	0	1	0	0	0	1	0	1	1	1	0
Run number	1	2		3	4	5			6	7	8			9

independent realizations of the same random variable, the distribution function of the run number $\tilde{r}$ is independent of the distribution of the underlying random variable $\tilde{b}'_{\Delta t}(0)$ (see equation (8.5)) but depends only on the number N of sample values. From a table of the run distribution, the interval can be found within which the run number falls with a given probability $1-p_1$: if the calculated value of the run number falls in this interval, then the assumption of stationarity can be accepted at the p_1 level of significance (Section 7.6.2). For example, Table A.9 in Appendix A.8 shows that, with $N = 14$ and equal number $N_0 = N_1 = N/2$ of 0's and 1's in the binary sequence, at the 5 % level the run number must be larger than 3 and smaller than 12: since in Table 8.1 $r = 9$, the test suggests to accept the stationarity of the process under investigation.

8.5 STATISTICAL ESTIMATION OF PARAMETERS OF STATIONARY RANDOM PROCESSES

Let $\dot{x}(t)$, $t \in (0, T)$, be a sample function from a random process known or accepted to be stationary. Then, the property of ergodicity (Appendix A5 and Section 2.3.2) must be assumed; in fact, it is necessary to get estimators of basic parameters by time-averaging (equations (A5.6) and (A5.13)) rather than by ensemble-averaging (equations (A5.2) and (A5.3)), because such estimation must be based on only one sample function. It is worth while to note, however, that ergodicity is often coupled with stationarity in at least two very common cases: Gaussian random processes (Section A6.2) with continuous power spectral density, and Gaussian Markov processes (Section A6.4). In general, weak ergodicity follows from weak stationarity under assumptions that are frequently encountered in practice (Section A5.1).

Thus, apart from unusual cases, estimation of basic parameters in stationary processes is performed by time-averaging over a single sample function.

8.5.1 Mean value of a stationary–ergodic random process

Consider the sample function $\dot{x}(t)$, $t \in (0, T)$. An estimator of the mean value of the underlying process $\tilde{x}(t)$ may be defined by (equation (A5.6)):

$$m_T = \frac{1}{T} \int_0^T \dot{x}(t)\, \mathrm{d}t \tag{8.3}$$

It is soon noted that m_T is a realization of a random variable $\tilde{m}_T$ as different realizations of $\tilde{x}$ are related to different realized sample functions.

The expected value of the estimator (8.3) is given by

$$\mathrm{E}[\tilde{m}_T] = \mathrm{E}\left[\frac{1}{T}\int_0^T \dot{x}(t)\,\mathrm{d}t\right] = \frac{1}{T}\int_0^T \mathrm{E}[\tilde{x}(t)]\,\mathrm{d}t = \mu_{x(t)} \tag{8.4}$$

proving that m_T is an unbiased estimate of $\mu_{\tilde{x}(t)}$ (Section 7.4.1).

Further, the definition of ergodic process (equation (A5.7)) implies that such an estimator is also consistent (see equation (7.20′)).

8.5.2 Mean square value of a stationary–ergodic process

Consider the following estimator of the mean square value $\mathrm{E}[\tilde{x}^2(t)]$ of the process (equations (A5.13) and (A5.5)):

$$b'_T(0) = \frac{1}{T}\int_0^T \dot{x}^2(t)\,\mathrm{d}t \tag{8.5}$$

Also $b'_T(0)$ is a random variable, $\dot{x}(t)$ being a realization of the random function $\tilde{x}(t)$. Then it is soon proved that

$$\mathrm{E}[\tilde{b}'_T(0)] = \mathrm{E}[\tilde{x}^2(t)] \tag{8.6}$$

showing that $b'_T(0)$ is an unbiased estimator. The estimator is also defined by equation (A5.14) written with $b'_T(0)$ instead of $b_T(0)$.

8.5.3 Autocorrelation function of a stationary–ergodic process

If the data exist for time $T+v$‡, instead of only for time T, the following estimator of the autocorrelation function $B'(v)$ can be introduced:

$$b'_T(v) = \frac{1}{T}\int_0^T \dot{x}(t)\dot{x}(t+v)\,\mathrm{d}t \tag{8.7}$$

It can be proved to be an unbiased estimator and, by the relevant equation (A5.14), a consistent estimator. One should expect an improvement of efficiency of the estimation when irregularly or randomly spaced data are used. In this case, in fact, information on the autocorrelation function is distributed more 'evenly' on the time axis. However, very sophisticated statistical techniques have been developed for this purpose.

8.5.4 Spectral density estimate of a stationary–ergodic process with zero mean value

Assume, for simplicity and without loss of generality, that $\mu_x(t) = 0$. Consider the following estimator of the actual ordinate $G'^{0}_{x(t)}(v)$ of the one-sided spectral density

‡ Otherwise, T must be replaced by $T-v$ in equation (8.7).

at the frequency ν:

$$g_T'^0(\nu) = \frac{1}{Tb_r}\int_0^T \dot{x}_f^2(t, \nu, b_r)\,dt \tag{8.8}$$

$\dot{x}_f(t, \nu, b_r)$ being the result of filtering (Section 8.3) the sample function $\dot{x}(t)$ through a narrow-band filter, whose bandwidth is b_r and is centred on the frequency ν of interest. The quantity b_r is also named the *resolution* bandwidth.

The estimator $g_T'^0(\nu)$ is the realization of the random variable $\tilde{g}_T'^0(\nu)$ in an analogous way to previous cases. It can be proved that

$$E[\tilde{g}_T'^0(\nu)] = \frac{1}{b_r}\int_I G_{x(T)}'^0(\nu)\,d\nu \tag{8.9}$$

I being the interval $(\nu - b_r/2,\ \nu + b_r/2)$.

Thus, unless b_r can be made very small,

$$E[\tilde{g}_T'^0(\nu)] \neq G_{x(t)}'^0(\nu) \tag{8.10}$$

and the estimator (8.8) is a biased estimator. The bias of the estimate, $b[\tilde{g}_T'^0(\nu)]$ (equation (7.22)), can be approximated as follows

$$B[\tilde{g}_T'^0(\nu)] \simeq \frac{b_r^2}{24}\left.\frac{d^2 G_{x(t)}'^0(\nu')}{d\nu'^2}\right|_{\nu'=\nu} \tag{8.11}$$

It turns out that the bias is proportional to the '*peakedness*' of the power spectrum. When the actual spectral density exhibits very pronounced peaks, the error of the estimator (8.8) is large; therefore, it is more suited for smooth spectral densities.On the other hand, the variance of the estimator is yielded by

$$\mathrm{Var}[\tilde{g}_T'^0(\nu)] \simeq \frac{[G_{x(t)}'^0(\nu)]^2}{Tb_r} \tag{8.12}$$

Therefore, the total quadratic error $\mathscr{E}_n$ (equation (7.21)) is given by

$$\mathscr{E}_n \simeq \frac{[G_{x(t)}'^0(\nu)]^2}{Tb_r} + \frac{b_r^2}{24}\left.\frac{d^2 G_{x(t)}'^0(\nu')}{d\nu'^2}\right|_{\nu'=\nu} \tag{8.13}$$

proving that a reduction in the filter bandwidth also reduces the bias term, but increases the pure random error.

An alternative approach to estimate the spectral density of a random process makes use of the finite Fourier transform. Assume that a sample record $\dot{x}(t)$ of unlimited length is available for a stationary, ergodic, Gaussian process $\tilde{x}(t)$, with zero mean value $\mu_{x(t)} = 0$. The power spectrum of $\tilde{x}(t)$ is given by equation (2.21) where $\dot{x}_F(\nu, T)$ (equation 2.25″) is the finite Fourier transform of $\dot{x}(t)$.

An estimator for $G_{\tilde{x}(t)}'^0(\nu)$ can be defined as

$$\tilde{g}_T'^0(\nu) = \frac{2}{T}[\dot{x}_F(\nu, T)]^2 \tag{8.14}$$

which is a realization of the random variable

$$\tilde{g}'^{0}_{T}(\nu) = \frac{2}{T}\,[\tilde{x}_{\mathrm{F}}(\nu, T)]^2 \tag{8.15}$$

with

$$\tilde{x}_{\mathrm{F}}(\nu, T) = \int_0^T \tilde{x}(t)\,\mathrm{e}^{-2\pi i \nu t}\,\mathrm{d}t \tag{8.16}$$

$\tilde{x}(t)$ being the random process.

It can be proved that the ratio of the random variable (8.15) to the actual value of $G'^{0}_{x(t)}(\nu)$ is a χ^2 variable with two degrees of freedom, apart from a numerical coefficient. Briefly

$$\tilde{c} = \frac{\tilde{g}'^{0}_{T}(\nu)}{G'^{0}_{x(t)}(\nu)} = \frac{\tilde{\chi}^2_2}{2} \tag{8.17}$$

The c variable is therefore independent of the record length T and hence if T is increased the error is not reduced, and the estimator (8.14) is inconsistent.

In order to define a consistent estimator, assume that n sample records of equal length T' are available, $\dot{x}_1(t), \ldots, \dot{x}_n(t)$, and estimate, for each sample function, the value of g'^{0}_{T}. Put

$$g'^{0}_{T'i}(\nu) = \frac{2}{T'}\left|x_{ki}(\nu, T')\right|^2 \tag{8.18}$$

Then consider the *averaged estimator*

$$g'^{0}_{T'}(\nu) = \frac{1}{n}\sum_{i=1}^{n} g'^{0}_{T_i}(\nu) \tag{8.19}$$

that is, the realization of the random variable

$$\tilde{g}'^{0}_{T}(\nu) = \frac{1}{n}\sum_{i=1}^{n} \tilde{g}'^{0}_{T'i}(v) = \frac{1}{n}\,\frac{G'^{0}_{x(t)}(\nu)}{2}\sum_{i=1}^{n}\tilde{\chi}^2_{2i} \tag{8.20}$$

Since a sum of χ^2 variables is still a χ^2 variable with number of degrees of freedom equal to the sum of the degrees of freedom of the addenda, one obtains the random quotient

$$\tilde{c} = \frac{\tilde{g}'^{0}_{T}(\nu)}{G'^{0}_{x(t)}(\nu)} = \frac{1}{2n}\,\tilde{\chi}^2_{2n} \tag{8.21}$$

The situation here is somewhat different than in equation (8.17), since the number n of sample records enters into the definition of the error quotient c. One gets

$$\begin{aligned} \mathrm{E}[\tilde{c}] &= 1 \\ \mathrm{Var}[\tilde{c}] &= 1/n \end{aligned} \tag{8.22}$$

showing that the coefficient of variation of c (equation (7.23))

$$e_c = \frac{\mathrm{Var}[\tilde{c}]}{\mathrm{E}[\tilde{c}]} = \frac{1}{n} \tag{8.23}$$

tends to zero as n increases.

Finally, it is obvious that the collection of n sample records $\dot{x}_1(t), \ldots, \dot{x}_n(t)$ can be looked at as a partition of an original record $\dot{x}(t)$ of length $T = nT'$. Noting that the resolution bandwidth b_r is equal to $1/T'$ one can assess that the standard error of the averaged estimation is given by

$$e_c = \frac{1}{Tb_r} \tag{8.24}$$

Thus, the random error via Fourier transform is quite similar to that obtained via filtering (equation (8.13)). The same happens for the bias error.

8.5.5 Probability density estimate of a stationary–ergodic process

Consider a sample record $\dot{x}(t)$ of length T from a stationary ergodic process $\tilde{x}(t)$. Moreover let

$$\phi(x, \Delta x) = \text{Prob}[(x - \tfrac{1}{2}\Delta x) < \tilde{x}(t) < (x + \tfrac{1}{2}\Delta x)] \tag{8.25}$$

denote the probability that $\tilde{x}(t)$ falls in the interval $[(x - \tfrac{1}{2}\Delta x) < \tilde{x} \leqslant (x + \tfrac{1}{2}\Delta x)]$. The quantity ϕ can be estimated by

$$\ddot{\phi}(x, \Delta x) = t_x/T \tag{8.26}$$

t_x being the total time spent by $\dot{x}(t)$ in the above-mentioned interval. Such an estimator is a realization of the random variable

$$\tilde{\phi}(x, \Delta x) = \tilde{t}_x/T \tag{8.27}$$

From the ergodicity assumption,

$$\text{E}[\tilde{\phi}(x, \Delta x)] = \lim_{T \to \infty} \frac{\tilde{t}_x}{T} \tag{8.28}$$

It follows that the estimator (8.26) is not affected by bias.

Once an unbiased estimator for $\phi(x, \Delta x)$ has been defined, consider the following estimator (equation (2.14)) for the probability density function $p_{\tilde{x}}(x)$ (Section A2.1)

$$\Pi(x) = \ddot{p}(x) = \frac{\ddot{\phi}(x, \Delta x)}{\Delta x} = \frac{t_x}{T\Delta x} \tag{8.29}$$

which is a realization of the random variable

$$\tilde{\Pi}(x) = \frac{\ddot{\phi}(x, \Delta x)}{\Delta x} \tag{8.30}$$

The expected value of the estimator is

$$\text{E}[\tilde{\Pi}(x)] = \frac{1}{\Delta x} \int_{x - \frac{1}{2}\Delta x}^{x + \frac{1}{2}\Delta x} p_{\tilde{x}}(x)\,\text{d}x \neq p_{\tilde{x}}(x) \tag{8.31}$$

thus proving that the estimator is biased.

In order to evaluate the mean square error of the estimate, note that it can be proved that

$$\mathrm{Var}[\tilde{\Pi}(x)] \simeq \frac{c^2 P_x(x)}{2bT\Delta x} \tag{8.32}$$

$$B[\tilde{p}(x)] \simeq \frac{(\Delta x)^2}{24} \frac{\mathrm{d}^2 p(x)}{\mathrm{d}x^2} \tag{8.33}$$

with b being the bandwidth of the sample record and c a constant mostly ranging in the neighbourhood of unity.

The normalized total quadratic error is given by

$$\mathscr{E}_p^{\,2} \simeq \frac{c^2}{2bT\Delta x} + \frac{(\Delta x)^4}{576} \left(\frac{\mathrm{d}^2 p(x)/\mathrm{d}x^2}{p(x)} \right)^2 \tag{8.34}$$

showing that the same conflict pointed out in the determination of spectral density with regard to the filter bandwidth, now arises with regard to the window magnitude Δx.

8.6 NON-STATIONARY STOCHASTIC DATA

Estimation of statistical properties of stochastic processes is a formidable task when the stationarity assumption does not hold (see Fig. 8.6). First, the possibility of working on only one sample function fails because ergodicity implies stationarity: in other words ensemble-averaging is needed, which requires that a *number of sample functions* are available. Nevertheless, in some particular cases, time-averaging may still be applied successfully, in an approximate fashion.

Generally speaking, for a non-stationary stochastic process, the first-order and higher-order probability density functions (equation (A5.1a)) are time-varying functions. For instance, the first-order probability density function is $p_x(x, t)$ (equation (2.14)) where x is the considered amplitude of the process at time t, and t is a given instant of the interval on which the process is developed. Statistical evaluation of the function would require that all possible pairs of values (x, t) are considered. Difficulties grow in an obvious manner when higher-order functions are to be calculated.

Additional assumptions usually help, and make analysis of non-stationary time series possible. If, in fact, the property of normality can be attached to the process, the time-dependent first-order density function can be characterized by simply specifying the time-dependent expected value and mean square value. At the same time, the correlation structure of the process should also be considered.

If the process is normal, the problem is solved by the autocovariance function and/or by the generalized power spectral density function.

The problem is thus much more tractable if there are reasons to believe that the probability density function is constant in form and depends on a small number of parameters, say the mean value and the variance.

In particular, estimators of mean value, variance, autocovariance function and

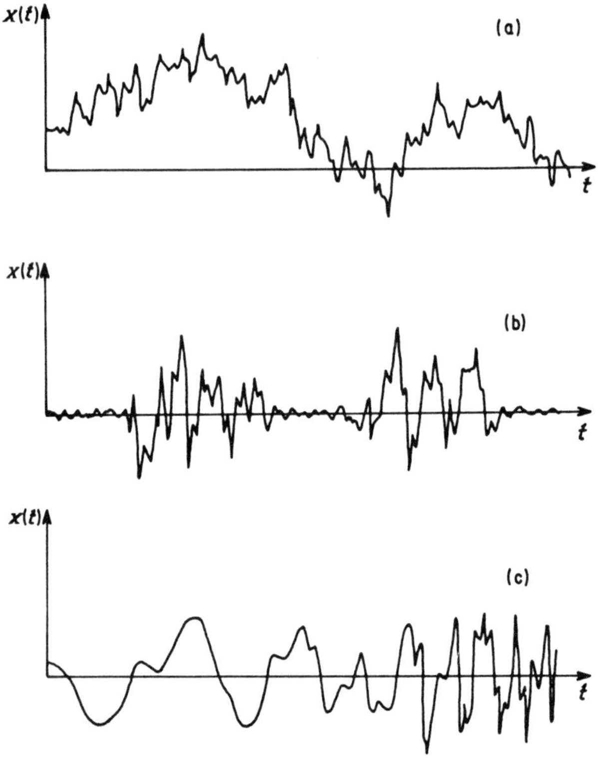

Fig. 8.6 Examples of non-stationary data: (a) time-varying mean value; (b) time-varying mean square value; (c) time-varying frequency structure [1].

spectral density functions will be dealt with in the following, bearing in mind that their knowledge is sufficient for Gaussian processes. Throughout the treatment it will be assumed that a collection of sample records $\dot{x}_i(t)$ $(i = 1, \ldots, n)$, is available on the interval $(0, T)$.

8.6.1 Mean value and variance of a non-stationary process

The following estimator of the mean value at time t

$$m_n(t) = \frac{1}{n} \sum_{i=1}^{n} \dot{x}_i(t) \tag{8.35}$$

is unbiased and consistent, since equation (A3.65) holds.

According to the development of Section 7.2.6 (equations (7.11) and (7.13)) and Section 7.4.1, suitable estimators of the variance at time t can also be introduced.

8.6.2 Autocorrelation function

An estimator for the autocorrelation function defined by equation (A5.3′) is given by

$$b'_n(t_1, t_2) = \frac{1}{n} \sum_{i=1}^{n} \dot{x}_i(t_1)\dot{x}_i(t_2) \tag{8.36}$$

This is a realization of the random variable

$$\tilde{b}'_n(t_1, t_2) = \frac{1}{n} \sum_{i=1}^{n} \tilde{x}_i(t_1)\tilde{x}_i(t_2) \tag{8.37}$$

The expected value is

$$\mu_{b'_n(t_1,t_2)}(t_1, t_2) = \frac{1}{n} \sum_{i=1}^{n} \mathrm{E}[\tilde{x}_i(t_1)\tilde{x}_i(t_2)] = \Sigma'_{x(t)}(t_1, t_2) \tag{8.38}$$

and hence the considered estimator is unbiased.

8.6.3 Spectral density of non-stationary processes

It is soon realized that spectral density cannot be defined for a non-stationary process in the same way and with the same evident physical interpretation as for a stationary process.

However, the generalized power spectral density function $G_{x(t)}(\nu_1, \nu_2)$ defined by equation (A5.54) can be introduced as the double Fourier transformation of the autocorrelation function.

It is not easy to attach a physical meaning to such a function that is, in general, a complex-valued function. However, it should be noted that $G_{x(t)}(\nu_1, \nu_2)$ is to be calculated by Fourier transform on the autocorrelation function. Time-averaging is not allowed, except in special instances.

8.7 SPECIAL NON-STATIONARY PROCESSES

In some cases, physical evidence suggests that a particular structure can be given to the non-stationary process. In the following, some special cases of *separable processes*, which allow great simplifications in statistical analysis will be illustrated. Such processes are also used in earthquake simulation (see for example Section 9.2.4).

8.7.1 Random disturbance

Assume that there are reasons to believe that the non-stationary random process can be given the form

$$\tilde{x}(t) = \Lambda(t) + \tilde{y}(t) \tag{8.39}$$

with $\Lambda(t)$ a deterministic function and $\tilde{y}(t)$ a stationary stochastic process with zero mean value. It is obvious, then, that

$$\mu_{x(t)}(t) = \Lambda(t) \tag{8.40}$$

Estimating the expected value of $\tilde{x}(t)$ then coincides with determining the deterministic component of the process. This can be accomplished by low-pass filtering, or by trend removal, or even by time-averaging.

As for the mean square value, noting that

$$\tilde{x}^2(t) = \Lambda^2(t) + 2\Lambda(t)\tilde{y}(t) + \tilde{y}^2(t) \tag{8.41}$$

and ensemble-averaging, one obtains

$$\mathrm{E}[\tilde{x}^2(t)] = \Lambda^2(t) + \mathrm{E}[\tilde{y}^2(t)] \tag{8.42}$$

Also $\mathrm{E}[\tilde{y}^2(t)]$ can be determined by time-averaging.

With reference to the autocorrelation function, note that

$$\tilde{x}(t_1)\tilde{x}(t_2) = \Lambda(t_1)\Lambda(t_2) + \Lambda(t_1)\tilde{y}(t_1) + \Lambda(t_2)\tilde{y}(t_2) + \tilde{y}(t_1)\tilde{y}(t_2) \tag{8.43}$$

Then, by ensemble-averaging

$$\mathrm{E}[\tilde{x}(t_1)\tilde{x}(t_2)] = \Lambda(t_1)\Lambda(t_2) + \mathrm{E}[\tilde{y}(t_1)\tilde{y}(t_2)] \tag{8.44}$$

Hence, also for the autocorrelation function it is possible to refer to time-averages on the difference $\tilde{x}(t) - \mu_{x(t)}(t)$, thus obtaining the autocorrelation function of the stationary disturbance.

Similar remarks apply to the spectral density, that is the Fourier transform of the autocorrelation function.

8.7.2 Stationary random factor

Consider now a non-stationary process $\tilde{x}(t)$ with constant zero mean, which can be given the form

$$\tilde{x}(t) = \Lambda(t)\,\tilde{y}(t) \tag{8.45}$$

where $\Lambda(t)$ is a deterministic function and $\tilde{y}(t)$ a stationary process with zero mean value and unit variance. The expected value of $\tilde{x}^2(t)$ is given by

$$\mathrm{E}[\tilde{x}^2(t)] = \mathrm{Var}[\tilde{x}(t)] = \Lambda^2(t) \tag{8.46}$$

A biased estimator for $\Lambda^2(t)$ can be obtained by short-time-averaging. Consider, in fact, the following estimator for $\Lambda^2(t)$:

$$\check{\Lambda}^2(t) = \frac{1}{\Delta t}\int_{t-\frac{1}{2}\Delta t}^{t+\frac{1}{2}\Delta t} \check{x}^2(t)\,\mathrm{d}t \tag{8.47}$$

$\check{x}(t)$ being, as usual, a sample function. The random value of the estimate would be

$$\tilde{\Lambda}^2(t) = \frac{1}{\Delta t}\int_{t-\frac{1}{2}\Delta t}^{t+\frac{1}{2}\Delta t} \tilde{x}^2(t)\,\mathrm{d}t \tag{8.48}$$

whose expected value can be obtained by ensemble-averaging

$$\mathrm{E}[\tilde{\Lambda}^2(t)] = \frac{1}{\Delta t}\int_{t-\frac{1}{2}\Delta t/2}^{t+\frac{1}{2}\Delta t} \Lambda^2(t)\,\mathrm{d}t \neq \mathrm{E}[\tilde{x}^2(t)] \tag{8.49}$$

The bias can be negligible if $\Lambda^2(t)$ is a smooth function and Δt is a sufficiently small interval of time.

The autocorrelation function is given by

$$\begin{aligned}\Sigma'_{x(t)}(t_1, t_2) &= \mathrm{E}[\tilde{x}(t_1)\tilde{x}(t_2)]\\ &= \mathrm{E}[\Lambda(t_1)\tilde{y}(t_1)\Lambda(t_2)\tilde{y}(t_2)]\\ &= \Lambda(t_1)\Lambda(t_2)B'_{y(t)}(|t_1 - t_2|)\end{aligned} \tag{8.50}$$

The autocorrelation function can then be investigated by considering the stationary process

$$\tilde{y}(t) = \frac{\tilde{x}(t)}{\Lambda(t)} \tag{8.51}$$

and then by time-averaging to discover statistical properties of $\tilde{y}(t)$.

8.7.3 Piecewise-separable processes

Finally put

$$\tilde{x}(t) = \Lambda_i(t)\tilde{y}_i(t) \qquad t \in (t_{i-1}, t_i) \tag{8.52}$$

i.e. assume that there are reasons to believe that in every interval (t_{i-1}, t_1), $i = 1, \ldots, n$, the stochastic process under consideration is the product of a deterministic function $\Lambda_i(t)$ and a stationary process $\tilde{y}_i(t)$ that may be different in every interval.

If a sample record $\grave{x}(t)$ of $\tilde{x}(t)$ is available, it can be divided into a number of sections, each relevant to a given interval. Hence, $\grave{x}(t)$ can be divided by $\Lambda_i(t)$ in the relevant interval, and the resultant record

$$\grave{y}_i(t) = \frac{\grave{x}(t)}{\Lambda_i(t)} \tag{8.53}$$

can be treated as a sample record from the stationary process $\tilde{y}_i(t)$. Time-averaging also applies in this case, as usual for stationary records.

The procedure has been successfully applied to study evolutionary properties of spectral density of earthquakes.

8.8 A NOTE ON RANDOM SPACE FIELDS

Previous sections were devoted to random functions of time. In particular, satisfactory estimators were analysed under the assumption that a wide-sense ergodic process is observed periodically at points with constant spacing.

However, in many problems related to structural engineering, the quantity of interest is an uncertain function of space coordinates: single or local-average observations are taken at discrete points. Such a problem, for example, arises in the estimation of the spatial variation of soil properties for geotechnical application, but also in the quality control of construction materials and/or the investigation of the spatial distribution of vertical loads (dead and live loads, snow action on structures, . . .).

On principle, such estimation procedures would require finite sampling grids, i.e. observations performed at points that form a regular (or almost regular) grid. However, engineering problems often require the knowledge of large-scale trends rather than the analysis of local random variations, and sometimes the engineer's aim is only to detect areas where the quantity of interest is likely to be greater (or lower) than a fixed threshold. In these cases, a grid with few sample points may be sufficient: the sampling can then be repeated until the result is achieved. The choice of the best exploration strategy becomes very important and involves sequential decision processes where past experience and engineering judgement act simultaneously. Models for the design of sampling networks, which have recently been proposed [2, 3], follow the Bayesian approach, but their presentation would be well beyond the scope of this book.

REFERENCES

1. Bendat, J. S. and Piersol, A. G. (1971) *Random Data: Analysis and Measurement Procedures*, Wiley Interscience, New York.
2. Veneziano, D. and Faccioli, E. (1975) Bayesian design of optimal experiments for the estimation of soil properties. *Proc. 2nd Int. Conf. on Applications of Statistics and Probability in Soil and Structural Engineering* (ICASP 2), Aachen, vol. II.
3. Veneziano, D. and Queiroz, C. S. (1977) *Two Problems of Spatial Sampling*, Research Report R 77-29, Department of Civil Engineering, M.I.T.

CHAPTER 9

Numerical methods and applications

This chapter discusses how complete statistical investigations, physical experiments and/or analytical solutions can be replaced by numerical procedures. The policy for approaching structural problems under random uncertainties by simulation is then discussed: a number of specific techniques are then presented and illustrated by detailed numerical examples.

9.1 SIMULATION AND MONTE CARLO TECHNIQUES

Analytical solution procedures are either unavailable or inefficient for many structural engineering problems whose parameters are random variables or functions. Some of these cases are listed in Chapter 6; they especially concern non-linear (in geometry and/or material) problems, dynamic analyses and damage accumulation. More generally, probabilistic closed-form solutions are never available when the structural problem must be approached by numerical, rather than analytical, procedures.

In these cases, two ways can be followed to obtain information on structural response randomness or on system reliability:

(1) introduce rough, inaccurate idealizations of the mechanical behaviour to which practicable analytical solutions correspond;
(2) repeat sampling from the statistical population of random quantities and simulate numerically a statistical sample of responses.

The rapid development of computers and computing procedures had led in many engineering fields to a widespread acceptance of simulation techniques. This is also the case in structural engineering, not only because of the difficulties encountered in the rigorous solution of many problems under uncertainty, but also because in many cases it is practically impossible to obtain sufficient statistical information from physical experiments.

Simulation methods essentially assume that the statistics of the input random quantities are known: then, each experiment consists of choosing a set of values of input quantities, performing a deterministic structural analysis and thus obtaining a set of output (response) quantities. The experiment is repeated n times, and the response statistics are finally calculated from the sample of responses.

An example of simulation is the estimation of integrals of the type (equation (A3.1a))‡

$$I = \int_{-\infty}^{\infty} g(x)\,\mathrm{d}P_x(x) = \int_{-\infty}^{\infty} g(x)\,p_x(x)\,\mathrm{d}x \tag{9.1}$$

by the random sample average

$$m_{g(x)} = \frac{1}{n}\,[g(\dot{x}_1) + \ldots + g(\dot{x}_n)] \tag{9.2}$$

$\tilde{x}$ being a random variable with cumulative distribution function (CDF) $P_x(x)$, $\dot{x}_1, \ldots, \dot{x}_n$ a sample of $\tilde{x}$ (Section 7.2.1), and $g(x)$ a deterministic function. The techniques making use of (9.2) for solving (9.1) are known as Monte Carlo techniques§.

The variance of the estimate (9.2) of (9.1) is (equation (A3.65))

$$s^2_{m_{g(x)}} = \frac{1}{n}\, s^2_{g(x)} \tag{9.3}$$

and obviously decreases with n, while

$$\mathrm{E}[\tilde{m}_{g(x)}] = \mathrm{E}[g(\tilde{x})] = I \tag{9.4}$$

i.e. $m_g(\tilde{x})$ is an unbiased estimate of I (equation (A3.65))

Some special techniques were set up in order to improve the efficiency of such a computation (Section 9.1.1). They are of some interest also in simulating the result of equations different from (9.1). In this context, other special techniques have been devised to reduce and/or simplify the computations needed for each experiment (Section 9.5).

9.1.1 Special Monte Carlo techniques

As the computational effort is proportional to the number of *experiments* n, the main aim of special techniques is to reduce the variance (9.3) of the estimate for a given number n, while preserving the relation (9.4).

‡ For simplicity, only one-dimensional formulae are presented in the following treatment.

§ In engineering applications, the words 'Monte Carlo techniques' are often used to indicate the whole set of simulation techniques. To avoid misunderstanding, this expression is not used in the following sections.

(a) Stratified sampling

Divide the range of the random variable $\tilde{x}$ into a number of conveniently chosen class intervals (Section 7.2.4)

$$\zeta_{i-1} = (x^i - \tfrac{1}{2}\Delta x^i) < x \leqslant (x^i + \tfrac{1}{2}\Delta x^i) = \zeta_i$$

and assume that the probabilities

$$p_i = \text{Prob}\,[\zeta_{i-1} < x \leqslant \zeta_i] = P_x(\zeta_i) - P_x(\zeta_{i-1}) \tag{9.5}$$

associated with the class intervals are known *a priori.*

Let n_i be the class frequencies relevant to the sample $(\dot{x}_1, \dot{x}_2, \ldots, \dot{x}_n)$, and x_{ij} the jth sample value $(j = 1, \ldots, n_i)$ falling into the ith class: the *stratified sample average*

$$m_{g(x)\,\text{str}} = \sum_i p_i \frac{1}{n_i}\left(\sum_{j=1}^{n_i} g(x_{ij})\right) \tag{9.6}$$

provides an unbiased estimate of I, with (see equation (A3.65))

$$\begin{aligned} s^2_{m_{g(\tilde{x})\,\text{str}}} &= \sum_i \frac{p_i^2}{n_i} s^2_{g(\tilde{x}_{ij})} \\ &= \sum_i \frac{p_i^2}{n_i^2}\left[\frac{1}{p_i}\int_{\zeta_{i-1}}^{\zeta_i} g^2(x)\,\mathrm{d}P_x(x) - \left(\frac{1}{p_i}\int_{\zeta_{i-1}}^{\zeta_i} g(x)\,\mathrm{d}P_x(x)\right)^2\right] \end{aligned} \tag{9.7}$$

(b) Correlated samples

In a random sample, the individual sample values x_j should be stochastically independent (see Section 7.2). However, if the values x_j are correlated, only equation (9.3) must be replaced by (see equation (A3.28))

$$s^2_{m_{g(x)\,\text{corr}}} = \frac{1}{n} s^2_{g(x)} + 2\sum_{k=1}^{n}\sum_{j=k+1}^{n} \text{Cov}\,[g(x_k), g(x_j)] \tag{9.8}$$

If one introduces a negative correlation between pairs $g(x_k)$, $g(x_j)$, negative covariance terms are produced in equation (9.8) and the variance of the estimate can be reduced without biasing it.

(c) Importance sampling

Importance sampling techniques attempt to estimate the integral (9.1) by the sample average $m_{g(y)}/\phi(\tilde{y})$, $\tilde{y}$ being a random variable with PDF

$$p_y(x) = \phi(x)\frac{\mathrm{d}P_x(x)}{\mathrm{d}x} = \phi(x)p_x(x) \tag{9.9}$$

Such an estimate is unbiased (equation (A3.2)).

The function ϕ ( ) is selected so that

$$\text{Var}\,[g(\tilde{y})/\phi(\tilde{y})] = \text{E}[(g(\tilde{y})/\phi(\tilde{y}) - I)^2] \tag{9.10}$$

is small and $\int_{-\infty}^{\infty} p_y(x)\,dx = 1$. Note that, if $\phi(x) = g(x)/I$, the variance of equation (9.10) becomes zero, but this would require knowledge of the unknown quantity I; in this way, however, a partial *a priori* knowledge of results can be judiciously exploited.

9.1.2 Random-number generators

In order to obtain a sample from the population of a random function, the first step is to generate random numbers with uniform distribution. At the present time, any computing machine (micro- or macro-computer) makes use of its own structural characteristics to pursue this goal in the best way, and hence there is no generally accepted program. However, most programs generating random numbers are founded on the so-called *congruential methods.* They generate random (integer) numbers u_i comprised between 0 and a given non-negative (integer) modulus m. For this purpose, the procedure starts with any $0 \leqslant u_0 < m$ and computes successive values

$$u_i = [c_1 u_{i-1} + c_2]_{\text{mod}\, m} \qquad i = 1, 2, \ldots; 0 < c_1 < m; 0 \leqslant c_2 < m \tag{9.11}$$

where 'modulo-m addition' is an (abstract) addition‡ (for which the identity is 0 and the inverse means to change sign) whose result is replaced by the remainder of its division by the modulus m. On a binary computing machine, the modulus m is conveniently selected equal to 2^n, n being the computer word length. Further, if $c_2 = 0$, the generator is called a *multiplicative congruential generator*; otherwise it is a *mixed congruential generator.*

The sequences obtained in this manner are not truly random but may have 'pseudo-random' properties, such as uniform distribution between 0 and m, zero correlation between different u_i, random appearing runs of odd and even numbers, etc. However, even with zero correlation, samples $(\grave{u}_1, \grave{u}_2, \ldots, \grave{u}_n)$ taken from such a sequence will not be stochastically independent. Alternatively, true random samples are often generated by analogue-to-digital conversion of true analogue noise.

The pseudo-random integers u_i are usually transformed into numbers $y_i = u_i/m$, uniformly distributed in the range (0, 1).

9.2 SIMULATION OF RANDOM QUANTITIES

9.2.1 Random variables

The basic aspect of the simulation techniques is to yield a set of random numbers for every design random variable x_i. These numbers must be stochastically independent and they must form a sample of a given distribution.

‡ The theoretical basis for this topic is 'congruence theory', which is beyond the scope of this book. The reader is referred to a modern algebra text.

Using any computer, standard library programs (Section 9.1.2) yield random numbers, uniformly distributed, $y_j \in (0, 1)$. A second algorithm, able to change a number in (0, 1) into the value of a random variable with non-uniform distribution, must then be applied (equations (A3.38) and (A3.39)).

Indicate by $\tilde{y}$ the random variable underlying the numbers y_j: obviously (Fig. 9.1)

$$p_y(y) = \begin{cases} 1 & \text{when } y \in (0, 1) \\ 0 & \text{otherwise.} \end{cases}$$

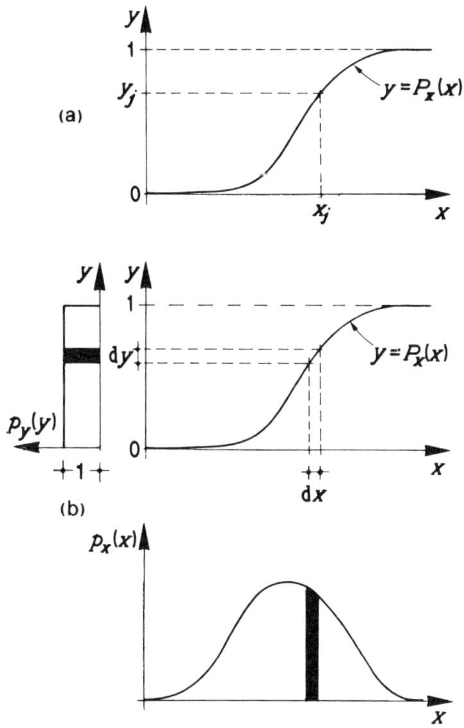

Fig. 9.1 Generation of random numbers of given CDF.

Let $\tilde{x}$ be the random variable to be simulated, and $P_x(x)$ its CDF. Relating $\tilde{y}$ and $\tilde{x}$ by the deterministic law

$$\tilde{y} = P_x(x) = \int_{-\infty}^{x} p_x(x')\,\mathrm{d}x' \tag{9.12}$$

and applying the general relationship between the PDFs of two functionally dependent random variables (equation (A3.32)), it is immediate (Fig. 9.1) that the numbers

$$\grave{x}_j = P_x^{-1}(\grave{y}_j) \tag{9.13}$$

associated to each $\dot{y}_j$ yielded by the random number generator, form a sample of the random variable $\tilde{x}$ with CDF $P_x(x)$.

9.2.2 Random functions: frequency domain models

Efficient numerical simulation procedures have been proposed in recent years for Gaussian random functions $\tilde{x}(t)$. These procedures are generally restricted to the stationary case, but generalization to non-stationary Gaussian processes and fields is sometimes possible.

(a) Stationary random functions

Consider a stationary Gaussian random function $\tilde{x}(t)$ and assume that the spectral density function $G_{x(t)}(\omega)$ (equation (A5.46)) has negligible magnitude outside the range $(-\bar{\omega}, \bar{\omega})$ (lower and upper cut-off frequencies). Then, lump the area under the power spectral density function into small finite areas $G_{x(t)}(\omega)\Delta\omega$ at equal frequency intervals $\Delta\omega = 2\bar{\omega}/N$ (Fig. 9.2) and introduce the sum of harmonic functions

$$\dot{x}_N(t) = \sum_{i=0}^{N} [2G_{x(t)}(-\bar{\omega} + i\Delta\omega)\Delta\omega]^{1/2} \cos[(-\bar{\omega} + i\Delta\omega)t + \dot{\phi}_i] \quad (9.14)$$

whose amplitude squared and divided by 2 is equal to the corresponding lumped

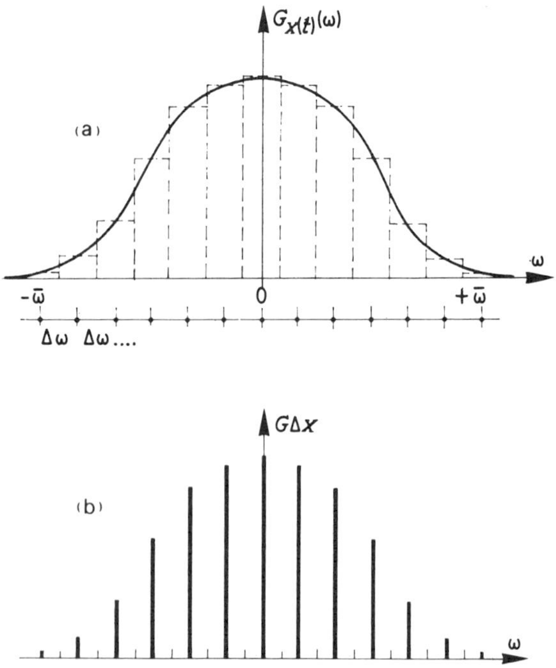

Fig. 9.2 Lumping of the area under a power spectral density function.

area of the spectral density (cf. sine wave in Table 2.3). If ϕ_i is a realization of a random phase angle $\tilde{\phi}_i$, uniformly distributed over $(0, 2\pi)$, equation (9.14) denotes a realization of an ergodic periodic random function $\tilde{x}_N(t)$, with period $T = \pi/\bar{\omega}$, zero mean (due to the circumstance that $\mathrm{E}[\cos\{(-\bar{\omega}+i\Delta\omega)t+\tilde{\phi}_i\}] = 0$ for all i), and spectral distribution function (equation (A5.43)) with concentrated increments $\Delta V_{x(t)}(-\bar{\omega}+i\Delta\omega) = G_{x(t)}(-\bar{\omega}+i\Delta\omega)\Delta\omega$ evaluated at the frequencies $(-\bar{\omega}+i\Delta\omega)$ (Fig. 9.2(b)).

A different form for equation (9.14) is founded on the symmetry of the spectral density function $G_{x(t)}(i\Delta\omega) = G_{x(t)}(-i\Delta\omega)$. In fact, when N is an odd number, one can write (9.14) in the form

$$\begin{aligned}\dot{x}_N(t) &= \sum_{i=0}^{(N-1)/2}\left[4G_{x(t)}\left(\frac{\Delta\omega}{2}+i\Delta\omega\right)\Delta\omega\right]^{1/2}\cos\left[\left(\frac{\Delta\omega}{2}+i\Delta\omega\right)t+\dot{\phi}_i\right]\\ &= \sum_{i=0}^{(N-1)/2}\left[2G^0_{x(t)}\left(\frac{\Delta\omega}{2}+i\Delta\omega\right)\Delta\omega\right]^{1/2}\cos\left[\left(\frac{\Delta\omega}{2}+i\Delta\omega\right)t+\dot{\phi}_i\right]\end{aligned} \tag{9.14}$$

where $G^0_{x(t)}(\ )$ denotes the one-sided power spectral density (equation (A5.52)). Note that equation (9.14′), used with summation starting with $i = h$, where $(\Delta\omega/2 + h\Delta\omega) = \omega_1$, permits one to operate on the ranges $(-\bar{\omega}, \omega_1)$ and $(\omega_1, \bar{\omega})$ instead of $(-\bar{\omega}, \bar{\omega})$.

For N an even number, the recursive point where ω is calculated is $(i\Delta\omega)$, summation starts with $i = 1$ and stops with $i = N/2$, and it is necessary to add to the r.h.s. of equation (9.14′) the contribution of the frequency $\omega = 0$.

Equation (9.14) can also be written in a form that is suitable for the application of the fast Fourier transform (FFT) algorithm‡:

$$\begin{aligned}\dot{x}_N(t) &= \mathrm{Re}\left(\sqrt{2}\sum_{j=0}^{N}[G_{x(t)}(-\bar{\omega}+j\Delta\omega)\Delta\omega]^{1/2}\,\mathrm{e}^{\mathrm{i}\dot{\phi}_j}\exp[\mathrm{i}(-\bar{\omega}+j\Delta\omega)t]\right)\\ &= \mathrm{Re}\left(\sqrt{2}\sum_{j=0}^{N}\dot{x}(j)\exp[\mathrm{i}(-\bar{\omega}+j\Delta\omega)t]\right)\end{aligned} \tag{9.14″}$$

where Re( ) indicates the real part and i is the imaginary unit.

If each $G_{x(t)}(-\bar{\omega}+i\Delta\omega)$ in (9.14) (or (9.14′)) is replaced by realizations from a non-negative random sequence with expected value $G_{x(t)}(-\bar{\omega}+i\Delta\omega)$, a realization of a random function, non-ergodic in the mean (Section A5.1), with the same second-moment properties as $\tilde{x}_N(t)$ is obtained. Non-ergodicity in the variance function (Section A5.1) can also be introduced by a fully random generation of the coefficients of the cosine function in equation (9.14) or (9.14′).

‡ The FFT is an efficient algorithm to calculate the discrete Fourier transform of a finite sequence in the form

$$z(j) = \sum_{k=0}^{N-1} Z(k)\mathrm{e}^{\mathrm{i}(2\pi/N)jk} \qquad j = 0, 1, \ldots, N-1$$

in which $Z(k)$ is a given sequence of complex Fourier coefficients.

As N tends to infinity, $\tilde{x}_N(t)$ tends to a stationary non-periodic random function $\tilde{x}_\infty(t)$ with zero mean and continuous spectral density function $G_{x(t)}(\omega)$ in $(-\bar{\omega}, \bar{\omega})$. As no single term dominates in the r.h.s. of equation (9.14), $\tilde{x}_\infty(t)$ is Gaussian because of the central limit theorem (Appendix A4). A more rapid convergence can be obtained by introducing variable intervals $\Delta\omega_j$ such that

$$G_{x(t)}\left(-\bar{\omega}+\sum_{i=1}^{j}\Delta\omega_i\right)\Delta\omega_j = s_{x(t)}^2/N = \text{constant}$$

In this case, for $N = 500$, the complementary CDFs, $1-P_{x(t)}(x)$, of the actual Gaussian process and the generated function were demonstrated to differ by about 1%, 2% and 5% for $x = 3s_x$, $4s_x$ and $5s_x$ respectively.

The covariance and the mean power spectral density function of the process of which equation (9.14) is a realization, have been proved to approach the associated target functions $B_{x(t)}(v)$ and $G_{x(t)}(\omega)$ as $1/N^2$.

The previous approach can be generalized to processes with non-differentiable spectral distribution function $V_{x(t)}(\omega)$ (equation (A5.43)) [1]. In this case in fact, one writes‡

$$V_{x(t)}(\omega) = V_{x_0(t)}(\omega) + V_{x_1(t)}(\omega) \tag{9.15}$$

where $V_{x_0(t)}$ is a differentiable spectral distribution of a random process $\tilde{x}_0(t)$, while $V_{x_1(t)}$, relevant to $\tilde{x}_1(t)$, corresponds to a discrete distribution of mean power on the real line (Section 2.4.4). For $\tilde{x}_0(t)$ and $\tilde{x}_1(t)$ Gaussian and independent, the original random function is given by

$$\tilde{x}(t) = \tilde{x}_0(t) + \tilde{x}_1(t) \tag{9.16}$$

Here, $\tilde{x}_0(t)$ can be simulated as $\dot{x}_{0_N}(t)$ from equation (9.14), with $G_{x_0(t)}(\omega) = \mathrm{d}V_{x_0(t)}(\omega)/\mathrm{d}\omega$. Further, let $\omega_k^\circ (k = 1, 2, \ldots)$ be the frequencies at which finite non-zero mean power masses, $\Delta V_{x_1(t)}(\omega_k^\circ)$, are concentrated. Random processes with the same second moments of $\tilde{x}_1(t)$ and with distribution closer to the normal are obtained as

$$\dot{x}_{1\mathbf{m}_k}(t) = \sum_k \frac{\sqrt{2}}{m_k}\Delta V_{x_1(t)}(\omega_k^0)\sum_{i=1}^{m_k}\cos(\omega_k^0 t + \dot{\phi}_{ki}) \tag{9.17}$$

where the kth frequency component is replaced by the sum of m_k harmonic functions with identical frequency but independent, uniformly distributed phases. Every m_k must be selected proportional to $\Delta V_{x_1(t)}(\omega_k^0)$, so that no single term dominates in the summation of equation (9.17) and $x_{1\mathbf{m}_k}(t)$ is nearly Gaussian. In fact, as $m_k \to \infty$ for all k and $N \to \infty$, the sum $\dot{x}_{0_N}(t) + \dot{x}_{1\mathbf{m}_k}(t)$ approaches a Gaussian random process $\tilde{x}_\infty(t)$ with the first- and second-moment properties of $\tilde{x}(t)$.

Extension of the above procedures to random fields $\tilde{\mathbf{x}}(t)$ in $\mathscr{R}^n$ is immediate. Also when the parameter space $\mathbf{t}$ is multidimensional, some procedures are available in the specialized literature [1], to which the reader is referred.

‡ By virtue of Wold's decomposition theorem.

(b) Non-stationary random functions

Consider now a non-stationary Gaussian random function $\tilde{z}(t)$ described by its *evolutionary mean power spectral density function* $G_{z(t)}(t, \omega)$ which is defined by

$$G_{z(t)}(t, \omega) = [D(t, \omega)]^2 G_{x(t)}(\omega) \tag{9.18}$$

with $D(t, \omega)$ an amplitude-modulating function that varies slowly with t, and $G_{x(t)}(\omega)$ the spectral density of a stationary r.p. $\tilde{x}(t)$.

Note that for fixed t, $G_{z(t)}(t, \omega)$ is the spectral decomposition of the variance

$$s^2_{z(t)}(t) = \int_{-\infty}^{\infty} G_{z(t)}(t, \omega)\,d\omega \tag{9.19}$$

Remarking that the evolutionary process can be interpreted as a process 'built up' by orthogonal oscillations with time-varying random amplitudes, one can introduce the function, analogous to (9.14)

$$\tilde{z}_N(t) = \sum_{i=0}^{N} [2G_{x(t)}(t, -\bar{\omega} + i\Delta\omega)\Delta\omega]^{1/2} \cos[(-\bar{\omega} + i\Delta\omega)t + \tilde{\phi}_i] \tag{9.20}$$

that, in the limit as $N \to \infty$, has the second-moment properties of $\tilde{z}(t)$.

In the special case of decomposable non-stationarity (equation (2.9)), $D(t, \omega)$ is not a function of ω and equation (9.18) becomes

$$G_{z,(t)}(t, \omega) = [D(t)]^2 G_{x(t)}(\omega) \tag{9.21}$$

where $D(t)$ is a deterministic function ($g(t)$ in equation (2.9)). In this case one can simulate $\tilde{z}(t)$ by first introducing $G_{x(t)}(\omega)$ in equation (9.14), then generating $\dot{x}_N(t)$ and forming the product $D(t)\dot{x}_N(t)$.

9.2.3 Random functions: time domain models

A more intuitive approach to the simulation of random processes consists of operating directly in the time domain, without introducing spectral measures. As seen in Chapter 3 the relevant time domain models are most often either in the form of filtered Poisson processes (Section A6.3.3) or in forms derived from the filtered white-noise process (Section A5.3).

(a) Filtered Poisson processes

The filtered Poisson process can be represented as (equation (A6.43))

$$\tilde{x}(t) = \sum_{m=1}^{\tilde{n}(t)} X(t, \tilde{\tau}_m, \tilde{y}_m) \tag{9.22}$$

where $\tilde{n}(t)$ is a Poisson process with intensity λ; $\tilde{y}_1, \tilde{y}_2, \ldots$ is a sequence of independent random variables identically distributed as a random variable $\tilde{y}$; $X(t, \tau, y)$ is the 'pulse' function representing the value at time t of a signal starting at time $\tau (X(t, \tau, y) = 0$ for $t \leqslant \tau)$; and $\tilde{\tau}_m$ denotes the mth waiting time relevant to the Poisson process $\tilde{n}(t)$.

Sample functions of this process can be generated by making direct use of equation (9.22), i.e. by simulating the values of the random variables $\tilde{n}(t)$, $\tilde{\tau}_m$, $\tilde{y}_m$. Moreover, such a process is easily made non-stationary by assuming, for example, that the mean arrival rate of the Poisson process is a function of time (equation (A6.31)).

(b) Filtered white-noise processes

For white-noise $\tilde{\upsilon}(t)$, the spectral density is a non-zero constant; hence all frequencies contribute with equal intensity to the mean square value, which is infinite. Moreover, since the autocorrelation function is the Fourier transform of a constant spectral density, it is possible to write (equation (A5.47b))

$$B(v) = 2\pi G_0 \partial(v) \tag{9.23}$$

v being the time lag and $\partial(v)$ the Dirac function. Equation (9.23) clearly expresses a particular feature of white-noise: its values at different times are uncorrelated. This circumstance suggests the idea that a sample function from a Gaussian white-noise process can be generated as a sequence of independent Gaussian ordinates, with equal mean and variance, at equally spaced points.

To simulate a Guassian white-noise process, the following procedure can be used [2]. First, sample a sequence of statistically independent random numbers $y_1, y_2, y_3, \ldots, y_n$ uniformly distributed over the range (0, 1) (Section 9.1.2). Then, generate a sequence of pairs of statistically independent random numbers $a_1, a_2, \ldots, a_n$ using the relations:

$$\begin{aligned} a_i &= (-2 \ln y_i)^{1/2} \cos(2\pi y_{i+1}) \\ a_{i+1} &= (-2 \ln y_i)^{1/2} \sin(2\pi y_{i+1}) \end{aligned} \qquad i = 1, 3, 5, \ldots \tag{9.24}$$

The random variables $\tilde{a}_i$ can be shown to possess a Gaussian distribution with zero mean and unit variance.

A sample function $\grave{\upsilon}_{\Delta t}(t)$ is then established by assigning the values $a_1, a_2, \ldots, a_n$ to n successive ordinates spaced at equal intervals Δt along a time abscissa, and assuming linear variation over each interval. The initial ordinate $\grave{\upsilon}(0) = a_0$ is usually put equal to zero (Fig. 9.3).

In the limit as $\Delta t \to 0$, this process can be proved to become Gaussian white-noise (Section A5.3) of intensity $G_0^1 = \Delta t/2\pi$ over the infinite frequency range

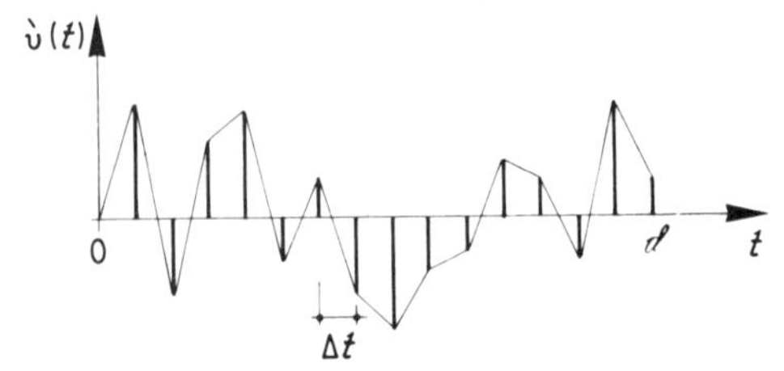

Fig. 9.3 Sample function generated for a Gaussian white-noise process $\upsilon(t)$.

$-\infty < \omega < \infty$. Hence the above procedure yields approximate sample functions of a Gaussian white-noise $\tilde{\upsilon}(t)$ (equation (A5.47)) of any given G_0, by multiplying each ordinate y_i by the normalization factor $\sqrt{(2\pi G_0/\Delta t)}$ [2].

A filtered white-noise process is a process $\tilde{x}(t)$ which satisfies the following equation

$$\ell[\tilde{x}(t)] = \tilde{\upsilon}(t) \tag{9.25}$$

in which $\ell[\quad]$ is a linear operator and $\tilde{\upsilon}(t)$ is Gaussian white-noise. Simulation of $\tilde{x}(t)$ is therefore immediate once $\tilde{\upsilon}(t)$ has been simulated; in particular, if $\ell[\quad]$ is a linear differential operator with constant coefficients, one can easily evaluate the frequency response function (equation (A5.41b)) $A(\omega)$ associated with the operator and hence, by equation (A5.73)

$$G_{x(t)} = |A(\omega)|^2 G_0 \tag{9.26}$$

Some algebra, then, permits one to derive the mean value, the covariance function, etc., of the filtered white-noise process.

Finally, sample functions of a non-stationary random process $\tilde{Z}(t)$ can be generated directly in the time domain in the case of decomposable (amplitude-modulated) non-stationarity

$$\tilde{Z}(t) = D(t)\tilde{x}(t) \tag{9.27}$$

where $D(t)$ is a deterministic function of time that varies slowly in its definition range, and $\tilde{x}(t)$ is a stationary process (cf. Section 8.7.2).

9.2.4 A model for the simulation of earthquake accelerograms

The procedure of the previous sections can be applied in the artificial generation of ground motion accelerograms, in particular following an approach first proposed by Ruiz and Penzien [2].

According to this model, the ground on which the structure is founded is idealized as a more or less soft soil layer resting on an underlying bedrock (Fig. 9.4)

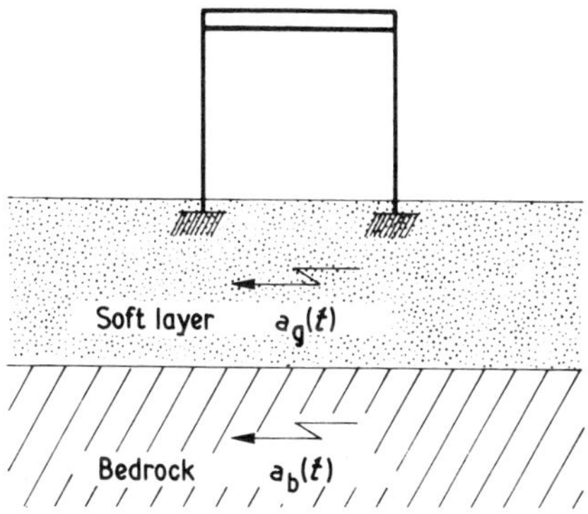

Fig. 9.4 Idealization of the ground on which the structure is founded.

Seismic waves travelling in the earth's crust hit the bedrock surface, hence are filtered by the soil layer and finally envelope the structure foundations.

The bedrock vibration is modelled by a non-stationary Gaussian process that is defined as the product of a white-noise process $\tilde{\upsilon}(t)$ multiplied by a deterministic function of time $\chi(t)$. Thus, a sample function $\grave{a}_b(t)$ of the bedrock acceleration $\tilde{a}_b(t)$ is obtained multiplying the shape function $\chi(t)$ by a sample function $\grave{\upsilon}(t)$ generated, according to the procedure of the previous section, over a time interval $(0, \mathscr{d})$ of given length $\mathscr{d}$ (Fig. 9.3). Let $G_{0\upsilon}$ be the relevant power spectral density function, which can be assumed equal to $\Delta t/2\pi$.

Next, filtering of the bedrock acceleration by the soil layer should be accounted for. The action of the layer is assimilated to a second-order linear filter, so that the random ground acceleration $\tilde{a}_g(t)$ at the foundation level is given by

$$\tilde{a}_g(t) = -2\xi\omega_0\dot{\tilde{z}}(t) - \omega_0^2(t)\tilde{z}(t) \tag{9.28}$$

where $\tilde{z}(t)$ is the solution of the differential equation

$$\ddot{\tilde{z}} + 2\xi\omega_0\dot{\tilde{z}} + \omega_0^2\tilde{z} = -\tilde{a}_b(t) \tag{9.29}$$

The parameters ω_0 and ξ may be thought of as characteristic ground frequency and characteristic damping ratio, respectively.‡ Some authors [2] have suggested to use, in the case of firm ground conditions, when specific experimental data are

‡ The basis for estimation of such quantities is given by recorded accelerograms near the site under investigation on the occasion of past earthquakes. Let $\grave{a}_k(t)$ $(k = 1, \ldots, n)$ be such *real* accelerograms, and let n be large enough that manipulation of such data makes sense from a statistical point of view (say $n \geqslant 4$, at least). An empirical estimate of the filter parameters can be performed from equation (9.32) by the following procedure.

Take the moduli of both sides of equation (9.32)

$$|\tilde{a}_{gF}(\omega)|^2 = |A(\omega)|^2|\tilde{a}_{bF}(\omega)|^2$$

and consider the intensity shape function $\chi(t)$ (equation (9.40)); put $a^{\chi}(t) = \chi(t)a(t)$ and denote by $G_{0\upsilon}^a$ the actual value of the white-noise spectral density: by averaging one gets

$$\mathrm{E}[|\tilde{a}_{gF}^{\chi}(\omega)|^2] = |A(\omega)|^2 2\pi G_{\upsilon}^a \int_{-\infty}^{\infty} \chi^2(t)\,\mathrm{d}t = C|A(\omega)|^2 \tag{9.29a}$$

C being the constant average value of the square modulus of $\tilde{a}_{bF}^{\chi}(\omega)$. Equation (9.29a) can be proved starting from the evolutionary autocorrelation function (8.50)

$$\Sigma'_{a^{\chi}}(t, v) = \chi(t - v/2)\chi(t + v/2)B'_a(v) \simeq \chi^2(t)B'_a(v) \tag{9.29b}$$

and considering its Fourier transform (9.21)

$$G'_{a^{\chi}}(t, \omega) = \chi^2(t)G'_a(\omega)$$

i.e. the evolutionary mean power spectral density function, for which (2.26) must be generalized to

$$\int_{-\infty}^{\infty} G(t, \omega)\,\mathrm{d}t = \mathrm{E}[|a_{gF}^{\chi}(t)|^2]$$

Equation (9.29a) indicates that the square modulus of the frequency response function has a similar behaviour to the square mean value of the ground acceleration.

Coming to the problem of estimating the parameters ω_0 and ξ, the l.h.s. of equation (9.29a) can be estimated by computing first the Fourier transform $\grave{a}_{kF}(\omega)$ of each seismogram $\grave{a}_k(t)$, and then lacking, the values $\omega_0 = 15.6 \text{ rad s}^{-1}$ and $\xi = 0.6$; other numerical values should be chosen for different soil conditions.

Since at the beginning of the quake the soil is assumed to be at rest, $\tilde{a}_g(t)$ is given by Duhamel's integral (see Section A5.3 and Chapter 4)

$$\tilde{a}_g(t) = \int_{-\infty}^{\infty} \tilde{a}_b(\tau) h(t-\tau)\,d\tau \tag{9.30}$$

$h(t)$ being the unit impulse response of the filter (equation (4.89)), which in this case is given by[§]

$$h(t) = \begin{cases} (\omega_0^2/\omega_d)\,e^{-\omega_0 t}\sin(\omega_d t + 2\phi) & t > 0 \\ 0 & t \leqslant 0 \end{cases} \tag{9.31}$$

with $\omega_d = \omega_0(1-\xi^2)$ and $\sin\phi = \xi$. .

The Fourier transform $\tilde{a}_{gF}(\omega)$ of the ground acceleration process $\tilde{a}_g(t)$ can be obtained from the Fourier transform $\tilde{a}_{bF}(\omega)$ of the input bedrock acceleration $\tilde{a}_b(t)$ through the complex frequency response function, $A(\omega)$ (equation (A5.42))

$$\tilde{a}_{gF}(\omega) = A(\omega)\tilde{a}_{bF}(\omega) \tag{9.32}$$

with (equation (4.101))

$$A(\omega) = \frac{1 + 2i\xi\omega/\omega_0}{1 - (\omega/\omega_0)^2 + 2i\xi\omega/\omega_0} \tag{9.33}$$

---

averaging the relevant squared moduli

$$E[|\tilde{a}^{\chi}_{gF}(\omega)|^2] \simeq \frac{1}{n}\sum_{k=1}^{n} |\dot{a}_{kF}(\omega)|^2$$

On the other hand, for any pair of values ξ and ω_0,

$$|A(\omega)|^2 = \frac{1 + 4\xi^2(\omega/\omega_0)^2}{[1-(\omega/\omega_0)^2]^2 + 4\xi^2(\omega/\omega_0)^2}$$

and from equations (9.29a) and (9.29b) the best-fitting values of ξ and ω_0 can be obtained once C is known. For the estimation of C, note that from equation (A5.66)

$$s^2_{a^{\chi}(t)}(t) = \int_{-\infty}^{\infty} 2\pi G_{0\upsilon}\chi^2(\tau)[h(t-\tau)]^2\,d\tau \tag{9.29c}$$

$h(\ )$ being the filter response function.

Denote now by $h_F^2(\omega)$, $\chi'_F(\omega)$ and $s^2_{aF}(\omega)$ the Fourier transforms of, respectively, $h^2(t)$, $2\pi G_{0\upsilon}\chi^2(t)$ and $s_a^2(t)$. The Fourier transform of equation (9.29c) gives

$$\chi'_F(\omega) = s^2_{aF}(\omega)/h_F^2(\omega)$$

and as $h_F^2(t)$ and $s^2_{aF}(\omega)$ can be calculated, the quantity $2\pi\chi^2(t)G_{0\upsilon}$ is estimated by inverse Fourier transformation of $\chi'_F(\omega)$.

The variance $s_a^2(t)$ of the ground acceleration is estimated from real seismograms, by

$$s_a^2(t) \simeq \frac{1}{n}\sum_{k=1}^{n} \dot{a}_k^2(t)$$

[§] Note that both equations (9.31) and (9.33) are somewhat different from the analogous (4.89) and (4.101), as the l.h.s. of the relevant differential equation is different. Note also the different meanings of the functions: in Chapter 4 $h(t)$ and $A(\omega)$ were the response functions in terms of *displacements*, while here they are the response functions in terms of *accelerations*.

Thus, the power spectral density function for the filtered process $\tilde{a}_g(t)$ is (equation (A5.73))

$$G_{a_g(t)}(\omega) = |A(\omega)|^2 G_{0\upsilon} \qquad -\infty < \omega < \infty \tag{9.34}$$

If the filter parameters ξ and ω_0 are known, a sample ground accelerogram $\grave{a}_g(t)$ can therefore be calculated from each available sample of bedrock acceleration $\grave{a}_b(t)$. For this purpose, one solves equation (9.29) with deterministic r.h.s. $-\grave{a}_b(t)$ and makes use of (9.28).

The filter described above attenuates the higher-frequency components and amplifies those in the neighbourhood of $\omega = \omega_0$. However, it does not change the amplitude as $\omega \to 0$, and hence strong singularities are present at $\omega = 0$. These undesirable singularities can be removed by passing $\grave{a}_g(t)$ through a second filter (Section 8.3) which greatly attenuates the very-low-frequency components. Such a filter is defined by [2]

$$|A_f(\omega)|^2 = \frac{(\omega/\omega_1)^4}{[1-(\omega/\omega_1)^2]^2 + 4\xi_1^2(\omega/\omega_1)^2} \tag{9.35}$$

so that the output process $\tilde{a}_{gf}(t)$ has the power spectral density function (Fig. 9.5):

$$G_{a_{gf}(t)}(\omega) = |A_f(\omega)|^2 \, |A(\omega)|^2 \, G_{0\upsilon} \qquad -\infty < \omega < \infty \tag{9.36}$$

In equation (9.35) ω_1 and ξ_1 are selected to give the desired filter characteristics: usually, the values $\omega_1 = 0.897 \text{ rad s}^{-1}$ and $\xi_1 = 1/\sqrt{2}$ can be assumed [3,4].

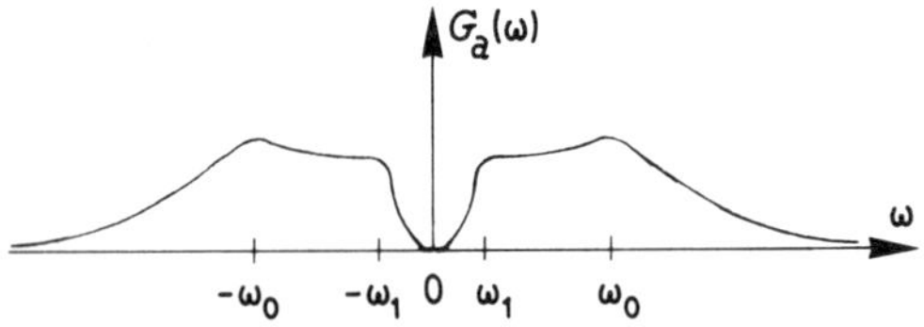

Fig. 9.5 Power spectral density function of twice filtered stationary white-noise $\tilde{a}_{gf}(t)$.

Note that this filter can be defined through the differential equation

$$\ddot{y} + 2\xi_1\omega_1\dot{y} + \omega_1^2 y = -\grave{a}_g(t) \tag{9.37}$$

of which equation (9.35) expresses the modulus of the frequency response function, and then putting

$$\grave{a}_{gf}(t) = \ddot{y}(t) \tag{9.38}$$

It is worth noting that one can also operate in the frequency domain by using equation (9.14) with $\omega = \omega_0$. The function $G_{x(t)} = G_{a_{gf}(t)}$ can be expressed in the form of equation (9.36), in which case the summation in equation (9.14′) should be started with $i = h$ where $(\Delta\omega/2 + h\Delta\omega)$ or $h\Delta\omega$ is equal to ω_1 for N odd or even respectively. It can also be expressed in the form of equation (9.34) and in this case summation starts with $i = 0$ or 1, respectively. Obviously, equation (9.14′) permits any arbitrary form for $G_{a_{gf}(t)}$ without posing difficulty.

At this stage, the intensity shape function $\chi(t)$ has still to be defined in order to complete the ground motion modelling. Usually, the estimated intensity function (cf. footnote on page 352) exhibits sharp irregularities (Fig. 9.6) which should not be present, reasonably, in the underlying earthquake ground motion process, and can be assumed to be due essentially to roughness in analysis and to the small number of manipulated earthquakes.

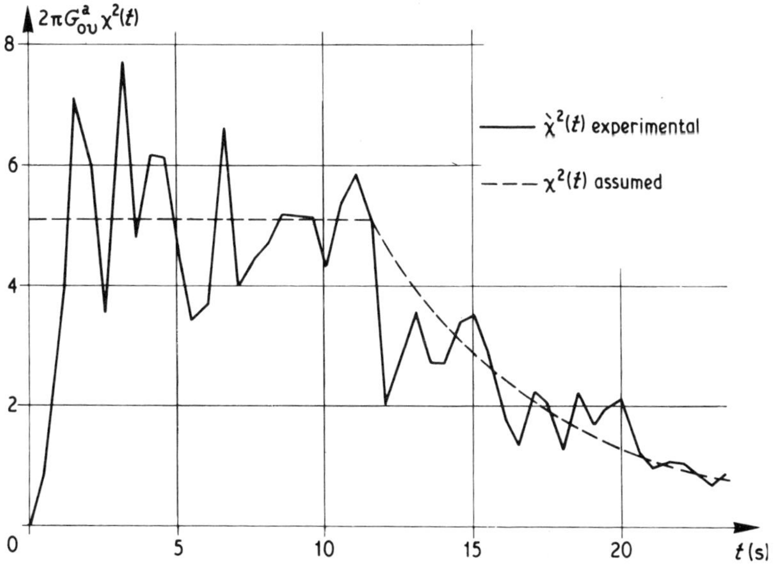

Fig. 9.6 Sketch of estimated intensity shape function $\chi(t)$.

Therefore, a smooth function $\chi(t)$, depending on three parameters, is generally assumed (Fig. 9.7):

$$\chi(t) = \begin{cases} (t/t_1)^2 & \text{if } t \leqslant t_1 \\ 1 & \text{if } t_1 < t \leqslant t_2 \\ e^{-c(t-t_2)} & \text{if } t > t_2 \end{cases} \tag{9.39}$$

Thus, the random processes $\tilde{a}_b(t)$ and $\tilde{a}_g(t)$ are assumed stationary in the interval (t_1, t_2). The parameters t_1, t_2, c in (9.39) can be estimated by fitting $\chi(t)$ to the estimated function in the least-squares sense.

When one lacks direct data, one can assume [2]: $t_1 = 1.21$ s, $t_2 = 11.5$ s and $c = 0.155\ \mathrm{s}^{-1}$ for $\mathscr{d} = 15$–20 s. Sometimes, $t_1 = 0$ is used (Fig. 9.6).

The seismicity of a site is often defined by the peak acceleration statistics, rather than the quantity $G_{0\upsilon}\chi^2(t)$. In this case, one can simulate the accelerogram by assuming any value for $G_{0\tilde{v}}$, on the condition that the expected peak acceleration of the simulated ground motion satisfies some conditions. It is expected that the maximum value $\tilde{a}_p$ of $\tilde{a}_{gf}(t)$ will occur in the stationarity interval (t_1, t_2). Therefore, the problem is to relate the maximum value of a Gaussian stationary process in an interval of length $t_2 - t_1$ to its fundamental parameters. As n is the number of random numbers y_j that are sampled for simulating an earthquake of typical duration d, the same number n of values $\grave{a}_{gf}(t_j)$ is obtained after executing simulation. The relevant standard deviation (equation (A5.49)) is

$$s_{a_s} = s_{a_{gf}(t)} = B_{a_{gf}(t)}(0) = \int_{-\infty}^{\infty} G_{a_{gf}(t)}(\omega)\,d\omega \tag{9.40}$$

Since n is very large, the expected number $N(a_0)$ of ordinates $\grave{a}_{gf}(t_j)$ exceeding any fixed values a_0 is given, empirically, by the product of n itself times the probability that the random quantity $\tilde{a}_{gf}(t)$ exceeds a_0 for any given t:

$$E[N(a_0)] = n \,\text{Prob}[\tilde{a}_{gf}(t) \geqslant a_0 \,|\, t] \tag{9.41}$$

The expected maximum value μ_{a_p} of the simulated sample functions corresponds to

$$E[N(\mu_{a_p})] = 1 \tag{9.42}$$

and hence

$$\text{Prob}[\tilde{a}_{gf}(t) \geqslant \mu_{a_p} | t] = 1/n \tag{9.43}$$

With n of order 1000, the expected peak value of the acceleration in the simulated process can be approximately set equal to three times the standard deviation s_{a_s} of the random variable $\tilde{a}_{gf}(t_j)$. In fact, noting that $E[\tilde{a}_{gf}(t)] = 0$ and putting $\mu_{a_p} = 3s_{a_s}$, one gets, from the tables of the Gaussian distribution (Table A.5),

$$\text{Prob}[\tilde{a}_{gf}(t) \geqslant \mu_{a_p} | t] = 1.35 \times 10^{-3} \simeq 1/1000$$

It follows that, if $\bar{a}_p$ is the actual peak acceleration that must be reproduced by the simulated earthquake, it is sufficient to multiply the available function by the amplitude factor

$$Y = \bar{a}_p / \mu_{a_p} \tag{9.44}$$

Finally, note that the randomness of peak acceleration $\bar{a}_p$ and earthquake duration $\tilde{d}$ should also be introduced in the simulation procedure. These are in fact random variables whose characteristics can be determined either by direct inspection of available data or by seismic risk analysis. Possibly one may make use of equations (3.40) and (3.39).

9.2.5 Random functions: a non-stationary 'data-based' model

The intrinsic lack of ergodicity in the non-stationary process implies that the reconstruction of its probabilistic nature requires a large number of sample functions (Chapter 8; Section 8.6). In contrast, when dealing with earthquake accelerations, wind-induced pressures, dynamic flight loads, etc., one is fortunate when one has a few, and not just one, sample functions. In order to overcome

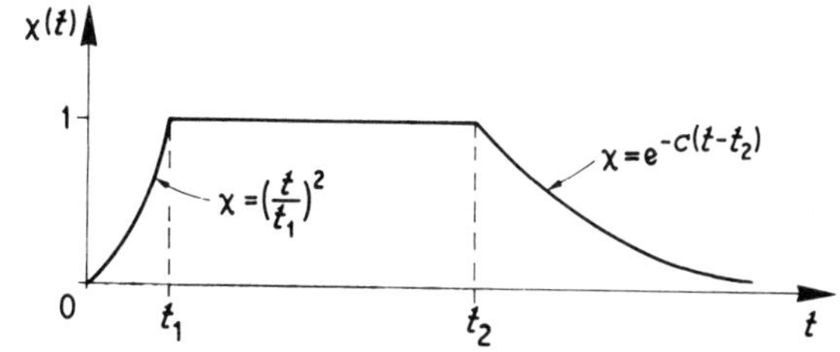

Fig. 9.7 Intensity function $\chi(t)$ usually assumed for non-stationary process $a(t)$.

this difficulty, another more sophisticated frequency domain model has been proposed [5, 6].

Let $x_0(t)$ be an observed record of duration d and $X_0(\omega)$ the Fourier transform of $x_0(t)$, multiplied by 2π, so that

$$x_0(t) = \frac{1}{2\pi}\int_{-\infty}^{\infty} X_0(\omega)\exp(\mathrm{i}\omega t)\,\mathrm{d}\omega$$

$$X_0(\omega) = 2\pi x_{0F}(\omega) = \int_{-\infty}^{\infty} x_0(t)\exp(-\mathrm{i}\omega t)\,\mathrm{d}t \tag{9.45}$$

Now introduce the phase angle $\phi_0(\omega)$ of $X_0(\omega)$ defined by

$$\phi_0(\omega) = -\phi_0(-\omega) = \tan^{-1}\left(\frac{\mathrm{Im}[X_0(\omega)]}{\mathrm{Re}[X_0(\omega)]}\right) \qquad \text{if } \mathrm{Re}[X_0(\omega)] \neq 0$$

$$\phi_0(\omega) = \pm\pi/2 + 2k\pi \qquad (k = 0, \pm 1, \ldots) \qquad \text{otherwise} \tag{9.46}$$

where Re[ ] = real part and Im[ ] = imaginary part. Then, as

$$X_0(\omega) = |X_0(\omega)|\exp[\mathrm{i}\phi_0(\omega)]$$

$x_0(t)$ can be expressed as

$$x_0(t) = \frac{1}{2\pi}\int_{-\infty}^{\infty} |X_0(\omega)|\exp\{\mathrm{i}[\omega t + \phi_0(\omega)]\}\,\mathrm{d}\omega \tag{9.47}$$

Construct now a non-stationary random process $\tilde{x}(t)$ so that

$$\tilde{x}(t) = \frac{1}{2\pi}\int_{-\infty}^{\infty} |X_0(\omega)|\exp\{\mathrm{i}[\omega t + \phi_0(\omega) + \tilde{\phi}(\omega)]\}\,\mathrm{d}\omega \tag{9.48}$$

where

$$\tilde{\phi}(\omega) = \tilde{\Lambda}\,\mathrm{sgn}(\omega) \tag{9.49}$$

with

$$\mathrm{sgn}(\omega) = \begin{cases} -1 & \text{for } \omega < 0 \\ 0 & \text{for } \omega = 0 \\ 1 & \text{for } \omega > 0 \end{cases}$$

and $\tilde{\Lambda}$ a random variable. A sample function of $\tilde{x}(t)$ can be generated by assigning a sample value of $\tilde{\Lambda}$ in equation (9.49) hence of $\tilde{\phi}$ in equation (9.48). Putting $\Lambda = 0$, one obtains the observed record (9.47).

Remember that in Section 9.2.2(a) the squared amplitudes X^2 of a set of discrete harmonics have been put in the form

$$X_i^2 = 2G_{x(t)}(-\bar{\omega} + i\Delta\omega)\Delta\omega \tag{9.50}$$

and hence, for a stationary random process, as $d \to \infty$ one formally obtains from equation (9.48)

$$\frac{1}{2\pi} X_0(\omega) = [2G_{x(t)}(\omega)\,\mathrm{d}\omega]^{1/2} \qquad -\infty < \omega < \infty \tag{9.51}$$

If only positive values of ω are considered, the amplitude has to be doubled to take into account the negative axis and the one-sided power spectral density (equation (A5.52)) can be introduced. Thus, for even N equation (9.48) can be reduced to

$$\tilde{x}(t) \simeq \sum_{i=1}^{n} [2G^{\circ}_{x(t)}(i\Delta\omega)\Delta\omega]^{1/2} \cos[i\Delta\omega t + \tilde{\phi}(i\Delta\omega)] + (G^{\circ}_{x(t)}(0)\Delta\omega)^{1/2} \cos\tilde{\phi}(0) \tag{9.52}$$

where $n = N/2$ and $\phi_0(\omega)$ has been absorbed into $\tilde{\phi}(i\Delta\omega)$. Comparison of equation (9.52) with equation (9.14′) indicates the essential difference between stationary and non-stationary cases. More precisely, the function $\tilde{\phi}(\omega) = -\tilde{\phi}(-\omega)$ is white-noise for $\omega > 0$ in the stationary case, while in the non-stationary case it is a random variable multiplied by the sign function (equation (9.49)) and, hence, its sample is not a function of ω but a constant for $\omega > 0$.

As $\phi_0(\omega) + \tilde{\phi}(\omega)$ is an odd function, equation (9.48) can also be written as

$$\tilde{x}(t) = \frac{1}{\pi}\int_0^{\infty} |X_0(\omega)| \cos[\omega t + \phi_0(\omega) + \tilde{\Lambda}]\,\mathrm{d}\omega \tag{9.53}$$

or equivalently

$$\tilde{x}(t) = x_0(t)\cos\tilde{\Lambda} - x_{0\mathrm{H}}(t)\sin\tilde{\Lambda} \tag{9.54}$$

where

$$x_0(t) = \frac{1}{\pi}\int_0^{\infty} |X_0(\omega)| \cos[\omega t + \phi_0(\omega)]\,\mathrm{d}\omega \tag{9.55}$$

is the observed record (equation (9.45)) and

$$x_{0\mathrm{H}}(t) = \frac{1}{\pi}\int_0^{\infty} |X_0(\omega)| \sin[\omega t + \phi_0(\omega)]\,\mathrm{d}\omega \tag{9.56}$$

can be proved to be the Hilbert transform of $x_0(t)$ (equation (A7.53)), which can also be written in the equivalent form

$$x_{0\mathrm{H}}(t) = \frac{1}{\pi}\int_{-\infty}^{\infty} \frac{x_0(\tau)}{t-\tau}\,\mathrm{d}\tau \tag{9.56′}$$

The process given by equation (9.54) is known as a *data-based non-stationary process of the first kind*. Its realizations can be obtained from equation (9.54) by simulating a value $\grave{\Lambda}$ (one value for each realization) for the random variable $\tilde{\Lambda}$. The Hilbert transform might be zero outside the range $(0, d)$; however, that does not occur if the record $x_0(t)$ oscillates rapidly and more or less symmetrically with respect to the baseline within $(0, d)$.‡

Equation (9.54) can also be written using the alternative expression

$$\tilde{x}(t) = \xi_0(t)\cos[\tilde{\Lambda} + \theta_0(t)] \tag{9.57}$$

where $\xi_0(t)$ is the envelope function (see equation (A7.52))

$$\xi_0(t) = \sqrt{[x_0^2(t) + x_{0H}^2(t)]} \tag{9.58}$$

and

$$\theta_0(t) = \tan^{-1}[x_{0H}(t)/x_0(t)] \tag{9.59}$$

The random variable $\tilde{\Lambda}$ is generally assumed to be distributed in accordance either with the uniform distribution in $(-m\pi, m\pi)$ or with the Gaussian distribution with mean value μ_Λ and standard deviation s_Λ. Note that, because of the periodicity of $\exp(\mathrm{i}\Lambda)$, its distribution function with normally distributed $\tilde{\Lambda}$ approaches, as $s_\Lambda \to \infty$, the distribution obtained under the assumption of $\tilde{\Lambda}$ being uniformly distributed.

Under either the assumption that $\tilde{\Lambda}$ is uniformly distributed in $(-m\pi, m\pi)$ (i.e. with $\mu_\Lambda = 0$) or the assumption that $\tilde{\Lambda}$ is Gaussian with $\mu_\Lambda = 0$ and $s_\Lambda \to \infty$, the following characteristics of the random function $\tilde{x}(t)$ are obtained:

$$\mu_{x(t)}(t) = 0 \tag{9.60}$$

$$s_{x(t)}(t) = \tfrac{1}{2}[x_0^2(t) + x_{0H}^2(t)] \tag{9.61}$$

$$\mathrm{E}[x(t_1)x(t_2)] = \tfrac{1}{2}[x_0(t_1)x_0(t_2) + x_{0H}(t_1)x_{0H}(t_2)] \tag{9.62}$$

$$G_{x(t)}(\omega_1, \omega_2) = \begin{cases} x_0(\omega_1)x_0^\dagger(\omega_2) & \omega_1\omega_2 \geqslant 0 \\ 0 & \text{otherwise} \end{cases} \tag{9.63}$$

$$P_{x(t)}(x) = \begin{cases} (1/\pi)[x_0^2(t) - x^2]^{1/2} & x < x_0(t) \\ 0 & \text{otherwise} \end{cases} \tag{9.64}$$

where† indicates the complex conjugate.

Extensions to multivariate one-dimensional processes are also possible [6].

9.3 CRITICAL APPRAISAL OF SIMULATION TECHNIQUES

From a theoretical point of view, the true value of any parameter describing a random quantity can only be achieved (see Section 7.3) when the number n of experiment's tends to ∞. However, simulation techniques are generally

‡ A *data-based non-stationary process of the second kind* has been suggested in case this condition is not satisfied by the available record. However, the reader is referred to the literature [5] for this second model, whose analytical derivation is more complex.

considered to yield sufficiently accurate results, with reference to *central* values (i.e. values in the neighbourhood of the average) of the studied distributions, in about 10^2–10^3 experiments. In contrast, in structural engineering, because of the high reliability required, very low values of probability and the *tails* of the PDFs of the response quantities have to be investigated: such *tails* are defined with good accuracy for n not less than 10^6–10^9. Such computational effort could be justified only if the object of the analysis were an expensive element of common and repeated use, but in most cases the large number of experiments (10^6–10^9) needed for an accurate reliability estimation by simulation makes such a procedure practically impossible, either because the number of structures or structural elements which will actually be produced is very low (often, unique structures are produced in civil engineering) or because the market cost of the single mechanical component does not justify such a sophisticated analysis. Thus, the basic problem of investigating the randomness of structural responses by numerical procedures appears to be without practical solution, unless a *hybrid* (or *mixed*) approach is adopted, as described in Section 9.4.

9.3.1 An example of simulation: the 'European' column curves

The study of a compressed column with random *imperfections* can be tackled by the methods presented in Chapter 6, but a more accurate and realistic analysis that does not neglect the residual stresses in the column can be performed only by using simulation techniques. For this purpose, one can employ any of the computer codes that were written for the *deterministic numerical 'simulation'* ‡ of the steel column behaviour. After the parameters measuring the initial out-of-straightness, the residual stresses and the yielding stress have been chosen by random sampling (Section 9.2) as independent input data, the deterministic calculations can be performed to determine the strength of the relevant column with given geometry and slenderness. By repeating this process with different artificially yielded data, a sample of size n of the column strength is obtained. Fig. 9.8 shows the results of samples of size 20 of the collapse stress σ_C (defined as the ratio of the load-carrying capacity of the compressed column to the nominal cross-sectional area) that were obtained by simulation for different slenderness values λ_B of a fixed I-section column; the statistics of the input data were obtained by direct measures on a population of columns, on which physical experiments were also performed [7].

Experimental and simulated results appear to be in good agreement. Some discrepancies appear only for slenderness values from 80 to 120, where the geometrical imperfections play a significant role, and can be justified by noting

‡ Note the different meaning that the word *simulation* has here. It characterizes in fact the deterministic study of a physical problem by means of an appropriate model that allows a numerical solution and that was proved to be accurate by comparison with the available experimental results. In contrast, *simulation* is generally employed in this book to indicate the repeated use of deterministic simulation algorithms in order to produce a statistical sample. However, in both cases this word denotes that laboratory experiments are substituted by numerical approaches.

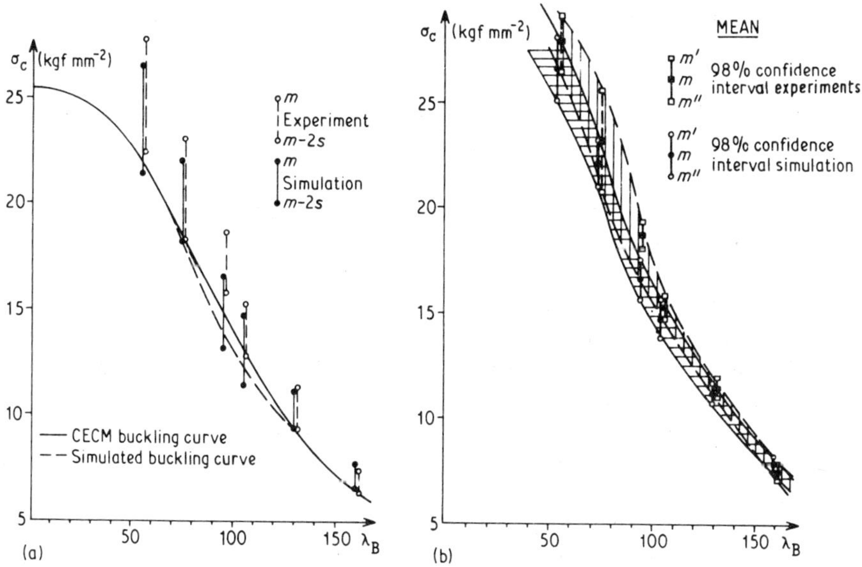

Fig. 9.8 Collapse stress of compressed steel column: comparison of experimental and numerical simulation results (sample size, $n=20$; m, sample mean; s, sample variance) [7].

that if the initial out-of-straightness is greater the residual stresses have generally lower values. In other words these two quantities are negatively correlated. However, stochastic independence was assumed in the simulation, and hence the actual strength was underestimated. On the whole, Fig. 9.8 shows that the numerical approach provides a good idealization of the physical behaviour and therefore, if the collapse stresses corresponding to very small probabilities were required, they could be yielded reliably by a large simulation ($n = 10^6 - 10^9$ for each λ_B).

Nevertheless, though the compressed steel column is certainly one of the most common elements in structural engineering, no national or international organization has deemed it suitable till now to finance such an investigation. In fact, the first simulation results confirmed the general accuracy of much less expensive analytical procedures, which were in the end used to derive column design curves. One of the key roles that simulation techniques can play in structural engineering thus becomes evident, namely to confirm the goodness and the accuracy of the results obtained by a simplified analytical model of the structural behaviour, rather than to provide results significant *per se*.

9.4 HYBRID SIMULATION PROCEDURES

Hybrid (or mixed) simulation consists of making use of simulation techniques only to yield samples of structural responses whose size is selected to ensure good accuracy (according to the theory of statistics in Chapter 7) on the estimation of the central values of the quantities of interest. An analytical‡ model is then used to translate statistical information into probabilistic statements and reliability evaluations.

In this way, simulation techniques can reasonably be employed to investigate the statistics of the structural responses that, in a deterministic context, can be studied only by numerical procedures. In such cases, in order to make operative the whole analysis process, the identification of the quantities whose randomnesses have to be simulated becomes a basic problem: a bad selection of these quantities, in fact, can involve either meaningless results or excessive analysis costs. For instance, if a structure is subjected to dynamic excitations only for comparatively short periods during its lifetime, a simulation of the whole time history is not necessary. A mixed approach that simulates the response to each single excitation and models in an analytical way the long-term structural behaviour can in fact be adopted. Section 9.4.1 deals with this particular problem as an example.

A more general example of very common use in structural reliability analysis is the estimation of the probability of attaining an ultimate limit state. Once both analytical form and parameter values of the relevant distribution functions have been chosen, the probability of failure is estimated by introducing, as a suitable probabilistic analytical model, the convolution integral (equation (2.68)). The properties of the function $P_R(\mathbf{W})$, i.e. the conditional probability of failure (CPF) under a given load $\mathbf{W}$, which depends on the randomness of the geometrical and mechanical structural properties, are investigated by simulation techniques whenever analytical procedure (cf. Chapters 4, 5 and 6) are not available.

Equation (2.68) is written in the load space, but a reliability problem can also be approached (Section 2.8.1) in the load-effect space or in the stress space (Fig. 2.7). In the last case, one must simulate the stress caused in a point of the structure by the random external actions and solve the convolution integral of the product of the relevant simulated PDF of the active stress by the CDF of the yield stress the analytical form of the latter is often known with acceptable accuracy. By contrast, if one operates in the load-effect space, one must simulate both functions whose product must be integrated. The PDF of the load effects and the CDF of the relevant resistant quantities, in fact, are both dependent on the random mechanical and geometrical characteristics of the structure.

9.4.1 An example of the hybrid approach: simulation of damage accumulation under a sequence of independent pulses

(a) Statement of the problem

In many instances during its lifetime, a structure is subjected not only to continuous load processes but also to random sequences of short-term excitations

‡ With reference to the estimation of the PDF of a fixed response parameter, available analytical models are the usual probability laws (Section A2.3) and the inferring procedures (Section 7.5) have to be employed to select the best-fitting one. (Note that the problem of choosing, from among several convenient mathematical forms, the distribution which best fits observed data affects all statistical information about the structural parameters, and therefore is not a specific limit of simulation.)

whose action is simultaneous with that of permanent and long-duration loads (Chapter 3). In these problems, the structure has usually been designed to high reliability under the constant and quasi-permanent loads, so that simulation is necessary only to check the reliability level when the random excitation acts.

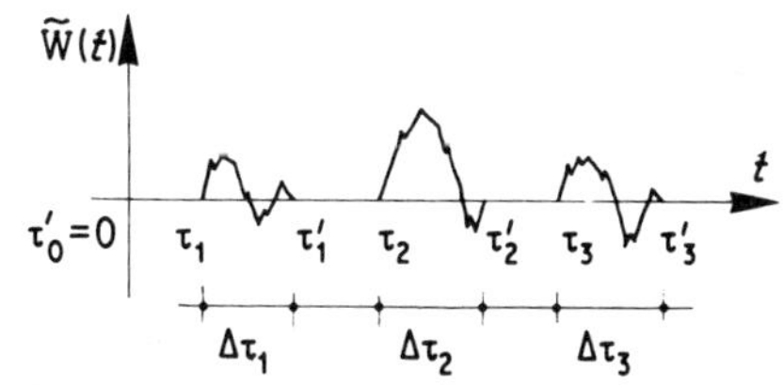

Fig. 9.9 Loading function as a sequence of random pulses.

Often, the random short-term-load sequence is well described by one scalar function $\tilde{W}(t)$, equal to zero in a number of finite random ranges (τ'_i, τ_{i+1}) (Fig. 9.9). If the pulses are assumed to occur according to a Poisson process with intensity λ (Section A6.3.2), $\tilde{W}(t)$ is a filtered Poisson process (equation (9.22)):

$$\tilde{W}(t) = \sum_{i=1}^{\tilde{n}(t)} \mathscr{W}(t, \tilde{\tau}_i, \tilde{\tau}'_i, \tilde{y}_{1i}, \tilde{y}_{2i}, \ldots) \tag{9.65}$$

where $\mathscr{W}(\quad)(\neq 0 \text{ for } \tau_i \leqslant t \leqslant \tau'_i)$ is the 'pulse' function; $\tilde{n}(t)$ denotes the random number of pulses at time t; $\tilde{y}_{1i}, \tilde{y}_{2i}, \ldots$ define the randomness of the ith pulse. The quantities $\tilde{y}_{1i}, \tilde{y}_{2i}, \ldots$ are generally supposed to be independent, identically distributed random variables. The same assumption can be made for the quantities $\Delta\tilde{\tau}_1 = \tilde{\tau}'_1 - \tilde{\tau}_1$, $\Delta\tilde{\tau}_2 = \tilde{\tau}'_2 - \tilde{\tau}_2, \ldots$.

When the load process acting on a structure is well described by equation (9.65), three types of problems can occur:

(1) The structure behaves elastically under every single excitation. This is the case of structures for which serviceability conditions determine the design. The analysis of the behaviour under a single pulse is sufficient for reliability purposes (Section 6.5).
(2) Each excitation may induce plastic deformations in the structure, but, whenever that occurs, the structure is repaired at the end of the pulse. Such a situation is sometimes assumed in earthquake engineering. Also in this case only the behaviour under a single pulse must be studied for reliability purposes.
(3) Each excitation may induce plastic deformations in the structure and no repair is performed. For these systems, a sequence of random dynamic or quasi-static loads involves the accumulation of residual deformations; at each instant, the local residual deformation is strictly related (and often identified) with the structural *damage*. In this case the behaviour during the whole time history must be investigated.

In order to study the randomness of a plastic deformation parameter (e.g. a residual deformation $\tilde{u}_r(t)$ at a given time t), one can simulate more (say n_h) load histories, by using equation (A6.37) for $n(t)$ and the concepts of Section 9.2.1, and perform n_h analyses whose result is a sample of size n_h of $u_r(t)$, from which its probability properties must be inferred. In this way the average number of deterministically analysed pulses is $n_h \mu_{n(t)}$. With the same computational effort a more accurate result can be achieved by using a hybrid approach, as anticipated: namely, the random properties of the structural response to each pulse are obtained by simulation, and those of the long-term response are derived analytically ‡.

(b) Solution for elastic–perfectly plastic structures

Assume first that plastic deformations do not influence the yielding level of the structure, i.e. assume an elastic–perfectly plastic behaviour of the structure and negligible geometrical effects.

If the duration $\Delta\tau_i$ of the ith pulse is much smaller than the mean inter-arrival time between two successive pulses, the residual deformation increment ψ_i due to the ith pulse may be considered constant at any time $t > \tau_i$ and equal to its actual value at $t = \tau_i + \Delta\tau_i$. In such a way, the total residual deformation $u_r(t)$ can be idealized as in Fig. 9.10(b) and both the input and output processes may be described as filtered Poisson processes. More precisely, at any time t the total residual displacement $\tilde{u}_r(t)$ may be idealized by

$$\begin{aligned}\tilde{u}_r(t) &= \sum_{i=1}^{\infty} \tilde{\psi}_i H(t - \tilde{\tau}_i) \\ &= \sum_{i=1}^{n(t)} \tilde{\psi}_i\end{aligned} \tag{9.66}$$

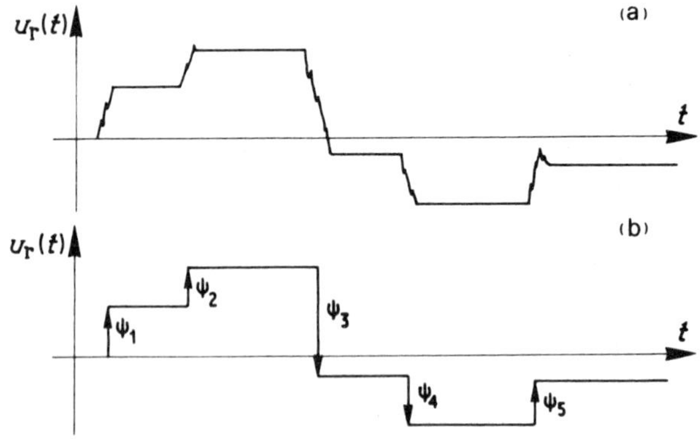

Fig. 9.10 Typical histories of residual deformation: (a) actual; (b) idealized.

‡ A different, fully analytical approach to tackle a similar problem was presented in Chapter 5, Sections 5.3.5 and 5.3.6: it yields only *bounds* on the response.

where the τ_i are the arrival times for the Poisson process of Fig. 9.9; $H(v_i)$ is the Heaviside step function, with $v_i = t - \tau_i$. The 'pulse' function of the filtered Poisson process is of the form:

$$U(v_i, \psi_i) = \psi_i H(v_i) = \begin{cases} \psi_i & \text{for } v_i \geq 0 \\ 0 & \text{otherwise} \end{cases} \tag{9.67}$$

and hence this process is also a compound Poisson process (equation (A6.38)).

Therefore the expected value of the total residual displacement at any time t may be expressed by (Section A6.3.3) [8].

$$\mathrm{E}[\tilde{u}_r(t)] = \lambda \int_0^t \mathrm{E}[U(v, \tilde{\psi})]\,\mathrm{d}v = \lambda\, t\, \mathrm{E}[\tilde{\psi}] \tag{9.68}$$

where

$$\mathrm{E}[U(v, \tilde{\psi})] = \int_{-\infty}^{+\infty} U(v, \psi)\, p_{\psi}(\psi)\,\mathrm{d}\psi \tag{9.69}$$

and the variance is given by

$$\mathrm{Var}[\tilde{u}_r(t)] = \lambda \int_0^t \mathrm{E}[U(v, \tilde{\psi})]\,\mathrm{d}v = \lambda\, t\, \mathrm{E}[\tilde{\psi}^2] \tag{9.70}$$

The 'characteristic function' $\Omega_{u_r(t)}(x, t)$ (i.e. the Fourier transform of the PDF; equation (A2.17) can be determined directly by equation (A6.39) as

$$\begin{aligned} \Omega_{u_r(t)}(x, t) &= \int_{-\infty}^{+\infty} \mathrm{e}^{\mathrm{i}u_r} p_{u_r(t)}(u_r, t)\,\mathrm{d}u_r \\ &= \exp\left(\lambda \int_0^t \mathrm{E}[\mathrm{e}^{\mathrm{i}xU(v,\tilde{\psi})} - 1]\,\mathrm{d}v\right) \\ &= \exp[\lambda\, t(\Omega_{\psi}(x) - 1)] \end{aligned} \tag{9.71}$$

where $\Omega_{\psi}(x)$ is the characteristic function for any of the $\tilde{\psi}_i$ random variables. Thus the problem of determining the probability distribution of the residual deformation $\tilde{u}_r(t)$ is completely ‡ solved once the probability density function $p_{\psi}(\psi)$ is known. This function can be easily inferred by a simulated sample of the plastic deformation $\tilde{\psi}$ at the end of a single loading pulse. If the size n of this sample is $n_h \mu_{n(t)}$, the computational effort is the same required by a time-history simulation, but the degree of precision achieved on the estimation of the parameters of $\tilde{\psi}$ is much better than that obtained for $u_r(t)$. Moreover, equation (9.71) solves the

‡ The inversion of the characteristic function can be made noting that the density function $p_{u_r(t)}(u_r, t)$ goes to zero quickly for large values of the argument. In fact, if $p_{u_r(t)}(u_r, t)$ is assumed to be zero outside finite limits $u_r = \pm L$ (where the value of L depends on the considered t), the complex Fourier series expansion for $p_{u_r(t)}(u_r, t)$ can be written as:

$$p_{u_r(t)}(u_r, t) = \sum_{k=-\infty}^{+\infty} c_k(t) \exp(-\mathrm{i}k\pi u_r/L)$$

where

$$c_k(t) \simeq \Omega_{u_r(t)}(k\pi/L, t)/2L$$

problem for any time t while the global simulation approach determines the sought quantities for an assigned value of the time.

By inverting (9.71), the structural reliability $\mathbb{R}(t)$, at any time t, can be calculated once given the permissible limit u_L for the total residual deformation u_r:

$$\mathbb{R}(t) = \int_{-u_L}^{+u_L} p_{u_r(t)}(u_r, t)\,du_r \tag{9.72}$$

Equation (9.72) holds if the limit state considered is assumed to be a serviceability limit state. By contrast, when the permanent deformation accumulation must be regarded as an ultimate limit state (for instance because of the limited ductility of the material considered), the reliability analysis must be performed by solving a first excursion probability problem (Appendix A7).

The procedure just described has been applied to the simple one-degree-of-freedom perfectly plastic structure drawn in Fig. 9.11. The assumed pulse loading function is $W(t) = Ma(t)$, where M is the mass of the beam and $a(t)$ is the ground acceleration. The limit (collapse) accelerations $\bar{a}$ and $-\bar{a}'$ are defined by the expression $\bar{a} = \bar{a}' = 4M_y/Mh$, where M_y is the limit moment in the plastic hinges and h the length of the columns. Under the assumption of rigid–plastic behaviour, the transverse displacement $u(t)$ of the beam is equal to the plastic deformation u_r and is described by the differential equations

$$\ddot{u} = \begin{cases} a - \bar{a} & \text{when } \dot{u} \geqslant 0 \\ a + \bar{a}' & \text{when } \dot{u} \leqslant 0 \\ 0 & \text{when } \dot{u} = 0 \text{ and } \bar{a}' \leqslant a \leqslant \bar{a} \end{cases}$$

The results obtained from the deterministic analysis of 100 artificially yielded accelerograms, followed by application of the analytical model of equation (9.71) are summarized in Fig. 9.11, together with the analogous results derived for a

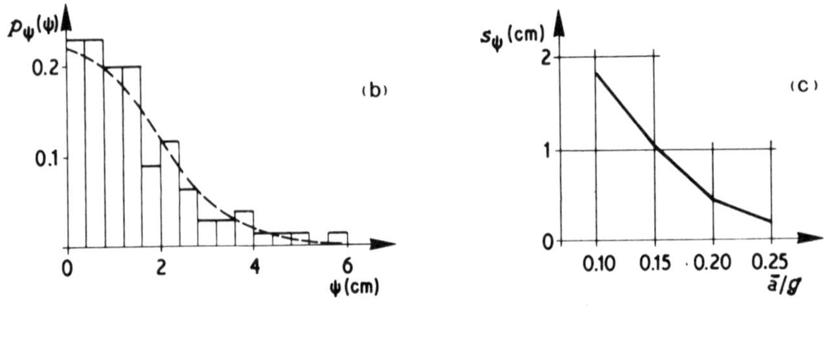

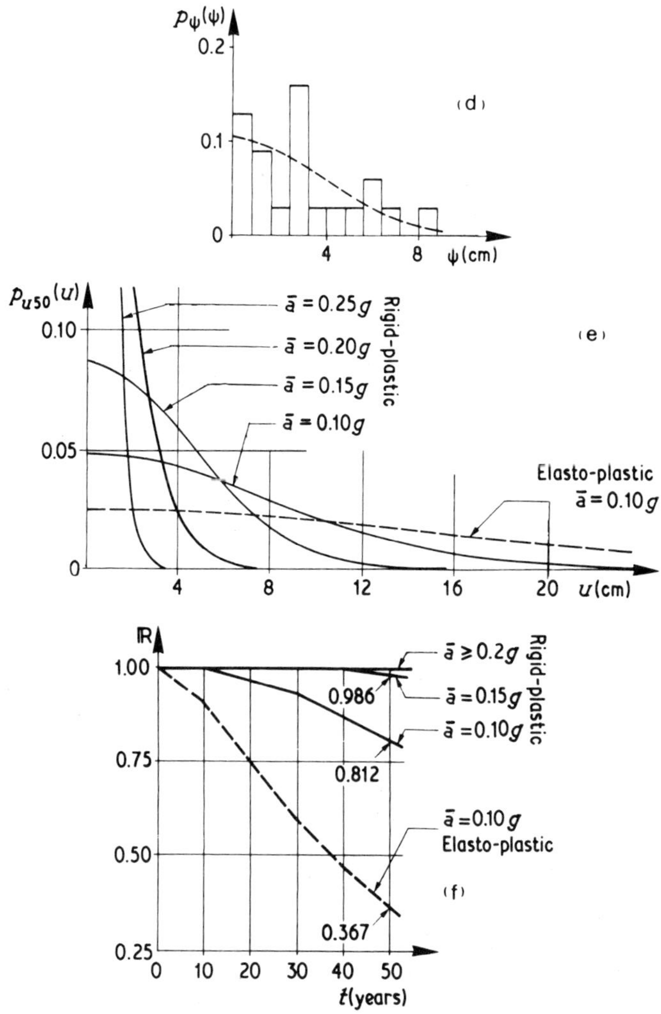

Fig. 9.11 (a) Considered single-degree-or-freedom rigid–perfectly plastic structure. (b) Histogram of the permanent displacement $\tilde{\psi}$ at the end of a single pulse (and its fitting normal distribution) for $\bar{a} = 0.1\,g$. (c) Relationship between the standard deviation of $\tilde{\psi}$ and the limit acceleration $\bar{a}$. (d) Histogram of $\tilde{\psi}$ in the elastic–plastic case ($k/M = 160\,\mathrm{s}^{-2}$. (e) PDFs $p_{u50}\,(u)$ for different values of the limit acceleration $\bar{a}$; mean arrival rate of pulses $\lambda = 0.4$/year. (f) Variation in time of the structural reliability; $\lambda = 0.4$/year (from Casciati *et al.* [9]).

smaller sample ($n = 20$) under the assumption of elastic–perfectly plastic behaviour.

(c) Solution for structures whose strength depends on the deformation

Consider now a structure whose load-carrying capacity varies with the deformation because of geometric non-linearities and/or (linear) hardening of the material. In this case the inelastic deformation variation due to each excitation depends on the inelastic deformation at the arrival time of the pulse. Therefore, Markov process idealizations (Section A6.4) are considered and developed.

Again let $u_r(t)$ denote the relevant plastic displacement (on which the carrying capacity of the structure now depends) and assume that failure occurs when the absolute maximum value of $u_r(t)$ during an excitation (peak inelastic displacement) reaches a limit value u_L, determined by ductility requirements or/and geometric non-linearities. After the failure time t_f, the fictitious state

$$u_r(t) = u_L \qquad \text{for } t > t_f \tag{9.73}$$

may be given to the structure.

The simplest model to describe the stochastic process $\tilde{u}_r(t)$ is a homogeneous Markov chain, discrete in both state and time (Section A1.4). Let both the reference time interval $(0, T)$ and the displacement range $(-u_L, u_L)$ be discretized as a set of m equally spaced times $t_1 = 0 < t_2 < \ldots < t_m$ and n $u_{r1} < u_{r2} < \ldots < u_{rn}$. The initial state is the undeformed one and the limit states $\pm u_L$ are two absorbing barriers (equation (A1.36)). The 'absolute probabilities' at times t_k are given by the vector (equation (A1.30))

$$\mathbf{p}_{(k)} = (\mathbf{p}_{(1)}\mathbf{P}_{(k-1)})^{\mathrm{T}} = \mathbf{p}_{(1)}[\mathbf{P}]^{(k-1)} \tag{9.74}$$

whose jth component $p_{(k)\,j}$ denotes the unconditional probability that the Markov chain is in the state j at time t_k and $\mathbf{p}_{(1)}$ defines the initial conditions. The sum of the probabilities of being in the absorbing states represents the probability that the structure fails at time $t = t_k$.

In equation (9.74) $\mathbf{P}$ is the one-step transition probability matrix of the considered homogeneous Markov chain. The element P_{ij} $(i,j = 1, \ldots, n)$ denotes the conditional probability that the Markov chain, in the state i at time $t = 0$, will move after one step to state j. The condition $\sum_j P_{ij} = 1$, for all i, follows from the definition.

The ith row of the matrix $\mathbf{P}$ can be evaluated for the considered mechanical problem by simulating a $(0, t_2)$ time history with the initial condition $u_r = u_{ri}$. If, during any simulated excitation, the inelastic deformation reaches either permissible limit $\pm u_L$, afterwards the structure always remains in the corresponding ultimate state.

A more realistic approach consists of modelling the stochastic process $u_r(t)$ as a homogeneous continuous-time discrete-state Markov process. In this case, the absolute probabilities at time t are the solutions of the differential linear equation system (equation (A6.61)):

$$\frac{\partial \mathbf{p}(t)}{\partial t} = \boldsymbol{\Xi}^{\mathrm{T}}\mathbf{p}(t) \tag{9.75}$$

In (9.75), $\boldsymbol{\Xi}$ is defined by equations (A6.56) and (A6.57) and denotes the transition intensity matrix. In order to evaluate the elements of this matrix, it is necessary to define first a matrix $\boldsymbol{\Pi}^0$ whose element Π_{ij}^0 $(i,j = 1, \ldots, n)$ represents the

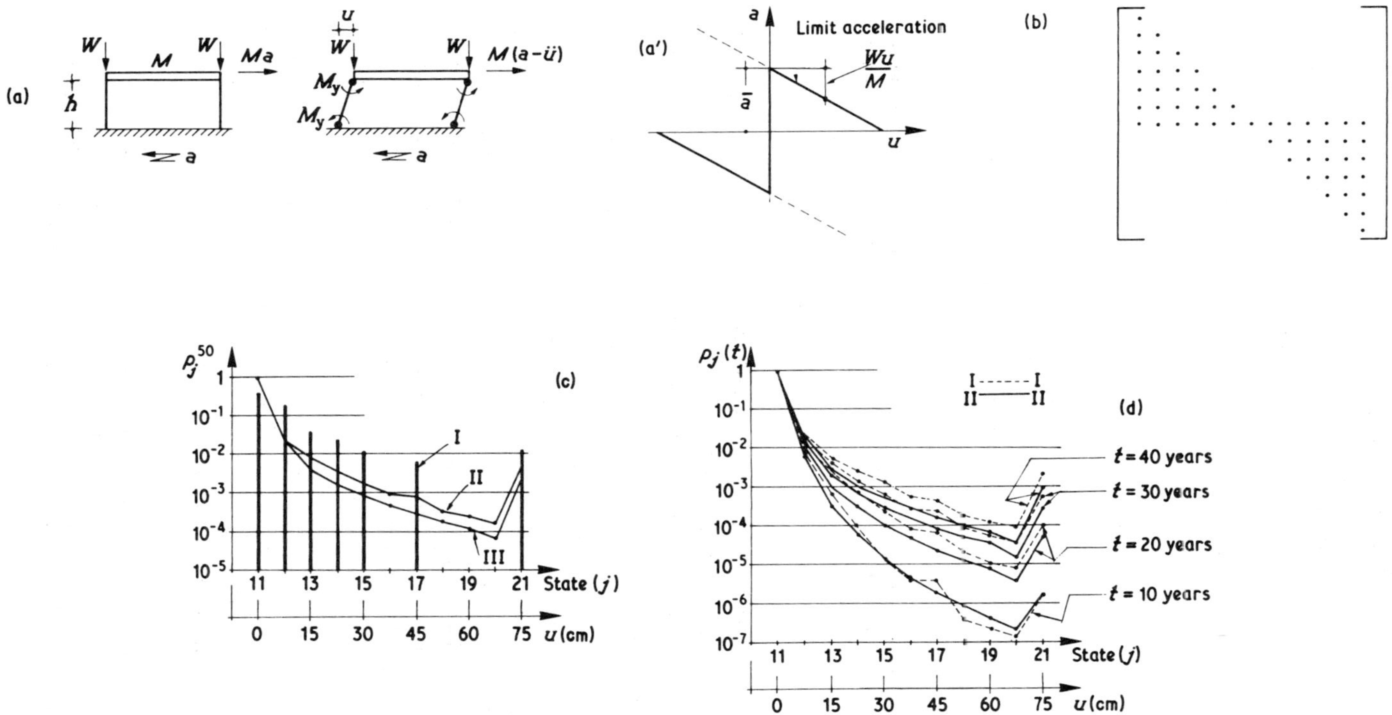

Fig. 9.12 (a) Considered single-degree-of-freedom rigid–perfectly plastic structure. (a′) Variation of limit acceleration with displacement. (b) Shape of the probability matrix **P** and of the conditional probability matrix **Π°** obtained in the considered numerical example. (c) The probability mass function $p_j(t)$ obtained for t = 50 yr and mean arrival rate of pulses λ = 0.2/yr using (I) a simulation method, (II) the discrete-time Markov model, (III) the continuous-time Markov model. (d) The probability mass functions $p_j(t)$ at different times t with λ = 0.2/year and (I) discrete-time Markov model, (II) continuous-time Markov model (after Casciati *et al* [9a]).

conditional probability that, if a ground motion occurs during $(0, h)$, the system moves from u_{ri} to u_{rj}. The matrix $\mathbf{\Pi}^0$ has generally non-zero diagonal elements, because the structure may return to the same state after the motion due to the excitation. The ith row of $\mathbf{\Pi}^0$ can be evaluated by simulating the structural response to a single excitation, taking into account the presence of the absorbing barriers [10].

Once one knows $\mathbf{\Pi}^0$, if the pulse arrivals are assumed to be Poisson events with intensity λ, one obtains

$$\begin{aligned} \Xi_{ii} &= -\lambda(1-\Pi_{ii}^0) \\ \Xi_{ij} &= \lambda(1-\Pi_{ii}^0)\frac{\Pi_{ij}^0}{1-\Pi_{ii}^0} = \lambda\Pi_{ij}^0 \end{aligned} \tag{9.76}$$

It is worth noting that the matrix $\mathbf{\Pi}^0$, which is to be evaluated by simulation methods, does not depend on the arrival rate λ of the ground motions. Therefore the influence of this parameter on the solution can be analysed without evaluating again the statistics of the structural response. Furthermore, the continuity of the mechanical behaviour would suggest a Markov process continuous in both state and time to be a more correct stochastic model. The absolute probabilities at time t are the solution of the Chapman–Kolmogorov equation (equation (A6.68)):

$$\frac{\partial p(u_r, t)}{\partial t} = -p(u_r, t)\int_{u'_r} \Xi(u_r, u'_r)\,du'_r + \int_{u'_r} p(u'_r, t)\,\Xi(u'_r, u_r)\,du_r \tag{9.77}$$

where $\Xi(u, u')$ is the transition probability density. However, this approach involves several computational difficulties in the solution of (9.77) and the approximations that must be introduced for the evaluation of the density function make it unreliable.

In conclusion, the homogeneous continuous-time discrete-state Markov process appears to be the most suitable model for analysing the stochastic process $\tilde{u}_r(t)$ representing the inelastic deformation of a structure subject to a sequence of load excitations. However, all the above approaches are founded on the same idealizations. The clumps of yield level crossings due to each loading pulse are automatically treated as 'points in time' by assuming that the structural response to a whole single excitation is instantaneous. Moreover 'absorbing states' are introduced to define the involved mode of structural failure (i.e. ductility capacity overcome or collapse for gravity loads). Thus, the 'peak inelastic deformation' during a single excitation is taken into account.

These procedures have again been applied to the single-degree-of-freedom frame of Fig. 9.11(a), but now the effective strength is reduced with the deformation because of the presence of the gravity loads W (Fig. 9.12(a)) [9a]: collapse occurs when this strength vanishes. Numerical results are presented in Fig. 9.12‡.

‡ Note explicitly that the illustrative examples of Figs 9.11 and 9.12 correspond to very simple governing mechanical relationships. In contrast, this sort of procedure is worth application to real problems when the mathematical complexity of the mechanical aspects requires (or suggests) that the (random) structural response to a single excitation be investigated numerically.

It is of interest to note the shape of the probability matrix **P** and of the conditional probability matrix $\mathbf{\Pi}^0$ obtained in the numerical example considered (Fig. 9.12(b)): from the mechanical point of view this shape means that the plastic deformations, once formed, cannot regress. However, this behaviour is not justified by the dynamic governing relations. Therefore the correct conclusion is that plastic deformations, once formed, have a very small probability of regressing.

9.5 PARAMETRIC TECHNIQUES IN SIMULATION

In simulating the response of a random structure, a great deal of the computational effort is spent in repeating the deterministic procedure of solution for each simulated structure. Although the adoption of a hybrid approach allows one to limit the number of such *experiments* to 10^2–10^3, this effort can still be excessive.

If the relation $u_j(x_i, \zeta_h)$ between the relevant response parameters u_j, the (random) design variables $\tilde{x}_i$ and the deterministic parameters ζ_h were known in closed form throughout the range of variability of the $\tilde{x}_i$, computations would be much easier.

In many other cases, the form of the relationship $u_j = u_j(x_i, \zeta_h)$ varies when some parameters are changed, but the mechanical model allows one to find these relationships and to follow their changes: in other words, a sensitivity analysis and a parameter study can be performed. In these cases, the computational effort can be much reduced by avoiding repeating the full solution of the structural problem for each simulated structure, i.e. by applying a *parametric technique* along the following steps:

(1) Sample a first set of random numbers $\grave{x}_i$. The analysis of the structure defined by these values of the design variables (first simulated structure), by assumption, allows one to obtain in closed form the relationship

$$\grave{u}_j = u_j(\grave{x}_i, \zeta_h^*) \tag{9.78}$$

where u_j is the relevant sample response quantity, and ζ_h^* the deterministic parameter holding in the *experimental point* $\mathbf{Q}(\grave{x}_i)$.

(2) Perform a parameter study and determine in the x_i space *a decision region* $\mathscr{A}$, whose points correspond to structures having the same behaviour as the first simulated structure (i.e. such that equation (9.78) holds). Then find the quickest operation sequence able to obtain the value u_j' of a new structure from the first one (even if the corresponding point does not belong to the decision region $\mathscr{A}$).

(3) Perform new trials using the results specified in (2).

(4) A histogram of the obtained values approximates the true PDF of $\tilde{u}_j$.

The difference between the usual simulation technique and the parametric one is shown diagrammatically in Fig. 9.13, from which one realizes that the computational efficiency of the parametric procedure depends on the selection of the first experiment $\mathbf{Q}$.

Some applications of this approach will be illustrated in the next sections: the examples that will be developed aim both at clarifying the computational aspects and at pointing out some general results.

9.5.1 Plastic collapse analysis

Limit analysis (cf. Chapter 5) is an example in which parametric techniques find a very appropriate use: in fact, each *collapse mechanism* corresponds to a well determined analytical relation between the load parameter(s) and the design variables (cf. equation (9.79) below), and the simulated structure can fail according to either one or the other mechanism.

In the following, two numerical methods for probabilistic limit analysis will be presented. The first is a logically trivial extension of linear programming, the second is a parametric technique.

(a) Stochastic programming

For simplicity, assume that the loads depend on one scalar parameter α (equation (5.8)) and neglect the 'fixed' load $\mathbf{W}_0$.

As evident from equation (5.7), the kinematically sufficient load factor α_{Kj}, corresponding to the collapse according to the jth mechanism, can be put in the form

$$\alpha_{Kj} = \frac{\boldsymbol{\chi}_j^{\mathrm{T}} \boldsymbol{\lambda}_{\mathrm{p}j}}{\mathbf{W}_{\mathrm{a}}^{\mathrm{T}} \mathbf{u}_j} \tag{9.79}$$

where the vector $\boldsymbol{\chi}_j$ depends on the mechanism and the local strengths, $\boldsymbol{\lambda}_{\mathrm{p}j}$ are the positive plastic multipliers, $\mathbf{u}_j$ is a displacement vector and $\mathbf{W}_{\mathrm{a}}$ is the reference load vector. Remembering that the vectors $\boldsymbol{\lambda}_{\mathrm{p}}$ and $\mathbf{u}$ must be compatible (equations (5.3) and (5.4)) and normalizing the displacements by the constraint $\mathbf{W}_{\mathrm{a}}^{\mathrm{T}}\mathbf{u} = 1$, the collapse load factor α_{p} can be expressed, through the kinematic theorem, by the constrained minimization problem (cf. equation (5.9))

$$\alpha_{\mathrm{p}} = \min \alpha_{\mathrm{K}} = \min \boldsymbol{\chi}^{\mathrm{T}} \boldsymbol{\lambda}_{\mathrm{p}} \tag{9.80a}$$

$$\mathbf{W}_{\mathrm{a}}^{\mathrm{T}}\mathbf{u} = 1 \tag{9.80b}$$

$$\mathbf{Q}\mathbf{u} - \mathbf{H}^{\mathrm{T}} \boldsymbol{\lambda}_{\mathrm{p}} = 0 \tag{9.80c}$$

$$\boldsymbol{\lambda}_{\mathrm{p}} \geqslant 0 \tag{9.80d}$$

whence, if $\mathbf{u}$ is eliminated from (9.80b) and (9.80c),

$$\alpha_{\mathrm{p}} = \min \boldsymbol{\chi}^{\mathrm{T}} \boldsymbol{\lambda}_{\mathrm{p}} \tag{9.81a}$$

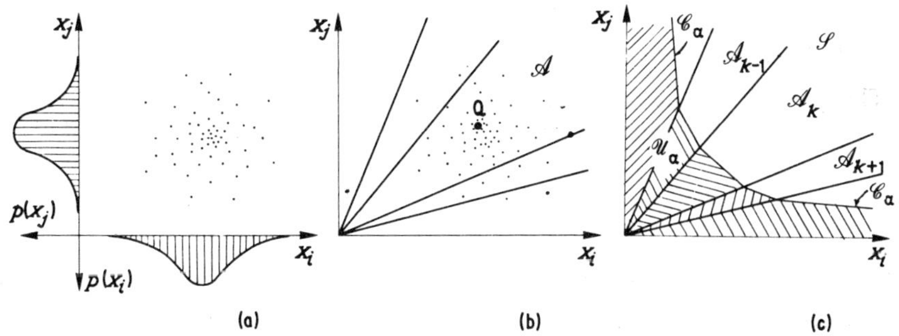

Fig. 9.13 Usual simulation procedure: the structural analysis is performed in every point $(\dot{x}_i, \dot{x}_j)$ yielded by a numerical experiment. (b) Parametic simulation: the structural analysis is developed completely only in **Q**, the first simulated structure; the same structural solution holds for all points in the decision region $\mathscr{A}$. Otherwise, a simplified procedure making use of the knowledge of the behaviour in **Q** is employed. (c) Decision regions corresponding to each collapse mechanism of a rigid–plastic structure, and example failure surface $\mathscr{C}_\alpha$ for a given α, in the form suggested by Ang and Amin [10] (cf. Figs. 1.5 and 1.6).

$$\mathbf{C}\boldsymbol{\lambda} \geqslant 0 \tag{9.81b}$$

$$\boldsymbol{\lambda} \geqslant 0 \tag{9.81c}$$

Constrained mathematical optimization problems like (9.81) can be solved by a well known numerical algorithm, called linear programming [11].

The natural extension of linear programming, when the strength parameters χ_j, and therefore the components of the vector $\boldsymbol{\lambda}_{\mathrm{p}}$, are random variables, is the *wait-and-see* stochastic programming procedure, which consists of:

(1) Using (9.81) and *parametric* linear programming, determine in the space of the design parameters x_i the decision regions $\mathscr{A}_k$ corresponding to each relevant collapse mechanism. It can be proved that these regions are *cones* with vertices in the axis origin (Fig. 9.13(c)).
(2) Fix a value of the load factor α.
(3) Specify the surface $\mathscr{C}_\alpha$, separating *safe* and *unsafe* regions for the given α. This surface is a plane in each of the decision regions $\mathscr{A}_k$ because it corresponds to a linear relation (9.79).
(4) Compute the conditional probability of collapse $P_{\mathrm{R}}(\alpha)$ for the fixed α (equation (5.53)) by the integral.

$$P_{\mathrm{R}}(\alpha) = \int_{\mathscr{U}(\alpha)} p(x_1, x_2, \ldots, x_n)\, \mathrm{d}x_1\, \mathrm{d}x_2 \ldots \mathrm{d}x_n \tag{9.82}$$

where $\mathscr{U}(\alpha)$ is the relevant unsafe region, and $p(x_1, x_2, \ldots, x_n)$ the JPDF of the design parameters x_i. If, as usual, the latter are assumed to be stochastically independent of each other, as well known,

$$p(x_1, x_2, \ldots, x_n) = p_1(x_1)\, p_2(x_2) \ldots p_n(x_n) \tag{9.83}$$

(5) Choose a new value of α, and repeat steps (3) and (4). The CPF $P_R(\alpha)$ is thus determined by points.

This procedure is subject to the same criticism as all procedures based on the kinematic approach only (Section 5.2.3). Moreover, note that each integral (9.82) must be performed numerically, with an effort that becomes unbearable as soon as the number of design variables x_i exceeds a few units.

(b) Parametric simulation

A frequency function $F_R(\alpha)$ approximating $P_R(\alpha)$ can be obtained by simulation of the x_i, whose PDFs are assumed known and independent. To each simulated point $\mathbf{Q}(x_1, x_2, \ldots, x_n)$ the value of α_p given by (9.81) is associated. If this operation is repeated a sufficient number n of times, a reliable histogram $F_R(\alpha)$ can be constructed.

The complete solution of the minimization problem (9.81) for each simulated point can be avoided thanks to the already described parametric technique. In fact, the first experiment yields a collapse load factor α_{p1} and a collapse kinematism; a parameter study allows one to define in the x_i space the decision region $\mathscr{A}$ whose points have the same collapse mechanism as the first simulated structure (Fig. 9.13 (b)). Once this decision region $\mathscr{A}$ is known, the points that in the x_i space represent the other simulated structures may be classified according to their belonging or not belonging to $\mathscr{A}$. For the first ones the collapse load factor α_{pj} is now a known linear combination of the x_i (equation (9.79)), whereas for the other points the new coefficients in the expression of α_{pj} may be determined by performing one or more further pivotal steps of the linear programming algorithm.

For a limit analysis problem, the first simulated structure (point $\mathbf{Q}$ in Fig. 9.13 (b)) may conveniently be substituted by the *expected structure* (i.e. the structure defined by the average values of the x_i parameters): in this way, in fact, the decision region $\mathscr{A}$ will contain most simulated points.

In order to illustrate the actual procedure in detail, consider a plane frame and introduce the following assumptions:

(1) Geometry changes do not influence the equilibrium equations (small displacements).
(2) Flexural load effects predominate (i.e. shear, axial force, axial shortening and instability effects can be neglected).
(3) Inelastic deformations concentrate in single sections (critical sections) and give rise to rotations θ (plastic hinge hypothesis).
(4) The moment–hinge rotation relationship is rigid–perfectly plastic.

Thus, the frame can be discretized in perfectly rigid elements connected by N critical sections where flexural deformations concentrate.

Let M_{yj}^+, $-M_{yj}^-$ and $\theta_j = \theta_j^+ - \theta_j^-$ $(j = 1, \ldots, N;\ \theta_j^+, \theta_j^- \geqslant 0)$ denote respectively the positive and negative full plastic moments and the plastic rotation at the jth critical section. Further, let $X_i = X_i^+ - X_0 (X_i^+, X_0 \geqslant 0)$ be the ith one of the redundancy parameters that must be introduced to make the structure statically determinate. Then the static theorem permits one to write the following linear programming problem:

$$\alpha_p = \max \alpha_S \tag{9.84a}$$

$$\alpha_S \mathbf{M}_a + \mathbf{M}^0 \mathbf{X}^+ \leqslant \mathbf{M}_y^+ \tag{9.84b}$$

$$-\alpha_S \mathbf{M}_a - \mathbf{M}^0 \mathbf{X}^+ \leqslant \mathbf{M}_y \tag{9.84c}$$

$$\alpha_S \geqslant 0 \qquad \mathbf{X}^+ \geqslant 0 \tag{9.84d}$$

where α_S denotes a statically admissible load factor and $(\mathbf{X}^+)^T = (X_1^+, X_2^+, \ldots, X_m^+, X_0)$. Moreover, in (9.84), $\mathbf{M}_a$ is the vector of length N of the bending moments at the N critical sections calculated on the statically determinate structure under the external actions and $\mathbf{M}^0$ is the $N \times (m+1)$ matrix whose ith column denotes the bending moments on the statically determinate structure due to the ith entry of the vector $\mathbf{X}^+$ for $i = 1, \ldots, m$ and

$$M^0_{j(m+1)} = -\sum_{i=1}^{m} M^0_{ji}$$

Then, the linear programming properties allow one to write the kinematic theorem as the dual of problem (9.84): thus (9.80) becomes, for the problem under analysis,

$$\alpha_p = \min \{ \mathbf{M}_y^+)^T \boldsymbol{\theta}^+ + (\mathbf{M}_y^-) \boldsymbol{\theta}^- \} \tag{9.85a}$$

$$(\mathbf{M}_a)^T \boldsymbol{\theta}^+ - (\mathbf{M}_a)^T \boldsymbol{\theta}^- \geqslant 1 \tag{9.85b}$$

$$(\mathbf{M}^0)^T \boldsymbol{\theta}^+ - (\mathbf{M}^0)^T \boldsymbol{\theta}^- \geqslant 0 \tag{9.85c}$$

$$\boldsymbol{\theta}^+ \geqslant 0 \qquad \boldsymbol{\theta}^- \geqslant 0 \tag{9.85d}$$

Problem (9.85) expresses that α_p is equal to the minimum of the values that the dissipated energy assumes in different compatible configurations (9.80c) for which the external work is 1 (equation (9.80b)).

The solution procedure of a linear programming problem involves that the constraints are written in the form of a system of simultaneous linear equations by introducing the so-called *slack* variable Y_i. Therefore, the system of constraints of problem (9.85) becomes a set of $r = m + 1$ equations in $k = 2N + m + 1$ unknowns:

$$\mathbf{D} \mathbf{z} = \mathbf{e} \tag{9.86}$$

where

$$\mathbf{D} = \left[\begin{array}{c|c} (\mathbf{M}_a)^T & \\ \hline (\mathbf{M}^0)^T & -\mathbf{I} \end{array} \right] \qquad \mathbf{e}^T = \{1, 0, 0, \ldots, 0\}$$

$$\mathbf{z}^T = \{\theta_1^+, \ldots, \theta_N^+, \theta_1^-, \ldots, \theta_N^-, Y_1, \ldots, Y_{m+1}\} \tag{9.87}$$

The objective function is

$$\mathbf{c}^{\mathrm{T}}\mathbf{z} = \{(\mathbf{M}_y^+)^{\mathrm{T}}, (\mathbf{M}_y^-)^{\mathrm{T}}, 0, \ldots, 0\}\mathbf{z} \tag{9.88}$$

and $\mathbf{c}$ is called the *price* vector.

The solution algorithm is iterative and each step consists of putting equal to zero the value of $2N$ unknown variables selected according to assigned decisional rules. Thus, one must solve a sequence of linear systems

$$\mathbf{D}^{\mathrm{s}}\mathbf{z}_{\mathrm{B}} = \mathbf{e} \tag{9.89}$$

where $\mathbf{D}^{\mathrm{s}}$ is a matrix that contains r linearly independent columns $\mathbf{d}_j$ $(j = 1, \ldots, k)$ of $\mathbf{D}$ and $\mathbf{z}_{\mathrm{B}}$ is a vector with the non-zero entries of $\mathbf{z}$ and denotes a basic solution. Note that any column $\mathbf{d}_j$ of $\mathbf{D}$ can be expressed as a linear combination of the column of $\mathbf{D}^{\mathrm{s}}$ and hence vectors $\mathbf{y}_j$ $(j = 1, \ldots, k)$ exist so that

$$\mathbf{y}_j = (\mathbf{D}^{\mathrm{s}})^{-1}\mathbf{d}_j \tag{9.90}$$

The procedure stops if the basic solution of the system (9.89) satisfies the optimum condition

$$\mathbf{z}_{\mathrm{B}}^0 = (\mathbf{D}^{\mathrm{s}0})^{-1}\mathbf{e} \tag{9.91}$$

so that

$$\alpha_{\mathrm{p}} = (\mathbf{c}_{\mathrm{B}}^0)^{\mathrm{T}}\mathbf{z}_{\mathrm{B}}^0 \tag{9.92}$$

where $\mathbf{c}_{\mathrm{B}}^0$ is a vector that contains the coefficients in equation (9.88) of the non-zero entries $\mathbf{z}$. For indices h of these non-zero components of $\mathbf{z}$ lower that $(2N + 1)$, the slack variable Z_h^0 of the dual problem (9.84) must be zero, while the other slack variables must be positive. In fact, if a critical section becomes a plastic hinge and undergoes rotation, there the bending moment has attained the relevant full plastic value. The linear programming properties ensure this correspondence between the solution of problem (9.85) and the solution of problem (9.84).

Suppose now that $\mathbf{M}_y$ has random entries of assigned PDFs; then also α_{p} is a random variable and its PDF is unknown. In order to calculate the approximation $F_{\mathrm{R}}(\alpha)$ to the CPF $P_{\mathrm{R}}(\alpha)$ (which coincides, for monoparametric load, to the CDF of α_{p}), one can make use of the following parametric simulation method. Consider the average structure defined by the mean values $\mu_{\mathbf{M}_y}$ of the full plastic moments at the N critical sections and determine the relevant quantities $\mathbf{c}_{\bar{\mathrm{R}}}$, $\mathbf{D}_{\bar{\mathrm{R}}}^{\mathrm{s}0}$, $\mathbf{z}_{\mathrm{B}\bar{\mathrm{R}}}^0$ (equation (9.91)), $\mathbf{c}_{\mathrm{B}\bar{\mathrm{R}}}^0$, $\alpha_{\mathrm{p}\bar{\mathrm{R}}}$ (equation (9.92)) and $\mathbf{z}_{\bar{\mathrm{R}}}^0$. Then, any simulated vector $\mathbf{M}_y$ can be expressed in the form

$$\grave{\mathbf{M}}_y = \mu_{\mathbf{M}_y} + \phi\Delta\grave{\mathbf{M}}_y \tag{9.93}$$

where $\Delta\mathbf{M}_y$ is an increment vector and ϕ is a scalar non-negative quantity. Therefore, the price vector becomes

$$\grave{\mathbf{c}} = \mu_{\mathrm{c}} + \phi\grave{\mathbf{g}} \tag{9.94}$$

with

$$\grave{\mathbf{g}}^{\mathrm{T}} = \{\Delta\grave{\mathbf{M}}_y^{\mathrm{T}}, 0, \ldots, 0\}$$

and the average-structure solution $\mathbf{z}_{\mathrm{B}\bar{\mathrm{R}}}^0$ can no longer be optimal for the new problem. In fact the solution $\mathbf{z}^0_{\mathrm{B\bar{R}}}$ is optimal if and only if all the entries of the vector $\mathbf{Z}$ of the slack variables of the dual problem remain non-negative. As

$$\dot{Z}_j = \dot{Z}^0_{\bar{\mathrm{R}}j} + \phi((\dot{\mathbf{g}}_{\mathrm{B}})^{\mathrm{T}}\mathbf{y}_j - \dot{g}_j) = \dot{Z}^0_{\bar{\mathrm{R}}j} + \phi f_j \tag{9.95}$$

where $\dot{\mathbf{g}}_{\mathrm{B}}$ contains the elements of $\mathbf{g}$ relevant to the entries of $\mathbf{c}_{\bar{\mathrm{R}}}$ in $\mathbf{c}^0_{\mathrm{B\bar{R}}}$, two cases can occur:

(1) All f_j are non-negative: it follows that the basic solution $\mathbf{z}^0_{\mathrm{B\bar{R}}}$ is also optimal for any ϕ.
(2) Otherwise, $\mathbf{z}^0_{\mathrm{B\bar{R}}}$ is optimal if and only if:

$$0 \leqslant \phi \leqslant \phi_{\mathrm{cr}} = -\frac{Z^0_{\mathrm{Rcr}}}{f_{\mathrm{cr}}} = \min_j \left(-\frac{Z^0_{\bar{\mathrm{R}}j}}{f_j}\right) \qquad \text{for } f_j < 0 \tag{9.96}$$

If the actual value of ϕ is lower than ϕ_{cr}, one can obtain directly $\alpha_{\mathrm{p}} = (\mathbf{c}_{\mathrm{B}})^{\mathrm{T}}\mathbf{z}^0_{\mathrm{B\bar{R}}}$. In the space of the quantities M_{yj}, the inequality $\phi(\mathbf{g}) \leqslant \phi_{\mathrm{cr}}(\mathbf{g})$ defines the decisional region $\mathscr{A}_1$ whose points define structures that collapse by the same mechanism as the average structure.

If $\phi > \phi_{\mathrm{cr}}$, the solution $\mathbf{z}_{\mathrm{B\bar{R}}}$ is no longer optimal, but it remains a basic solution (equation (9.89)) for the new problem: thus, the re-search for the initial basic solution (that is generally the phase of the linear programming algorithm that requires more computational effort) is avoided. The new optimal solution can be obtained by few pivotal steps from $\mathbf{z}_{\mathrm{B\bar{R}}}$.

(c) Examples of probabilistic limit analyses by parametric simulation

The results of some examples are now presented; each of them aims to emphasize some different aspects of probabilistic limit analysis. The first permits one to perform a precision–efficiency comparison between some different approaches, while the other examples show the way to derive the CPF of a structure subject to more independent loadings and point out some properties of this CPF and its dependence on the structural redundancy that can be easily generalized.

The comparison between different approaches is performed with reference to the simple portal frame of Fig. 9.14(a). The structure is divided into 40 rigid elements connected by 41 critical sections, where potential plastic hinges can form (8 in each column, 5 in each 'joint', 15 in the cross beam; Fig. 9.14(b)).

The relevant inelastic rotation–bending moment relationship is drawn in Fig. 9.14(c), where the randomness of the plastic bending moment $\tilde{M}_{\mathrm{y}}$ is also described by the relevant histogram‡.

Calculations were first developed under the assumption that the plastic bending moments relevant to the critical sections belonging to the same column, joint or beam are strictly correlated. Thus, five independent parameters have to be simulated and the results are drawn in Fig. 9.14(e): there, $F_{\mathrm{E}}(\alpha)$ is the CDF of the

‡ As the actual CPF of the structural carrying capacity must be inferred from the results of the simulated experiments, this operation can be avoided on the input data.

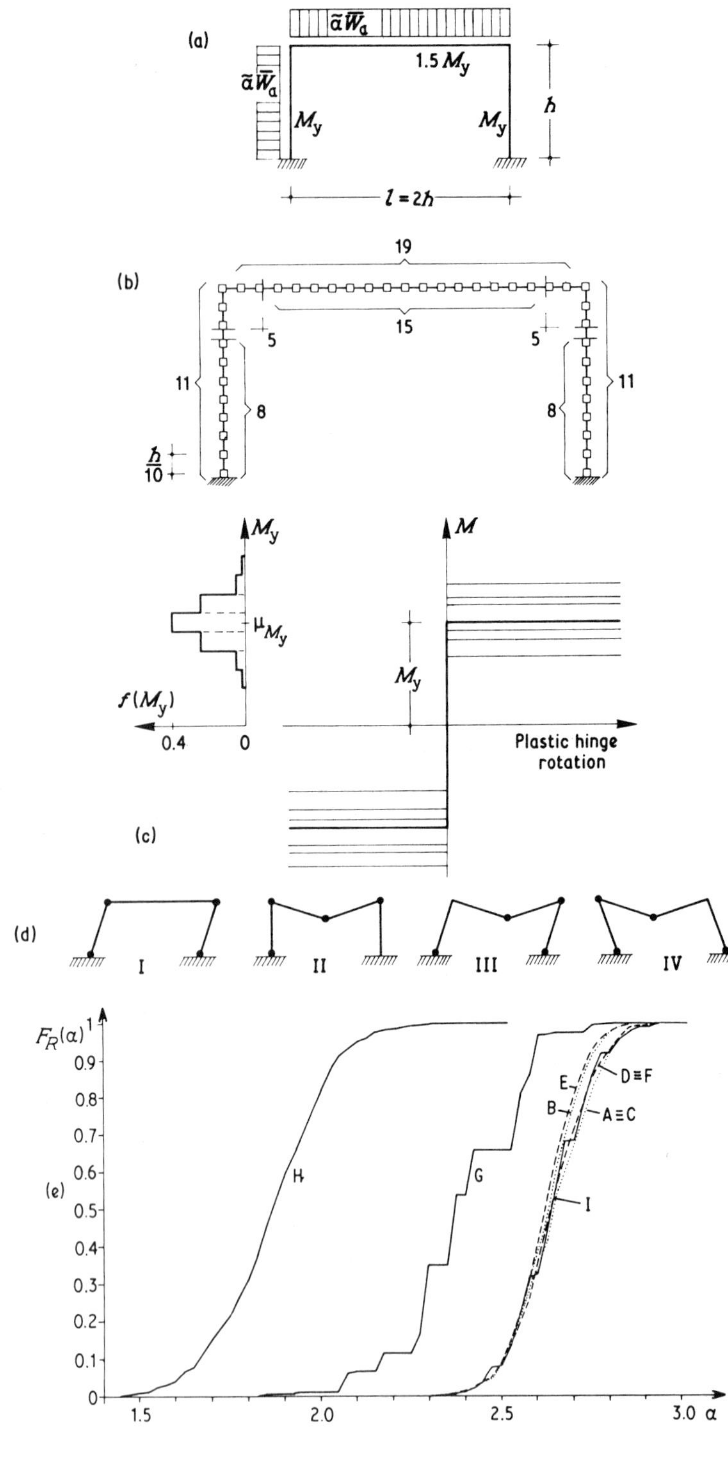
(a)
α̃W̄a
1.5 My
My
My
h
l = 2h
(b)
19
15
5
5
11
11
8
8
h/10
My
M
μMy
My
f(My)
0.4
0
Plastic hinge rotation
(c)
(d)
I
II
III
IV
FR(α)
1
0.9
0.8
0.7
0.6
0.5
0.4
0.3
0.2
0.1
0
(e)
H
G
E
B
D ≡ F
A ≡ C
I
1.5
2.0
2.5
3.0 α

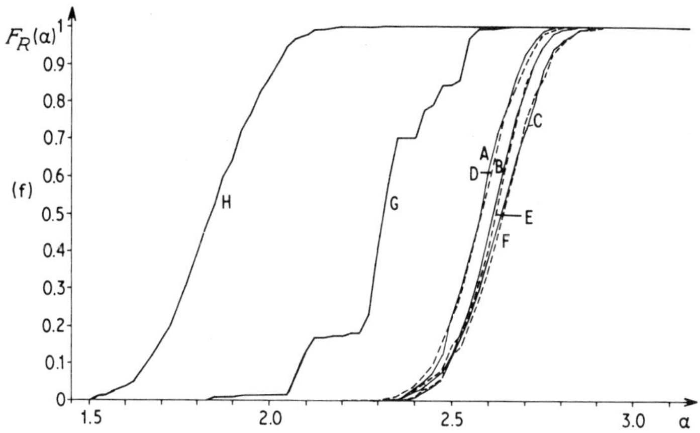

Fig. 9.14 First example of limit analysis by simulation (from [11]). (a) Example frame; (b) discretization; (c) constitutive law and plastic moment randomness; (d) plastic collapse mechanisms for calculations; (e) results for the correlated case (f) results for the uncorrelated case: A, D = $F_R(\alpha)$: B, E = $\Sigma F_{Ki}(\alpha)$: C, F = max $F_{Ki}(\alpha)$: G = static upper bound (Chapter 5): H = $F_E(\alpha)$: I = stochastic programming; first sample A, B, C (n = 500): second sample D, E, F (n = 500).

load factor α_E for which the elastic limit state is attained; α_{Ki} is the kinematic load factor relevant to the ith collapse mechanism presented in Fig. 9.14(d) and ($\Sigma_i F_{Ki}(\alpha)$ and (max $F_{Ki}(\alpha)$) denote the bounds expressed by equations (1.8), (1.6) and (5.17). The static upper bound (Chapter 5) is calculated on the basis of three different equilibrated moment diagrams at collapse.

It is worth noting the accuracy of the simulation approach for the central values of the distribution, while the tails are fully ignored: if required, they must be inferred. Further, the computer-time comparison (Table 9.1) confirms, for this example, the efficiency that can be achieved making use of the parametric technique of simulation.

Fig. 9.14(f) shows the results for uncorrelated resistance. In this case all the curves translate towards the left according to the considerations discussed in

Table 9.1 Comparison of the computer time required by different approaches to the estimation of $P_R(\alpha)$ for the problem of Fig. 9.14. The time required by 'used' simulation is put equal to 1; the ratio between the simulation times required by the two cases is 2.11.

	Methods				
Number of parameters	*Stochastic programming*	*'Usual' simulation*	*Parametric simulation*	*Kinematic bounds*	*Static upper bound*
5 (Fig. 9.14(e))	10.53	1	0.32	0.23	0.03
41 (Fig. 9.14(f))	—	1	0.73	0.61	0.02

Section 1.4. Moreover, $\Sigma\, F_{Ki}(\alpha)$ becomes a lower bound to the 'exact' solution because only the above four independent mechanisms were considered instead of the complete set of possible ones (cf. Section 5.2.4(a)).

The example of Fig. 9.15 concerns a structure subject to two simultaneous random actions $\tilde{\alpha}\bar{W}_a$ and $\tilde{\alpha}_g\bar{W}_g$. The reliability analysis of such a problem requires the calculation of a two-dimensional approximate CPF $F_R(\alpha, \alpha_g)$. This estimate can be obtained by simulating the function $F_R(\alpha, \zeta\alpha)$ for different values of

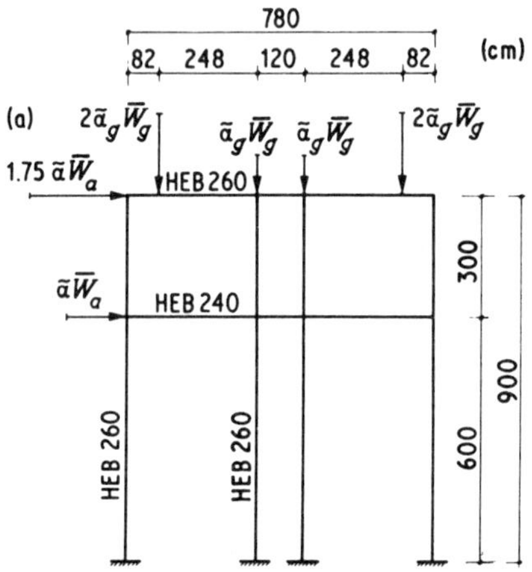

$\bar{W}_a = 0.4\ t$ | $\bar{W}_g = 12\ t$
$\tilde{\alpha}$ = extreme type II | $\tilde{\alpha}_g$: normal
$\mu_\alpha = 1$ | $\mu_{\alpha g} = 1$
c.o.v. $\delta_\alpha = 0.1$ to 0.7 | c.o.v. $\delta_{\alpha g} = 0.05$

yield stress: nomal; c.o.v. 0.05

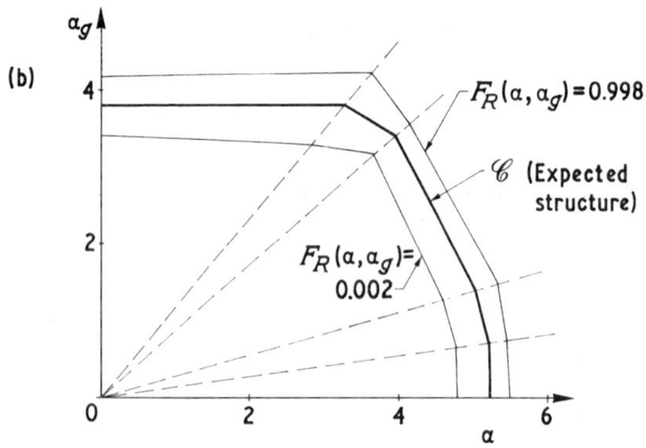

(c)

$\zeta = \frac{\alpha_g}{\alpha}$	$\frac{m_{R(\zeta)}}{\alpha_{pR(\zeta)}}$	$\frac{s_{R(\zeta)}}{m_{R(\zeta)}}$
0.00	0.997	2.32
0.13	0.995	2.38
0.27	0.998	2.33
0.83	0.998	2.14
1.14	0.997	3.51
∞	0.991	3.50

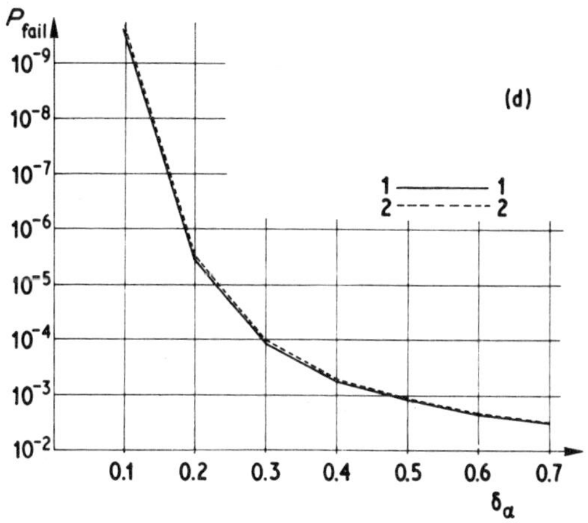

Fig. 9.15 Second example of limit analysis by simulation (from [12]), (a) Structural diagram; (b) contour lines of estimated CPF $F_R(\alpha, \alpha_g)$ (sample size (500); (c) table of numerical results in terms of the carrying capacity R_p of the structure in function of the ratio $\zeta = \alpha_g/\alpha$ (d) comparison of the obtained probabilities of failure: 1, bivariate convolution integral; 2, univariate convolution integral.

$\zeta = \alpha_g/\alpha$. In order to obtain $F_R(\alpha, \zeta\alpha)$ for given ζ, note that this function is the CDF of the carrying capacity $\tilde{\alpha}_p$ of the structure subject to the loads $\tilde{\alpha}\bar{W}_a$, $\tilde{\alpha}\zeta\bar{W}_g$ so that equation (9.85) is still able to solve a deterministic experiment.

However, if the action of the only load $\tilde{\alpha}_g\bar{W}_g$ causes failure with a very small probability, one can also write equation (2.68) with reference only to $\tilde{\alpha}W_a$ on condition that the estimation $F_R(\alpha)$ of $P_R(\alpha)$ takes into account the load $\tilde{\alpha}_g\bar{W}_g$ as a quantity that increases or decreases the local resistance (Fig. 9.15(d)). With reference to the plastic collapse limit state, in particular, the presence of this vertical load $\tilde{\alpha}_g\bar{W}_g$ influences the value of the moment due to $\tilde{\alpha}\bar{W}_a$ for which the plastic hinge forms at a given critical section. In other words, equation (9.84) can be written considering $\mathbf{M}_a$ due only to $\tilde{\alpha}\bar{W}_a$, if the bending moment $\tilde{\mathbf{M}}_g = \tilde{\alpha}_g\mathbf{M}\bar{W}_g$ involved by $\tilde{\alpha}_g\bar{W}_g$ on the statically determinate frame is subtracted from (added to) the l.h.s. of equation (9.84b) (or (9.84 c)). It follows that the use of a simulation approach requires, as input data of each experiment, both the simulated yielding stresses at the critical sections and a simulated value of the load $\tilde{\alpha}_g$.

Fig. 9.15(c) shows that the coefficient of variation of the carrying capacity is higher where the number of plastic hinges involved by the collapse mechanism is smaller, i.e. where the vertical loads are prevalent for the analysed structure.

The example of Fig. 9.16, in which several random vertical loads $\tilde{W}_{gi}$ are taken into account on the 1.h.s. of equations (9.84b) and (9.84c) confirms the indications

(a)

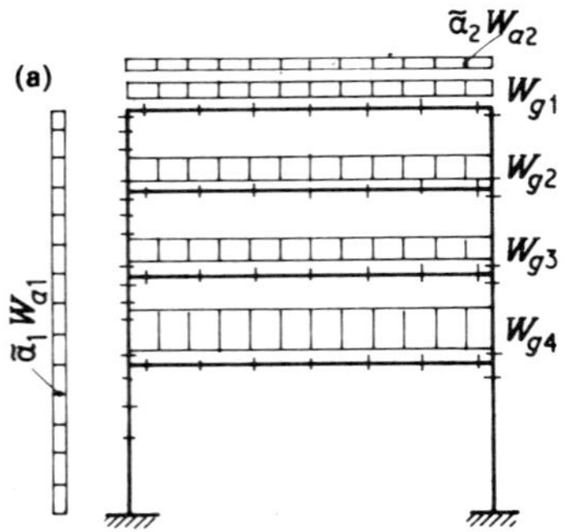

(b)

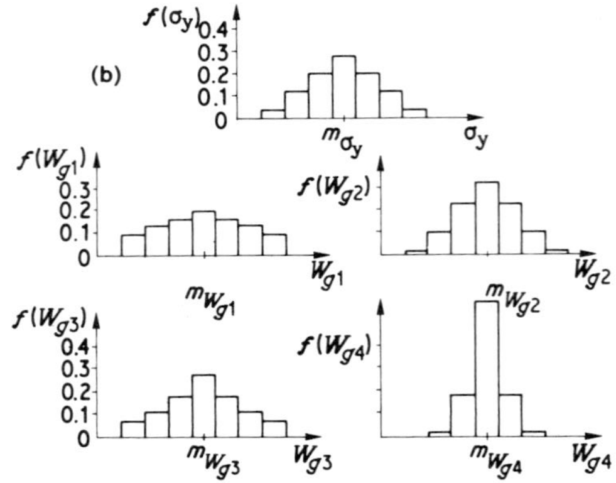

(c)

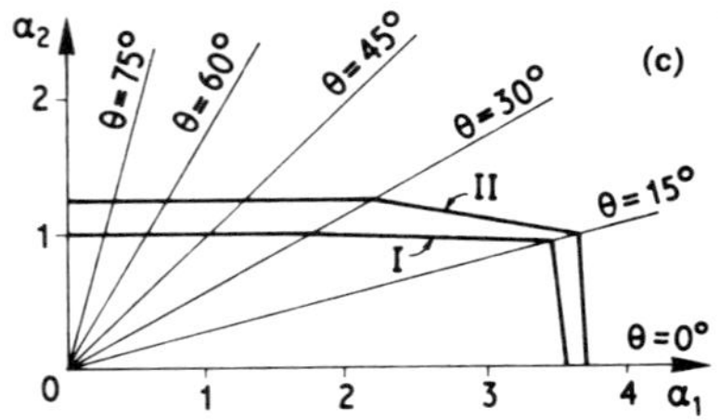

	Mean m	c.o.v. s/m
Yield stress σ_y	2400 kg cm^{-2}	0.0180
Dead loads W_{g1}	1.2 kg cm^{-1}	0.1801
W_{g2}	2.0 kg cm^{-1}	0.1265
W_{g3}	2.0 kg cm^{-1}	0.1619
W_{g2}	2.8 kg cm^{-1}	0.0755

(d)

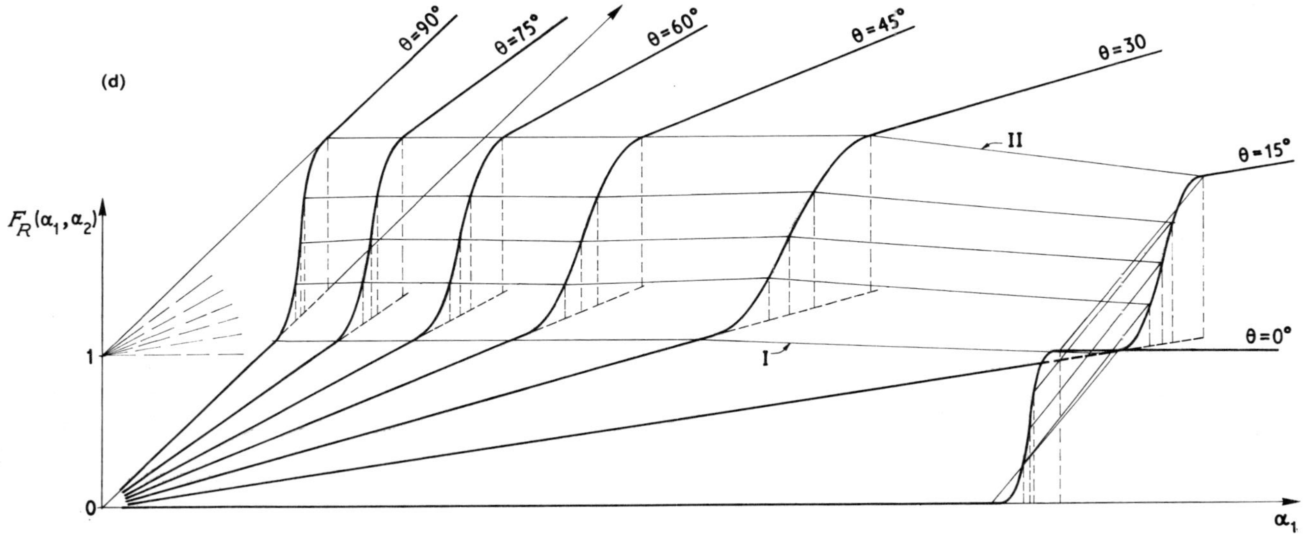

Fig. 9.16 Third example of limit analysis (after [13]). (a) Diagram of example structure; (b) discretization and definition of the relevant random parameters; (c, d) estimated CPF, taking into account the loads $\tilde{W}_{gi}$ on the l.h.s. of (9.84 b) and (9.84 c) [((c) plan; (d) assonometry].

$\theta = \tan^{-1}(\alpha_2/\alpha_1)$; line I: $F_R \simeq 0$; line II: $F_R \simeq 1$.

of Fig. 9.15(c). The estimation $F_R(\alpha_1, \alpha_2)$ of the CPF $P_R(\alpha_1, \alpha_2)$ is drawn in Fig. 9.16(b): it was obtained by solving problem (9.85) for seven different values of the ratio $\zeta = \alpha_2/\alpha_1$. Line I connects the lower values and line II the upper values of the carrying capacity that were simulated. The width of the range between these two lines is a rough measure of the standard deviation of the carrying capacity along the straight line defined by ζ. Thus, Fig. 9.16(c) shows that the coefficient of variation decreases when ζ increases from 0 to $0.268 = \tan 15°$, when the more probable mechanism is the shear mechanism of the frame. Then, for $\zeta \geqslant 0.577 = \tan 30°$ the most probable mechanism involves only the top of the structure and hence a lower degree of redundancy than that of the original frame influences the structural behaviour: this yields a higher value of the coefficient of variation of the carrying capacity, which remains approximately constant in this range. In fact, from equations (A3.26) and (A3.27), one obtains, for uncorrelated plastic moments, that the coefficient of variation δ_{α_K} of the kinematic load factor $\tilde{\alpha}_{Ki} = \mathbf{c}_{Bi}^T \mathbf{z}_{Bi}$ is given by

$$\delta_{\alpha_{Ki}} = \frac{s_{\alpha_{Ki}}}{\mu_{\alpha_{Ki}}} = \frac{\left(\sum_j z_{B_{ij}}^2 s_{M_yj}^2\right)^{1/2}}{\sum_j z_{B_{ij}} \mu_{M_yj}}$$

$$= \frac{\left(\sum_j z_{Bij}^2 \delta_{M_yj}^2 \mu_{M_yj}^2\right)^{1/2}}{\sum_j z_{Bij} \mu_{M_yj}} \tag{9.97}$$

$$= \frac{\left[\sum_j z_{Bij}^2 \delta_{M_yj}^2 \Lambda_j^2 \left(\mu_{M_y}^{\max}\right)^2\right]^{1/2}}{\sum_j z_{Bj} \Lambda_j \mu_{M_y}^{\max}}$$

with $\mu_{M_y}^{\max} = \max_j \mu_{M_yj}$ and hence $\Lambda_j < 1$.

Then, if $\delta_{M_yj} = \delta_{M_y} =$ constant,

$$\delta_{\alpha_{Ki}} = \frac{\left(\sum_j z_{Bij}^2 \Lambda_j^2\right)^{1/2}}{\sum_j z_{Bij} \Lambda_j} \delta_{M_y} = \chi_i \delta_{M_y} \tag{9.98}$$

As $\mathbf{z}_{Bi}$ denotes the vector of the plastic rotations at incipient collapse, $\mathbf{z}_{Bi} \ll 1$ and hence $\chi_i < 1$: further, as the number of plastic hinges involved by the collapse mechanism increases, χ_i decreases.

Therefore, the coefficient of variation of a kinematic load factor $\tilde{\alpha}_{Ki}$ is lower than that of the full plastic moment δ_{M_y}, in the case that such value is common to all critical sections, and decreases as the involved redundancy degree increases. If δ_{M_yj} is not constant, introduce the maximum value $\delta_{M_y}^{\max}$ and so equation (9.97) becomes

$$\delta_{\alpha_{Ki}} \leqslant \frac{\left(\sum_j z_{Bij}^2 \Lambda_j^2\right)^{1/2}}{\sum_j z_{Bij}\Lambda_j} \delta_{M_y}^{\max} \leqslant \delta_{M_y}^{\max} \tag{9.99}$$

If some load parameters are taken into account in the l.h.s. of the constraints of (9.84), then the l.h.s. of (9.99) should take into account their randomness. Therefore $\delta_{M_y}^{\max}$ must be replaced by $\delta_{M,W_g}^{\max}$, i.e. by the maximum value of the coefficients of variation of the left-hand sides of (9.85b).

Then, (9.85a) states that the collapse load factor $\tilde{\alpha}_p$ is the $\min_i \tilde{\alpha}_{Ki}$. It follows that the higher values calculated on the ith mechanism are generally cut off by the values of load factors computed on different collapse kinematisms and hence (see also Section 5.2.6)

$$\mu_{\alpha_p} \lesssim \min_i \mu_{\alpha_{Ki}} \qquad s_{\alpha_p} \lesssim \min_i s_{\alpha_{Ki}} \tag{9.100}$$

Therefore, no general conclusion can be drawn on the coefficient of variation δ_{α_p}. Nevertheless, numerical experience showed that

$$s_{\alpha_p}/m_{\alpha_p} \leqslant \delta_{M_y, W_g}^{\max} \tag{9.101}$$

and that the r.h.s. of (9.101) decreases as the actual redundancy degree and/or the number of possible mechanisms increase. Moreover, as the actual redundancy degree involved by the plastic collapse mechanism increases, the estimate m_{α_p} tends, in the numerical experiments performed, to the load factor of the average structure $\alpha_{p\bar{R}}$, i.e. (equation (A3.26′)) to $\min_i \mu_{\tilde{\alpha}_{Ki}}$.

Both these circumstances are confirmed by another example where five structures of different redundancy degree were considered (Fig. 9.17) [14]. The second structure is the one of Fig. 9.16 subject to the permanent loads W_{gi} and to the horizontal action $\tilde{\alpha} W_a$: problem I is obtained by adding another span, while problems III, IV and V are derived from problem II by making statically determinate one, two or three storeys respectively. The analyses were developed taking into account the axial force–bending moment interaction profile shown in Fig. 9.17(a), i.e. the second basic assumption (item (2)) of the previous subsection 9.5.1(b) is partially released. For this purpose, the constraints (9.84b) and (9.84c) become

$$\begin{aligned} \alpha_S(M_{ai}/\tilde{M}_{yi}^+ + N_{ai}/\tilde{N}_{yi}^+) + (M_{ij}^0/\tilde{M}_{yi}^+ + N_{ij}^0/\tilde{N}_{yi}^+) &\leqslant 1 \\ \alpha_S(M_{ai}/\tilde{M}_{yi}^+ - N_{ai}/\tilde{N}_{yi}^-) + (M_{ij}^0/\tilde{M}_{yi}^+ - N_{ij}^0/\tilde{N}_{yi}^-) &\leqslant 1 \\ -\alpha_S(M_{ai}/\tilde{M}_{yi}^- + N_{ai}/\tilde{N}_{yi}^-) - (M_{ij}^0/\tilde{M}_{yi}^- + N_{ij}^0/\tilde{N}_{yi}^-) &\leqslant 1 \\ -\alpha_S(M_{ai}/M_{yi}^- - N_{ai}/N_{yi}^+) - (M_{ij}^0/\tilde{M}_{yi}^- - N_i/\tilde{N}_{yi}^+) &\leqslant 1 \end{aligned} \tag{9.102}$$

where $\tilde{N}_{yi}^+$ and $-\tilde{N}_{yi}$ are the carrying capacities of an element in tension and in compression respectively. It is worth noting that the effect of the axial force

(a)

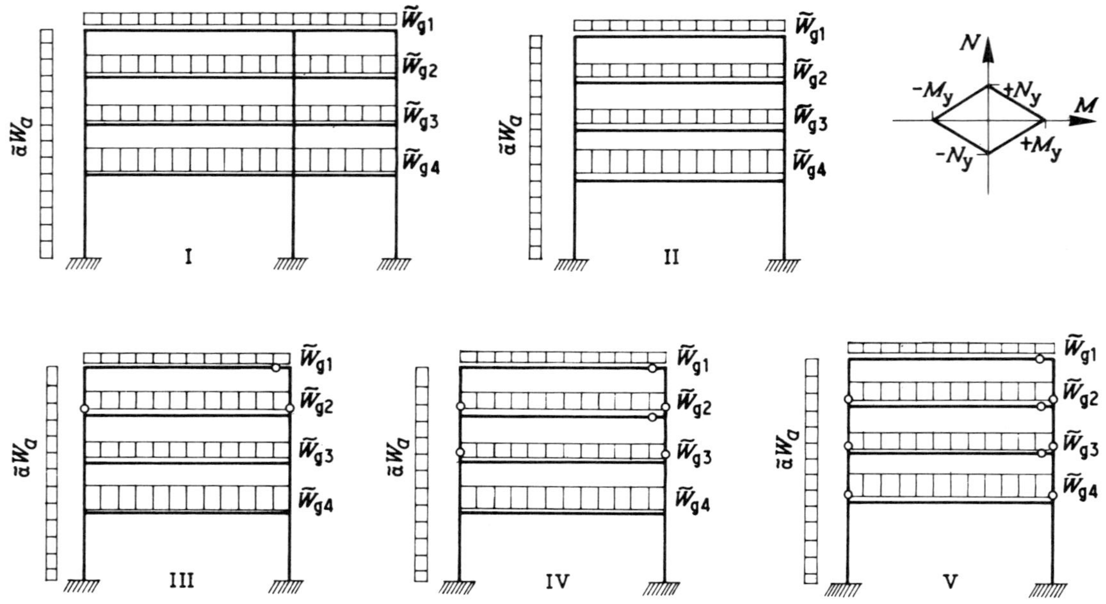

$\tilde{W}_{g1}$
$\tilde{W}_{g2}$
$\tilde{W}_{g3}$
$\tilde{W}_{g4}$
$\tilde{\alpha}W_a$
I
II
III
IV
V
N
M
$-M_y$
$+M_y$
$+N_y$
$-N_y$

Sample Size n	Structure	I	II	III	IV	V
	$\alpha_{p\bar{R}}$	5.286	3.577	3.542	1.540	0.859
500	$\frac{m_{\alpha_p}}{\alpha_{p\bar{R}}}$	0.99967	0.99967	0.99075	0.98972	0.98269
	$\frac{s_{\alpha_p}}{m_{\alpha_p}}$	0.0059	0.0078	0.0148	0.0895	0.1125
250	$\frac{m_{\alpha_p}}{\alpha_{p\bar{R}}}$	0.99972	0.99960	0.98976	0.98693	0.98354
	$\frac{s_{\alpha_p}}{m_{\alpha_p}}$	0.0062	0.0078	0.0153	0.0895	0.1116
100	$\frac{m_{\alpha_p}}{\alpha_{p\bar{R}}}$	0.99944	1.00058	0.98820	0.98654	0.9804
	$\frac{s_{\alpha_p}}{m_{\alpha_p}}$	0.0061	0.0080	0.0164	0.0845	0.1126
0	$\frac{m_{\alpha_p}}{\alpha_{p\bar{R}}}$	0.99911	1.00226	0.98863	0.99052	0.9926
	$\frac{s_{\alpha_p}}{m_{\alpha_p}}$	0.0060	0.0081	0.0181	0.0903	0.1002

(b)

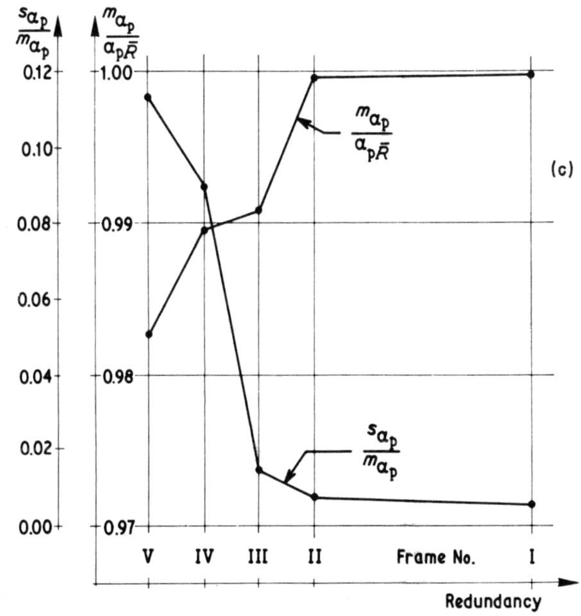

(c)

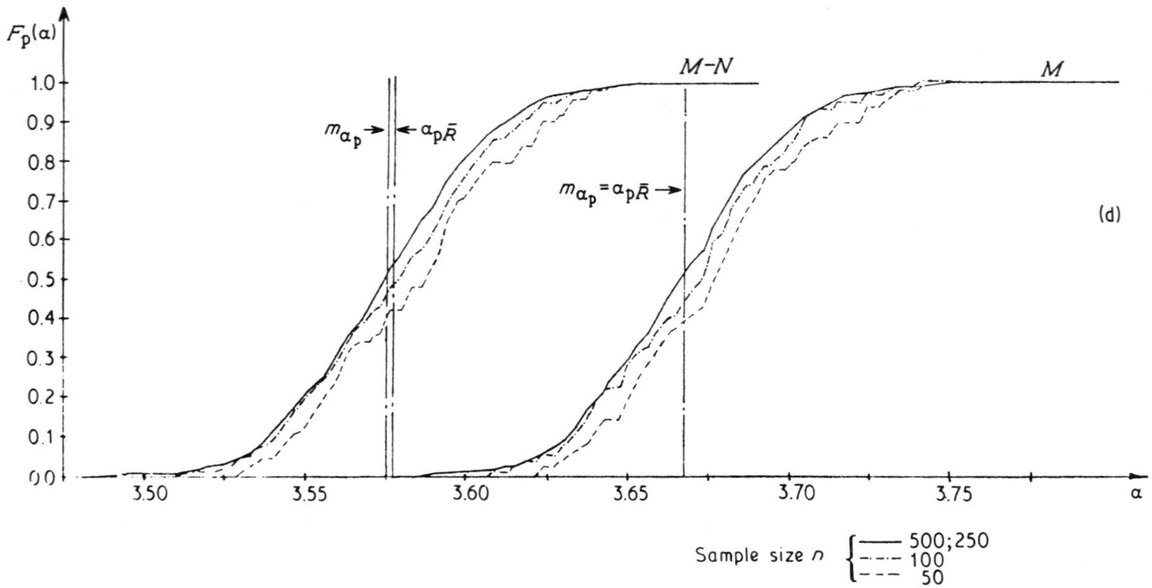

Fig. 9.17 Comparison of results of limit analysis of parametric simulation applied to five four-storey frames (from [14]). (a) Example frames; (b) effect of sample size; (c) effect of degree of redundancy; (d) M–N interaction effect: function $F_p(\alpha) = F_R(\alpha)$ of structure II simulated by considering only the bending moment (M) or taking into account the axial force-bending moment interaction (M–N).

translate the curves $F_R(\alpha)$ towards the left, but the coefficient of variation of $\tilde{\alpha}_p$ maintains approximately the same value (Fig. 9.17(d)).

Finally, this example also shows the dependence of the estimation on the sample size n. Particularly, the estimates of μ_{α_p} and s_{α_p} are satisfactory from $n = 50$ (Fig. 9.17(b)), while the convergence on the frequencies is only achieved for $n \geqslant 250$ (Fig. 9.17(d)).

(d) Shakedown

In case of *incremental collapse* (Section 5.3), the limit condition is still described by a linear programming problem. More precisely, the static (*shakedown*) theorem (see Section 5.3.1) permits one to write (equations (5.103) and (5.106))

$$\begin{aligned} &\tilde{\alpha}_{pV} = \max \tilde{\alpha}_{SV} \\ &\tilde{\alpha}_{SV}\, \mathbf{M}_{Ea}^{max} + \mathbf{M}^0 \mathbf{X}_V^+ \leqslant \tilde{\mathbf{M}}_y^+ \\ &-\tilde{\alpha}_{SV}\, \mathbf{M}_{Ea}^{min} - \mathbf{M}^0 \mathbf{X}_V^+ \leqslant \tilde{\mathbf{M}}_y^- \\ &\tilde{\alpha}_{SV} \geqslant 0 \qquad \mathbf{X}_V \geqslant 0 \end{aligned} \tag{9.103}$$

where $\mathbf{M}_{Ea}^{max}$ and $-\mathbf{M}_{Ea}^{min}$ are the vectors of the maximum and minimum bending moments at the critical sections and $\mathbf{X}_V^+$ is a vector of self-equilibrated redundancies with the same properties of the vector $\mathbf{X}^+$ in (9.84).

The parametric simulation approach summarized in equation (9.96) can also be employed in order to obtain an estimate $F_{RV}(\alpha)$ of the CPF $P_{RV}(\alpha)$ relevant to the incremental collapse. Fig. 9.18 presents the results obtained for the structure of Fig. 9.16(a) and compares them with the ones relevant to the limit states defined by the first plasticization and by the 'static' plastic collapse.

(e) Other limit states described by constrained optimization problems

The problems treated in this section (9.5.1) are solved by a constrained mathematical optimization: whenever the structural behaviour can be modelled in a similar manner, the analysis of the corresponding parametric problem permits one to perform many experiments without solving over again the governing equations [12]. Besides linear programming, the complementarity linear programming and/or quadratic programming is often employed in structural engineering for the solution of elasto–plastic problems: in these cases, an extension of this parametric technique can also be adopted [12].

9.5.2 Deflections at incipient collapse

Another problem in which parametric techniques are usefully applied is the determination of the deflections at incipient collapse of elastic–perfectly plastic structures subjected to monotonically increasing quasi-static loads.

This determination is important for a more complete analysis of the limit state connected with plastic collapse, and is reported here also as the starting point for the approximate determination of deflections under time-varying random loads (Section 9.6.1).

(a) The deterministic problem

As is well known, the deflections at incipient collapse for a deterministic elastic–perfectly plastic structure, without using step-by-step techniques, can be evaluated if some assumptions are introduced in addition to the usual hypotheses of simple plastic theory.

The structure considered must be initially free from stresses, and the parameter defining the magnitude of the acting loads increases monotonically up to the collapse. Moreover, the effects of strain-hardening and of the spread of plastic zones along the structural members are assumed to be negligible (so that the bending moment–curvature relation may be assumed elastic–perfectly plastic). Further, plastic hinge rotations, once initiated, increase monotonically with the load parameter.

Under these assumptions, when the collapse load is reached, all except one of the plastic hinges (apart from exceptional cases of simultaneous formation of two or more hinges) have formed and undergone rotation. The value of these rotations can be computed by using the technique of elastic structural analysis, if the last-formed plastic hinge is known.

Several methods have been formulated in order to determine the position of the last-formed plastic hinge, but difficulties arise for the determination of the plastic hinges which, at incipient collapse, do not appear in the mechanism when failure is partial. In this case, however, it is better to employ the step-by-step procedures presented in Section 9.6.2.

Excluding the case of partial failure, the virtual work approach may be advantageous especially in view of a probabilistic analysis. At incipient collapse, under the assumption that the mechanism is of the complete type, the real deformation state of a structure with m redundancies is defined by the rotations θ_j ($j = 1, \ldots, m+1$) in $m+1$ plastic hinges and by the elastic curvatures M/EJ. This deformation state can be combined by the equation of virtual work with m independent sets of self-stressing moments M_i^0 to give a system of m linear equations in $m+1$ unknowns‡:

$$\sum_{j=1}^{m+1} M_i^0 \theta_j + \int_L M_i^0 \frac{M}{EJ} \mathrm{d}l = 0 \qquad i = 1, \ldots, m \tag{9.104}$$

‡ When the collapse is partial, that is when part of the frame remains indeterminate, there will again be $m+1$ unknown quantities, some of which are values of the redundancies while the rest are hinge rotations.

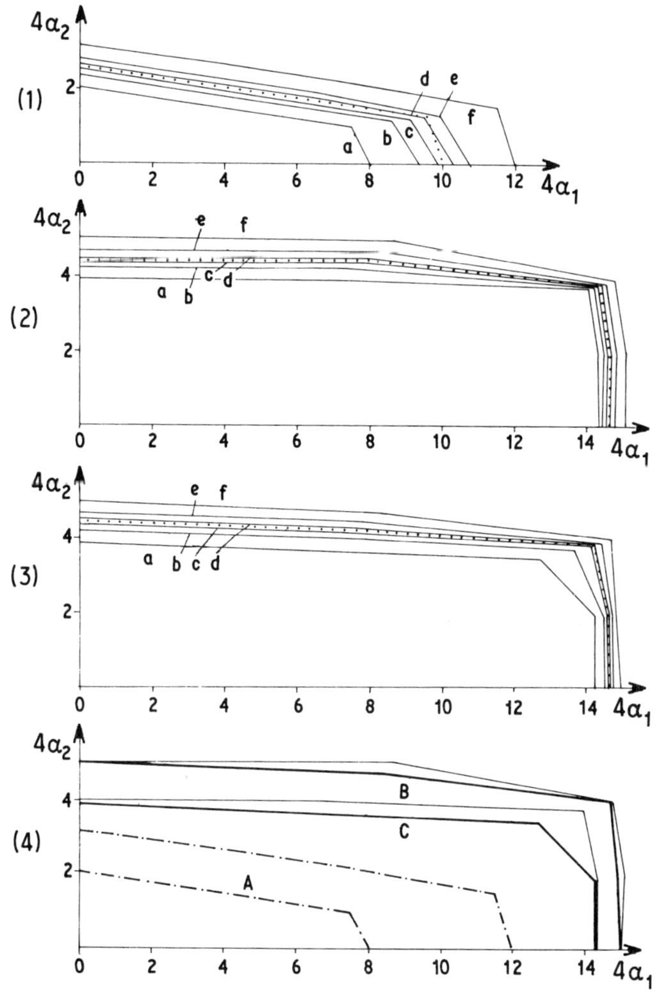

Fig. 9.18 Comparison of reliability analyses of frame in Fig. 9.16 with respect to different limit states (from Casciati and Sacchi [13]).

(1), (2), (3): Level curves of the estimated probabilities of failure F_i:
(1) first plasticization; F_E;
(2) static plastic collapse; F_R;
(3) incremental collapse; F_{RV}.
lines a: $F_i = 0.002$; lines b: $F_i = 0.2$; lines c: $F_i = 0.4$; lines d: $F_i = 0.6$; lines e: $F_i = 0.8$; lines f: $F_i = 0.998$.
dotted lines: boundary of safe region of expected structure.

(4) Limit positions of the level curves:
A: first plasticization (F_E);
B: static plastic collapse (F_R);
C: incremental collapse (F_{RV}).

Moreover, the plastic hinge rotations, in addition to equation (9.104), must satisfy the condition that their signs and those of the corresponding full plastic moments must be the same for any hinge. In matrix notation, the problem becomes for a discretized structure:

$$\mathbf{\Lambda\theta} = \mathbf{\Upsilon M}_{\mathrm{y}} + \Xi\alpha_{\mathrm{p}} + \mathbf{c}\,\mathbf{W}_g \tag{9.105a}$$

$$\mathbf{M}_{\mathrm{y}}\mathbf{\theta}^{\mathrm{T}} \geqslant 0 \tag{9.105b}$$

where $\mathbf{\Lambda}$ is the $m \times (m+1)$ matrix of the coefficients of the plastic rotations (vector $\mathbf{\theta}$), $\mathbf{\Upsilon}$ is an $m \times (m+1)$ matrix, Ξ is a vector of length m, and $\mathbf{c}$ is a matrix with columns of length m whose number coincides with the number of permanent loads. The right-hand side of (9.105a) covers the integrals of (9.104) and indicates that these integrals depend linearly on the full plastic moments in the plastic hinges formed, on the live load value α_{p} for which the structure fails, and on the dead load values $\mathbf{W}_g$.

Solving the system (9.105), the last-formed plastic hinge, i.e. the critical section where the plastic rotation θ_0 is zero at incipient collapse, is determined and the other plastic rotations are then known: $\mathbf{\theta}_{\mathrm{B}} = \mathbf{D}^{-1}(\mathbf{\Upsilon M}_{\mathrm{y}} + \Xi\alpha_{\mathrm{p}} + \mathbf{c}\mathbf{W}_g)$. The $m \times m$ matrix $\mathbf{D}$ contains the m linearly independent columns of $\mathbf{\Lambda}$ in which the coefficients of θ_0 do not appear. Moreover, the plastic rotations $\mathbf{\theta}_{\mathrm{B}}$ satisfy the condition $\mathbf{M}_{\mathrm{yB}}\mathbf{\theta}_{\mathrm{B}}^{\mathrm{T}} \geqslant 0$, where $\mathbf{M}_{\mathrm{yB}}$ is the vector containing the coefficients of $\mathbf{\theta}_{\mathrm{B}}$ in the constraints (9.105b).

Once the state of the frame has been found, the deflection at a particular point in the structure may be determined using the virtual work equation again:

$$1\,u = \sum_j M'\theta_j + \int_L M' + \frac{M}{EJ}\,\mathrm{d}l \tag{9.106}$$

where M' is any distribution of bending moments which is in equilibrium with the unit load applied at the particular point where u is to be computed.

(b) The problem for a random structure

If the mechanical characteristics of each critical section are random quantities with given CDF, in the previous formulae the full plastic moment vector $\mathbf{M}_{\mathrm{y}}$ and the α_{p} value assume a random nature. The deflection u at incipient collapse becomes, therefore, a random variable of unknown CDF, which also depends on the (possibly random) dead load $\tilde{\mathbf{W}}_g$ and on the flexural rigidity EJ. In the following, EJ is assumed to be deterministic.

In order to obtain this CDF, it is useful to observe that a different collapse mechanism and a different last-formed plastic hinge correspond to different samples of the $\tilde{\mathbf{M}}_{\mathrm{y}}$ vector and the $\tilde{\mathbf{W}}_g$ values. However, in the space of these random variables a set $\mathcal{A}_k$ of points having the same mechanism of collapse (*decision region*) may be identified (Section 9.5.1(b)); it was already stated that each $\mathcal{A}_k$ is a cone with the vertex at the origin. Furthermore, in every decision region $\mathcal{A}_k$ there is a set $\mathcal{G}_{kj}$ of points having the same last-formed hinge: again it

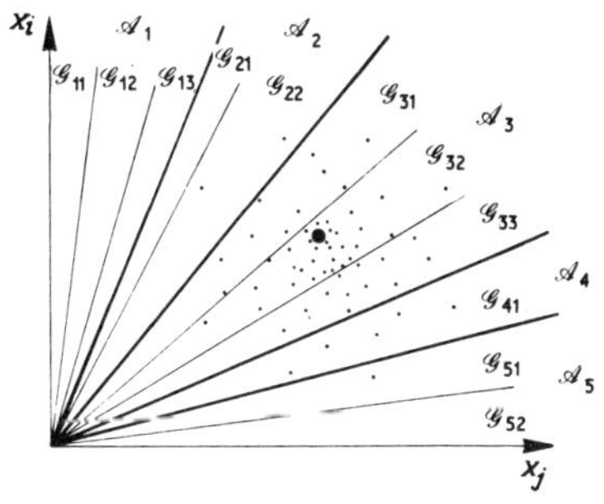

Fig. 9.19 Each region $\mathscr{A}_k$ corresponding to the same collapse mechanism is divided into subregions $\mathscr{G}_{kj}$ corresponding to the same last-formed hinge.

can be proved that each $\mathscr{G}_{kj}$ is a cone with the vertex at the origin (Fig. 9.19).

A rigorous probabilistic description of the structural geometry at collapse would involve specification of all possible regions $\mathscr{A}_k$ and $\mathscr{G}_{kj}$. Their knowledge, in fact, allows one to determine the true distribution function of any deflection parameter u, given by

$$P_u(u) = \sum_k \sum_j \int_{\mathscr{G}_{kju}} p(M_{y1}, \ldots, M_{yn}, W_g) \mathrm{d}M_{y1}, \ldots, \mathrm{d}M_{yn} \, \mathrm{d}W_g \qquad (9.107)$$

or in general terms [15]

$$P_u(u) = \sum_k \sum_j \int_{\mathscr{G}_{kju}} p_i(x_i) \, \mathrm{d}x_1, \ldots, \mathrm{d}x_n \qquad (9.108)$$

where $p_i(x_i)$ is the JPDF of the design parameters $\tilde{x}_i$; $k = 1, \ldots, h$, with h the number of the regions $\mathscr{A}_k$ in the x_i space; $i = 1, \ldots, n_k$, with n_k the number of regions $\mathscr{G}_{kj}$ in the kth region $\mathscr{A}_k$ and $\mathscr{G}_{kju} = (\mathscr{G}_{kj} \cap \Theta)$, Θ being the half-plane in the x_i-space in which the displacement is smaller than u.

The beam of Fig. 9.20 is a very simple example, in which $\mathscr{A}_k$ is unique and divided into two subregions $\mathscr{G}_{k1}$ and $\mathscr{G}_{k2}$.

However, the use of equation (9.108) is subject to the same computational difficulties already discussed with reference to stochastic programming, and the development of approximate techniques seems more promising. In this case, for each of the simulated structures, it is necessary to determine the last-formed plastic hinge and then to obtain the parameters of the deformation state as linear combinations of the independent random design parameters.

In order to make the procedure more efficient, a parametric study is possible, but the sensitivity analysis of the solution can be performed only with reference to a given kinematism. The cones $\mathscr{G}_{kj}$ are the relevant decision regions that allow one to apply a parametric simulation method.

(a)

(b)

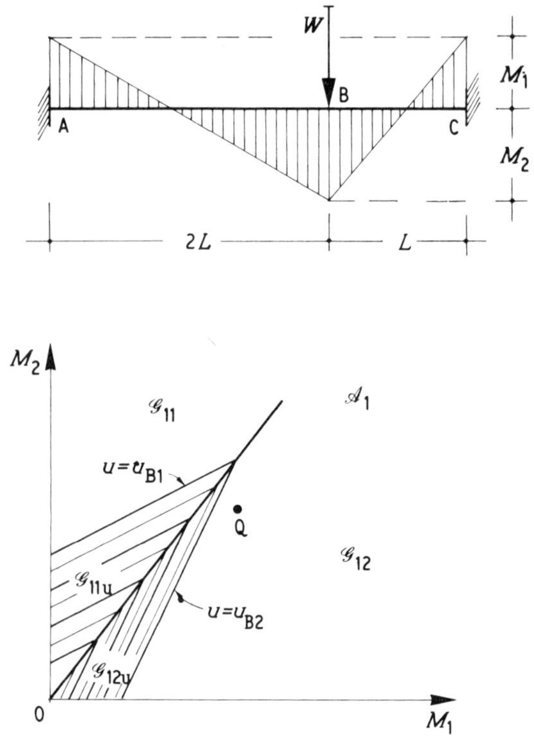

Fig. 9.20 (a) Example elastic–plastic beam and bending moment diagram (from Casciati [15]). (b) Regions corresponding to formation of last hinge at the built-in ends ($\mathscr{G}_{11}$) or under the force ($\mathscr{G}_{12}$).

9.6 SIMULATION OF RESIDUAL DEFORMATIONS UNDER AN ARBITRARY LOADING HISTORY

Another problem, most important in practice, in which numerical approaches can find application, is the computation of residual deformations under an arbitrary loading history.

In fact, a structure can be put out of effective service before ultimate plastic collapse by the accumulation of plastic deformations (Section 6.1.3). Therefore, in a complete structural reliability analysis it may be necessary to evaluate the probability $\mathrm{Prob}[\tilde{u}_r > u_L \mid T]$ that the inelastic deformation $\tilde{u}_r$ attains a given permissible limit u_L during the structural lifetime T.

A similar problem, including inertia forces but with the load history limited to a succession of independent pulses, was discussed in Section 9.4.1. In this section, in contrast, inertia forces will be neglected; moreover, in Section 9.6.2 no limitation is posed on the load history.

9.6.1 A simplified treatment

Assume that the stochastic load on the structure cannot change sign; then

$$\text{Prob}[\tilde{u}_r > u_L | T] = \text{Prob}[\tilde{W}_{max} > \phi^{-1}(u_L) | T] \tag{9.109}$$

where $\tilde{W}_{max}$ is the largest load value in the time interval $(0, T)$ and $u_r = \phi(W_{max})$ is the relationship between the residual deformation and the maximum load.

In order to avoid step-by-step techniques, the following upper bound to the probability (9.109) can be introduced. With reference to Fig. 9.21 the $\phi(0)$ function is idealized with the straight line $\phi'(W_{max})$ joining the points representing the elastic limit and the plastic collapse. Since $W_L = \phi'^{-1}(u_L) \leqslant \phi^{-1}(u_L)$, it is possible to write

$$\text{Prob}[\tilde{u}_r > u_L | T] \leqslant \text{Prob}[\tilde{W}_{max} \geqslant \tilde{W}_L | T] \tag{9.110}$$

Since

$$\tilde{W}_L = \frac{(\tilde{W}_p - \tilde{W}_E)}{\tilde{u}_p} u_L + \tilde{W}_E \tag{9.111}$$

the PDF of $\tilde{W}_L$ may be easily estimated after the distribution laws of the parameters defining the elastic limit and the plastic collapse are obtained by simulation (cf. Section 9.5.2).

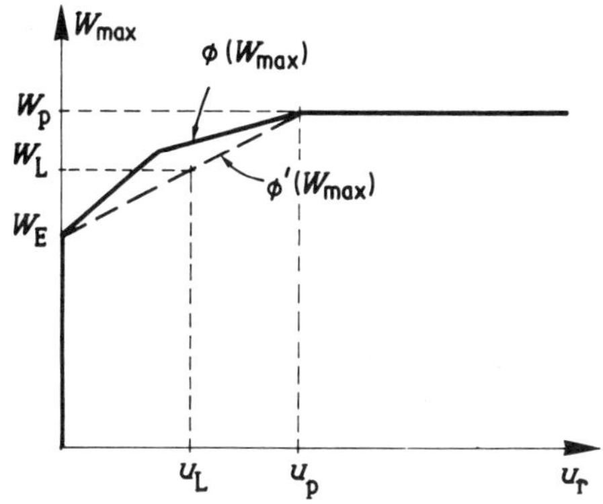

Fig. 9.21 Determination of residual displacement: approximate load–permanent displacement relationships.

9.6.2 A step-by-step procedure

With reference to a plane frame, introduce the hypotheses of Section 9.5.1(b) and modify item (4): the moment–deformation relationships are assumed to be elastic–perfectly plastic and non-holonomic (Fig. 9.22).

Consider then a plane frame discretized in perfectly elastic elements connected

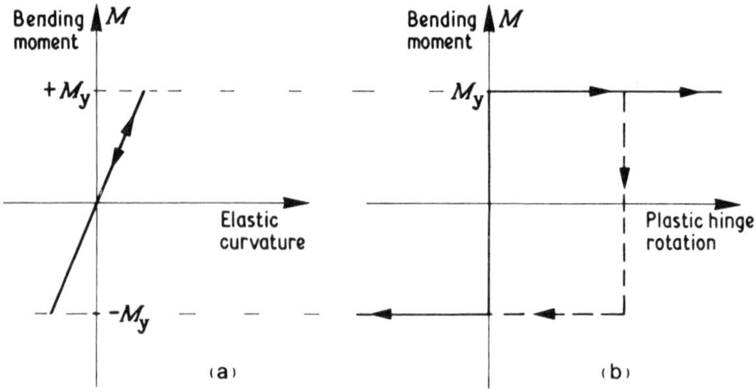

Fig. 9.22 Idealized elastic–perfectly plastic moment–deformation relationships: (a) moment vs elastic curvature; (b) moment vs plastic hinge rotation.

by N critical sections. Moreover let α_0 and $\boldsymbol{\theta}_0$ denote the load factor and the inelastic rotations that define a known initial solution.

Under the above assumptions, the hinge-by-hinge elastic–perfectly plastic analysis, as the load factor α increases (or decreases), may be performed by solving the following parametric linear programming (PLP) problem:

$$\max\{-(\mathbf{X}^T\Delta\boldsymbol{\theta}^+ + \mathbf{Y}^T\Delta\boldsymbol{\theta}^- + \mathbf{x}^T\mathbf{z}^+ + \mathbf{y}^T\mathbf{z}^-\} \tag{9.112a}$$

$$\mathbf{Z}\Delta\boldsymbol{\theta}^+ - \mathbf{Z}\Delta\boldsymbol{\theta}^- + \mathbf{z}^+ = \mathbf{M}_y^+ - \alpha_0\mathbf{M}_E - \mathbf{Z}\boldsymbol{\theta}_0 - \Delta\alpha\,\mathbf{M}_E \tag{9.112b}$$

$$-\mathbf{Z}\Delta\boldsymbol{\theta}^+ + \mathbf{Z}\Delta\boldsymbol{\theta}^- + \mathbf{z}^- = \mathbf{M}_y^- + \alpha_0\mathbf{M}_E + \mathbf{Z}\boldsymbol{\theta}_0 + \Delta\alpha\,\mathbf{M}_E \tag{9.112c}$$

$$\Delta\boldsymbol{\theta}^+ \geqslant 0 \qquad \Delta\boldsymbol{\theta}^- \geqslant 0 \qquad \mathbf{z}^+ \geqslant 0 \qquad \mathbf{z}^- \geqslant 0 \tag{9.112d}$$

In (9.112) $\mathbf{M}_E$ denotes the vector of the elastic bending moments in the N critical sections; $\mathbf{M}_y^+$ and $-\mathbf{M}_y^-$ are the vectors of the positive and negative yielding moments; $\mathbf{z}^+$ and $\mathbf{z}^-$ are vectors of *slack variables*. The $N \times N$ matrix $\mathbf{Z}$ is the matrix of the influence coefficients. The vectors $\Delta\boldsymbol{\theta}^+$ and $\Delta\boldsymbol{\theta}^-$ denote the positive and negative increments, respectively, of the inelastic rotations at the N critical sections; the actual inelastic rotations $\boldsymbol{\theta}$ are given by $\boldsymbol{\theta} = \boldsymbol{\theta}_0 + (\Delta\boldsymbol{\theta}^+ - \Delta\boldsymbol{\theta}^-)$.

The orthogonality relations due to the constitutive law are satisfied by defining the components of the vectors $\mathbf{X}$, $\mathbf{Y}$, $\mathbf{x}$ and $\mathbf{y}$ in the following way: $X_i = 0$, $Y_i = 1$, $x_i = 1$, $y_i = 0$ at the positive plastic hinges; $X_i = 1$, $Y_i = 0$, $x_i = 0$, $y_i = 1$ at the negative plastic hinges; $X_i = 1$, $Y_i = 1$, $x_i = 0$ and $y_i = 0$ at the remaining sections.

The objective function (9.112a) attain a zero-valued maximum for the actual solution of the problem. This solution can be expressed in the form (9.91):

$$\{\Delta\boldsymbol{\theta}_B^{+T}, \Delta\boldsymbol{\theta}_B^{-T}, \mathbf{z}_B^{+T}, \mathbf{z}_B^{-T}\}^T = \mathbf{d} + \mathbf{D}\,\Delta\alpha + \mathbf{c}\{\mathbf{z}_{NB}^{+T}, \mathbf{z}_{NB}^{-T}, \Delta\boldsymbol{\theta}_{NB}^{+T}, \Delta\boldsymbol{\theta}_{NB}^{-T}\}^T \tag{9.113}$$

where the suffix B individuates the basic variables and the suffix NB the non-basic variables (T = transpose of the whole vector). Note that for the arbitrary jth critical section $\Delta\theta_j^+$ (or $\Delta\theta_j^-$) is non-basic if z_j^+ (or z_j^-) is basic and vice versa, according to the material constitutive law.

Therefore, each step of this incremental structural analysis may be summarized in the following two phases:

(1) phase I: determine the critical value $\Delta\alpha_c$ to which the load factor α can be increased (or decreased) before the initial *basic* solution (9.113) will no longer be *optimal*;
(2) phase II: find the new stress–strain solution of (9.112) for $\alpha = \alpha_0 + \Delta\alpha_c$.

From the initial solution α_0, $\boldsymbol{\theta}_0$, phase I is solved by performing a sensitivity analysis of the solution to changes in the requirement vector of (9.112). In fact the critical value $\Delta\alpha_c$, for which a new plastic hinge forms or a local unloading occurs at the jth critical section, and the index j can be evaluated by using the parametric linear programming rules (Section 9.5.1(b)). Then, phase II consists of the application of the pivot algorithm. For $\Delta\alpha = \Delta\alpha_c$ the price and the requirement vectors of (9.112) must be changed in order to take into account the actual plastic hinges and the new initial condition respectively. For such a modified LP problem (9.112), the inelastic rotation increments $\Delta\boldsymbol{\theta} = \Delta\boldsymbol{\theta}^+ - \Delta\boldsymbol{\theta}^-$ are evaluated by performing one *pivotal* step.

The actual bending moments $\mathbf{M}$ at the N critical sections can be calculated, for any load factor $\alpha = \alpha_0 + \Delta\alpha$ ($\Delta\alpha < \Delta\alpha_c$), by the expression

$$\mathbf{M} = (\alpha_0 + \Delta\alpha)\mathbf{M}_E + \mathbf{Z}(\boldsymbol{\theta}_0 + \Delta\boldsymbol{\theta}) \tag{9.114}$$

Then the displacements at given points of the structure can be determined by using the techniques of elastic structural analysis (9.110). The load–deflection relationship is approximately linear between the formation of two successive plastic hinges. The procedure discussed above is repeated until the final solution, for which $\Delta\alpha$ cannot be increased (or decreased) without losing feasibility, is obtained.

By the approach summarized in (9.112) any piecewise-linear loading path can be analysed and no problem arises if more independent loading conditions act simultaneously on the structure. However, for the sake of simplicity one monotonically increasing load parameter is considered in the following.

In problem (9.112), the matrix $\mathbf{Z}$ and the vector $\mathbf{M}_E$ may have random elements because of the geometry randomness. The uncertainties on the material strength affect the yielding moment vector $\mathbf{M}_y^+$ and $\mathbf{M}_y^-$. Furthermore the requirement vector may also depend on a random self-stress distribution. However, at least for steel structures, the geometry randomness is generally negligible and the structural response (i.e. inelastic rotations, displacements, etc.) is mainly influenced by the uncertainties on the requirement vector of (9.112).

Because of the numerical nature of the relationship between the data and the structural response, the statistics of the latter can be evaluated only by using simulation techniques. To this purpose an improved simulation technique can be formulated as follows.

The analysis of the structure defined by the average values of the random design parameters is first performed by solving problem (9.112): thus the relevant relationship $\boldsymbol{\theta}(\alpha)$ between the inelastic rotations and the load factor is determined and can be stored. Then a first simulated ensemble of the random variables is obtained by selecting a series of random numbers each of which is associated with a particular value of the random variable of interest. The structure defined by these values of the design parameters has, in the elastic branch, the same solution ($\boldsymbol{\theta} = 0$) as the average structure. Performing phase I of the deterministic analysis, the position of the first-formed plastic hinge is determined. If this plastic hinge coincides with the first one of the average structure, phase II of the analysis is avoided by using the corresponding stored solution. This phase is avoidable for every branch of the piecewise-linear relationship $\boldsymbol{\theta}(\alpha)$ as long as the found hinge is the same as that which forms in the average structure. Otherwise, the stress–strain solution is to be calculated by the pivot algorithm.

The whole procedure is repeated for all the simulated structures [16].

In order to explain the computational aspects of this technique, consider problem (9.112) in the compact form:

$$\max(\boldsymbol{\zeta}^{T}\mathbf{x}) \tag{9.115a}$$

$$\boldsymbol{\Upsilon}\mathbf{x} \leqslant \boldsymbol{\Xi} + \Delta\alpha\,\mathbf{c} \tag{9.115b}$$

$$\mathbf{x} \geqslant 0 \tag{9.115c}$$

where the elements of the vector $\boldsymbol{\Xi}$ are random quantities and the price vector $\boldsymbol{\zeta}$ changes as soon as a new plastic hinge forms or a local unloading occurs. For a given price vector the basic variables $\mathbf{x}_B$ are determined by (equation 9.113)

$$\mathbf{x}_B = \mathbf{D}^{-1}(\boldsymbol{\Xi} + \Delta\alpha\mathbf{c}) \tag{9.116}$$

where $\mathbf{D}$ is the optimal basis matrix (equation (9.91)). Equation (9.116) holds for $\Delta\alpha$ in the range $(0, \Delta\alpha_c)$; $\Delta\alpha_c$ is the smallest load increment for which a basic variable x_j becomes zero. If x_j denotes a slack variable, a new plastic hinge forms at the jth critical section for $\Delta\alpha = \Delta\alpha_c$; otherwise, a local unloading occurs. Therefore for $\Delta\alpha \geqslant \Delta\alpha_c$ the price vector must be changed.

Once the elastic–plastic analysis of the average structure is performed, the inverse matrix $\mathbf{D}^{-1}$, the index vector defining the variables in the basis and the index j of the variable leaving the basis for $\Delta\alpha = \Delta\alpha_c$ are stored for each branch of the piecewise-linear law $\boldsymbol{\theta}(\alpha)$. Then the incremental solution of any simulated structure can be directly obtained by putting the corresponding stored matrix $\mathbf{D}^{-1}$ in (9.115), as long as the order in which the plastic hinges form is the same as the average structure. From a computational point of view, this control is to verify that the variable leaving the basis is the same for the simulated and the average structure.

By this parametric analysis, phase II of the deterministic structural analysis (the phase that requires the more cumbersome computations) is avoided, as long as possible, for any simulated structure. However, the computational advantage depends on the characteristics of the considered problem. For instance, for steel structures, the material strength is defined by small coefficients of variation and therefore changes in the order in which plastic hinges form rarely occur: the procedure is hence very efficient.

The results of an example are presented in Fig. 9.23. Thc frame was discretized in eight perfectly elastic elements and 14 critical sections were considered.

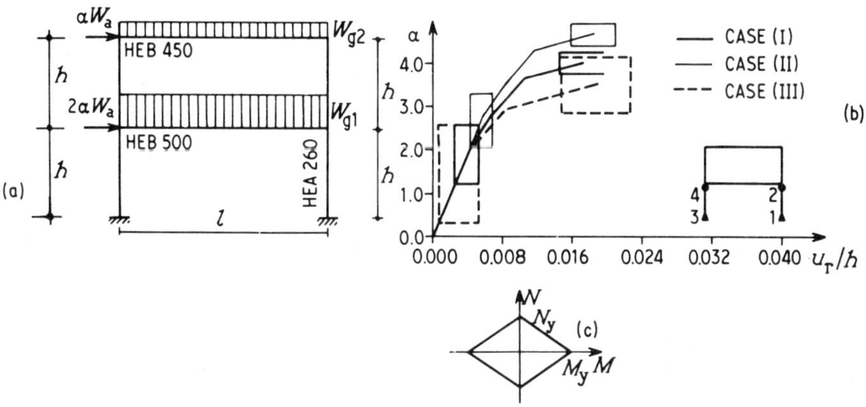

Fig. 9.23 (a) Analysed structural problem: $W_a = 1.275$ t; $E[\tilde{W}_{g1}] = 3.6$ t m^{-1}; $E[\tilde{W}_{g2}] = 1.5$ t m^{-2}. (b) Behaviour of the average structure and variability of the relevant structural response parameter: (I) axial force effects considered, (II) axial force effects neglected, (III) as (I) but with weak connections. (c) Assumed axial force–bending moment interaction profile (from Casciati and Faravelli [16]). Considered critical sections: top beam, A, B; lower beam, C, D; bottom of columns, E, F.

The yielding stresses σ of the material at the critical sections were assumed to be uncorrelated, normally distributed random variables with the same mean value $E[\tilde{\sigma}_y]$ and the same coefficient of variation $\delta_{\sigma_y} = 5\%$. A first analysis was performed assuming that the connections do not influence the structural behaviour. That is generally obtained by designing welded connection with stiffeners, capable of developing the fixed-end plastic moment of the supported beam and with adequate rotation capacity. The analysis was performed for l_k equal to twice the cross-section depth d_k in the equation

$$\bar{\theta}_k = \Lambda_k \theta_k^E = \Lambda_k M_{yk} l_k/(EJ)_k \tag{9.117}$$

that expresses the permissible limit $\bar{\theta}_k$ to the inelastic rotation at the kth critical section. Equation (9.117) is obtained assuming the angle $\bar{\theta}_k$ to be uniformly distributed over the length l_k of partial plastification and computing the average curvature in this length. Thus, at any critical section, $\bar{\theta}_k$ may be assumed to be proportional to the elastic rotation Λ_k of the plastic zone and subject to the full plastic moment M_{yk}. The coefficients Λ_k were assumed to be uncorrelated, uniformly distributed over the interval (6, 10). The analysis was also performed for $l_k = d_k$ in equation (9.117) in order to investigate a less ductile local behaviour.

Then, more economical connections were considered and the structural analysis was repeated. Particularly, the beam-to-column joints (sections A, B, C and D of Fig. 9.23(a)) were designed to be capable of developing 80% of the fixed-end plastic moment of the supported beam as expected value. The considered c.o.v. of the connection strength was 20%. The rotation capacity at the kth joint is again obtained by equation (9.117) where $l_k = d_k$ and the coefficients Λ_k were assumed to be uncorrelated, uniformly distributed over the interval (Λ^l, Λ^u).

Results were calculated for both brittle joints ($\Lambda^l = 0.5$, $\Lambda^u = 2$) and ductile joints ($\Lambda^l = 2$, $\Lambda^u = 6$). Moreover, the expected value of the strength of the column anchorages (sections E and F of Fig. 9.23) was taken equal to 80% of the column plastic moment with c.o.v. = 20%. For these connections the values $\Lambda^l = 2$ and $\Lambda^u = 6$ were considered.

The analysis was also performed assuming that the connections in sections A and B are capable of developing, as expected value, 80% of the column plastic moment instead of 80% of the fixed-end plastic moment of the supported beam (case III′).

Figs 9.24 and 9.25 present the results obtained by the discussed parametric simulation approach. There α_p, α_E, α_d and α_D denote the limit states defined by the plastic collapse, the first plasticization, the ultimate ductility and the excessive residual deformation (i.e. $u_{ri} \geqslant u_{Li} = \varepsilon i h_h$, ε being a given constant, i the floor number, and u_{ri} its permanent horizontal displacement.)

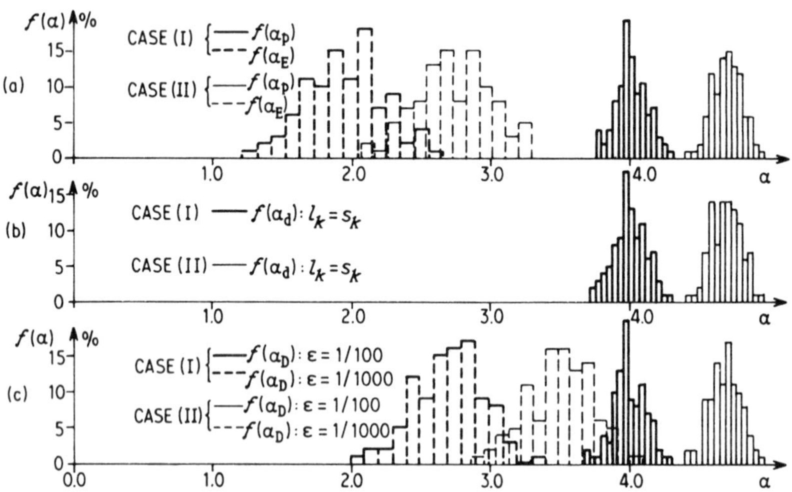

Fig 9.24 Frame of Fig. 9.23 (normal connections): histograms of the load parameter at attainment of several limit states (s_k is d_k in the text) (from Casciati and Faravelli [16]).

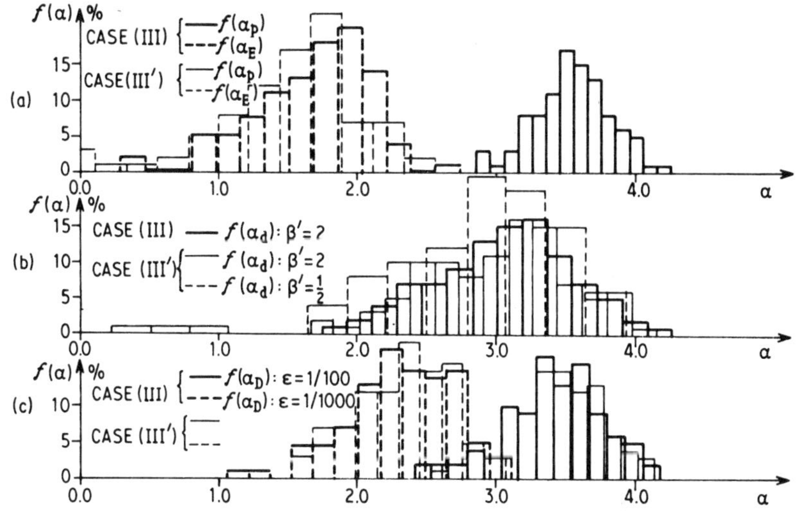

Fig. 9.25 Frame of Fig. 9.23 (weak connections): histograms of the load parameter at attainment of several limit states (β' is Λ^1 in text) (from Casciati and Faravelli [16]).

REFERENCES

1. Veneziano, D. (1978) *Random Processes for Engineering Application* (Lecture notes), Department of Civil Engineering, M.I.T.
2. Clough, R. A. and Penzien, J. (1975) *Dynamics of Structures*, McGraw-Hill, New York.
3. Penzien, J. (1978) Non-linear stochastic response of structures to strong earthquake ground motions. *First Int. Seminar on Advances in Vibrations*, London.
4. Penzien, J. (1978) Stochastic modelling of three-dimensional earthquake ground motions. *First Int. Seminar on Advances in Vibrations*, London.
5. Mitsuma, H. (1979) *Application of Data-Based Nonstationary Random Processes to Engineering Mechanics Problems*. Ph.D. Thesis in Engineering Sciences, Columbia University, New York.
6. Shinozuka, M., Ishikawa, H. and Mitsuma, H. (1979) Data-based nonstationary random processes. *Proc. ASCE Specialty Conf. on Probabilistic Mechanics and Structural Reliability*, Tucson, Arizona, pp. 39–42.
7. Strating, J. and Vos, H. (1973) Computer simulation of the E.C.C.S. buckling curve using a Monte Carlo method. *Heron*, **19** (2).
8. Casciati, F., Faravelli, L. and Gobetti, A. (1979) Permanent deformations of rigid-plastic structures subject to random dynamic loads. *Engineering Structures*, **1** (3), 139–44.
9. Casciati, F., Faravelli, L. and Gobetti, A. (1978) Residual displacement for rigid plastic structures subject to stochastic ground accelerations. *Proc. 6th European Conf. Earthquake Engineering* (6. ECEE), Dubrovnik, Yugoslavia, vol. 2, pp. 31–38

9a. Casciati, F., Faravelli, L. and Gobetti, A. (1979) Reliability analysis of frames with geometric non-linearity subject to stochastic ground accelerations. *Proc. 3rd Int. Conf. on Applications of Statistics and Probability in Soil and Structural Engineering* (ICASP 3) Sydney, vol. 2, pp. 756–64.

10. Ang, A. H. S. and Amin, M. (1970) *Probabilistic Structural Mechanics and Engineering* (Course Notes), University of Illinois, Urbana, USA.

11. Casciati, F. and Sacchi, G. (1974) On the reliability theory of the structures. *Meccanica*, **9** (4) 281–98.

12. Casciati, F. and Faravelli, L. (1977) La sicurezza strutturale nei confronti degli stati limite descrivibili come problemi di ottimizzazione vincolata (in Italian). *Costruzioni Metalliche*, **29** (3), 153–8.

13. Casciati, F. and Sacchi, G. (1975) On the longevity of structures having random resistance. *Proc. 2nd Int. Conf. on Applications of Statistics and Probability in Soil and Structural Engineering* (ICASP 2), Aachen, vol. 1, pp. 101–14.

14. Casciati, F., Faravelli, L. and Sacchi, G. (1977) Etude probabiliste des structures par programmation paramétrique (in French). *Bulletin Technique de la Suisse Romande*, **14,** 181–6.

15. Casciati, F. (1978) A probabilistic approach to the deformation analysis of elastic-plastic frames. *Journal of Structural Mechanics*, **6** (1), 45–60.

16. Casciati, F. and Faravelli, L. (1980) Elasto-plastic analysis of random structures by simulation methods; in *Simulation of Systems '79* (eds L. Dekker, G. Savastano, and G. G. Vansteenkiste), North-Holland, Amsterdam, pp. 497–506.

CHAPTER 10

Probabilistic methods and structural codes

This chapter contains a discussion of reliability-based codes of practice, in the light of what has been illustrated in the previous chapters. It is not intended to illustrate or summarize any specific code (for most of which satisfactory handbooks exist, like the 1977 CIRIA Report [1]), but rather to highlight some significant problems (e.g. earthquake design spectra, selection of target reliability in Level 2 methods, calibration of partial safety factors in Level 1 approaches). The numerical values included in the treatment are to be considered as examples, not as a guide to code applications.

10.1 CODES OF PRACTICE

A code of practice is a state of the art expressed in simple and practical (though often somewhat arbitrary or empirical) terms for easy application by designers or control authorities. Such codes have mainly been developed during the last 30 or 40 years to provide authoritative guidance to designers and constructors and a basis of compliance for building specifications. Unfortunately, different committees have prepared, often without any common basis, different codes for the various structural materials and types of construction or for the loading. The consequent anomalies and discrepancies mean that different levels of safety and performance result from the use of different codes, and such a circumstance sometimes makes it impossible to compare the convenience of alternative designs, especially if different structural materials are used.

On the other hand, during the last 20 years, the belief that absolute safety can be provided has been recognized to be mistaken; whatever design method is employed, in fact, loads and strengths are uncertain variables. Thus, the introduction of the reliability concept provided an all-embracing criterion of structural adequacy by which structural design codes can be rationalized [1] and the whole design process simplified. Obviously, lack of data on the uncertainty of the various design parameters does not invalidate a method which can take account of the best information available. Even so, the better the statistical data, the more accurate will be the result.

10.2 PARAMETERS INTRODUCED BY RELIABILITY-BASED CODES

Statistical data are required on the input and system quantities (Chapter 1) in order to develop reliability calculations. The problem is often simplified by the assumption that the reliability during an assigned period is sufficient for code purposes, instead of the whole time function. This period is called the *design life of the structure* during which the integrity of the structure or structural element must be assured. The design life is usually a conventionally chosen interval of time (e.g. 50 years) whose value can be varied according to the building destination (e.g. 100 years for hospitals and schools; 30 years for factory buildings).

For common practice, then, the code must first of all provide values for the parameters that describe input and system quantities, to be used at least when direct data on the specific problem at hand are unavailable or unreliable; otherwise, the code may remain unapplicable on the whole.

Some typical examples of code prescriptions are discussed in the following.

10.2.1 System quantities

The system quantities of interest in structural engineering can be of mechanical or geometrical nature. Elasticity moduli, yield and ultimate stresses, and hardening parameters are of the first type and generally contribute to define the material constitutive law. Length and out-of-straightness of members, dimensions of cross-sections, and position of bars in reinforced concrete are the main geometrical quantities and contribute to define the structure.

Codes of practice generally neglect uncertainties in the values of the latter quantities, as statistical data on the past cannot be applied explicitly to the future without objections by the contractor. Therefore, the current trend is to introduce larger-than-actual variabilities in resistance mechanical parameters in order to allow implicitly for the uncertainty in geometrical quantities. For instance, the coefficient of variation of the steel yield stress σ_y, statistically estimated as $\delta = 2\%$, is generally assigned equal to 5%; thus the effects of dimensional tolerances of members are considered indirectly.

When the geometrical uncertainty must be specified explicitly, as in the case of the maximum out-of-straightness in view of instability calculations, the codes generally prescribe a conventional deterministic value (unavoidable imperfection) that must be taken into account in the analysis. Typically, the unavoidable out-of-straightness of steel columns accounting also for end load eccentricities is taken to be of the order of one-thousandth of the column length.

Among mechanical quantities, elastic moduli show comparatively small random variations and are therefore assumed deterministic in most codes. In particular, the elastic modulus of concrete is usually taken proportional to the square root of the *characteristic strength.*

Thus, only the uncertainty in the yield and/or ultimate strength has to be specified. With regard to the strength of ductile materials, the probability laws are generally assumed to be normal or log-normal (so that negative values are avoided) and are sometimes truncated in order to allow for the quality control cut-off. For materials failing in a brittle manner (i.e. essentially according to a *weakest-link* type of mechanism; Section 1.5), there are good theoretical reasons for fitting the data to a Weibull distribution (Section 6.3.3). The coefficients of variation are assigned in a fictitious way that takes into account other uncertainties (such as dimensional tolerances) or they, as well as the mean values, have to be determined by experimental tests, whose design is indicated by the codes and is founded on the principles of statistics.

10.2.2 Input quantities

(a) General aspects

The types of loading generally covered by the codes are summarized in Table 3.4. For support settlements, fire or thermal effects, and other special types of actions, there are not consistent assessments, mainly because of lack of data.

A loading code, after a classification of the actions (for instance the one of Section 3.2), should contain some numerical indications. For loads constant or nearly so throughout the structure design life, a normal distribution is often adopted so that the average value μ and the standard deviation are sufficient to define the self-weight or the imposed dead load.

Long-duration loads like sustained live loading on buildings are generally described by the distribution of the maximum intensity in 50 years and the relevant number of independent repetitions

$$r = 50\,\lambda^l \tag{10.1}$$

where λ^l is the yearly mean rate of load changes. The spatial variability is generally taken into account by equivalent uniform distributed loads (EUDL, Section 3.3.4), whose intensity can be related to the influence area of the structural element under investigation: the greater the area, the smaller is the value of the EUDL to be introduced in the calculations. Further, suitable *concentration factors* should be associated with very small value of the influence area [2].

(b) Short-duration actions

Short-duration actions require, in addition to the distribution of the maximum intensity in 50 years and the mean rate of arrival, at least the knowledge of the duration of each excitation in order to obtain a satisfactory time description; more details are necessary for the spatial variability. Often, the random distribution of intensity and spatial variability is not considered explicitly, so that probabilistic handling by the code user is avoided. Consider, for instance, the wind velocity in any direction. The code establishes the wind profile on a vertical line (Section 3.5.1, equation (3.55)) and a *reference velocity* v_0^* which is typically (Section 3.5.3) the velocity averaged on an interval of 10 min at $z^* = 10$ m above ground in a plain site, corresponding to a return period $T_r = 50$ yr[‡].

The *design velocity* v_d is then established in the form [3]

$$v_d = \zeta_1 \zeta_2 \zeta_3 v_0^* \tag{10.2}$$

rather than by equation (3.62). In equation (10.2)

(1) ζ_1 is a topography coefficient, depending on the profile at the site, varying in the range 0.9–1.1;
(2) ζ_2 is a coefficient depending on the roughness of the terrain surrounding the site, on the building's size (stiffness and damping) and on the variation of strong winds with heights;
(3) ζ_3 is a coefficient depending on return period T of the maximum action chosen for structural calculations.

The coefficient ζ_2 contains the relatively finer time factor (Section 3.2) relevant to the wind action (the *gust factor* ζ in equation (3.62)). This factor takes into account the load structure interaction and, though many codes [3, 4] specify its value explicitly, a satisfactory definition needs more investigation. The other contributions to ζ_2 and the value of ζ_1 are relatively finer space factors and appear well established at the present time, while estimation of the temporal factor ζ_3 can be performed according to the results of Section 3.5.4.

After the design velocity has been established, the *wind pressure* can be calculated (equation (3.63)) and the unit load w on unit area of a solid surface exposed to the wind action is finally obtained through equation (3.64) by means of a suitable *exposure coefficient* c_p.

Exposure coefficients are also introduced in other cases, e.g. for defining the snow load on roofs. No snow load is usually considered on roofs with slopes larger than 60° and no obstacle to snow gliding, but for intermediate slope values, as well as for particular roof shapes, a suitable exposure coefficient [5] is introduced: this coefficient has to be applied to the uniformly distributed snow load w_y on horizontal flat roofs and is taken equal to 1 for slopes not larger than 25°. In turn, w_y is generally assumed to be 80% of the snow load on ground (Section 3.6)

$$w_y = 0.8 w_{gy} \tag{10.3}$$

The coefficient in (10.3) takes account of the snow drifting with the wind, which reduces the total amount of snow on roofs. Equation (10.3) is relevant to wind

‡ Definition of the reference wind speed on the basis of the return period offers consistency with regard to the effect of the wind loads on the structural safety, in the sense that the influence of different wind climates on structural safety can be neutralized by fixing the return period and by the appropriate CDF of the wind velocity for the site under consideration.

velocities in the range from 4 to 6 m s^{-1}: for faster winds, the correlations between wind and snow should be accounted for explicitly.

It is worth noting, from a philosophical viewpoint, that snow loading can be reduced by a suitable design of the roof, by manual removal of the snow (that is generally the case for accessible sites, such as balconies), or, in some special instances (like air-supported structures) by continuous heating. It follows that the spatial distribution of the load can be modified artificially and unexpectedly: hence, the most unfavourable load distribution should be considered for structural analysis purpose.

10.2.3 Seismic actions and response

A completely different description of the action characterizes the building codes for seismic regions. Inelastic response history analyses would be an impractical requirement in the design of most buildings: therefore, simpler methods of analysis are prescribed, namely elastic statical calculations under fictitious equivalent lateral forces, or at most elastic dynamical calculations founded on modal analysis (Section 4.3.4) and, possibly, empirical corrections to account for inelastic behaviour.

(a) *Modal analysis and response spectra*

In elastic modal analysis, the structure is regarded as an ensemble of several single-degree-of-freedom elastic systems with pulsation ω_i and damping ξ_i. Then the maximum modal displacements $u_{i\mathrm{m}}$ are related to the relevant maximum global displacement u_m by empirical relations. The most usual expression makes use of the sum of the squared modal responses:

$$u_\mathrm{m} = \left(\sum_i u_{i\mathrm{m}}^2\right)^{1/2} \tag{10.4}$$

in which only the lower few modes need to be included in the summation. Equation (10.4) has no theoretical support but is founded on equation (4.159).

In this context, knowledge of the functional dependence of $u_{i\mathrm{m}}$ on the modal pulsation ω_i, damping ξ_i and accelerogram $a(t)$ becomes of interest in view of practical applications. From equation (4.90), the displacement of the ith single-degree-of-freedom elastic system with subcritical damping can be expressed by

$$u_i(t) = \frac{1}{\omega_i} v_i(t) \tag{10.5}$$

where

$$v_i(t) = \frac{1}{\sqrt{(1-\xi_i^2)}} \int_0^t a(\tau) \exp\left[-\xi_i\omega_i(t-\tau)\right] \sin\left[\omega_i\sqrt{(1-\xi_i^2)}\,(t-\tau)\right] \mathrm{d}\tau \tag{10.6}$$

has the dimension of a velocity and is called *pseudo-velocity*. Further, the ratio between the force on each system (equal to the displacement times the stiffness) and the relevant mass is given by (equations (4.68) and (10.5))

$$a_i(t) = \frac{K_i}{M_i} u_i(t) = \omega_i^2 u_i(t) = \omega_i v_i(t) \tag{10.7}$$

and is called *pseudo-acceleration.* Note that the pseudo-velocity is a measure of the energy absorption in each one-degree-of-freedom system and the pseudo-acceleration a measure of its elastic reaction.

The maximum value of $v(t)$ is generally plotted in terms of the system natural frequency ν_i or period $T_i = 1/\nu_i$ or pulsation $\omega_i = 2\pi\nu_i$, one curve for each chosen value of ξ_i: the results are the (pseudo-) *velocity spectra* S_v (Section 3.4.3)‡. Often the pseudo-velocity spectrum is plotted in logarithmic scale: then, according to (10.5) and (10.7), ascending and descending 45° lines correspond to constant maximum values of displacement and pseudo-acceleration respectively: the three spectra can thus be represented in a single figure (Fig. 10.1).

The response spectra for an actual earthquake motion are quite irregular, but the local irregularities depend on the specific record and attenuate as the damping increases; therefore, it is customary to simplify the spectra into a piecewise-linear (trapezoidal) shape. For example, the line A in Fig. 10.1, that defines the *standard response spectrum,* commonly adopted by seismic codes, was derived by smoothing the results obtained after a statistical handling of a large amount of available records; in the same figure line B gives the maximum ground motion for comparison with the maximum acceleration of the single degree of freedom.

When, in a seismic region, a sufficient number of earthquake records are available, they can be normalized to the same intensity value in order to obtain a sample of response spectra on which statistics can be estimated. Fig. 3.13 shows the mean pseudo-velocity spectra obtained in this way for the western United States. Otherwise, local seismicity in the nearby seismic sources has to be simulated according to the discussion of Section 3.4.2. Then, after statistical handling of the result, the mean response spectrum, or, more generally, the spectral ordinates corresponding to given probabilities of exceedance for a given magnitude and focal distance, can be obtained.

Modern seismic codes usually assign a *design spectrum,* by the use of which modal analysis is performed deterministically. In a reliability-based code, the *design spectrum* is the one which is supposed to have a given probability of exceedance in the region: it can be obtained as the mean of the analysed sample plus k standard deviations (e.g. $k = 1$) [6]. The ordinates of the design spectrum can also be obtained by multiplying the maximum ground motion displacement velocity and acceleration by suitable amplification factors that depend on the damping of the system (Table 10.1). The amplification factor of the acceleration has to be applied up to 8 Hz, while for frequencies larger than 33 Hz, the factor becomes 1; a linear interpolation is introduced in the range 8–33 Hz. Line B in

‡ Pseudo-velocity and pseudo-acceleration do not coincide with the actual system velocity and acceleration, although their maximum values are usually very close: therefore the spectra are commonly identified with each other.

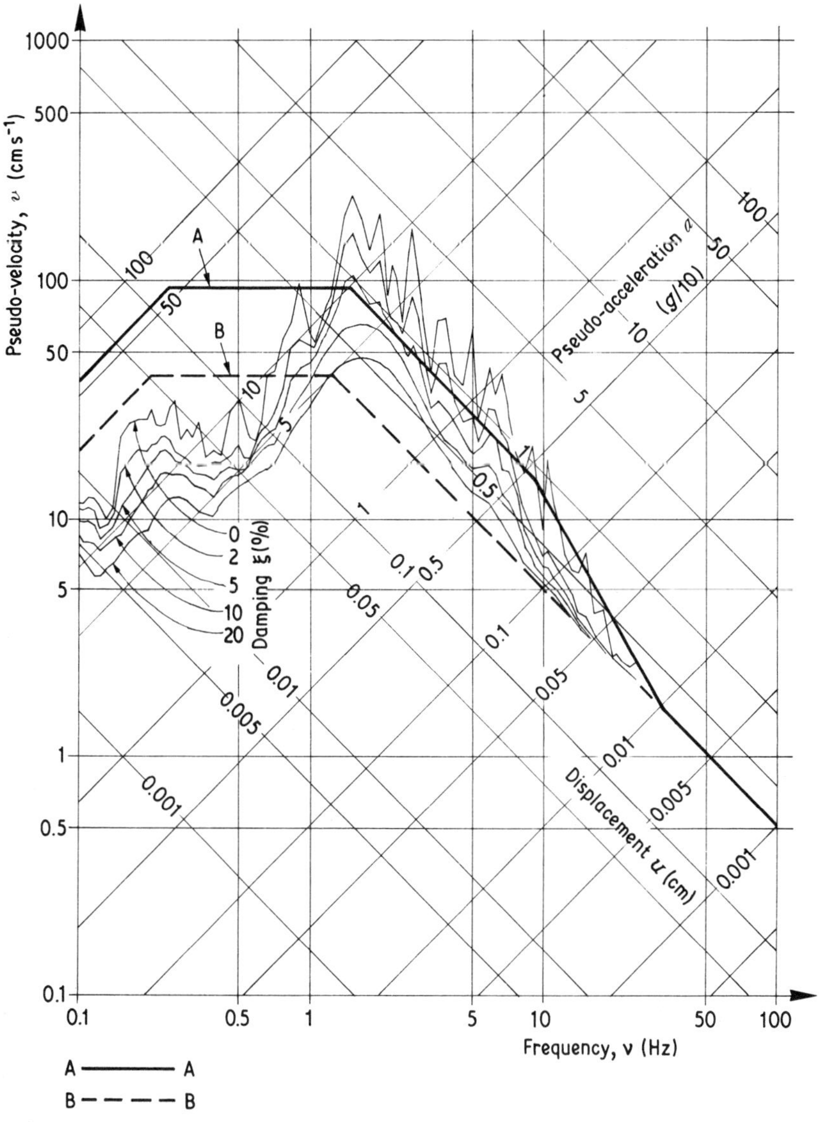

Fig. 10.1 Response spectra of Friuli earthquake of 6 May 1976 recorded at Tolmezzo; E–W horizontal component. Line A, standard response spectrum ($\xi_i = 5\%$, $a_g = 0.33g$); line B, corresponding spectrum of maximum ground motion. (after ENEL (1976) *Terremoto del Friuli, Maggio* 1976: *Spettri di risposta* (in Italian), ENEL, Rome.)

Fig. 10.1 is the ground motion corresponding to the standard design response spectrum and, when statistical data are not available, one can employ this diagram suitably scaled to the appropriate peak ground acceleration. For this purpose, note that standard spectra may be either velocity-normalized (in which

Table 10.1 Amplification factors for average-plus-one-standard-deviation spectra of horizontal elastic maximum standard ground motion (from Duff [7]).

Damping	*Amplification factors*		
% *critical*	a	v	u
0.5	5.10	3.84	3.04
1	4.38	3.38	2.73
2	3.66	2.92	2.42
3	3.24	2.64	2.24
5	2.71	2.30	2.01
7	2.36	2.08	1.85
10	1.99	1.84	1.69
20	1.26	1.37	1.38

case the maximum ground motion for $a_g = 1g$ is given by $v_g = 26\,\text{in s}^{-1}$ and $u_g = 15$ in) or acceleration-normalized (in which case the maximum ground motion for $a = 1g$ is given by $v_g = 48$ in s^{-1} and $u_g = 36$ in). For the first type, a least-squares fitted straight line in logarithmic scale (Section 7.5.3) of the response vs modified Mercalli intensity is plotted in Fig. 10.2: a set of amplification factors slightly different from those in Table 10.1 should be derived for this case. For acceleration-normalized spectra, a very simplified rule consists in assuming u_g and v_g proportional to a_g: in Fig. 10.1, for instance, a_g is taken equal to $0.33g$ and hence $v_g = 48 \times 0.33$ in $\text{s}^{-1} = 15.84$ in $\text{s}^{-1} \simeq 40$ cm s^{-1} and $u_g = 36 \times 0.33$ in $= 11.88$ in $\simeq 30$ cm. Then the standard design response spectrum is derived making use of the factors of Table 10.1 for the appropriate value of damping (in particular $\xi = 5\%$, in Fig. 10.1).

(b) *Modified spectra for inelastic response*

Consider now the case in which a simple mass–spring oscillator deforms inelastically according to a relation as in Fig. 10.3(a) and idealize this relation as elastic–perfectly plastic, with u_u a value for the maximum displacement selected as a useful limit.

Remembering that the pseudo-acceleration a is defined as the force divided by the mass, the maximum pseudo-acceleration of an elastic system for which $u_{\max} = u_\text{u}$ is given by equation (10.7):

$$a_{\max} = \omega^2 u_{\max} = \omega^2 u_\text{u} \tag{10.8}$$

while for an inelastic system and small excursions into the inelastic range

$$a_{\max} = \omega^2 u_\text{y} = \omega^2 u_\text{u}/\zeta \tag{10.8$'$}$$

where ζ is the *ductility factor*, i.e. the ratio

$$\zeta = u_\text{u}/u_\text{y} \tag{10.9}$$

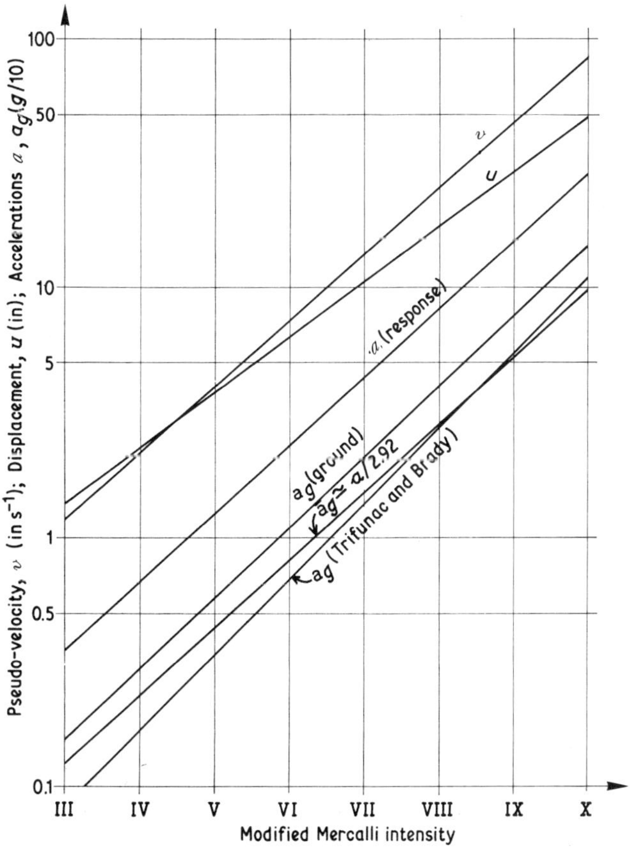

Fig. 10.2 Logarithmic plot of mean plus one standard deviation of the peak horizontal response for a 5% damping system and of peak ground acceleration vs Modified Mercalli intensity (from Duff [7]).

Then, in the zones of the spectrum corresponding to larger displacements or velocities (I and II in Fig. 10.3(b)), one can assume equal displacements for elastic and inelastic systems, while the accelerations are in the ratio ζ to 1. Vice versa, in the high-frequency zone (V in Fig. 10.3(b)), the maximum accelerations can be assumed to be the same, while the displacement of the inelastic system is ζ times that of the elastic system. Finally, in the high-acceleration zone (III in Fig. 10.3(b)), the total absorbed energy is assumed to be the same, for elastic and inelastic systems, thus obtaining a ratio $\sqrt{(2\zeta - 1)}$ between the accelerations. Fig. 10.3 shows how *two* inelastic spectra (lines 2 and 3) are thus constructed from *one* elastic spectrum (line 1).

Inelastic spectra explain why code design spectra generally have ordinates too small with respect to probabilistic assessment: a structure designed for $0.1\,g$, for instance, can be safe also for earthquakes with spectrum $0.4g$ if it is able to provide a ductility factor $\zeta = 4$. However, the direct use of inelastic spectra in modal analysis of multi-degree-of-freedom structures, as sometimes suggested [8], is not only incorrect from a theoretical viewpoint, but has been shown to produce unconservative designs [9].

(c) Equivalent static forces

As already stated, codes of practice generally also offer the possibility of making use of an equivalent lateral force procedure, instead of modal analysis. Although the building is idealized in the same manner, this procedure involves less computational effort because it requires only the fundamental period, and not periods and shapes of the higher vibration modes. Moreover, the vibration period T of the building can be estimated (better if underestimated so that the design is generally conservative) by simple formulae which involve only a general description of the building type and its overall dimensions. Then, the magnitude of lateral forces is also given by simple formulae appropriate for buildings with regular distribution of mass and stiffness over height. Formulae of the following type are usually recommended [6]:

$$W_i = \frac{a(T)}{g}\left(\sum_{j=1}^{N} W_{gj}\right) W_{gi} h_i^C \left(\sum_{j=1}^{N} W_{gj} h_j^C\right)^{-1} \tag{10.10}$$

where W_i is the equivalent lateral force at the ith floor level; W_{gj} is the contribution to the total building weight by the jth floor; h_i is the height of the ith storey; N is the number of floors; $a(T)$ is the pseudo-acceleration corresponding to the estimated fundamental period and the appropriate damping determined from the design (either elastic or inelastic) spectrum; and C is a coefficient defined as follows:

$$C = \begin{cases} 1 & \text{for } T \leqslant 0.5\,\text{s} \\ \dfrac{T+1.5}{2} & \text{for } 0.5 < T < 2.5\,\text{s} \\ 2 & \text{for } T \geqslant 2.5\,\text{s} \end{cases} \tag{10.11}$$

(d) Three-dimensional response

Planar models appropriate for each of two orthogonal lateral directions must be analysed separately whether by modal analysis or an equivalent lateral force procedure. The results of the two analyses have then to be combined without neglecting the vertical component of the motion and the effects of torsional responses in the structure that arise from eccentricity in the mass and stiffness distributions. However, buildings with large eccentricities, as well as buildings with close values of natural frequencies of the lower modes and coincident centres of mass and stiffness, exhibit generally coupled lateral–torsional motions and independent analyses for the two lateral directions may not suffice. In such cases, a modal analysis allowing for three degrees of freedom per floor (two translational and one torsional motions) should be employed.

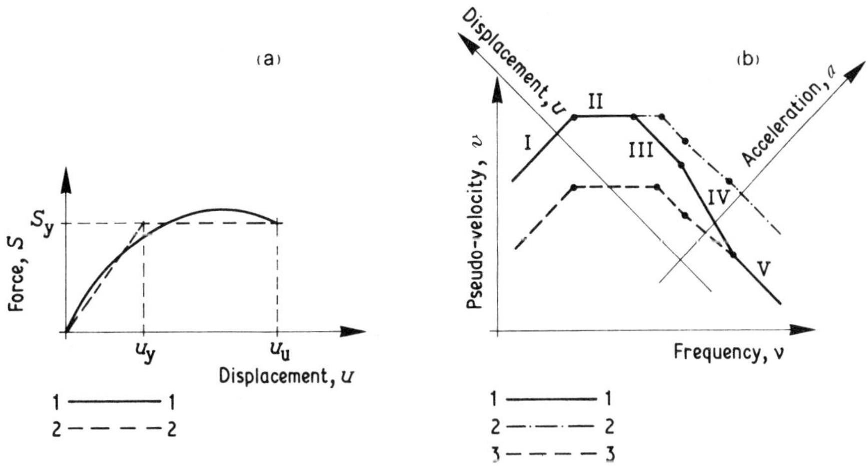

Fig. 10.3 (a) Inelastic force–displacement relationship: 1, actual (————); 2, idealized, elastic–perfectly plastic (– – – – –). (b) Earthquake design spectra (qualitative): 1, elastic spectrum, accelerations or displacements (————); 2, inelastic displacement spectrum (—·—·—); 3, inelastic acceleration spectrum (– – – – –).

10.3 RELIABILITY ANALYSIS IN A CODE FRAMEWORK

It is well known that one of the main objective of structural engineering is the definition of a suitable scalar quantity in order to compare easily and consistently different design solutions: the definitions of *yield criteria* and relevant *equivalent stresses* proposed in the past were aimed at this purpose.

At the present time, the structural reliability P_{suc} (or its complement $P_{fail} = 1 - P_{suc}$) have been recognized to be very appropriate to evaluate a structural design. Indeed, most reliability-based codes tend to identify the aim of structural design with 'the achievement of acceptable probabilities that the structure being designed will not become unfit for the use for which it is required during some reference period and having regard to its intended life' [10].

With this approach, if Υ_i, $i = 1, \ldots, N$, denote the possible solutions of a structural design problem, they can be accepted if

$$P_{fail}(\Upsilon_i) \leqslant \bar{P}_{fail} \tag{10.12}$$

where $\bar{P}_{fail}$ is the assigned reference value. Among the acceptable designs, the safest one is the design for which P_{fail} is minimum, while the most economical solution in terms of initial cost is generally the design for which P_{fail} is closest to $\bar{P}_{fail}$.

Thus, the main tasks of building codes reduce to (a) providing the methods of reliability analysis capable of computing P_{fail} starting from the available data on the design parameters, and (b) selecting a limit value $\bar{P}_{\text{fail}}$ for the calculated probability of failure.

10.3.1 Reliability analysis methods

In Section 2.8 the bases of reliability analysis have been illustrated. If the design Υ_i is obtained by selecting the values $\mathbf{x}_i$ for the design parameters $\mathbf{x}$, the unsafe region $\mathscr{U}$ in the space of the random quantities $\mathbf{Q}$ relevant to the problem being analysed depends on this choice: so does the probability of failure (equation (2.56)). Thus to every design Υ_i one can associate a value of the probability of failure $P_{\text{fail}}(\Upsilon_i)$, given by the appropriate convolution integral (Section 2.8) or by simulation (Chapter 9): this probability must satisfy inequality (10.12) for the design to be acceptable.

Such methods of checking the safety of structures, often denoted as *Level* 3 reliability approaches, tacitly assume that all uncertainties in design are contained in the JPDF $p_{\mathbf{Q}}$ and that this JPDF is known. However, in structural problems the latter condition is seldom satisfied, because of the usual scarcity of data: in many cases only mean and variance of each $\tilde{Q}_j$ may be known with any confidence. Moreover, the relevant mathematical functions may be highly non-linear in the basic variables, or given only by numerical relations, so that the evaluation of $P_{\text{fail}}(\Upsilon_i)$ becomes impracticable, even if statistical information were sufficient. Very clearly, Level 3 methods can be applied to special and/or important structures (including structures for which there is little or indirect past experience), but are not suitable for normal design purposes, because of theoretical and numerical difficulties.

Sections 10.4 and 10.5 will be devoted to the description of simplified reliability methods that at present are usually classified into two categories or *levels*; namely (Section 2.8.3):

(1) *Level* 2, at which safety is checked only at selected point(s) on the failure boundary, and inequality (10.12) is substituted by a comparison on a *reliability index*.
(2) *Level* 1, at which the appropriate value(s) of reliability is (are) provided by a number of partial safety factors related to the *characteristic values* of the basic variable $\mathbf{Q}$.

However, before going into details of these methods, the other task of structural codes (i.e. specifying the reference value $\bar{P}_{\text{fail}}$ that appears on the r.h.s. of inequality (10.12)) is briefly discussed.

10.3.2 Selection of the target reliability

The selection of the target reliability involves investigations at higher levels than the third one [11], investigations in fields that are at the limit (and sometimes well beyond) the jurisdiction of the engineering profession. Moreover, a general consensus on the criteria for selecting an objective function to optimize does not exist at the present time (cf. Chapter 6, and especially Sections 6.7.3 and 6.7.4).

Therefore, the acceptable level of risk of attaining an ultimate limit state can only be assessed with regard to social criteria, but it must be also kept in mind (Section 1.7) that such low probabilities are significant only as conventional, comparative values.

A proposed formula aimed at obtaining consistent target values $\bar{P}''_{\text{fail}}$ of the probability of failure of any structure due to any cause in its design life is [1]

$$\bar{P}''_{\text{fail}} = \frac{10^{-4}\zeta_s T}{L} \tag{10.13}$$

where ζ_s is a social criterion factor (Table 10.2), T is the design life of the structure in years and L is the average number of people within or near to the structure during the period of the risk.

Table 10.2 Social criterion factor ζ_s (from CIRIA [1])

Nature of structure	ζ_s
Places of public assembly, dams	0.005
Domestic, office, trade and industry	0.05
Bridges	0.5
Towers, masts, offshore structures	5

Nevertheless, attempts to make use of present probabilistic approaches, founded on theoretical probabilistic concepts alone, may lead to unsatisfactory results on the partial safety factor calibration. In fact, the methods that have been presented in Chapter 6 in order to calculate the *theoretical* probability of failure P_{fail} of a structure are concerned with the random nature of loads and resistance. But they generally neglect some sources of failure, such as error or negligence, that can be introduced during the design, erection and use of the construction. Therefore, the actual probability of failure P''_{fail} is much larger than the *theoretical* probability of failure P_{fail}: moreover, an increase in P_{fail} does not necessarily result in a comparable increase in P''_{fail}. All this question is very open at the present state of knowledge, and the results of a few current researches will be presented in Chapter 11 (Section 11.5). In the meantime, very crude, rule-of-thumb relationships between P''_{fail} and P_{fail} must be accepted for design purposes, such as for instance [1]

$$P''_{\text{fail}} = 10P_{\text{fail}} \tag{10.14}$$

Introducing equation (10.14) into (10.13), the target value of the *theoretical* probability of failure becomes

$$\bar{P}_{\text{fail}} \simeq \frac{10^{-5}\zeta_s T}{L} \tag{10.15}$$

In most of this book, the theoretical probability of failure P_{fail} is the only contribution considered: the target reliability level expressed by equation (10.15) should therefore be assumed. For instance, if the design life T is assumed equal to 50 years and the usual value 0.05 (Table 10.1) is given to the social criterion factor ζ_s, then

$$\bar{P}_{\text{fail}} = \frac{2.5 \times 10^{-5}}{L} \simeq 10^{-5}\text{–}10^{-7} \qquad (10.15')$$

10.4 LEVEL 2 RELIABILITY METHODS

10.4.1 Performance function and reliability index β

Following Sections 1.6 and 2.8, let $\mathscr{U}$, $\mathscr{S}$ and $\mathscr{C}$ be respectively the unsafe region, the safe region and the surface separating the two regions in the space of the n basic random variables $\tilde{\mathbf{Q}}$. Assume the random variables $\tilde{\mathbf{Q}}$ to be stochastically independent‡.

It is in general possible to characterize a function $Z = Z(\mathbf{Q})$ such that

$$\begin{aligned} Z(\mathbf{Q}) &> 0 \qquad \text{when } \mathbf{Q} \in \mathscr{U} \\ Z(\mathbf{Q}) &= 0 \qquad \text{when } \mathbf{Q} \in \mathscr{C} \\ Z(\mathbf{Q}) &< 0 \qquad \text{when } \mathbf{Q} \in \mathscr{S} \end{aligned} \qquad (10.16)$$

i.e. such that

$$\begin{aligned} \{Z(\mathbf{Q}) > \mathbf{0}\} &\Rightarrow \{\text{success}\} \\ \{Z(\mathbf{Q}) < \mathbf{0}\} &\Rightarrow \{\text{failure}\} \end{aligned} \qquad (10.16')$$

The choice of the function $Z(\mathbf{Q})$, usually called *performance function* of the structure, is not unique. For instance, in the most simple case of two positive-definite random variables $\tilde{R}$ and $\tilde{S}$, and condition of success defined by equation (2.59), the performance function can be identified with any of the following

$$Z_1 = R - S \qquad (10.17a)$$

$$Z_2 = R^2 - S^2 \qquad (10.17b)$$

$$Z_3 = \ln(R/S) \qquad (10.17c)$$

etc.

In general, however, the performance function is not linear in the basic variables $\mathbf{Q}$. A linear approximation can be obtained by taking only the first two

‡ Most of the relationships in this section hold only for uncorrelated random variables $\tilde{\mathbf{Q}}$. If the variables $\tilde{\mathbf{Q}}$ are correlated, they can be replaced by the uncorrelated variables $\tilde{\mathbf{Q}}^*$ derived from the former by the relation

$$\tilde{\mathbf{Q}}^* = \mathbf{U}^{\mathrm{T}}\tilde{\mathbf{Q}}$$

where $\mathbf{U}$ is the matrix of the eigenvectors of the covariance matrix $\boldsymbol{\Sigma}_{\mathbf{Q}}$: $\boldsymbol{\Sigma}_{\mathbf{Q}}$ and the diagonal covariance matrix of the uncorrelated variables, $\boldsymbol{\Sigma}_{\mathbf{Q}^*}$, have the same eigenvalues, coincident with the variances $s^2_{\mathbf{Q}^*}$.

Note that for Gaussian random variables, uncorrelation implies independence.

terms of a power-series expansion in a points $\mathbf{Q}_L$ (*linearization point*)‡

$$Z = Z(\mathbf{Q}) = C_0 + C_j Q_j \qquad j = 1, \ldots, n \tag{10.18}$$

where, indicating by $\partial Z_L/\partial Q_j$ the derivatives of Z in $\mathbf{Q}_L$,

$$C_j = \frac{\partial Z_L}{\partial Q_j} \tag{10.18'}$$

$$C_0 = Z(\mathbf{Q}_L) - C_j Q_{L_j} \tag{10.18''}$$

If $\tilde{Q}_j$ are random variables and equation (10.18) holds, equations (A3.26′) and (A3.27′) yield immediately the mean and standard deviation of $\tilde{Z} = Z(\tilde{\mathbf{Q}})$:

$$\mu_Z = C_0 + C_j \mu_{Q_j} = Z(\mathbf{Q}_L) + C_j(\mu_{Q_j} - Q_{L_j}) \tag{10.18'''}$$

$$s_Z = \sqrt{(C_j^2 s_{Q_j}^2)} \tag{10.18^{iv}}$$

From the third inequality of (10.16)

$$P_{\text{fail}} = \text{Prob}\{Z < 0\} = \int_{-\infty}^{0} p_Z(Z)\,\mathrm{d}Z \tag{10.19}$$

where $p_Z(Z)$ is the PDF of the random variable $\tilde{Z} = Z(\tilde{\mathbf{Q}})$ (Fig. 10.4(a)). For a given type of PDF, P_{fail} is a decreasing function of the ratio

$$\beta = \mu_Z / s_Z \tag{10.20}$$

which is the reciprocal of the c.o.v. of $\tilde{Z}$, and can be taken as a measure of safety (or *reliability index*).

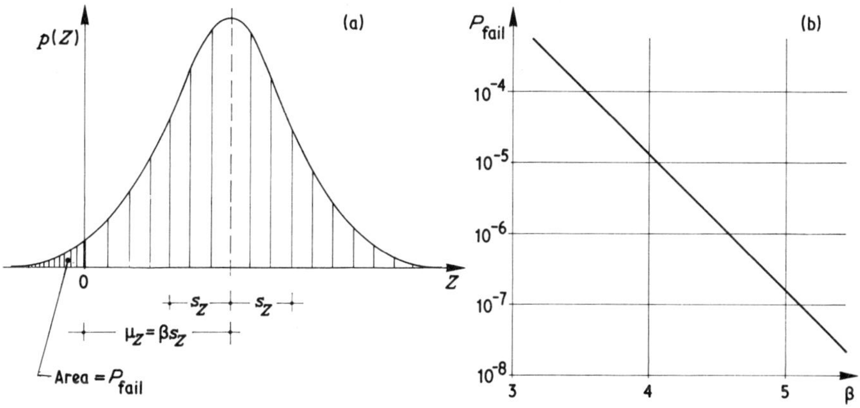

Fig. 10.4 (a) Definition of the reliability index β from the PDF of the performance function Z. (b) Approximate log–linear relationship P_{fail} vs β.

‡ In the following formulae, summation over the repeated index j is implied ($j = 1, 2, \ldots, n$).

If $Z(\mathbf{Q})$ is linear, and all the variables $\tilde{Q}_j$ are normally distributed, so is $\tilde{Z}$, and the relationship between P_{fail} and β is given by

$$P_{\text{fail}} = P_{\text{fail}}^{N} = \Phi(-\beta) \tag{10.21}$$

where $\Phi(\ )$ is the standardized normal distribution function; otherwise, equation (10.21) yields only an approximate estimate of β. In particular, if $\tilde{Q}_j$ are normally distributed but $Z(\mathbf{Q})$ is not linear, the true probability of failure for a given β is bounded by [12]

$$0 \leqslant P_{\text{fail}} \leqslant P_{\text{fail}}^{N} \tag{10.22}$$

when the safe region is concave (Fig. 10.5(a)), while if the safe region is convex (Fig. 10.5(c))

$$P_{\text{fail}}^{N} \leqslant P_{\text{fail}} \leqslant P_{\text{fail}}^{\chi} \tag{10.23}$$

where

$$P_{\text{fail}}^{\chi} = 1 - \chi_n^2(\beta^2) \tag{10.23$'$}$$

and $\chi_n^2(\ )$ is the cumulative χ^2 distribution with n degrees of freedom, n being the number of basic variables $\tilde{Q}_j$.

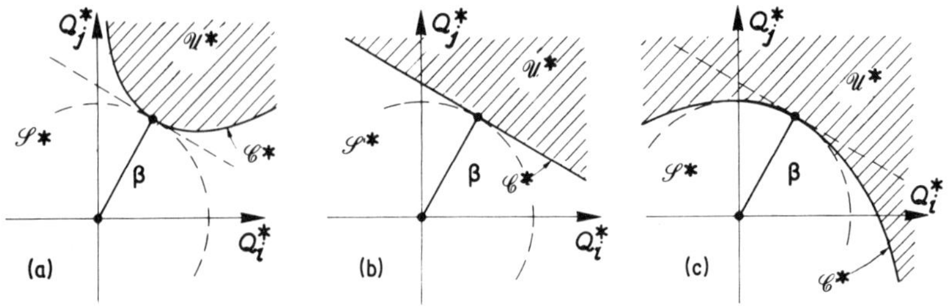

Fig. 10.5 Definition of the reliability index β in the space of standardized variables Q_j^*.

It is easy to show that, for any unbounded distribution, in the range of very low probabilities, the relationship between P_{fail} and β is of the type

$$P_{\text{fail}} = \Lambda e^{-D\beta} \tag{10.24}$$

with Λ and D appropriate constants. In particular, the following relationship has been proposed (Fig. 10.4(b))

$$P_{\text{fail}} = 460\,e^{-4.3\beta} \qquad \text{or} \qquad \beta = 0.23\,(6.13 - \ln P_{\text{fail}}) \tag{10.24$'$}$$

which can be used to choose a target reliability index β from a target reliability $\bar{P}_{\text{fail}}$ [13].

10.4.2 Alternative second-moment methods

The level 2 methods which neglect the actual distribution of the involved random variables are also denoted as *second-moment methods*, because they take account of first and second moments (averages and standard deviations) only.

(a) *Mean value first-order second-moment methods*

In the simplest second-moment methods, the linearization in equation (10.18) is performed in the mean point of the random variables $\tilde{Q}_j$:

$$\mathbf{Q}_L = \mu_{\mathbf{Q}} \tag{10.25}$$

In this way, however, for non-linear performance functions, the error caused by linearization can be quite significant, because it generally increases with increasing distance from the linearization point, and the mean point must be well within the safe region.

(b) *Advanced second-moment methods*

The above remark suggests that one chooses $\mathbf{Q}_L$ on the failure boundary $\mathscr{C}$; further, a correct selection for $\mathbf{Q}_L$ should be the point $\mathbf{Q}_P$ of the failure boundary associated with the maximum density of probability of failure, i.e. the point where the derivatives of the integral (2.56) are zero. Then, introducing the second inequality of (10.16), equation (10.18‴) becomes‡

$$\mu_Z = \sum_{j=1}^{n} C_j(\mu_{Q_j} - Q_{P_j}) \tag{10.26'}$$

Equation (10.18^{iv}) can be written in the form

$$s_Z = \sum_{j=1}^{n} \alpha_j C_j s_{Q_j} \tag{10.26''}$$

with

$$\alpha_j = \frac{C_j s_{Q_j}}{\left(\sum_{k=1}^{n} C_k^2 s_{Q_k}^2\right)^{1/2}} \tag{10.26''}$$

From equation (10.20)

$$\mu_Z - \beta s_Z = 0 \tag{10.20'}$$

Introducing equations (10.26′) and (10.26″) and rearranging

$$\sum_{j=1}^{n} C_j(\mu_{Q_j} - Q_{P_j} - \beta\alpha_j s_{Q_j}) = 0 \tag{10.27}$$

The coordinates of the sought linearization point $\mathbf{Q}_P$ are then derived as those§ for which all the terms in parentheses in (10.27) vanish, i.e.

$$Q_{P_j} = \mu_{Q_j} - \alpha_j \beta s_{Q_j} \tag{10.28}$$

The proof that equation (10.28) gives the point of $\mathscr{C}$ of maximum density of probability of failure will be presented in the next section. Here note that equation (10.28) can only be solved by an iterative procedure that involves the

‡ The implicit summation convention is *not* used in the following formulae.
§ Occasionally, no single point can be found where the density of the probability of failure is a maximum, but such cases seldom occur in practice.

following steps: guess a value of β and a vector $\mathbf{Q}_P^0$ for computing $C_j(\mathbf{Q}_P^0)$ and hence α_j; compute $\mathbf{Q}_P^1$ from equation (10.28) and iterate until convergence on $\mathbf{Q}_P$ is achieved; evaluate $Z = Z(\mathbf{Q}_P)$ and modify till $Z = 0$.

10.4.3 The 'minimum-distance' method

Now project the regions $\mathscr{U}$ and $\mathscr{S}$ in the space of the *standardized* variables $\mathbf{Q}^*$, defined by

$$Q_j^* = \frac{Q_j - \mu_{Q_j}}{s_{Q_j}} \tag{10.29}$$

and let

$$Z^* = Z^*(\mathbf{Q}^*) = Z(\boldsymbol{\mu}_Q + \boldsymbol{\Sigma}_Q \mathbf{Q}^*) = 0 \tag{10.30}$$

be the equation, in the space of the standardized variables, of the failure boundary $\mathscr{C}^*$, i.e. the boundary between the projections $\mathscr{S}^*$ and $\mathscr{U}^*$ of the safe and unsafe regions $\mathscr{S}$ and $\mathscr{U}$. Level 2 reliability methods can also be regarded as the ones assuming as a safety measure the smallest distance β of the surface (10.30) from the origin in the space of the standardized variables $\mathbf{Q}^*$ (Fig. 10.5).

It is possible to show that such a geometrical approach to level 2 reliability methods is fully equivalent to the approach of Section 10.4.2(a) for linear performance functions and of Section 10.4.2(b) for non-linear performance functions.

The proof is presented with reference to the latter case. After the standardization of the variables $\mathbf{Q}$, equations (10.28) and (10.26‴) become

$$Q_{Pj}^* = -\beta\alpha_j \tag{10.28'}$$

and

$$\alpha_j = \frac{c_j}{\left[\sum_{k=1}^{n} (c_k)^2\right]^{1/2}} \tag{10.26^{iv}}$$

where $c_j = \partial Z^*/\partial Q_j^*|_{\mathbf{Q}_P^*}$. Thus, the α_j are the direction cosines of the perpendicular from the origin to the hyperplane which is tangential to the failure boundary at the point $\mathbf{Q}_P$ and β may be interpreted as the distance to the origin from the point $\mathbf{Q}_P$ which is therefore the point of $\mathscr{C}$ closest to the origin. On the other hand, as all points at the same distance from the origin have equal joint probability densities, the point on $\mathscr{C}$ with the highest density of probability of failure is the one which is closest to the origin and, hence, coincides with $\mathbf{Q}_P$ [14]. Such a geometrical approach allows one to write the solving problem in the form

$$\begin{aligned} \beta &= \min\left(\sum_{j=1}^{n} Q_j^{*2}\right)^{1/2} \\ Z^*(\mathbf{Q}^*) &= 0 \end{aligned} \tag{10.31}$$

so that, if the safety domain boundary does not admit differentiation, the reliability problem can be solved by using specific minimization techniques. The whole procedure summarized in Section 10.4.2(b) is the well known 'direction cosine' algorithm for solving equation (10.31). When the derivatives c_j can be computed in an analytical way, this algorithm is generally the one associated with the highest computational efficiency.

10.4.4 Approximate full-distribution methods

If a particular problem involves random variables which are definitely non-normal, direct application of the above procedures may lead to meaningless results.

From a theoretical viewpoint, any random variable $\tilde{Q}$ with CDF $P_{\tilde{Q}}(Q)$ can be replaced by a normal standardized variable u, the two variables being related by the equality of probabilities

$$P_{\tilde{Q}}(Q) = \Phi(u) \tag{10.32}$$

and hence

$$u = \Phi^{-1}(P_{\tilde{Q}}(Q)) \tag{10.32'}$$

However, some simplified procedures have been proposed for Level 2 reliability calculations. Of these, two appear meaningful [1]: the first one replaces the variables $\tilde{Q}_j$ by normal variables having the same mean μ_{Q_j} and the same fractile of an assigned order p; the second makes use of a Taylor expansion of (10.32') in the design point and neglects the non-linear terms.

10.4.5 Criticism of Level 2 reliability methods

The measure β is generally regarded as the *best* reliability index that can practically be introduced into a structural code. However β has some undesirable features:

(1) β is not necessarily the same whenever the probability of failure P_{fail} of the analysed system or the relevant Chebyshev bounds are the same.
(2) β does not account adequately for the geometry of the safety domain; in fact, for equal β's, P_{fail} in the case of Fig. 10.5 (b) is likely to be larger than in Fig. 10.5 (a) and smaller than in Fig. 10.5 (c). Moreover, P_{fail} is definitely affected by the number n of relevant random variables, as shown by equation (10.23) in the case of convex safe region‡.

‡ To take account of this *dimension effect*, it has been suggested [15] that the prescribed index β_n should be varied with n according to the law

$$\chi_n^2\,(\beta_n^2) = \text{constant } (n)$$

Table 10.3 shows the example of the set of $\bar{\beta}_n$ corresponding to $\beta_1 = 3$.

Table 10.3 Dimension-biased reliability index

$n =$	1	2	4	6	8	10	20	30
$\beta_n =$	3.00	3.46	4.02	4.47	4.85	5.20	6.48	7.48

(3) β cannot be employed if the mean vector and the covariance matrix of the basic variables **Q** are totally or partially unknown.
(4) β cannot be modified for cases of partial knowledge of the distribution functions of the basic variables.
(5) β is not defined as a function of time and therefore cannot account for stochastic variations of the parameters involved in structural analysis.

With reference to the last point, it is worth noting in particular that present Level 2 reliability methods treat the load combination problem under the two following assumptions [14]:

(A) for each time-varying load, the design life T can be subdivided into a number of elementary time intervals such that the peak values of load in consecutive time intervals can be considered to be independent;
(B) each load remains constant at its peak value throughout its own elementary time interval.

The latter assumption is clearly conservative, and the whole treatment leads to the performance of reliability analysis by considering the load combinations given in Table 3.5 for both short- and long-duration loads.

Recently, a new reliability index that overcomes the difficulties involved by the use of β has been proposed [16]. However, before its adoption by structural code, this index must be verified and discussed with reference to practical problems as has been done in the last decade for the index β.

10.5 LEVEL 1 RELIABILITY METHODS

10.5.1 General formats

Level 1 reliability methods require first that the relevant variables be distinguished into *action* quantities, say $W_j (j = 1, \ldots, n)$, and *resistance* quantities, say $\chi_l (l = 1, \ldots, N)$. Then, the performance requirement is expressed by an inequality of the type

$$\mathscr{G}_1[\gamma_i^f, W_{kj}] \leqslant \mathscr{G}_2[\gamma_h^m, \chi_{kl}] \tag{10.33}$$

where $\mathscr{G}_1$ and $\mathscr{G}_2$ are appropriate functions (usually, very simple *linear* combinations); γ_i^f and γ_h^m are the so-called *partial safety factors* ($i = 1, \ldots, m$; $h = 1, \ldots, r$) whose number and values are specified by the code; W_{kj} and χ_{kl} are the *characteristic values* of the corresponding *action* and *resistance* parameters. The *characteristic value* is defined as the appropriate *fractile* (equations (2.70) and (A2.24)) of the relevant random variable

$$\chi_{kl}: \quad \text{Prob}[\tilde{\chi}_k < \chi_{kl}] = p \tag{10.34}$$

$$W_{kj}: \quad \text{Prob}[\tilde{W}_j > W_{kj}] = p \tag{10.35}$$

$$\text{or} \quad \text{Prob}[\tilde{W}_j < W_{kj}] = 1 - p \tag{10.35'}$$

The value of p is typically set at 0.05; so W_{kj} is the 95% fractile of W_j, and χ_{kl} the 5% fractile of χ_l.

In the case of random processes $\tilde{W}_j = \tilde{W}_j(t)$ and/or $\tilde{\chi}_l = \tilde{\chi}_l(t)$, the definitions (10.34) and (10.35) of the characteristic values can be extended by reading them as the probability that the event in square brackets is verified *at least once* during the *design life* of the structure (or the probability that the event in square brackets in equation (10.35′) is verified *at all times* throughout the design life).

The structural reliability is controlled through the values of the *partial safety factors*; therefore, different values are prescribed for different limit states.

This type of approach, at the basis of many recent codes, is in particular prescribed by two general documents on structural safety [10, 17], the only difference between the two lying in the number and meaning of the factors γ. The ISO document [17] prescribes seven safety factors ($m = 5$; $r = 2$). The coefficient γ_1^m covers the possible reduction in strength of the material with respect to that obtained from tests, while γ_2^m covers possible weaknesses arising from any other causes: in the European recommendations [10] these two coefficients are reduced to one, denoted by γ_m. On the l.h.s., two coefficients concern the loads: the first (γ_{f1} in the European recommendations) covers unfavourable deviations and the second (ψ_0 in European recommendations) takes account of the reduced probability that various time-varying loads will all be simultaneously at their characteristic value. A third factor (γ_{f3} in the European recommendations) is related to possible adverse modifications of the load effects due to incorrect or inaccurate modelling of the structure. The last two coefficients of the ISO document (which the European recommendations unify into one indicated by γ_n) respectively take account of the nature of the structure and its behaviour, and of the severity of the consequences of attaining the considered limit state.

The partial safety factors are thus reduced to five in the European recommendations; the factors of the latter document will be discussed in the following.

Note that to use Level 1 codes, no probabilistic calculation (nor, in the limit, any knowledge of probability or statistics) is required. In other words, Level 1 codes entrust all probabilistic calculations onto the code writers, who in particular have to choose on probabilistic bases the values of the partial safety factors. In principle, these values should be calibrated by higher-level procedures, but a special remark applies to the factors γ_n and γ_{f3}. The first is not used explicitly in the critical inequality (10.33), but only to modify γ_m or γ_{f1}: it corresponds to different selections of the target reliability levels, depending on the problem being analysed. Thus, its value must not be derived from higher levels of reliability analysis, but on the basis of the considerations developed in Section 10.3.2. Analogously γ_{f3} should take into account the difference between *actual* and *theoretical* probability of failure (Section 10.3.2): its value corresponds to different estimates of the coefficient in equation (10.14). The target reliability level resulting from these considerations will then be employed for the *calibration* of γ_m, γ_{f1} and ψ_0.

10.5.2 Calibration of the partial safety factors

The partial safety factors might be obtained by calibration with existing standards, but in this case non-homogeneous levels of reliability would be achieved in the areas of different material technologies. Further, the values of the partial safety factors would be different from code to code and that would cause much confusion to the user.

On the other hand, if a constant set of γ_m, γ_{f1} and ψ_0 is prescribed, the associated reliability deviates from the target reliabilities for some design situations. It is in principle possible to select one single set of partial safety factors that minimizes the extent of this deviation. But, the resistance factor γ_m depends on the material and limit state under investigation, while the load factors γ_{f1} and ψ_0 are independent of those aspects: they only depend on the loads. Therefore, γ_{f1} and ψ_0 can also be calibrated from the knowledge of the load properties, for different structural situations each of which is characterized by the same value of γ_m. In other words, a better policy appears to be the calibration of the load factors for different resistance situations, γ_m being given *a priori*. In the following, γ_m will be taken equal to 1.

In general, an optimal set of load factors can be selected by [18].

(1) defining some function to measure an overall *distance* between the target reliability and the reliability associated with the proposed load factor set for the different design situations; possibly, these design situations may be affected by suitable weights [1], allowing for their expected recurrence in future practice;
(2) selecting γ_{f1} and ψ_0 so as to minimize this function (*economy requirement*).

However, although it is possible to select the objective function so as to penalize unconservatism heavily, such an approach does not exclude designs with reliability lower than the fixed target level. That is why a recent proposal [19] associates with the economy requirement a *safety requirement*: only design solutions with reliability equal to, or greater than, the target value are admissible and have to be considered in the calibration procedure.

The introduction of this safety requirement involves that the second step of the previous procedure must be modified by restricting the class of design solutions on which the objective function must be minimized.

Let Λ denote the definition field of the basic quantities $\mathbf{Q}$ relevant to the group of structures to which the partial safety factors are to be applied. Calibration can be formulated as a constrained minimization problem:

$$\begin{aligned} &\min \int_{\Lambda} [\bar{P}_{\text{fail}} - P_{\text{fail}}(\mathbf{Q}, \gamma_{f1}, \psi_0)]\,\mathrm{d}\mathbf{Q} \\ &\bar{P}_{\text{fail}} - P_{\text{fail}}(\mathbf{Q}, \gamma_{f1}, \psi_0) \geqslant 0 \qquad \text{for all } \mathbf{Q} \in \Lambda \end{aligned} \tag{10.36}$$

However, when different structural typologies have to be considered, the shapes of the corresponding failure boundaries may be very different from each other. Therefore, as clearly illustrated by the example in Fig. 10.6, the same set of partial safety factors may correspond to very different values of the probability of failure (just as in Level 2 approaches, it would happen for the reliability index β: cf. Fig. 10.5). Hence, two possibilities arise: namely either each calibration is performed with reference to a very small group Λ of structures with a well defined shape of the safety domain, or idealized resistance properties are introduced for all structures of a wide group. In the first case, many sets of partial safety factors are derived: each of them can be applied to the corresponding class of structures but this may cause confusion to the code user. In contrast, the second way requires that the partial safety factors ensure an actual designed structure more conservative than the idealized one; however, it has the great asset of proposing load factors independent of materials, technologies and limit states. In fact an idealization of the resistance characteristics consists of assuming an identical CPF shape for all structures; the parameters of this function and the load properties then become the only quantities of interest for calibration purposes.

Fig. 10.6 illustrates also a related circumstance to be outlined. Only one check is unable to ensure that inequality (10.33) is verified in all possible directions in the load space, as it should, and therefore does not warrant a sufficient reliability, even from a limited Level 1 viewpoint. Therefore, most Level 1 codes require that inequality (10.33) be repeatedly employed for different load combinations, defined by different values of the relevant load safety factors (Fig. 10.7 (a)).

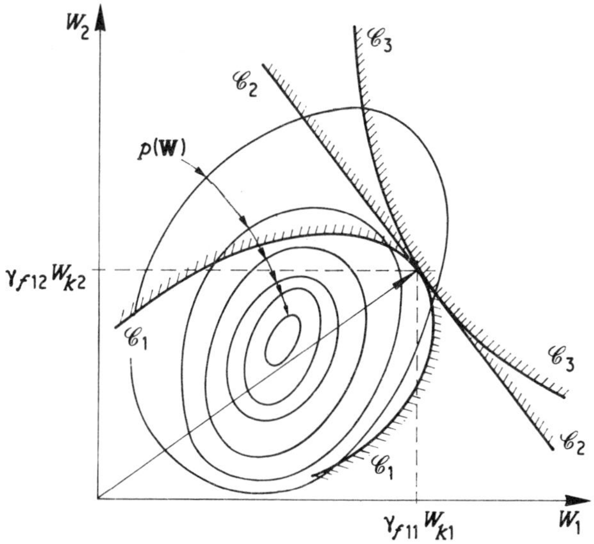

Fig. 10.6 Different deterministic safety domains associated with different P_{fail} satisfying the same Level 1 relation (for structures with random resistance, the safety domain is replaced by the contour lines of the CPF).

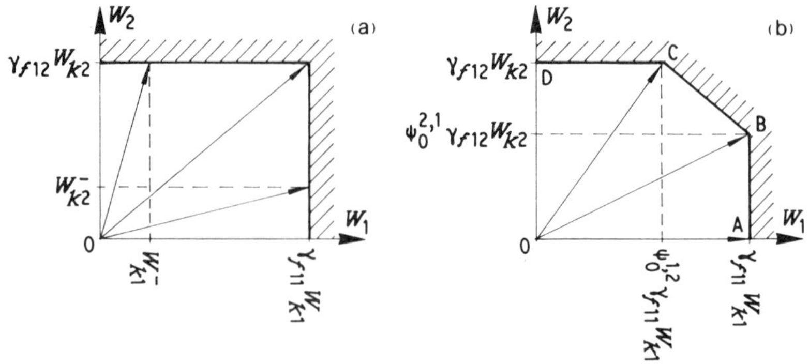

Fig. 10.7 (a) Checks required by present codes for time-constant actions (W_{k1}^- and W_{k2}^- denote the characteristic values to be introduced when the load action is favourable for safety). (b) Checks required for time-varying actions.

Therefore two tasks are entrusted by a Level 1 philosophy to the load partial safety factors γ_{f1} and ψ_0 being calibrated:

(1) they must ensure that in the load space, along the directions of interest, the contour lines of the conditional probability of failure (CPF) of all relevant structures are sufficiently distant from the origin of the axes;
(2) they must define these directions of interest.

If the policy of introducing idealized resistance properties is adopted, the second task coincides with the definition of the load conditions by whose check one ensures that the actual structure is more conservative than the idealized one needed to obtain the fixed target reliability level.

This structure must be selected to have a CPF attaining significant values at a sufficiently large distance from the origin of the axes in order to satisfy the first aim: the parameters of this CPF must be determined as the one for which the optimal solution of problem (10.36) is achieved in the subset Λ° of Λ that contains only the variables defining the loads and the CPF.

If the CPF depends on one parameter only, this assumes the meaning of a conventional resistance and will be denoted by R. In this case there is a one-to-one correspondence

$$\bar{P}_{\text{fail}} \Leftrightarrow \bar{R} \tag{10.37a}$$

such that

$$\{(\bar{P}_{\text{fail}} - P_{\text{fail}}) \geqslant 0\} \Leftrightarrow \{(R - \bar{R}) \geqslant 0\} \tag{10.37b}$$

Hence, remembering that the variables of interest are the $\mathbf{Q}^\circ$ defined in $\Lambda^\circ \subset \Lambda$, (10.36) becomes

$$\begin{aligned} &\min \int_{\Lambda^\circ} [R(\mathbf{Q}^\circ, \gamma_{f1}, \psi_0) - \bar{R}] \mathrm{d}\mathbf{Q} \\ &R(\mathbf{Q}^\circ, \gamma_{f1}, \psi_0) - \bar{R} \geqslant 0 \end{aligned} \tag{10.38}$$

Once the shape of the CPF function is stated, the r.h.s. of (10.37b) can be derived, for instance by a trial-and-error procedure, making use of the basic equations presented in Sections 2.8 and 3.7. Moreover, if the calibration is performed separately for different shapes of the CPF, the conventional resistance R does not depend on the shape but only on the values and the properties of the actions. Thus, one can express R as a function of the characteristic value of the loads multipied by factors ϕ related to the relevant safety factors. Further, a suitable discretization of the subset Λ° can be introduced and (10.38) reduces to a mathematical programming problem:

$$\begin{gathered} \min \sum [R(\phi_i W_{ki}) - \bar{R}] \\ R(\phi_i W_{ki}) \geqslant \bar{R} \end{gathered} \tag{10.39}$$

It has recently been suggested [20] to use, for calibration purposes, CPFs with piecewise-linear contours which allows one to obtain the results directly in accordance with current code trends (Fig. 10.7), as will be shown in the following.

Introduce in the load space an appropriated oblique hyperplane ζ. Then, a suitable CPF idealization is defined by the relation

$$P_R(\mathbf{W}) = \max_i [P_i(R_i), P_\chi(\chi)] \tag{10.40}$$

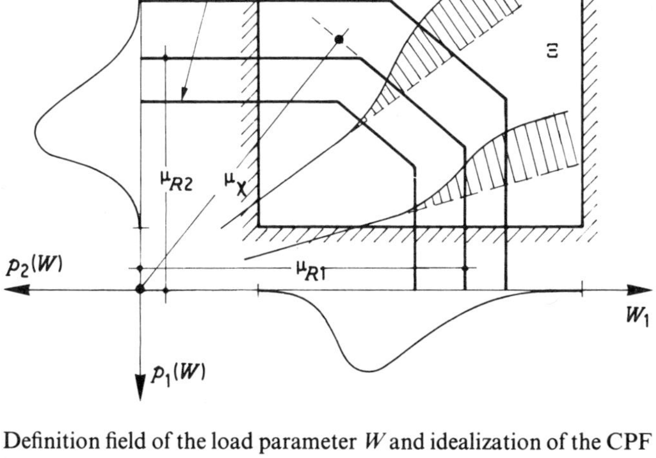

Fig. 10.8 Definition field of the load parameter W and idealization of the CPF (two-load case).

where $P_i(R_i)$ denotes the normal CDF with mean μ_{R_i} and assigned coefficient of variation δ_R. The parameter χ is the distance from the origin of the hyperplane parallel to ζ in $\mathbf{W}$ and $P_\chi(\chi)$ is the normal CDF with mean μ_χ and c.o.v. δ_χ. In equation (10.38), μ_{R_i} is selected as the mean value required by a structure subjected only to the ith action in order to present a fixed probability of failure $\bar{P}_{fW}$. Then, the only free parameter to define this function is the mean value μ_χ, or possibly the characteristic value χ_k: they assume the meaning of mean or characteristic value of a conventional resistance R.

Once problem (10.39) is solved and the optimum values of the factors ϕ_i^{opt} are derived, a successive step is to state the values of γ_{f1} and ψ_0 that ensure that the actual safety domain is, according to the Level 1 philosophy, outside the contour line $\bar{P}_R(\mathbf{W}) = 5\%$ defined by the optimal values ϕ_i^{opt}. This step can be performed, if the actual safety domain is convex, by easy geometrical considerations (Fig. 10.9).

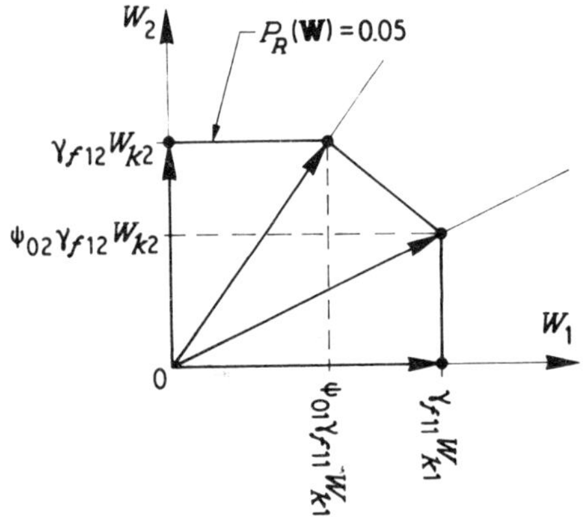

Fig. 10.9 Definition of the check directions necessary to provide safety.

The numerical computations that have been developed show few advantages with respect to the solution of Fig. 10.7 (a) when permanent loads are considered, but the whole procedure does appear to yield significant results in the calibration of the load combination factor ψ_0 [21], as the γ_{f1} are obtained directly when the value $\bar{P}_{fW}$ is assigned.

A final remark appears of interest. If the same target reliability level is required under the different load combinations, a rigorous calibration of the load factors should lead to different values of the safety coefficients for each combination. As such a circumstance does not meet the requirement of not confusing the code user, different target reliability levels have to be employed as the number of loads in combination increases. For instance, the NBS [18] proposed $\bar{P}^0_{fail} = 1.35 \times 10^{-3}$ ($\bar{\beta} = 3$) for structures or structural elements subjected to permanent loads only; $\bar{P}^{v}_{fail} = 6.21 \times 10^{-3}$ ($\bar{\beta} = 2.5$) and $\bar{P}^{e}_{fail} = 4.01 \times 10^{-2}$ ($\bar{\beta} = 1.75$) in the presence of wind and seismic action respectively. These values, derived from current practice, appear to be too high in comparison with the values presented in Section 10.3.2; nevertheless they can be employed in the presented calibration procedure, in which $\bar{P}_{fW}$ and $\bar{P}_{fail}$ are to be regarded as different target values of the probability of failure related to different design situations.

REFERENCES

1. Construction Industry Research and Information Association (CIRIA) (1977) *Rationalisation of Safety and Serviceability Factors in Structural Codes*, Report 63, CIRIA, London.
2. Köning, G. and Marten, K. (1977) Nutzlasten in Bürogebäuden (in German), *Die Bautechnik*, **7**, 165–70.
3. ECCS-CECM (1978) *Recommendation pour le Calcul de l'Action du Vent sur les Constructions* (in French), ECCS–CECM, Paris.
4. British Standards Institution (BSI), Council for Code of Practice (1972) *Code Basic Data for the Design of Buildings*, Chap. V, Loading (Part 2, Wind loads), BSI, London.
5. International Standards Organization (ISO) (1974) *Snow Load on Roofs*, TC 98/SEC 3/WG 1, ISO.
6. Chopra, A. K. and Newmark, N. M. (1980) Analysis; in *Design of Earthquake Resistant Structures* (ed. E. Rosenblueth), Pentech Press, London.
7. Duff, C. G. (1980) Design ground response spectra; in *Reference Seismic Ground Motions in Nuclear Safety Assessment*, OECD, Paris.
8. Bertero, V. V. and Kamil, H. (1975) Non-linear seismic design of multistorey frames. *Canadian Journal of Civil Engineering*, **2**(4), 504–16.
9. Casciati, F., Faravelli, L. and Gobetti, A. (1979) Reliability of seismic resistant frames designed by inelastic spectra. *U.S. Earthquake Engineering Conf.*, Stanford, pp. 553–62.
10. Joint Committee on Structural Safety (1976) Common unified rules for different types of constructions and materials. *CEB Bull. d'Info.*, **116 E.**
11. Sexsmith, R. G. and Lind, N. C. (1977) Policies for selection of target safety levels. *Proc. 2nd Int. Conf. on Structural Safety and Reliability* (ICOSSAR '77), Munich, pp. 149–61.
12. Veneziano, D. (1974) *Contributions to Second-Moment Reliability Theory*, Dept of Civil Engineering MIT, Struct. Publ. No. 389.
13. Rosenblueth, E. (1972) Code Specification on Safety and Serviceability, in *Planning and Design of Tall Building*, TC 10, ASCE-IABSE.
14. Rackwitz, R. (1976) Practical probabilistic approach to design. *CEB Bull. d'Information*, **112**, 13–89.
15. Ditlevsen, O. (1978) *Generalized Second-Moment Reliability Index*, DCAAM Rep. No. 144.
16. Veneziano, D. (1979) New index of reliability. *J. Engrg. Mech. Div., Proc. ASCE,* **105** (EM2), 277–96.
17. International Standards Organization (ISO) (1973) *General Principles for the Verification of the Safety of Structures*, ISO 2394.

18. National Bureau of Standards (NBS) (1980) *Development of a Probability Based Load Criterion for American National Standard A58*, NBS Special Publ. 577.
19. Casciati, F. and Faravelli, L. (1980) Calibration of the load enhancement factors. *Engineering Structures*, **2**(2), 113–22.
20. Faravelli, L. (1982) A proposal on load combination for level I formats. *Engineering Structures*, **4**, 197–206.
21. Ballio, G., Casciati, F. and Faravelli, L. (1981) Combination values of the actions for level I verifications: their evaluation via fully probabilistic methods. *Proc. 3rd Int. Conf. on Structural Safety and Reliability* (ICOSSAR '81), Trondheim, pp. 627–37.

CHAPTER 11

Approximations, errors and open questions in probabilistic structural engineering

Engineers' calculations are always performed on much-simplified models of the real world. Safety factors and other empirical devices have always been used to warrant a satisfactory degree of safety notwithstanding such approximations: the introduction of probabilistic approaches, as so far illustrated in this book, has done much in order to make such devices more rational, but not all problems have been solved, and a great deal of empiricism is still necessary to overcome uncertainties in practical applications.

In view of future developments, this concluding chapter discusses some general questions and open problems related to the modelling of structural systems.

11.1 STRUCTURAL MODELS AND 'PERTURBATIONS'

In the practice of technical problems, following either a deterministic or a probabilistic approach, the evaluation of structural response under given loading is performed under simplifying assumptions, like typically linear elastic behaviour, small displacements, conservative forces and so on. Thus, the mathematical model is greatly simplified compared to the much greater complexity of the real structure.

Similarly, the loading introduced in the calculations is always much simpler than the actual load (as already discussed many times throughout this book) and most often includes only the actions that are directly related to the essential functionality demand by the structure (for instance, in a column, the load transferred by the floors, the occupants and the furniture, but not the one deriving from the initial out-of-straightness).

This corresponds to the requirement that both the structural model and the pattern of loading could be easily manipulated; in particular, that an analytical or numerical solution could be found for the characteristic equation of the structure.

The consequence of the above circumstance is that a possibly large set exists containing actions or load effects that are not explicitly taken into account in the design procedure, but that actually occur on the structure, either because some loads are not included in the forecasted load conditions, or because the actual structure differs significantly from the mathematical model, or even because approximations are introduced in the calculations. These differences are usually neglected with respect to the *main effects*, and therefore called *perturbations* (or, more simply though less expressively, *errors*).

Practical structural analysis usually ignores perturbations, implicitly relying on the assumption that relative deviations of the actual structural response from the calculated one (*output*) are of the same order as the differences between the actual and idealized structure and loading (*input*).

Nevertheless, the real situation does not always correspond to such expectation, and many times the designer is faced with the question whether the equilibrium configuration that he is analysing is sensibly modified by perturbations. In other words, the question arises of the *reliability* of the results of the chosen mechanical model with respect to possible perturbations.

In classical structural mechanics, this sort of problem arises in connection with the stability/instability of the equilibrium configuration. The well known example of the compressed slender column is very effective from this point of view, in that it is the typical case where, for example, a small bending of the axis (the *perturbation*) may cause a disproportionate increase in the displacements and stresses, if the normal force approaches the *critical* values.

As noted by Bolotin [1], the concept of stable and/or unstable equilibrium is very important in structural engineering, just as a tool to take under control the sensitivity of the structure to unknown perturbations. The introduction of probabilistic methods allows one to make explicit the use that the structural engineers really make of the concept of stability. It can be argued, in fact, that a parameter exists measuring the *degree of stability* of an equilibrium configuration of a mechanical system, and that the same parameter also measures the *probability* that the system does not tend to take, in the actual situation, a configuration significantly different from the calculated (compatible) one. As stressed by Bolotin, this relation between the degree of stability and the quoted probability justifies the reason of stability analysis in the structural design.

11.1.1 A simple example: one-degree-of-freedom model

This discussion can be made clear by the following elementary example [2].

Let the actual structure be a rigid column, hinged at the bottom end, elastically restrained at the top end, and loaded by a well determined axial load W and, in the real situation, by a small random moment $\tilde{M}$ (Fig. 11.1) with zero mean ($\mu_M = 0$). All other quantities are deterministic.

However, in the calculations, the moment $\tilde{M}$ is neglected, so that the analytical relations are greatly simplified, while, being very small by assumption, the moment $\tilde{M}$ is assumed not to influence response and strength of the column significantly.

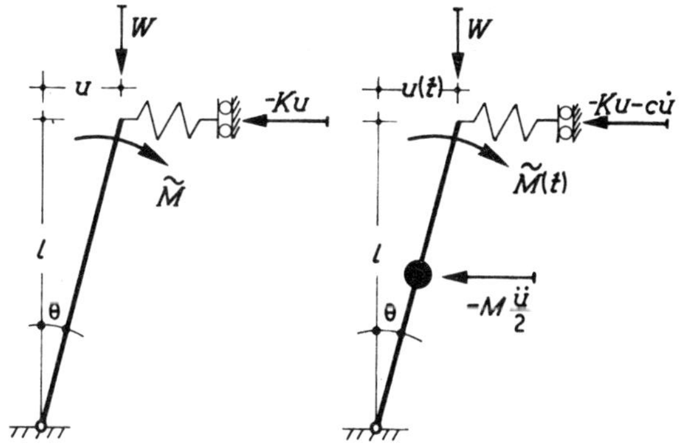

Fig. 11.1 A rigid column, the actual situations: (a) static; (b) dynamic.

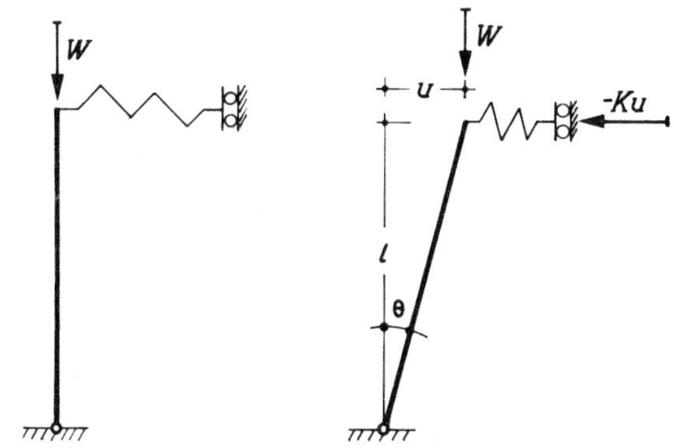

Fig. 11.2 A rigid column, the *perfect* model used in the calculations.

Thus, the *calculation model* is the *perfect column* in Fig. 11.2, of which $\tilde{M}$ is the perturbation‡.

(a) Static analysis

The solution of the model of Fig. 11.2 for $W \neq Kl$ is trivial: the equilibrium configuration is defined by $\theta = 0$, and only normal force is present in the column. On the contrary, analysis of the actual structure with the perturbation (Fig. 11.1(a)) yields the following equilibrium equation (within the small displacement assumptions, i.e. $\sin\theta = \theta$; $\cos\theta = 1$):

$$W\,l\tilde{\theta} - K\,l^2\tilde{\theta} + \tilde{M} = 0 \tag{11.1}$$

‡ A similar procedure is often followed in the analysis of elastic columns, plates, etc., where the *imperfections* are not included in the calculation model and a *buckling* (or *bifurcation*) load is sought.

whence

$$\tilde{\theta} = \frac{\tilde{M}}{Kl^2 - Wl} \tag{11.2}$$

Equation (11.2) points out that the actual equilibrium configuration of the column deviates from the position $\theta = 0$: the amount of the deviation is a random variable proportional to the intensity of the perturbation ($\tilde{M}$).

Assume now that stability analysis is performed for the *perfect* model in Fig. 11.2. Let U_0 be the total energy of the system, given by

$$U_0 = -\frac{Wl\theta^2}{2} + \frac{Kl^2\theta^2}{2} \tag{11.3}$$

and consider the second derivative of the energy calculated at the point $\theta = 0$, i.e. at the considered equilibrium configuration

$$q_s = \left.\frac{d^2U_0}{d\theta^2}\right|_{\theta=0} = Kl^2 - Wl = \left(1 - \frac{W}{W_c}\right)W_c l \tag{11.4}$$

where

$$W_c = Kl \tag{11.5}$$

is the *critical load* (or *buckling load*, or *bifurcation load*) of the column.

Introducing q_s, equation (11.2) can be written

$$\tilde{\theta} = \tilde{M}/q_s \tag{11.2'}$$

It is well known that if $q_s > 0$ the considered configuration $\theta = 0$ is stable, and if $q_s < 0$ the same configuration is not stable. Apart from special instances, the case $q_s = 0$ denotes transition from stable to unstable states.

Equation (11.2′), however, shows that the simple statement of stability ($q_s > 0$) is not sufficient from the point of view of safety or functionality, in that the occurrence of small perturbations may cause large rotations of the column, with untolerable elongation of the top spring, unless q_s is sufficiently large (i.e. W is sufficiently less than W_c).

Roughly speaking, the larger is q_s, the *more stable* is the structure. This justifies the idea of denoting by *degree of stability* the second derivative of the total energy of the *perfect* model (without perturbations).

Let us now evaluate the uncertainty affecting the equilibrium configuration of the real structure with respect to the solution ($\theta = 0$) of the *perfect* model.

Remembering first that by assumption $\mu_M = 0$, one gets immediately

$$\mu_\theta = 0 \tag{11.6}$$

i.e. the solution of the *perfect* model is *exact in the mean*. Then, a good measure of the uncertainty of this solution is the variance of the random angle $\tilde{\theta}$. From equations (10.2′) and (A3.5b) it is easy to get

$$s_\theta^2 = s_M^2/q_s^2 \tag{11.7}$$

s_M^2 being the variance of the perturbation. Equation (11.7) shows that the larger is q_s (i.e. the more stable the theoretical equilibrium configuration), the smaller is the uncertainty affecting the real structure and consequently the *more reliable is the perfect model* (Fig. 11.2) introduced in the calculations. Thus, the introduction of a safety factor with respect to the critical load is a tool to keep q_s large and then to reduce uncertainties of the main configuration $\theta = 0$.

Stability analysis, then, takes the significance of a test for *credibility* of the structural model, thus establishing a definite relationship between stability and uncertainty of a given configuration.

(b) Dynamic analysis

The same problem can also be treated with a dynamic perturbation.

Assume (Fig. 11.1(b)) that the perturbing moment is a random function of time $\tilde{M}(t)$, described by Gaussian stationary white-noise stochastic process, with average and autocorrelation function given by

$$\mu_M(t) = 0 \qquad B_M(v) = \Lambda\, \partial(v) \tag{11.8}$$

Λ being a positive constant and $\partial(\ )$ the Dirac delta-function. Let M be the mass of the rod and c the damping of the spring, so that the equation of motion of the rod is

$$\tfrac{1}{4}Ml^2\ddot{\theta} + cl^2\dot{\theta} + K\theta l^2 = W\theta l + \tilde{M}(t) \tag{11.9}$$

i.e. in standard form (cf. equation (4.69))

$$\ddot{\theta} + 2\xi\omega_0\dot{\theta} + \omega_0^2\theta = w\theta + \tilde{y}(t) \tag{11.10}$$

with

$$\omega_0^2 = \frac{4K}{M} \qquad \xi = \frac{2c}{M\omega_0} \qquad w = \frac{4W}{Ml} \qquad \tilde{y} = \frac{4\tilde{M}}{Ml} \tag{11.10$'$}$$

The steady-state response $\tilde{\theta}^s(t)$ can be calculated by the methods presented in Section 4.3.

The impulse response function $h(v)$ is in fact the same as in equation (4.89), with the only difference that, as can be easily verified, ω_d is here given by

$$\omega_d = \omega_0\sqrt{[1 - (w/\omega_0^2) - \xi^2]} \tag{11.11}$$

In order to calculate the properties of the steady-state response, equations (4.96) and (4.97) can be applied with

$$B_w(v) = \frac{4\Lambda}{Ml^2}\,\partial(v) \tag{11.12}$$

One gets immediately

$$\mu_\phi = 0 \tag{11.13}$$

and after integration

$$B_\theta^s(\tau) = \frac{4\Lambda e^{-\xi\omega_0\tau}}{Ml^2\,\omega_d^2}(C_1\cos\omega_d\tau + C_2\sin\omega_d\tau) \tag{11.14}$$

with

$$C_1 = \frac{\omega_d^3}{4\xi^3\omega_0^3 + 4\xi\omega_0\omega_d^2} \qquad C_2 = \frac{\omega_d^3}{4\xi^2\omega_0^2 + 4\xi\omega_d^2} \tag{11.14'}$$

Because of linearity, the output process $\tilde{\theta}(t)$ is also a Gaussian process as the input $\tilde{y}(t)$. Hence the structure response is completely characterized by equations (11.13) and (11.14).

In particular, consider the *state vector* $(\theta, \dot{\theta})$, which describes the condition of the system at each instant.

Following the same procedure as in Section 4.2.1(d), the point properties of the state vector are given by

$$B^s_{\theta,\dot{\theta}}(\tau) = -\dot{B}^s_{\dot{\theta}}(\tau) = \xi\omega_0 B^s_{\theta}(\tau) + \frac{e^{-\xi\omega_0\tau}}{\omega_d^2}\frac{4\Lambda}{Ml^2}(C_1\omega_d \sin\omega_d + C_2\omega_d\cos\omega_d\tau) \tag{11.15a}$$

$$B^s_{\dot{\theta}}(\tau) = -\ddot{B}^s_{\theta}(\tau) = 2\xi\omega_0\dot{B}^s_{\theta}(\tau) + (\xi^2\omega_0^2 + \omega_d^2)B^s_{\theta}(\tau) \tag{11.15b}$$

whence

$$\begin{aligned} s_\theta^2 &= B^s_\theta(0) = \frac{4\Lambda}{Ml^2}\frac{\omega_d}{4\xi^3\omega_0^3 + 4\xi\omega_0\omega_d^2} \\ \Sigma_{\theta,\dot{\theta}} &= \mathrm{Cov}\,(\tilde{\theta}, \dot{\theta}) = 0 \\ s_{\dot{\theta}}^2 &= B^s_{\dot{\theta}}(0) = \frac{4\Lambda}{Ml^2}\frac{\omega_d}{4\xi\omega_0} \end{aligned} \tag{11.16}$$

Since, as already noted, the output process is Gaussian, the instantaneous JPDF is given by

$$p(\theta, \dot{\theta}) = \frac{1}{2\pi s_\theta s_{\dot{\theta}}} \exp\left[-\left(\frac{\theta^2}{2s_\theta^2} + \frac{\dot{\theta}^2}{2s_{\dot{\theta}}^2}\right)\right] \tag{11.17}$$

Rearranging equations (11.10′), one can write

$$\frac{\theta^2}{2s_\theta^2} = \frac{8c}{\Lambda M\omega_d}U_0 \tag{11.18}$$

$$\frac{\dot{\theta}^2}{2s_{\dot{\theta}}} = \frac{8c}{\Lambda M\omega_d}Q_0 \tag{11.19}$$

where U_0 is given by equation (11.3) and

$$Q_0 = \frac{Ml^2}{8}\dot{\theta}^2 \tag{11.20}$$

is the kinetic energy of the rod. Introducing equations (11.18) and (11.19), equation (11.17) becomes

$$p(\theta, \dot{\theta}) = \frac{1}{2\pi s_\theta s_{\dot{\theta}}} \exp\left(-\frac{8c}{M\Lambda\omega_d}(U_0 + Q_0)\right) \tag{11.21}$$

proving that the instantaneous JPDF of the components of the state vector is larger, for a given value of the main load W, on trajectories where the total energy $U_0 + Q_0$ is minimum, i.e. for *stable trajectories*. The narrower is the *valley* of the total energy functional, the larger the probability to find the structure close to the path of minimum energy, in analogy to the previous static example.

11.1.2 Concluding remark

The conclusions reached for the simple model of Section 11.1.1 are quite general, as proved by Bolotin [1] in a rather different way. It can be stated therefore that in both static and dynamic problems, the stability of the solution of the (deterministic) unperturbed structural model is strictly related to the reliability of the model, and that a large probability of occurrence of the calculated configuration is the counterpart of stability with respect to the *perturbations* neglected in the calculations.

11.2 PROBABILISTIC INTERPRETATION OF THE CONCEPT OF STABILITY

In investigating the reliability of a deterministic structural system versus a random perturbation, Section 11.1 has shown a first strict relation between probability and stability. But another question also arises in probabilistic mechanics, which will be discussed in this section.

11.2.1 Statement of the problem

The classical definitions of *stability* and *instability* lead to a sharp qualification of a state of (static or dynamic) equilibrium. The stability of mechanical systems is investigated with respect to the whole set of possible system quantities and/or forcing functions. If only one possibility of *instability* is found to exist, the equilibrium is defined as *unstable*, and the structure classified as unable to withstand the considered loads.

On the contrary, taking a probabilistic point of view, one can imagine cases in which, despite the *possibility* that some form of unstable behaviour is activated, the *probability* that this happens is very low: also, it may happen that the structure always tends to return to the main configuration after some excursion.

The importance of these considerations from a designer's point of view is self-evident: whence, the opportunity of a probabilistic approach also in stability problems. This leads essentially to weaker definitions of stability than the classical ones: in fact, to assess that a system has zero (or practically zero) probability of following an unstable path is less restrictive than the requirement that such paths do not exist, although the two statements may sound almost equivalent to an engineer's ear.

In view of the importance of these questions, probabilistic definitions of stability are reported below, although they do not yet find much application in technical practice.

11.2.2 Probabilistic definitions of stability

Consider stability in the Liapunov sense. Let the system input/output quantities $\mathbf{Y} = (Y_1, Y_2, \ldots, Y_m)$ describe the reasons causing a deviation of the system from the equilibrium configuration $\mathbf{u}_0 = \mathbf{0}$. Then the unperturbed equilibrium is called stable in the time interval T relative to the response output parameters $\mathbf{u} = (u_1, u_2, \ldots, u_n)$, if for any small positive number ζ a positive number ε exists so that in the whole interval $(0, T)$

$$\{\|\mathbf{Y}\| \leqslant \varepsilon\} \Rightarrow \{|\mathbf{u}| < \zeta \qquad (11.22)$$

where $|\ \ |$ and $\|\ \ \|$ denote suitable vector norms.

In other words, the unperturbed equilibrium is called stable if the output can be made smaller than any assigned quantity by decreasing the system quantities that cause deviation of equilibrium.

Assume now that the above condition is not satisfied, i.e. that there exists at least one perturbation vector $\mathbf{Y}$ causing a response $\mathbf{u}$ bounded from below for any value of ε: classically, one should conclude that equilibrium is not stable and try to modify the system to ensure safety. On the contrary, further investigation is possible in the probabilistic context. The next step, in fact, is to ask for which and how many $\mathbf{Y}$ vectors the structure exhibits an unstable behaviour, and how large is the probability of occurrence of such vectors. It is possible that such probability is zero, or so small that it may be considered not significant. This is the reason to turn to the probabilistic stability criteria: perturbations existing in real structures are random in nature, so that the response output vector $\mathbf{u}$ should be regarded as a stochastic field. The following probabilistic definitions of stability can be formulated [3]:

(1) The solution $\mathbf{u}(t) = \mathbf{0}$ is stable in probability if for any $\zeta > 0$ and $\chi > 0$ there is an $\varepsilon(\zeta, \chi) > 0$ such that the inequality $\|\mathbf{Y}\| < \varepsilon$ implies that

$$\mathrm{Prob}\left[\sup_{0 < t < \infty} |\tilde{\mathbf{u}}(t)| < \zeta\right] > 1 - \chi \qquad (11.23)$$

(2) The solution $\mathbf{u}(t) = \mathbf{0}$ is asymptotically stable in probability if for any $\zeta_1 > 0$

$$\lim_{t \to \infty} \mathrm{Prob}[|\tilde{\mathbf{u}}(t)| < \zeta_1] = 1 \qquad (11.24)$$

(3) The solution $\mathbf{u}(t) = \mathbf{0}$ is stable in probability on the finite time interval $(0, T)$ if for any $\zeta > 0$ and $\chi > 0$ there is an $\varepsilon(\zeta, \chi) > 0$ such that the inequality $\|\mathbf{Y}\| < \varepsilon$ implies

$$\text{Prob}\left[\sup_{0 < t \leqslant T} |\tilde{\mathbf{u}}(t)| < \zeta\right] > 1 - \chi \tag{11.25}$$

(4) The solution $\mathbf{u}(t) = \mathbf{0}$ is stable with respect to the mean norm $\mathrm{E}[|\tilde{\mathbf{u}}|]$ if for any $\zeta > 0$ there is an $\varepsilon(\zeta) > 0$ such that $\|\mathbf{Y}\| < \varepsilon$ implies

$$\sup_{0 < t < \infty} \mathrm{E}[|\mathbf{u}|] < \zeta \tag{11.26}$$

(5) The solution $\mathbf{u}(t) = \mathbf{0}$ is stable with respect to the mean number of the crossings $\tilde{n}(t)$ of the sphere $|\mathbf{u}| = \zeta$ (Section A7.4.1) if for any $\zeta > 0$ there is an $\varepsilon(\zeta) > 0$ such that the inequality $\|\mathbf{Y}\| < \varepsilon$ implies

$$\lim_{t \to \infty} \mathrm{E}[\tilde{n}(t)] < \zeta \tag{11.27}$$

(6) When definition (11.27) is not applicable (for example, if the disturbed motion is a stationary random process) one must introduce the rate of positive crossings $\nu^+(t)$

$$\nu^+(t) = \lim_{\Delta t \to 0} \frac{\mathrm{E}[\tilde{n}(t + \Delta t)] - \mathrm{E}[\tilde{n}(t)]}{\Delta t} \tag{11.28}$$

and replace (11.27) by

$$\sup_{0 < t < \infty} \nu^+(t) < \zeta \tag{11.27$'$}$$

Some of the definitions presented above are connected to each other. For example, when the solution is stable with respect to the mean norm, it is stable in probability (while the inverse statement does not hold in general); stability with respect to the mean number of crossings implies stability in probability (Appendix A4); and so on.

11.2.3 Some remarks

Examples of areas where application of the probabilistic concept of stability can be most useful are all the problems where it is not possible to exclude *a priori* that loads causing instability may occur, although they can be very rare. Long bay cable bridges prone to aerodynamic flutter may be a typical example.

Of course, the definitions presented again lead to a sharp, yes-or-no qualification of the equilibrium, as in the classical, deterministic approach: the underlying assumption is of course the elastic response of the structure. A more complete probabilistic approach should introduce some form of *graduated* evaluation of the response, from either the point of view of safety (probability of attaining a condition in which the structure *collapses* or is *damaged*) and/or serviceability (intervals during which the response is beyond an *acceptable threshold*). More elementary problems of this type have been tackled in various places in this book, but their connections with the concept of *stability*, although they may be clear in principle, require further research in order to be useful in applications.

11.3 ERRORS IN STRUCTURAL ANALYSIS INVOLVED BY APPROXIMATIONS

11.3.1 Basic ideas

As already stated, the probabilistic point of view underlines and clarifies the interactions between the solution of mathematical problems involved by structural design and the use that can be made of the results of computations. Indeed, mathematical results are never employed in a direct form, but always manipulated, and often neglected in the details, by the engineer's judgement that enters into the procedure of structural analysis and design as a decisive factor.

As a matter of fact, the occurrence of perturbations in the real structure often makes futile any effort to give pure rigorous solutions to the mathematical model. Therefore, *exact* mathematics is often a uselessly severe requirement, while the model used in the calculations is only a *conventional* description of the expected reality. Nevertheless, errors in analysis may be decisive in causing structural malfunctions, and the control of allowable approximations should be required and founded on well defined rational criteria.

In general, one of the main obstacles found in the judgement of errors in calculations is the difficulty of getting bounds on *local* mistakes in stresses, displacements, etc.: indeed, comparisons are only possible on the basis of some *global* measure of structural response or performance.

A possible approach to the question is provided by the probabilistic theory of structural safety: since the assessment of reliability is a central objective of a structural design, the analysis of the propagation of numerical errors on the safety indices or on the probability of failure can give useful criteria, in order to judge the admissibility of approximations, or to calibrate increments of the safety coefficients necessary to neutralize the effect of errors on structural safety.

In this way, the probabilistic approach might really give a decisive impact for an *engineering* evaluation of errors introduced by approximations: unfortunately, and somewhat surprisingly, there is not much work on the subject at the present stage, despite the interest that the matter should have in practical applications.

Some studies have been performed by Lind [4], who explored random sources of errors in calculations, and by Baratta [5], who analysed the possibility of combining non-random mathematical error estimates of currently adopted numerical methods with probabilistic description of the structural environment. The latter study is summarized in the following, in order to give an idea of the possible results of such investigations.

The problem is simplified by assuming that the structural model is adequate to represent the real structure; moreover in the calculations the actual random time-varying loads are replaced by an *n*-dimensional *design load* $\tilde{\mathbf{W}}_D$, constant in time, and are applied once, at the beginning of the structure's lifetime.

Let $\mathbf{W}$ be any possible load vector, $\mathbf{u}$ the structure response vector, and $\mathbf{K}$ the characteristic operator of the structure, so that, neglecting dynamic effects, the response equation is

$$\mathbf{Ku} = \mathbf{W} \tag{11.29}$$

and assume that such equation has only one solution for any $\mathbf{W}$.

Let $\bar{\mathbf{u}}(\mathbf{W})$ be an approximate solution of equation (11.29), and put

$$\bar{\bar{\mathbf{W}}} = \mathbf{K}\bar{\bar{\mathbf{u}}}(\mathbf{W}) = \mathbf{DW} \tag{11.30}$$

$\bar{\bar{\mathbf{W}}}$ is called the *emerging load* associated with $\mathbf{W}$. It is assumed that an approximate solution can be found for every possible $\mathbf{W}$.

The norm of the difference between $\mathbf{D}$ and the identity operator $\mathbf{I}$ can be identified with the *numerical error* of the approximation

$$\mathscr{E}_1 = |\mathbf{D} - \mathbf{I}| \tag{11.31}$$

Assume also that the reliability of the structure is fully defined by one factor β (*reliability index*, cf. Section 10.4), and that $\bar{\beta}$ is the minimum prescribed value of this index. If $\mathscr{S}$ is the success region of the structure in the space of possible loads, the design load $\mathbf{W}_D$ used in the calculations should be defined by the assumption that, coupling exact structural analysis with correct design rules and building practice, if the structure resists $\mathbf{W}_D$ it will also resist any foreseable loads, except possibly those with very low probability of occurrence.

In symbols

$$\{\mathbf{W}_D \in \mathscr{S}\} \Rightarrow \{\beta \geqslant \bar{\beta}\} \tag{11.32}$$

(Note that in the present treatment the circumstance is neglected that design sometimes requires consideration of two or more load conditions.)

Generally speaking, approximate procedures do not allow one to test whether the l.h.s. of implication (11.32) is true. They can only say if the design load falls in an approximate success domain $\bar{\bar{\mathscr{S}}}$, more or less different from $\mathscr{S}$. It is however obvious that

$$\{\mathbf{W}_D \in \bar{\bar{\mathscr{S}}}\} \Rightarrow \{\beta \geqslant \bar{\beta}] \tag{11.32'}$$

A significant *global* measure of the error is given by the coefficient γ^* that must be applied to the design load in order to neutralize errors in the solution procedures as regards the satisfactory performance of the structure with respect to the reliability index, i.e. the coefficient defined by

$$\{\gamma^* \mathbf{W}_D \in \mathscr{S}\} \Rightarrow \{\beta \geqslant \bar{\beta}\} \tag{11.33}$$

In the following, γ^* will be called the *additional safety factor*, while the *reliability error* is defined by the position

$$\mathscr{E} = \gamma^* - 1 \tag{11.34}$$

To calculate $\mathscr{E}$ for a given procedure may be very difficult. However, an upper bound is given by [5]

$$\mathscr{E} \leqslant \mathscr{E}_u = \Delta \frac{1 + \bar{\beta}_n \delta_W^0}{\bar{\beta}_n \delta_W^0} \tag{11.35}$$

where

$$\Delta = \frac{r\mathscr{E}_1}{1 - r\mathscr{E}_1} \tag{11.36}$$

$$\delta_W^0 = \frac{1}{|\mathbf{\Sigma}_W^{-\frac{1}{2}} \boldsymbol{\mu}_W|} \tag{11.37}$$

In turn, $\mathscr{E}_1$ is the *numerical error* defined by equation (11.31), μ_W the average vector of the loads, and

$$r = |\mathbf{\Sigma}_W^{-\frac{1}{2}}|\,|\mathbf{\Sigma}_W^{-\frac{1}{2}}| \tag{11.38}$$

the *condition number* of their covariance matrix $\mathbf{\Sigma}_W$.

Thus, Δ is a measure of the difference between the emerging and the originating loads (called *relative numerical error* in the following), δ_W^0 essentially measures the coefficient of variation of the service loads, while $\bar{\beta}_n$ is the prescribed index biased by the dimension effect (n is the order of vectors **u**, **W**, etc.) (Section 10.4.5).

This upper bound has some significant properties that can reasonably be supposed to be properties of the reliability error $\mathscr{E}$ also: thus it can be stated that

(1) The reliability error does *not* depend on the error in *local* effects, but only on a global estimate of the error between applied and emerging loads.
(2) $\mathscr{E}$ is a decreasing function of the product $\bar{\beta}_n \delta_W^0$, i.e. it is smaller when the loads are affected by large uncertainties (larger δ_W^0) and it is smaller when high reliability (larger $\bar{\beta}_n$) is required for the structures.
(3) $\mathscr{E}$ is proportional to the relative numerical error Δ.
(4) If $\bar{\beta}_n \delta_W^0 = 0$, $\mathscr{E}$ is infinitely large unless the relative numerical error $\Delta = 0$. In other words, approximations are not allowed if no margin of safety exists for the structure, or if the design load coincides deterministically with the service load. This is probably due to the upper-bounding procedure used to obtain $\mathscr{E}_u$ independently of analysis of local effects; in such a case, analysis of the propagation of the error on local effects cannot be avoided.

It is interesting to note that in the particular case that all the components of the actual load vector are stochastically independent and have the same expected value and variance, the upper bound becomes

$$\mathscr{E}_{\mathrm{u}} = \frac{1+C_{\mathrm{n}}\delta_{\mathrm{w}}}{C_{\mathrm{n}}\delta_{\mathrm{w}}} \tag{11.39}$$

δ_{w} being the common value of the coefficient of variation of each single load component, and C_{n} a coefficient slightly dependent on n and on the prescribed safety index $\bar{\beta}$. Typical plots of the ratio $\mathscr{E}_{\mathrm{u}}/\Delta$ versus δ_{w} for a number of values of n and a given $\bar{\beta}$ are shown in Fig. 11.3.

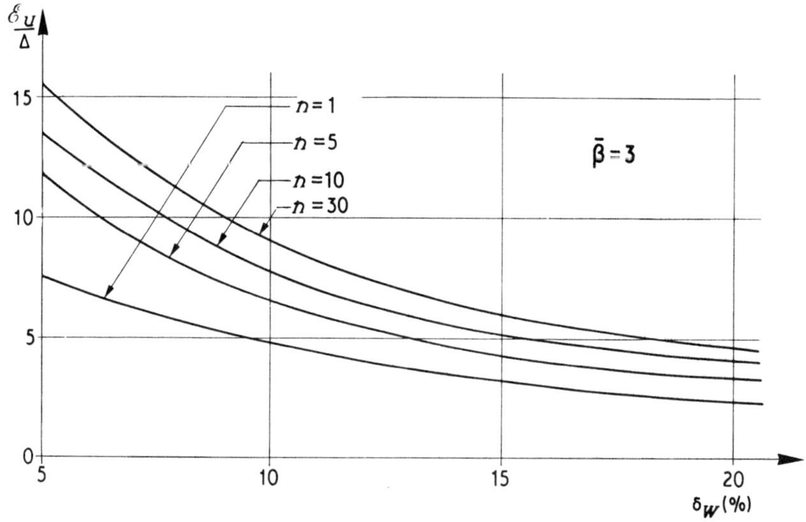

Fig. 11.3 Relationship between reliability error and coefficient of variation of the loads, for prescribed reliability index $\bar{\beta}=3$ and different dimensions n of the problem (from Baratta [5]).

It can further be shown that, under the same conditions as above, an evaluation (not necessarily an upper bound) of the reliability error is provided by

$$\mathscr{E} \simeq \Delta \frac{\gamma^0}{\gamma^0 - 1} \tag{11.40}$$

γ^0 being the *central safety factor*, defined by

$$\gamma^0 = \frac{|\mathbf{W}_{\mathrm{D}}|}{|\boldsymbol{\mu}_{\mathrm{w}}|} \tag{11.41}$$

Fig. 11.4 shows plots of the *additional* safety factor $\gamma^* = 1+\mathscr{E}_{\mathrm{u}}$ (equations (11.33) and (11.34)) versus the *central* safety factor γ° for a number of values of the relative numerical error Δ. equation (11.36).

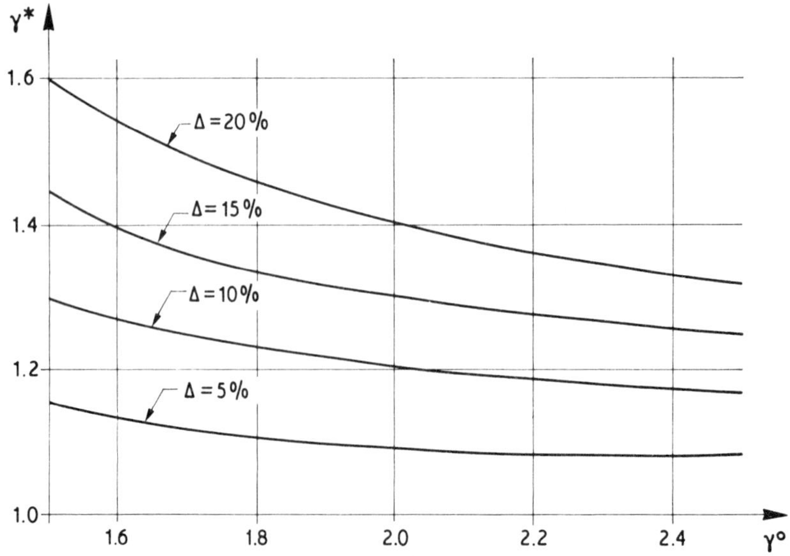

Fig. 11.4 Relationship between additional and central safety factor for different values of the relative numerical error Δ (from [5]).

11.3.2 Some applications

In the preceding section, it has been shown that the evaluation of the reliability error is a rather simple task, provided that the relative numerical error between actual and emerging loads Δ is available.

Let us now look at two examples showing how the matter can be tackled in practical instances.

(a) Iterative methods for solution of linear equations

Consider a discrete linearly elastic structure: then equation (11.29) reduces to a system of linear equations (equation (4.1)), with **K** an $n \times n$ non-singular matrix. Let this system of linear equations be solved by an iterative procedure stopped after a small number h of iterations: the consequent error is sought. Any iterative solution of linear equations is founded on a decomposition of the matrix **K** of the type

$$\mathbf{K} = \mathbf{X} + \mathbf{Y} \tag{11.42}$$

with **X** non-singular. Then, it can be proved [5] that a well approximated value for Δ is given by

$$\Delta = \frac{\Theta^h}{1 - \Theta^h} \tag{11.43}$$

with Θ the maximum eigenvalue of the matrix $\mathbf{Z}^T\mathbf{Z}$, **Z** being the product $-\mathbf{Y}\mathbf{X}^{-1}$. Note that in structural problems, Θ is always smaller than unity. In Fig. 11.5, equation (11.43) is plotted for several values of Θ. For instance, if

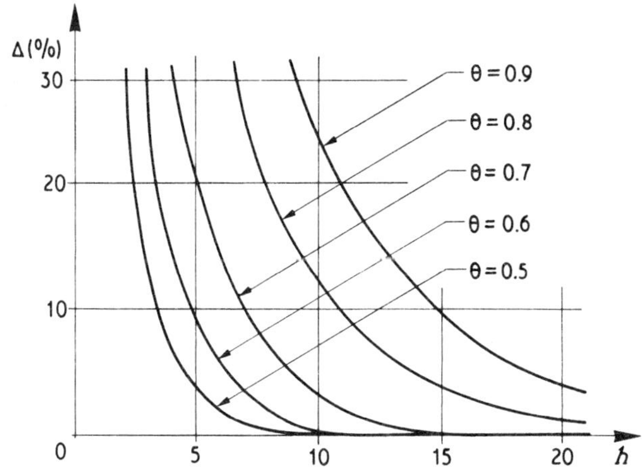

Fig. 11.5 Error vs number of iterations in solution of linear equations (from [5]).

$\Theta = 0.9$ and the procedure is stopped after 15 iterations, $\Delta = 10\%$. Then, from Fig. 11.4, assuming $\gamma^\circ = 2$, an additional safety factor $\gamma^* = 1.2$ should be applied to the design loads.

(b) Errors in the evaluation of the stiffness matrix

Consider again a linear elastic structure. Let **K** be the actual stiffness matrix (unknown, by assumption) and $\bar{\mathbf{K}}$ a calculated matrix. In this case, then, the error source lies in the process of constructing the stiffness matrix, while it is assumed that an exact mathematical procedure is employed in solving the equations.

Let $\bar{\mathbf{K}}$ be an approximation of **K** in the sense that

$$\mathbf{K} = \bar{\mathbf{K}} + \zeta \mathbf{e} \tag{11.44}$$

ζ being a number smaller than unity, and **e** an *error matrix* such that

$$|\mathbf{e}| = |\bar{\mathbf{K}}| \tag{11.45}$$

It can be seen [5] that in this case the operator **D** is defined by

$$\mathbf{D} = \zeta \mathbf{e} \bar{\mathbf{K}}^{-1} \tag{11.46}$$

and the relative numerical error by

$$\Delta = \frac{\zeta r'}{1 - \zeta r'} \tag{11.47}$$

with $r' = |\bar{\mathbf{K}}|\,|\bar{\mathbf{K}}^{-1}|$ the condition number of the approximate stiffness matrix $\bar{\mathbf{K}}$. Then, applying equation (11.40), the reliability error is estimated by

$$\mathscr{E} \simeq \frac{\gamma^0}{\gamma^0 - 1} \frac{\zeta r'}{1 - \zeta r'} \tag{11.48}$$

11.4 PROBLEMS IN THE DISCRETIZATION OF UNCERTAIN STRUCTURES

As well known (cf. Section 4.1) many structural problems are studied on a *discretized* model: for instance, a frame is divided into linear elements (*beams* and *columns*), a plate into *finite elements*, and so on. When the mechanical and geometrical properties are random, so will be those of each element. And unless the properties in each point of the structure are perfectly correlated with each other, the study of the random structure would require not only the probabilistic properties of the single elements, but also the cross-properties between different elements. In the technical literature, this question (which was anticipated in Section 4.1) has so far received little attention, mostly paid to the determination of the structural *strength* (cf. Section 1.5): the discretization of a random field in a more general structural problem is illustrated below on an elementary example

11.4.1 Elongation of an elastic tie

(a) Formulation of the problem

Consider the elastic rod in Fig. 11.6, subjected to a well determinated tensile load W; each elementary length $\mathrm{d}x$ of the rod elongates by the quantity

$$\mathrm{d}\Delta l = \frac{W}{EA}\mathrm{d}x = \varepsilon \mathrm{d}x \tag{11.49}$$

The quantity EA (*elongation stiffness*) may vary stochastically along the rod axis (Fig. 11.6(a)). Put‡

$$\tilde{e}(x) = \frac{1}{E\tilde{A}(x)} \tag{11.50}$$

and assume that $\tilde{e}(x)$ is a segment from a stochastic process with expected value $\mu_e(x)$ and autocovariance function $\Sigma_e(x_1, x_2)$.

Consider first the bar as a continuous system, and calculate the properties of the total elongation Δl. Integrating equation (11.49) gives

$$\widetilde{\Delta l} = \int_0^l \frac{W}{E\tilde{A}}\mathrm{d}x = \int_0^l W\tilde{e}(x)\,\mathrm{d}x \tag{11.51}$$

It turns out that $\widetilde{\Delta l}$ is a random variable with expected value

$$\mu_{\Delta l} = W\int_0^l \mu_e(x)\,\mathrm{d}x \tag{11.52}$$

‡ The random variations of Young's modulus E are usually negligible with respect to those of the cross-section A.

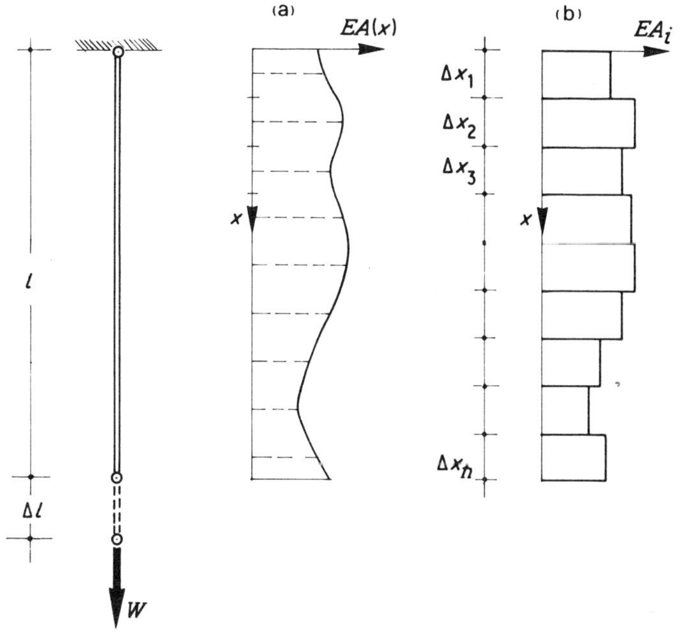

Fig. 11.6 Tie with random elongation stiffness *EA*.

and variance

$$s_{\Delta l}^2 = W^2 \int_0^l \int_0^l \Sigma_e(x_1, x_2)\,dx_1\,dx_2 \tag{11.53}$$

Assume now that continuous solutions are not tractable, as it may be for more complicated structural patterns.

A possible simplification, usual in most engineering problems, is to convert the original structure into a discretized one.

Subdivision of the rod into a chain of small constant-stress/constant-strain elements is equivalent to assuming a piecewise-constant $\widetilde{EA}(x)$ (Fig. 11.6(b)): the values of $\tilde{e}_i = 1/\widetilde{EA}_i$ in the different elements constitute a set of n random variables whose statistical properties should be related to the underlying random process $\tilde{e}(x)$.

Let $\tilde{\varepsilon}_i$, the strain in the ith element, be defined by averaging the actual strain on the length $\Delta x_i = x_i - x_{i-1}$

$$\tilde{\varepsilon}_i = \frac{1}{\Delta x_i} \int_{x_{i-1}}^{x_i} \tilde{\varepsilon}(x)\,dx = \frac{W}{\Delta x_i} \int_{x_{i-1}}^{x_i} \tilde{e}(x)\,dx \tag{11.54}$$

The probabilistic properties of $\tilde{\varepsilon}_i$ can be obtained by the theory of integration of stochastic processes, namely by equations (A5.16) to (A5.20), whence it is easy to write

$$\mu_{\varepsilon_i} = \frac{W}{\Delta x_i} \int_{x_{i-1}}^{x_i} \mu_e(x)\, dx \tag{11.55a}$$

$$s^2_{\varepsilon_i} = \frac{W^2}{\Delta x_i^2} \int_{x_{i-1}}^{x_i} \int_{x_{i-1}}^{x_i} \Sigma_e(x_1, x_2)\, dx_1\, dx_2 \tag{11.55b}$$

$$\mathrm{Cov}[\tilde{\varepsilon}_i, \tilde{\varepsilon}_j] = \frac{W^2}{\Delta x_i \Delta x_j} \int_{x_{j-1}}^{x_j} dx_1 \int_{x_{i-1}}^{x_i} \Sigma_e(x_1, x_2)\, dx_2 \tag{11.55c}$$

The statistics of $e_i = 1/EA_i$ in the single elements are then derived from the above formulae remembering that

$$\mu_{\varepsilon_i} = W\mu_{e_i} \qquad s^2_{\varepsilon_i} = W^2 s^2_{e_i} \qquad \mathrm{Cov}[\tilde{\varepsilon}_i, \tilde{\varepsilon}_j] = W^2\, \mathrm{Cov}[\tilde{e}_i, \tilde{e}_j] \tag{11.56}$$

whence

$$\mu_{e_i} = \frac{1}{\Delta x_i} \int_{x_{i-1}}^{x_i} \mu_e(x)\, dx \tag{11.57a}$$

$$s^2_{e_i} = \frac{1}{\Delta x_i^2} \int_{x_{i-1}}^{x_i} \int_{x_{i-1}}^{x_i} \Sigma_e(x_1, x_2)\, dx_1\, dx_2 \tag{11.57b}$$

$$\mathrm{Cov}[\tilde{e}_i, \tilde{e}_j] = \frac{1}{\Delta x_i \Delta x_j} \int_{x_{j-1}}^{x_j} dx_1 \int_{x_{i-1}}^{x_i} \Sigma_e(x_1, x_2)\, dx_2 \tag{11.57c}$$

Equations (11.57) yield an evaluation of the probabilistic properties of the sequence of random variables $\tilde{e}_1, \ldots, \tilde{e}_n$ into which the stochastic process $\tilde{e}(x)$ has been broken.

It is hardly questionable, from the above formulae, that in general such variables should be correlated, at least in a Markov sense (i.e. e_i correlated to $\tilde{e}_{i-1}$ and $\tilde{e}_{i+1}$).

However, remember that, as already stated (Chapters 2 and 4), often the autocovariance function $\Sigma_e(x_1, x_2)$ tends to become smaller and smaller as $x_1 - x_2$ increases, so that for $|x_1 - x_2|$ larger than some limit d_{ext} (the *extinction length*, Section 2.4.3) it is reasonably possible to set $\Sigma_e(x_1, x_2) = 0$. Thus, making Δx rather large, the integrand in equation (11.57c) tends to vanish, and the properties of different elements tend to become uncorrelated, thus making the handling of the structural equations much simpler.

Unfortunately, the larger is Δx, the worse is the capability of the discretized model to approach the behaviour of the original structure. Therefore, when accurate calculations are considered necessary, random quantities describing mechanical properties in every finite element should be properly correlated.

Of course, the evaluation of these correlations is reliable only when appropriate and sufficient data are available. If on the contrary this is not the case, as in many practical problems, it is preferable either to choose the size of the elements such that the assumption of stochastic independence is acceptable, or, at the other extreme, to assume perfect correlation between the properties of the single elements. The latter approach can be followed when the uncertainty of the relevant property is larger *among* elements of the same population than *along* each element: this may be assumed, for instance, if the rod of Fig. 11.6 is drawn at random out of a batch from a mill, of which only *nominal* properties are known *a priori*. In this case, the stochastic processes of the more general approach are reduced to random variables, and the discretization does not pose any problem connected with the randomness of the mechanical properties.

(b) A numerical application

With reference to the tie in Fig. 11.6, let $\tilde{e}(x)$ $(x \in 0, l)$ be a segment from a stationary random process with mean value

$$\mu_e = 1.5158 \times 10^{-7}\ \mathrm{kg}^{-1}$$

variance

$$s_e^2 = 2.2975 \times 10^{-16}\ \mathrm{kg}^{-2}$$

and autocovariance function

$$\Sigma_e(x_1, x_2) = s_e^2 \exp(-\zeta |x_1 - x_2|)$$

Assume that the extinction length d_{ext} corresponds to

$$\Sigma_e(x_1, x_2) = 0.01\, s_e^2$$

and put

$$l = 1000\ \mathrm{cm}$$
$$\zeta = 4.6052 \times 10^{-2}\ \mathrm{cm}^{-1}$$

Hence

$$d_{\mathrm{ext}} = 100\ \mathrm{cm} = 0.1l$$

Put finally

$$W = 6280\ \mathrm{kg}$$

and let us calculate the statistics of the elongation $\widetilde{\Delta l}$.

From equations (11.52) and (11.53) one gets

$$\mu_{\Delta l} = W \mu_e l = 0.95192\ \mathrm{cm}$$

$$\begin{aligned} s_{\Delta l}^2 &= W^2 \int_0^l \int_0^l s_e^2 \exp(-\zeta |x_1 - x_2|)\,\mathrm{d}x_1\,\mathrm{d}x_2 \\ &= \frac{2W^2 s_e^2}{\zeta^2}(\zeta l + \mathrm{e}^{-\zeta l} - l) \\ &= 3.84968 \times 10^{-4}\ \mathrm{cm}^2 \end{aligned}$$

$$s_{\Delta l} = 1.96206 \times 10^{-2}\ \mathrm{cm}$$

If the tie is discretized in n finite segments, by equation (11.54), one gets, for constant normal force in every cross section of the tie,

$$\widetilde{\Delta l} = W \sum_{i=1}^{n} \tilde{e}_i \Delta x_i$$

The discretization made following equations (11.50) to (11.57) yields the exact result under constant normal force, for any n, and introduces approximations only under variable normal force. In particular, the statistics of the random variables denoting the mean value of $\tilde{e}(x)$ in every segment Δx_i are given by

$$\mu_{ei} = \mu_e$$

$$\mathrm{Cov}[\tilde{e}_i, \tilde{e}_j] = \frac{s_e^2}{\zeta^2 \Delta x_i \Delta x_j} [\exp(-\zeta x_{i-1}) - \exp(-\zeta x_i)][\exp(\zeta x_j) - \exp(\zeta x_{j-1})] \qquad i > j$$

$$s_{ei}^2 = \frac{2 s_e^2}{\zeta^2 (\Delta x_i)^2} [\zeta \Delta x_i + \exp(-\zeta \Delta x_i) - 1]$$

while the statistics of $\widetilde{\Delta l}$ would be expressed by

$$\mu_{\Delta l} = W \Sigma_{\mu_{ei}} \Delta x_i = W \mu_e l$$

$$s_{\Delta l}^2 = \sum_{i=1}^{n} \sum_{j=1}^{n} W^2 \Delta x_i \Delta x_j \mathrm{Cov}[\tilde{e}_i, \tilde{e}_j]$$

It is worth while to note that if the random variables are assumed independent (i.e. $\mathrm{Cov}[\tilde{e}_i, \tilde{e}_j] = I_{ij} s_{ei}^2$) the following results are obtained as an approximation $s_{\Delta l}^0$ for the standard deviation:

n	$\Delta x_i = l/n$ (cm)	d_{ext} (cm)	$s_{\Delta l}^0$ (cm)	$s_{\Delta l}$ (cm)
5	200	100	1.873×10^{-2}	1.962×10^{-2}
10	100	100	1.758×10^{-2}	1.962×10^{-2}
20	50	100	1.548×10^{-2}	1.962×10^{-2}
40	25	100	1.264×10^{-2}	1.962×10^{-2}
80	12.5	100	0.971×10^{-2}	1.962×10^{-2}

Thus it can be argued that the assumption of statistical independence is consistent only if the distance between adjacent elements is larger than the extinction length.

A simpler way to discretize the random process is to identify the random variables $\tilde{e}_i$ with the values that the process $\tilde{e}(x)$ takes in the centroids of the elements. Let x_i $(i = 1, \ldots, n)$ be the abscissa of the centroid of the ith element. The covariance matrix of the $\tilde{e}_i$ would be given in this case by

$$\mathrm{Cov}[\tilde{e}_i, \tilde{e}_j] = s_e^2 \exp(-\zeta |x_i - x_j|)$$

In this case the variance of $\widetilde{\Delta l}$

$$s_{\Delta l}^{*2} = \sum_{i=1}^{n} \sum_{j=1}^{n} W^2 \Delta x_i \Delta x_j \mathrm{Cov}[\tilde{e}_i, \tilde{e}_j]$$

does not coincide with $s_{\Delta l}^2$, not even for constant normal force. Convergence of $s_{\Delta l}^*$ to $s_{\Delta l}$ by increasing n can be shown by the following numerical results:

n	$\Delta x_i = l/n$	d_{ext}	$s^*_{\Delta l}$	$s_{\Delta l}$
5	200	100	4.257×10^{-2}	1.962×10^{-2}
10	100	100	3.037×10^{-2}	1.962×10^{-2}
20	50	100	2.341×10^{-2}	1.962×10^{-2}
40	25	100	2.070×10^{-2}	1.962×10^{-2}
80	12.5	100	1.990×10^{-2}	1.962×10^{-2}

Also in this case it is interesting to consider results deriving from neglecting correlations in the $\tilde{e}_i$. In this case, the standard deviation

$$s'^2_{\Delta l} = \sum_{i=1}^{n} W^2 (\Delta x_i)^2 s^{*2}_{ei}$$

has the following behaviour

n	Δx_i	d_{ext}	$s'_{\Delta l}$	$s^*_{\Delta l}$
5	200	100	4.257×10^{-2}	4.257×10^{-2}
10	100	100	3.010×10^{-2}	3.037×10^{-2}
20	50	100	2.128×10^{-2}	2.341×10^{-2}
40	25	100	1.505×10^{-2}	2.070×10^{-2}
80	12.5	100	1.064×10^{-2}	1.990×10^{-2}

The above results suggest that:

(1) Discretization is best performed by taking mean values of the random process on the finite elements.
(2) Proper correlations should be accounted for. When statistical independence is assumed, increasing the number of elements, i.e. improving the structural model, is in contrast with improving the approximation in the statistics of results.
(3) The correlation matrix of random variables in the finite elements can be built up by simply calculating covariances corresponding to single points in different elements, but very poor convergence is achieved.

11.5 GROSS ERRORS

In the last few years, it has become evident that the largest responsibility for most actual structural failures is the occurrence of errors that cause very significant differences of the observed failure frequency from the theoretical failure probability.

Recently, the study of available statistical surveys has suggested the following tentative indications [6]:

(1) the failure frequency is much larger than that established by neglecting gross errors;
(2) failure is almost always associated with human errors;
(3) most often more than one human error is associated with failure;
(4) human errors are also present in structures that do not fail.

Human errors occur with much larger frequency than large random variations of the material strength or of the applied loads: this is the logical conclusion of the above statements. In fact, actual observed failures seldom appear to be caused by insufficient member dimension or reinforcement: the choice of the latter, within normal practice, appears to influence only the *theoretical* failure probability (Section 10.3.2) [7].

In order to take into account the above features of observed structural failures, and in particular to fit the unexpectedly high frequency of failure to theoretical probabilities through the introduction of high probabilities of occurrence of human errors, a model has been proposed [6] by which the production of any building is looked at as the result of sequence of *activities*: the architect's, the engineer's, the contractor's and the inspector's. Each of the above operators is subject to error with his own probability of occurrence, with the condition that the architect's, engineer's or contractor's errors cannot have practical consequences if not associated with the inspector's error. Besides human errors, the probability of random variations of loads and strengths can also be taken into account by the model.

An example of a set of *credible* values attributed to the above probabilities is shown in Table 11.1, where a difference of about one order of magnitude has been adopted between human errors and abnormal variations in strength or loads.

Table 11.1 Probabilities of occurrence of errors as assumed in the example (from Lind [6]).

Event	*Probability*
A = Architect's error	0.1
E = Engineer's error	0.4
C = Contractor's error	0.5
I = Inspector's error	0.1
S = Excess load	0.02
M = Weak material	0.02
U = User error	0.02

The events in the table are assumed to be independent. Moreover, it is assumed that, as stated under point (4) above, the presence of a single error or abnormal deviation does not imply immediately the failure of the structure: it is to be thought, rather, that a collapse occurs as the result of the coincidence of two or more errors. Let m be the number of human errors having occurred during design and construction of the building and n the number of abnormal deviations in the structure–environment properties (the last three events in Table 11.1). A reasonable criterion for failure, taking into account points (1)–(4), is

$$m + n \geqslant 3 \tag{11.58}$$

The resulting probabilities for the single and joint events leading to failure are in Table 11.2. The symbols AI, EI, CI denote simultaneous occurrence of the inspector's error, together with the architect's (A), engineer's (E) and contractor's (C) respectively: by virtue of the assumed independence, introducing the values of Table 11.1,

$$\begin{aligned} \text{Prob}[AI] &= 0.1 \times 0.1 = 0.01 \\ \text{Prob}[EI] &= 0.4 \times 0.1 = 0.04 \\ \text{Prob}[CI] &= 0.5 \times 0.1 = 0.05 \end{aligned} \tag{11.59}$$

Table 11.2 Probabilities of joint occurrence of errors ($\times 10^6$) (from Lind [6]).

	(a)	(b)	(c)	(d)	(e)	(f)
$n =$	0	1	2	3	*Total*	*Failure*
1 $AI \cap EI \cap CI$	19	1	—	—	20	20
2 $AI \cap EI$	358	22	—	—	380	22
3 $AI \cap CI$	452	28	1	—	480	29
4 $EI \cap CI$	1864	114	2	—	1980	116
5 AI	8584	526	11	—	9120	11
6 EI	35408	2168	44	—	37620	44
7 CI	44725	2738	56	—	47520	56
8 No error	849783	52028	1068	8	902880	8
9 Total	941192	57624	1176	8	1000000	
10 Failures	19	165	114	8		306

The probability $P(\ )$ of occurrence of each possible combination of human errors is then easily obtained. For instance

$$\begin{aligned} P(AI \cap EI \cap CI) &= \text{Prob}[AI]\ \text{Prob}[EI]\ \text{Prob}[CI] \\ &= 0.01 \times 0.04 \times 0.05 = 2 \times 10^{-5} \end{aligned} \tag{11.60a}$$

$$\begin{aligned} P(AI \cap CI) &= \text{Prob}[AI]\ \text{Prob}[CI]\ (1 - \text{Prob}[EI]) \\ &= 0.01 \times 0.05 \times (1 - 0.04) = 4.80 \times 10^{-4} \end{aligned} \tag{11.60b}$$

$$\begin{aligned} P(AI) &= \text{Prob}[AI]\ (1 - \text{Prob}[EI])\ (1 - \text{Prob}[CI]) \\ &= 0.01\,(1 - 0.04)\,(1 - 0.05) = 9.12 \times 10^{-3} \end{aligned} \tag{11.60c}$$

Such values are reported in column (e) of Table 11.2, while columns (a), (b), (c) and (d) show the probabilities that each combination of human errors occurs in coincidence with a number of abnormal deviations $n = 0, 1, 2, 3$ respectively, as will be soon explained.

First, remembering the last three rows of Table 11.1,

$$\begin{aligned}\text{Prob}[\tilde{n} = 0] &= [1 - P(S)]\,[1 - P(M)]\,[1 - P(U)]\\ &= (0.98)^3 = 0.941\,192 \end{aligned} \tag{11.61a}$$

$$\begin{aligned}\text{Prob}[\tilde{n} = 1] &= P(S)[1 - P(M)]\,[1 - P(U)] + P(M)[1 - P(S)]\,[1 - P(U)]\\ &\quad + P(U)[1 - P(M)]\,[1 - P(S)]\\ &= 3 \times 0.02 \times 0.98^2 = 0.057\,624 \end{aligned} \tag{11.61b}$$

$$\begin{aligned}\text{Prob}[\tilde{n} = 2] &= P(S)P(M)[1 - P(U)] + P(S)P(U)\,[1 - P(M)]\\ &\quad + P(M)P(U)[1 - P(S)] = 0.001\,176 \end{aligned} \tag{11.61c}$$

$$\text{Prob}[\tilde{n} = 3] = P(S)P(M)P(U) = 0.000\,008 \tag{11.61d}$$

These values are reported in row 9 of Table 11.2.

Consider now the numbers in any of the rows 1 to 8, for instance those in row 3. These are given by

column (a):

$$\begin{aligned}P(AI \cap CI)\,\text{Prob}[\tilde{n} = 0] &= 4.80 \times 10^{-4} \times 0.941\,192\\ &= 4.52 \times 10^{-4} = 452 \times 10^{-6}\end{aligned}$$

column (b):

$$P(AI \cap CI)\,\text{Prob}[\tilde{n} = 1] = 2.77 \times 10^{-5} = 28 \times 10^{-6} \tag{11.62}$$

column (c):

$$P(AI \cap CI)\,\text{Prob}[\tilde{n} = 2] = 8.45 \times 10^{-7} = 1 \times 10^{-6}$$

column (d):

$$P(AI \cap CI)\,\text{Prob}[\tilde{n} = 3] = 3.84 \times 10^{-9} \simeq 0$$

The examined row 3 corresponds to the occurrence of two human errors ($m = 2$; AI and CI). Hence the structure fails, according to condition (11.58), when at least one abnormal deviation occurs ($n \geqslant 1$). Summing up the probabilities in columns (b), (c) and (d) (corresponding to $n = 1$, 2 and 3), one gets a total frequency of failures (29×10^{-6}) that can be associated to simultaneous occurrence of architect's and contractor's errors. This value, i.e. the sum of the probabilities on the right of the heavy line, is reported in column (f) for each row 1 to 8.

The above calculations are, as already noted, only a first crude attempt, based on the empirically estimated probabilities of Table 11.1: nevertheless the results are in good agreement with observations and are an encouraging premise to investigate the question further.

Table 11.2 underlines that the probability of failure in the absence of gross errors is rather small if compared to the overall probability of failure (about 8/306), thus confirming that mainly gross errors should be prevented in order to reduce the observed frequency of failure.

Thus, the question is how the probability of gross errors can be reduced. The matter has been approached recently [7], under the assumption that gross errors are affected by the choice of the structural layout, that should enter into the design procedure as a decisive step. Therefore, a criterion to measure the *proneness to failure* of any structural layout must be sought, in order to get elements for decision. It has been suggested that the question could be tackled by polyvalent logics and by the modern theory of *fuzzy sets*. The fuzzy set approach is a generalization of probabilistic methods that may be very attractive in perspective. At present, however, this technique is in very rapid evolution, and cannot be treated here.

In any case, while the present operative procedures of probabilistic structural engineering and the related codes of practice do not allow for explicit inclusion of gross errors (which are indirectly accounted for by raising the target reliability: cf. Section 10.3.2), it is fair to forecast that this question will be the object of much research in the near future, and perhaps *the* new thing in the next generation of codes.

REFERENCES

1. Bolotin, V. V. (1966) Statistical aspects in the theory of structural stability; in *Dynamic Stability of Structures* (ed. G. Hermann), Pergamon Press, New York.
2. Augusti, G., Baratta, A. and Casciati, F. (1979) Structural response under random uncertainties. *Proc. 3rd Int. Conf. on Applications of Statistics and Probability in Soil and Structural Engineering* (ICASP 3), Sydney, vol 3, pp. 280–328.
3. Bolotin, V. V. (1972) Reliability theory and stochastic stability; in *Stability* (ed. H. H. Leipholz), University of Waterloo Press, Ontario, pp. 385–442.
4. Lind, N. C. (1976) Approximate analysis and economics of structures. *J. Struct. Div., Proc. ASCE*, **102** (ST6).
5. Baratta, A. (1980) Engineering error evaluation in approximate structural analysis. *Meccanica*, **15**(4), 206–20.
6. Lind, N. C. (1979) Optimization, cost–benefit analysis, specifications. *Proc. 3rd Int. Conf. on Applications of Statistics in Soil and Structural Engineering* (ICASP 3), Sydney, vol. 3, pp. 373–84.
7. Ditlevsen, O. (1980) *Formal and Real Structural Safety: Influence of Gross Errors.* IABSE Proceedings, **P36**, pp. 185–204.

PART THREE

Appendices

A1

Elements of probability theory

These Appendices cover the mathematical background of probabilistic methods. However, too rigorous proofs would be beyond the scope of this book.

A1.1 BASIC AXIOMS

With reference to any random phenomenon, its *sample description space* $\mathscr{S}$ is the complete set of descriptions of all possible outcomes of the phenomenon.

An *event* E is any subset of the sample description space ($E \subset \mathscr{S}$). An event E has occurred if the observation outcome of the random phenomenon has a sample description that is a member of E. The space $\mathscr{S}$ is a subset of itself and is called the *certain event*.

Technical reasons do not usually permit all subsets of $\mathscr{S}$ to be events. Rather, one adopts a family $\mathscr{F}$ of subsets of $\mathscr{S}$ with the following properties: (a) $\mathscr{S} \in \mathscr{F}$; (b) $E^c \in \mathscr{F}$, E^c being the complement of any set $E \in \mathscr{F}$ (i.e. the event that E does not occur); (c) $\bigcup_{n=1}^{\infty} E_n \in \mathscr{F}$, $\bigcup_{n=1}^{\infty} E_n$ being the union of any sequence of sets E_1, $E_2, \ldots$ belonging to $\mathscr{F}$.‡

The smallest family of subsets of $\mathscr{S}$ with the properties (a), (b) and (c), and also containing as members all the sets in which one expects to be interested, is often sufficient to develop the mathematical theory of probability. Thus, if $\mathscr{S}$ is the real

‡ Note that the *intersection* ($E_1 \cap E_2$) of two events is defined as consisting of the descriptions that belong to both E_1 and E_2, whereas the *union* ($E_1 \cup E_2$) is defined as consisting of the descriptions that belong to at least one of the events E_1 and E_2. The following relations are properties of the operations $\cup$ and $\cap$ for any events E_1, E_2 and E_3:

Commutative law

$$E_1 \cup E_2 = E_2 \cup E_1 \qquad E_1 \cap E_2 = E_2 \cap E_1$$

Associative law

$$E_1 \cup (E_2 \cup E_3) = (E_1 \cup E_2) \cup E_3 \qquad E_1 \cap (E_2 \cap E_3) = (E_1 \cap E_2) \cap E_3$$

Distributive law

$$E_1 \cap (E_2 \cup E_3) = (E_1 \cap E_2) \cup (E_1 \cap E_3)$$
$$E_1 \cup (E_2 \cap E_3) = (E_1 \cup E_2) \cap (E_1 \cap E_3)$$

Idempotency law

$$E_1 \cup E_1 = E_1 \qquad E_1 \cap E_1 = E_1$$

Morgan's laws

$$(E_1 \cup E_2)^c = E_1^c \cap E_2^c \qquad (E_1 \cap E_2)^c = E_1^c \cup E_2^c$$

line, one adopts as $\mathscr{F}$ the family $\mathscr{B}$ of Borel sets‡, where $\mathscr{B}$ is defined as the smallest family of sets of real numbers that possesses properties (a), (b) and (c) and contains as members all intervals.

Then one defines for each event $E \in \mathscr{F}$ a non-negative real number that is denoted by Prob[E] and is called the 'probability that E will occur' (i.e. the probability that an observed outcome of the random phenomenon is a member of E). Thus, in a sample description space $\mathscr{S}$, *probability* is a function that assigns a non-negative real number to every event E; the function Prob[ ] must satisfy three axioms:

Axiom 1

$$\text{Prob}[E] \geqslant 0, \text{ for any } E$$

Axiom 2

$$\text{Prob}[\mathscr{S}] = 1, \ \mathscr{S} \text{ being the certain event}$$

Axiom 3

For any sequence of events $E_1, E_2, \ldots, E_n, \ldots$ which are mutually exclusive (that is $(E_j \cap E_k) = \varnothing$ for any $j \neq k$, where $\varnothing$ denotes the *empty set*)

$$\text{Prob}\left[\bigcup_{j=1}^{\infty} E_j\right] = \sum_{j=1}^{\infty} \text{Prob}[E_j] \tag{A1.1}$$

A *probability space* $\mathscr{S}^{\mathrm{P}}$ is a sample description space on which a probability function (satisfying axioms 1, 2 and 3) has been defined.

The following formulae may be proved to hold for any E_1 and E_2 on a probability space:

Probability of the impossible event

$$\text{Prob}[\varnothing] = 0 \tag{A1.2}$$

Probability of the difference of two events

$$\text{Prob}[E_1 \cap E_2^c] = \text{Prob}[E_1] - \text{Prob}[E_1 \cap E_2] \tag{A1.3}$$

Probability of the complement of an event

$$\text{Prob}[E_1^c] = 1 - \text{Prob}[E_1] \tag{A1.4}$$

Probability of the union of two events

$$\text{Prob}[E_1 \cup E_2) = \text{Prob}[E_1] + \text{Prob}[E_2] - \text{Prob}[E_1 \cap E_2] \tag{A1.5}$$

Inequality for the probability of a subevent

$$\text{Prob}[E_1 \cap E_2^c) = \text{Prob}[E_1] - \text{Prob}[E_2] \qquad \text{if } E_2 \subset E_1 \tag{A1.6}$$

$$\text{Prob}[E_2] \leqslant \text{Prob}[E_1] \qquad \text{if } E_2 \subset E_1 \tag{A1.7}$$

‡ A *Borel set* is a set, bounded or not, that can be obtained from the open and closed sets by a finite or infinite sequence of unions, intersections and/or complementations.

Probability that 1 of 2 events will occur

$$\text{Prob}[(E_1 \cap E_2^c) \cup (E_2 \cap E_1^c)] = \text{Prob}[E_1] + \text{Prob}[E_2] - 2\,\text{Prob}[E_1 \cap E_2] \tag{A1.8}$$

Moreover, let E_1 and E_2 be two events on a probability space; the *conditional probability* of the event E_2, given the event E_1, denoted by $\text{Prob}[E_2|E_1]$, is defined by

$$\text{Prob}[E_2|E_1] = \frac{\text{Prob}[E_2 \cap E_1]}{\text{Prob}[E_1]} \qquad \text{if } \text{Prob}[E_1] \geqslant 0 \tag{A1.9}$$

while, if $\text{Prob}[E_1] = 0$, $\text{Prob}[E_2|E_1]$ is undefined.

The conditional probability can be interpreted as the probability that E_2 will occur, under the assumption that E_1 has occurred. In other words $\text{Prob}[E_2|E_1]$ represents our re-evaluation of the probability of E_2 in the light of the information that E_1 has occurred.

Then, if $E_1 \Rightarrow E_2$, i.e. if $\text{Prob}[E_2|E_1] = 1$, equation (A1.9) implies

$$\text{Prob}[E_2] = \frac{\text{Prob}[E_1 \cap E_2]}{\text{Prob}[E_1|E_2]} \geqslant \text{Prob}[E_1 \cap E_2] = \text{Prob}[E_1] \tag{A1.9$'$}$$

A1.2 INDEPENDENT EVENTS

If $\text{Prob}[E_1] > 0$, the event E_2 is said to be *stochastically independent* of the event E_1 if and only if

$$\text{Prob}[E_2|E_1] = \text{Prob}[E_2] \tag{A1.10}$$

Then, if both E_1 and E_2 have positive probability, from equation (A1.9)

$$\text{Prob}[E_1 \cap E_2] = \text{Prob}[E_2|E_1]\,\text{Prob}[E_1]$$

and from equation (A1.10)

$$\text{Prob}[E_1 \cap E_2] = \text{Prob}[E_2]\,\text{Prob}[E_1] \tag{A1.11}$$

Therefore the events E_1 and E_2 are stochastically independent if and only if equation (A1.11) holds. It should be noted that two mutually exclusive events E_1 and E_2 are independent if and only if $\text{Prob}[E_1]\,\text{Prob}[E_2] = 0$, which is so if and only if either E_1 or E_2 has probability zero.

Extensions of equations (A1.9), (A1.10) and (A1.11) to n events $E_1, E_2, \ldots, E_n$ are immediate. Further n families of events ($\mathscr{F}_1, \mathscr{F}_2, \ldots, \mathscr{F}_n$, whose members are events on some sample description space $\mathscr{S}$) are said to be independent if any set of n events $E_1, E_2, \ldots, E_n$ (where E_1 is selected from $\mathscr{F}_1$, E_2 is selected from $\mathscr{F}_2, \ldots$) is independent in the sense that satisfies the relation:

$$\text{Prob}[E_1 \cap E_2 \cap \ldots \cap E_n] = \text{Prob}[E_1]\,\text{Prob}[E_2] \ldots \text{Prob}[E_n] \tag{A1.12}$$

Suppose now that each description in a probability space is an n-tuple. More explicitly, let the described random phenomenon be defined as consisting of n

trials, and let $\mathscr{F}_k$, $k = 1, 2, \ldots, n$ be the family of events on $\mathscr{S}$ that depends on the kth trial. The n trials are independent if the families of events $\mathscr{F}_1, \mathscr{F}_2, \ldots, \mathscr{F}_n$ are independent, i.e. if equation (A1.12) holds for every set of events $E_1, E_2, \ldots, E_n$, such that (for $k = 1, 2, \ldots, n$) E_k depends only on the kth trial.

An experiment which has only two possible outcomes (success $\mathscr{s}$ and failure $\mathscr{f}$, with $\text{Prob}[\mathscr{s}] = p$, $\text{Prob}[\mathscr{f}] = q$ and $p+q = 1$) is called a *Bernoulli trial* with sample description space $\mathscr{S} = \{\mathscr{s}, \mathscr{f}\}$. Consider now the sample description space $\mathscr{S}_n$ of independent repeated Bernoulli trials $E_1, E_2, \ldots, E_n$: it contains 2^n descriptions each of n-tuple with $\mathscr{s}$ or $\mathscr{f}$ as single member. Then, for the relevant probability space $\mathscr{S}_n^{\mathrm{P}}$:

$$\text{Prob}[E_1 \cap E_2 \cap \ldots \cap E_n] = p^k q^{n-k} \tag{A1.13}$$

where k is the number of successes $\mathscr{s}$ among the components of the description.

The event 'k successes in n trials' can happen in as many ways as k letters can be distributed among n places: that is, there are $\binom{n}{k}$ descriptions containing k successes. The probability $\mathscr{b}(k; n, p)$ that n independent repeated Bernoulli trials, with probabilities p for success and $q = 1-p$ for failure, will result in k successes and $n-k$ failures ($k = 0, \ldots, n$) is given by

$$\mathscr{b}(k; n, p) = \binom{n}{k} p^k q^{n-k} \tag{A1.14}$$

and is called the *binomial law*.

The binomial law may also be approximated, under certain circumstances, by the Poisson probability law. A random phenomenon whose $\mathscr{S}$ consists of all the integers from 0 onwards ($\mathscr{S} = \{0, 1, 2, \ldots\}$), with a probability function defined in terms of a parameter $\lambda > 0$ by

$$\eta(k; \lambda) = \mathrm{e}^{-\lambda} \frac{\lambda^k}{k!} \qquad k = 0, 1, 2, \ldots \tag{A1.15}$$

is said to obey the *Poisson probability law*.

Since

$$\lim_{n \to \infty} \binom{n}{k} p^k (1-p)^{n-k} = \mathrm{e}^{-np} \frac{(np)^k}{k!}$$

for large values of n equation (A1.14) is approximated by equation (A1.15) with $\lambda = np$. (In practice, this approximation is used if $p \leqslant 0.1$.)

Consider next the independent trials of an experiment with r (several) possible outcomes and let p_k ($k = 1, \ldots, r$, $\sum p_k = 1$) be the probability that at each trial the kth outcome will occur. Corresponding to the binomial law it is possible to introduce the *multinomial law*: the probability that in n trials the first outcome will occur k_1 times, the second outcome will occur k_2 times, ..., the rth outcome will occur k_r times, for any $k_j > 0$ with $\sum_{j=1}^{r} k_j = n$, is given by

$$\mathscr{b}(k_1, k_2, \ldots, k_r; n, p_1, p_2, \ldots, p_r) = \frac{n!}{k_1!\, k_2! \ldots k_r!} p_1^{k_1} p_2^{k_2} \ldots p_r^{k_r} \tag{A1.16}$$

A1.3 DEPENDENT TRIALS

Consider an event E that may be written as the intersection ($E = E_1 \cap E_2 \cap E_3 \cap \ldots \cap E_n$) of dependent events $E_1, E_2, \ldots, E_n$; the probability function is expressed by the multiplicative rule

$$\text{Prob}[E] = \text{Prob}[E_1]\text{Prob}[E_2|E_1] \ldots \text{Prob}[E_n|E_1, E_2, \ldots, E_{n-1}] \quad \text{(A1.17)}$$

Consequently, to know the value Prob$[E]$ for any event E it suffices to know Prob$[E_1]$ for any event E_1 depending on the first trial and, for $k = 2, \ldots, n$, the conditional probability Prob$[E_k|E_1, E_2, \ldots, E_{k-1}]$ of any event E_k depending on the kth trial, given any events $E_1, E_2, \ldots, E_{k-1}$ depending on the 1st, 2nd, ..., $(k-1)$th trial respectively.

Assume that $E_1, E_2, \ldots, E_n$ are n mutually exclusive events, with Prob$[E_j] > 0$, and are also *exhaustive* (that is $\bigcup_{j=1}^{n} E_j = \mathscr{S}$ is the certain event). By using equation (A1.17), the unconditional probability Prob$[E^\circ]$ for any event E° may be expressed in terms of the conditional probabilities Prob$[E^\circ|E_1], \ldots$, Prob$[E^\circ|E_n]$ and the unconditional probabilities Prob$[E_1], \ldots$, Prob$[E_n]$:

$$\text{Prob}[E^\circ] = \text{Prob}[E^\circ|E_1]\,\text{Prob}[E_1] + \ldots + \text{Prob}[E^\circ|E_n]\,\text{Prob}[E_n] \quad \text{(A1.18)}$$

An interesting consequence of equation (A1.18) is called *Bayes's theorem*:

$$\text{Prob}[E_i|E^\circ] = \frac{\text{Prob}[E^\circ \cap E_i]}{\text{Prob}[E^\circ]} = \frac{\text{Prob}[E^\circ|E_i]\,\text{Prob}[E_i]}{\sum_{j=1}^{n} \text{Prob}[E^\circ|E_j]\,\text{Prob}[E_j]} \quad \text{(A1.19)}$$

If the event E_i is regarded as a 'cause', equation (A1.19) expresses the probability that the event E°, which has occurred, is the result of the cause E_i. However, one will rarely know all the r.h.s. probabilities.

Moreover, a sequence of trials of an experiment which has only two possible outcomes ($\mathscr{s}$ or $\mathscr{f}$) are defined as *Markov dependent Bernoulli trials* if

$$\text{Prob}[E_{k+1}|E_k, E_{k-1}, \ldots, E_1] = \text{Prob}[E_{k+1}|E_k] \quad \text{(A1.20)}$$

In other words, if at the kth trial the conditional probability of the outcome of the next trial does not depend on what has happened in past trials but only on what is happening at the present time (the trials have no memory). If Prob $(\mathscr{s}, \mathscr{s}) = \text{Prob}[E_{k+1} \equiv \mathscr{s}|E_k \equiv \mathscr{s}]$,[‡] Prob $(\mathscr{f}, \mathscr{s}) = \text{Prob}[E_{k+1} \equiv \mathscr{s}|E_k \equiv \mathscr{f}]$, Prob $(\mathscr{f}, \mathscr{f}) = \text{Prob}[E_{k+1} \equiv \mathscr{f}|E_k \equiv \mathscr{f}]$ and Prob $(\mathscr{s}, \mathscr{f}) = \text{Prob}[E_{k+1} \equiv \mathscr{f}|E_k \equiv \mathscr{s}]$ are quantities independent of k, the Markov dependent Bernoulli trials are said to be *repeated*. The above conditional probabilities, in addition to the quantity Prob$[E_1 \equiv \mathscr{s}]$, suffice to specify the probability of any event for Markov dependent repeated Bernoulli trials (note that for independent Bernoulli trials the probability p of success at each trial is required). In fact, for $k = 2, 3, \ldots, n$,

$$\text{Prob}[E_k] = \text{Prob}[E_{k-1}]\,\text{Prob}[E_k|E_{k-1}] + \text{Prob}[E^c_{k-1}]\,\text{Prob}[E_k|E^c_{k-1}]$$

‡ $E_k \equiv \mathscr{s}$ stands for 'the outcome of the kth event (trial) is $\mathscr{s}$.'

and hence

$$\begin{aligned}\text{Prob}\,[E_k \equiv \mathscr{s}] &= \text{Prob}[E_{k-1} \equiv \mathscr{s}]\,\text{Prob}(\mathscr{s},\mathscr{s}) + \text{Prob}[E_{k-1} \equiv \mathscr{f}]\,\text{Prob}(\mathscr{f},\mathscr{s})\cdot \\ &= \text{Prob}\,[E_{k-1} \equiv \mathscr{s}]\,\{\text{Prob}\,(\mathscr{s},\mathscr{s}) - [1 - \text{Prob}\,(\mathscr{f},\mathscr{f})]\} \\ &\quad + \{1 - \text{Prob}\,(\mathscr{f},\mathscr{f})\}\end{aligned} \tag{A1.21}$$

Then, under the assumption

$$|\text{Prob}\,(\mathscr{s},\mathscr{s}) + \text{Prob}\,(\mathscr{f},\mathscr{f}) - 1| < 1 \tag{A1.22}$$

equation (A1.21) gives

$$\begin{aligned}\text{Prob}\,[E_k \equiv \mathscr{s}] &= \left(\text{Prob}\,[E_1 \equiv \mathscr{s}] - \frac{1 - \text{Prob}\,(\mathscr{f},\mathscr{f})}{2 - \text{Prob}\,(\mathscr{s},\mathscr{s}) - \text{Prob}\,(\mathscr{f},\mathscr{f})}\right)\{\text{Prob}\,(\mathscr{s},\mathscr{s}) \\ &\quad + \text{Prob}\,(\mathscr{f},\mathscr{f}) - 1\}^{k-1} + \frac{1 - \text{Prob}\,(\mathscr{f},\mathscr{f})}{2 - \text{Prob}\,(\mathscr{s},\mathscr{s}) - \text{Prob}\,(\mathscr{f},\mathscr{f})}\end{aligned} \tag{A1.23}$$

In many cases the probability $\text{Prob}\,[E_1 \equiv \mathscr{s}]$ is not known and only the conditional probabilities can be computed (in the case that equation (A1.22) holds):

$$\begin{aligned}\text{Prob}\,[E_k \equiv \mathscr{s} | E_1 \equiv \mathscr{s}] &= \frac{1 - \text{Prob}\,(\mathscr{s},\mathscr{s})}{2 - \text{Prob}\,(\mathscr{s},\mathscr{s}) - \text{Prob}\,(\mathscr{f},\mathscr{f})}\{\text{Prob}\,(\mathscr{s},\mathscr{s}) \\ &\quad + \text{Prob}\,(\mathscr{f},\mathscr{f}) - 1\}^{k-1} + \frac{1 - \text{Prob}(\mathscr{f},\mathscr{f})}{2 - \text{Prob}(\mathscr{s},\mathscr{s}) - \text{Prob}(\mathscr{f},\mathscr{f})}\end{aligned} \tag{A1.24}$$

$$\begin{aligned}\text{Prob}[E_k \equiv \mathscr{s} | E_1 \equiv \mathscr{f}] &= \frac{1 - \text{Prob}\,(\mathscr{f},\mathscr{f})}{2 - \text{Prob}\,(\mathscr{s},\mathscr{s}) - \text{Prob}\,(\mathscr{f},\mathscr{f})}\{\text{Prob}\,(\mathscr{s},\mathscr{s}) \\ &\quad + \text{Prob}\,(\mathscr{f},\mathscr{f}) - 1\}^{k-1} + \frac{1 - \text{Prob}\,(\mathscr{f},\mathscr{f})}{2 - \text{Prob}\,(\mathscr{s},\mathscr{s}) - \text{Prob}\,(\mathscr{f},\mathscr{f})}\end{aligned} \tag{A1.25}$$

By interchanging the role of $\mathscr{s}$ and $\mathscr{f}$, one obtains similarly $\text{Prob}\,[E_k \equiv \mathscr{f} | E_1 \equiv \mathscr{s}]$ and $\text{Prob}\,[E_k \equiv \mathscr{f} | E_1 \equiv \mathscr{f}]$.

Further equation (A1.22) implies that

$$\lim_{k \to \infty} |\text{Prob}\,(\mathscr{s},\mathscr{s}) + \text{Prob}\,(\mathscr{f},\mathscr{f}) - 1|^{k-1} = 0$$

and hence for large values of k equation (A1.23) involves, by interchanging the role of $\mathscr{s}$ and $\mathscr{f}$,

$$\begin{aligned}\text{Prob}[E_k \equiv \mathscr{s}] &= \frac{1 - \text{Prob}\,(\mathscr{f},\mathscr{f})}{2 - \text{Prob}\,(\mathscr{s},\mathscr{s}) - \text{Prob}\,(\mathscr{f},\mathscr{f})} \\ \text{Prob}\,[E_k \equiv \mathscr{f}] &= \frac{1 - \text{Prob}\,(\mathscr{s},\mathscr{s})}{2 - \text{Prob}\,(\mathscr{s},\mathscr{s}) - \text{Prob}\,(\mathscr{f},\mathscr{f})}\end{aligned} \tag{A1.26}$$

In other words, after a large number of Markov dependent repeated Bernoulli trials have been performed, one is in a state of *statistical equilibrium* in the sense that the probability $\text{Prob}\,[E_k \equiv \mathscr{s}]$ of success on the kth trial is the same for all large values of k and indeed is functionally independent of the initial conditions, represented by Prob $[E_1 \equiv \mathcal{s}]$.

A1.4 MARKOV CHAINS

Consider now n trials of an experiment with r possible outcomes $\mathcal{s}_1, \mathcal{s}_2, \ldots, \mathcal{s}_r$ with $r > 2$. For $k = 1, 2, \ldots, n$ and $j = 1, 2, \ldots, r$, let $E_k^{(j)}$ be the event that on the kth trial outcome $\mathcal{s}_j$ occurred. The trials are defined as *Markov dependent* if, for any $k \in (1, n)$ and integers $j_1, j_2, \ldots, j_k \in (1, r)$, the events $E_k^{(j_k)}$ satisfy the condition:

$$\text{Prob}[E_k^{(j_k)} \mid E_{k-1}^{(j_{k-1})}, \ldots, E_1^{(j_1)}] = \text{Prob}[E_k^{(j_k)} \mid E_{k-1}^{(j_{k-1})}] \tag{A1.27}$$

Using an intuitively meaningful language, one speaks of observing at n times the 'state of a system' which has r possible states, numbered $1, 2, \ldots, r$, instead of speaking of n trials of an experiment with r possible outcomes. Let $E_k^{(j)}$ be the event that the system is in state j at time k. If equation (A1.27) holds, the system is a 'Markov chain' with r possible states. In other words, the *Markov chain* is a system without memory of the past, as the conditional probability of transition from one's present state to any other states does not depend on how one arrived in one's present state.

If the conditional probability $\text{Prob}(i, j) = \text{Prob}[E_k^{(j)} \mid E_{k-1}^{(i)}]$ is independent of k, the Markov chain is said to be *homogeneous* (which corresponds to the notion of Markov dependent repeated trials with r possible outcomes).

The transition probabilities $\text{Prob}(i, j)$ are best exhibited in the form of an $r \times r$ matrix called the *transition probability matrix*:

$$\mathbf{P} = \begin{bmatrix} \text{Prob}\ (1,1) & \text{Prob}\ (1,2) & \ldots & \text{Prob}\ (1,r) \\ \text{Prob}\ (2,1) & \text{Prob}\ (2,2) & \ldots & \text{Prob}\ (2,r) \\ \vdots & \vdots & & \vdots \\ \text{Prob}\ (r,1) & \text{Prob}\ (r,2) & \ldots & \text{Prob}\ (r,r) \end{bmatrix} \tag{A1.28}$$

If one defines the product $\mathbf{x} = \mathbf{yz}$ of two matrices by the usual row–column multiplication rule ($x_{ij} = y_{ik} z_{kj}$), the m-step transition probability matrix $\mathbf{P}_{(m)}$ of a homogeneous Markov chain, whose element lying at the intersection of the ith row and jth column is given by $\text{Prob}[E_{k+m}^{(j)} \mid E_k^{(i)}]$, may be expressed concisely by

$$\mathbf{P}_{(m)} = \mathbf{P}\mathbf{P}_{(m-1)} = \mathbf{P}_{(m-1)}\mathbf{P} \qquad m = 2, 3, \ldots \tag{A1.29}$$

Moreover, the unconditional probability vector $\mathbf{p}_{(k)}$, whose jth element expresses that at time k the Markov chain is in state j, is given, for $k \geqslant 2$, in terms of the initial unconditional probability vector $\mathbf{p}_{(1)}$

$$\mathbf{p}_{(k)} = (\mathbf{p}_{(1)}^{\mathrm{T}} \mathbf{P}_{(k-1)})^{\mathrm{T}} = (\mathbf{p}_{(1)}^{\mathrm{T}} \mathbf{P})^{(k-1)})^{\mathrm{T}} = (\mathbf{P})^{(k-1)})^{\mathrm{T}} \mathbf{p}_{(1)} \tag{A1.30}$$

In order to determine the $(k-1)$-step transition probability matrix, equation

(A1.30) shows that numerical results can be obtained by using the theory of eigenvalues and eigenvectors of matrices (Section 4.3).

A matrix **x** is defined as 'stochastic' if the sum of entries in any row $\sum_{j=1}^{r} x_{ij} = 1)$ and 'doubly stochastic' if in addition the sum of the entries in any column is equal to 1. The transition probability matrix **P** of a Markov chain is stochastic.

A homogeneous Markov chain with r states is *ergodic* if numbers $\pi_1, \pi_2, \ldots, \pi_r$ exist such that for any states i and j

$$\lim_{m \to \infty} \text{Prob}[E_{k+m}^{(j)} \mid E_k^{(i)}] = \pi_j \tag{A1.31}$$

i.e. as m tends to ∞, the elements of the m-step transition probability matrix $\mathbf{P}_{(m)}$ tend to a limit that depends only on the final state j and not on the initial state. Then equation (A1.31) implies

$$\lim_{k \to \infty} \text{Prob}[E_k^{(j)}] = \pi_j \tag{A1.32}$$

The vector π is called the *stationary probability vector*; its elements represent the probabilities of being in the various states after one has achieved the statistical equilibrium.

With reference to the conditions under which a homogeneous Markov chain is ergodic, the following theorems are stated:

(1) A homogeneous Markov chain with transition probability matrix **P** is ergodic if all entries of $\mathbf{P}_{(m)}$ are positive.
(2) A homogeneous Markov chain is ergodic if a state i exists such that $\text{Prob}[E_k^{(i)} \mid E_{k-1}^{(i)}] > 0$ and all states communicate.

Two states *communicate* if one can reach i from j and also j from i. In order to obtain the stationary probability vector $\boldsymbol{\pi}$, it may be shown that the solution of the system

$$\pi = \mathbf{P}^{\mathrm{T}}\pi \tag{A1.33}$$

satisfying the conditions

$$\pi_j \geqslant 0 \quad \text{for} \quad j = 1, 2, \ldots, r \qquad \sum_{j=1}^{r} \pi_j = 1 \tag{A1.34}$$

is unique and necessarily satisfies equations (A1.31) and (A1.32). Moreover if **P** is doubly stochastic

$$\pi_1 = \pi_2 = \ldots = \pi_r = 1/r \tag{A1.35}$$

An important class of non-ergodic Markov chains groups the chains possessing *absorbing states* (or barriers). A state j is said to be absorbing if $\text{Prob}(j, i) = 0$ for all $i \neq j$ (i.e. $\text{Prob}(j, j) = 1$) so that it is impossible to leave an absorbing state. If j is an absorbing state in a Markov chain with states $\{1, 2, \ldots, r\}$, then the *absorption conditional probabilities* $\Pi_j(i) = \text{Prob}[j \mid E_1^{(i)}]$ of ever arriving at the absorbing state j, given that one started from state i, $i = 1, \ldots, r$, are given by the vector $\mathbf{\Pi}_j$ that is unique solution to the following system of equations

$$\begin{aligned} &\Pi_j(j) = 1 & & \\ &\Pi_j(i) = 0 & &\text{if } j \text{ cannot be reached from } i \\ &\mathbf{\Pi}_j = \mathbf{P}\mathbf{\Pi}_j & &\text{otherwise} \end{aligned} \tag{A1.36}$$

A2

Random variables

A2.1 DEFINITIONS

A *numerical-valued random phenomenon* is a random phenomenon whose sample description space is the set $\mathscr{R}$, consisting of all real numbers from $-\infty$ to ∞. The set $\mathscr{R}$ may be represented geometrically by a real line: to every point of the line there corresponds a real number and vice versa.

A *random variable* $\tilde{x}$ is a real-valued function defined on a sample description space $\mathscr{S}$ with a probability function Prob[ ] defined on family $\mathscr{F}$ of subsets of $\mathscr{S}$. Further, for every Borel set B of real numbers, the set $\{S: \tilde{x}(s) \in B\}$ must be an event in $\mathscr{F}$.

The *probability function* of a random variable $\tilde{x}$ ‡, denoted by $\mathrm{Prob}_{\tilde{x}}[\ \ ]$, is a function defined for every Borel set B of real numbers by

$$\mathrm{Prob}_{\tilde{x}}[B] = \mathrm{Prob}[\{S: \tilde{x}(s) \in B\}] = \mathrm{Prob}[\tilde{x} \in B] \qquad \text{(A2.1)}$$

and it means the probability that an observed value of $\tilde{x}$ will be in B.

Two random variables $\tilde{x}$ and $\tilde{y}$ are *identically distributed* if their probability functions are equal: $\mathrm{Prob}_{\tilde{x}}[B] = \mathrm{Prob}_{\tilde{y}}[B]$ for all Borel sets B.

The *probability law* of $\tilde{x}$ is a probability function that coincides with $\mathrm{Prob}_{\tilde{x}}[\ \ ]$. The probability law of a random variable can always be specified by stating the *distribution function* (or cumulative distribution function (CDF)) $P_{\tilde{x}}(x)$ of the random variable $\tilde{x}$, defined for any real number by

$$P_{\tilde{x}}(x) = \mathrm{Prob}[\tilde{x} \leqslant x] \qquad \text{(A2.2)}$$

The distribution function must be non-decreasing ($P_{\tilde{x}}(a) \leqslant P_{\tilde{x}}(b)$ if $a < b$) and furthermore:

$$\lim_{x \to -\infty} P_{\tilde{x}}(x) = 0 \qquad \lim_{x \to +\infty} P_{\tilde{x}}(x) = 1 \qquad \lim_{x \to b+} P_{\tilde{x}}(x) = P_{\tilde{x}}(b) \qquad \text{(A2.3)}$$

‡ Throughout these appendices, as well as throughout the whole book, a superimposed tilde ($\tilde{}$) indicates *randomness* of the relevant quantity. It is sometimes omitted for simplicity if not confusing.

A random variable $\tilde{x}$ is said to be *discrete* or *continuous*, depending on whether its distribution function $P_{\tilde{x}}(x)$ is discrete or continuous.

A random variable $\tilde{x}$ is discrete if there exists a function $p_{\tilde{x}}^0(x)$, called the *probability mass function* (PMF) of $\tilde{x}$, in terms of which the probability function $\text{Prob}_{\tilde{x}}[\ \ ]$ may be expressed as a sum (see Fig. A.1(a)); for any Borel set B of real numbers

$$\text{Prob}_{\tilde{x}}[B] = \text{Prob}[\tilde{x} \in B] = \sum_{\substack{\text{over all} \\ x^+ \text{ in } B}} p_{\tilde{x}}^0(x) \tag{A2.4}$$

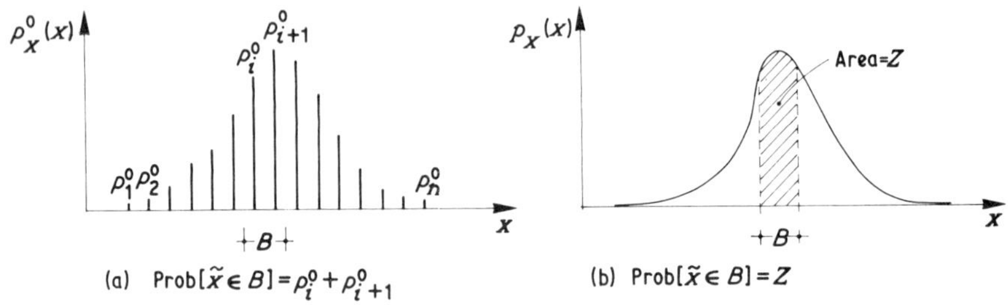

Fig. A.1 (a) Discrete random variable; (b) continuous random variable.

where x^+ denotes a value of x in B such that $p_{\tilde{x}}^0(x^+) > 0$. The real number x^+ is called a *probability mass point* (or a 'spectral point') of the random variable $\tilde{x}$ and

$$\sum_{\substack{\text{over all} \\ x^+}} p_{\tilde{x}}^0(x^+) = 1 \tag{A2.5}$$

$$P_{\tilde{x}}(x') = \sum_{\substack{\text{over all} \\ x^+ \leqslant x'}} p_{\tilde{x}}^0(x^+) \tag{A2.6}$$

It then follows that, for any real number x,

$$p_{\tilde{x}}^0(x) = \text{Prob}[\tilde{x} = x] = P_{\tilde{x}}(x) - \lim_{a \to x-} P_{\tilde{x}}(a) \tag{A2.7}$$

The distribution function $P_{\tilde{x}}(x)$ of a discrete random variable $\tilde{x}$ is a piecewise-constant or 'step' function with positive increments.

A random variable $\tilde{x}$ is continuous if there exists a function $p_{\tilde{x}}(x)$, called the *probability density function* (PDF) of $\tilde{x}$, in terms of which $\text{Prob}_{\tilde{x}}[\ \ ]$ may be expressed as an integral (see Fig. A.1(b)); for any Borel set B

$$\text{Prob}_{\tilde{x}}[B] = \text{Prob}[\tilde{x} \in B] = \int_B p_{\tilde{x}}(x)\,\mathrm{d}x \tag{A2.8}$$

Moreover, from

$$P_{\tilde{x}}(x) = \int_{-\infty}^{x} p_{\tilde{x}}(x')\,\mathrm{d}x' \qquad -\infty < x < \infty \tag{A2.9}$$

it follows that

$$p_{\tilde{x}}(x) = \frac{\mathrm{d}}{\mathrm{d}x} P_{\tilde{x}}(x) \tag{A2.10}$$

at all real numbers x at which the derivative exists.

A2.2 EXPECTATIONS OF A RANDOM VARIABLE

The *expectation* (or 'mean') of a random variable $\tilde{x}$, denoted by $\mu_{\tilde{x}} = \mathrm{E}[\tilde{x}]$, is defined (when it exists) by

$$\mu_{\tilde{x}} = \mathrm{E}[\tilde{x}] = \int_{-\infty}^{\infty} x \,\mathrm{d}P_{\tilde{x}}(x) \tag{A2.11}$$

where the integral is a Stieltjes integral‡. The expectation of $\tilde{x}$ is said to exist if the improper integral of equation (A2.11) is absolutely convergent ($\mathrm{E}[|\tilde{x}|] < \infty$). Further, if $\tilde{x}$ is a continuous random variable specified by its PDF,

$$\mu_{\tilde{x}} = \mathrm{E}[\tilde{x}] = \int_{-\infty}^{\infty} x p_{\tilde{x}}(x)\,\mathrm{d}x \tag{A2.11'}$$

whereas, if $\tilde{x}$ is a discrete random variable specified by its PMF,

$$\mu_{\tilde{x}} = \mathrm{E}[\tilde{x}] = \sum_{\text{over all } x^+} x p_{\tilde{x}}^0(x^+) \tag{A2.11''}$$

The *nth moment* $\mathrm{E}[\tilde{x}^n]$ of a random variable $\tilde{x}$ (for any integer n) is the expectation of the nth power of $\tilde{x}$. The *variance* $s_{\tilde{x}}^2$ of $\tilde{x}$ is defined as its second central moment

$$s_{\tilde{x}}^2 = \mathrm{Var}[\tilde{x}] = \mathrm{E}[(\tilde{x} - \mathrm{E}[\tilde{x}])^2] = \mathrm{E}[\tilde{x}^2] - \mathrm{E}^2[\tilde{x}] \tag{A2.12}$$

The *standard deviation* $s_{\tilde{x}}$ of $\tilde{x}$ is defined as the positive square root of its variance:

$$s_{\tilde{x}} = +\sqrt{\mathrm{Var}[\tilde{x}]} \tag{A2.13}$$

and the *coefficient of variation* (c.o.v.) $\delta_{\tilde{x}}$ is the ratio between the standard deviation and the expectation:

$$\delta_{\tilde{x}} = s_{\tilde{x}}/\mu_{\tilde{x}} \tag{A2.14}$$

‡ Given a continuous function $g(x)$, a distribution function $P(x)$ and a half-open interval $(a, b]$ on the real line, the Stieltjes integral of $g(x)$ with respect to $P(x)$ over $(a, b]$ is defined as follows.

Consider a partition of $(a, b]$ into n subintervals (x_{i-1}, x_i), $a = x_0 < x_1 < \ldots < x_n = b$. Further, consider a point x_i' in each subinterval, so that $x_{i-1} < x_i' < x_i$; then

$$\int_{a+}^{b} g(x)\,\mathrm{d}P(x) = \lim_{n\to\infty} \sum_{i=1}^{n} g(x_i')[P(x_i) - P(x_{i-1})]$$

in which the limit is taken over all partitions of (a, b), as the maximum length of subinterval tends to 0.

Furthermore the *moment generating function* of a random variable $\tilde{x}$, denoted by $\Psi_{\tilde{x}}(\ )$, is defined for every real number u by

$$\Psi_{\tilde{x}}(u) = \mathrm{E}(\mathrm{e}^{u\tilde{x}}] \tag{A2.15}$$

which, if it is finite for $|u| < U$, $U > 0$, possesses the power-series expansion (valid for $|u| < U$):

$$\Psi_{\tilde{x}}(u) = 1 + \mathrm{E}[\tilde{x}]u + \mathrm{E}[\tilde{x}^2]\frac{u^2}{2!} + \ldots + \mathrm{E}[\tilde{x}^n]\frac{u^n}{n!} + \ldots \tag{A2.16}$$

Then the *characteristic function* $\Omega_{\tilde{x}}(\ )$ of $\tilde{x}$ is defined, for any real number u, by

$$\Omega_{\tilde{x}}(u) = \mathrm{E}[\mathrm{e}^{\mathrm{i}u\tilde{x}}] \tag{A2.17}$$

(with $\mathrm{i} = \sqrt{(-1)}$ = unit imaginary number). As there is a one-to-one correspondence between characteristic function and distribution function, to specify the $\mathrm{Prob}_{\tilde{x}}[\ ]$ of $\tilde{x}$ it suffices to state its characteristic function. Various inversion formulae may be stated; for any continuous random variable $\tilde{x}$ one can prove that

$$\mathrm{Prob}_{\tilde{x}}[\tilde{x} = x] = \lim_{U \to \infty} \frac{1}{2U} \int_{-U}^{U} \mathrm{e}^{-\mathrm{i}ux}\Omega_{\tilde{x}}(u)\,\mathrm{d}u \qquad -\infty < x < \infty \tag{A2.18}$$

and, if $\Omega_{\tilde{x}}(\ )$ is absolutely integrable, then $\tilde{x}$ is continuous and its probability density function is given by

$$p_{\tilde{x}}(x) = \frac{1}{2\pi} \int_{-\infty}^{\infty} \mathrm{e}^{-\mathrm{i}ux}\Omega_{\tilde{x}}(u)\,\mathrm{d}u \qquad -\infty < x < \infty \tag{A2.19}$$

Equation (A2.19) expresses that $p_{\tilde{x}}(\ )$ is the Fourier transform of $\Omega_{\tilde{x}}(\ )$ and conversely that

$$\Omega_{\tilde{x}}(u) = \int_{-\infty}^{\infty} \mathrm{e}^{\mathrm{i}ux}\, p_{\tilde{x}}(x)\,\mathrm{d}x \qquad -\infty < u < \infty \tag{A2.20}$$

Recalling that

$$\mathrm{e}^{\mathrm{i}ux} = \cos(ux) + \mathrm{i}\sin(ux) \tag{A2.21}$$

for any random variable, $\mathrm{E}[|\mathrm{e}^{\mathrm{i}u\tilde{x}}|] = \mathrm{E}[1]$ and consequently $\Omega_{\tilde{x}}(\ )$ always exists. Moreover note that

$$\mathrm{E}[\tilde{x}^k] = \frac{\mathrm{d}^k}{\mathrm{d}u^k}\Psi_{\tilde{x}}(0) = \frac{1}{\mathrm{i}^k}\frac{\mathrm{d}^k}{\mathrm{d}u^k}\Omega_{\tilde{x}}(0) \tag{A2.22}$$

In addition to the expected value, two other measures of location of the probability law are provided by the median and the mode. The *median* of a probability law is defined as a number $m_{e\tilde{x}}$ such that

$$\lim_{x \to m_{e\tilde{x}}^-} P_{\tilde{x}}(x) = P_{\tilde{x}}(m_{e\tilde{x}} - o) \leqslant \tfrac{1}{2} \leqslant P_{\tilde{x}}(m_{e\tilde{x}} + o) = \lim_{x \to m_{e\tilde{x}}^+} P_{\tilde{x}}(x) \tag{A2.23}$$

or, for a continuous probability law

$$\int_{-\infty}^{m_{e\tilde{x}}} p_{\tilde{x}}(x)\,\mathrm{d}x = 0.5 \tag{A2.23'}$$

For a continuous probability law, a *mode* is defined as a number $m_{o\tilde{x}}$ at which $p_{\tilde{x}}(x)$ has a relative maximum. Hence a probability law may be 'unimodal', 'bimodal', and so on.

Measures of dispersion are the variance, the standard deviation and the coefficient of variation. Furthermore, the *p fractile* (or quantiles of order p) of a probability law are values x_{kp} of x such that

$$\text{Prob}\,[\tilde{x} \leqslant x_{kp}] = p \qquad 0 < p < 1 \qquad \text{(A2.24)}$$

Note that the median is the 0.5 fractile.

A2.3 PROBABILITY LAWS

The probability laws may be classified on the basis of similar functional forms. The relevant tables are collected in Appendix A.8. Tables A.1 and A.2 present some usual probability laws and their properties; the rest of Appendix A.8 contains numerical tables of the CDF and/or PDF connected with the binomial, Poisson, normal (Gaussian), extreme type I, chi-square, Student's t and run distributions.

A number of modifications to the more common distributions are frequently adopted:

(1) *Shifted distributions* are obtained by shift of the origin.
(2) *Transformed distributions* are obtained by changing the independent variable; e.g. if the logarithm of the variable $\tilde{x}$ of interest is normally distributed, $\tilde{x}$ is log-normally distributed.
(3) *Truncated distributions* are obtained by truncating a basic distribution $p_{\tilde{x}}(x)$ below x_0:

$$p_{\tilde{y}}(x) = \begin{cases} 0 & x < x_0 \\ \zeta p_{\tilde{x}}(x) & x \geqslant x_0 \end{cases}$$

where $\zeta = 1/(1 - P_{\tilde{x}}(x_0))$ is the renormalizing factor.
(4) *Censored distributions* coincide with the parent distribution $p_{\tilde{x}}(x)$ except that the probability above (and/or below) x_0 is assumed to be lumped at x_0:

$$\text{Prob}\,[\tilde{y} > x_0] = 0$$
$$\text{Prob}\,[\tilde{y} = x_0] = \text{Prob}[\tilde{x} \geqslant x_0] = 1 - P_{\tilde{x}}(x_0)$$

A2.4 JOINTLY DISTRIBUTED RANDOM VARIABLES

In many cases the result of a random experiment is not expressed by a single quantity but by a family of simultaneously observed quantities.

Several random variables $\tilde{x}_1, \tilde{x}_2, \tilde{x}_3, \ldots, \tilde{x}_n$ are said to be *jointly distributed* if they are defined as functions on the same probability space $\mathscr{S}^{\text{P}}$. The *joint probability function* of n jointly distributed random variables is defined for every Borel set of n-tuples of real numbers by

$$\text{Prob}_{\tilde{x}_1, \tilde{x}_2, \ldots, \tilde{x}_n}[B] = \text{Prob}[\{S \in \mathscr{S}: (\tilde{x}_1(S), \tilde{x}_2(S), \ldots, \tilde{x}_n(S)) \in B\}] \quad \text{(A2.25)}$$

and represents the probability that the n-tuple $(\tilde{x}_1, \tilde{x}_2, \ldots, \tilde{x}_n)$ of observed values of the random variables will lie in the set B.

To specify the joint probability function of $\tilde{x}_1, \tilde{x}_2, \ldots, \tilde{x}_n$, it suffices to specify the *joint distribution function* that for any real numbers $x_1, x_2, \ldots, x_n$, is given by

$$P_{\tilde{x}_1, \tilde{x}_2, \ldots, \tilde{x}_n}(x_1, x_2, \ldots, x_n) = \text{Prob}[\tilde{x}_1 \leqslant x_1, \tilde{x}_2 \leqslant x_2, \ldots, \tilde{x}_n \leqslant x_n] \quad \text{(A2.26)}$$

or the *joint characteristic function*, defined for all real numbers $u_1, u_2, \ldots, u_n$ by

$$\Omega_{\tilde{x}_1, \tilde{x}_2, \ldots, \tilde{x}_n}(u_1, u_2, \ldots, u_n) = \mathrm{E}[\exp\{\mathrm{i}\,(u_1 \tilde{x}_1 + \ldots + u_n \tilde{x}_n)\}] \quad \text{(A2.27)}$$

Then the *joint probability density function* (if the derivative below exists) is given by

$$p_{\tilde{x}_1, \tilde{x}_2, \ldots, \tilde{x}_n}(x_1, x_2, \ldots, x_n) = \frac{\partial^n}{\partial x_1 \partial x_2 \ldots \partial x_n} P_{\tilde{x}_1, \tilde{x}_2, \ldots \tilde{x}_n}(x_1, x_2, \ldots, x_n) \quad \text{(A2.28)}$$

and the *joint probability mass function* is given by

$$p^0_{\tilde{x}_1, \tilde{x}_2, \ldots, \tilde{x}_n}(x_1, x_2, \ldots, x_n) = \text{Prob}\,[\tilde{x}_1 = x_1, \tilde{x}_2 = x_2, \tilde{x}_3 = x_3, \ldots, \tilde{x}_n = x_n] \quad \text{(A2.29)}$$

A *discrete joint probability law* is specified by its probability mass function: for any Borel set B of n-tuples

$$\text{Prob}_{\tilde{x}_1, \tilde{x}_2, \ldots, \tilde{x}_n}[B] = \sum_{\substack{\text{over all} \\ (x_1, x_2, \ldots, x_n)^+ \\ \text{in } B}} p^0_{\tilde{x}_1, \tilde{x}_2, \ldots, \tilde{x}_n}(x_1, x_2, \ldots, x_n)^+ \quad \text{(A2.30)}$$

where $(x_1, x_2, \ldots, x_n)^+$ means a point $(x_1, x_2, \ldots, x_n)$ in B such that $p^0_{\tilde{x}_1, \tilde{x}_2, \ldots, \tilde{x}_n}(x_1, x_2, \ldots, x_n)^+ > 0$.

A *continuous joint probability law* is specified by its probability density function: for any Borel set B of n-tuples

$$\text{Prob}_{\tilde{x}_1, \tilde{x}_2, \ldots, \tilde{x}_n}[B] = \iint_B \cdots \int p_{\tilde{x}_1, \tilde{x}_2, \ldots, \tilde{x}_n}(x_1, x_2, \ldots, x_n)\,\mathrm{d}x_1\,\mathrm{d}x_2 \ldots \mathrm{d}x_n \quad \text{(A2.31)}$$

Finally the *marginal probability law* of each of the random variables $\tilde{x}_1, \tilde{x}_2, \ldots, \tilde{x}_n$ may be obtained from the joint probability law.

In the continuous case, for any $k = 1, 2, \ldots, n$ and any fixed number x_k

$$p_{\tilde{x}_k}(x_k) = \int_{-\infty}^{\infty} \mathrm{d}x_1 \ldots \int_{-\infty}^{\infty} \mathrm{d}x_{k-1} \int_{-\infty}^{\infty} \mathrm{d}x_{k+1} \ldots \int_{-\infty}^{\infty} \mathrm{d}x_n\, p_{\tilde{x}_1, \tilde{x}_2, \ldots, \tilde{x}_n}(x_1, \ldots, x_{k-1}, x_k, x_{k+1}, \ldots, x_n) \quad \text{(A2.32)}$$

and an analogous formula may be written in the discrete case.

Introduce now the vector notation $\tilde{\mathbf{x}}^{\mathrm{T}} = \{\tilde{x}_1, \tilde{x}_2, \ldots, \tilde{x}_n\}$. One defines the *mean vector*

$$\mathrm{E}[\tilde{\mathbf{x}}] = \boldsymbol{\mu}_{\mathbf{x}} = \{\mu_{\tilde{x}_1}, \mu_{\tilde{x}_2}, \ldots, \mu_{\tilde{x}_n}\}^{\mathrm{T}} \tag{A2.33}$$

and for $i,j = 1, \ldots, n$, the *covariances*

$$\mathrm{Cov}[\tilde{x}_i, \tilde{x}_j] = \mathrm{E}[(\tilde{x}_i - \mu_{\tilde{x}_i})(\tilde{x}_j - \mu_{\tilde{x}_j})] \tag{A2.34}$$

Then the array

$$\boldsymbol{\Sigma} = \begin{bmatrix} \mathrm{Cov}[\tilde{x}_i, \tilde{x}_j] \\ i, j = 1, \ldots, n \end{bmatrix} = \mathrm{E}[(\tilde{\mathbf{x}} - \boldsymbol{\mu})(\tilde{\mathbf{x}} \quad \boldsymbol{\mu})^{\mathrm{T}}] \tag{A2.35}$$

is called the *covariance matrix* of $\tilde{\mathbf{x}}$, being the multivariate analogue to the variance of a random variable. The matrix $\boldsymbol{\Sigma}$ is symmetric and positive-semidefinite, i.e.

$$\boldsymbol{\Sigma} = \boldsymbol{\Sigma}^{\mathrm{T}} \qquad \mathbf{y}^{\mathrm{T}} \boldsymbol{\Sigma} \mathbf{y} \geqslant 0 \tag{A2.36}$$

for any given vector $\mathbf{y}$.

Moreover the *correlation coefficient* ρ of two jointly distributed random variables is defined by

$$\rho_{\tilde{x}_i, \tilde{x}_j} = \frac{\mathrm{Cov}[\tilde{x}_i, \tilde{x}_j]}{s_{\tilde{x}_i} s_{\tilde{x}^j}} \tag{A2.37}$$

Two jointly distributed random variables $\tilde{x}_i$ and $\tilde{x}_j$ with finite second moments are said to be *uncorrelated* if their correlation coefficient vanishes.

Finally the joint probability density function of jointly normally distributed random variables $\tilde{x}_1, \tilde{x}_2, \ldots, \tilde{x}_n$ in terms of their correlation coefficient $\rho_{\tilde{x}_i, \tilde{x}_j}$ and variances $s^2_{\tilde{x}_i}$ is

$$p_{\tilde{x}_1 \cdot \tilde{x}_2, \ldots, \tilde{x}_n}(\tilde{x}_1, \tilde{x}_2, \ldots \tilde{x}_n)$$

$$= \frac{1}{(2\pi)^{n/2} s_{\tilde{x}_1} s_{\tilde{x}_2} \ldots s_{x_n} |\rho|^{1/2}} \exp\left[-\frac{1}{2} \sum_{i,j=1}^{n} \left(\frac{x_i - \mu_{\tilde{x}_i}}{s_{\tilde{x}_i}} \right) \rho^{ij} \left(\frac{x_j - \mu_{\tilde{x}_j}}{s_{\tilde{x}_j}} \right) \right] \tag{A2.38}$$

where $|\rho|$ is the determinant of the matrix ρ, whose members are the $\rho_{\tilde{x}_i, \tilde{x}_j}$, and ρ^{ij} is the element of the inverse matrix $\boldsymbol{\rho}^{-1}$.

Jointly distributed random variables $\tilde{x}_1, \tilde{x}_2, \ldots, \tilde{x}_n$ are said to be *independent* if and only if any of the following equivalent statements is true:

(1) *Criterion in terms of probability functions*: for all Borel sets $B_1, B_2, \ldots, B_n$ of real numbers

$$\mathrm{Prob}[\tilde{x}_1 \in B_1, \ldots, \tilde{x}_n \in B_n] = \mathrm{Prob}[\tilde{x}_1 \in B_1] \ldots \mathrm{Prob}[\tilde{x}_n \in B_n] \tag{A2.39a}$$

(2) *Criterion in terms of distribution functions*: for all real numbers $x_1, x_2, \ldots, x_n$

$$P_{\tilde{x}_1, \tilde{x}_2, \ldots, \tilde{x}_n}(x_1, x_2, \ldots, x_n) = P_{\tilde{x}_1}(x_1) \ldots P_{\tilde{x}_n}(x_n) \tag{A2.39b}$$

(3) *Criterion in terms of characteristic functions*: for all real numbers $u_1, u_2, \ldots, u_n$

$$\Omega_{\tilde{x}_1, \tilde{x}_2, \ldots, \tilde{x}_n}(x_1, x_2, \ldots, x_n) = \Omega_{\tilde{x}_1}(u_1) \ldots \Omega_{\tilde{x}_n}(u_n) \tag{A2.39c}$$

If two random variables are independent, their covariance and correlation coefficient are zero. In fact, independence implies that

$$p_{\tilde{x}_i,\tilde{x}_j}(x_i, x_j) = p_{\tilde{x}_i}(x_i)\, p_{\tilde{x}_j}(x_j) \tag{A2.40}$$

and hence the integral in equation (A2.34) becomes simply the product $\mathrm{E}[\tilde{x}_i - \mu_{\tilde{x}_i}]\mathrm{E}[\tilde{x}_j - \mu_{\tilde{x}_j}]$, both factors of which are zero. So if the variables are independent they are uncorrelated. However, the converse of this last statement is not true.

A2.5 CONDITIONAL PROBABILITY AND CONDITIONAL EXPECTATION

Consider the *conditional distribution function*

$$P_{\tilde{y}|\tilde{x}}(y|x) = \mathrm{Prob}[\tilde{y} \leqslant y | \tilde{x} = x] \tag{A2.41}$$

of a random variable $\tilde{y}$, given (the value of) a random variable $\tilde{x}$, and the *conditional expectation*

$$\mu_{\tilde{y}|\tilde{x}}(x) = \mathrm{E}[\tilde{y}|x = x] = \int_{-\infty}^{\infty} y \, \mathrm{d}P_{\tilde{y}|\tilde{x}}(y|x) \tag{A2.42}$$

of a random variable $\tilde{y}$ given a random variable $\tilde{x}$.

If $\tilde{x}$ is discrete (conditioning by a discrete random variable), one obtains from equation (A1.9)

$$P_{\tilde{y}|\tilde{x}}(y|x) = \frac{\mathrm{Prob}[\tilde{y} \leqslant y \cap \tilde{x} = x]}{\mathrm{Prob}[\tilde{x} = x]} \tag{A2.43}$$

If $\tilde{x}$ and $\tilde{y}$ are jointly discrete, then the *conditional probability mass function* is

$$p^0_{\tilde{y}|\tilde{x}}(y|x) = \mathrm{Prob}[\tilde{y} = y | \tilde{x} = x] = \frac{p^0_{\tilde{x},\tilde{y}}(x, y)}{p^0_{\tilde{x}}(x)} \tag{A2.44}$$

if x is such that $p^0_{\tilde{x}}(x) > 0$. Hence the conditional distribution function may be expressed by equation (A2.6) and the conditional expectation by equation (A2.11″).

Moreover let $\tilde{x}$ and $\tilde{y}$ be jointly distributed random variables. Suppose that $\mu_{\tilde{y}}$ is finite and that $\tilde{x}$ is discrete. Then the following fundamental properties of the conditional expectation $\mu_{\tilde{y}|\tilde{x}}(x)$ hold:

$$\mu_{\tilde{y}} = \sum_{\text{over } x^+} \mu_{\tilde{y}|\tilde{x}}(x^+) p_{\tilde{x}}(x^+) \tag{A2.45}$$

$$\mu_{\tilde{y}|\tilde{x}}(x^+) = \mu_{\tilde{y}} \qquad \text{if } \tilde{x} \text{ and } \tilde{y} \text{ are independent} \tag{A2.46}$$

where x^+ denotes the values of x such that $p^0_{\tilde{x}}(x) > 0$.

Finally for discrete random variables $\tilde{x}$ and $\tilde{y}$, any event E and set B, the following properties of the conditional probability hold:

$$\text{Prob}[E] = \sum_{\text{over } x^+} \text{Prob}[E|\tilde{x} = x^+]\, p^0_{\tilde{x}}(x^+) \qquad \text{(A2.47)}$$

$$\text{Prob}[\tilde{y} \in B|\tilde{x} = x] = \text{Prob}[\tilde{y} \in B] \qquad \text{if } \tilde{x} \text{ and } y \text{ are independent} \qquad \text{(A2.48)}$$

If $\tilde{x}$ and $\tilde{y}$ are continuous random variables (conditioning by a continuous random variable) and have joint probability density function $p_{\tilde{x},\tilde{y}}(x, y)$, the *conditional probability density function* $p_{\tilde{x}|\tilde{y}}(\ \ |\ \ )$ of $\tilde{y}$ given $\tilde{x}$ is defined for all y and for all x such that $p_{\tilde{x}}(x) > 0$ by

$$p_{\tilde{y}|\tilde{x}}(y|x) = \frac{p_{\tilde{x},\tilde{y}}(x, y)}{p_{\tilde{x}}(x)} = \frac{p_{\tilde{x},\tilde{y}}(x, y)}{\displaystyle\int_{-\infty}^{\infty} p_{\tilde{x},\tilde{y}}(x, y)\, \mathrm{d}y} \qquad \text{(A2.49)}$$

Then the conditional expection is given by equation (A2.11) and the relevant property is

$$\mu_{\tilde{y}} = \int_{-\infty}^{\infty} \mu_{\tilde{y}|\tilde{x}}(x)\, \mathrm{d}P_{\tilde{x}}(x) \qquad \text{(A2.50)}$$

The *conditional variance* of $\tilde{y}$, given $\tilde{x}$, is defined, assuming $\text{E}[\tilde{y}^2] < \infty$, by

$$\begin{aligned} s^2_{\tilde{y}|\tilde{x}}(x) &= \text{E}[\{\tilde{y} - \mu_{\tilde{y}|\tilde{x}}(x)\}^2|\tilde{x} = x] \\ &= \int_{-\infty}^{\infty} \{y - \mu_{\tilde{y}|\tilde{x}}(x)\}^2\, p_{\tilde{y}|\tilde{x}}(y|x)\, \mathrm{d}y \end{aligned} \qquad \text{(A2.51)}$$

and its relevant property is

$$s^2_{\tilde{y}} = \text{E}[\text{Var}[\tilde{y}|\tilde{x} = x]] + \text{Var}[\text{E}[\tilde{y}|\tilde{x} = x]] \qquad \text{(A2.52)}$$

i.e. the variance of $\tilde{y}$ is equal to the mean of the conditional variance plus the variance of the conditonal mean.

From equation (A2.52) it follows that, for any random variable $\tilde{x}$ and any constant c (see equation (A3.4)),

$$\text{E}[(\tilde{x} - c)^2] = s^2_{\tilde{x}} + (\mu_{\tilde{x}} - c)^2 \qquad \text{(A2.53)}$$

Finally, let $\tilde{x}, \tilde{y}, \tilde{z}, \tilde{q}$ be random variables and c a real number. Then the following assertions hold with probability 1:

$\mu_{\tilde{y}\|\tilde{x}}(x)$	is unique	(A2.54)
$\mu_{\tilde{y}\|\tilde{x}}(x) = \mu_{\tilde{y}}$	if $\tilde{x}$ and $\tilde{y}$ are independent	(A2.55)
$\mu_{c\|\tilde{x}}(x) = c$		(A2.56)
$\mu_{\tilde{y}\|\tilde{x}}(x) \geqslant 0$	if $\tilde{y} \geqslant 0$	(A2.57)
$\mu_{\tilde{z}\|\tilde{x}}(x) \leqslant \mu_{\tilde{y}\|\tilde{x}}(x) \leqslant \mu_{\tilde{q}\|\tilde{x}}(x)$	if $\tilde{z} \leqslant \tilde{y} \leqslant \tilde{q}$	(A2.58)

A2.6 CHEBYSHEV'S INEQUALITY

For any probability law with finite mean $\mu_{\tilde{x}}$ and finite variance $s_{\tilde{x}}^2$, introduce the quantity $P(h)$, for any $h > 0$, as the probability assigned to the interval $\{x: \mu_{\tilde{x}} - hs_{\tilde{x}} < x \leqslant \mu_{\tilde{x}} + hs_{\tilde{x}}\}$ by the probability law

$$P(h) = P_{\tilde{x}}(\mu_{\tilde{x}} + hs_{\tilde{x}}) - P_{\tilde{x}}(\mu_{\tilde{x}} - hs_{\tilde{x}}) \tag{A2.59}$$

Note that $P(h) = \Phi(h) - \Phi(-h)$ for the normal probability law‡; $P(h) = 1 - e^{-(1+h)}$ if $h \geqslant 1$ and $P(h) = e^{-1}(e^h - e^{-h})$ for $h \leqslant 1$ for the exponential law; and $P(h) = h/\sqrt{3}$ if $h < \sqrt{3}$ for the uniform distribution. For the other common probability laws, one cannot so readily evaluate $P(h)$, but it is possible to obtain a lower bound for it, which does not depend on the probability law under consideration.

Chebyshev's inequality: For any distribution function $P_{\tilde{x}}(x)$ and any $h \geqslant 0$

$$P(h) = P_{\tilde{x}}(\mu_{\tilde{x}} + hs_{\tilde{x}}) - P_{\tilde{x}}(\mu_{\tilde{x}} - hs_{\tilde{x}}) \geqslant 1 - 1/h^2 \tag{A2.60}$$

The proof is omitted. A graph of the function $P(h)$ is given in Fig. A.2.

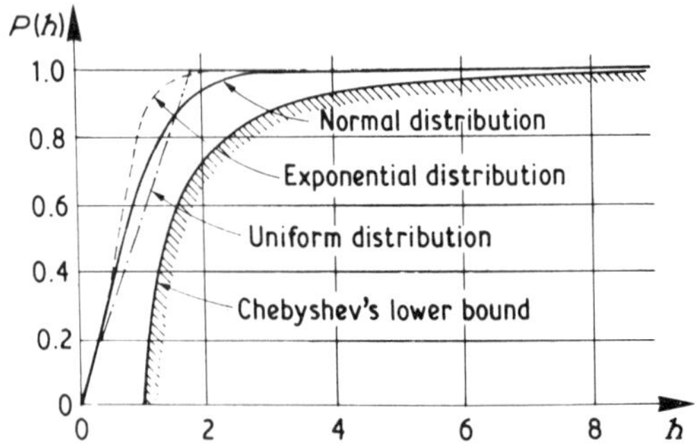

Fig. A.2 Graphs of the function $P(h)$ (equation (A2.60)) for several CDFs.

Chebyshev's inequality may be also reformulated as follows: $P(h)$ is the probability that an observed value of a numerical random phenomenon will lie in an interval centred at the mean and of length $2h$ standard deviations. So Chebyshev's inequality for a random variable $\tilde{x}$ with finite mean and variance may be written for any $h > 0$ and any $\varepsilon > 0$:

$$\begin{aligned} &\text{Prob}[|\tilde{x} - \mu_{\tilde{x}}| \leqslant hs_{\tilde{x}}] \geqslant 1 - 1/h^2 \\ &\text{Prob}[|\tilde{x} - \mu_{\tilde{x}}| \geqslant hs_{\tilde{x}}] \leqslant 1/h^2 \\ &\text{Prob}[|\tilde{x} - \mu_{\tilde{x}}| \leqslant \varepsilon] \geqslant 1 - s_{\tilde{x}}^2/\varepsilon^2 \\ &\text{Prob}[|\tilde{x} - \mu_{\tilde{x}}| > \varepsilon] \leqslant s_{\tilde{x}}^2/\varepsilon^2 \end{aligned} \qquad \text{A2.61)}$$

‡ Φ denotes the distribution function of a normal variable with mean value 0 and standard deviation 1.

A3

Functions of random variables

A3.1 EXPECTATION OF A FUNCTION

Given a random variable $\tilde{y}$, which arises as a Borel function‡ of a random variable $\tilde{x}$ so that $\tilde{y} = g(\tilde{x})$ for some Borel function $g(\ )$, the expectation $\mu_{\tilde{y}} = \mathrm{E}[g(\tilde{x})]$ is given, as seen already, by

$$\begin{aligned} \mu_{\tilde{y}} &= \mathrm{E}_{\tilde{y}}[\tilde{y}] = \mathrm{E}_{\tilde{y}}[g(\tilde{x})] = \int_{-\infty}^{\infty} y\,\mathrm{d}P_{g(\tilde{x})}(y) \\ &= \mathrm{E}_{\tilde{x}}[g(\tilde{x})] = \int_{-\infty}^{\infty} g(x)\,\mathrm{d}P_{\tilde{x}}(x) \end{aligned} \qquad \text{(A3.1)}$$

Equation (A3.1) holds if $\mathrm{E}_{\tilde{y}}[g(\tilde{x})]$ and $\mathrm{E}_{\tilde{x}}[g(\tilde{x})]$ exist and is a direct consequence of the fact that the integrals used to define expectations are required to be absolutely convergent.

Further, if $\tilde{x}$ is a continuous random variable it is specified by its PDF

$$\mu_{\tilde{y}} = \mathrm{E}_{\tilde{x}}[g(\tilde{x})] = \int_{-\infty}^{\infty} g(x)\,p_{\tilde{x}}(x)\,\mathrm{d}x \qquad \text{(A3.1')}$$

whereas if $\tilde{x}$ is a discrete random variable it is specified by its PMF

$$\mu_{\tilde{y}} = \mathrm{E}_{\tilde{x}}[g(\tilde{x})] = \sum_{\substack{\text{over all}\\ x^+}} p^0_{\tilde{x}}(x^+)g(x^+) \qquad \text{(A3.1'')}$$

where x^+ denotes a value of x such that $p_{\tilde{x}}(x^+) > 0$.

An extension of equation (A3.1) implies that for any Borel function $\phi(\ )$

$$\mathrm{E}[\phi(\tilde{y})] = \mathrm{E}_{\tilde{x}}[\phi(g(\tilde{x}))] \qquad \text{(A3.2)}$$

‡ A real function $g(x)$ of a real variable x, defined on $\mathscr{R}^1$, is called a Borel function if, for any Borel set $B \in \mathscr{R}^1$, $g^{-1}(B)$ is a Borel set of $\mathscr{R}^1$. Note that all continuous functions are Borel functions. In these appendices the word 'function' (unless otherwise qualified) will mean a Borel function and the word 'set' (of real members) will mean a Borel set.

if either of these expectations exists. Thus one can find the mean and variance of

$$\tilde{y} = g(\tilde{x}) \tag{A3.3}$$

without actually finding the probability law of $\tilde{y}$. In fact, as immediate consequence of equation (A3.2), equation (A2.12) involves

$$s_{\tilde{y}}^2 = \mathrm{Var}[g(\tilde{x})] = \mathrm{E}_{\tilde{x}}[g^2(\tilde{x})] - \mathrm{E}_{\tilde{x}}^2[g(\tilde{x})] \tag{A3.4}$$

Therefore if C and D are real numbers and $\tilde{x}$ is a random variable with finite mean and variance, then

$$\mathrm{E}[C\tilde{x} + D] = C\mu_{\tilde{x}} + D \tag{A3.5a}$$

$$\mathrm{Var}[C\tilde{x} + D] = C^2 s_{\tilde{x}}^2 \tag{A3.5b}$$

$$s_{(C\tilde{x}+D)} = |C| s_{\tilde{x}} \tag{A3.5c}$$

Moreover for any functions $g(x)$, $g_1(x)$ and $g_2(x)$, whose expectations exist,

$$|\mathrm{E}[g(\tilde{x})]| \leqslant \mathrm{E}[|g(\tilde{x})|] \tag{A3.6a}$$

$$\mathrm{E}[g_1(\tilde{x})] \leqslant \mathrm{E}[g_2(\tilde{x})] \qquad \text{if } g_1(x) \leqslant g_2(x) \text{ for all } x \tag{A3.6b}$$

A3.2 EXPECTATION OF A FUNCTION OF JOINTLY DISTRIBUTED RANDOM VARIABLES

Consider n jointly distributed random variables $\tilde{x}_1, \tilde{x}_2, \ldots, \tilde{x}_n$. For any Borel function $g(x_1, x_2, \ldots, x_n)$ of n real variables, the expectation may be expressed in terms of the joint probability law of $\tilde{x}_1, \ldots, \tilde{x}_n$:

$$\mathrm{E}[g(\tilde{x}_1, \tilde{x}_2, \ldots, \tilde{x}_n)] = \int_{-\infty}^{\infty}\int_{-\infty}^{\infty}\cdots\int_{-\infty}^{\infty} g(x_1, x_2, \ldots, x_n) p_{\tilde{x}_1,\tilde{x}_2,\ldots,\tilde{x}_n}(x_1, x_2, \ldots, x_n)\,\mathrm{d}x_1 \ldots \mathrm{d}x_n \tag{A.3.7}$$

for jointly continuous variables, and

$$\mathrm{E}[g(\tilde{x}_1, \tilde{x}_2, \ldots, \tilde{x}_n)] = \sum_{\substack{\text{over all} \\ (x_1, \ldots x_n)^+}} g(x_1, x_2, \ldots, x_n)^+ p^0_{\tilde{x}_1,\tilde{x}_2,\ldots,\tilde{x}_n}(x_1, x_2, \ldots, x_n)^+ \tag{A3.7'}$$

where $(x_1, x_2, \ldots, x_n)^+$ is such that $p^0_{\tilde{x}_1,\tilde{x}_2,\ldots\tilde{x}_n}(x_1, x_2, \ldots, x_n)^+ > 0$.

Also, for any Borel function $\phi(\ )$ of one real variable and $\tilde{y} = (\tilde{x}_1, x_2, ..\tilde{x}_n)$

$$\mathrm{E}[\phi(\tilde{y})] = \mathrm{E}_{\tilde{x}_1,\tilde{x}_2,\ldots,\tilde{x}_n}[\phi(g(\tilde{x}_1, \tilde{x}_2, \ldots, \tilde{x}_n))] \tag{A3.8}$$

A very important property possessed by the operation of expectation of a random variable is its *linearity property*. Let $\tilde{x}_1$ and $\tilde{x}_2$ be two jointly distributed random variables with finite expectations; then the sum $(\tilde{x}_1 + \tilde{x}_2)$ has a finite expectation given by

$$\mathrm{E}[\tilde{x}_1 + \tilde{x}_2] = \mu_{\tilde{x}_1} + \mu_{\tilde{x}_2} \tag{A3.9}$$

The variance of the sum of two random variables is

$$\mathrm{Var}[\tilde{x}_1+\tilde{x}_2] = s^2_{\tilde{x}_1} + s^2_{\tilde{x}_2} + 2\,\mathrm{Cov}[\tilde{x}_1,\tilde{x}_2] \qquad \text{(A3.10)}$$

where the covariance (see equation (A2.34)) can be proved to be equal to the *product moment* $\mathrm{E}[\tilde{x}_1, \tilde{x}_2]$ minus the product of the means; in symbols

$$\mathrm{Cov}[\tilde{x}_1,\tilde{x}_2] = \mathrm{E}[\tilde{x}_1,\tilde{x}_2] - \mu_{\tilde{x}_1}\mu_{\tilde{x}_2} \qquad \text{(A3.11)}$$

For any two functions $g_1(\ )$ and $g_2(\ )$, the linearity property of the operation of taking expectations implies that

$$\mathrm{E}[g_1(\tilde{x}_1,\tilde{x}_2)+g_2(\tilde{x}_1,\tilde{x}_2)] = \mathrm{E}[g_1(\tilde{x}_1,\tilde{x}_2)] + \mathrm{E}[g_2(\tilde{x}_1,\tilde{x}_2)] \qquad \text{(A3.12)}$$

if the expectations on the r.h.s. exist. However, it is not true that a similar relation holds for products.

If the random variables $\tilde{x}_1$ and $\tilde{x}_2$ are independent, then for any two Borel functions $g_1(\ )$ and $g_2(\ )$ of one real variable, the product moment $\mathrm{E}[g_1(\tilde{x}_1)g_2(\tilde{x}_2)]$ is equal to the product of their means, i.e.

$$\mathrm{E}[g_1(\tilde{x}_1)g_2(\tilde{x}_2)] = \mathrm{E}[g_1(\tilde{x}_1)]\,\mathrm{E}[g_2(\tilde{x}_2)] \qquad \text{(A3.13)}$$

Particularly two random variables $\tilde{x}_1$ and $\tilde{x}_2$ which satisfy

$$\mathrm{E}[\tilde{x}_1\tilde{x}_2] = \mu_{\tilde{x}_1}\mu_{\tilde{x}_2} \qquad \text{(A3.14)}$$

are said to be *uncorrelated*. In fact, from equation (A3.11), it follows that $\mathrm{Cov}[\tilde{x}_1, \tilde{x}_2] = 0$ and hence the correlation coefficient $\rho_{\tilde{x}_1,\tilde{x}_2}$ defined by equation (A2.37) becomes zero. Therefore, two random variables that are independent are uncorrelated, whereas the converse is not true in general (Section A2.4).

If and only if $\tilde{x}_1$ and $\tilde{x}_2$ are uncorrelated, from equation (A3.10) it follows that

$$\mathrm{Var}[\tilde{x}_1+\tilde{x}_2] = s^2_{\tilde{x}_1} + s^2_{\tilde{x}_2} \qquad \text{(A3.15)}$$

Furthermore, for any two jointly distributed random variables $\tilde{x}_1$ and $\tilde{x}_2$ with finite second moments it is possible to prove:

Schwarz's inequality

$$|\mathrm{E}[\tilde{x}_1\tilde{x}_2]|^2 \leqslant \mathrm{E}[\tilde{x}_1^2]\,\mathrm{E}[\tilde{x}_2^2] \qquad \text{(A3.16)}$$

Cauchy's inequality

$$|\mathrm{Cov}[\tilde{x}_1, \tilde{x}_2]| \leqslant s_{\tilde{x}_1}s_{\tilde{x}_2} \qquad \text{(A3.17)}$$

A3.3 CONDITIONAL EXPECTATION OF A FUNCTION OF RANDOM VARIABLES

Let $\tilde{x}$ and $\tilde{y}$ be jointly distributed random variables and $g(x, y)$ a function of two variables. Suppose that $\mu_{\tilde{y}}$ is finite and that $\tilde{x}$ is discrete. Then one can prove that

$$\mathrm{E}[g(\tilde{x},\tilde{y})|\tilde{x}=x] = \mathrm{E}[g(x,\tilde{y})|\tilde{x}=x] \qquad \text{(A3.18)}$$

for all x such that $p^0_{\tilde{x}}(x) > 0$.

If $\tilde{x}$ and $\tilde{y}$ are continuous jointly distributed random variables, with reference to the product of $\tilde{y}$ and any function $g(\tilde{x})$, the following property holds if $\mathrm{E}[g(\tilde{x})\tilde{y}]$ is finite:

$$\mathrm{E}[g(\tilde{x})\tilde{y}|\tilde{x}=x] = g(x)\,\mathrm{E}[\tilde{y}|\tilde{x}=x] \tag{A3.19}$$

Combining equations (A3.19) and (A2.50) it follows that

$$\mathrm{E}[g(\tilde{x})\tilde{y}] = \mathrm{E}[g(\tilde{x})\mu_{\tilde{y}|\tilde{x}}(\tilde{x})] \tag{A3.20}$$

and equation (A3.20) can be shown to provide an axiomatic definition of the conditional expectation.

Finally, let $\tilde{x}$, $\tilde{y}$, $\tilde{z}$ and $\tilde{q}$ be random variables, c a real number and $g(x)$ any function of one variable. Then the following assertions hold with probability 1:

$$\mathrm{E}[g(\tilde{x})|\tilde{x}=x] = g(x) \tag{A3.21}$$

$$\mathrm{E}[c\tilde{y}|\tilde{x}=x] = c\mu_{\tilde{y}|\tilde{x}}(x) \tag{A3.22}$$

$$\mathrm{E}[g(\tilde{x})\tilde{y}|\tilde{x}=x] = g(x)\mu_{\tilde{y}|\tilde{x}}(x) \tag{A3.23}$$

$$\mathrm{E}[\tilde{z}+\tilde{q}|\tilde{x}=x] = \mu_{\tilde{z}|\tilde{x}}(x)+\mu_{\tilde{q}|\tilde{x}}(x) \tag{A3.24}$$

$$|\mathrm{E}[\tilde{y}|\tilde{x}=x]|^r \leqslant \{\mathrm{E}[|y||\tilde{x}=x]\}^r \leqslant \mathrm{E}[|y|^r|\tilde{x}=x] \qquad \text{if } r \geqslant 1 \tag{A3.25}$$

A3.4 EXPECTATIONS OF SUMS OF RANDOM VARIABLES

Let $\tilde{x}_1, \tilde{x}_2, \ldots, \tilde{x}_n$ be n jointly distributed random variables. From the linearity properties of the expectation operation, one obtains the following formulae for the mean, mean square and variance of the sum:

$$\mathrm{E}\left[\sum_{k=1}^{n} \tilde{x}_k\right] = \sum_{k=1}^{n} \mu_{\tilde{x}_k} \tag{A3.26}$$

$$\mathrm{E}\left[\left(\sum_{k=1}^{n} \tilde{x}_k\right)^2\right] = \sum_{k=1}^{n} \mathrm{E}[x_k^2] + 2\sum_{k=1}^{n}\sum_{j=k+1}^{n} \mathrm{E}[\tilde{x}_k\tilde{x}_j] \tag{A3.27}$$

$$\mathrm{Var}\left[\sum_{k=1}^{n} \tilde{x}_k\right] = \sum_{k=1}^{n} s_{\tilde{x}_k}^2 + 2\sum_{k=1}^{n}\sum_{j=k+1}^{n} \mathrm{Cov}[\tilde{x}_k, \tilde{x}_j] \tag{A3.28}$$

Equation (A3.26) simplifies if the random variables $\tilde{x}_1, \tilde{x}_2, \ldots, \tilde{x}_n$ are uncorrelated:

$$\mathrm{Var}\left[\sum_{k=1}^{n} \tilde{x}_k\right] = \sum_{k=1}^{n} s_{\tilde{x}_k}^2 \qquad \text{if } \mathrm{Cov}[\tilde{x}_k, \tilde{x}_j] = 0 \text{ for } k \neq j \tag{A3.29}$$

Then, if the random variables are independent, for any real number u, the following formula for the moment-generating function of their sum holds:

$$\Psi_{(\tilde{x}_1+\tilde{x}_2+\ldots+\tilde{x}_n)}(u) = \Psi_{\tilde{x}_1}(u)\Psi_{\tilde{x}_2}(u)\ldots\Psi_{\tilde{x}_n}(u) \tag{A3.30}$$

Equations (A3.26)–(A3.30) are useful for finding the mean and variance of a random variable $\tilde{y}$ (without knowing the probability law of $\tilde{y}$) if one can represent

$\tilde{y}$ as a sum of random variables $\tilde{x}_1, \tilde{x}_2, \ldots, x_n$, the mean, variance and covariances of which are known. For instance, consider the number of successes $\tilde{z}_n$ in n independent repeated Bernoulli trials with probability p of success at each trial. It is a random variable and, from equations (A3.26), (A3.29) and (A3.30), the parameters of its probability law are

$$\mu_{\tilde{z}_n} = np \qquad s^2_{\tilde{z}_n} = np(1-p) \qquad \Psi_{\tilde{z}_n}(u) = [p(e^u - 1) + 1]^u \tag{A3.31}$$

More generally, for the vector case $\mathbf{y} = \mathbf{D} + \mathbf{C}\tilde{\mathbf{x}}$ in which $\mathbf{D}$ is a given vector and $\mathbf{C}$ is a given matrix, using linearity of expectation, one obtains

$$\boldsymbol{\mu}_{\tilde{\mathbf{y}}} = \mathrm{E}[\tilde{\mathbf{y}}] = \mathrm{E}[\mathbf{D} + \mathbf{C}\tilde{\mathbf{x}}] = \mathbf{D} + \mathbf{C}\boldsymbol{\mu}_{\tilde{\mathbf{x}}} \tag{A3.26$'$}$$

and

$$\boldsymbol{\Sigma}_{\tilde{\mathbf{y}}} = \mathrm{E}[\tilde{\mathbf{y}} - \boldsymbol{\mu}_{\tilde{\mathbf{y}}})(\tilde{\mathbf{y}} - \boldsymbol{\mu}_{\tilde{\mathbf{y}}})^{\mathrm{T}}] = \mathbf{C}\boldsymbol{\Sigma}_{\tilde{\mathbf{x}}}\mathbf{C}^{\mathrm{T}} \tag{A3.27$'$}$$

A3.5 THE PROBABILITY LAW OF A FUNCTION OF A RANDOM VARIABLE

Consider a random variable $\tilde{y}$ which arises as a function of another random variable $\tilde{x}$, so that for some Borel function $g(\ )$: $\tilde{y} = g(\tilde{x})$. Once $P_{\tilde{x}}(x)$ is known, the basic problem is to find the distribution function $P_{\tilde{y}}(y) = \mathrm{Prob}[\tilde{x}{:}g(x) \leqslant y]$ and/or the probability density function $p_{\tilde{y}}(y)$ or the probability mass function $p^0_{\tilde{y}}(y)$ in the case in which these functions exist.

Let $\tilde{x}$ be a continuous random variable. If $y = g(x)$ is differentiable for all x, and either $\mathrm{d}g/\mathrm{d}x > 0$ for all x or $\mathrm{d}g/\mathrm{d}x < 0$ for all x, then $\tilde{y} = g(\tilde{x})$ is a continuous random variable with probability density function given by

$$p_{\tilde{y}}(y) = \begin{cases} p_{\tilde{x}}(g^{-1}(y))\left|\dfrac{\mathrm{d}}{\mathrm{d}y}(g^{-1}(y))\right| & \text{if } a < y < b \\ 0 & \text{otherwise} \end{cases} \tag{A3.32}$$

in which a and b, with $a < b$, are defined as $\lim_{x\to-\infty} g(x)$ and $\lim_{x\to+\infty} g(x)$ respectively or vice versa.

If the function $y = g(x)$ is not monotonic (increasing or decreasing) the following formula for $p_{\tilde{y}}(y)$ at the real number y must be employed:

$$P_{\tilde{y}}(y) = \mathrm{Prob}[\{x{:}g(x) \leqslant y\}] \tag{A3.33}$$

Equation (A3.33) applied to the function $y = Cx + D$ with $C > 0$ and $-\infty < D < \infty$ provides

$$P_{C\tilde{x}+D}(y) = \mathrm{Prob}[C\tilde{x} + D \leqslant y] = \mathrm{Prob}\left[\tilde{x} \leqslant \frac{y-D}{C}\right] = P_{\tilde{x}}\left(\frac{y-D}{C}\right) \tag{A3.34a}$$

$$p_{C\tilde{x}+D}(y) = \frac{\mathrm{d}}{\mathrm{d}y}P_{C\tilde{x}+D}(y) = \frac{1}{C}p_{\tilde{x}}\left(\frac{y-D}{C}\right) \quad \text{if } \tilde{x} \text{ is continuous} \tag{A3.34b}$$

$$p^0_{C\tilde{x}+D}(y) = p^0_{\tilde{x}}\left(\frac{y-D}{C}\right) \quad \text{if } \tilde{x} \text{ is discrete} \tag{A3.34c}$$

Further examples are summarized in the following relations:

(1) $\tilde{y} = \tilde{x}^2$:

$$P_{\tilde{x}^2} = \begin{cases} P_{\tilde{x}}(\sqrt{y}) - P_{\tilde{x}}(-\sqrt{y}) + p^0_{\tilde{x}}(-\sqrt{y}) & \text{for } y \geqslant 0 \\ 0 & \text{for } y < 0 \end{cases} \tag{A3.35a}$$

$$p_{\tilde{x}^2} = \begin{cases} [p_{\tilde{x}}(\sqrt{y}) + p_{\tilde{x}}(-\sqrt{y})]/(2\sqrt{y}) & \text{for } y \geqslant 0 \\ 0 & \text{for } y < 0 \end{cases} \quad \text{if } \tilde{x} \text{ is continuous} \tag{A3.35b}$$

$$p^0_{\tilde{x}^2} = \begin{cases} p^0_{\tilde{x}}(\sqrt{y}) + p^0_{\tilde{x}}(-\sqrt{y}) & \text{for } y \geqslant 0 \\ 0 & \text{for } y < 0 \end{cases} \quad \text{if } \tilde{x} \text{ is discrete} \tag{A3.35c}$$

(2) $\tilde{y} = |\tilde{x}|$:

$$p_{|\tilde{x}|}(y) = \begin{cases} p_{\tilde{x}}(y) + p_{\tilde{x}}(-y) & \text{for } y > 0 \\ 0 & \text{for } y < 0 \end{cases} \tag{A3.36}$$

(3) $\tilde{y} = \sqrt{|\tilde{x}|}$:

$$p_{\sqrt{|\tilde{x}|}}(y) = \begin{cases} 2y[p_{\tilde{x}}(y^2) + p_{\tilde{x}}(-y^2)] & \text{for } y > 0 \\ 0 & \text{for } y < 0 \end{cases} \tag{A3.37}$$

Moreover, in terms of the inverse function $P_{\tilde{x}}^{-1}(\ )$ to the distribution function $P_{\tilde{x}}(\ )$ of the random variable $\tilde{x}$, it is possible to state the following theorem that plays an important role in several applications.

Let $\tilde{z}_1, \tilde{z}_2, \ldots, \tilde{z}_n$ be independent random variables, each uniformly distributed over the interval 0 to 1. The random variables

$$\tilde{x}_1 = P_{\tilde{x}}^{-1}(\tilde{z}_1) \qquad \tilde{x}_2 = P_{\tilde{x}}^{-1}(\tilde{z}_2) \quad \ldots \quad \tilde{x}_n = P_{\tilde{x}}^{-1}(\tilde{z}_n) \tag{A3.38}$$

are then a random sample of the random variable $\tilde{x}$. Conversely, if $\tilde{x}_1, \tilde{x}_2, \ldots, \tilde{x}_n$ are a random sample of the random variable $\tilde{x}$ and if the distribution function $P_{\tilde{x}}(\ )$ is continuous, the random variables

$$\tilde{z}_1 = P_{\tilde{x}}(\tilde{x}_1) \qquad \tilde{z}_2 = P_{\tilde{x}}(\tilde{x}_2) \quad \ldots \quad \tilde{z}_n = P_{\tilde{x}}(\tilde{x}_n) \tag{A3.39}$$

are a random sample of the random variable $\tilde{z} = P_{\tilde{x}}(x)$, which is uniformly distributed on the interval 0 to 1.

The transformation of a random variable $\tilde{x}$ into a uniformly distributed random variable $\tilde{z} = P_{\tilde{x}}(\tilde{x})$ is called the *probability integral transformation.*

Further, for discrete variables $\tilde{x}$ and $\tilde{y}$, function $g(x, y)$ and Borel set B, the following property of the conditional probability holds:

$$\text{Prob}[g(\tilde{x}, \tilde{y}) \in B \,|\, \tilde{x} = x^+] = \text{Prob}[g(x, \tilde{y}) \in B \,|\, x = x^+] \tag{A3.40}$$

where x^+ denotes the values of x such that $p^0_{\tilde{x}}(x) > 0$.

A3.6 THE PROBABILITY LAW OF A FUNCTION OF RANDOM VARIABLES

Let $\tilde{y}$ be a random variable which arises as a function $\tilde{y} = g(\tilde{x}_1, \tilde{x}_2, \ldots, \tilde{x}_n)$ of n jointly distributed random variables $\tilde{x}_1, \tilde{x}_2, \ldots, \tilde{x}_n$ with joint probability function $\text{Prob}_{\tilde{x}_1,\tilde{x}_2,\ldots,\tilde{x}_n}[B]$. For any real number y

$$P_{\tilde{y}}(y)=\text{Prob}[\{\tilde{x}_1,\tilde{x}_2,\ldots,\tilde{x}_n\}:g(x_1,x_2,\ldots,x_n)\leqslant y] \tag{A3.41}$$

Some formulae that are consequences of the above basic statement are presented in the following where two jointly continuous random variables $\tilde{x}_1$ and $\tilde{x}_2$ with joint probability density function $p_{\tilde{x}_1,\tilde{x}_2}(x_1,x_2)$ are considered.

Let $\tilde{y}=\tilde{x}_1+\tilde{x}_2$, then

$$P_{\tilde{y}}(y)=\int_{-\infty}^{\infty}\mathrm{d}x_1\int_{-\infty}^{y-x_1}\mathrm{d}x_2\,p_{\tilde{x}_1,\tilde{x}_2}(x_1,x_2)=\int_{-\infty}^{\infty}\mathrm{d}x_2\int_{-\infty}^{y-x_2}\mathrm{d}x_1\,p_{\tilde{x}_1,\tilde{x}_2}(x_1,x_2) \tag{A3.42}$$

$$\begin{aligned}p_{\tilde{y}}(y)&=\int_{-\infty}^{\infty}\mathrm{d}x_1\,p_{\tilde{x}_1,\tilde{x}_2}(x_1,y-x_1)\\&=\int_{-\infty}^{\infty}\mathrm{d}x_2\,p_{\tilde{x}_1,\tilde{x}_2}(y-x_2,x_2)^{\ddagger}\end{aligned} \tag{A3.43}$$

If $\tilde{x}_1$ and $\tilde{x}_2$ are independent, then equation (A3.43) becomes

$$\begin{aligned}p_{\tilde{y}}(y)&=\int_{-\infty}^{\infty}\mathrm{d}x_1\,p_{\tilde{x}_1}(x_1)p_{\tilde{x}_2}(y-x_1)\\&=\int_{-\infty}^{\infty}\mathrm{d}x_2\,p_{\tilde{x}_1}(y-x_2)p_{\tilde{x}_2}(x_2)\end{aligned} \tag{A3.44}$$

It is worth noting that the mathematical operation involved in equation (A3.44) arises in many branches of mathematics. Let $p_1(\ )$, $p_2(\ )$ and $p_3(\ )$ be three functions which are such that

$$p_3(y)=\int_{-\infty}^{\infty}p_1(x)p_2(y-x)\mathrm{d}x=p_1(x)\circledast p_2(y-x) \tag{A3.45}$$

The function $p_3(y)$ is called the *convolution* of $p_1(\ )$ and $p_2(\ )$. Hence $p_{(\tilde{x}_1+\tilde{x}_2)}(y)$ is the convolution of the probability density functions of the random variables $\tilde{x}_1$ and $\tilde{x}_2$.

Moreover the probability density functions of the difference, product and

‡ If $\tilde{x}_1$ and $\tilde{x}_2$ are jointly discrete, then

$$p_{\tilde{y}}^{0}(y)=\sum_{\substack{\text{over all}\\x^+}}p_{\tilde{x}_1,\tilde{x}_2}^{0}(x^+,y-x^+)=\sum_{\substack{\text{over all}\\x^*}}p_{\tilde{x}_1,\tilde{x}_2}^{0}(y-x^*,x^*) \tag{A3.43'}$$

where x^+ and x^* denote the values of x such that $p_{\tilde{x}_1,\tilde{x}_2}^{0}(x^+,y-x^+)>0$ and $p_{\tilde{x}_1,\tilde{x}_2}^{0}(y-x^*,x^*)>0$ respectively.

quotient of two jointly continuous random variables are

$$p_{(\tilde{x}_1-\tilde{x}_2)}(y) = \int_{-\infty}^{\infty} dx_1\, p_{\tilde{x}_1,\tilde{x}_2}(x_1, x_1 - y)$$
$$= \int_{-\infty}^{\infty} dx_2\, p_{\tilde{x}_1,\tilde{x}_2}(y + x_2, x_2) \tag{A3.46}$$

$$p_{\tilde{x}_1\tilde{x}_2}(y) = \int_{-\infty}^{\infty} dx_1 \frac{1}{|x_1|}\, p_{\tilde{x}_1,\tilde{x}_2}(x_1, y/x_1)$$
$$= \int_{-\infty}^{\infty} dx_2 \frac{1}{|x_2|}\, p_{\tilde{x}_1,\tilde{x}_2}(y/x_2, x_2) \tag{A3.47}$$

$$p_{(\tilde{x}_1/\tilde{x}_2)}(y) = \int_{-\infty}^{\infty} dx_2\, p_{\tilde{x}_1,\tilde{x}_2}(yx_2, x_2) \tag{A3.48}$$

while the distribution function of $\tilde{y} = (\tilde{x}_1^2 + \tilde{x}_2^2)^{1/2}$ is better expressed by means of polar coordinates ($x_1 = R\cos\theta$; $x_2 = R\sin\theta$):

$$P_{\tilde{y}}(y) = \int_0^{2\pi} d\theta \int_0^y R\, dr\, p_{\tilde{x}_1,\tilde{x}_2}(R\cos\theta, R\sin\theta) \tag{A3.49}$$

$$p_{\tilde{y}}(y) = \begin{cases} y\int_0^{2\pi} d\theta\, p_{\tilde{x}_1,\tilde{x}_2}(y\cos\theta,\ y\sin\theta) & \text{for } y > 0 \\ 0 & \text{for } y < 0 \end{cases} \tag{A3.50}$$

$$p_{(\tilde{x}_1^2+\tilde{x}_2^2)}(y) = \begin{cases} \frac{1}{2}\int_0^{2\pi} d\theta\, p_{\tilde{x}_1,\tilde{x}_2}(\sqrt{y}\,\cos\theta,\ \sqrt{y}\sin\theta) & \text{for } y > 0 \\ 0 & \text{for } y < 0 \end{cases} \tag{A.351}$$

Now take two independent random variables, $\tilde{x}_1$ and $\tilde{x}_2$, and refer to problem of finding the probability distribution of their sum. Equations (A3.42)–(A3.33) can be used to prove that:

(1) If $\tilde{x}_1$ and $\tilde{x}_2$ are normally distributed, then $(\tilde{x}_1 + \tilde{x}_2)$ is normally distributed with

$$\mu_{(\tilde{x}_1+\tilde{x}_2)} = \mu_{\tilde{x}_1} + \mu_{\tilde{x}_2} \qquad s_{(\tilde{x}_1+\tilde{x}_2)} = \sqrt{(s_{\tilde{x}_1}^2 + s_{\tilde{x}_2}^2)} \tag{A3.52}$$

(2) If $\tilde{x}_1$ obeys a binomial probability law with parameters n_1 and p and $\tilde{x}_2$ obeys the same distribution with n_2 and p, then $(\tilde{x}_1 + \tilde{x}_2)$ obeys a binomial probability law with parameters p and

$$n = n_1 + n_2 \tag{A3.53}$$

Analogous statements are true for Poisson ($\lambda = \lambda_1 + \lambda_2$), Cauchy ($a = a_1 + a_2$, $b = b_1 + b_2$)‡ and gamma (λ, $r = r_1 + r_2$) probability laws.

‡ The Cauchy distribution is described by

$$P_{\tilde{x}}(x) = \frac{1}{\pi a}\,\frac{1}{1 + [(x-b)/a]^2} \tag{A3.53$'$}$$

More generally, a distribution is said to be *stable* if it is not changed by a linear transformation. The normal distribution (Section A6.2) is an example of a stable distribution.

Finally, a geometrical method for finding the probability law of a function $\tilde{y}$ of several random variables can be obtained if there is a function of a real variable denoted by $p_g(\ )$ such that

$$p_{\tilde{x}_1, \tilde{x}_2, \ldots, \tilde{x}_n}(x_1, x_2, \ldots, x_n) = p_g(y) \tag{A3.54}$$

with $g(x_1, x_2, \ldots, x_n) = y$ a suitable continuous function of $x_1, x_2, \ldots, x_n$. In this case, for any real number y

$$p_{\tilde{y}}(y) = p_g(y)\frac{\mathrm{d}V_g(y)}{\mathrm{d}y} \tag{A3.55}$$

where $V_g(y)$ is the volume within the surface with equation $g(x_1, x_2, \ldots, x_n) = y$ that coincides with a contour line of the JPDF of the variables $\tilde{x}_1, \tilde{x}_2, \ldots, \tilde{x}_n$:

$$V_g(y) = \underset{\{(x_1,x_2,\ldots,x_n):\ g(x_1,x_2,\ldots,x_n) \leqslant y\}}{\iint\cdots\int} \mathrm{d}x_1\,\mathrm{d}x_2\ldots\mathrm{d}x_n \tag{A3.56}$$

A3.7 THE JOINT PROBABILITY LAW OF FUNCTIONS OF RANDOM VARIABLES

The problem of obtaining the joint probability law of several random variables which arise as functions is not different from those previously considered, but details are more complicated. Hence developments are limited to an often-used particular formula.

Let $\tilde{y}_1, \tilde{y}_2, \ldots, \tilde{y}_n$ be n random variables which arise as functions of n jointly continuous random variables $\tilde{x}_1, \tilde{x}_2, \ldots, \tilde{x}_n$, whose joint probability density function is continuous at all but a finite number of points in the $(x_1, x_2, \ldots, x_n)$ space:

$$\tilde{y}_1 = g_1(\tilde{x}_1, \tilde{x}_2, \ldots, \tilde{x}_n) \qquad \tilde{y}_2 = g_2(\tilde{x}_1, \tilde{x}_2, \ldots, \tilde{x}_n)$$
$$\ldots \quad \tilde{y}_n = g_n(\tilde{x}_1, \tilde{x}_2, \ldots, \tilde{x}_n) \tag{A3.57}$$

Further, let $g_1(\ ), g_2(\ ), \ldots, g_n(\ )$ have continuous partial derivative at all points $(x_1, x_2, \ldots, x_n)$ and be such that the Jacobian

$$J(x_1, x_2, \ldots, x_n) = \begin{vmatrix} \frac{\partial g_1}{\partial x_1} & \frac{\partial g_1}{\partial x_2} & \cdots & \frac{\partial g_1}{\partial x_n} \\ \frac{\partial g_2}{\partial x_1} & \frac{\partial g_2}{\partial x_2} & \cdots & \frac{\partial g_2}{\partial x_n} \\ \vdots & \vdots & & \vdots \\ \frac{\partial g_n}{\partial x_1} & \frac{\partial g_n}{\partial x_2} & \cdots & \frac{\partial g_n}{\partial x_n} \end{vmatrix} \neq 0 \tag{A3.58}$$

at all points $(x_1, x_2, \ldots, x_n)$.

Let Λ denote the set of points $(y_1, y_2, \ldots, y_n)$ such that the n equations

$$y_i = g_i(x_1, x_2, \ldots, x_n) \qquad i = 1, \ldots, n \tag{A3.59}$$

possess at least one solution $(x_1, x_2, \ldots, x_n)$

$$x_i = g^{-1}(y_1, y_2, \ldots, y_n) \qquad i = 1, \ldots, n \tag{A3.60}$$

Then the random variables $\tilde{y}_1, \tilde{y}_2, \ldots, \tilde{y}_n$ defined by equation (A3.57) are jointly continuous with a joint probability density function given by

$$p_{\tilde{y}_1, \tilde{y}_2, \ldots, \tilde{y}_n}(y_1, y_2, \ldots y_n) = \begin{cases} p_{\tilde{x}_1, \tilde{x}_2, \ldots, \tilde{x}_n}(x_1, x_2, \ldots, x_n) \,|J(x_1, x_2, \ldots, x_n)|^{-1} \\ \qquad \text{if } (y_1, y_2, \ldots, y_n) \in \Lambda \\ 0 \qquad \text{if } (y_1, y_2, \ldots, y_n) \notin \Lambda \end{cases} \tag{A3.61}$$

A3.8 SUMS OF INDEPENDENT RANDOM VARIABLES

The problem of addition of independent random variables consists of finding, either exactly or approximately, the probability law of a random variable that arises as the sum of n independent random variables $\tilde{x}_1, \tilde{x}_2, \ldots, \tilde{x}_n$ whose joint probability law is known. Let

$$\tilde{z}_n = \tilde{x}_1 + \tilde{x}_2 + \ldots + \tilde{x}_n \tag{A3.62}$$

be this sum; also the arithmetic mean

$$\tilde{m}_n = \tilde{z}_n / n \tag{A3.63}$$

is a random variable.

If $\tilde{x}_1, \tilde{x}_2, \ldots, \tilde{x}_n$ are identically distributed as $\tilde{x}$, the following expressions for the mean, variance and moment-generating function of $\tilde{z}_n$ and $\tilde{m}_n$ are derived from equations (A3.26), (A3.29) and (A3.30):

$$\mu_{\tilde{z}_n} = n\mu_{\tilde{x}} \qquad s^2_{\tilde{z}_n} = ns^2_{\tilde{x}} \qquad \Psi_{\tilde{z}_n}(u) = [\Psi_{\tilde{x}}(u)]^n \tag{A3.64}$$

$$\mu_{\tilde{m}_n} = \mu_{\tilde{x}} \qquad s^2_{\tilde{m}_n} = (1/n)s^2_{\tilde{x}} \qquad \Psi_{\tilde{m}_n}(u) = [\Psi_{\tilde{x}}(u/n)]^n \tag{A3.65}$$

Equations (A3.64) and (A3.65) are very useful in the analysis of sample averages. A *sample of size n* of a random variable $\tilde{x}$ is a set of n jointly distributed random variables $\tilde{x}_1, \tilde{x}_2, \ldots, \tilde{x}_n$ whose individual probability laws coincide with that of $\tilde{x}$. The sample is a *random sample* if the random variables $\tilde{x}_1, \tilde{x}_2, \ldots, \tilde{x}_n$ are independent. Then the random variable $\tilde{m}_n$ defined by equation (A3.63) is called the *sample average*, while the *sample variance* is defined by

$$s^2_n = \frac{1}{n}\sum_{k=1}^{n}(\tilde{x}_k - \tilde{m}_n)^2 = \frac{1}{n}\sum_{k=1}^{n}\tilde{x}^2_k - \tilde{m}^2_n \tag{A3.66}$$

The sample average $\tilde{m}_n$ is a random variable with mean variance and moment-generating function given by equation (A3.65). It is worth noting that the variance of the sample mean $\tilde{m}_n$ tends to zero as the sample size tends to infinity.

The techniques for computing these quantities, the distribution function $P_{\tilde{m}_n}(\ )$ and the relevant probability function, as well as the study of the behaviour of large sample sizes provide the framework for many of the concepts of statistics (Chapter 7).

The sum of independent random variables also arises in considering the description of the motion of a particle that is located at the point 0 on a certain straight line and suffers random displacements along this line in the form of a series of steps $\tilde{x}_1, \tilde{x}_2, \ldots, \tilde{x}_n$. The size $\tilde{x}_k$ of the kth step is assumed to be a random variable with a known probability law. The particle is executing a *random walk* along the line and its position after n steps $\tilde{z}_n$ is a random variable that arises as the sum of the n steps $\tilde{x}_1, \tilde{x}_2, \ldots, \tilde{x}_n$ (see equation (A3.62)).

The problem of addition of independent random variables is generally solved by the use of characteristic functions (see equation (A2.17)). In fact the probability law of a sum of independent random variables has as its characteristic function the product of the characteristic functions of the individual random variables:

$$\Omega_{\tilde{z}_n}(u) = \Omega_{\tilde{x}_1}(u) \ldots \Omega_{\tilde{x}_n}(u) \tag{A3.67}$$

One may be able to invert the characteristic function of $\tilde{z}_n$ in order to obtain its probability density function (see equation (2.19)). Otherwise the moments of the random variable $\tilde{z}_n$ that exist may be obtained by equation (A2.22).

A4

Sequences of random variables

A4.1 CONVERGENCE OF A SEQUENCE OF RANDOM VARIABLES

Consider a sequence of jointly distributed random variables $\tilde{x}_1, \tilde{x}_2, \ldots, \tilde{x}_n$ defined on the same probability space $\mathscr{S}^{\mathrm{P}}$ and let $\tilde{x}$ be another random variable defined on the same $\mathscr{S}^{\mathrm{P}}$.

The sequence of random variables $\tilde{x}_n$ *converges to* $\tilde{x}$ *with probability* 1 if

$$\text{Prob}\left[\lim_{n\to\infty} \tilde{x}_n = \tilde{x}\right] = 1 \qquad \text{(A4.1)}$$

This convergence implies a 'probability statement simultaneously' about all but a finite number of members of the sequence $\tilde{x}_1, \tilde{x}_2, \ldots, \tilde{x}_n$; given any positive numbers ε and ∂, an integer N exists such that

$$\text{Prob}[|\tilde{x}_N - \tilde{x}| < \varepsilon, \quad |\tilde{x}_{N+1} - \tilde{x}| < \varepsilon, \ldots] > 1 - \partial \qquad \text{(A4.2)}$$

The sequence $\tilde{x}_1, \tilde{x}_2, \ldots, \tilde{x}_n$ *converges in mean square* to the random variable $\tilde{x}$ if

$$\lim_{n\to\infty} \mathrm{E}[(\tilde{x}_n - \tilde{x})^2] = 0 \qquad \text{(A4.3)}$$

The sequence $\tilde{x}_1, \tilde{x}_2, \ldots, \tilde{x}_n$ *converges in probability* to the random variable $\tilde{x}$ if for every positive number ε.

$$\lim_{n\to\infty} \text{Prob}[|\tilde{x}_n - \tilde{x}| > \varepsilon] = 0 \qquad \text{(A4.4)}$$

The convergence differs from the first one as it implies only simultaneous probability statements about each of all but a finite number of members of the sequence $\tilde{x}_1, \tilde{x}_2, \ldots, \tilde{x}_n$; given any positive numbers ε and ∂, an integer N exists such that

$$\text{Prob}[|\tilde{x}_N - \tilde{x}| < \varepsilon] > 1 - \partial \qquad \text{Prob}[|\tilde{x}_{N+1} - \tilde{x}| < \varepsilon] > 1 - \partial \quad \ldots \qquad \text{(A4.5)}$$

If convergence in mean square holds, then Chebyshev's inequality (equation

(A2.61)) for $\varepsilon > 0$

$$\text{Prob}[|\tilde{x}_n - \tilde{x}| > \varepsilon] \leqslant \frac{1}{\varepsilon^2} \text{E}[|\tilde{x}_n - \tilde{x}|^2] \tag{A4.6}$$

implies convergence in probability.

Moreover if the sequence $\tilde{x}_1, \tilde{x}_2, \ldots, \tilde{x}_n$ converges with probability 1, it converges to $\tilde{x}$ in probability. In fact the first convergence involves that for every $\varepsilon > 0$

$$\lim_{N \to \infty} \text{Prob}\left[\sup_{n \geqslant N} |\tilde{x}_n - \tilde{x}| > \varepsilon\right] = 0 \tag{A4.7}$$

while the second convergence implies equation (A4.4). But if $|\tilde{x}_N - \tilde{x}| > \varepsilon$, then $\sup_{n \geqslant N} |\tilde{x}_n - \tilde{x}| > \varepsilon$ and hence

$$\text{Prob}[|\tilde{x}_N - \tilde{x}| > \varepsilon] \leqslant \text{Prob}\left[\sup_{n \geqslant N} |\tilde{x}_n - \tilde{x}| > \varepsilon\right] \tag{A4.8}$$

Thus equation (A4.7) implies equation (A4.4).

However, without additional conditions, convergence in probability implies neither convergence in mean square nor convergence with probability 1. Further, the last one neither implies nor is implied by convergence in mean square. But if a sequence $\tilde{x}_1, \tilde{x}_2, \ldots, \tilde{x}_n$ converges in mean square to 0 in such a way that

$$\sum_{n=1}^{\infty} \text{E}[\tilde{x}_n^2] < \infty \tag{A4.9}$$

then it follows that the sequence converges to 0 with probability 1.

A4.2 THE LAW OF LARGE NUMBERS

A rough statement of the law of large numbers, first established by Bernoulli in 1713, expresses that the sample average (equation (A3.63)) of a random sample $\tilde{x}_1, \tilde{x}_2, \ldots, \tilde{x}_n$ of a random variable $\tilde{x}$ is approximately equal to the ensemble mean $\mu_{\tilde{x}}$ with a probability that can be made as close to 1 as desired by taking a large enough sample.

A more precise mathematical statement is the following. A sequence of jointly distributed random variables $\tilde{x}_1, \tilde{x}_2, \ldots, \tilde{x}_n$, with finite mean, is said to obey the strong law of large numbers, the weak law or the quadratic mean *law of large numbers* if (see equation (A3.63)).

$$\tilde{\zeta}_n = \tilde{z}_n/n - \mu_{\tilde{z}}/n \tag{A4.10}$$

converges to 0 as n tends to ∞, with probability 1, in probability or in quadratic mean respectively.

More explicitly, the sequence $\tilde{x}_1, \tilde{x}_2, \ldots, \tilde{x}_n$ obeys the *strong law of large numbers* if to every pair $\varepsilon > 0$ and $\partial > 0$ there corresponds an integer N such that

$$\text{Prob}[|\zeta_N| < \varepsilon, |\zeta_{N+1}| < \varepsilon, \ldots] > 1 - \partial \tag{A4.11}$$

while the same sequence obeys the *weak law of large numbers* if for every $\varepsilon > 0$

$$\lim_{n \to \infty} \text{Prob}[|\zeta_n| > \varepsilon] = 0 \tag{A4.12}$$

If the variables $\tilde{x}_1, \tilde{x}_2, \ldots, \tilde{x}_n$ are mutually independent and identically distributed, the finiteness of the common mean $\mu_{\tilde{x}}$ implies the respect of the strong law of large numbers. For not necessarily identically distributed random variables with finite means, the weak law of large numbers holds whenever $\tilde{x}_1, \tilde{x}_2, \ldots, \tilde{x}_n$ are uniformly bounded ($|\tilde{x}_k| < c$ for all k, where c is a constant). Also, a sufficient (but not necessary) condition is that the sample variance (equation (A3.66) divided by n, s^2/n, converges to 0 as n tends to ∞. This is a direct consequence of the Chebyshev inequality (A2.61). More generally, for some $\xi > 0$, the condition

$$\lim_{n \to \infty} \frac{1}{n^{1+\xi}} \sum_{k=1}^{n} \text{E}[|\tilde{x}_k - \mu_{\tilde{x}_k}|^{1+\xi}] = 0 \tag{A4.13}$$

is the Markov condition for the validity of the weak law of large numbers.

Consider now the case of dependent random variables $\tilde{x}_k$ with finite means and finite variances and introduce the covariance between the nth summand $\tilde{x}_n$ and the nth sample mean $\tilde{m}_n$ (equation (A3.63)):

$$\text{Cov}[\tilde{m}_n, \tilde{x}_n] = \frac{1}{n} \sum_{k=1}^{n} \text{E}[\tilde{x}_k \tilde{x}_n] \tag{A4.14}$$

Then a sequence of jointly distributed random variables $\tilde{x}_1, \tilde{x}_2, \ldots, \tilde{x}_n$ with zero mean and uniformly bounded variance (i.e. $s_{\tilde{x}}^2 \leqslant c$ for all n) obeys the quadratic law of large numbers if and only if

$$\lim_{n \to \infty} \text{Cov}[\tilde{m}_n, \tilde{x}_n] = 0 \tag{A4.15}$$

while it obeys the strong law of large numbers if positive constants C and D exist such that for all integers n

$$|\text{Cov}[\tilde{m}_n, \tilde{x}_n]| \leqslant C/n^D \tag{A4.16}$$

Consider now n independent repeated Bernoulli trials (see equations (A1.13) and (A1.14)) with probability p for success. Let 1 and 0 be the success and the failure respectively. Thus the sum $\tilde{z}_n$ of n outcomes coincides with the number of successes in n trials and hence equation (A1.14) and Table A.1 involve that $\mu_{\tilde{z}_n} = np$. Therefore, in this case, the weak law of large numbers (A4.12) becomes

$$\lim_{n \to \infty} \text{Prob}\left[\left|\frac{\tilde{z}_n}{n} - p\right| > \varepsilon\right] = 0 \tag{A4.17}$$

i.e. as n increases, the probability that the average number of successes deviates from p by more than any given ε tends to zero. This statement serves as a basis for the intuitive notion of probability as a measure of relative frequencies.

Moreover, the strong law of large numbers (A4.11) for Bernoulli trials, as formulated by Cantelli in 1917, implies that

$$\text{Prob}\left[\lim_{n\to\infty}\left|\frac{\tilde{x}_n}{n}-p\right|=0\right]=1 \tag{A4.18}$$

i.e. the frequency of successes is bounded to stay near p if the number of trials is increased, whereas the weak law only states that for every sufficiently large fixed n the average $\tilde{x}_n/n$ is likely to be near p.

A4.3 CONVERGENCE IN DISTRIBUTION OF A SEQUENCE OF RANDOM VARIABLES

For $n = 1, 2, \ldots$, let $\tilde{x}_n$ be a random variable with distribution function $P_{\tilde{x}_n}(\ )$ and characteristic function $\Omega_{x_n}(\ )$ and let $\tilde{x}$ be a random variable with $P_{\tilde{x}}(\ )$ and $\Omega_{\tilde{x}}(\ )$. The law of $\tilde{x}_n$ converges to the law of $\tilde{x}$

$$\mathscr{L}(\tilde{x}_n) \to \mathscr{L}(\tilde{x}) \tag{A4.19}$$

if any one of the following equivalent statements holds:

(1) For every bounded continuous function $g(\ )$ of a real variable, there is convergence of the expectation $\mathrm{E}[g(\tilde{x}_n)]$ to $\mathrm{E}[g(\tilde{x})]$:

$$\lim_{n\to\infty} \mathrm{E}[g(\tilde{x}_n)] = \mathrm{E}[g(\tilde{x})] \tag{A4.20}$$

(2) For every continuous function, there is convergence of the distribution functions

$$\lim_{n\to\infty} P_{g(\tilde{x}_n)}(y) = P_{g(\tilde{x})}(y) \tag{A4.21}$$

at every real number y at which $P_{g(\tilde{x})}(\ )$ is continuous.

(3) At every real number, there is convergence of the characteristic functions

$$\lim_{n\to\infty} \Omega_{\tilde{x}_n}(u) = \Omega_{\tilde{x}}(u) \tag{A4.22}$$

(4) At every real number x that is a point of continuity of $P_{\tilde{x}}(\ )$, there is convergence of the distribution functions

$$\lim_{n\to\infty} P_{\tilde{x}_n}(x) = P_{\tilde{x}}(x) \tag{A4.23}$$

(5) At every two points x_1 and x_2, with $x_1 < x_2$, at which $P_{\tilde{x}}$ is continuous, there is convergence of the probability functions over the interval x_1 to x_2

$$\lim_{n\to\infty}\left(P_{\tilde{x}_n}(x_2) - P_{\tilde{x}_n}(x_1)\right) = P_{\tilde{x}}(x_2) - P_{\tilde{x}}(x_1) \tag{A4.24}$$

These results involve that the one-to-one correspondence between distribution and characteristic functions is continuous in the sense that a sequence of distribution functions $P_n(\quad)$ converges to a distribution function $P(\quad)$ at all points of continuity of $P(\quad)$ if and only if the sequence of characteristic functions $\Omega_n(u)$ converges at each real number u to the characteristic function $\Omega(\quad)$ of $P(\quad)$. This is stated by the *continuity theorem of probability theory*: consider a sequence of distribution functions $P_n(\quad)$ with characteristic functions $\Omega_n(u)$; in order that a function $P(\quad)$ exists such that

$$\lim_{n \to \infty} P_n(x) = P(x) \tag{A4.25}$$

at all points x, which are continuity points of $P(x)$, it is necessary and sufficient that a function $\Omega(u)$, continuous at $u = 0$, exists such that

$$\lim_{n \to \infty} \Omega_n(u) = \Omega(u) \tag{A4.26}$$

at all real u.

A4.4 THE CENTRAL LIMIT THEOREM

A rough statement of the central limit theorem, already surmised by Laplace and Gauss in the early 1800s, expresses that the sum of a large number of independent identically distributed random variables with finite means and variances, normalized to have mean zero and variance 1, is approximately normally distributed. For this purpose, given any random variable $\tilde{x}$ with finite mean and variance, its *standardization* $\tilde{x}$ is the random variable

$$\tilde{x}^* = (\tilde{x} - \mu_{\tilde{x}})/s_{\tilde{x}} \tag{A4.27}$$

The 'standardization' $\tilde{x}^*$ is a dimensionless random variable with $\mu_{\tilde{x}^*} = 0$ and $s^2_{\tilde{x}^*} = 1$.

A more precise mathematical statement of the central limit theorem is the following. A sequence of jointly distributed random variables $\tilde{x}_1, \tilde{x}_2, \ldots, \tilde{x}_n$, with finite mean and variance, obeys the classical central limit theorem if the sequence $\tilde{\varkappa}_1, \tilde{\varkappa}_2, \ldots, \tilde{\varkappa}_n$ of the standardization of their sum

$$\tilde{\varkappa}_n = \tilde{x}_1 + \tilde{x}_2 + \ldots + \tilde{x}_n \tag{A4.28}$$

converges in distribution to a random variable that is normally distributed with mean 0 and variance 1.

From the practical viewpoint, the following two versions of the central limit theorem are of interest.

(1) Let $\tilde{x}_1, \tilde{x}_2, \ldots, \tilde{x}_n$ be independent identically distributed as the random variable $\tilde{x}$ with finite mean $\mu_{\tilde{x}}$ and standard deviations $s_{\tilde{x}}$. Then

$$\tilde{\varkappa}_n^* = \frac{(\tilde{x}_1 + \tilde{x}_2 + \ldots + \tilde{x}_n) - n\mu_{\tilde{x}}}{\sqrt{n}\, s_{\tilde{x}}} \tag{A4.29}$$

for $n = 1, 2, \ldots$ defines a sequence that converges in distribution to a random variable that is normally distributed with mean 0 and variance 1.

(2) Let $\tilde{x}_1, \tilde{x}_2, \ldots, \tilde{x}_n$ be independent random variables with finite mean $\mu_{\tilde{x}_k}$ and finite $(2+\xi)$ central moment $E[|\tilde{x}_k - \mu_{\tilde{x}_k}|^{2+\xi}]$. Then the central limit theorem holds if

$$\lim_{n \to \infty} \frac{1}{s_{\tilde{x}_n}^{2+\xi}} \sum_{k=1}^{n} E[|\tilde{x}_k - \mu_{\tilde{x}_k}|^{2+\xi}] = 0 \tag{A4.30}$$

with

$$s_{\tilde{x}_n}^2 = \sum_{k=1}^{n} s_{\tilde{x}_k}^2$$

Equation (A4.30) is called the *Liapunov condition* for the validity of the central limit theorem.

Equation (A4.29) for n independent repeated Bernoulli trials becomes the De Moivre–Laplace limit theorem: for fixed $z_1 < z_2$

$$\lim_{n \to \infty} \text{Prob}[np + z_1\sqrt{(npq)} \leqslant \tilde{x}_n < np + z_2\sqrt{(npq)}] = \Phi(z_2) - \Phi(z_1) \tag{A4.31}$$

where $\Phi(\ )$ denotes the normal distribution function.

A5

Random processes

A5.1 DESCRIBING A RANDOM PROCESS

A *random* (or *stochastic*) *process* is a family of random variables $\{\tilde{x}(t), t \in T\}$ indexed by a parameter t varying in an index set T.

For practical purposes it is adequately represented by some finite number of ordinates (say n). Hence a way of describing a stochastic process is to specify either the joint distribution function of the n random variables $\tilde{x}(t_1), \tilde{x}(t_2), \ldots, \tilde{x}(t_n)$, with $t_1, t_2, \ldots, t_n$ in T,

$$P_{\tilde{x}(t_1), \tilde{x}(t_2), \ldots, \tilde{x}(t_n)}(x_1, x_2, \ldots, x_n) = \text{Prob}[\tilde{x}(t_1) \leqslant x_1, \tilde{x}(t_2) \leqslant x_2, \ldots, \tilde{x}(t_n) \leqslant x_n] \quad \text{(A5.1a)}$$

or their joint characteristic function

$$\Omega_{\tilde{x}(t_1), \tilde{x}(t_2), \ldots, \tilde{x}(t_n)}(u_1, u_2, \ldots, u_n) = \text{E}[\exp\{\text{i}(u_1 \tilde{x}(t_1) + u_2 \tilde{x}(t_2) + \ldots + u_n \tilde{x}(t_n))\}] \quad \text{(A5.1b)}$$

A simpler, although less complete, description of stochastic processes consists of analysing the behaviour of their first and second moments, when they exist and are finite. In the last case the *mean value function* $\mu_{\tilde{x}(t)}$ is defined for all t in T by

$$\mu_{\tilde{x}(t)}(t) = \text{E}[\tilde{x}(t)] \quad \text{(A5.2)}$$

and the *covariance kernel* or *autocovariance function* $\Sigma_{\tilde{x}(t)}(t_1, t_2)$ is defined for all t_1 and t_2 in T by

$$\Sigma_{\tilde{x}(t)}(t_1, t_2) = \text{Cov}[\tilde{x}(t_1), \tilde{x}(t_2)] \quad \text{(A5.3)}$$

Further, for two random processes $\tilde{x}(t)$ and $\tilde{y}(t)$, the cross-covariance function $\Sigma_{\tilde{x}(t), \tilde{y}(t)}(t_1, t_2)$ is defined for all t_1 and t_2 in T by

$$\Sigma_{\tilde{x}(t), \tilde{y}(t)}(t_1, t_2) = \text{Cov}[\tilde{x}(t_1), \tilde{y}(t_2)] \quad \text{(A5.4)}$$

Instead of the covariance function, one can describe random processes by means of the correlation functions; particularly the *autocorrelation function* is

$$\Sigma'_{\tilde{x}(t)}(t_1, t_2) = \mathrm{E}[\tilde{x}(t_1)\tilde{x}(t_2)] = \Sigma_{\tilde{x}(t)}(t_1, t_2) + \mu_{\tilde{x}(t)}(t_1)\mu_{\tilde{x}(t)}(t_2) \qquad \text{(A5.3')}$$

while the *cross-correlation function* is

$$\Sigma'_{\tilde{x}(t),\tilde{y}(t)}(t_1, t_2) = \mathrm{E}[\tilde{x}(t_1)\tilde{y}(t_2)] = \Sigma_{\tilde{x}(t),\tilde{y}(t)}(t_1, t_2) + \mu_{\tilde{x}(t)}(t_1)\mu_{\tilde{y}(t)}(t_2) \qquad \text{(A5.4')}$$

Stochastic processes can be divided into two broad classes: *stationary* and *evolutionary*. Intuitively, a stationary process is one whose distribution remains the same as time progresses; an evolutionary process is one which is not stationary. For a more formal definition of stationary, consider a 'linear index set' T, i.e. a set that has the property that the sum $(t+v)$ of any two members t and v of T also belongs to T. Then, a stochastic process $\{\tilde{x}(t), t \in T\}$ is said to be:

(1) *strictly stationary of order* k, where k is a given positive integer, if the k-dimensional random vectors $(\tilde{x}(t_1), \ldots, \tilde{x}(t_k))$ and $(\tilde{x}(t_1+v), \ldots, \tilde{x}(t_k+v))$ are identically distributed for any k points $t_1, \ldots, t_k$ in T and any v in T;
(2) *strictly stationary* if for any integer k is strictly stationary of order k;
(3) *weakly stationary* (or *covariance stationary*) if it possesses finite second moments and the covariance kernel $\Sigma_{\tilde{x}(t)}(t_1, t_2)$ is a function only of the absolute difference $|t_1 - t_2| = v$, in the sense that there exists a function $B(v)$ such that for all t_1 and t_2 in T

$$B_{\tilde{x}(t)}(v) = \Sigma_{\tilde{x}(t)}(t_1, t_2) \qquad \text{(A5.5)}$$

The function $B_{\tilde{x}(t)}(v)$ is called the 'covariance function' (or autocovariance function) of the stationary $\{\tilde{x}(t), t \in T\}$.

In a similar way, one can define the cross-covariance function $B_{\tilde{x}(t),\tilde{y}(t)}(v)$ for stationary random processes. Further, according to equation (A5.3'), one can define the *autocorrelation function* of the stationary process $\tilde{x}(t)$ as

$$B_{\tilde{x}(t)}(v) = \Sigma'_{\tilde{x}(t)(t_1, t_2)} = B_{\tilde{x}(t)}(v) - \mu^2_{\tilde{x}(t)} \qquad \text{(A5.5')}$$

The stationary process is sometimes called *homogeneous*, in particular if t is a space parameter.

Finally consider the sample (or time) means (see equation (A3.63):

$$\tilde{m}_T = \begin{cases} \dfrac{1}{T}\displaystyle\sum_{t=1}^{T} \tilde{x}(t) & \text{If } \tilde{x} \text{ is a discrete-parameter stochastic process} \\ \dfrac{1}{T}\displaystyle\int_0^T \tilde{x}(t)\,\mathrm{d}t & \text{if } \tilde{x} \text{ is a continuous-parameter stochastic process} \end{cases} \qquad \text{(A5.6)}$$

formed from increasingly larger samples. The sequence of sample means is said to be *ergodic* if

$$\lim_{T \to \infty} \mathrm{Var}[\tilde{m}_T] = 0 \qquad \text{(A5.7)}$$

i.e. $\mathrm{Var}[\tilde{m}_T]$ converges in mean square (equation (A4.3)) to 0. If $\tilde{x}$ is a discrete parameter stochastic process, from equation (A3.28) $\mathrm{Var}[\tilde{m}_T]$ is given by

$$\mathrm{Var}[\tilde{m}_T] = \frac{1}{T^2} \sum_{t_1=1}^{T} \sum_{t_2=1}^{T} \Sigma_{\tilde{x}(t)}(t_1, t_2) \tag{A5.8}$$

while if $\tilde{x}$ is a continuous parameter stochastic process, from equation (A5.22) it follows that

$$\mathrm{Var}[\tilde{m}_T] = \frac{2}{T^2} \int_0^T \mathrm{d}t_2 \int_0^{t_2} \Sigma_{\tilde{x}(t)}(t_1, t_2)\, \mathrm{d}t_1 \tag{A5.9}$$

Then, for processes with bounded variance, equation (A5.7) is satisfied if (but not only if)

$$\lim_{|t_1 - t_2| \to \infty} \Sigma_{\tilde{x}(t))}(t_1, t_2) = 0 \tag{A5.10}$$

Then, in general, a stochastic process is said to be *ergodic in the mean* if sample averages formed from an observed record of the process may be used as an approximation to the corresponding averages.

If $\tilde{x}(t)$ is a weakly stationary process, the condition (A5.7) becomes

$$\begin{aligned} &\lim_{T\to\infty} \frac{1}{T} \sum_{v=0}^{T-1} B_{\tilde{x}(t)}(v) = 0 \\ &\lim_{T\to\infty} \frac{1}{T} \int_0^T B_{\tilde{x}(t)}(v)\, \mathrm{d}v = 0 \end{aligned} \tag{A5.11}$$

for discrete and continuous parameter processes respectively and this convergence occurs (but the condition is not necessary) if

$$\lim_{v\to\infty} B_{\tilde{x}(t)}(v) = 0 \tag{A5.12}$$

for processes with bounded variance. The convergence to 0 of equation (A5.12) is equivalent to continuity of the spectral distribution function $V_{\tilde{x}(t)}(\omega)$ (equation (A5.43)) at $\omega = 0$ when the spectral representation of $\tilde{x}(t)$ is employed (see Section A5.3).

Further let $\tilde{x}(t)$ be a continuous parameter process stationary up to fourth moments and consider the sample means of the random process $\tilde{\zeta}_v(t) = [\tilde{x}(t+v) - \mu_{\tilde{x}(t)}][\tilde{x}(t) - \mu_{\tilde{x}(t)}]$:

$$\tilde{b}_T(v) = \frac{1}{T} \int_0^T \zeta_v(t)\, \mathrm{d}t = \frac{1}{T} \int_0^T [\tilde{x}(t+v) - \mu_{\tilde{x}(t)}][\tilde{x}(t) - \mu_{\tilde{x}(t)}]\, \mathrm{d}t \tag{A5.13}$$

formed from increasingly larger samples. The sequence of $\tilde{b}_T(v)$ is ergodic if

$$\lim_{T\to\infty} \mathrm{Var}[\tilde{b}_T(v)] = 0 \tag{A5.14}$$

with v given. Then if equation (A5.14) holds for all $v \geqslant 0$, the stochastic process $\tilde{x}(t)$ is said to be *ergodic in the covariance function.*

If all higher-order moments of $\tilde{x}(t)$ equal the corresponding time averages, $\tilde{x}(t)$ is *ergodic in the strict sense.*

A5.2 INTEGRATION AND DIFFERENTIATION OF RANDOM PROCESSES

A5.2.1 Integration

Consider the random variable

$$\tilde{z} = \int_a^b \tilde{x}(t)\,\mathrm{d}t = \lim_{\max(t_k - t_{k-1}) \to 0} \sum_{k=1}^{n} \tilde{x}(t_k)(t_k - t_{k-1}) \tag{A5.15}$$

defined as the integral of a stochastic process over an interval $a < t \leqslant b$, where different convergence methods may be considered (see Section A4.1). The following is mainly concerned with convergence in mean square (equation (A4.3)).

Let $\{\tilde{x}(t), t \geqslant 0\}$ be a continuous parameter stochastic process with finite second moments, whose mean value function $\mu_{\tilde{x}(t)}(t)$ and covariance kernel $\Sigma_{\tilde{x}(t)}(t_1, t_2)$ are continuous functions of t_1 and t_2. Then for any $b > a \geqslant 0$, the integral $\tilde{z}$ is well defined as the limit in mean square of the usual approximating sums of the form of equation (A5.15). Then for any non-negative real numbers a, b, a' and b' the following formulae hold:

$$\mu_{\tilde{z}} = \int_a^b \mu_{\tilde{x}(t)}(t)\,\mathrm{d}t \tag{A5.16}$$

$$\mathrm{E}[\tilde{z}^2] = \int_a^b \int_a^b \mathrm{E}[\tilde{x}(t_1)\tilde{x}(t_2)]\,\mathrm{d}t_1\,\mathrm{d}t_2 \tag{A5.17}$$

$$s_{\tilde{z}}^2 = \int_a^b \int_a^b \Sigma_{\tilde{x}(t)}(t_1, t_2)\,\mathrm{d}t_1\,\mathrm{d}t_2 = 2\int_a^b \mathrm{d}t_2 \int_a^{t_2} \Sigma_{\tilde{x}(t)}(t_1, t_2)\,\mathrm{d}t_1{}^{\ddagger} \tag{A5.18}$$

$$\mathrm{E}\left[\int_a^b \tilde{x}(t_1)\,\mathrm{d}t_1 \int_{a'}^{b'} \tilde{x}(t_2)\,\mathrm{d}t_2\right] = \int_a^b \mathrm{d}t_1 \int_{a'}^{b'} \mathrm{d}t_2\,\mathrm{E}[\tilde{x}(t_1)\tilde{x}(t_2)] \tag{A5.19}$$

$$\mathrm{Cov}\left[\int_a^b \tilde{x}(t_1)\,\mathrm{d}t_1, \int_{a'}^{b'} \tilde{x}(t_2)\,\mathrm{d}t_2\right] = \int_a^b \mathrm{d}t_1 \int_{a'}^{b'} \mathrm{d}t_2\,\Sigma_{\tilde{x}(t)}(t_1, t_2) \tag{A5.20}$$

Thus, for a stochastic process $\tilde{x}(t)$, $t \geqslant 0$, which is being observed continuously over an interval $0 \leqslant t \leqslant T$, equations (A5.16) and (A5.17) provide the mean and variance of the sample mean $\tilde{m}$ defined by equation (A5.6):

$$\mu_{\tilde{m}_T} = \frac{1}{T}\int_0^T \mu_{\tilde{x}(t)}(t)\,\mathrm{d}t \tag{A5.21}$$

$$s_{\tilde{m}_T}^2 = \frac{2}{T^2}\int_0^T \mathrm{d}t_2 \int_0^{t_2} \mathrm{d}t_1\,\Sigma_{\tilde{x}(t)}(t_1, t_2) \tag{A5.22}$$

Furthermore equations (A5.16) and (A5.17) can be generalized for the weighted

‡ This is true because $\Sigma_{\tilde{x}(t)}(t_1, t_2)$ is a symmetric function, $\Sigma(t_1, t_2) = \Sigma(t_2, t_1)$.

integral

$$\tilde{z}_h = \int_{-\infty}^{\infty} h(t)\tilde{x}(t)\,dt \tag{A5.23}$$

where $h(t)$ is a given weight function. In this case

$$\mu_{\tilde{z}_h} = \int_{-\infty}^{\infty} h(t)\mu_{\tilde{x}(t)}(t)\,dt \tag{A5.24}$$

$$s^2_{\tilde{z}_h} = \int_{-\infty}^{\infty}\int_{-\infty}^{\infty} h(t_1)h(t_2)\Sigma_{x(t)}(t_1, t_2)\,dt_1\,dt_2 \tag{A5.25}$$

A5.2.2 Differentiation

Let $\{\tilde{x}(t), t \geqslant 0\}$ be a stochastic process with finite second moments. The derivative $\dot{\tilde{x}}(t)$ is defined by

$$\tilde{y}(t) = \dot{\tilde{x}}(t) = \lim_{v \to \infty} \frac{\tilde{x}(t+v) - x(t)}{v} \tag{A5.26}$$

where the limit is in the sense of convergence in mean square. A sufficient condition for the existence of the r.h.s. limit in equation (A5.26) is that the function $\mu_{\tilde{x}(t)}(t)$ is differentiable and the mixed second derivative $\partial^2(\Sigma_{\tilde{x}(t)}(t_1, t_2))/\partial t_1 \partial t_2$ exists and is continuous. Under these conditions, the following formulae hold:

$$\mu_{\tilde{y}(t)}(t) = \frac{d}{dt}(\mu_{\tilde{x}(t)}(t)) = \dot{\mu}_{\tilde{x}(t)}(t) \tag{A5.27}$$

$$\text{Cov}[\tilde{y}(t_1), \tilde{y}(t_2)] = \frac{d}{dt_1}\frac{d}{dt_2}\text{Cov}[\tilde{x}(t_1), \tilde{x}(t_2)] = \frac{\partial^2}{\partial t_1 \partial t_2}(\Sigma_{\tilde{x}(t)}(t_1, t_2)) \tag{A5.28}$$

$$\text{Cov}[\tilde{y}(t_1), \tilde{x}(t_2)] = \frac{d}{dt_1}\text{Cov}[\tilde{x}(t_1), \tilde{x}(t_2)] = \frac{\partial}{\partial t_1}(\Sigma_{\tilde{x}(t)}(t_1, t_2) \tag{A5.29}$$

A stochastic process $\{\tilde{x}(t), t \geqslant 0\}$ is said to be *differentiable in mean square* if $\tilde{y}(t) = \dot{\tilde{x}}(t)$ exists as a limit in mean square for all $t \geqslant 0$.

The derivative process $\{\dot{\tilde{x}}(t), t \geqslant 0\}$ of a covariance stationary process $\{\tilde{x}(t), t \geqslant 0\}$ is covariance stationary with

$$B_{\dot{\tilde{x}}(t)}(v) = -\ddot{B}_{\tilde{x}(t)}(v) \tag{A5.30}$$

A5.2.3 Stochastic differential equations

Consider the linear differential equation with constant coefficients a_j ($j = 1, \ldots, n$):

$$a_n \frac{\partial^n \tilde{y}(t)}{\partial t^n} + a_{n-1}\frac{\partial^{n-1}\tilde{y}(t)}{\partial t^{n-1}} + \ldots + a_0 \tilde{y}(t) = \tilde{x}(t) \tag{A5.31}$$

$$\tilde{y}(0) = y_{00} \qquad \dot{\tilde{y}}(0) = y_{01} \quad \ldots \quad \left.\frac{\partial^{n-1}\tilde{y}(t)}{\partial t^{n-1}}\right|_{t=0} = y_{0(n-1)} \tag{A5.32}$$

where $\tilde{x}(t)$ is a random process with mean value function $\mu_{\tilde{x}(t)}(t)$ and covariance kernel $\Sigma_{\tilde{x}(t)}(t_1 t_2)$ and equation (A5.32) denotes given initial conditions. Then the mean value function of $\tilde{y}(t)$ satisfies the differential equation

$$a_n \frac{\partial^n}{\partial t^n} \mu_{\tilde{y}(t)}(t) + \ldots + a_0 \mu_{\tilde{y}(t)} = \mu_{\tilde{x}(t)}(t)$$

$$\mu_{\tilde{y}(t)}(0) = y_{00} \qquad \frac{\partial}{\partial t} \mu_{\tilde{y}(t)}(t) \bigg|_{t=0} = y_{01} \quad \ldots \tag{A5.33}$$

The cross-covariance $\Sigma_{\tilde{x}(t),\, \tilde{y}(t)}(t_1, t_2)$ between $\tilde{x}(t)$ and $\tilde{y}(t)$ satisfies the differential equation

$$a_n \frac{\partial^n}{\partial t_2^n} \Sigma_{\tilde{x}(t),\, \tilde{y}(t)}(t_1, t_2) + \ldots + a_0 \Sigma_{\tilde{x}(t),\, \tilde{y}(t)}(t_1, t_2) = \Sigma_{\tilde{x}(t)}(t_1, t_2)$$

$$\Sigma_{\tilde{x}(t),\, \tilde{y}(t)}(t_1, 0) = 0 \qquad \frac{\partial}{\partial t_2} \Sigma_{\tilde{x}(t),\, \tilde{y}(t)}(t_1, t_2) \bigg|_{t_2=0} = 0 \quad \ldots \tag{A5.34}$$

and the covariance kernel $\Sigma_{\tilde{y}(t)}(t_1, t_2)$ of $\tilde{y}(t)$ satisfies the differential equation

$$a_n \frac{\partial^n}{\partial t_1^n} \Sigma_{\tilde{y}(t)}(t_1, t_2) + \ldots + a_0 \Sigma_{\tilde{y}(t)}(t_1, t_2) = \Sigma_{\tilde{x}(t),\, \tilde{y}(t)}(t_1, t_2)$$

$$\Sigma_{\tilde{y}(t)}(0, t_2) = 0 \qquad \frac{\partial}{\partial t_1} \Sigma_{\tilde{y}(t)}(t_1, t_2) \bigg|_{t_1=0} = 0 \quad \ldots \tag{A5.35}$$

A5.2.4 Chebyshev's inequality for stochastic processes

Let $\{\tilde{x}(t), a \leqslant t \leqslant b\}$ be a stochastic process which is differentiable in mean square. Then an upper bound for the probability

$$\text{Prob}[E] = \text{Prob}[|\tilde{x}(t) - \mu_{\tilde{x}(t)}(t)| \leqslant \varepsilon \quad \text{for all } t\!: a \leqslant t \leqslant b] \tag{A5.36}$$

is given by the inequality

$$\text{Prob}[E] \geqslant 1 - \frac{s_{\tilde{x}(a)}^2 + s_{\tilde{x}(b)}^2}{2\varepsilon^2} + \frac{1}{\varepsilon^2} \int_a^b s_{\tilde{x}(t)} s_{\dot{\tilde{x}}(t)} \mathrm{d}t \tag{A5.37}$$

A5.3 DEPENDENT RANDOM PROCESSES AND HARMONIC ANALYSIS OF RANDOM PROCESSES

Consider a 'black box' which in response to an input $\{x(t), -\infty < t < \infty\}$ yields an output represented by $\{y(t), -\infty < t < \infty\}$. The black box is said to be a 'time-invariant linear filter' if it can be regarded as converting an input to an output and satisfies the following two conditions:

(1) the output relevant to the superposition of two inputs is the superposition of the relevant output;
(2) the delay in the input by a fixed time only involves the delay in the output by the same time.

Examples of time-invariant linear filters are linear combinations, the combination of derivatives and integral operators. When the output function is expressed by ($v = t_2 - t_1$)

$$y(t_2) = \int_{-\infty}^{t_2} h(t_2 - t_1)x(t_1)\,dt_1 = \int_0^{\infty} h(v)x(t_2 - v)\,dv \qquad \text{(A5.38)}$$

the function $h(\ )$ is called the *impulse response function.*

A filter can be described by an explicit formula for the output $y(t) = g(x(t))$ or in terms of their *frequency response.* Consider the function

$$x(t) = X e^{i\omega t} \qquad \text{(A5.39)}$$

for some real number ω. It is called the complex harmonic with amplitude X and frequency ω, while the functions

$$x(t) = X \cos \omega t \qquad \text{and} \qquad x(t) = X \sin \omega t \qquad \text{(A5.40)}$$

are called real harmonics with amplitude X and frequency ω (remember that $e^{i\omega t} = \cos \omega t + i \sin \omega t$).

It is possible to show that if a complex harmonic of unit amplitude, frequency ω is applied to a time-invariant linear filter, then the corresponding output (if it exists) $y(t)$ is a complex harmonic of amplitude $A(\omega)$ and frequency ω, i.e.

$$y(t) = A(\omega)e^{i\omega t} \qquad \text{(A5.41a)}$$

where the *frequency response function* of the time-invariant linear filter defined by equation (A5.38) is expressed by

$$A(\omega) = \int_0^{\infty} e^{-i\omega v} h(v)\,dv \qquad \text{(A5.41b)}$$

This function is the most intuitively meaningful way of describing a filter because for those inputs which possess a Fourier transform and which give rise to an output possessing a Fourier transform the effect of the filtering operation can be expressed very simply by

$$\begin{aligned} y_F(\omega) &= \frac{1}{2\pi}\int_{-\infty}^{\infty} e^{-i\omega t} y(t)\,dt \\ &= \frac{1}{2\pi}\left(\int_0^{\infty} dv\, h(v)e^{-i\omega t}\right)\int_{-\infty}^{\infty} dt_2\, e^{-i\omega t_2}x(t_2) \\ &= A(\omega)x_F(\omega) \end{aligned} \qquad \text{(A5.42)}$$

In other words, the Fourier transform of the output is equal to the product of the frequency response function of the filter and the Fourier transform of the input.

However, no meaning can be attached to the integral defining the Fourier transform of a stochastic process. In fact, their sample functions are non-periodic undamped functions and hence do not satisfy the usual assumptions in the usual theory of Fourier integrals. Nevertheless, it is possible to conceive a different approach that consists of searching for the contribution of each frequency ω to the content of the process: such an approach is called the *harmonic analysis of stochastic processes.* The essential of this theory is stated in the following with reference to continuous parameter time series $\{\tilde{x}(t), t \geqslant 0\}$ that are covariance stationary.

If the stochastic process is ergodic in the covariance function (i.e. equation (A5.14) holds), it may be shown that there exists a bounded non-decreasing function of a real variable ω, $V_{\tilde{x}(t)}(\omega)$, called the *spectral distribution function,*

$$B_{\tilde{x}(t)}(v) = \int_{-\infty}^{\infty} e^{iv\omega} \, dV_{\tilde{x}(t)}(\omega) \qquad -\infty < v < \infty \tag{A5.43}$$

For any frequencies $\omega_1 < \omega_2$, $V_{\tilde{x}(t)}(\omega_2) - V_{\tilde{x}(t)}(\omega_1)$ may be interpreted as a measure of the contribution of the frequency band ω_1 to ω_2 to the 'content' of the stochastic process. If $V_{\tilde{x}(t)}(\omega)$ is everywhere differentiable, its derivative $G_{\tilde{x}(t)}(\omega)$ is called the *spectral density function* of the stochastic process:

$$B_{\tilde{x}(t)}(v) = \int_{-\infty}^{\infty} e^{iv\omega} G_{\tilde{x}(t)}(\omega) \, d\omega \tag{A5.44}$$

A sufficient condition for equation (A5.44) to hold is that

$$\int_{-\infty}^{\infty} |B(v)| \, dv < \infty \tag{A5.45}$$

Then an explicit formula for the spectral density function is given by

$$G_{\tilde{x}(t)}(\omega) = \frac{1}{2\pi} \int_{-\infty}^{\infty} e^{-iv\omega} B_{\tilde{x}(t)}(v) \, dv \tag{A5.46}$$

The spectral density function derives its importance from the fact that the following statement may be proved: the spectral density function of the output stochastic process is equal to the product of the square of the modulus of the frequency response function of the filter by the spectral density function of input stochastic process (see Section A5.5 (equation (A5.73)).

It is worth noting that a covariance stationary stochastic process $\{\upsilon(t), t \geqslant 0\}$ is said to be a *white-noise* if it possesses a constant spectral density function: for some $G_0 > 0$

$$G_{\tilde{x}(t)}(\omega) = G_0 \qquad \text{for all } \omega \tag{A5.47a}$$

The covariance function is then, in terms of the Dirac delta-function $\partial(v)$

$$B_{\tilde{x}(t)}(v) = 2\pi G_0 \partial(v) \tag{A5.47b}$$

Let $\tilde{x}(t)$ and $\tilde{y}(t)$ be jointly stationary random processes. Then equation (A5.46) is generalized by introducing the *cross-spectral density function* $G_{\tilde{x}(t),\tilde{y}(t)}(\omega)$ defined by

$$G_{\tilde{x}(t),\tilde{y}(t)}(\omega) = \frac{1}{2\pi}\int_{-\infty}^{\infty} B_{\tilde{x}(t),\tilde{y}(t)}(v)\,\mathrm{e}^{-\mathrm{i}\omega v}\mathrm{d}v \qquad \text{(A5.48)}$$

Moreover, equations analogous to equations (A5.46) and (A5.48) can be written starting from equations (A5.3′), (A5.4′) and (A5.5′). In this way a different notion of spectral density function is introduced; it will be marked by a prime:

$$G'_{\tilde{x}(t)}(\omega) = \frac{1}{2\pi}\int_{-\infty}^{\infty} B'_{\tilde{x}(t)}(v)\,\mathrm{e}^{-\mathrm{i}\omega v}\mathrm{d}v \qquad \text{(A5.46′)}$$

$$G'_{\tilde{x}(t),\tilde{y}(t)}(\omega) = \frac{1}{2\pi}\int_{-\infty}^{\infty} B'_{\tilde{x}(t),\tilde{y}(t)}(v)\,\mathrm{e}^{-\mathrm{i}\omega v}\,\mathrm{d}v \qquad \text{(A5.48′)}$$

For real $\tilde{x}(t)$, substitution of $v = 0$ in the inverse of the analogue to equation (A5.44) yields

$$\mathrm{E}[\tilde{x}^2(t)] = B'_{\tilde{x}(t)}(0) = \int_{-\infty}^{\infty} G'_{\tilde{x}(t)}(\omega)\,\mathrm{d}\omega \qquad \text{(A5.49)}$$

and for real $\tilde{x}(t)$ and $\tilde{y}(t)$, similarly

$$\mathrm{E}[\tilde{x}(t)\tilde{y}(t)] = B'_{\tilde{x}(t),\tilde{y}(t)}(0) = \int_{-\infty}^{\infty} G'_{\tilde{x}(t),\tilde{y}(t)}(\omega)\,\mathrm{d}\omega \qquad \text{(A5.50)}$$

Some alternative ensemble spectral densities can be found in the literature. In particular $G_{\tilde{x}(t)}(\omega)$ is often called the *two-sided spectral density in x-units²/radians per second* while, introducing the frequency $\nu = \omega/2\pi$,

$$G^{\nu}_{\tilde{x}(t)}(\nu) = 2\pi G_{\tilde{x}(t)}(2\pi\nu) \qquad \text{(A5.51)}$$

is called the *two-sided spectral density in x-units²/cycles per second.*

Further, the *one-sided spectral densities* are often defined only for non-negative frequencies:

$$G^{0}_{\tilde{x}(t)}(\omega) = 2G_{\tilde{x}(t)}(\omega) \qquad \text{for } \omega \geqslant 0 \qquad \text{(A5.52)}$$

$$G^{0\nu}_{\tilde{x}(t)}(\nu) = 2G^{\nu}_{\tilde{x}(t)}(\nu) \qquad \text{for } \nu \geqslant 0 \qquad \text{(A5.53)}$$

Similar definitions also apply to cross-spectral densities.

Finally, as the spectral density of stationary data can be defined by the Fourier transform of a stationary covariance function, given the covariance $\Sigma_{\tilde{x}(t)}(t_1, t_2)$ of a non-stationary random process (see equation (A5.3)), the *double frequency generalized power spectral density function* of the process is defined by

$$G_{\tilde{x}(t)}(\omega_1,\omega_2) = \frac{1}{4\pi^2}\int_{-\infty}^{\infty}\int_{-\infty}^{\infty} \Sigma_{\tilde{x}(t)}(t_1,t_2)\exp[-\mathrm{i}(\omega_1 t_1 - \omega_2 t_2)]\,\mathrm{d}t_1\,\mathrm{d}t_2 \qquad \text{(A5.54)}$$

Similarly one can define the double Fourier transformation relation

$$G_{\tilde{x}(t),\tilde{y}(t)}(\omega_1,\omega_2) = \frac{1}{4\pi^2}\int_{-\infty}^{\infty}\int_{-\infty}^{\infty} \Sigma_{\tilde{x}(t),\tilde{y}(t)}(t_1,t_2)\exp[-\mathrm{i}(\omega_1 t_1 - \omega_2 t_2)]\,\mathrm{d}t_1\,\mathrm{d}t_2 \qquad \text{(A5.55)}$$

A5.4 OPERATIONS ON RANDOM PROCESSES

Let $\tilde{x}(t)$ and $\tilde{y}(t)$ be two random processes and consider a third random process $\tilde{z}(t)$ defined, for real C and D, by

$$\tilde{z}(t) = C\tilde{x}(t) + D\tilde{y}(t) \tag{A5.56}$$

Then the relevant covariance functions are given by

$$\Sigma_{\tilde{x}(t),\tilde{z}(t)}(t_1,t_2) = C\Sigma_{\tilde{x}(t)}(t_1,t_2) + D\Sigma_{\tilde{x}(t),\tilde{y}(t)}(t_1,t_2) \tag{A5.57}$$

$$\Sigma_{\tilde{z}(t),\tilde{x}(t)}(t_1,t_2) = C\Sigma_{\tilde{x}(t)}(t_1,t_2) + D\Sigma_{\tilde{y}(t),\tilde{x}(t)}(t_1,t_2) \tag{A5.58}$$

$$\begin{aligned}\Sigma_{\tilde{z}(t)}(t_1,t_2) = C^2\Sigma_{\tilde{x}(t)}(t_1,t_2) \\ + D^2\Sigma_{\tilde{y}(t)}(t_1,t_2) + CD[\Sigma_{\tilde{x}(t),\tilde{y}(t)}(t_1,t_2) \\ + \Sigma_{\tilde{y}(t),\tilde{x}(t)}(t_1,t_2)]\end{aligned} \tag{A5.59}$$

For stationary random processes, the same relations can apply to the covariance functions $B_{\tilde{x}(t),\tilde{z}(t)}(v)$, $B_{\tilde{z}(t),\tilde{x}(t)}(v)$ and $B_{\tilde{z}(t)}(v)$, while the corresponding spectral densities are

$$G_{\tilde{x}(t),\tilde{z}(t)}(\omega) = CG_{\tilde{x}(t)}(\omega) + DG_{\tilde{x}(t),\tilde{y}(t)}(\omega) \tag{A5.60}$$

$$G_{\tilde{z}(t),\tilde{x}(t)}(\omega) = CG_{\tilde{x}(t)}(\omega) + DG_{\tilde{y}(t),\tilde{x}(t)}(\omega) \tag{A5.61}$$

$$G_{\tilde{z}(t)} = C^2G_{\tilde{x}(t)}(\omega) + D^2G_{\tilde{y}(t)}(\omega) + CD[G_{\tilde{x}(t),\tilde{y}(t)}(\omega) + G_{\tilde{y}(t),\tilde{x}(t)}(\omega)] \tag{A5.62}$$

Consider now a real linear system with real input $x(t)$ and output $y(t)$ given by

$$y(t_2) = \int_{-\infty}^{\infty} h^*(t_2,t)x(t)\,dt = \int_{-\infty}^{\infty} h(t_2,v)x(t_2-v)\,dv \tag{A5.38$'$}$$

where the Green's function is the system response to a unit impulse $\partial(t_2-t_1)$ applied at time t, and $h(t_2,v) = h^*(t_2,t_2-v)$ with $v = t_2-t$.

If $\tilde{x}(t)$ is a random process and $E[\tilde{x}^2(t)]$ and $E[\tilde{y}^2(t)]$ exist, then

$$\mu_{\tilde{y}(t)}(t_2) = \int_{-\infty}^{\infty} h^*(t_2,t_1)\mu_{\tilde{x}(t)}(t)\,dt \tag{A5.63}$$

$$\Sigma_{\tilde{x}(t),\tilde{y}(t)}(t_1,t_2) = \Sigma_{\tilde{y}(t),\tilde{x}(t)}(t_2,t_1) = \int_{-\infty}^{\infty} h^*(t_2,t)\Sigma_{\tilde{x}(t)}(t_1,t)\,dt \tag{A5.64}$$

$$\begin{aligned}\Sigma_{\tilde{y}(t)}(t_1,t_2) &= \int_{-\infty}^{\infty} h^*(t_2,t)\Sigma_{\tilde{y}(t),\tilde{x}(t)}(t_1,t)\,dt \\ &= \int_{-\infty}^{\infty} h^*(t_1,t)\Sigma_{\tilde{y}(t),\tilde{x}(t)}(t_2,t)\,dt\end{aligned} \tag{A5.65}$$

$$E[\tilde{y}^2(t_2)] = \int_{-\infty}^{\infty} h^*(t_2,t)\Sigma_{\tilde{y}(t),\tilde{x}(t)}(t_2,t)\,dt + \mu^2_{\tilde{y}(t)}(t_2) \tag{A5.66}$$

Note that these equations hold for a filter that is not time-invariant; in fact the impulse response function $h^*(t_2,t)$ is a function of both t_2 and t.

Moreover, if the input $\tilde{x}(t)$ is stationary and the system is time-invariant, then the system output $\tilde{y}(t)$ is also stationary. The process $\tilde{y}(t)$ will be ergodic if this is true for $\tilde{x}(t)$.

In this case

$$h^*(t_2, t) = h(t_2 - t) = h(v) \tag{A5.38'}$$

$h(t_2 - t_1)$ being defined by equation (A5.38), and equations (A5.63)–(A5.66), for real $\tilde{x}(t)$ and $\tilde{y}(t)$, reduce in this case to

$$\mu_{\tilde{y}(t)} = \mu_{\tilde{x}(t)} \int_{-\infty}^{\infty} h(v)\,\mathrm{d}v \tag{A5.67}$$

$$B_{\tilde{x}(t),\tilde{y}(t)}(v_2) = B_{\tilde{y}(t),\tilde{x}(t)}(-v_2) = \int_{-\infty}^{\infty} h(v_1) B_{\tilde{x}(t)}(v_2 - v_1)\,\mathrm{d}v_1 \tag{A5.68}$$

$$B_{\tilde{y}(t)}(v_2) = \int_{-\infty}^{\infty} h(v_1 - v_2) B_{\tilde{y}(t),\tilde{x}(t)}(-v_1)\,\mathrm{d}v_1 \tag{A5.69}$$

$$\mathrm{E}[\tilde{y}^2(t)] = \int_{-\infty}^{\infty} h(v_1) B_{\tilde{y}(t),\tilde{x}(t)}(-v_1)\,\mathrm{d}v_1 + \mu^2_{\tilde{y}(t)} \tag{A5.70}$$

where $h(t_2 - t_1) = 0$ for $t_2 - t_1 < 0$ in physical applications.

Equations (A5.68) and (A5.69) are often called the *Wiener–Lee relations*, while equations (A5.64) and (A5.65) are the *generalized Wiener–Lee relations*.

Further equations (A5.68) and (A5.69) can be expressed in terms of spectral densities according to equation (A5.42):

$$G_{\tilde{x}(t),\tilde{y}(t)}(\omega) = A(\omega) G_{\tilde{x}(t)}(\omega) \tag{A5.71}$$

$$G_{\tilde{y}(t),\tilde{x}(t)}(\omega) = A^{\dagger}(\omega) G_{\tilde{x}(t)}(\omega) \tag{A5.72}$$

$$G_{\tilde{y}(t)} = A(\omega) G_{\tilde{x}(t),\tilde{y}(t)}(\omega) = |A^{\dagger}(\omega)|^2 G_{\tilde{x}(t)}(\omega) \tag{A5.73}$$

where $A^{\dagger}(\omega)$ is the complex conjugate of $A(\omega)$.

Finally, in the special case of stationary white-noise input (equation (A5.47)):

$$B_{\tilde{x}(t),\tilde{y}(t)}(v) = 2\pi G_0 h(v) \tag{A5.74}$$

$$B_{\tilde{y}(t)}(v) = 2\pi G_0 \int_{-\infty}^{\infty} h(v')h(v' + v)\,\mathrm{d}v' = 2\pi G_0 \phi_h(v) \tag{A5.75}$$

$$\mathrm{E}[\tilde{y}^2(t)] = 2\pi G_0 \int_{-\infty}^{\infty} h^2(v')\,\mathrm{d}v' + \mu^2_{\tilde{y}(t)} = 2\pi G_0 \phi_h(0) + \mu^2_{\tilde{y}(t)} \tag{A5.76}$$

Consider now a generating random process $\tilde{x}(t)$ and a single-valued non-linear measurable function $g(x)$. Then $\tilde{y}(t) = g(\tilde{x}(t))$ is a new random process produced by a zero-memory operation on $\tilde{x}(t)$. The probabilistic properties of $\tilde{y}(t)$ can be obtained by the methods of Appendix A3 and in particular, for real variables, one obtains

$$\Sigma_{\tilde{y}(t)}(t_1,t_2) = \int_{-\infty}^{\infty}\int_{-\infty}^{\infty} y(x_1)y(x_2)p_{\tilde{x}(t_1),\tilde{x}(t_2)}(x_1,x_2)\,\mathrm{d}x_1\,\mathrm{d}x_2 - \mu_{\tilde{y}(t)}(t_1)\mu_{\tilde{y}(t)}(t_2) \tag{A5.77}$$

where $x_1 = x(t_1)$ and $x_2 = x(t_2)$.

A5.5 MULTIDIMENSIONAL RANDOM FIELDS

The theory of *random fields* $\tilde{x}(t)$, with $t = \{t_1, t_2, \ldots, t_k\}^T$ a k-dimensional vector parameter, is a direct generalization of the theory of ordinary random processes developed in this Appendix (A5).

Further, a random field is said to be 'isotropic' in the strict or the broad sense if its distribution functions or its first two moments are invariant with respect to rigid-body motion and with respect to minor reflections. Isotropy implies homogeneity (i.e. stationarity) while the reverse is true only for unidimensional random process.

A6

Mathematical models for random process analysis

A6.1 SPECIAL CLASSES OF RANDOM PROCESSES

A6.1.1 Constant-in-time process (random variable)

If each realization of the random process $\tilde{x}(t)$ is identically equal to a random parameter $\tilde{y}$ constant in time, the distribution function $P_{\tilde{y}}(y)$ determines uniquely the stationary (but not ergodic) random process $\tilde{x}(t)$. If $\mathrm{E}[\tilde{y}^2]$ exists then

$$\mu_{\tilde{x}(t)} = \mu_{\tilde{y}} \tag{A6.1}$$

$$B_{\tilde{x}}(v) = \mathrm{E}[\tilde{y}^2] - \mu_{\tilde{y}}^2 \tag{A6.2}$$

A6.1.2 General periodic processes

A *general periodic process* is defined by the Fourier series

$$\tilde{x}(t) = \tilde{c}_0 + \sum_{k=1}^{\infty} [\tilde{c}_{1k} \cos(k\omega_0 t) + \tilde{c}_{2k} \sin(k\omega_0 t)] \tag{A6.3}$$

with real random coefficients $\tilde{c}_0$, $\tilde{c}_{1k}$, $\tilde{c}_{2k}$ assuming that the series converges in mean square (equation (A4.3)).

Equation (A6.3) denotes a weakly stationary random process if and only if

$$\mu_{\tilde{c}_{1k}} = \mu_{\tilde{c}_{2k}} = 0 \tag{A6.4}$$

$$\mathrm{E}[\tilde{c}_{1k}^2] = \mathrm{E}[\tilde{c}_{2k}^2] \tag{A6.5}$$

$$\mathrm{E}[\tilde{c}_0\tilde{c}_{1k}] = \mathrm{E}[\tilde{c}_0\tilde{c}_{2k}] = \mathrm{E}[\tilde{c}_{1k}\tilde{c}_{2k}] = 0 \tag{A6.6}$$

$$\mathrm{E}[\tilde{c}_{1i}\tilde{c}_{1k}] = \mathrm{E}[\tilde{c}_{2i}\tilde{c}_{2k}] \qquad i \neq k \tag{A6.7}$$

Equations (A6.4)–(A6.7) then imply

$$\mu_{\tilde{x}(t)} = \mu_{\tilde{c}_0} \tag{A6.8}$$

$$B'_{\tilde{x}(t)}(v) = B_{\tilde{x}(t)}(v) + \mu^2_{\tilde{x}(t)} = \mathrm{E}[\tilde{c}_0^2] + \frac{1}{2}\sum_{k=1}^{\infty} \mathrm{E}[c_{1k}^2 + c_{2k}^2]\cos(k\omega_0 v) \tag{A6.9}$$

A particular form of equation (A6.3) is the *random-phase periodic process*:

$$\tilde{x}(t) = c_0 + \sum_{k=1}^{\infty} \{c_{1k}\cos[k(\omega_0 t + \tilde{\Phi})] + c_{2k}\sin[k(\omega_0 t + \tilde{\Phi})]\} \tag{A6.10}$$

where $\tilde{\Phi}$ is uniformly distributed between 0 and 2π. The process is stationary and ergodic with

$$\mu_{\tilde{x}(t)} = c_0 \tag{A6.11}$$

$$B_{\tilde{x}(t)}(v) = \frac{1}{2}\sum_{k=1}^{\infty} (c_{1k}^2 + c_{2k}^2)\cos(k\omega_0 v) \tag{A6.12}$$

$$2\pi G_{\tilde{x}(t)}(\omega) = \frac{1}{4}\sum_{k=1}^{\infty} (c_{1k}^2 + c_{2k}^2)[\partial(\omega - k\omega_0) + \partial(\omega + k\omega_0)] \tag{A6.13}$$

A function $x(t)$ is said to be *band-limited* between $\omega = 0$ and $\omega = 2\pi b$ if and only if its Fourier transform exists and equals zero for $|\omega| > 2\pi b$. The parameter b if called the *bandwidth* associated with $x(t)$. A stochastic process $\tilde{x}(t)$ is band-limited between $\omega = 0$ and $\omega = 2\pi b$ if and only if $G'_{\tilde{x}(t)}(\omega)$ (see equation (A5.46′)) exists and equals zero for $|\omega| > 2\pi b$.

A6.2 NORMAL PROCESSES

A stochastic process $\{\tilde{x}(t), t \in T\}$ is said to be a *normal process* if for any integer n and any subset $\{t_1, t_2, \ldots, t_n\}$ of T the n random variables $\tilde{x}(t_1), \ldots, \tilde{x}(t_n)$ are jointly normally distributed (see equation (A2.39)) in the sense that their joint characteristic function is given, for any real numbers $u_1, u_2, \ldots, u_n$, by

$$\Omega_{\tilde{x}(t_1),\tilde{x}(t_2),\ldots,\tilde{x}(t_n)}(u_1, u_2, \ldots, u_n) = \exp\left(\mathrm{i}\sum_{j=1}^{n} u_j \mu_{\tilde{x}(t_j)} - \frac{1}{2}\sum_{j,k=1}^{n} u_j u_k \mathrm{Cov}[\tilde{x}(t_j), \tilde{x}(t_k)]\right) \tag{A6.14}$$

It is worth noting that knowledge of the mean value function $\mu_{\tilde{x}(t)}(t)$ and the covariance kernel $\mathrm{Cov}[\tilde{x}(t_1), \tilde{x}(t_2)] = \Sigma_{\tilde{x}(t)}(t_1, t_2)$ implies knowledge of the complete probability law of the process.

Any stochastic process derived by means of linear operations from a normal process is itself a normal process. That is an immediate consequence of the following theorem, which generalizes equation (A3.52). Let $\tilde{y}_1, \ldots, \tilde{y}_m$ be linear combinations, by the coefficients c_{ij} ($i = 1, \ldots, m$; $j = 1, \ldots, n$), of n jointly normally distributed random variables $\tilde{x}_1, \ldots, \tilde{x}_n$. Then $\tilde{y}_1, \ldots, \tilde{y}_m$ are jointly normally distributed with

$$\mu_{\tilde{y}_j} = c_{j1}\mu_{\tilde{x}_1} + \ldots + c_{jn}\mu_{\tilde{x}_n} \tag{A6.15}$$

$$\mathrm{Cov}[\tilde{y}_j, \tilde{y}_k] = \sum_{i,l=1}^{n} c_{ji}c_{kl}\,\mathrm{Cov}[\tilde{x}_i, \tilde{x}_l] \tag{A6.16}$$

Further let $\tilde{z}_1, \tilde{z}_2, \ldots, \tilde{z}_n$ be a sequence of normally distributed random variables which converges in mean square to a random variable $\tilde{z}$. Then $\tilde{z}$ can be shown to be normally distributed.

When non-linear operations on normal processes are considered, the following theorem may be useful. Let $\{\tilde{x}(t), t \geqslant 0\}$ be a normal process with zero means and covariance kernel $\Sigma_{\tilde{x}(t)}(t_1, t_2)$. Then for any non-negative numbers t_1, t_2 and v:

$$\mathrm{E}[\tilde{x}^2(t)] = \Sigma_{\tilde{x}(t)}(t, t) \tag{A6.17}$$

$$\mathrm{Var}[\tilde{x}^2(t)] = 2\Sigma^2_{\tilde{x}(t)}(t, t) \tag{A6.18}$$

$$\mathrm{Cov}[\tilde{x}^2(t_1), \tilde{x}^2(t_2)] = 2\Sigma^2_{\tilde{x}(t)}(t_1, t_2) \tag{A6.19}$$

$$\mathrm{E}[\tilde{x}(t)\tilde{x}(t+v)] = \Sigma_{\tilde{x}(t)}(t, t+v) \tag{A6.20}$$

$$\mathrm{Var}[\tilde{x}(t)\tilde{x}(t+v)] = \Sigma_{\tilde{x}(t)}(t,t)\Sigma_{\tilde{x}(t)}(t+v, t+v) + \Sigma^2_{\tilde{x}(t)}(t, t+v) \tag{A6.21}$$

$$\begin{aligned}\mathrm{Cov}[\tilde{x}(t_1)\tilde{x}(t_1+v), \tilde{x}(t_2)\tilde{x}(t_2+v)] &= \Sigma_{\tilde{x}(t)}(t_1, t_2)\Sigma_{\tilde{x}(t)}(t_1+v, t_2+v) \\ &\quad + \Sigma_{\tilde{x}(t)}(t_1, t_2+v)\Sigma_{\tilde{x}(t)}(t_1+v, t_2)\end{aligned} \tag{A6.22}$$

Note that, if $\tilde{x}(t)$ is normal and stationary, the fourth moment can be expressed in terms of only $B_{\tilde{x}(t)}(v)$ and $\mu_{\tilde{x}}$ and the condition for ergodicity in the covariance (equation (A5.14)) becomes

$$\lim_{T\to\infty} \frac{1}{T}\int_0^T [B(v)]^2\,\mathrm{d}v = 0 \tag{A6.23}$$

Finally a stochastic process $\{\tilde{x}(t),\ t \in T\}$ with finite second moment whose probability law depends on some parameter η is said to be asymptotically normal as that parameter tends to a given limit, if, for every finite set of indices $t_1, t_2, \ldots, t_n \in T$, the joint characteristic function of the standardized random variables

$$\tilde{x}^*(t) = \frac{\tilde{x}(t) - \mu_{\tilde{x}(t)}(t)}{s_{\tilde{x}(t)}t} \tag{A6.24}$$

satisfies (for every $u_1, \ldots, u_n$) the relation

$$\lim_{\eta\to\infty} \log \Omega_{\tilde{x}^*(t_1), \ldots, \tilde{x}^*(t_n)}(u_1, \ldots, u_n) = -\frac{1}{2}\sum_{j,k=1}^{n} u_j \rho_{jk} u_k \tag{A6.25}$$

In equation (A6.25) ρ_{ik} is the correlation between $\tilde{x}(t_j)$ and $\tilde{x}(t_k)$:

$$\rho_{jk} = \mathrm{Cov}[\tilde{x}(t_j), \tilde{x}(t_k)]/(s_{\tilde{x}(t_j)}s_{\tilde{x}(t_k)}) \tag{A6.26}$$

A6.3 COUNTING PROCESSES AND POISSON PROCESSES

A6.3.1 Counting processes

An integer-valued process $\{\tilde{n}(t),\ t \geqslant 0\}$ which counts the number of points occurring in an interval is said to be a *counting process*. Let $\tilde{w}_1 < \tilde{w}_2 < \ldots < \tilde{w}_n$ be the times at which events of specified character have occurred; $\tilde{w}_j$ is often called the *waiting time* to the jth event. The random variables

$$\tilde{\tau}_1 = \tilde{w}_1, \qquad \tilde{\tau}_2 = \tilde{w}_2 - \tilde{w}_1, \quad \ldots, \quad \tilde{\tau}_n = \tilde{w}_n - \tilde{w}_{n-1} \tag{A6.27}$$

are called the successive inter arrival times. Among counting processes the Poisson process plays a prominent part.

A6.3.2 Poison processes

An integer-valued process $\{\tilde{n}(t), t \geqslant 0\}$ is said to be a *Poisson process* with mean rate (or intensity) λ if the process has stationary independent increments and for any times t_1 and t_2 (with $t_1 < t_2$) the number $\tilde{n}(t_2) - \tilde{n}(t_1)$ is Poisson distributed (see Section A2.3) with mean $\lambda(t_2 - t_1)$. Hence, for $k = 0, 1, 2, \ldots$

$$\mathrm{Prob}[\tilde{n}(t_2) - \tilde{n}(t_1) = k] = \mathrm{e}^{-\lambda(t_2 - t_1)} \frac{[\lambda(t_2 - t_1)]^k}{k!} \tag{A6.28}$$

$$\mathrm{E}[\tilde{n}(t_2) - \tilde{n}(t_1)] = \lambda(t_2 - t_1) \tag{A6.29}$$

$$\mathrm{Var}[\tilde{n}(t_2) - \tilde{n}(t_1)] = \lambda(t_2 - t_1) \tag{A6.30}$$

It is worth noting that the parameter λ represents the mean rate of occurrence for unit time of the events being counted.

A counting process $\{\tilde{n}(t), t \geqslant 0\}$ that satisfies the following five assumptions is a Poisson process:

(1) zero is the initial time for counting events; thus $\tilde{n}(0) = 0$;
(2) the process $\tilde{n}(t)$ has independent increments;
(3) in any interval, there is a positive probability that an event will occur, but it is not certain that an event will occur;
(4) in sufficiently small intervals, at most one event can occur;
(5) the counting process $\tilde{n}(t)$ has stationary increments (i.e. the random variables $\tilde{n}(t_2) - \tilde{n}(t_1)$ and $\tilde{n}(t_2 + v) - \tilde{n}(t_1 + v)$ are identically distributed).

If assumption (2) is dropped, the generalized Poisson process is obtained, while if assumption (5) is dropped, the non-homogeneous Poisson processes are introduced with intensity function $\lambda(t)$ given by

$$\lambda(t) = \frac{\mathrm{d}}{\mathrm{d}t} \mu_{\tilde{n}(t)}(t) \tag{A6.31}$$

It is worth noting that for a Poisson process at time t the probability of a jump within an ensuing short time interval between t and $t+v$ equals $\lambda v + o(v)$ while the probability of more than one jump is $o(v)$‡.

Moreover consider the sequence of waiting times $\tilde{w}_1, \tilde{w}_2, \ldots$ (see equation (A6.27)) corresponding to a counting process $\{\tilde{n}(t),\ t \geqslant 0\}$. For any $t > 0$ and $n = 1, 2, \ldots$

$$\tilde{n}(t) \leqslant n \qquad \text{if and only if } \tilde{w}_{n+1} > t \tag{A6.32}$$

$$\tilde{n}(t) = n \qquad \text{if and only if } \tilde{w}_n \leqslant t \text{ and } \tilde{w}_{n+1} > t \tag{A6.33}$$

$$\text{Prob}[\tilde{n}(t) \leqslant n] = \text{Prob}[\tilde{w}_{n+1} > t] = 1 - P_{\tilde{w}_{n+1}}(t) \qquad n = 0, 1, \ldots \tag{A6.34}$$

$$\begin{aligned}\text{Prob}[\tilde{n}(t) = n] &= \text{Prob}[\tilde{w}_n \leqslant t] - \text{Prob}[\tilde{w}_{n+1} \leqslant t] \\ &= P_{\tilde{w}_n}(t) - P_{\tilde{w}_{n+1}}(t) \qquad n = 1, 2, \ldots \end{aligned} \tag{A6.35}$$

$$\text{Prob}[\tilde{n}(t) = 0] = 1 - \text{Prob}[\tilde{w}_1 \leqslant t] = 1 - P_{\tilde{w}_1}(t) \tag{A6.36}$$

Then, if $\{\tilde{n}(t), t \geqslant 0\}$ is a Poisson process with mean rate λ, it is possible to prove that

$$\text{Prob}[\tilde{w}_n > t] = \text{Prob}[\tilde{n}(t) < n] = \sum_{k=0}^{n-1} \mathrm{e}^{-\lambda t} \frac{(\lambda t)^k}{k!} \tag{A6.37}$$

and hence $\tilde{w}_n$ obeys the gamma probability law with parameters n and λ (see Table A.2). Further, the successive inter-arrival times $\tilde{\tau}_1, \tilde{\tau}_2, \ldots$ of events of Poisson type of intensity λ are independent identically distributed random variables obeying an exponential probability law with mean $1/\lambda$ (see Table A.2). Finally, under the condition that $\tilde{n}(T) = k$, the k times $\tilde{w}_1 < \tilde{w}_2 < \tilde{w}_3 < \ldots < \tilde{w}_k$ in the interval 0 to T at which events occur are random variables that, considered as unordered, are distributed independently and uniformly in the interval 0 to T.

A6.3.3 Random processes generated by a Poisson process

A stochastic process $\{\tilde{x}(t), t \geqslant 0\}$ is said to be a *compound Poisson process* if it can be represented, for $t \geqslant 0$, by

$$\tilde{x}(t) = \sum_{m=1}^{\tilde{n}(t)} \tilde{y}_m \tag{A6.38}$$

in which $\{\tilde{n}(t), t \geqslant 0\}$ is a Poisson process and is independent of the sequence $\{\tilde{y}_m, m = 1, 2, \ldots\}$, which is a family of independent random variables identically distributed as a random variable $\tilde{y}$. It is possible to show that a compound Poisson process $\{\tilde{x}(t),\ t \geqslant 0\}$ has stationary independent increments and characteristic function, for any $t \geqslant 0$,

$$\Omega_{\tilde{x}(t)}(u) = \exp\{\lambda t[\Omega_{\tilde{y}}(u) - 1]\} \tag{A6.39}$$

‡ $o(v)$ stands for a quantity such that $o(v)/v \to 0$ as $v \to 0$.

Then, if $E[\tilde{y}^2] < \infty$, the moments of $\tilde{x}(t)$ are given by

$$\mu_{\tilde{x}(t)}(t) = \lambda t \mu_{\tilde{y}} \tag{A6.40}$$

$$s^2_{\tilde{x}(t)}(t) = \lambda t \, E[\tilde{y}^2] \tag{A6.41}$$

$$\mathrm{Cov}[\tilde{x}(t_1), \tilde{x}(t_2)] = \lambda E[\tilde{y}^2] \min(t_1, t_2) \tag{A6.42}$$

Furthermore a stochastic process $\{\tilde{x}(t), t \geqslant 0\}$ is said to be a *filtered Poisson process* (FPP) if it can be represented, for $t \geqslant 0$, by

$$\tilde{x}(t) = \sum_{m=1}^{\tilde{n}(t)} U(t, \tilde{w}_m, \tilde{y}_m) \tag{A6.43}$$

where $\{\tilde{n}(t), t \geqslant 0\}$ is a Poisson process with intensity λ; $\tilde{y}_1, \tilde{y}_2, \ldots$ is a sequence of independent random variables, identically distributed as a random variable $\tilde{y}$ and independent of $\{\tilde{n}(t), t \geqslant 0\}$. The function $U(t, \tilde{w}_m, \tilde{y}_m)$ is called the *response function*, where $\tilde{w}_m$ denotes the mth waiting time relevant to the Poisson process $\{\tilde{n}(t), t \geqslant 0\}$. A typical response function is

$$U(t, w, y) = U(t - w, y) = \begin{cases} y & \text{if} \quad t - w \geqslant 0 \\ 0 & \text{if} \quad t - w < 0 \end{cases} \tag{A6.44}$$

which corresponds to the compound Poisson process defined by equation (A6.38). It is possible to show that, for the filtered Poisson process defined by equation (A6.43), for any positive number t and real number u,

$$\Omega_{\tilde{x}(t)}(u) = \exp\left(\lambda \int_0^t E[\exp\{iuU(t,w,\tilde{y})\} - 1] \, dw \right) \tag{A6.45}$$

and for any $t_2 > t_1 \geqslant 0$ and real numbers u_1 and u_2

$$\Omega_{\tilde{x}(t_1),\tilde{x}(t_2)}(u_1, u_2) = \exp\left(\lambda \int_0^{t_1} E[\exp\{i[u_1 U(t_1,w,\tilde{y}) + u_2 U(t_2,w,\tilde{y})]\} - 1] \, dw \right.$$
$$\left. + \lambda \int_{t_1}^{t_2} E[\exp\{i[u_2 U(t_2,w,\tilde{y})\} - 1] \, dw \right) \tag{A6.46}$$

Then if $E[U^2(t,w,\tilde{y})] < \infty$ for all w, the moments of $\tilde{x}(t)$ are given by

$$\mu_{\tilde{x}(t)}(t) = \lambda \int_0^t E[U(t, w, \tilde{y})] \, dw \tag{A6.47}$$

$$s^2_{\tilde{x}(t)}(t) = \lambda \int_0^t E[U^2(t, w, \tilde{y})] \, dw \tag{A6.48}$$

$$\mathrm{Cov}[\tilde{x}(t_1), \tilde{x}(t_2)] = \lambda \int_0^{\min(t_1, t_2)} E[U(t_1,w,\tilde{y}) \, U(t_2,w,\tilde{y})] \, dw \tag{A6.49}$$

Finally a stochastic process $\{\tilde{x}(t), -\infty < t < \infty\}$ tends to be stationary when t is large, if $U(t - \tilde{w}_m, \tilde{y}_m)$ is essentially zero except for small $(t - \tau)$; in this case one may write

$$\tilde{x}(t) = \sum_{m=-\infty}^{\infty} U(t - \tilde{w}_m, \tilde{y}_m) \tag{A6.50}$$

where the times $\{\tilde{w}_m\}$ are assumed to occur in accord with a Poisson process with intensity λ and $\tilde{y}_1, \tilde{y}_2, \ldots$ is assumed to be a sequence of independent identically distributed random variables. In the same way that equations (A6.45), (A6.47)–(A6.49) are proved, one may prove 'Campbell's theorem', consisting of the same relations but with the integrals over the interval $-\infty$ to $+\infty$.

The class of random pulse sequences considered in this section contains also one type of *shot noise*, i.e. a r.p. with zero mean and

$$\Sigma(t_1, t_2) = D(t_1)\partial(t_2 - t_1) \tag{A6.50$'$}$$

where $D(t_1)$ is the *intensity function* of the shot noise. This type of shot noise is represented by

$$\tilde{x}(t) = \sum_{m=1}^{n(t)} y_m \partial(t - w_m) \tag{A6.50$''$}$$

and

$$D(t) = \mathrm{E}[\tilde{y}^2]\lambda(t) \tag{A6.50$'''$}$$

A6.4 MARKOV PROCESSES

A discrete-parameter stochastic process $\{\tilde{x}(t), t = 0, 1, 2, \ldots\}$ or a continuous-parameter stochastic process $\{\tilde{x}(t), t \geqslant 0\}$ is said to be a *Markov process* if, for any set of n time points $t_1 < t_2 < \ldots < t_n$, the conditional distribution of $\tilde{x}(t_n)$ for given values of $\tilde{x}(t_1), \ldots, \tilde{x}(t_{n-1})$ depends only on $\tilde{x}(t_{n-1})$:

$$\begin{aligned}\operatorname{Prob}[\tilde{x}(t_n) \leqslant x_n | \tilde{x}(t_1) = x_1, \ldots, \tilde{x}(t_{n-1}) = x_{n-1}]\\ = \operatorname{Prob}[\tilde{x}(t_n) \leqslant x_n | \tilde{x}(t_{n-1}) = x_{n-1}]\end{aligned} \tag{A6.51}$$

Intuitively equation (A6.51) means that, given the present of the process, the future is independent of its past.

A real number x is said to be a *state* of a stochastic process $\{\tilde{x}(t), t \in T\}$ if there exists a time t in T such that $\operatorname{Prob}[x - v < \tilde{x}(t) < x + v] > 0$ for every $v > 0$. The set of possible values of a stochastic process is then called its 'state space'. Markov processes, as well as all stochastic processes, can be classified depending on the discrete or continuous character of the parameter t and of the state x (see Fig. A.3).

A Markov process is described by a *transition probability function* $\operatorname{Prob}[E, t | x, t_0]$, which represents the conditional probability that the state of the system will belong to the set E at time t, given that at time t_0 ($t_0 < t$) the system is in state x. The Markov process is said to be *homogeneous* (or to have 'stationary transition probabilities') if $\operatorname{Prob}[E, t | x, t_0|$ depends on t and t_0 only through the difference $t - t_0$.

A Markov process whose state space is discrete is called a *Markov chain*. In order to specify a discrete parameter Markov chain $\{\tilde{x}(t), t = 1, 2, \ldots\}$, equation (A5.1a), requires knowledge of

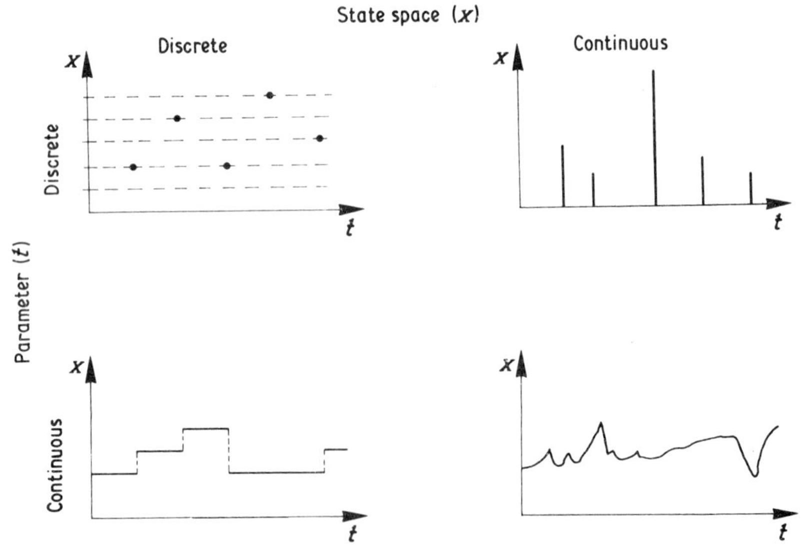

Fig. A.3 Sample realizations of a random process.

$$\text{Prob}[\tilde{x}(t_1) = k_1, \tilde{x}(t_2) = k_2, \ldots, \tilde{x}(t_n) = k_n]$$
$$p^0_{\tilde{x}(t_1)}(x_1) p^0_{\tilde{x}(t_2)|\tilde{x}(t_1)}(x_1, x_2) \ldots p^0_{\tilde{x}(t_n)|\tilde{x}(t_{n-1})}(x_{n-1}, x_n) \quad \text{(A6.52)}$$

and hence for all times $t_j \geqslant t_k > 0$ and states x_j and x_k knowledge of the probability mass function

$$p^0_{\tilde{x}(t_j)}(x_j) = \text{Prob}[\tilde{x}(t_j) = x_j] \quad \text{(A6.53)}$$

and of the probability mass function

$$p^0_{\tilde{x}(t_j)|\tilde{x}(t_k)}(x_k, x_j) = \text{Prob}[\tilde{x}(t_j) = x_j | \tilde{x}(t_k) = x_k] \quad \text{(A6.54)}$$

that is, the 'transition probability function'. Then if the Markov chain is homogeneous, equation (A6.54) depends only on the difference $t_j - t_k$; thence the n-step transition probability function (matrix) and the unconditional probability vector are given by equations (A1.29) and (A1.30) respectively. (See also the properties developed in Section A1.4.)

A fundamental relation satisfied by the function (A6.54) is the so-called *Chapman–Kolmogorov equation*: for any times $t_j > t_l > t_k \geqslant 0$ and states x_j and x_k

$$p^0_{x(t_j)|\tilde{x}(t_k)}(x_k, x_j) = \sum_{\substack{\text{all} \\ \text{states } i}} p^0_{\tilde{x}(t_j)|\tilde{x}(t_i)}(x_i, x_j)\, p^0_{\tilde{x}(t_i)|\tilde{x}(t_k)}(x_k, x_i) \quad \text{(A6.55)}$$

Similarly, if $\{\tilde{x}(t), t \geqslant 0\}$ is a continuous-parameter Markov chain, to specify the probability law of $\{\tilde{x}(t), t > 0\}$ it suffices to state for all times $t_j \geqslant t_k \geqslant 0$ and states x_j and x_k equations (A6.53) and (A6.54). Then the transition probability function (A6.54) satisfies (A6.55).

However, for a continuous-parameter Markov chain, the role of the one-step transition probabilities (see Section A1.4) is played by the *transition intensities:*

$$\Xi_{kk}(t) = -c_k(t) = -\lim_{v\to 0} \frac{1 - p^0_{\tilde{x}(t+v)|\tilde{x}(t)}(x_k, x_k)}{v} \tag{A6.56}$$

$$\Xi_{kj}(t) = c_k(t)\Pi_{kj}(t) = \lim_{v\to 0} \frac{p^0_{\tilde{x}(t+v)|\tilde{x}(t)}(x_k, x_j)}{v} \qquad j \neq k \tag{A6.57}$$

In equation (A6.56) $c_k(t)v + o(v)$ represents the probability that a change of state occurs between t and $t+v$. In equation (A6.57) $\Pi_{kj}(t)$ is continuous in t and, for every fixed t, j,

$$\sum_j \Pi_{kj}(t) = 1 \quad \Rightarrow \quad \sum_j \Xi_{kj}(t) = 0 \tag{A6.58}$$

The coefficient $\Pi_{kj}(t)$ can be interpreted as the conditional probability that, if a change from x_k occurs between t and $t+v$, this change takes the system from x_k to x_j.

Then *Kolmogorov's forward equation*

$$\frac{\partial}{\partial t_j} p^0_{\tilde{x}(t_j)|\tilde{x}(t_k)}(x_k, x_j) = +\Xi_{jj}(t_j) p^0_{\tilde{x}(t_j)|\tilde{x}(t_k)}(x_k, x_j) + \sum_{i\neq j} p^0_{\tilde{x}(t_i)|\tilde{x}(t_k)}(x_k, x_i)\Xi_{ij}(t_j)$$

$$\Rightarrow \frac{\partial}{\partial t_j}\mathbf{p}^0 = \mathbf{p}^0\Xi \tag{A6.59}$$

can be stated by equation (A6.55). The forward equation holds if one makes the assumption that, for fixed j, the passage to the limit in equation (A6.57) is uniform with respect to k. Equation (A6.59) involves differentiation with respect to the later time t_j, while *Kolmogorov's backward equation*

$$-\frac{\partial}{\partial t_k} p^0_{\tilde{x}(t_j)|\tilde{x}(t_k)}(x_k, x_j) =$$

$$+\Xi_{kk}(t_k) p^0_{\tilde{x}(t_j)|\tilde{x}(t_k)}(x_k, x_j) + \sum_{i\neq k} p^0_{\tilde{x}(t_j)|\tilde{x}(t_i)}(x_i, x_j)\Xi_{ki}(t_k)$$

$$\Rightarrow \frac{\partial}{\partial t_k}\mathbf{p}^0 = -\Xi\mathbf{p}^0 \tag{A6.60}$$

involves differentiation with respect to the earlier time t_k and holds without any additional assumption on the passage to the limit.

Moreover the role of equation (A1.30) on the unconditional probability vector $\mathbf{p}(t)$ for homogeneous continuous-parameter Markov chains is played by the differential linear equation system

$$\frac{\partial \mathbf{p}(t)}{\partial t} = (\mathbf{p}^{\mathrm{T}}(t)\Xi)^{\mathrm{T}} = \Xi^{\mathrm{T}}\mathbf{p}(t) \tag{A6.61}$$

where the diagonal elements of the matrix Ξ are given by equation (A6.56) and the relevant non-diagonal elements are defined by equation (A6.57): these coefficients become constant because of the homogeneous assumption. Then the theory of linear simultaneous ordinary differential equations with constant coefficients permits one to determine the formal solution

$$\mathbf{p}(t) = \exp(\Xi^{\mathrm{T}} t)\,\mathbf{p}(1) \tag{A6.62}$$

whose numerical values can be obtained by using the theory of eigenvalues and eigenvectors of matrices.

Finally if a homogeneous Markov process, continuous in both state and time, is considered, equation (A6.61) becomes

$$\frac{\partial p(x,t)}{\partial t} = -p(x,t)\int_{x'} \Xi(x,x')\,\mathrm{d}x + \int_{x'} p(x',t)\,\Xi(x',x)\,\mathrm{d}x' \tag{A6.63}$$

where $\Xi(x,x')$ is the transition probability density, whose estimation generally involves several computational difficulties. In this case, let $p_{x(t)|x(0)}(x,t,x_0) = p(x,t,x_0)$ be the conditional probability density of $\tilde{x}(t)$ at t given $\tilde{x}(0) = x_0$. Then, the forward Kolmogorov equation is

$$\frac{1}{2}\frac{\partial^2}{\partial x^2}(c_1(x,t)p(x,t,x_0)) - \frac{\partial}{\partial x}(c_2(x,t)p(x,t,x_0)) = \frac{\partial p(x,t,x_0)}{\partial t} \tag{A6.59$'$}$$

The functions $c_1(x,t)$ and $c_2(x,t)$ in equation (A6.59′) (often called *Fokker-Planck diffusion equation*) are defined as follows:

$$c_2(x,t) = \lim_{\Delta t \to 0} \frac{1}{\Delta t}\,\mathrm{E}[\{\tilde{x}(t+\Delta t) - \tilde{x}(t)\} \,|\, \tilde{x}(t) = x]$$

$$c_1(x,t) = \lim_{\Delta t \to 0} \frac{1}{\Delta t}\,\mathrm{Var}[\{\tilde{x}(t+\Delta t) - \tilde{x}(t)\} \,|\, \tilde{x}(t) = x]$$

A7

Crossing and extremes of random processes

A7.1 STOCHASTIC EXTREMAL PROBLEM

The stochastic extremal problem concerns the knowledge of the probability distribution of the maximum $\tilde{y}(t_1, t_2)$ of a random process $\tilde{x}(t)$ during the interval from t_1 to t_2.

If the interval from t_1 to t_2 is very large, the maximum $\tilde{y}(t_1, t_2)$ can be regarded as the maximum of the maxima $\tilde{y}(t_i, t_j)$, with $(t_j - t_i) \leqslant (t_2 - t_1)$, $t_i < t_j$, $t_1 \leqslant t_i$ and $t_2 \geqslant t_j$. These maxima can be independent identically distributed random variables, and so one can make use of the extreme distribution theory (see Section A7.2).

However, when the above approach does not hold or the probability distribution of $\tilde{y}(t_i, t_j)$, is not known *a priori*, one must estimate $P_{\tilde{y}(t_1,t_2)}(\cdot)$ from the knowledge of the stochastic process properties. In this case the basic 'first excursion' problem must be investigated to obtain results on the extremes of random processes:

$$\text{Prob}[\tilde{y}(t_1, t_2) \leqslant a] = 1 - \text{Prob}[\tau_a \leqslant t] \tag{A7.1}$$

where τ_a is the time to the first crossing above a given level a.

A7.2 EXTREME VALUE DISTRIBUTIONS

If the variable $\tilde{y}$ is the maximum of n random variables $\tilde{x}_1, \tilde{x}_2, \ldots, \tilde{x}_n$, then

$$P_{\tilde{y}}(y) = \text{Prob}[\tilde{y} \leqslant y] = \text{Prob}[\text{all } n \text{ of the } \tilde{x}_i \leqslant y] \tag{A7.2}$$

If the $\tilde{x}_i$ are independent

$$\begin{aligned} P_{\tilde{y}}(y) &= \text{Prob}[\tilde{x}_1 \leqslant y]\,\text{Prob}[\tilde{x}_2 \leqslant y] \ldots \text{Prob}[\tilde{x}_n \leqslant y] \\ &= P_{\tilde{x}_i}(y) P_{\tilde{x}_2}(y) \ldots P_{\tilde{x}_n}(y) \end{aligned} \tag{A7.3}$$

and in the special case where all the $\tilde{x}_i$ are identically distributed

$$P_{\tilde{y}}(y) = P^n_{\tilde{x}}(x) \tag{A7.4a}$$

$$p_{\tilde{y}}(y) = \frac{\mathrm{d}}{\mathrm{d}y} P_{\tilde{y}}(y) = nP^{n-1}_{\tilde{x}}(y)p_{\tilde{x}}(y) \tag{A7.4b}$$

If the conditions of stochastic independence and identical distribution hold among the $\tilde{x}_i$, then in many cases the shape of the distribution of $\tilde{y}$ is relatively insensitive to the exact shape of the distribution of the $\tilde{x}_i$ and, as n grows, limiting forms of the distribution of $\tilde{y}$ can be found. Three cases have been studied in some detail and will be presented in the following sections.

A7.2.1 Type I distribution

Let the common distribution of the $\tilde{x}_i$ be unlimited in the positive direction and its upper tail fall off in an exponential manner. When $n \to \infty$, the limiting cumulative distribution function and the limiting probability density function of the largest value $\tilde{y}$ of n independent $\tilde{x}_i$ tend to

$$P_{\tilde{y}}(y) = \exp(-\mathrm{e}^{-c(y-u)}) \qquad -\infty \leqslant y \leqslant \infty \tag{A7.5}$$

$$p_{\tilde{y}}(y) = c\exp[-c(y-u) - \mathrm{e}^{-c(y-u)}] \qquad -\infty \leqslant y \leqslant \infty \tag{A7.6}$$

The constants c and u are related to the mean and standard deviation of $\tilde{y}$ by

$$\mu_{\tilde{y}} = u + \gamma/c \simeq u + 0.577/c \tag{A7.7}$$

$$s_{\tilde{y}} = \pi/(\sqrt{6}c) \simeq 1.282/c \tag{A7.8}$$

where $\gamma \simeq 0.577$ is the so-called Euler–Mascheroni constant.

In an analogous manner, for the minimum value $\tilde{z}$ of n independent variables with a common unlimited distribution in the negative direction and an exponential-like lower tail, one obtains in the limit for $n \to \infty$

$$P_{\tilde{z}}(z) = 1 - \exp(-\mathrm{e}^{c(z-u)}) \qquad -\infty \leqslant z \leqslant \infty \tag{A7.9}$$

$$p_{\tilde{z}}(z) = c\exp[c(z-u) - \mathrm{e}^{c(z-u)}] \qquad -\infty \leqslant z \leqslant \infty \tag{A7.10}$$

$$\mu_{\tilde{z}} = u - \gamma/c \tag{A7.11}$$

$$s_{\tilde{z}} = \pi/(\sqrt{6}c) \tag{A7.12}$$

A7.2.2 Type II distribution

In this case the underlying variables has a distribution limited at zero on the left, while the tail on the right is of the form

$$p_{\tilde{x}}(x) = 1 - \zeta(1/x)^c \tag{A7.13}$$

The asymptotic distribution of $\tilde{y}$ (i.e. the limit distribution of the largest of n

independent $\tilde{x}_i$ with $n + \infty$) is

$$P_{\tilde{y}}(y) = \exp[-(u/y)^c] \qquad y \geqslant 0 \qquad \text{(A7.14)}$$

$$p_{\tilde{y}}(y) = \frac{c}{u}\left(\frac{u}{y}\right)^{c+1} \exp\left[-\left(\frac{u}{y}\right)^c\right] \qquad y \geqslant 0 \qquad \text{(A7.15)}$$

$$\mu_{\tilde{y}} = u\Gamma\left(1 - \frac{1}{c}\right) \qquad c > 1 \qquad \text{(A7.16)}$$

$$s_{\tilde{y}}^2 = u^2\left[\Gamma\left(1 - \frac{2}{c}\right) - \Gamma^2\left(1 - \frac{1}{c}\right)\right] \qquad c > 2 \qquad \text{(A7.17)}$$

$$\frac{s_{\tilde{y}}^2}{\mu_{\tilde{y}}^2} = \frac{\Gamma(1 - 2/c)}{\Gamma^2(1 - 1/c)} - 1 \qquad c > 2 \qquad \text{(A7.18)}$$

In equations (A7.16)–(A7.18) $\Gamma(\;)$ denotes the gamma function: $\mu_{\tilde{y}}/u$ and $s_{\tilde{y}}/\mu_{\tilde{y}}$ versus parameters u are plotted in Fig. A.4.

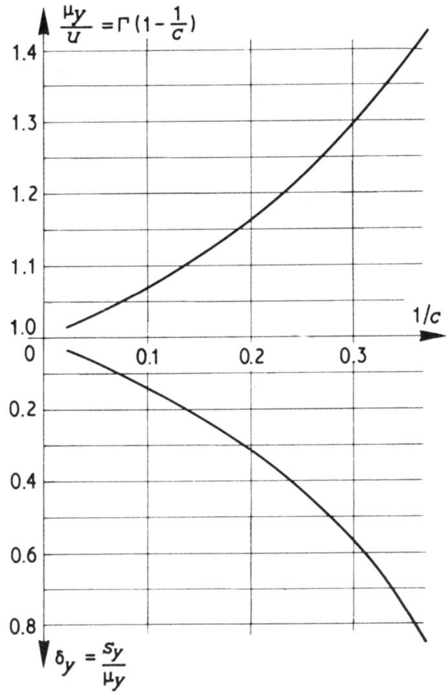

Fig. A.4 Relation between type II extreme distribution parameters.

A7.2.3 Type III distribution

This type of extreme value distributions arises from underlying distributions limited in the tail of interest, i.e. above an upper cut-off w for the maxima, and below a lower cut-off ε for the minima.

Thus, let in the vicinity of w the common CDF of the underlying variables be defined by

$$P_x(x) = 1 \qquad x \geqslant w$$
$$p_x(x) = 1 - c(w-x)^k \qquad x \leqslant w$$

with $k > 0$. The asymptotic distribution of $\tilde{y}$, the largest $\tilde{x}_i$ for $n \to \infty$, is

$$P_{\tilde{y}}(y) = \exp\left[-\left(\frac{w-y}{w-u}\right)^c\right] \qquad y \leqslant w \tag{A7.19}$$

$$p_{\tilde{y}}(y) = \frac{c}{w-u}\left(\frac{w-y}{w-u}\right)^{c-1} \exp\left[-\left(\frac{w-y}{w-u}\right)^c\right] \qquad y \leqslant w \tag{A7.20}$$

Symmetrically, let in the vicinity of ε

$$P_x(x) = 0 \qquad x \leqslant \varepsilon$$
$$p_x(x) = c(x-\varepsilon)^k \qquad x \geqslant \varepsilon$$

The asymptotic distribution of $\tilde{z}$, the smallest $\tilde{x}_i$ for $n \to \infty$, is

$$P_{\tilde{z}}(z) = 1 - \exp\left[-\left(\frac{z-\varepsilon}{u-\varepsilon}\right)^c\right] \qquad z \geqslant \varepsilon \tag{A7.21}$$

$$p_{\tilde{z}}(z) = \frac{c}{u-\varepsilon}\left(\frac{z-\varepsilon}{u-\varepsilon}\right)^{c-1} \exp\left[-\left(\frac{z-\varepsilon}{u-\varepsilon}\right)^c\right] \qquad z \geqslant \varepsilon \tag{A7.22}$$

$$\mu_{\tilde{z}} = \varepsilon + (u-\varepsilon)\Gamma\left(1+\frac{1}{c}\right) \tag{A7.23}$$

$$s_{\tilde{z}}^2 = (u-\varepsilon)^2\left[\Gamma\left(1+\frac{2}{c}\right) - \Gamma^2\left(1+\frac{1}{c}\right)\right] \tag{A7.24}$$

A7.3 RETURN PERIOD

Let $\tilde{x}(t)$ be a random process whose CDF is $P_{\tilde{x}(t)}(x)$. Let $\tilde{y}_1 = \tilde{y}(0, 1)$ be the maximum value that the process attains in a unit of time (e.g. a year). Let the unit time be so long that it is possible to conceive that the maximum values during unit times are stochastically independent. Thus it can be thought that whenever a unit time elapses, an experiment is performed and an independent value of $\tilde{y}_1$ is observed. Let $\tilde{y}_n = \tilde{y}(0, n)$: it is soon realized that the probability $P_{\tilde{y}_n}(x)$ that no value larger than x occurs in n trials is given by (see equation (A7.4a))

$$P_{\tilde{y}_n}(x) = \{P_{\tilde{y}_1}(x)\}^n \tag{A7.25}$$

Further, the probability that a value of $\tilde{y}_1$ larger than x occurs at the nth trial is given by the geometrical law (Table A.1) with $p = 1 - P_{\tilde{y}_1}(x)$:

$$p^0_{\tilde{n}(x)}(n) = \text{Prob}[\tilde{y}_1 > x, n] = \{1 - P_{\tilde{y}_1}(x)\}\{P_{\tilde{y}_1}(x)\}^{n-1} \tag{A7.26}$$

Hence, the mean number $\mu_{\tilde{n}}$ of trials required to exceed at least once the considered value x is given by

$$\mu_{\tilde{n}(x)}(x) = \frac{1}{1 - P_{\tilde{y}_1}(x)} \tag{A7.27}$$

and is usually denoted by $T(x)$ when the unit of time is the year. In this case it is often referred to as the *return period* of the value x.

It is interesting to note that some quantitative features of the probability distribution of the number of trials to exceed the pre-fixed value x are directly related to the return period. For instance, the variance of $\tilde{n}$ is given by (Table A.1):

$$s^2_{\tilde{n}(x)}(x) = \frac{P_{y_1}(x)}{[1 - P_{y_1}(x)]^2} = T^2(x) - T(x) \tag{A7.28}$$

while the probability that exceeding occurs before the return period is given by $1 - P_{\tilde{y}_n}(x)$ (equation (A7.25)), i.e.

$$\begin{aligned} P_{\tilde{n}(x)}[\mu_{\tilde{n}(x)}(x) = T(x)] &= \text{Prob}[\tilde{y} > x, \text{ before } T(x)] \\ &= 1 - \{1 - [1 - P_{y_1}(x)]\}^{\mu_{\tilde{n}(x)}(x)} \\ &\simeq 1 - [1 - 1/T(x)]^{T(x)} \end{aligned} \tag{A7.29}$$

where $P_{\tilde{n}(x)}(n)$ is the CPF of $\tilde{n}(x)$. For large T (say $T > 10$, i.e. for large values of x, provided that $\lim_{x \to -\infty} T(x) = 1$), approximately

$$\begin{aligned} P_{\tilde{n}(x)}(n) &= 1 - \{1 - [1 - P_{y_1}(x)]\}^n \\ &\simeq 1 - \exp\{-n[1 - P_{y_1}(x)]\} \\ &= 1 - \exp[-n/T(x)] \end{aligned} \tag{A7.30}$$

and the distribution of the trials tends to be independent of the particular shape of $P_{\tilde{y}_1}(x)$. In particular the probability that exceeding occurs before the return period tends to

$$P_{\tilde{n}(x)}(T(x)) \simeq 1 - 1/\mathrm{e} = 0.63\,^{\ddagger} \tag{A7.31}$$

a value that does not depend on the return period itself.

The comparison of equations (A7.27) and (A7.30) permits one to make some basic remarks. The value x of the considered random process can be regarded as the argument of both the distribution functions $P_{y_1}(x)$ and $P_{y_n}(x)$:

$$x = P^{-1}_{y_1}(x) \qquad \text{and} \qquad x = P^{-1}_{y_n}(x)$$

‡ For integer T, one obtains from equation (A7.29)

$$(1 - 1/T)^T = \sum_{j=0}^{T} \binom{T}{j}\left(\frac{1}{T}\right)^j = 1 - T\left(\frac{1}{T}\right) + \frac{T(T-1)}{2}\left(\frac{1}{T}\right)^2 - \frac{T(T-1)(T-2)}{2.3}\left(\frac{1}{T}\right)^3$$

$$\simeq 1 - \frac{u}{1!} + \frac{u^2}{2!} - \frac{u^3}{3!} \simeq \mathrm{e}^{-u}$$

where $u = m(1/m) = 1$, and the approximation holds for large T.

Therefore, two return periods $T(x)$ and $T_n(x)$ (equation (A7.27)) can be associated with x and they have generally different values:

$$T(x) = \frac{1}{1 - P_{y_1}(x)} \neq \frac{n}{1 - P_{y_n}(x)} = T_n(x) \qquad \text{(A7.31')}$$

as the quantities are different from a conceptual viewpoint, the nature of the trial being different (e.g. let $P_{y_1}(x) = 70\%$ and $n = 3$, then, $T(x) = 3.33$ years and $T_n(x) = 4.56$ years).

However, for large T, equation (A7.30) provides a relationship between $T(x)$ and $P_{y_n}(x)$:

$$T(x) \simeq \frac{n}{-\ln[1 - P_{\tilde{n}(x)}(x)]} = \frac{n}{-\ln[P_{y_n}(x)]} \qquad \text{(A7.32)}$$

(e.g. let $P_{y_1}(x) = 95\%$ and $n = 3$, then, $P_{y_n}(x) = 85.74\%$ and $T = 20$ years, both by equation (A7.27) and (A7.32)). Moreover, for $1 - P_{y_n}(x) \ll 1$, as $-\ln(P_{y_n}(x)) \simeq 1 - P_{y_n}(x)$, one obtains

$$T(x) \simeq \frac{n}{1 - P_{y_n}(x)} \qquad \text{(A7.32')}$$

(e.g. let $P_{y_1}(x) = 99.5\%$ and $n = 3$, then, $P_{y_n}(x) = 98.5\%$ and $T = 200$ years, both by equation (A7.27) and (A7.32').

Knowledge of the return period most often offers all the informations needed for rational design. In fact, let $\bar{P}$ be an admissible value of the probability that the unfavourable event occurs before N unit times. Hence from equation (A7.30) it is possible to find the reference return period

$$\bar{T}(x) = -N/\ln \bar{P} \simeq -N/(1 - \bar{P}) \qquad \text{(A7.32'')}$$

The latter equality holds if $\bar{P} \ll 1$. Equation (A7.32'') denotes the return period of exceeding the design parameter x^∇ that is required in order that the desired value of $\bar{P}$ be effective. If the relation between the return period and the limit value of the variable is available, or if the CDF of $\tilde{y}_1$ is known, it is possible to decide the design value x of the parameter $\tilde{x}$; by putting $\Pi(x) = 1 - P_{\tilde{y}_1}(x)$

$$x^\nabla = \Pi^{-1}[-(\ln \bar{P})/N] \qquad \text{(A7.33)}$$

A7.4 CROSSING TIMES AND LOCAL MAXIMA OF DIFFERENTIABLE PROCESSES

Let $\{\tilde{x}(t), t > 0\}$ be a differentiable stochastic process in continuous time, with continuous derivative $\{\dot{\tilde{x}}(t), t > 0\}$. Also let a be a given 'threshold' (or 'barrier') level; in this case one can consider a differentiable time-dependent barrier $a(t)$.

The stochastic process is said to *cross from below* at time t if $\tilde{x}(t) = a$ and $\dot{\tilde{x}}(t) > 0$ and to *cross from above* if $\tilde{x}(t) = a$ and $\dot{\tilde{x}}(t) < 0$. Then it is possible to define the following random quantities (Fig. A.5):

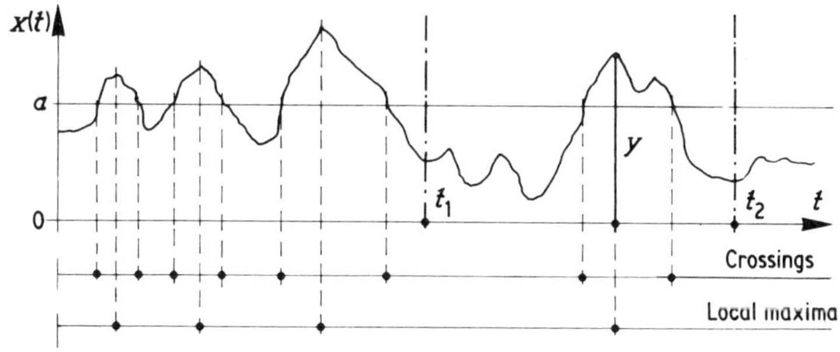

Fig. A.5 Realization of $\tilde{x}(t)$ and relevant associated random quantities.

(1) the point process of crossing times;
(2) the point process of local maxima;
(3) the random variable $\tilde{y}$ defined as the maximum of $\tilde{x}(t)$ in the time interval t_1 to t_2.

A7.4.1 Point process of crossing times

Let $\tilde{n}_a(t_1\ t_2)$, with $t_1 < t_2$, denote the process that counts the number of crossing events of level a. It can be conceived as the integral of the impulse process

$$\tilde{\zeta}_a(t) = \sum_i H(t - t_{a_i}) \tag{A7.34}$$

in which t_{a_i} is the ith crossing time and $H(\ \ )$ is the Heaviside step function; thence

$$\tilde{n}_a(t_1, t_2) = \int_{t_1}^{t_2} \zeta_a(t)\,\mathrm{d}t \tag{A7.35}$$

For the process defined by equation (A7.34) it is possible to find the moments of any order; in particular

$$\begin{aligned}\mu_{\tilde{\zeta}_a(t)}(t) = v_a(t) &= \int_{-\infty}^{\infty}\int_{-\infty}^{\infty} |\dot{x}|\,H(x-a)\,p_{\tilde{x}(t),\,\dot{\tilde{x}}(t)}(x, \dot{x})\,\mathrm{d}x\,\mathrm{d}\dot{x}\\ &= p_{\tilde{x}(t)}(a)\int_{-\infty}^{\infty} |\dot{x}|\,p_{\dot{\tilde{x}}(t)|\tilde{x}(t)}(\dot{x}, a)\,\mathrm{d}\dot{x}\\ &= p_{\tilde{x}(t)}(a)\,\mathrm{E}[\,|\dot{\tilde{x}}(t)|\,|\,\tilde{x}(t) = a]\end{aligned} \tag{A7.36}$$

and allowing for different processes $\tilde{x}_1(t)$ and $\tilde{x}_2(t)$ and for different threshold levels a_1 and a_2

$$\begin{aligned}&\mathrm{E}[\tilde{\zeta}_{a_1}(t_1)\tilde{\zeta}_{a_2}(t_2)]\\ &= p_{\tilde{x}_1(t_1),\,\tilde{x}_2(t_2)}(a_1, a_2)\int_{-\infty}^{\infty}\int_{-\infty}^{\infty} |\dot{x}_1\dot{x}_2|\\ &\quad\times p_{\dot{\tilde{x}}_1(t_1),\,\dot{\tilde{x}}_2(t_2)|\tilde{x}_1(t_1),\,\tilde{x}_2(t_2)}(\dot{x}_1, \dot{x}_2, a_1, a_2)\,\mathrm{d}\dot{x}_1\,\mathrm{d}\dot{x}_2\\ &= p_{\tilde{x}_1(t_1),\,\tilde{x}_2(t_2)}(a_1, a_2)\mathrm{E}[\,|\dot{\tilde{x}}_1(t_1)\dot{\tilde{x}}_2(t_2)|\tilde{x}_1(t_1) = a, \tilde{x}_2(t_2) = a_2\,]\end{aligned} \tag{A7.37}$$

In equation (A7.36), which corresponds to Rice's well known result, $\nu_a(t)$ is the *mean crossing rate of $\tilde{x}(t)$ with respect to level a.* Then, from equations (A7.36) and (A7.37) one obtains

$$\mathrm{E}[\tilde{n}_a(t_1, t_2)] = \int_{t_1}^{t_2} \nu_a(t)\,\mathrm{d}t \tag{A7.38}$$

$$\mathrm{E}[\tilde{n}_{a_1}(t_1, t_2)\tilde{n}_{a_2}(t_3, t_4)] = \int_{t_1}^{t_2} \mathrm{d}\tau_1 \int_{t_3}^{t_4} \mathrm{d}\tau_2\, \mathrm{E}[\tilde{\zeta}_{a_1}(\tau_1)\tilde{\zeta}_{a_2}(\tau_2)] \tag{A7.39}$$

If the *upcrossing process $\tilde{n}_a^+(t_1, t_2)$* is considered, with the associated impulse process $\tilde{\zeta}_a^+(t)$, the following results are obtained:

$$\mu_{\tilde{\zeta}_a^+(t)}(t) = \nu_a^+(t) = p_{\tilde{x}(t)}(a) \int_0^{\infty} \dot{x}\, p_{\dot{\tilde{x}}(t)|\tilde{x}(t)}(\dot{x}, a)\,\mathrm{d}\dot{x} \tag{A7.40}$$

$$\mathrm{E}[\tilde{\zeta}_{a_1}^+(t_1)\tilde{\zeta}_{a_1}^+(t_2)] = p_{\tilde{x}_1(t_1),\,\tilde{x}_2(t_2)}(a_1, a_2) \int_0^{\infty}\int_0^{\infty} \dot{x}_1 \dot{x}_2$$
$$p_{\dot{\tilde{x}}(t_1),\,\dot{\tilde{x}}(t_2)|\tilde{x}(t_1),\,x(t_2)}(\dot{x}_1, \dot{x}_2, a_1, a_2)\,\mathrm{d}\dot{x}_1\,\mathrm{d}\dot{x}_2 \tag{A7.41}$$

If $\tilde{x}(t)$ is a stationary Gaussian stochastic process, equation (A7.36) and (A7.38) become (see equation (A5.30))

$$\nu_a(t) = \nu_a = \frac{1}{\pi}\frac{s_{\dot{\tilde{x}}}}{s_{\tilde{x}}} \exp\left(-\frac{a^2}{2s_{\tilde{x}}^2}\right) = \frac{1}{\pi}\sqrt{\left(-\frac{\ddot{B}(0)}{B(0)}\right)} \exp\left(-\frac{a^2}{2B(0)}\right) \tag{A7.42}$$

and

$$\nu_a^+(t) = \nu_a^-(t) = \tfrac{1}{2}\nu_a \tag{A7.43}$$

where $\nu_a^-(t)$ is the mean downcrossing rate.

The preceding results can be generalized to vector-valued stochastic processes, random-field stochastic processes with discontinuous realization (e.g. some filtered Poisson processes) and stochastic processes with discrete time. In particular, for generalized filtered Poisson processes $\tilde{x}(t)$ with triangular response function, of equal small base width and random height $\tilde{y}(t)$, the mean crossing rate of $\tilde{x}(t)$ is given by

$$\nu_a^+(t) = \nu_a^-(t) = [1 - P_{\tilde{y}(t)}(a, t)]\,\lambda(t) \tag{A7.44}$$

where λ is the intensity of the basic Poisson process.

A7.4.2 Point process of local maxima

A (local) maximum of a twice-differentiable process $\tilde{x}(t)$ occurs whenever $\dot{\tilde{x}}(t) = 0$ and $\ddot{\tilde{x}}(t) < 0$. Thence it corresponds to a zero-downcrossing event of $\dot{\tilde{x}}(t)$ and their occurrence characteristics can be obtained by the relevant procedure.

However, the quantity of interest is the distribution of the local maxima of $\tilde{x}(t)$. If $\tilde{x}(t)$ is stationary and Gaussian with zero mean, the probability density function of the magnitude $x_{\max}(t)$ of the peak of $\tilde{x}(t)$ that occurs at time t reduces to

$$p_{\tilde{x}_{\max}(t)}(x) = \frac{x}{c_0}\exp\left(-\frac{x^2}{2c_0}\right) \qquad 0 < x < \infty \tag{A7.45}$$

for a narrow-band process. In equation (A7.45) (cf. equation (A5.44)).

$$c_0 = \int_{-\infty}^{\infty} G(\omega)\, \mathrm{d}\omega = B_{\tilde{x}}(0) = s_{\tilde{x}}^2 \tag{A7.46}$$

and $G(\omega)$ is the spectral density function of $\tilde{x}(t)$. Equation (A7.45) characterizes the *Rayleigh density function.* By contrast, if $\tilde{x}(t)$ is a wide-band process, the probability density of equation (A7.45) is well approximated by the normal density of $\tilde{x}(t)$.

In intermediate cases, the distribution is neither Rayleigh nor normal.

A7.5 FIRST PASSAGE TIME DISTRIBUTION

The exact probability distribution $P_{\tilde{\tau}_a}(t)$ of the first-passage time (i.e. the first time at which $x(t) = a$) is known only in few cases concerning Markov process. Thence many approximate methods have been proposed to tackle theoretical difficulties. They may be classified as:

(1) methods based on the mean upcrossing rate of $\tilde{x}(t)$ or of an envelope process $\tilde{\xi}(t)$;
(2) methods based on the random process of peaks of $\tilde{x}(t)$ or of $\tilde{\xi}(t)$;
(3) methods that replace $\tilde{x}(t)$ with an equivalent first-order Markov process;
(4) numerical procedures (e.g. Monte Carlo simulation).

The first method is the most common approach and can be applied to a wide variety of Gaussian processes. It will be discussed in the following. For methods of type (4), see Chapter 9.

The random process $\tilde{x}(t)$ is assumed to be stationary with zero mean and unit variance. Then, if a is a number larger than 1, the distribution function $P_{\tilde{\tau}}(t)$ can be approximated through functions of the type (see Fig. A.6).

$$P_{\tilde{\tau}_a}(t) \simeq 1 - P\mathrm{e}^{-ct} \qquad 0 \leqslant P \leqslant 1,\ c \geqslant 0 \tag{A7.47}$$

The different methods (1) and (2) correspond to different choices of P and c.

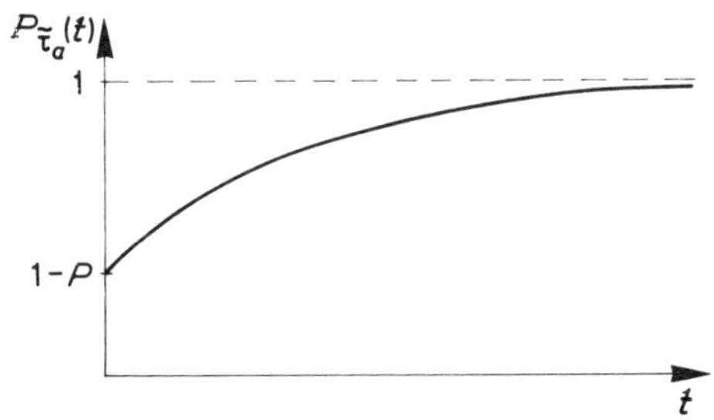

Fig. A.6 Approximate function for $P_{\tau_a}(t)$.

A7.5.1 Methods based on the mean upcrossing rate

An approximation to the first-passage time distribution follows from the assumption that the upcrossings are Poisson events with mean rate ν_a^+. With reference to Gaussian processes, this assumption becomes exact asymptotically as a tends to ∞ and involves

$$P_{\tilde{\tau}_a}(t) = 1 - \exp(-\nu_a^+ t) \tag{A7.48}$$

where ν_a^+ is given by equation (A7.36).

Generalization of equation (A7.48) can be obtained by introducing the non-homogeneous Poisson process (see equation (A6.31)), so that, for a general random process, one can write

$$P_{\tilde{\tau}_a}(t) = 1 - \exp\left(-\int_0^t \nu_a^+(t')\,\mathrm{d}t'\right) \tag{A7.49}$$

For the filtered Poisson process for which equation (A7.44) holds, equation (A7.49) involves

$$P_{\tilde{\tau}_a}(t) = 1 - \exp\left(-[1 - P_{\tilde{w}}(a)]\int_0^t \lambda(t')\,\mathrm{d}t'\right) \tag{A7.50}$$

and, from equation (A7.1)

$$P_{\tilde{y}(0,t)}(t) = \exp\left(-[1 - P_{\tilde{w}}(a)]\int_0^t \lambda(t')\,\mathrm{d}t'\right) \tag{A7.51}$$

For $a > 3$, the Poisson assumption is generally accurate for wide-band processes.

In narrow-band processes the upcrossing points are not independent of each other: rather they tend to cluster. For these processes it is suitable to define an envelope process $\tilde{\xi}(t)$, that is a random process for which $\tilde{\xi}(t) \geqslant \tilde{x}(t)$ for all t; $\tilde{\xi}(t) \simeq \tilde{x}(t)$ near the peaks of $\tilde{x}(t)$ and $\tilde{\xi}(t)$ is smoother than $\tilde{x}(t)$.

For instance, let the spectral decomposition of $\tilde{x}(t)$ (stationary, with zero mean and finite variance) be

$$\tilde{x}(t) = \int_0^\infty [\cos(\zeta t)\,\mathrm{d}\tilde{u}_1(\zeta) + \sin(\zeta t)\,\mathrm{d}\tilde{u}_2(\zeta)]$$

Then the following envelope process can be introduced:

$$\tilde{\xi}(t) = \sqrt{[\tilde{x}^2(t) + \tilde{x}_{\mathrm{H}}^2(t)]} \tag{A7.52}$$

where $\tilde{x}_{\mathrm{H}}(t)$ is the Hilbert transform of $\tilde{x}(t)$, i.e.

$$x_{\mathrm{H}}(t) = \int_0^\infty [\sin(\zeta t)\,\mathrm{d}\tilde{u}_1(\zeta) - \cos(\zeta t)\,\mathrm{d}\tilde{u}_2(\zeta)] \tag{A7.53}$$

with

$$\begin{aligned} \tilde{u}_1(\zeta) &= \lim_{T\to\infty}\frac{1}{\pi}\int_{-T}^{T}\frac{\sin(\zeta t)}{t}\,\tilde{x}(t)\,\mathrm{d}t \\ \tilde{u}_2(\zeta) &= \lim_{T\to\infty}\frac{1}{\pi}\int_{-T}^{T}\frac{1-\cos(\zeta t)}{t}\,\tilde{x}(t)\,\mathrm{d}t \end{aligned} \tag{A7.54}$$

Once the process $\tilde{\xi}(t)$ is defined the mean upcrossing rate of the envelope can be obtained and then equation (A7.48) employed.

In all previous developments, the parameter P of equation (A7.47) was assumed equal to 1: that is true under the assumption that $\tilde{x}(0) < a$. Otherwise it is necessary to introduce in equations (A7.48)–(A7.51)

$$P = 1 - P_{\tilde{x}(0)}(a) \tag{A7.55}$$

as a factor to the exponential.

In the case that P is smaller than 1, then the previous values of c must be divided by P in order to take into account the smaller expected value of $\tilde{\tau}_a$ given that the process starts below the level a.

A8

Tables of probability functions

Tables of probability functions appear on the following pages.

Table A.1 Discrete probability laws.

Probability law	Parameters	Probability mass function $p^0(\ )$
Bernoulli	$0 \leqslant p \leqslant 1$ $q = 1 - p$	$p^0(x) = \begin{cases} p & x = 1 \\ q & x = 0 \\ 0 & \text{otherwise} \end{cases}$
Binomial	$n = 1, 2, \ldots$ $0 \leqslant p \leqslant 1$	$p^0(x) = \begin{cases} \binom{n}{x} p^x q^{n-x} & x = 0, 1, 2 \ldots, n \\ 0 & \text{otherwise} \end{cases}$
Poisson	$\lambda > 0$	$p^0(x) = \begin{cases} e^{-\lambda}\lambda^x/x! & x = 0, 1, 2, \ldots \\ 0 & \text{otherwise} \end{cases}$
Geometric	$0 \leqslant p \leqslant 1$	$p^0(x) = \begin{cases} pq^{x-1} & x = 1, 2, \ldots \\ 0 & \text{otherwise} \end{cases}$
Negative binomial (Pascal)	$r > 0$ $0 \leqslant p \leqslant 1$	$p^0(x) = \begin{cases} \binom{r+x-1}{x} p^r q^x = \binom{-r}{x} p^r(-q)^x & x = 0, 1, 2, \ldots \\ 0 & \text{otherwise} \end{cases}$
Hypergeometric	$N = 1, 2, \ldots$ $n = 1, 2, \ldots, N$ $p = 0, \frac{1}{N}, \frac{2}{N}, \ldots, 1$	$p^0(x) = \begin{cases} \dfrac{\binom{Np}{x}\binom{Nq}{n-x}}{\binom{N}{n}} & x = 0, 1, \ldots, n \\ 0 & \text{otherwise} \end{cases}$

Mean $\mu_{\tilde{x}} = \mathrm{E}[\tilde{x}]$	Variance $s_{\tilde{x}}^2 = \mathrm{Var}[\tilde{x}]$	Moment-generating function $\Psi_{\tilde{x}}(u) = \mathrm{E}[e^{u\tilde{x}}]$	Characteristic function $\Omega_{\tilde{x}}(u) = \mathrm{E}[e^{iu\tilde{x}}]$
p	pq	$pe^u + q$	$pe^{iu} + q$
np	npq	$(pe^u + q)^n$	$(pe^{iu} + q)^n$
λ	λ	$\exp[\lambda(e^u - 1)]$	$\exp[\lambda(e^{iu} - 1)]$
$\frac{1}{p}$	$\frac{q}{p^2}$	$\frac{pe^u}{1 - qe^u}$	$\frac{pe^{iu}}{1 - qe^{iu}}$
$\frac{rq}{p} = r\zeta$	$\frac{rq}{p^2} = r\zeta\Lambda$	$\left(\frac{p}{1 - qe^u}\right)^r = (\Lambda - \zeta e^u)^{-r}$	$\left(\frac{p}{1 - qe^{iu}}\right)^r = (\Lambda - \zeta e^{iu})^{-r}$
with $\zeta = \frac{q}{p}$;	$\Lambda = \frac{1}{p}$		
np	$npq\left(\frac{N-n}{N-1}\right)$	–	–

Table A.2 Continuous probability laws

Probability law	Parameters	Probability density function $p(\ )$
Uniform over interval a to b	$-\infty < a < b < \infty$	$p(x) = \begin{cases} \dfrac{1}{b-a} & a < x < b \\ 0 & \text{otherwise} \end{cases}$
Normal (Gaussian)†	$-\infty < \mu < \infty$ $s > 0$	$p(x) = \dfrac{1}{s\sqrt{(2\pi)}} \exp\left[-\dfrac{1}{2}\left(\dfrac{x-\mu}{s}\right)^2\right]$
Exponential	$\lambda > 0$	$p(x) = \begin{cases} \lambda e^{-\lambda x} & x > 0 \\ 0 & x < 0 \end{cases}$
Gamma	$r > -1$ $\lambda > 0$	$p(x) = \begin{cases} \dfrac{\lambda}{\Gamma(r)} (\lambda x)^{r-1} e^{-\lambda x} & x > 0 \\ 0 & x \leqslant 0 \end{cases}$
χ^2 with n degrees of freedom ‡	$n > 0$	$p(x) = \begin{cases} \dfrac{1}{2^{n/2}\Gamma(n/2)} x^{(n/2)-1} e^{-x/2} & x > 0 \\ 0 & x \leqslant 0 \end{cases}$
F with m, n degrees of freedom ‡	$n, m > 0$	$p(x) = \begin{cases} \dfrac{\Gamma[(m+n)/2]}{\Gamma(m/2)\Gamma(n/2)} \left(\dfrac{m}{n}\right)^{m/2} \dfrac{x^{(m/2)-1}}{[1+(m/n)x]^{(m+n)/2}} & x > 0 \\ 0 & x \leqslant 0 \end{cases}$

† The log-normal distribution is easily derived from the normal as follows.

Let $\tilde{x}$ be a normally distributed r.v. with mean μ_x and variance s_x^2. The variable $\tilde{u} = e^x$ is log-normally distributed, with mean $\mu_u = \exp(\mu_x + \frac{1}{2}s_x^2)$ and variance $s_u^2 = \mu_u^2(e^{s_x^2} - 1)$; note that $\tilde{u} \geqslant 0$.

Conversely, if μ_u and s_u are known,

$$s_x^2 = \ln\left\{\left(\frac{s_u}{\mu_u}\right)^2 + 1\right\}; \qquad \mu_x = \ln((\mu_u) - \tfrac{1}{2}s_x^2$$

Further, if $\tilde{x}$ and $\tilde{y}$ are jointly normally distributed with coefficient of correlation ρ_{xy}, one can introduce the r.v. $\tilde{u} = e^x$ and $v = e^y$, that are lognormally distributed with

$$\rho_{uv} = \exp(\rho_{xy}s_x s_y - 1)/[(e^{s_x^2} - 1)(e^{s_y^2} - 1)]^{1/2}$$

Conversely

$$\rho_{xy} = \ln\{1 + \rho_{uv}[(e^{s_x^2} - 1)(e^{s_y^2} - 1)]^{1/2}\}/(s_x s_y).$$

Mean $\mu_{\tilde{x}} = \mathrm{E}[\tilde{x}]$	Variance $s_{\tilde{x}}^2 = \mathrm{Var}[\tilde{x}]$	Moment-generating function $\Psi_{\tilde{x}}(u) = \mathrm{E}[\mathrm{e}^{u\tilde{x}}]$	Characteristic function $\Omega_{\tilde{x}}(u) = \mathrm{E}[\mathrm{e}^{\mathrm{i}u\tilde{x}}]$
$\frac{a+b}{2}$	$\frac{(b-a)^2}{12}$	$\frac{\mathrm{e}^{ub}-\mathrm{e}^{ua}}{u(b-a)}$	$\frac{\mathrm{e}^{\mathrm{i}ub}-\mathrm{e}^{\mathrm{i}ua}}{\mathrm{i}u(b-a)}$
μ	s^2	$\exp(u\mu+\frac{1}{2}u^2s^2)$	$\exp(\mathrm{i}u\mu-\frac{1}{2}u^2s^2)$
$\frac{1}{\lambda}$	$\frac{1}{\lambda^2}$	$\frac{\lambda}{\lambda-u}=\left(1-\frac{u}{\lambda}\right)^{-1}$	$\left(1-\frac{\mathrm{i}u}{\lambda}\right)^{-1}$
$\frac{r}{\lambda}$	$\frac{r}{\lambda^2}$	$\left(1-\frac{u}{\lambda}\right)^{-r}$	$\left(1-\frac{\mathrm{i}u}{\lambda}\right)^{-r}$
n	$2n$	$(1-2u)^{-n/2}$	$(1-2\mathrm{i}u)^{-n/2}$
$\frac{n}{n-2}$ if $n>2$	$\frac{2n(m+n-2)}{m(n-2)^2(n-4)}$ if $n>4$	–	–

† The *gamma function* is defined by

$$\Gamma(n+1) = \int_0^\infty x^n \mathrm{e}^{-x}\,\mathrm{d}x.$$

Note that $\Gamma(\frac{1}{2}) = \sqrt{\pi}$; while if $n = 0, 1, 2, \ldots, \Gamma(n+1) = n!$

Table A.3 Binomial probabilities: table of $\binom{n}{k}p^k(1-p)^{n-k}$.

		p							
n	*k*	0.01	0.05	0.10	0.20	0.30	0.3333	0.40	0.50
2	0	0.9801	0.9025	0.8100	0.6400	0.4900	0.4444	0.3600	0.2500
	1	0.0198	0.0950	0.1800	0.3200	0.4200	0.4444	0.4800	0.5000
	2	0.0001	0.0025	0.0100	0.0400	0.0900	0.1111	0.1600	0.2500
3	0	0.9703	0.8574	0.7290	0.5120	0.3430	0.2963	0.2160	0.1250
	1	0.0294	0.1354	0.2430	0.3840	0.4410	0.4444	0.4320	0.3750
	2	0.0003	0.0071	0.0270	0.0960	0.1890	0.2222	0.2880	0.3750
	3	0.0000	0.0001	0.0010	0.0080	0.0270	0.0370	0.0640	0.1250
4	0	0.9606	0.8145	0.6561	0.4096	0.2401	0.1975	0.1296	0.0625
	1	0.0388	0.1715	0.2916	0.4096	0.4116	0.3951	0.3456	0.2500
	2	0.0006	0.0135	0.0486	0.1536	0.2646	0.2963	0.3456	0.3750
	3	0.0000	0.0005	0.0036	0.0256	0.0756	0.0988	0.1536	0.2500
	4	0.0000	0.0000	0.0001	0.0016	0.0081	0.0123	0.0256	0.0625
5	0	0.9510	0.7738	0.5905	0.3277	0.1681	0.1317	0.0778	0.0312
	1	0.0480	0.2036	0.3280	0.4096	0.3602	0.3292	0.2592	0.1562
	2	0.0010	0.0214	0.0729	0.2048	0.3087	0.3292	0.3456	0.3125
	3	0.0000	0.0011	0.0081	0.0512	0.1323	0.1646	0.2304	0.3125
	4	0.0000	0.0000	0.0004	0.0064	0.0284	0.0412	0.0768	0.1562
	5	0.0000	0.0000	0.0000	0.0003	0.0024	0.0041	0.0102	0.0312
6	0	0.9415	0.7351	0.5314	0.2621	0.1176	0.0878	0.0467	0.0156
	1	0.0571	0.2321	0.3543	0.3932	0.3025	0.2634	0.1866	0.0938
	2	0.0014	0.0305	0.0984	0.2458	0.3241	0.3292	0.3110	0.2344
	3	0.0000	0.0021	0.0146	0.0819	0.1852	0.2195	0.2765	0.3125
	4	0.0000	0.0001	0.0012	0.0154	0.0595	0.0823	0.1382	0.2344
	5	0.0000	0.0000	0.0001	0.0015	0.0102	0.0165	0.0369	0.0938
	6	0.0000	0.0000	0.0000	0.0001	0.0007	0.0014	0.0041	0.0156
8	0	0.9227	0.6634	0.4305	0.1678	0.0576	0.0390	0.0168	0.0039
	1	0.0746	0.2793	0.3826	0.3355	0.1977	0.1561	0.0896	0.0312
	2	0.0026	0.0515	0.1488	0.2936	0.2965	0.2731	0.2090	0.1094
	3	0.0001	0.0054	0.0331	0.1468	0.2541	0.2731	0.2787	0.2188
	4	0.0000	0.0004	0.0046	0.0459	0.1361	0.1707	0.2322	0.2734
	5	0.0000	0.0000	0.0004	0.0092	0.0467	0.0683	0.1239	0.2188
	6	0.0000	0.0000	0.0000	0.0011	0.0100	0.0171	0.0413	0.1094
	7	0.0000	0.0000	0.0000	0.0001	0.0012	0.0024	0.0079	0.0213
	8	0.0000	0.0000	0.0000	0.0000	0.0001	0.0002	0.0007	0.0039
10	0	0.9044	0.5987	0.3487	0.1074	0.0282	0.0173	0.0060	0.0010
	1	0.0914	0.3151	0.3874	0.2684	0.1211	0.0867	0.0403	0.0098
	2	0.0042	0.0746	0.1937	0.3020	0.2335	0.1951	0.1209	0.0439
	3	0.0001	0.0105	0.0574	0.2013	0.2668	0.2601	0.2150	0.1172
	4	0.0000	0.0010	0.0112	0.0881	0.2001	0.2276	0.2508	0.2051
	5	0.0000	0.0001	0.0015	0.0264	0.1029	0.1366	0.2007	0.2461
	6	0.0000	0.0000	0.0001	0.0055	0.0368	0.0569	0.1115	0.2051
	7	0.0000	0.0000	0.0000	0.0008	0.0090	0.0163	0.0425	0.1172
	8	0.0000	0.0000	0.0000	0.0001	0.0014	0.0030	0.0106	0.0439
	9	0.0000	0.0000	0.0000	0.0000	0.0001	0.0003	0.0016	0.0098
	10	0.0000	0.0000	0.0000	0.0000	0.0000	0.0000	0.0001	0.0010

Reprinted from Parzen [A.1] by permission of John Wiley and Sons Ltd.

Table A.4 Poisson probabilities: table of $e^{-\lambda}\lambda^x/x!$.

λ	x = 0	1	2	3	4	5	6	8	10	12	14	16	18	20
0.1	0.9048	0.0905	0.0045	0.0002										
0.2	0.8187	0.1637	0.0164	0.0011	0.0001									
0.3	0.7408	0.2222	0.0333	0.0033	0.0002									
0.4	0.6703	0.2681	0.0536	0.0072	0.0007	0.0001								
0.5	0.6065	0.3033	0.0758	0.0126	0.0016	0.0002								
0.6	0.5488	0.3293	0.0988	0.0198	0.0030	0.0004								
0.7	0.4966	0.3476	0.1217	0.0284	0.0050	0.0007	0.0001							
0.8	0.4493	0.3595	0.1438	0.0383	0.0077	0.0012	0.0002							
0.9	0.4066	0.3659	0.1647	0.0494	0.0111	0.0020	0.0003							
1.0	0.3679	0.3679	0.1839	0.0613	0.0153	0.0031	0.0005							
1.1	0.3329	0.3662	0.2014	0.0738	0.0203	0.0045	0.0008							
1.2	0.3012	0.3614	0.2169	0.0867	0.0260	0.0062	0.0012							
1.3	0.2725	0.3543	0.2303	0.0998	0.0324	0.0084	0.0018	0.0001						
1.4	0.2466	0.3452	0.2417	0.1128	0.0395	0.0111	0.0026	0.0001						
1.5	0.2231	0.3347	0.2510	0.1255	0.0471	0.0141	0.0035	0.0001						
1.6	0.2019	0.3230	0.2584	0.1378	0.0551	0.0176	0.0047	0.0002						
1.7	0.1827	0.3106	0.2640	0.1496	0.0636	0.0216	0.0061	0.0003						
1.8	0.1653	0.2975	0.2678	0.1607	0.0723	0.0260	0.0078	0.0005						
1.9	0.1496	0.2842	0.2700	0.1710	0.0812	0.0309	0.0098	0.0006						
2.0	0.1353	0.2707	0.2707	0.1804	0.0902	0.0361	0.0120	0.0009						
3.0	0.0498	0.1494	0.2240	0.2240	0.1680	0.1008	0.0504	0.0081	0.0008	0.0001				
4.0	0.0183	0.0733	0.1465	0.1954	0.1954	0.1563	0.1042	0.0298	0.0053	0.0006				
5.0	0.0067	0.0337	0.0842	0.1404	0.1755	0.1755	0.1462	0.0653	0.0181	0.0034	0.0005			
6.0	0.0025	0.0149	0.0446	0.0892	0.1339	0.1606	0.1606	0.1033	0.0413	0.0113	0.0022	0.0003		
7.0	0.0009	0.0064	0.0223	0.0521	0.0912	0.1277	0.1490	0.1304	0.0710	0.0264	0.0071	0.0014	0.0002	
8.0	0.0003	0.0027	0.0107	0.0286	0.0573	0.0916	0.1221	0.1396	0.0993	0.0481	0.0169	0.0045	0.0009	0.0002
9.0	0.0001	0.0011	0.0050	0.0150	0.0337	0.0607	0.0911	0.1318	0.1186	0.0728	0.0324	0.0109	0.0029	0.0006
10.0	0.0000	0.0005	0.0023	0.0076	0.0189	0.0378	0.0631	0.1126	0.1251	0.0948	0.0521	0.0217	0.0071	0.0019

Table A.5 CDF of the standardized normal distribution:

$$\Phi(x^*) = \Phi(u) = \frac{1}{\sqrt{(2\pi)}} \int_{-\infty}^{u} e^{-\frac{1}{2}y^2} \, dy; \qquad \tilde{x}^* = \frac{\tilde{x} - \mu_x}{s_x}.$$

(*a*) $x^* \leqslant 1.99$ (*includes PDF, in italic*)

x^*	0.00	0.01	0.02	0.03	0.04	0.05	0.06	0.07	0.08	0.09
0.0	0.5000	0.5040	0.5080	0.5120	0.5160	0.5199	0.5239	0.5279	0.5319	0.5359
	0.3989	*0.3989*	*0.3989*	*0.3988*	*0.3986*	*0.3984*	*0.3982*	*0.3980*	*0.3977*	*0.3973*
0.1	0.5398	0.5438	0.5478	0.5517	0.5557	0.5596	0.5636	0.5675	0.5714	0.5753
	0.3970	*0.3965*	*0.3961*	*0.3956*	*0.3951*	*0.3945*	*0.3939*	*0.3932*	*0.3925*	*0.3918*
0.2	0.5793	0.5832	0.5871	0.5910	0.5948	0.5987	0.6026	0.6064	0.6103	0.6141
	0.3910	*0.3902*	*0.3894*	*0.3885*	*0.3876*	*0.3867*	*0.3857*	*0.3847*	*0.3836*	*0.3825*
0.3	0.6179	0.6217	0.6255	0.6293	0.6331	0.6368	0.6406	0.6443	0.6480	0.6517
	0.3814	*0.3802*	*0.3790*	*0.3778*	*0.3765*	*0.3752*	*0.3739*	*0.3725*	*0.3712*	*0.3697*
0.4	0.6554	0.6591	0.6628	0.6664	0.6700	0.6736	0.6772	0.6808	0.6844	0.6879
	0.3683	*0.3668*	*0.3653*	*0.3637*	*0.3621*	*0.3605*	*0.3589*	*0.3572*	*0.3555*	*0.3538*
0.5	0.6915	0.6950	0.6985	0.7019	0.7054	0.7088	0.7123	0.7157	0.7190	0.7224
	0.3521	*0.3503*	*0.3485*	*0.3467*	*0.3448*	*0.3429*	*0.3410*	*0.3391*	*0.3372*	*0.3352*
0.6	0.7257	0.7291	0.7324	0.7357	0.7389	0.7422	0.7454	0.7486	0.7517	0.7549
	0.3332	*0.3312*	*0.3292*	*0.3271*	*0.3251*	*0.3230*	*0.3209*	*0.3187*	*0.3166*	*0.3144*
0.7	0.7580	0.7611	0.7642	0.7673	0.7703	0.7734	0.7764	0.7794	0.7823	0.7852
	0.3123	*0.3101*	*0.3079*	*0.3056*	*0.3034*	*0.3011*	*0.2989*	*0.2966*	*0.2943*	*0.2920*
0.8	0.7881	0.7910	0.7939	0.7967	0.7995	0.8023	0.8051	0.8078	0.8106	0.8133
	0.2897	*0.2874*	*0.2850*	*0.2827*	*0.2803*	*0.2780*	*0.2756*	*0.2732*	*0.2709*	*0.2685*
0.9	0.8159	0.8186	0.8212	0.8238	0.8264	0.8289	0.8315	0.8340	0.8365	0.8389
	0.2661	*0.2637*	*0.2613*	*0.2589*	*0.2565*	*0.2541*	*0.2516*	*0.2492*	*0.2468*	*0.2444*
1.0	0.8413	0.8438	0.8461	0.8485	0.8508	0.8531	0.8554	0.8577	0.8599	0.8621
	0.2420	*0.2396*	*0.2371*	*0.2347*	*0.2323*	*0.2299*	*0.2275*	*0.2251*	*0.2227*	*0.2203*

1.1	0.8643	0.8665	0.8686	0.8708	0.8729	0.8749	0.8770	0.8790	0.8810	0.8830
	0.2179	*0.2155*	*0.2131*	*0.2107*	*0.2083*	*0.2059*	*0.2036*	*0.2012*	*0.1989*	*0.1965*
1.2	0.8849	0.8869	0.8888	0.8907	0.8925	0.8944	0.8962	0.8980	0.8997	0.9015
	0.1942	*0.1919*	*0.1895*	*0.1872*	*0.1849*	*0.1826*	*0.1804*	*0.1781*	*0.1758*	*0.1736*
1.3	0.9032	0.9049	0.9066	0.9082	0.9099	0.9115	0.9131	0.9147	0.9162	0.9177
	0.1714	*0.1691*	*0.1669*	*0.1647*	*0.1626*	*0.1604*	*0.1582*	*0.1561*	*0.1539*	*0.1518*
1.4	0.9192	0.9207	0.9222	0.9236	0.9251	0.9265	0.9279	0.9292	0.9306	0.9319
	0.1497	*0.1476*	*0.1456*	*0.1435*	*0.1415*	*0.1394*	*0.1374*	*0.1354*	*0.1334*	*0.1315*
1.5	0.9332	0.9245	0.9357	0.9370	0.9382	0.9394	0.9406	0.9418	0.9429	0.9441
	0.1295	*0.1276*	*0.1257*	*0.1238*	*0.1219*	*0.1200*	*0.1182*	*0.1163*	*0.1145*	*0.1127*
1.6	0.9452	0.9463	0.9474	0.9484	0.9495	0.9505	0.9515	0.9525	0.9535	0.9545
	0.1109	*0.1092*	*0.1074*	*0.1057*	*0.1040*	*0.1023*	*0.1006*	*0.0983*	*0.0973*	*0.0957*
1.7	0.9554	0.9564	0.9573	0.9582	0.9591	0.9599	0.9608	0.9616	0.9625	0.9633
	0.0940	*0.0925*	*0.0909*	*0.0893*	*0.0878*	*0.0863*	*0.0848*	*0.0833*	*0.0818*	*0.0804*
1.8	0.9641	0.9649	0.9656	0.9664	0.9671	0.9678	0.9686	0.9693	0.9699	0.9706
	0.0789	*0.0775*	*0.0761*	*0.0748*	*0.0734*	*0.0721*	*0.0707*	*0.0694*	*0.0681*	*0.0669*
1.9	0.9713	0.9719	0.9726	0.9732	0.9738	0.9744	0.9750	0.9756	0.9761	0.9767
	0.0656	*0.0644*	*0.0632*	*0.0619*	*0.0608*	*0.0596*	*0.0584*	*0.0573*	*0.0562*	*0.0551*

(*b*) $x^* \geqslant 2{,}0$

2.0	0.9772	0.9778	0.9783	0.9788	0.9793	0.9798	0.9803	0.9808	0.9812	0.9817
2.1	0.9821	0.9826	0.9830	0.9834	0.9838	0.9842	0.9846	0.9850	0.9854	0.9857
2.2	0.9861	0.9864	0.9868	0.9871	0.9875	0.9878	0.9881	0.9884	0.9887	0.9890
2.3	0.9893	0.9896	0.9898	0.9901	0.9904	0.9906	0.9909	0.9911	0.9913	0.9916
2.4	0.9918	0.9920	0.9922	0.9925	0.9927	0.9929	0.9931	0.9932	0.9934	0.9936
2.5	0.9938	0.9940	0.9941	0.9943	0.9945	0.9946	0.9948	0.9949	0.9951	0.9952
2.6	0.9953	0.9955	0.9956	0.9957	0.9959	0.9960	0.9961	0.9962	0.9963	0.9964
2.7	0.9965	0.9966	0.9967	0.9968	0.9969	0.9970	0.9971	0.9972	0.9973	0.9974
2.8	0.9974	0.9975	0.9976	0.9977	0.9977	0.9978	0.9979	0.9979	0.9980	0.9981
2.9	0.9981	0.9982	0.9982	0.9983	0.9984	0.9984	0.9985	0.9985	0.9986	0.9986
3.0	0.9987	0.9987	0.9987	0.9988	0.9988	0.9989	0.9989	0.9989	0.9990	0.9990
3.1	0.9990	0.9991	0.9991	0.9991	0.9992	0.9992	0.9992	0.9992	0.9993	0.9993

Table A.5 (*Contd.*)

x^*	0.00	0.01	0.02	0.03	0.04	0.05	0.06	0.07	0.08	0.09
3.2	0.9993	0.9993	0.9994	0.9994	0.9994	0.9994	0.9994	0.9995	0.9995	0.9995
3.3	0.9995	0.9995	0.9995	0.9996	0.9996	0.9996	0.9996	0.9996	0.9996	0.9997
3.4	0.9997	0.9997	0.9997	0.9997	0.9997	0.9997	0.9997	0.9997	0.9997	0.9998

(c) $1-\Phi(x^*)$; $2.32 \leqslant x^* \leqslant 6.36$

x^*	2.32	2.50	3.00	3.09
$1-\Phi(x^*)$	1.00E − 2†	6.21E − 3	1.35E − 3	1.00E − 3
x^*	3.50	3.72	4.00	4.27
$1-\Phi(x^*)$	2.13E − 4	1.00E − 4	3.17E − 5	1.00E − 5
x^*	4.50	4.75	5.00	5.20
$1-\Phi(x^*)$	3.30E − 6	1.00E − 6	2.90E − 7	1.00E − 7
x^*	5.50	5.61	6.00	6.36
$1-\Phi(x^*)$	1.90E − 8	1.00E − 8	1.00E − 9	1.00E − 10

† $1.00\,E-2 \equiv 1.00 \times 10^{-2}$.

Table A.6 Values of the distribution: $P_y(y) = \exp(-e^{-y})$ (extreme type I (maxima) distribution).

$P_{\hat{y}}(y)$	y	$P_{\hat{y}}(y)$	y	$P_{\hat{y}}(y)$	y
0.0000	$-\infty$	0.500	0.367	$0.9550 = 1-5\,E-3$	5.296
0.0001	-2.220	0.550	0.514	$0.9980 = 1-2\,E-3$	6.214
0.0010	-1.933	0.600	0.672	$0.9990 = 1-1\,E-3$	6.907
0.0050	-1.667	0.650	0.842	$0.9995 = 1-5\,E-4$	7.601
0.0100	-1.527	0.700	0.031	$0.9998 = 1-2\,E-4$	8.517
0.0200	-1.364	0.750	1.246	$0.9999 = 1-1\,E-4$	9.210
0.0500	-1.097	0.800	1.500	$1-5\,E-5$	9.903
0.1000	-0.834	0.850	1.817	$1-2\,E-5$	10.820
0.1500	-0.640	0.900	2.250	$1-1\,E-5$	11.513
0.2000	-0.476	0.950	2.970	$1-1\,E-6$	13.815
0.2500	-0.327	0.980	3.902	$1-1\,E-7$	16.118
0.3000	-0.186	0.990	4.600	$1-1\,E-8$	18.421
0.4000	0.087	0.992	4.824	$1-1\,E-9$	20.723

After [A.2].

Table A.7 Values of the χ_n^2 variable distributed according to the χ^2 distribution.

	p_{χ^2}									
n	0.005	0.010	0.025	0.050	0.100	0.90	0.95	0.975	0.99	0.995
1	0.000039	0.00016	0.00098	0.0039	0.0158	2.71	3.84	5.02	6.63	7.88
2	0.0100	0.0201	0.0506	0.103	0.211	4.61	5.99	7.38	9.21	10.60
3	0.0717	0.155	0.216	0.352	0.584	6.25	7.81	9.35	11.34	12.84
4	0.207	0.297	0.484	0.711	1.06	7.78	9.49	11.14	13.28	14.86
5	0.412	0.554	0.831	1.15	1.61	9.24	11.07	12.83	15.09	16.75
6	0.676	0.872	1.24	1.64	2.20	10.64	12.59	14.45	16.81	18.55
7	0.989	1.24	1.69	2.17	2.83	12.02	14.07	16.01	18.48	20.28
8	1.34	1.65	2.18	2.73	3.49	13.36	15.51	17.53	20.09	21.96
9	1.73	2.09	2.70	3.33	4.17	14.68	16.92	19.02	21.67	23.59
10	2.16	2.56	3.25	3.94	4.87	15.99	18.31	20.48	23.21	25.19
11	2.60	3.05	3.82	4.57	5.58	17.28	19.68	21.92	24.73	26.76
12	3.07	3.57	4.40	5.23	6.30	18.55	21.03	23.34	26.22	28.30
13	3.57	4.11	5.01	5.89	7.04	19.81	22.36	24.74	27.69	29.82
14	4.07	4.66	5.63	6,57	7.79	21.06	23.68	26.12	29.14	31.32
15	4.60	5.23	6.26	7.26	8.55	22.31	25.00	27.49	30.58	32.80

16	5.14	5.81	6.91	7.96	9.31	23.54	26.30	28.85	32.00	34.27
17	5.70	6.41	7.56	8.67	10.08	24.77	27.59	30.19	33.41	35.72
18	6.26	7.01	8.23	9.39	10.86	25.99	28.87	31.53	34.81	37.16
19	6.84	7.63	8.91	10.12	11.65	27.20	30.14	32.85	36.19	38.58
20	7.43	8.26	9.59	10.85	12.44	28.41	31.41	34.17	37.57	40.00
21	8.03	8.90	10.28	11.59	13.24	29.62	32.67	35.48	38.93	41.40
22	8.64	9.54	10.98	12.34	14.04	30.81	33.92	36.78	40.29	42.80
23	9.26	10.20	11.69	13.09	14.85	32.01	35.17	38.08	41.64	44.18
24	9.89	10.86	12.40	13.85	15.66	33.20	36.42	39.36	42.98	45.56
25	10.52	11.52	13.12	14.61	16.47	34.38	37.65	40.65	44.31	46.93
26	11.16	12.20	13.84	15.38	17.29	35.56	38.88	41.92	45.64	48.29
27	11.81	12.88	14.57	16.15	18.11	36.74	40.11	43.19	46.96	49.64
28	12.46	13.56	15.31	16.93	18.94	37.92	41.34	44.46	48.28	50.99
29	13.12	14.26	16.05	17.71	19.77	39.09	42.56	45.72	49.59	52.34
30	13.79	14.95	16.79	18.49	20.60	40.26	43.77	46.98	50.89	53.67
40	20.71	22.16	24.43	26.51	29.05	51.81	55.76	59.34	63.69	66.77
60	35.53	37.48	40.48	43.19	46.46	74.40	79.08	83.30	88.38	91.95
120	83.85	86.92	91.58	95.70	100.62	140.23	146.57	152.21	158.95	163.65

From [A.3].

Table A.8 Values of the t_n variable distributed according the Studient's t distribution.

n	$p_{t_n}(t)$								
	0.750	0.875	0.900	0.950	0.975	0.9875	0.990	0.995	0.9975
1	1.00000	2.4142	3.078	6.3138	12.706	25.452	31.821	63.657	127.32
2	0.81650	1.6036	1.886	2.9200	4.3027	6.2053	6.965	9.9248	14.089
3	0.76489	1.4226	1.638	2.3534	3.1825	4.1765	4.541	5.8409	7.4533
4	0.74070	1.3444	1.533	2.1318	2.7764	3.4954	3.747	4.6041	5.5976
5	0.72669	1.3009	1.476	2.0150	2.5706	3.1634	3.365	4.0321	4.7733
6	0.71756	1.2733	1.440	1.9432	2.4469	2.9687	3.143	3.7074	4.3168
7	0.71114	1.2543	1.415	1.8946	2.3646	2.8412	2.998	3.4995	4.0293
8	0.70639	1.2403	1.397	1.8595	2.3060	2.7515	2.896	3.3554	3.8325
9	0.70272	1.2297	1.383	1.8331	2.2622	2.6850	2.821	3.2498	3.6897
10	0.69981	1.2213	1.372	1.8125	2.2281	2.6338	2.764	3.1693	3.5814
11	0.69745	1.2145	1.363	1.7959	2.2010	2.5931	2.718	3.1058	3.4966
12	0.69548	1.2089	1.356	1.7823	2.1788	2.5600	2.681	3.0545	3.4284
13	0.69384	1.2041	1.350	1.7709	2.1604	2.5326	2.650	3.0123	3.3725
14	0.69242	1.2001	1.345	1.7613	2.1448	2.5096	2.624	2.9768	3.3257
15	0.69120	1.1967	1.341	1.7530	2.1315	2.4899	2.602	2.9467	3.2860

16	0.69013	1.1937	1.337	1.7459	2.1199	2.4729	2.583	2.9208	3.2520
17	0.68919	1.1910	1.333	1.7396	2.1098	2.4581	2.567	2.8982	3.2225
18	0.68837	1.1887	1.330	1.7341	2.1009	2.4450	2.552	2.8784	3.1966
19	0.68763	1.1866	1.328	1.7291	2.0930	2.4334	2.539	2.8609	3.1737
20	0.68696	1.1848	1.325	1.7247	2.0860	2.4231	2.528	2.8453	3.1534
21	0.68635	1.1831	1.323	1.7207	2.0796	2.4138	2.518	2.8314	3.1352
22	0.68580	1.1816	1.321	1.7171	2.0739	2.4055	2.508	2.8188	3.1182
23	0.68531	1.1802	1.319	1.7139	2.0687	2.3979	2.500	2.8073	3.1040
24	0.68485	1.1789	1.318	1.7109	2.0639	2.3910	2.492	2.7969	3.0905
25	0.68443	1.1777	1.316	1.7081	2.0595	2.3846	2.485	2.7874	3.0782
26	0.68405	1.1766	1.315	1.7056	2.0555	2.3788	2.479	2.7787	3.0669
27	0.68370	1.1757	1.314	1.7033	2.0518	2.3734	2.473	2.7707	3.0565
28	0.68335	1.1748	1.313	1.7011	2.0484	2.3685	2.467	2.7633	3.0469
29	0.68304	1.1739	1.311	1.6991	2.0452	2.3638	2.462	2.7564	3.0380
30	0.68276	1.1731	1.310	1.6973	2.0423	2.3596	2.457	2.7500	3.0298
40	0.68066	1.1673	1.303	1.6839	2.0211	2.3289	2.323	2.7045	2.9712
60	0.67862	1.1616	1.292	1.6707	2.0003	2.2991	2.390	2.6603	2.9146
120	0.67656	1.1559	1.289	1.6577	1.9799	2.2699	2.358	2.6174	2.8599
∞	0.67449	1.1503	1.282	1.6449	1.9600	2.2414	2.326	2.5758	2.8070

From [A.3].

Table A.9 Run distribution: values of $r_{N,p'}$.

r: run number
$p_1 = 2p'$: level of significance of the test;
N: number of sample values;
$N_0 = N_1 = N/2$: number of 0 and 1 observations:
$\mu_r = N/2 + 1$; $s_r^2 = N(N-2)/[4(N-1)]$

	Prob $[\tilde{r} > r_{N,p'}]$					
N	0.99	0.975	0.95	0.05	0.025	0.01
10	2	2	3	8	9	9
12	2	3	3	10	10	11
14	3	3	4	11	12	12
16	4	4	5	12	13	13
18	4	4	6	13	14	15
20	5	6	6	15	15	16
22	6	7	7	16	16	17
24	7	7	8	17	18	18
26	7	8	9	18	19	20
28	8	9	10	19	20	21
30	9	10	11	20	21	22
32	10	11	11	22	22	23
36	11	12	13	24	25	26
40	13	14	15	26	27	28
50	17	18	19	32	33	34
60	21	22	24	37	39	40
70	25	27	28	43	44	46
80	30	31	33	48	50	51
90	34	36	37	54	55	57
100	38	40	42	59	61	63
110	43	45	46	65	66	68
120	47	49	51	70	72	74
130	52	54	56	75	77	79
140	56	58	60	81	83	85
150	61	63	65	86	88	90
160	65	68	70	91	93	96
170	70	72	74	97	99	101
180	74	77	79	102	104	107
190	79	82	84	107	109	112
200	84	86	88	113	115	117

After [A.4].

REFERENCES (APPENDICES)

The following are the main sources used for the text of the Appendices and the Tables:

A.1 Parzen, E. (1960) *Modern Probability Theory and its Applications*, Wiley, New York.
A.2 Ferry-Borges, J. and Castanheta, M. (1971) *Structural Safety*, LNEC, Lisbon.
A.3 Hadley, G. (1969) *Elementary Statistics*, Holden-Day, San Francisco.
A.4 Bendat, J. S. and Piersol, A. G. (1971) *Random Data: Analysis and Measurement Procedures*, Wiley-Interscience, New York.
A.5 Benjamin, J. R. and Cornell, C. A. (1970) *Probability, Statistics, and Decision for Civil Engineers*, McGraw-Hill, New York.
A.6 Feller, W. (1970) *An Introduction to Probability Theory and its Applications*, Wiley, New York.
A.7 Gumbel, E. J. (1967) *Statistics of Extremes*, Columbia University Press, New York.
A.8 Parzen, E. (1962) *Stochastic Processes*, Holden-Day, San Francisco.
A.9 Srinivasan, S. K. and Vasudevan, R. (1971) *Introduction to Random Differential Equations and Their Applications*, American Elsevier, New York.
A.10 Veneziano, D. (1978) *Random Processes for Engineering Application* (Lecture Notes), Department of Civil Engineering, M.I.T., Cambridge, Mass., USA.

Index